# Lecture Notes in Control and Information Sciences

Volume 484

This series reports new developments in the fields of control and information sciences—quickly, informally and at a high level. The type of material considered for publication includes:

1. Preliminary drafts of monographs and advanced textbooks
2. Lectures on a new field, or presenting a new angle on a classical field
3. Research reports
4. Reports of meetings, provided they are
   (a) of exceptional interest and
   (b) devoted to a specific topic. The timeliness of subject material is very important.

Indexed by EI-Compendex, SCOPUS, Ulrich's, MathSciNet, Current Index to Statistics, Current Mathematical Publications, Mathematical Reviews, IngentaConnect, MetaPress and Springerlink.

More information about this series at http://www.springer.com/series/642

Alexandre Mauroy · Igor Mezić ·
Yoshihiko Susuki
Editors

# The Koopman Operator in Systems and Control

## Concepts, Methodologies, and Applications

*Editors*
Alexandre Mauroy
Department of Mathematics
University of Namur
Namur, Belgium

Igor Mezić
Department of Mechanical Engineering
University of California Santa Barbara
Santa Barbara, CA, USA

Yoshihiko Susuki
Department of Electrical
and Information Systems
Osaka Prefecture University
Sakai, Japan

ISSN 0170-8643 ISSN 1610-7411 (electronic)
Lecture Notes in Control and Information Sciences
ISBN 978-3-030-35712-2 ISBN 978-3-030-35713-9 (eBook)
https://doi.org/10.1007/978-3-030-35713-9



Mathematics Subject Classification (2010): 37N35, 47D03, 47N70, 93B07, 93B28, 93C10, 93C57, 93D05



This Springer imprint is published by the registered company Springer Nature Switzerland AG
The registered company address is: Gewerbestrasse 11, 6330 Cham, Switzerland

*To our families, students, and collaborators.*
Alexandre Mauroy
Igor Mezić
Yoshihiko Susuki

# Preface

As an example of fruitful cross-fertilization between mathematics and engineering, nonlinear control theory has attracted considerable effort driven by the need to understand, predict, and control dynamical phenomena. Past decades have witnessed significant advances and achievements, with the promise to meet current needs including, among others, more severe (energetic and economic) constraints in industrial processes or innovative control designs in synthetic biology. Nonlinear control theory has, therefore, become instrumental in addressing real-world problems and challenges in emerging fields such as computational neuroscience and environmental science.

A few limitations and challenges, however, arose. In spite of intense research activity at a theoretical level, nonlinear system theory is not applied to real-world problems as much as hoped, and linear control theory prevails in real-world applications because it provides a systematic and general, easy to implement framework. This generally results in suboptimal solutions to nonlinear problems which, sooner or later, might not meet the growing demand for better understanding, higher performances, or improved accuracy. Another important challenge is related to the increasing complexity and diversity of systems, which cannot be described through classical systems identification and are not amenable to model-based control. At the same time, these systems generate big data that may be leveraged to describe them. In this context, there is an increasing need for innovative data-driven model-free methods in nonlinear control theory.

The chapters of this edited book demonstrate that the Koopman operator approach offers a promising way to address the aforementioned challenges. The Koopman operator provides a global description of nonlinear dynamical systems in terms of the evolution of "observable-functions" of the state space. Although this description seems to be overlooked in control theory at first sight, it has evocatively been considered for decades: for instance, it is reminiscent of Lyapunov functions characterization in stability analysis and to the use of value functions in optimal control. The most noticeable advantage of the Koopman approach relies on the fact that it turns a nonlinear system into a linear (but infinite-dimensional) system. At the cost of working in high-dimensional spaces providing a good approximation of the observable dynamics, systematic methods from linear control theory can be applied in a straightforward manner to controlled nonlinear systems. Moreover, the Koopman operator framework is amenable to data analysis through the so-called Koopman mode decomposition and is directly connected to efficient numerical algorithms such as Arnoldi-type methods and (extended) dynamic mode decomposition. Therefore, it provides data-driven methods for model-free control of nonlinear systems.

This book is motivated by the increasing interest in the Koopman operator framework in the control community and aims at providing a snapshot of the current research effort in this area. Due to intense research activity, it is undoubtedly not exhaustive, and new chapters could be added as these lines are written. For the novice reader, the book will provide a broad overview of the Koopman formalism in various topics of interest in nonlinear control theory. For the expert researcher in the field, it will open up to a wealth of research perspectives, with limitations to be overcome and open problems to be solved.

The volume consists of 20 chapters contributed by more than 40 authors with different perspectives and backgrounds, but with a common interest in the broad field of nonlinear control theory. The chapters are organized into three parts: I. Concepts, II. Methodologies, and III. Applications. Part I mainly investigates the connections between classical concepts in control theory and properties of the Koopman operator. It also provides basic material on the Koopman operator framework, both from theoretical and numerical points of view. Part II presents several Koopman-based methods developed in the context of nonlinear control. Most chapters in this part focus on numerical aspects, but theoretical aspects are also considered. Part III presents various applications of the Koopman operator framework in the broad context of control theory, ranging from nanopositioning to heat transfer in buildings. The content of each chapter is briefly described as follows.

- **Part I. Concepts**
  - *Chapter 1. Introduction to the Koopman operator.* In this introductory chapter, we provide an overview of the Koopman operator approach in the context of systems and control theory. We review the basic definitions and properties related to the Koopman operator, including its spectral properties. Approximations of the operator in finite-dimensional subspaces are also presented, as well as numerical methods allowing to obtain those approximations from data.
  - *Chapter 2. Stability.* A. Mauroy, I. Mezić, and A. Sootla demonstrate in this chapter that global stability properties can be obtained from the spectral properties of the Koopman operator. This yields a systematic method for stability analysis that mirrors classical linear stability analysis.
  - *Chapter 3. Observability and estimation.* Using the eigenfunctions of the Koopman operator, A. Surana defines the so-called Koopman canonical transform to propose a novel characterization of nonlinear observability.

This allows to use classical techniques for linear and bilinear systems and apply them to nonlinear constrained state estimation, possibly with non-convex state constraints.
- *Chapter 4. Controllability.* In this chapter, D. Goswami and D. Paley present sufficient conditions for bilinearizability of control-affine systems. Using the Koopman canonical transform and leveraging classical Lie-algebraic tools, they propose a reachability analysis for those systems.
- *Chapter 5. Interconnections.* The study of interconnected systems is crucial in control theory. In the case of cascaded linear systems, R. Mohr and I. Mezić show that the (principal) Koopman eigenvalues and eigenfunctions of an interconnected system are strongly connected to those of its decoupled subsystems, so that the dynamic properties of the interconnected system can be inferred from the properties of the decoupled subsystems. These results are also extended to nonlinear systems through topological conjugacies.
- *Chapter 6. Nonautonomous and random dynamical systems.* Control systems are generally nonautonomous and noisy. In this chapter, S. Maćešić and N. Črnjarić-Žic provide an extension of the Koopman operator framework to the case of nonautonomous and random dynamical systems. In both cases, they investigate the spectral properties of the Koopman operator and propose numerical methods to compute these properties from data.
- *Chapter 7. Numerical linear algebra perspective.* The Koopman operator framework is complemented with numerical methods that can be leveraged for data-driven control. In this chapter, Z. Drmač revisits the dynamic mode decomposition algorithm and studies its robustness through a numerical linear algebra perspective. An improved version of the algorithm is proposed based on his theoretical study. This chapter also provides a useful background for the second part of this volume.

- **Part II. Methodologies**
  - *Chapter 8. Data-driven control.* This chapter serves as an introduction to the second part of this volume. E. Kaiser, J. N. Kutz, and S. L. Brunton present a broad and detailed overview of data-driven control in the framework of the Koopman operator and its adjoint, the Perron–Frobenius operator. They review various problems such as optimal control, system identification, observability and controllability analysis, and sensor and actuator placement.
  - *Chapter 9. Prediction and model predictive control.* In this chapter, M. Korda and I. Mezić propose linear predictors for controlled nonlinear systems, which are obtained through a lifting method in the framework of the Koopman operator. They demonstrate that these predictors achieve superior performance than classical linear predictors and use them in the context of model predictive control to derive the so-called Koopman-MPC control scheme.
  - *Chapter 10. Model predictive control for partial differential equations.* S. Peitz and S. Klus use reduced order modeling based on the Koopman operator approach to develop two model predictive control schemes for

partial differential equations. In the first case, they consider a small number of autonomous systems associated with constant inputs and turn the control problem into a switching problem. In the second case, they obtain a bilinear surrogate model through linear interpolation between the reduced order models derived in the first approach.

- *Chapter 11. Optimal pulse-based control.* Motivated by biological applications, A. Sootla, G.-B. Stan, and D. Ernst study the problem of optimal convergence to an equilibrium with pulse-based control. Using an eigenfunction of the Koopman operator, they recast the optimal problem into a static optimization program, which they study and solve in the case of monotone systems.
- *Chapter 12. Feedback stabilization.* In this chapter, B. Huang, X. Ma, and U. Vaidya present a systematic, data-driven method to design stabilizing feedback controllers for control-affine systems. This method relies on the bilinear form of the dynamics obtained with the Koopman canonical transform and on the Lyapunov function based approach for the design of feedback controllers. They also propose a systematic convex optimization-based formulation for computing control Lyapunov functions.
- *Chapter 13. System identification.* A. Mauroy and J. Goncalves exploit the idea that a nonlinear dynamical system can be identified through its associated Koopman operator in a lifted space of observables. They present two dual methods for nonlinear system identification, which rely solely on linear regression techniques.
- *Chapter 14. Manifold learning and state estimation.* In this chapter, T. Shnitzer, R. Talmon, and J.-J. Slotine use a data-driven manifold learning technique based on diffusion maps to obtain a linear state-space representation of the dynamics, a method which is reminiscent of Koopman spectral analysis. Then they design two observers for this linear model by considering a linear contracting observer and a Kalman filter.
- *Chapter 15. Phase–amplitude reduction.* Phase–amplitude reduction of limit-cycling systems is directly connected to the eigenfunctions of the Koopman operator. S. Shirasaka, W. Kurebayashi, and H. Nakao review this framework and propose a bi-orthogonalization method to compute the phase–amplitude sensitivity of the system. They illustrate the approach by considering the optimal injection timing of a weak control input to suppress deviations from the limit cycle.

- **Part III. Applications**
  - *Chapter 16. Active learning and experimental applications.* In unknown and changing environments, active learning should be incorporated to ensure robust performance of the control process. In this chapter, T. A. Berrueta, I. Abraham, and T. Murphey investigate active learning methods in the

framework of the Koopman operator. They also illustrate active learning based on the Koopman operator with several experimental applications.

- *Chapter 17. Nanopositioning.* S. Zelenika, E. Kamenar, M. Korda, and I. Mezić apply Koopman-based model predictive control (Koopman-MPC) to the nanometric positioning control problem and demonstrate that Koopman-MPC provides good performance (i.e., small overshoot and settling time) compared with classical controllers.
- *Chapter 18. Heat transfer estimation in buildings.* Y. Kono, Y. Susuki, and T. Hikihara present a novel methodology for heat transfer modeling based on Koopman mode decomposition. This method allows them to estimate parameters of the advection and diffusion equations, such as the effective velocity and diffusivity, and is used with real data in the case of a building with atrium.
- *Chapter 19. Dynamics reconstruction in power grids.* In this chapter, Y. Susuki and K. Sako consider the data-driven analysis of voltage dynamics in power grids. Using extended dynamic mode decomposition with delay embedding, they reconstruct the full-state dynamics of a power grid using voltage–amplitude measurements of a single bus.
- *Chapter 20. Performance of consensus networks.* H. K. Mousavi, C. Somarakis, Q. Sun, and N. Motee propose a novel method to estimate the performance of nonlinear consensus networks. Based on Koopman mode decomposition and interpolation methods, the performance estimate is expressed in terms of Koopman modes and eigenvalues.

We believe that these 20 chapters illustrate the potential and effectiveness of the Koopman operator framework in various contexts, as a general approach to design systematic, data-driven techniques for analysis and control of nonlinear systems. While this framework is still in its infancy, the chapters provide many research perspectives and open problems—some of which are summarized in Chap. 1. Finally, we hope that this volume will serve as an invitation to contribute to the field and will stimulate further progress in Koopman based control theory.

| Namur, Belgium | Alexandre Mauroy |
| --- | --- |
| Santa Barbara, USA | Igor Mezić |
| Sakai, Japan | Yoshihiko Susuki |
| June 2019 | |

**Acknowledgements** We would like to thank all authors for their invaluable work and contributions. We also acknowledge our institutions and departments for their support: University of Namur, Belgium (naXys Institute and Department of Mathematics), University of California, Santa Barbara, USA (Department of Mechanical Engineering), and Osaka Prefecture University, Japan (Department of Electrical and Information Systems). We acknowledge financial support from: the Belgian National Fund for Scientific Research (FNRS); the National Science Foundation (NSF

CAREER ECCS-1454022); the US Defense Advanced Research Projects Agency; the US Air Force Office of Scientific Research (AFOSR FA9550-19-1-0004); the US Office of Naval Research (ONR YIP N00014-16-1-2645); the US Army Research Office (ARO); the Japan Science Agency, Core Research for Evolutional Science and Technology (JST-CREST) Program #JP-MJCR15K3; the Ministry of Education, Culture, Sports, Science and Technology—Japan (MEXT) KEKENHI Grant #15H03964.

# Contents

## Part I Concepts

# Part I
# Concepts

# Chapter 1
# Introduction to the Koopman Operator in Dynamical Systems and Control Theory

**Alexandre Mauroy, Yoshihiko Susuki and Igor Mezić**

**Abstract** This introductory chapter provides an overview of the Koopman operator framework. We present basic notions and definitions, including those related to the spectral properties of the operator. We also review numerical methods that are used to represent the operator on a finite-dimensional basis and to compute its spectral properties from data (DMD algorithm and Arnoldi-based methods). Finally, several extensions of the Koopman operator to input–output systems are reviewed.

## 1.1 Introduction

The idea of linearizing nonlinear systems with no approximation has always been attractive. Toward this aim two adjoint linear operators have emerged as powerful tools to study nonlinear systems: the Perron–Frobenius (or transfer) operator and the Koopman (or composition) operator, which traces back to the works by Koopman and von Neumann [26, 54]. Related infinite-dimensional methods such as Carleman embedding [8] have also been developed in a similar context. Although both operators have attracted considerable interest in statistical mechanics [12], the Perron–Frobenius approach has prevailed in nonlinear systems theory for a long time, enabling the development of set-oriented methods and the study of coherent sets in measure-preserving dynamical systems [9].

A. Mauroy (✉)
Department of Mathematics, University of Namur, Namur, Belgium
e-mail: alexandre.mauroy@unamur.be

Y. Susuki
Department of Electrical and Information Systems, Osaka Prefecture University, Osaka, Japan
e-mail: susuki@eis.osakafu-u.ac.jp

I. Mezić
Department of Mechanical Engineering, University of California Santa Barbara, Santa Barbara, CA, USA
e-mail: mezic@engineering.ucsb.edu



A. Mauroy et al. (eds.), *The Koopman Operator in Systems and Control*,
Lecture Notes in Control and Information Sciences 484,
https://doi.org/10.1007/978-3-030-35713-9_1

However, the last decade has witnessed a growing interest in the Koopman operator framework, which originated from the work [44, 48]. In that work, it was shown that the eigenfunctions of the Koopman operator reveal important global geometric properties of the underlying nonlinear system, such as ergodic and periodic partitions (see [7] for a review). Those results mainly focused on measure-preserving conservative systems inducing a unitary operator in $L^2$ space. They were further extended to nonconservative systems, in which case the Koopman operator is not unitary [40, 42, 51]. This suggested that the Koopman approach is particularly well suited to provide a novel approach to nonconservative systems, and paved the way for applications in nonlinear control theory, as already envisioned in [48] (see Conclusions).

Since then, the Koopman operator framework gained increased attention in the control community. Recent theoretical contributions have been obtained in the context of observability [65, 79], controllability analysis [15, 78], and stability [41]. Based on finite-dimensional approximations of the operator, novel methods have been proposed for model predictive control [27, 55], optimal control [23] (see also [71] for early related contributions), feedback stabilization [19], and identification [39, 60]. Moreover, several control-oriented applications have been considered, such as synthetic biology [64], power grids [29], robotics [1], and consensus [53].

Numerous methods have been proposed to compute the spectrum of the Koopman operator, its eigenfunctions, and the so-called Koopman modes directly from data (e.g., Arnoldi method [61] and DMD method [62]). They belong to two classes: those based on rigorous projections, such as (generalized) Laplace averages, and those based on the approximation of the underlying linear operator. These data-driven methods complement the aforementioned developments, making the Koopman operator approach not only applicable to a wealth of real-world problems (e.g., fluids dynamics [61], power grids [67], epidemiology [57], climatology [63], time-series classification [66] to list a few) but also very appealing to develop novel control schemes in the big data era. Namely, a notable property of the stability assessment and control approaches using the Koopman operator is that they can be pursued directly from data, i.e., they do not require an underlying model [27, 68].

The chapter is organized as follows. In Sect. 1.2, we present the Koopman operator and review a few general properties. Spectral properties of the operator are discussed and related to geometric properties of the underlying system in Sect. 1.3. Finite-dimensional representations and numerical methods are presented in Sect. 1.4. In Sect. 1.5, the framework is extended to the case of systems with inputs. Finally, perspectives and concluding remarks are given in Sects. 1.6 and 1.7, respectively. A preliminary version of the chapter can be found in [43].

## 1.2 The Koopman Operator

This section presents the definition of the Koopman operator in a general framework. The main properties of the operator and its dual are also reviewed.

### 1.2.1 Koopman Operator for Autonomous Discrete-Time and Continuous-Time Systems

The Koopman operator can be defined for discrete-time systems (nonlinear transformations) and continuous-time systems (semigroups of transformations). This is discussed in detail in this section.

#### 1.2.1.1 Discrete-Time Systems

Consider a discrete-time dynamical system defined by the nonlinear (non-singular) transformation $\mathbf{S} : X \to X$, where the state space $X$ is a finite-dimensional metric space (e.g., Euclidean space, or a manifold). The system is typically described through its orbits $\{\mathbf{S}^k(\mathbf{x})\}_{k=0}^{\infty}$, where $\mathbf{x}$ is an initial condition. Alternatively, one can focus on an output function $f : X \to \mathbb{C}$, also called *observable (function)*, and on its orbit $\{f(\mathbf{S}^k(\mathbf{x}))\}_{k=0}^{\infty}$. The evolution of all observables is given by the action of the *Koopman operator* associated with the system, defined next.

**Definition 1.1** (*Koopman operator (discrete-time)*) Consider a (Banach) space $\mathscr{F}$ of observables $f : X \to \mathbb{C}$. The Koopman operator $U_{\mathbf{S}} : \mathscr{F} \to \mathscr{F}$ associated with the map $\mathbf{S} : X \to X$ is defined through the composition

$$U_{\mathbf{S}} f = f \circ \mathbf{S} \quad \forall f \in \mathscr{F} .$$

△

Note that, in the remainder of this chapter, we will omit the subscript and denote the operator by $U$. We also remark that $\mathscr{F}$ should be closed under composition. Moreover, the operator could be defined for vector-valued observables $\mathbf{f} : X \to \mathbb{C}^m$ (with $m \geq 2$).

The Koopman operator offers two main advantages. First, it gives a *global* picture of the system, in contrast to the pointwise description in terms of orbits. Second, it provides a *linear* approach to the (nonlinear) system since it is a linear operator: one easily checks that $U(c_1 f_1 + c_2 f_2) = c_1 U f_1 + c_2 U f_2$ for all $f_1, f_2 \in \mathscr{F}$ and $c_1, c_2 \in \mathbb{C}$. However, unless the state space is a finite set, the Koopman operator is infinite-dimensional. This is the main challenge to overcome when dealing with the operator-theoretic viewpoint. The focus in [44, 48], and much of the follow-up work, was on resolving this issue by focusing on the spectral objects—eigenvalues, eigenfunctions and modes, to be defined below—and reducing the dimensionality by finite-dimensional projections onto eigenspaces.

#### 1.2.1.2 Continuous-Time Systems

Consider now a continuous-time system described by the one-parameter family of (non-singular) maps $\mathbf{S}^t : X \to X, t \in \mathbb{R}^+$, which generate the trajectories $\{\mathbf{S}^t(\mathbf{x})\}_{t=0}^{\infty}$. Similarly to the discrete-time case, one can focus on the evolution of observables

$f : X \to \mathbb{C}$ along the trajectories of the system, i.e., $\{f(\mathbf{S}^t(\mathbf{x}))\}_{t=0}^{\infty}$. The evolution of all observable functions is given by the one-parameter family of Koopman operators associated with the system.

**Definition 1.2** (*Koopman operators (continuous-time)*) Consider a (Banach) space $\mathcal{F}$ of observables $f : X \to \mathbb{C}$. The family of Koopman operators $U_{\mathbf{S}^t} : \mathcal{F} \to \mathcal{F}$ associated with the family of maps $\mathbf{S}^t : X \to X$, $t \in \mathbb{R}^+$, is defined through the composition

$$U^t_{\mathbf{S}^t} f = f \circ \mathbf{S}^t \quad \forall f \in \mathcal{F} .$$

△

In the following, we drop the subscript $\mathbf{S}^t$. Definition 1.2 is general and, in particular, does not assume any regularity property on $(\mathbf{S}^t)_{t \geq 0}$. For instance, the family of Koopman operators can be defined for switched systems and more generally for hybrid systems: see, e.g., [16, 55]. However, we will assume that $\mathbf{S}^t$ is a semigroup, which implies that $(U^t)_{t \geq 0}$ is a *Koopman semigroup of operators*. Indeed, we verify in this case that $U^0 f = f \circ \mathbf{S}^0 = f$ and

$$U^{t+s} f = f \circ \mathbf{S}^{t+s} = f \circ (\mathbf{S}^s \circ \mathbf{S}^t) = U^t U^s f ,$$

for all $t, s \geq 0$. Similarly to the discrete-time case, the Koopman semigroup of operators is linear, i.e., $U^t(c_1 f_1 + c_2 f_2) = c_1\, U^t f_1 + c_2\, U^t f_2$ for all $f_1, f_2 \in \mathcal{F}$, $c_1, c_2 \in \mathbb{C}$, and $t \geq 0$.

If the semigroup $(\mathbf{S}^t)_{t \geq 0}$ possesses some additional properties, the Koopman semigroup of operators can be strongly continuous [11], that is

$$\lim_{t \downarrow 0} \|U^t f - f\| = 0 ,$$

where $\|\cdot\|$ denotes the norm defined in $\mathcal{F}$. For instance, if $(\mathbf{S}^t)_{t \geq 0}$ is continuous with respect to $t$, then the semigroup $(U^t)_{t \geq 0}$ is strongly continuous in $\mathbb{L}^2(\Omega)$, where $\Omega \subset \mathbb{R}^n$ is an open forward-invariant set [11]. If $(\mathbf{S}^t)_{t \geq 0}$ is uniformly Lipschitz continuous with respect to $t$, then the semigroup $(U^t)_{t \geq 0}$ is strongly continuous in $C(K)$, where $K \subset \mathbb{R}^n$ is a compact forward-invariant set [5]. For strongly continuous semigroups $(U^t)_{t \geq 0}$, the limit

$$\lim_{t \downarrow 0} \frac{U^t f - f}{t} \triangleq Lf \quad \forall f \in \mathcal{D}$$

exists (in the strong sense) for all $f \in \mathcal{D}$, where $\mathcal{D}$ is a dense set in $\mathcal{F}$. This defines the infinitesimal (Koopman) generator $L : \mathcal{D} \to \mathcal{F}$ of the semigroup of operators, which provides the time derivative of observables along the trajectories. If the semigroup of maps $(\mathbf{S}^t)_{t \geq 0}$ is the flow induced by the dynamics[1] $\dot{\mathbf{x}} = \mathbf{F}(\mathbf{x}), \mathbf{x} \in \mathbb{R}^n$, the infinitesimal

[1] More generally, one could consider that the state $\mathbf{x}$ and the vector field $\mathbf{F}$ lie on a finite-dimensional manifold and its tangent bundle, respectively.

generator of the associated Koopman semigroup of operators is given by

$$Lf = \mathbf{F} \cdot \nabla f \tag{1.1}$$

(when it exists[2]), where $\cdot$ denotes the inner product and $\nabla$ denotes the gradient operator. In this case, the infinitesimal generator provides a direct and useful connection between the Koopman operator framework and the vector field of the underlying system.

## 1.2.2 Some Properties of the Koopman Operator

The Koopman operator is a linear operator that is characterized by remarkable properties. These properties are presented in this section for the discrete-time Koopman operator, but are also satisfied in the continuous-time setting. More details can be found in [5, 35].

The properties of the Koopman operator depend on the choice of the space $\mathscr{F}$ of observables. This choice is crucial and determined by the type of systems that are studied and the properties of those systems that should be captured. For measure-preserving dynamical systems, the Hilbert space of $L^2$ functions with respect to the invariant measure are usually considered (e.g., ergodic theory). For nonconservative systems which admit an attractor, more appropriate spaces are spaces of continuously differentiable functions [41], spaces of analytic functions [42], or generalized Hardy spaces [51].

Here are a few basic properties of the Koopman operator. The Koopman operator exists and is unique if the map $\mathbf{S}$ (or the flow $\mathbf{S}^t$) exists and is unique. It is generally bounded (and therefore continuous) in usual spaces. However, it is not compact (unless the state space is finite) and not self-adjoint. For conservative systems, it is a unitary[3] operator in $L^2$ space [35].

### 1.2.2.1 Positivity and Contractivity

Let $\mathscr{F}$ be a space of real-valued observables $f : X \to \mathbb{R}$ and $\mathscr{K}$ be the cone of positive functions, that is

$$\mathscr{K} = \{f \in \mathscr{F} \mid f(\mathbf{x}) \geq 0 \quad \forall \mathbf{x} \in X\}\,.$$

For all $f \in \mathscr{K}$, it is clear that $Uf = f \circ \mathbf{S} \geq 0$ so that $\mathscr{K}$ is invariant under the action of the Koopman operator. It follows that the Koopman operator is *positive* in Banach

[2] Some regularity property (e.g., $C^1$) is required on the vector field $\mathbf{F}$ to ensure that the semigroup is strongly continuous.

[3] This property explains the original use of the letter $U$ in the seminal paper [26].

spaces endowed with the order $\geq_{\mathscr{K}}$ induced by $\mathscr{K}$, that is $f_1 \leq_{\mathscr{K}} f_2 \Leftrightarrow f_2 - f_1 \in \mathscr{K}$ for all $f_1, f_2 \in \mathscr{F}$. These spaces are for instance $C(K)$ and $L^p(\Omega)$, $1 \leq p \leq \infty$. If in addition the underlying system preserves an order in the state space $X$ (i.e., monotone system), this property implies that the Koopman operator is positive with respect to a more specific cone in $\mathscr{F}$.

The Koopman operator is a positive *Markov operator*. Indeed, the function $\mathbf{1} \in \mathscr{K}$ defined by $\mathbf{1}(\mathbf{x}) = 1\ \forall \mathbf{x} \in X$ is a fixed point of the Koopman operator, that is $U\mathbf{1} = \mathbf{1}$. The Markov property and the positivity property on $L^\infty$ imply that the Koopman operator is a *contraction* with respect to the supremum norm:

$$\|Uf\|_\infty \leq \|f\|_\infty$$

or equivalently $\|U\|_\infty = \sup_f\{\|Uf\|_\infty : \|f\|_\infty = 1\} \leq 1$.

#### 1.2.2.2 Duality

According to the theory of linear operators on Banach spaces, the Koopman operator has a dual operator $U^*$ acting on the conjugate space $\mathscr{F}^*$ of bounded linear functionals $\xi : \mathscr{F} \to \mathbb{C}$. This dual operator satisfies

$$\xi(Uf) = (U^*\xi)(f) \qquad \forall f \in \mathscr{F},\ \xi \in \mathscr{F}^*\ . \tag{1.2}$$

A bounded linear functional $\xi$ defined on $\mathscr{F}^* = C(K)^*$ can be associated with a measure (Riesz theorem), which can in turn be associated with a density function $\rho \in L^1(K)$ if the measure is absolutely continuous (Radon–Nikodym theorem). In this case, we have

$$\xi(f) = \int_K f(\mathbf{x})\,\rho(\mathbf{x})\,d\mathbf{x}$$

and we can define a dual operator $P : L^1(K) \to L^1(K)$ such that

$$(U^*\xi)(f) = \int_K f(\mathbf{x})\,(P\rho)(\mathbf{x})\,d\mathbf{x}\ .$$

This operator is the so-called *Perron–Frobenius operator*, or *transfer operator*. From the duality (1.2), it is clear that

$$\int_K f(\mathbf{x})\,(P\rho)(\mathbf{x})\,dx = \int_K (Uf)(\mathbf{x})\,\rho(\mathbf{x})\,d\mathbf{x}\ .$$

Considering a $\sigma$-algebra $\mathscr{A}$ of subsets of $K$ and the characteristic function $f = \mathbf{1}_A$ of the set $A \subseteq \mathscr{A}$, defined by

$$\mathbf{1}_A(\mathbf{x}) = \begin{cases} 1 & \text{if } \mathbf{x} \in A \\ 0 & \text{otherwise} \end{cases},$$

we obtain

$$\int_A (P\rho)(\mathbf{x})\, dx = \int_{\mathbf{S}^{-1}(A)} \rho(\mathbf{x})\, dx \qquad \forall A \in \mathscr{A}\,. \tag{1.3}$$

This equality shows that the Perron–Frobenius operator is related to the *forward* propagation of densities along the orbits of the map $\mathbf{S}$, while the action of the Koopman operator can be seen as the *backward* propagation of observables.[4] The equality (1.3) can be used as a definition of the Perron–Frobenius operator in general integrable spaces of functions.

In the case of a measure-preserving map, the two operator-theoretic descriptions are equivalent. The invariant measure $\mu^*$ satisfying $\mu^*(\mathbf{S}^{-1})(A) = \mu^*(A)$ for all $A \subseteq \mathscr{A}$ can be used to define the space $\mathscr{F} = \mathscr{F}^* = L^2(\mu^*)$ equipped with the norm $\|f\|^2 = \int |f|^2\, d\mu^*$. In this space, the operators are *isometries*, i.e.,

$$\|Uf\| = \|Pf\| = \|f\| \qquad \forall f \in \mathscr{F}$$

and, more generally, they satisfy $UU^* = UP = I$, where $I$ is the identity operator (note that $P$ acts on densities defined with respect to the invariant measure). Moreover, the operators are *unitary*, i.e., $U = P^{-1}$, when they are invertible. These properties imply that the system can be equivalently described with both operators.

In the context of control, systems are usually not conservative (i.e., they are dissipative in the sense of [34]) so that they rarely preserve a non-singular measure. In this case, the two dual operator-theoretic descriptions are not equivalent and it is preferable to use the Koopman operator, which is characterized by a smoother behavior than the Perron–Frobenius operator, as illustrated below through its spectral properties. However, it should be noted that some works have successfully applied the Perron–Frobenius operator to specific control problems (e.g., optimal control [36], stability analysis [74], optimal stabilization [59]).

#### 1.2.2.3 Invariant Sets

There exists a direct connection between the fixed points of the Koopman operator $U$ and $\mathbf{S}$-*invariant* sets $A \subset K$ satisfying $\mathbf{S}^{-1}(A) = A$. This is summarized in the following result.

**Proposition 1.1** (Invariant sets) *Suppose that $U$ is the Koopman operator associated with the non-singular transformation $\mathbf{S} : X \to X$. Then, a set $A \subseteq X$ is $\mathbf{S}$-invariant if and only if $U\mathbf{1}_A = \mathbf{1}_A$. Moreover, if there exists a function $f$ such that $Uf = f$, then the sublevel sets*

---

[4] This duality is easily captured when one considers the Perron–Frobenius operator $P$ acting on measures. In this case, (1.3) implies that $P\mu(A) = \mu(\mathbf{S}^{-1}(A))$ for all $A \in \mathscr{A}$.

$$A_r = \{\mathbf{x} \in X : f(\mathbf{x}) \leq r\}$$

*are* $\mathbf{S}$*-invariant.*

The proof is straightforward and can be found in [35, Theorem 4.2.1]. The result is also valid for continuous-time systems. Since ergodic maps admit no invariant set (except the whole state space $X$ and sets of zero measure), fixed points of the Koopman operator should be constant almost everywhere. In the case of systems that admit equilibria, sublevel sets of the fixed points characterize the basin of attraction of those equilibria [48, 70].

Proposition 1.1 shows a clear connection between algebraic properties of the Koopman operator (i.e., fixed points) and geometric properties of the system (i.e., invariance). In fact, a fixed point of the operator is a specific eigenfunction associated with the eigenvalue 1. In the next section, we show that the interplay between the Koopman operator and the geometric properties of the underlying systems can be further investigated through the spectral properties of the operator.

## 1.3 Spectral Properties of the Koopman Operator

Since the Koopman operator is linear, it is natural to consider its spectrum and associated eigenfunctions. As illustrated in the work [44], the spectral properties of the Koopman operator are of paramount importance, as they reveal global properties of the underlying dynamical system.

### *1.3.1 Koopman Eigenvalues and Eigenfunctions*

The eigenfunctions and eigenvalues of the Koopman operator are called *Koopman eigenfunctions* and *Koopman eigenvalues* in short. For discrete-time systems, they are defined as follows.

**Definition 1.3** (*Koopman eigenfunction and eigenvalue (discrete time)*) An eigenfunction of the Koopman operator associated with the discrete-time map $\mathbf{S}$ is an observable $\phi_\mu \in \mathscr{F} \setminus \{0\}$ that satisfies

$$U\phi_\mu = \phi_\mu \circ \mathbf{S} = \mu\phi_\mu \ ,$$

where $\mu \in \mathbb{C}$ is the corresponding eigenvalue. △

*Example 1.1* (*Linear discrete-time system* [61]) Consider the linear transformation $\mathbf{S}(\mathbf{x}) = \mathbf{A}\mathbf{x}$, $\mathbf{x} \in \mathbb{R}^n$, where $\mathbf{A}$ is a matrix with eigenvalues $\mu_j$ and corresponding left eigenvectors $\mathbf{w}_j$. The spectrum of the Koopman operator contains the eigenvalues $\mu_j$ and the associated eigenfunctions are given by $\phi_{\mu_j}(\mathbf{x}) = \mathbf{w}_j^T\mathbf{x}$ where $^T$ denotes the transposed vector. (Note that $\mathscr{F}$ is chosen so that $\phi_{\mu_k} \in \mathscr{F}$.) We easily verify that

$$(U\phi_{\mu_j})(\mathbf{x}) = \phi_{\mu_j}(\mathbf{A}\mathbf{x}) = \mathbf{w}_j^T A\mathbf{x} = \mu_j \mathbf{w}_j^T \mathbf{x} = \mu_j\, \phi_{\mu_j}(\mathbf{x})\ .$$

These eigenfunctions are linear, and therefore capture the linearity of $\mathbf{S}$. △

The eigenfunctions described in the above example are associated with the eigenvalues of the matrix $\mathbf{A}$ and are called the *principal eigenvalues* [52] (see also Chap. 5). For nonlinear systems with a hyperbolic equilibrium, they are associated with the eigenvalues of the Jacobian matrix at the equilibrium. Similarly, we can define the eigenvalues and eigenfunctions of Koopman operators associated with continuous-time systems

**Definition 1.4** (*Koopman eigenfunction and eigenvalue (continuous time)*) An eigenfunction of the Koopman operator associated with the semigroup of maps $(\mathbf{S}^t)_{t\geq 0}$ is an observable $\phi_\lambda \in \mathscr{F} \setminus \{0\}$ that satisfies

$$U^t \phi_\lambda = \phi_\lambda \circ \mathbf{S}^t = e^{\lambda t}\phi_\lambda \quad \forall t \geq 0\ ,$$

where $\lambda \in \mathbb{C}$ is the corresponding eigenvalue. △

If the semigroup of operators $U^t$ is strongly continuous, the Koopman eigenfunctions and eigenvalues are equivalently defined through the equality $L\phi_\lambda = \lambda\phi_\lambda$. If $(\mathbf{S}^t)_{t\geq 0}$ is induced by the dynamics $\dot{\mathbf{x}} = \mathbf{F}(\mathbf{x})$, they can be computed with the eigenvalue equation

$$\mathbf{F} \cdot \nabla\phi_\lambda = \lambda\phi_\lambda\ . \tag{1.4}$$

*Example 1.2* (*Continuous-time system with a cubic nonlinearity*) Consider the dynamics $\dot{x} = F(x) = -x - x^3$, $x \in \mathbb{R}$. An eigenfunction of the associated Koopman operator, solution to (1.4), is given by

$$\phi_\lambda(x) = \frac{x}{\sqrt{1+x^2}}$$

and the corresponding eigenvalue is $\lambda = -1$. In this case, the eigenfunction is not linear, and thereby captures the nonlinearity of the system. △

The eigenvalues belong to the point spectrum $\sigma_p$ of the Koopman operator, which is generally an infinite set. If $\mu_1, \mu_2 \in \sigma_p$, then $\mu_1^{j_1}\mu_2^{j_2} \in \sigma_p$ for all $j_1, j_2 \in \mathbb{R}$ provided that the associated eigenfunction $\phi_{\mu_1}^{j_1}\phi_{\mu_2}^{j_2}$ belongs to $\mathscr{F}$. It follows that the set of eigenfunctions can be seen as an algebra generated by the principal eigenfunctions. In the continuous-time setting, one has similarly that $\lambda_1, \lambda_2 \in \sigma_p$ implies $j_1\lambda_1 + j_2\lambda_2 \in \sigma_p$, provided that $\phi_{\lambda_1}^{j_1}\phi_{\lambda_2}^{j_2} \in \mathscr{F}$.

The Koopman operator may also admit a continuous spectrum.[5] Typically, the Koopman operator associated with chaotic systems has a non-empty continuous spectrum. A non-empty continuous spectrum can also be observed in the following example.

*Example 1.3* (*Continuous spectrum* [47]) Consider the dynamics

$$\dot{r} = 0$$
$$\dot{\theta} = r$$

with $r \in \mathbb{R}^+$ and $\theta \in \mathbb{S}$. In the $L^2(\mathbb{R}^+ \times \mathbb{S})$ space, the associated Koopman semigroup of operators has a point spectrum $\sigma_p = \{0\}$ and a continuous spectrum $\sigma_c = \mathrm{i}\mathbb{R} \setminus \{0\}$. The continuous spectrum is associated with generalized eigenfunctions of the form $e^{\mathrm{i}\theta}\,\delta(r - \overline{r})$ for all $\overline{r} > 0$. △

**Generalized Laplace Averages**

Koopman eigenfunctions can be obtained through the so-called *generalized Laplace averages* (GLA) [7, 45, 51]. For a discrete-time system, suppose that $\mu_1, \ldots, \mu_j$ are simple Koopman eigenvalues with $1 \geq |\mu_1| \geq |\mu_2| \geq \cdots \geq |\mu_j|$ and there is no other eigenvalue $\mu$ such that $|\mu| \geq |\mu_j|$. For a bounded, continuous observable $f$, the generalized Laplace average is given by

$$f_j^*(\mathbf{x}) = \lim_{n \to \infty} \frac{1}{K} \sum_{k=0}^{K-1} \mu_j^{-k} \left( f(\mathbf{S}^k(\mathbf{x})) - \sum_{i=1}^{j-1} \mu_i^k f_i^*(\mathbf{x}) \right).$$

Note that a similar definition can be given in the continuous-time case. The function $f_j^*$ is equal to the projection of $f$ onto the space spanned by the eigenfunction $\phi_{\mu_j}$[6] and is therefore an eigenfunction. In practice, it is difficult to compute the GLA numerically since it requires to multiply very small values with very large ones. However, dominant eigenfunctions (i.e., $\phi_{\mu_1}$) can be computed efficiently with GLA, yielding Fourier averages [40] if $|\mu_1| = 1$ and Laplace averages [42] when $|\mu_1| < 1$ (see also Chaps. 11 and 15). There exist other numerical methods to compute the Koopman eigenfunctions, which are based on finite-dimensional approximations of the operator and also yield data-driven methods (Sect. 1.4).

---

[5]The spectrum of $U$ is composed of three parts:

- the *point spectrum* $\sigma_p$ is the set of values $\lambda \in \mathbb{C}$ such that $U - \lambda I$ is not injective;
- the *continuous spectrum* $\sigma_c$ is the set of values $\lambda \in \mathbb{C} \setminus \sigma_p$ such that the operator $U - \lambda I$ is not surjective and has a dense image;
- the *residual spectrum* $\sigma_r$ is the set of values $\lambda \in \mathbb{C} \setminus \sigma_p$ such that the operator $U - \lambda I$ is not surjective and does not have a dense image.

[6]More precisely, we have $f_j^* = v_j \phi_{\mu_j}$, where $v_j$ is the Koopman mode (see Sect. 1.3.3).

## 1.3.2 Relationship with Geometric Properties

In Sect. 1.2.2.3, it has been shown that the eigenfunctions of the Koopman operator associated with the eigenvalue $\mu = 1$ (discrete time), or equivalently $\lambda = 0$ (continuous time), are related to invariant sets. More precisely, the level sets of these eigenfunctions define a partition of invariant sets, called the ergodic partition [50]. Similar results can be obtained for other eigenfunctions. According to the associated eigenvalue, we distinguish two classes of eigenfunctions: *conservative eigenfunctions* and *dissipative eigenfunctions*. The former is related to properties of the dynamics on the attractor while the latter captures properties of the dynamics off the attractor.

### 1.3.2.1 Conservative Eigenfunctions

Conservative eigenfunctions are associated with eigenvalues $\mu = e^{\pm i\omega}$ (discrete time) or $\lambda = \pm i\omega$ (continuous time), with $\omega \in \mathbb{R}^+$. With a slight abuse of notation, we denote them by $\phi_{i\omega}$ for both discrete and continuous-time cases. These eigenfunctions capture the dynamics of conservative (measure-preserving) systems, or the dynamics of nonconservative systems (i.e., dissipative in the sense of [34]) on the attractor.

The absolute value of a conservative eigenfunction is constant along the orbits of the system, and thus on each ergodic subset of the state space. This implies that all the information on the dynamics is retained in the argument of the eigenfunction, which defines a phase coordinate $\theta = \angle\phi_{i\omega} : X \to \mathbb{S}$ in the state space. Thus, as shown in [44], the level sets of the eigenfunction yield a *periodic partition* of the state space.

**Proposition 1.2** (Periodic partition) *Suppose that the Koopman operator $U$ associated with the non-singular transformation $\mathbf{S} = X \to X$ admits an eigenfunction $\phi_{i\omega}$ with the eigenvalue $e^{i\omega}$, $\omega \in \mathbb{R}^+$. Then, the sets*

$$I_\theta = \{\mathbf{x} \in X : \angle\phi_{i\omega}(\mathbf{x}) = \theta\}\,, \qquad \theta \in \mathbb{S}$$

*define a periodic partition: $\mathbf{S}^k(I_\theta) = I_{\theta+k\omega}$ and, in particular, $\mathbf{S}^k(I_\theta) = I_\theta$ if $k = 2\pi/\omega$.*

*Proof* The result follows from the equality

$$\angle\phi_{i\omega}(\mathbf{S}^k(\mathbf{x})) = \angle(U^k\phi_{i\omega})(\mathbf{x}) = \angle\left(\phi_{i\omega}(\mathbf{x})\, e^{ik\omega}\right) = \theta + k\omega\,.$$

□

The result is summarized in Fig. 1.1a. A similar result also holds for continuous-time systems.

In the case of nonconservative systems, conservative eigenfunctions capture the asymptotic behavior of the system on the attractor. For continuous-time systems that

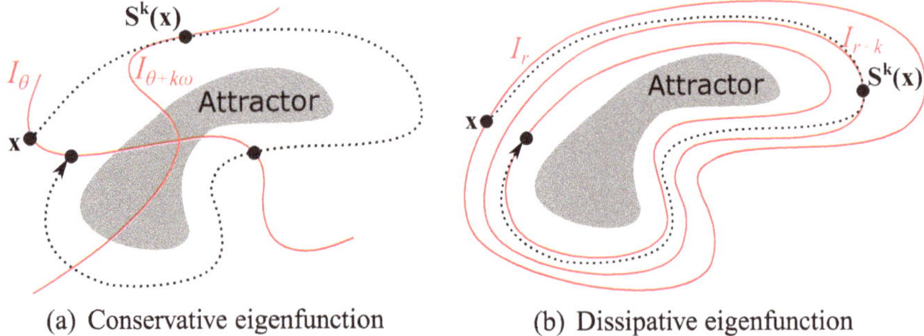


(a) Conservative eigenfunction (b) Dissipative eigenfunction

**Fig. 1.1** Level sets of Koopman eigenfunctions. **a** The level sets of conservative eigenfunctions yield a periodic partition of the state space. **b** The level sets of dissipative eigenfunctions yield a partition that captures the convergence toward the attractor

admit a limit cycle with fundamental frequency $\omega$, there exists an eigenfunction $\phi_{\mathrm{i}\omega}$ whose level sets correspond to the so-called *isochrons* [40].

#### 1.3.2.2 Dissipative Eigenfunctions

Dissipative eigenfunctions are associated with eigenvalues $\mu$ such that $|\mu| = e^s < 1$ (discrete time) or $\lambda$ such that $\mathrm{Re}\{\lambda\} = s < 0$ (continuous time), with $s \in \mathbb{R}^-$. With a slight abuse of notation, we denote them by $\phi_s$ for both discrete and continuous-time cases. The values of dissipative eigenfunctions decrease exponentially along the orbits of the system. For conservative measure-preserving systems, this implies that they should be equal to zero almost everywhere in each element of the ergodic partition. Indeed, Birkhoff's ergodic theorem yields the equality

$$\int_X \phi_s(\mathbf{x})\, d\mu^*(\mathbf{x}) = \lim_{K\to\infty} \frac{1}{K} \sum_{k=1}^{K} U^k \phi_s = 0\,.$$

In fact, dissipative eigenfunctions are associated with systems that are not conservative and capture the transient behavior of the system off the attractor. Their value is equal to zero (almost everywhere) on the attractor.

The absolute value of a dissipative eigenfunction defines an "amplitude coordinate" $r = |\phi_s| : X \to \mathbb{R}^+$ in the state space that complements phase coordinates related to conservative eigenfunctions. This coordinate provides a measure of the time needed to converge toward the attractor.

**Proposition 1.3** *Suppose that the Koopman operator $U$ associated with the non-singular transformation $\mathbf{S} = X \to X$ admits an eigenfunction $\phi_s$ with the eigenvalue $e^s$, $s \in \mathbb{R}^-$. Then, the sets*

$$I_r = \{\mathbf{x} \in X : |\phi_s(\mathbf{x})| = e^{rs}\}\,, \qquad r \in \mathbb{R}^+ \tag{1.5}$$

*define a partition of $X$ that satisfies $\mathbf{S}^k(I_r) = I_{r+k}$.*

*Proof* The result follows from the equality

$$|\phi_s(\mathbf{S}^k(\mathbf{x}))| = |(U^k\phi_s)(\mathbf{x})| = |\phi_s(\mathbf{x})\, e^{ks}| = e^{(r+k)s}\,.$$

□

The result is summarized in Fig. 1.1b. A similar result also holds for continuous-time systems.

**Equilibria and Limit Cycles**

For systems that admit an (hyperbolic) equilibrium or limit cycle, there exist continuous dissipative eigenfunctions associated with the eigenvalues of the linearized system at the equilibrium or with the Floquet exponents of the limit cycle [47], respectively. These eigenfunctions yield linearizing coordinates of the system [33] and capture the global stability properties of the attractor [41] (see also Chap. 2). In addition, their zero level sets are related to the stable and unstable manifolds of the attractor [46]. The partition (1.5) defined with the dissipative eigenfunction associated with the dominant (stable) eigenvalue yields a set of *isostables* that complements the set of isochrons [42]. These sets can be used to obtain action-angle representations of fixed point and limit-cycle dynamics.

### *1.3.3 Koopman Modes*

The (infinite set of) Koopman eigenfunctions can provide a complete basis of the space $\mathscr{F}$. For example, this holds when the dynamics are integrable and defined in a compact space. In this case, every observable that belongs to $\mathscr{F}$ can be expanded in terms of the eigenfunctions. More generally, this expansion leads to the notion of *Koopman modes*, which are isolated from the continuous spectrum [44, 61].

**Definition 1.5** (*Koopman mode expansion*) Suppose that the span of Koopman eigenfunctions $\{\phi_{\mu_j}\}_{j=1}^{\infty}$ densely fills the space $\mathscr{G} \subseteq \mathscr{F}$. The *Koopman mode expansion* of $f \in \mathscr{G}$ is given by

$$f = \sum_{j=1}^{\infty} v_j\, \phi_{\mu_j}\,. \tag{1.6}$$

The coefficients $v_j$ are the *Koopman modes* related to the observable $f$. △

This definition is valid for both discrete-time and continuous-time systems. Note also that if the observable $\mathbf{f}$ is vector-valued, then the Koopman mode $\mathbf{v}_j$ is a vector.

And more generally, if $f$ is a function from some configuration space to $\mathbb{C}$, then $v_j$ is a function [44].

The Koopman modes are defined with respect to Koopman eigenfunctions associated with specific dynamics, and therefore they also capture some information on these underlying dynamics. Using the Koopman mode expansion and the definition of Koopman eigenfunctions, we can write

$$U^k f = \sum_{j=1}^{\infty} v_j \, \mu_j^k \, \phi_{\mu_j} \tag{1.7}$$

for discrete-time systems, and similarly

$$U^t f = \sum_{j=1}^{\infty} v_j \, e^{\lambda_j t} \, \phi_{\lambda_j}$$

for continuous-time systems. It becomes clear that the modulus $|v_j|$ and argument $\angle v_j$ yield the amplitude and phase, respectively, of a specific oscillation mode in the evolution of the observable $f$. For the identity function $\mathrm{id}(\mathbf{x}) = \mathbf{x}$, the Koopman mode expansion provides the orbits of the system for all initial conditions (the dependence on the initial condition is given by the eigenfunctions). This is illustrated in the following examples.

*Example 1.4* (*Linear discrete-time system (continued)*) Consider again the linear transformation $\mathbf{S}(\mathbf{x}) = \mathbf{A}\mathbf{x}$ with $\mathbf{x} \in \mathbb{R}^n$ (Example 1.1). An orbit of the system is given by

$$\mathbf{S}^k(\mathbf{x}) = \sum_{j=1}^{n} \mathbf{v}_j \, (\mathbf{w}_j^T \mathbf{x}) \, \mu_j^k,$$

where $\mathbf{v}_j$ and $\mathbf{w}_j$ are the right and left eigenvectors of $\mathbf{A}$, respectively. This expression is the evolution of the identity function $\mathrm{id}(\mathbf{x}) = \mathbf{x}$ under the action of the Koopman operator. Comparing it with (1.7), we recover the fact that $\phi_{\mu_j}(\mathbf{x}) = \mathbf{w}_j^T \mathbf{x}$ and we observe that the Koopman modes associated with the identity function are the right eigenvectors $\mathbf{v}_j$. We note that the Koopman mode expansion is finite, owing to the linear dynamics and the specific observable id that is also linear in state. △

*Example 1.5* (*Continuous-time system with a cubic nonlinearity (continued)*) Consider again the dynamics $\dot{x} = F(x) = -x - x^3$, $x \in \mathbb{R}$, which induces the flow

$$S^t(x) = \frac{x e^{-t}}{\sqrt{1 + x^2 - x^2 e^{-2t}}} \, .$$

Using the change of variable $y = e^{-t}$ and computing the Taylor expansion of $xy/\sqrt{1 + x^2 - x^2 y^2}$ around $y = 0$, we obtain the expansion

$$S^t(x) = \sum_{\substack{j=1 \\ j \text{ odd}}}^{\infty} \frac{j!!(j-2)!!}{j!} \left( \frac{x}{\sqrt{1+x^2}} \right)^j y^j = \sum_{\substack{j=1 \\ j \text{ odd}}}^{\infty} \frac{j!!(j-2)!!}{j!} \left( \frac{x}{\sqrt{1+x^2}} \right)^j e^{-jt},$$

where $j!! = j(j-2)(j-4)\ldots 3$ denotes the double factorial. We recover the principal eigenfunction $\phi_\lambda(x) = x/\sqrt{1+x^2}$ associated with $\lambda = 1$ (Example 1.2) and its "harmonics" $\phi_\lambda^j$, $j \in \mathbb{N}$, which are also eigenfunctions by the algebra property (see Sect. 1.3.1). For an analytic observable $f$, the Taylor expansion around the origin yields the Koopman mode expansion

$$\begin{aligned}(U^t f)(x) = f \circ S^t(x) &= f(0) + f'(0) S^t(x) + \frac{f''(0)}{2} (S^t(x))^2 + \frac{f^{(3)}(0)}{3!} (S^t(x))^3 + \cdots \\ &= f(0) + f'(0) \frac{x}{\sqrt{1+x^2}} e^{-t} + \frac{f''(0)}{2} \left( \frac{x}{\sqrt{1+x^2}} \right)^2 e^{-2t} \\ &\quad + \left( \frac{f'(0)}{2} + \frac{f'''(0)}{3!} \right) \left( \frac{x}{\sqrt{1+x^2}} \right)^3 e^{-3t} + R(x)\end{aligned}$$

where $R(x)$ stands for the higher order terms (see also [13] for general formulae). It follows that the Koopman modes are given by

$$v_0 = f(0) \qquad v_1 = f'(0) \qquad v_2 = f''(0)/2 \qquad v_3 = f'(0)/2 + f'''(0)/3! \qquad \cdots$$

$\triangle$

In the above example, the Koopman mode expansion holds for analytic observables. In the case of other functional spaces such as $C^k$, the Koopman mode expansion may be finite, in which case a remainder $R(x)$ of the Taylor series accounts for the continuous spectrum (see [13] for more details).

The Koopman modes are related to the projections of $f$ onto the eigenfunctions: we have

$$\Pi_j f = v_j \phi_{\mu_j}$$

with the projection operator $\Pi_j$ defined by $\Pi_j \phi_{\mu_i} = \delta_{ij} \phi_{\mu_j}$ (where $\delta_{ij}$ is the Kronecker delta). This spectral projection can be obtained through the eigenfunctional $\xi_{\mu_j}$ of the dual operator $U^*$ to $f$, which satisfies $U^* \xi_{\mu_j} = \mu_j \xi_{\mu_j}$ and is normalized so that $\xi_{\mu_j}(\phi_{\mu_i}) = \delta_{ij}$. In particular, if $f$ admits the Koopman mode expansion (1.6), we have

$$\xi_{\mu_j}(f) = \sum_{i=1}^{\infty} v_i \, \xi_{\mu_j}(\phi_{\mu_i}) = v_j \,. \tag{1.8}$$

In the case of Example 1.5, it follows that $\xi_0(f) = f(0)$, $\xi_\lambda(f) = f'(0)$, $\xi_{2\lambda}(f) = f''(0)/2$, and $\xi_{3\lambda}(f) = f'(0)/2 + f'''(0)/3!$ (see also [13]). These eigenfunctionals only capture local properties in the vicinity of the equilibrium and the associated eigenfunctions of the Perron–Frobenius operator are not well defined (they are the

Dirac functions and its derivatives). This contrasts with the fact that Koopman eigenfunctions are smooth functions. From this perspective, it is much more appropriate to use the Koopman operator in the case of nonconservative systems.

We finally note that the Koopman modes can be computed from the time series generated by the system dynamics. These data-driven methods rely on finite-dimensional approximations of the Koopman operator. This is the focus of the next section.

## 1.4 Finite-Dimensional Approximations: Toward Numerical Methods

Since the Koopman operator is infinite-dimensional, it is necessary to consider a finite-dimensional approximation, as long as numerical methods are concerned. This approximation is at the basis of most data-driven methods to compute the spectral properties of the operator. It is also useful for prediction purposes and serves as a crucial step in numerical methods for control.

### *1.4.1 Finite-Dimensional Approximation of the Koopman Operator*

In this section, we derive the matrix representation of the Koopman operator projected onto a finite-dimensional subspace. We also consider the dual operator and the spectral properties of the finite-dimensional representation. Some developments are inspired from [10, 75].

#### 1.4.1.1 Representation of the Operator in a Finite-Dimensional Subspace

Consider an $N$-dimensional linear subspace $\mathscr{F}_N \subset \mathscr{F}$ spanned by the basis functions $\{\psi_j\}_{j=1}^N$. Consider also a projection operator $\Pi : \mathscr{F} \to \mathscr{F}_N$. For a given subspace $\mathscr{F}_N$ and projection operator $\Pi$, a finite-dimensional approximation of the Koopman operator $U$ is given by the compression

$$U_N = \Pi \, U\big|_{\mathscr{F}_N} : \mathscr{F}_N \to \mathscr{F}_N \, . \tag{1.9}$$

The projection operator can be decomposed as

$$\Pi = \Psi^T \Gamma \tag{1.10}$$

where the operator $\Gamma : \mathscr{F} \to \mathbb{C}^N$ can be seen as a bounded linear map[7] that yields the coordinates of $f$ in the basis $\{\psi_j\}_{j=1}^N$ and $\Psi^T : \mathbb{C}^N \to \mathscr{F}_N$ is given by $\Psi^T \mathbf{a} = (\psi_1 \cdots \psi_N)\,\mathbf{a}$. We note that

$$\Gamma \Psi^T = \mathbf{I}\,, \tag{1.11}$$

which implies $\Gamma f = \Gamma \Psi^T \Gamma f = \Gamma \Pi f$ so that $\Gamma$ is well defined for all $f \in \mathscr{F}$. In particular, if $\mathscr{F}$ is a Hilbert space and $\Pi$ is an orthogonal projection, then $\Gamma f = \mathbf{0}$ for all $f \perp \mathscr{F}_N$.

Injecting (1.10) in (1.9), we obtain

$$U_N f = \Pi U \Pi f = \Psi^T \Gamma U \Psi^T \Gamma f = \Psi^T \mathbf{U} \Gamma f \qquad f \in \mathscr{F}_N, \tag{1.12}$$

where

$$\mathbf{U} : \mathbb{C}^N \to \mathbb{C}^N\,, \qquad \mathbf{U}\,\mathbf{a} = \Gamma U \Psi^T \mathbf{a}$$

is the matrix representation of $U_N$. It is called the *Koopman matrix* of the system and corresponds to the action of Koopman operator on observables $f \in \mathscr{F}_N$ in the coordinates $\Gamma f$ related to the basis $\{\psi_j\}_{j=1}^N$. Since $\Psi^T \Gamma f = \Pi f = f$ for $f \in \mathscr{F}_N$, it also follows from the last equality of (1.12) that

$$\Gamma U f = \mathbf{U}\,\Gamma f \qquad f \in \mathscr{F}_N\,. \tag{1.13}$$

For $f = \psi_j$, we have $\Gamma \psi_j = \mathbf{e}_j$, where $\mathbf{e}_j$ is the $j$th unit vector, and (1.13) implies that

$$\mathbf{U}\,\mathbf{e}_j = \Gamma U \psi_j\,. \tag{1.14}$$

Hence the $j$th column of $\mathbf{U}$ contains the coordinates of $\Pi U \psi_j = U_N \psi_j$ in the basis of functions, i.e., $U_N \psi_j = \sum_{l=1}^N \mathbf{U}_{ij} \psi_i$

*Example 1.6* Consider the transformation $S(x) = 2x - x^2$, $x \in \mathbb{R}$. We choose the subspace $\mathscr{F}_4$ spanned by the basis functions $\{\psi_j\}_{j=1}^4 = \{x^{j-1}\}_{j=1}^4$ and the (non-orthogonal) projection that satisfies

$$\Pi x^j = \begin{cases} x^j & \text{if } j \leq 3 \\ 0 & \text{if } j > 3 \end{cases}. \tag{1.15}$$

Using (1.14) with $\Gamma x^j = \mathbf{e}_{j+1} \in \mathbb{R}^4$ for $j \leq 3$ and $(U\psi_j)(x) = \left(2x - x^2\right)^{j-1}$, we obtain the Koopman matrix

$$\mathbf{U} = \begin{pmatrix} 1 & 0 & 0 & 0 \\ 0 & 2 & 0 & 0 \\ 0 & -1 & 4 & 0 \\ 0 & 0 & -4 & 8 \end{pmatrix}.$$

[7] If $\mathscr{F}$ is a Hilbert space, one can consider inner products instead.

The finite-dimensional representation obtained here with the projection (1.15) corresponds to the finite section method. We note that the representation of $U\psi_j$ is cut off for $j = 3, 4$. △

The above developments can also be applied to the continuous-time setting. In this case, it suffices to replace $U$ by $U^t$ for some time $t > 0$. Equivalently, for continuous-time systems of the form $\dot{\mathbf{x}} = \mathbf{F}(\mathbf{x})$, one can compute the matrix representation $\mathbf{L}$ of the compression $L_N = \Pi L|_{\mathscr{F}_N}$ of the infinitesimal generator (1.1) (see also [4, 41, 71] and Example 1.7 below).

#### 1.4.1.2 Dual Formulation and Lifted Dynamics

The dual operator $U_N^* : \mathscr{F}_N^* \to \mathscr{F}_N^*$ (where $\mathscr{F}_N^*$ is the set of linear bounded functionals on $\mathscr{F}_N$) is defined by

$$\xi(U_N f) = (U_N^* \xi)(f) , \qquad \forall f \in \mathscr{F}_N , \xi \in \mathscr{F}_N^* .$$

Using (1.12) and linearity of $\xi$, we obtain

$$(U_N^* \xi)(f) = \xi(U_N f) = \xi(\Psi^T \mathbf{U} \Gamma f) = \xi(\Psi)^T \mathbf{U} \Gamma f \qquad f \in \mathscr{F}_N$$

with $\xi(\Psi)^T = (\xi(\psi_1) \cdots \xi(\psi_N))$ (this is the vector of moments if the basis functions are monomials). Replacing $f$ by $\Psi$ in the above equality and using (1.11), we obtain that the dual to (1.13) is given by

$$(U_N^* \xi)(\Psi) = \xi(\Psi)^T \mathbf{U} \Gamma \Psi = \xi(\Psi)^T \mathbf{U} \mathbf{I} = \mathbf{U}^T \xi(\Psi) . \tag{1.16}$$

As we would expect, the matrix representation of $U_N^*$ is the transpose $\mathbf{U}^T$. It corresponds to the action of the dual operator on linear bounded functionals $\xi \in \mathscr{F}_N^*$ expressed through their values $\xi(\Psi)$ (e.g., moments). Note that the values $\xi(\Psi)$ are the coordinates of the functional $\xi$ in a basis of functionals $\{\xi_i\}_{i=1}^N$ with $\xi_i(\psi_j) = \delta_{ij}$.

**Lifted Dynamics**

In the space of continuous functions, if we consider the evaluation functional $\xi_{\mathbf{x}}(f) = f(\mathbf{x})$ for some $\mathbf{x} \in X$, we obtain

$$\Pi \Psi(\mathbf{S}(\mathbf{x})) = (U_N \Psi)(\mathbf{x}) = (U_N^* \xi_{\mathbf{x}})(\Psi) = (\mathbf{U}^T \Psi)(\mathbf{x})$$

so that successive iterates of $\mathbf{U}^T$ provide an approximation of the values of $\Psi$ along an orbit of the system. Defining the *lifted state* $\mathbf{z} = \Psi(\mathbf{x})$ and neglecting the projection, we get the (approximate) linear *lifted dynamics*

$$\mathbf{z}(k+1) = \mathbf{A}\, \mathbf{z}(k) \tag{1.17}$$

with $\mathbf{A} = \mathbf{U}^T$, which is used in several numerical methods for prediction and control based on the Koopman operator. Note that the lifting space is typically chosen of dimension much larger than the dimension of the underlying attractor of the system.

In the continuous-time setting, we obtain similarly

$$\Pi\dot{\Psi}(\mathbf{x}) = (L_N \Psi)(\mathbf{x}) = (L_N^* \xi_{\mathbf{x}})(\Psi) = (\mathbf{L}^T \Psi)(\mathbf{x}) \, ,$$

which leads to the lifting dynamics

$$\dot{\mathbf{z}} = \mathbf{A}\,\mathbf{z} \tag{1.18}$$

with $\mathbf{A} = \mathbf{L}^T$.

*Remark 1.1* The output of the system can be considered as a particular observable $h$, that is $y = h(\mathbf{x})$. In this case, the lifted dynamics (1.17) or (1.18) are complemented with the equality

$$y = \mathbf{C}^T\,\mathbf{z}$$

where $\mathbf{C} = \Gamma h$. △

#### 1.4.1.3 Spectral Properties

The hope is that the eigenfunctions and eigenvalues of $U_N$, which we denote by $\tilde{\phi}_{\mu_j}$ and $\tilde{\mu}_j$ respectively, approximate those of $U$ (see [28] for theoretical results in the case of systems with an ergodic invariant measure on the attractor). We remark that $\tilde{\phi}_{\mu_j}$ is not necessarily a projection of $\phi_{\mu_j}$. In fact, we have for instance $\tilde{\phi}_{\mu_j} = \Pi\phi_{\mu_j} = \phi_{\mu_j}$ when $U$ commutes with $\Pi$, in which case the subspace $\mathscr{F}_N$ is invariant under the action of $U$. The equality $U_N\tilde{\phi}_{\mu_j} = \tilde{\mu}_j\tilde{\phi}_{\mu_j}$ with (1.10) and (1.12) implies that $\Psi^T\mathbf{U}\Gamma\tilde{\phi}_{\mu_j} = \tilde{\mu}_j\Psi^T\Gamma\tilde{\phi}_{\mu_j}$, or equivalently

$$\mathbf{U}\Gamma\tilde{\phi}_{\mu_j} = \tilde{\mu}_j\,\Gamma\tilde{\phi}_{\mu_j} \, .$$

Thus the eigenvalues of $U_N$ are the eigenvalues of the Koopman matrix $\mathbf{U}$ and the coordinates of the corresponding eigenfunctions in the basis of functions are the right eigenvectors of $\mathbf{U}$. In the case of continuous-time systems, the Koopman eigenvalues and eigenfunctions are obtained similarly by computing the eigenvalues and eigenvectors of $\mathbf{L}$. This is illustrated with the following example.

*Example 1.7* Consider the dynamics $\dot{x} = -x - x^3$, $x \in \mathbb{R}$. Similar to Example 1.6, we choose the subspace $\mathscr{F}_4$ spanned by the basis functions $\{\psi_j\}_{j=1}^4 = \{x^{j-1}\}_{j=1}^4$ and the same (non-orthogonal) projection. We compute the matrix representation $\mathbf{L}$ of the infinitesimal generator $(Lf)(x) = (-x - x^3)df/dx$. Using (1.14) (where $U$ is replaced by $L$) with $(L\psi_j)(x) = (j-1)(-x^{j-1} - x^{j+1})$, we obtain

$$\mathbf{L} = \begin{pmatrix} 0 & 0 & 0 & 0 \\ 0 & -1 & 0 & 0 \\ 0 & 0 & -2 & 0 \\ 0 & -1 & 0 & -3 \end{pmatrix}.$$

The eigenvalues of $\mathbf{L}$ are $0, -1, -2, -3$ and correspond to the exact Koopman eigenvalues (see Example 1.2). The right eigenvector associated with the eigenvalue $-1$ yields the eigenfunction $\tilde{\phi}_{-1}(x) = 0.894\,x - 0.0447\,x^3$. This is an approximation of the Koopman eigenfunction $\phi_{-1}$ which is mainly valid close to the origin. Note that we have considered here the Carleman realization of the Koopman operator since the basis consists of monomials [8]. △

The left eigenvectors of $\mathbf{U}$ are associated with the eigenfunctionals $\tilde{\xi}_{\mu_j}$ satisfying $U_N^* \tilde{\xi}_{\mu_j} = \tilde{\mu}_j \tilde{\xi}_{\mu_j}$: using (1.16) and evaluating the previous expression at $\Psi$, we have

$$\tilde{\xi}_{\mu_j}(\Psi)^T \mathbf{U} = \tilde{\mu}_j \, \tilde{\xi}_{\mu_j}(\Psi)^T .$$

The eigenvectors can be normalized so that

$$\tilde{\xi}_{\mu_j}(\tilde{\phi}_{\mu_i}) = \tilde{\xi}_{\mu_j}(\Psi)^T \Gamma \tilde{\phi}_{\mu_i} = \delta_{ij}$$

and according to (1.8), the left eigenvectors $\tilde{\xi}_{\mu_j}(\Psi)$ of $\mathbf{U}$ are the (approximate) Koopman modes of the vector-valued observable $\Psi$. Moreover, the Koopman modes of an observable $f = \Psi^T \Gamma f$ are approximated by

$$\tilde{v}_j = \tilde{\xi}_{\mu_j}(f) = \tilde{\xi}_{\mu_j}(\Psi)^T \Gamma f .$$

### *1.4.2 Data-Driven Methods*

The data-driven methods to obtain the spectral properties of the dynamics (e.g., Koopman mode decomposition and Koopman eigenfunctions) can be separated into those aiming to devise a finite-dimensional matrix approximation of the Koopman operator, described in Sect. 1.4.1, such as the *(extended) dynamic mode decomposition* (E)DMD [61, 62, 73, 75], and methods based on generalized Laplace averages (GLA) [7, 44, 45, 48, 51]. GLA methods do not provide an approximation of the operator, but first seek an approximation to eigenvalues and then use projection theorems to obtain eigenfunctions, and modes [3] (see also Sect. 1.3.1).

The original algorithms [61, 62] use observable data sampled along a trajectory of the system. The Arnoldi-type[8] algorithm presented in [61] constructs a companion matrix that attempts to approximate the Koopman operator, utilizing Krylov

[8]As noted in Chap. 7, the name *Arnoldi* is misplaced and the terminology *Krylov–Rayleigh–Ritz procedure* should have been preferred.

subspaces of the state space. Recent development of Arnoldi-type methods, called Hankel DMD, is the use of Krylov subspaces in observable space, where the basis functions are given by $\psi_j = U^j f$ $(j = 0, 1, \ldots)$ for some observable $f$ [3, 69]. In contrast, the original (SVD-enhanced) DMD method [62] uses linear basis functions $\psi_j(\mathbf{x}) = x_j$ in the state space. The EDMD method uses an extended basis and does not need to sample data along a trajectory, but can use one-step iteration [75]. We refer to Chap. 7 for a general overview of Arnoldi-type and SVD-enhanced DMD methods and their connections.

#### 1.4.2.1 Arnoldi-Type Method

Suppose we have the finite-length evenly sampled time series $\{\mathbf{x}_0, \mathbf{x}_1, \ldots, \mathbf{x}_{M-1}\}$, where $\mathbf{x}_k \in \mathbb{R}^n$ is the $k$th snapshot of the vector of observables $\mathbf{x}$ (i.e., $\mathbf{x}_k = \mathbf{S}^k(\mathbf{x}_0)$) and $M$ the number of available snapshots. Then, the so-called *empirical Ritz values* $\tilde{\mu}_j$ and *empirical Ritz vectors* $\tilde{\mathbf{v}}_j$ of the data are defined with the following algorithm:

1. Compute the constant vector $\mathbf{c} = (c_0, \ldots, c_{M-2})^T$ such that, for some $\mathbf{r} \in \mathbb{R}^n$ satisfying $\mathbf{r} \perp \operatorname{span}\{\mathbf{K}_{M-1}\}$, we have

$$\mathbf{r} = \mathbf{x}_{M-1} - \mathbf{K}_{M-1}\,\mathbf{c} \tag{1.19}$$

   with the *Krylov subspace* defined as

$$\mathbf{K}_{M-1} := (\mathbf{x}_0, \ldots, \mathbf{x}_{M-2}) \,. \tag{1.20}$$

   The vector $\mathbf{c}$ is given by

$$\mathbf{c} = \mathbf{K}_{M-1}^{+}\mathbf{x}_{M-1} \,, \tag{1.21}$$

   where $\mathbf{K}_{M-1}^{+}$ is the Moore–Penrose pseudoinverse of $\mathbf{K}_{M-1}$.
2. Define the matrix $\mathbf{C}_{M-1}$ as

$$\mathbf{C}_{M-1} := \begin{pmatrix} 0 & 0 & \cdots & 0 & c_0 \\ 1 & 0 & \cdots & 0 & c_1 \\ 0 & 1 & \cdots & 0 & c_2 \\ \vdots & \vdots & \ddots & \vdots & \vdots \\ 0 & 0 & \cdots & 1 & c_{M-2} \end{pmatrix},$$

   called the *companion matrix*. Then, denote its $M-1$ eigenvalues as the empirical Ritz values $\tilde{\mu}_1, \ldots, \tilde{\mu}_{M-1}$.
3. Define the *Vandermonde* matrix $\mathbf{T}$ using $\tilde{\mu}_j$ as

$$\mathsf{T} := \begin{pmatrix} 1 & \tilde{\mu}_1 & \tilde{\mu}_1^2 & \cdots & \tilde{\mu}_1^{M-2} \\ 1 & \tilde{\mu}_2 & \tilde{\mu}_2^2 & \cdots & \tilde{\mu}_2^{M-2} \\ \vdots & \vdots & \vdots & \ddots & \vdots \\ 1 & \tilde{\mu}_{M-1} & \tilde{\mu}_{M-1}^2 & \cdots & \tilde{\mu}_{M-1}^{M-2} \end{pmatrix} . \tag{1.22}$$

4. Define the empirical Ritz vectors $\tilde{\mathbf{v}}_j$ to be the columns of $\mathbf{V} := \mathbf{K}_{M-1}\mathbf{T}^{-1}$. Note that the columns of $\mathbf{T}^{-1}$ are the dual basis to the basis composed of the rows of $\mathbf{T}$. Each row of $\mathbf{T}$ can be interpreted as the set of values of an eigenfuction corresponding to the eigenvalue $\mu_j$ on the trajectory of the system in state space. Thus each column $j$ of $\mathbf{K}_{M-1}\mathbf{T}^{-1}$ is the projection of the vector of observables onto the eigenspace spanned by the eigenfunction associated with the eigenvalue $\mu_j$. If $\mu_j$ approximates the true eigenvalue of the Koopman operator, then $\mathbf{v}_j$ is the corresponding approximation to the Koopman mode.

It is shown in [61] that if all of the empirical Ritz values are nonzero and distinct, then the following decompositions of the data are obtained

$$\mathbf{x}_k = \sum_{j=1}^{M-1} \tilde{\mu}_j^k \tilde{\mathbf{v}}_j, \quad k = 0, \ldots, M-1, \qquad \mathbf{x}_{M-1} = \sum_{j=1}^{M-1} \tilde{\mu}_j^{M-1} \tilde{\mathbf{v}}_j + \mathbf{r} . \tag{1.23}$$

Comparing with (1.7), the empirical Ritz values $\tilde{\mu}_j$ and vectors $\tilde{\mathbf{v}}_j$ behave precisely in the same manner as the Koopman eigenfunctions $\mu_i$ and the terms $\phi_i(\mathbf{x}_0)\mathbf{v}_i$ containing the Koopman eigenfunctions and Koopman modes, but for the finite sum (1.23) instead of the infinite sum (1.7). If the data are generated by a continuous-time system with a sampling time $\Delta t$ (i.e., $\mathbf{x}_k = \mathbf{S}^{k\Delta t}(\mathbf{x}_0)$), continuous-time eigenvalues are given by $\lambda_k = \log(\mu_k)/\Delta t$.

Here, the matrix representation of the operator $U_N = \Pi U$ on the subspace spanned by observables $\mathbf{x}$, is given by the transpose of the companion matrix $\mathbf{C}_{M-1}^T$.[9]

Only the last column is obtained in (1.21) through the discrete orthogonal projection $\Pi U^M f$, where $\Pi : \mathscr{F} \to \mathscr{F}_N$ yields the least squares fit at the points $\mathbf{x}_k$, $k = 1, \ldots, M$:

$$\Pi_N g = \underset{\tilde{g} \in \text{span}\{\psi_1, \ldots, \psi_N\}}{\text{argmin}} \sum_{k=1}^{M} |\tilde{g}(\mathbf{x}_k) - g(\mathbf{x}_k)|^2 . \tag{1.24}$$

We also note that the Arnoldi algorithm requires that $n > M - 1$.

A variant of the Arnoldi algorithm is the Prony-type method (or Hankel DMD) which uses time delays and Krylov subspaces in observable space [3, 69]. Its advantage is that it prevents the rank deficiency observed in the original Arnoldi-type algorithm.

---

[9]Note that the eigenvectors of the transpose are contained in the columns of the transpose of the Vandermonde (VdM) matrix, due to the fact that the VdM matrix is determined by the eigenvalues, that are the same for the companion matrix and its transpose.

#### 1.4.2.2 (Extended) Dynamic Mode Decomposition

Consider a set of snapshot pairs of data $\{(\mathbf{x}_k, \mathbf{y}_k)\}_{k=1}^{M}$, where $M$ is the number of samples, such that $\mathbf{S}(\mathbf{x}_k) = \mathbf{y}_k$. For a given set of basis functions $\{\psi_1, \psi_2, \ldots, \psi_N\}$ (with $\psi_j \in \mathscr{F} \setminus \{0\}$), we can define the data matrices

$$\mathbf{P}_\mathbf{x} := \begin{pmatrix} \psi_1(\mathbf{x}_1) & \psi_2(\mathbf{x}_1) & \cdots & \psi_N(\mathbf{x}_1) \\ \psi_1(\mathbf{x}_2) & \psi_2(\mathbf{x}_2) & \cdots & \psi_N(\mathbf{x}_2) \\ \vdots & \vdots & \ddots & \vdots \\ \psi_1(\mathbf{x}_K) & \psi_2(\mathbf{x}_K) & \cdots & \psi_N(\mathbf{x}_K) \end{pmatrix} \quad \mathbf{P}_\mathbf{y} := \begin{pmatrix} \psi_1(\mathbf{y}_1) & \psi_2(\mathbf{y}_1) & \cdots & \psi_N(\mathbf{y}_1) \\ \psi_1(\mathbf{y}_2) & \psi_2(\mathbf{y}_2) & \cdots & \psi_N(\mathbf{y}_2) \\ \vdots & \vdots & \ddots & \vdots \\ \psi_1(\mathbf{y}_K) & \psi_2(\mathbf{y}_K) & \cdots & \psi_N(\mathbf{y}_K) \end{pmatrix}.$$

The Koopman matrix $\mathbf{U}$ associated with the basis functions $\{\psi_j\}_{j=1}^{N}$ is computed from data by

$$\mathbf{U} = \mathbf{P}_\mathbf{x}^{+} \mathbf{P}_\mathbf{y}$$

where $\mathbf{P}_\mathbf{x}^{+}$ denotes the Moore–Penrose pseudoinverse of $\mathbf{P}_\mathbf{x}$. This is the matrix representation of $\Pi U$, where $\Pi : \mathscr{F} \to \mathscr{F}_N$ is again the discrete orthogonal projection (1.24).

As explained in Sect. 1.4.1.3, the Koopman eigenvalues, eigenfunctions, and modes can be obtained from the eigenvalues, right eigenvectors, and left eigenvectors of $\mathbf{U}$, respectively (see [75] for details and [28] for theoretical convergence results).

Several extensions of the original method based on machine learning techniques have been developed; e.g., kernel-based EDMD [76], reproducing kernels [24], deep learning [37, 77].

The DMD method corresponds to the special case of EDMD with $\psi_j(\mathbf{x}) = x_j$. In this context, efficient numerical schemes have been proposed to obtain more accurate computations of the Koopman matrix (e.g., exact DMD [73]). An important body of work has also contributed to improve and generalize the DMD method with a great variety of extensions: total DMD [18], sparsity promoting DMD [22], compressed DMD [6], tensor-based DMD [25], Bayesian DMD [72], methods for selection of eigenvectors based on data-driven evaluation of residuals—DDMD RRR [10], DMD for control [56], DMD for interconnected control systems [17], DMD for nonautonomous systems [38, 49], among others. We refer to the monograph [32] for further details on DMD.

## 1.5 Koopman Operator for Systems with Inputs

Now we consider dynamical systems with inputs, in both discrete-time and continuous-time cases.

## 1.5.1 Discrete-Time Systems

A discrete-time system with a control input $\mathbf{u}$ is represented by the transformation $\mathbf{S}: X \times \mathbb{R}^p \to X$, $(\mathbf{x}, \mathbf{u}) \mapsto \mathbf{S}(\mathbf{x}, \mathbf{u})$. We denote by $\mathscr{U}$ the space of admissible control signals $\overline{\mathbf{u}}(\cdot) : \mathbb{N} \to \mathbb{R}^p$.

### 1.5.1.1 Koopman Operator

Following [27], we can consider an extended (infinite-dimensional) state space $X \times \mathscr{U}$ and define the Koopman operator for the skew product system

$$\big(\mathbf{S}(\mathbf{x}, \overline{\mathbf{u}}(0)), T\overline{\mathbf{u}}(\cdot)\big) : X \times \mathscr{U} \to X \times \mathscr{U}$$

as

$$(Uf)\big(\mathbf{x}, \overline{\mathbf{u}}(\cdot)\big) = f\big(\mathbf{S}(\mathbf{x}, \overline{\mathbf{u}}(0)), T\overline{\mathbf{u}}(\cdot)\big)$$

where $T : \mathscr{U} \to \mathscr{U}$ is the left shift operator, i.e., $(T\overline{\mathbf{u}})(k) = \overline{\mathbf{u}}(k+1)$, $k \in \mathbb{N}$. Note that the observables $f : X \times \mathscr{U} \to \mathbb{C}$ are defined on an infinite-dimensional domain.

Given that the dependence of the evolution on $X$ on the input $\overline{\mathbf{u}}$ is only through $\overline{\mathbf{u}}(0)$, we can consider instead the $(n+p)$-dimensional space $X \times \mathbb{R}^p$, where input variables $\mathbf{u}$ are interpreted as additional state variables. In this case, the Koopman operator is defined by

$$(Uf)(\mathbf{x}, \mathbf{u}) = f(\mathbf{S}(\mathbf{x}, \mathbf{u}), \mathbf{u}), \tag{1.25}$$

for all $f : X \times \mathbb{R}^p \to \mathbb{C}$ (see, e.g., [58]). We note that the whole sequence of control signal $\overline{\mathbf{u}}(\cdot)$ is not taken into account in this formulation. In fact, the value $\mathbf{u}$ plays the role of a parameter. Equivalently, one could define a family of Koopman operators parametrized by $\mathbf{u}$, where each operator is associated with the map $\mathbf{S}(\cdot, \mathbf{u})$:

$$(U_{\mathbf{u}} f)(\mathbf{x}) = f(\mathbf{S}(\mathbf{x}, \mathbf{u})) \ .$$

Several iterations of the Koopman operator therefore correspond to a sequence of operators $U_{\overline{\mathbf{u}}(0)} U_{\overline{\mathbf{u}}(1)} \cdots U_{\overline{\mathbf{u}}(k)}$ determined by the control input $\overline{\mathbf{u}}(\cdot)$.

### 1.5.1.2 Finite-Dimensional Approximation and Lifted Dynamics

In the case of a system with input, the Koopman operator (1.25) can be approximated in a finite-dimensional subspace spanned by basis functions $\psi(\mathbf{x}, \mathbf{u})$. The results presented in Sect. 1.4.1 still hold; one can compute a matrix approximation $\mathbf{U}$ of the operator, which leads to the lifted dynamics (1.17). In that case, the lifted state depends on the input in a nonlinear manner. However, if one chooses the basis functions

$$\{\psi_j(\mathbf{x})\}_{j=1}^{N} \cup \{u_j\}_{j=1}^{p} , \tag{1.26}$$

it is easy to see that we obtain the linear lifted dynamics (see [27])

$$\mathbf{z}(k+1) = \mathbf{A}\,\mathbf{z}(k) + \mathbf{B}\,\mathbf{u}(k)$$

with $\mathbf{z} = (\psi_1(\mathbf{x}) \ \cdots \ \psi_N(\mathbf{x}))^T$, $\mathbf{A} \in \mathbb{R}^{N\times N}$, and $\mathbf{B} = \mathbb{R}^{N\times p}$. An output can also be considered in this case (see Remark 1.1).

### *1.5.2 Continuous-Time Systems*

A continuous-time system with a control input $\overline{\mathbf{u}}$ is represented by a family of one-parameter maps $\mathbf{S}^t : X \times \mathscr{U} \to X$, $t \in \mathbb{R}^+$, where $\mathscr{U}$ is the space of admissible control signals $\overline{\mathbf{u}}(\cdot) : \mathbb{R}^+ \to \mathbb{R}^p$.

#### 1.5.2.1 Koopman Operator

Considering the extended state space $X \times \mathscr{U}$, we can define the family of Koopman operators

$$(U^t f)\big(\mathbf{x}, \overline{\mathbf{u}}(\cdot)\big) = f\big(\mathbf{S}^t(\mathbf{x}, \overline{\mathbf{u}}(\cdot)), T^t\overline{\mathbf{u}}(\cdot)\big),$$

for all $f : X \times \mathscr{U} \to \mathbb{C}$, where $T^t : \mathscr{U} \to \mathscr{U}$ is the left shift semigroup defined by $T^t\mathbf{u}(\tau) = \mathbf{u}(\tau + t)$. Note that $U^t$ has the semigroup property $U^\tau U^t f = U^{t+\tau} f$ since $\mathbf{S}^\tau(\mathbf{S}^t(\mathbf{x}, \overline{\mathbf{u}}(\cdot)), T^t\overline{\mathbf{u}}(\cdot)) = \mathbf{S}^{t+\tau}(\mathbf{x}, \overline{\mathbf{u}}(\cdot))$.

Alternatively, one can define the family of Koopman operators

$$(U^t f)\big(\mathbf{x}, \overline{\mathbf{u}}(\cdot)\big) = f\big(\mathbf{S}^t(\mathbf{x}, \overline{\mathbf{u}}(\cdot)), \overline{\mathbf{u}}(\cdot)\big)$$

where the input signal is kept fixed along a trajectory. In this case, $U^t$ does not have the semigroup property. Equivalently, this corresponds to a family of operators parametrized by $\overline{\mathbf{u}}(\cdot)$:

$$(U^t_{\overline{\mathbf{u}}(\cdot)} f)(\mathbf{x}) = f(\mathbf{S}^t(\mathbf{x}, \overline{\mathbf{u}}(\cdot))) \tag{1.27}$$

for all $f : X \to \mathbb{C}$, where we have dropped the dependence on $\overline{\mathbf{u}}$. If the flow $\mathbf{S}^t$ is induced by the dynamics $\dot{\mathbf{x}} = \mathbf{F}(\mathbf{x}, \overline{\mathbf{u}}(t))$ and if $g(t, \mathbf{x}) = f(\mathbf{S}^t(\mathbf{x}, \overline{\mathbf{u}}(\cdot)))$, we also obtain the dynamics of observables

$$\frac{\partial g}{\partial t}(t, \mathbf{x}) = \mathbf{F}\left(\mathbf{S}^t(\mathbf{x}, \overline{\mathbf{u}}(\cdot)), \overline{\mathbf{u}}(t)\right) \cdot \nabla f\left(\mathbf{S}^t(\mathbf{x}, \overline{\mathbf{u}}(\cdot))\right) , \tag{1.28}$$

or equivalently

$$\frac{\partial g}{\partial t}(t, \mathbf{x}) = U^t_{\overline{\mathbf{u}}(\cdot)} L_{\overline{\mathbf{u}}(t)} U^{-t}_{\overline{\mathbf{u}}(\cdot)} g\,(t, \mathbf{x}) \ ,$$

with the operator $L_{\overline{\mathbf{u}}(t)} = \mathbf{F}(\cdot, \overline{\mathbf{u}}(t)) \cdot \nabla$. Note that, in the non-autonomous case, the operators $U^t_{\overline{\mathbf{u}}(\cdot)}$ and $L_{\overline{\mathbf{u}}(t)}$ do not commute [2]. If the system dynamics are control affine, that is $\dot{\mathbf{x}} = \mathbf{F}(\mathbf{x}) + \mathbf{G}(\mathbf{x})\, u(t)$, with $u(\cdot) : \mathbb{R}^+ \rightarrow \mathbb{R}$, (1.29) is bilinear in $f$ and $u$ [4, 14, 30]:

$$\frac{\partial g}{\partial t}(t, \mathbf{x}) = \left(\mathbf{F}\left(\mathbf{S}^t(\mathbf{x}, \overline{\mathbf{u}}(\cdot))\right) + u(t)\,\mathbf{G}\left(\mathbf{S}^t(\mathbf{x}, \overline{\mathbf{u}}(\cdot))\right)\right) \cdot \nabla f\left(\mathbf{S}^t(\mathbf{x}, \overline{\mathbf{u}}(\cdot))\right) \ . \quad (1.29)$$

As in the discrete-time case, one can also consider an extended state space $X \times \mathbb{R}^p$. Moreover, under the zero-order hold assumption, we have $\mathbf{S}^t(\mathbf{x}, \overline{\mathbf{u}}(\cdot)) \approx \mathbf{S}^t(\mathbf{x}, \overline{\mathbf{u}}(\cdot) = \mathbf{u}(0))$ for $t$ sufficiently small. Thus we can define the Koopman semi-group of operators

$$(U^t f)(\mathbf{x}, \mathbf{u}) = f(\mathbf{S}^t(\mathbf{x}, \overline{\mathbf{u}}(\cdot) = \mathbf{u}), \mathbf{u}) \qquad t \ll 1 \quad (1.30)$$

for all $f : X \times \mathbb{R}^p \rightarrow \mathbb{C}$. For a fixed value $t$, we simply recover the discrete-time case.

#### 1.5.2.2 Finite-Dimensional Approximation and Lifted Dynamics

Similarly to the discrete-time case, the operators (1.27) and (1.30) can be approximated in finite-dimensional spaces (see Sect. 1.5.1.2).

The observable dynamics (1.29) can also be projected on a finite-dimensional space. If the system dynamics are control affine, (1.29) yields the bilinear lifted dynamics

$$\frac{d\mathbf{z}}{dt} = \mathbf{A}\,\mathbf{z} + \mathbf{B}\,\mathbf{z}\,u(t) \quad (1.31)$$

with $\mathbf{z} = \Psi(\mathbf{x})$ and where $\mathbf{A}^T = \mathbf{L} \in \mathbb{R}^{N \times N}$ and $\mathbf{B}^T \in \mathbb{R}^{N \times N}$ are the matrix representations of the operators $\mathbf{F} \cdot \nabla$ and $\mathbf{G} \cdot \nabla$, respectively, in the subspace. For more general dynamics, one can obtain the linear lifted dynamics

$$\dot{\mathbf{z}} = \mathbf{A}\,\mathbf{z} + \mathbf{B}\,\mathbf{u}(t)$$

with $\mathbf{z} = (\psi_1(\mathbf{x}) \ \cdots \ \psi_N(\mathbf{x}))^T$ if the subspace is spanned by the basis functions (1.26), which depend also linearly on $\mathbf{u}$.

## 1.6 Perspectives in Control Theory

While the Koopman operator framework is gaining increasing attention in the control community, there remain unexplored research areas and challenges. Below, we list a few open problems that have been raised in the context of the Koopman operator framework applied to control theory.

- **Input–output systems.** There have been so far some attempts to define the Koopman operator for systems with inputs (see Sect. 1.5). However, the spectral properties of the operator defined in this case are not well studied (see [38, 49] and Chap. 6 for preliminary results in the case of nonautonomous systems). A proper Koopman operator (spectral) theory for input–output systems could provide new insight on interconnected systems and complement classic results in input–output stability, absolute stability theory, and nonlinear controllability/observability.
- **Systems identification.** Provided that the Koopman operator is properly defined for input–output systems, it could be used to develop linear methods for nonlinear system identification in the classic sense. These methods might complement preliminary results obtained in the context of parameter estimation [39].
- **Geometric theory of nonlinear control.** Mature techniques from the geometric theory of nonlinear control [20], such as zero dynamics, feedback linearization, and output regulation, could be applied to the Koopman operator and reformulated in terms of its spectral properties. Connections to chronological calculus should also be investigated [2].
- **Hybrid systems.** The Koopman operator can also be defined in the case of hybrid systems, as long as a flow map is well defined. However, the operator has not been used so much in this context and could be considered in the case of control. Note also preliminary numerical results on the control of switched systems [55] (see also Chap. 10).
- **Stochastic systems.** In the case of stochastic systems inducing a flow $\mathbf{S}^t : X \times \Omega \to X$, where $\Omega$ is the probability space, the Koopman operator is easily generalized through the definition (see, e.g., [44])

  $$U^t f(\mathbf{x}) = \mathbb{E}[f(\mathbf{S}^t(\mathbf{x}, \omega)] \quad \mathbf{x} \in X, \omega \in \Omega$$

  where $\mathbb{E}$ is the mathematical expectation. This framework could be used in the context of control of nonlinear stochastic systems (see also Chap. 6).
- **Networked dynamical systems.** Large-scale networked dynamical systems such as multi-agent systems, communication and energy networks have attracted a lot of interest in the control community. Koopman theory for networked systems and associated DMD methods are reported in [17]. This direction is expected to yield computationally tractable design techniques for distributed control in order to shape complex dynamics occurring in the network (see also Chap. 20).
- **Theoretical framework for numerical methods.** All numerical methods based on the Koopman operator rely on finite-dimensional approximations of the (infinite-dimensional) operator. There is currently a lack of theoretical results on the estima-

tion of error bounds for these approximations (see, e.g., some preliminary results in [21] and also Chap. 7 for a linear algebra perspective) and of the rate of convergence of the algorithms (see, e.g., the recent work [31] in the context of approximation theory). These results are crucial to study the performance and validity of current and future Koopman-based control methods.
- **Comparison with classic methods and applications.** A thorough comparison of Koopman operator techniques with respect to classic nonlinear control techniques is missing, but necessary to measure the advantage of these techniques over standard methods. A large variety of real applications should also be considered to identify current needs and drive further theoretical development (see, e.g., Chaps. 16, 17, 18, 19, and 20).

## 1.7 Conclusion

The Koopman operator is a powerful tool that can be used to *linearize nonlinear systems in the large*. In the context of nonlinear control theory, it yields infinite-dimensional (bi)linear representations of systems (with inputs), which can be reduced to finite-dimensional (bi)linear approximations in a well-chosen set of basis functions. These approximations can be subsequently combined with classic (linear) techniques in the context of stability analysis (Chaps. 2 and 5), data-driven identification and approximation (Chaps. 7 and 13), analysis of observability/controllability (Chaps. 3, 4, and 14), model reduction (Chaps. 15 and 19), and controller design (Chaps. 8, 9, 10, 11, and 12), to list a few.

The Koopman operator framework is gaining increasing attention in the control community, but is still in its infancy. There also remain many unexplored research areas and open problems. The main challenge is to deal with the infinite-dimensional nature of the operator and the inherent approximations of numerical schemes, which is somehow the price to pay to develop linear methods for nonlinear systems.

## References

1. Abraham, I., De La Torre, G., Murphey, T.D.: Model-based control using Koopman operators. In: Proceedings of Robotics: Science and Systems XIII (2017)
2. Agrachev, A.A., Sachkov Y.: Control theory from the geometric viewpoint, vol. 87. Springer (2013)
3. Arbabi, H., Mezić, I.: Ergodic theory, dynamic mode decomposition and computation of spectral properties of the Koopman operator. SIAM J. Appl. Dyn. Syst. **16**(4), 2096–2126 (2017)
4. Banks, S.P.: On the generation of infinite-dimensional bilinear systems and Volterra series. Int. J. Syst. Sci. **16**(2), 145–160 (1985)
5. Bátkai, A., Fijavž, M.K., Rhandi, A.: Positive Operator Semigroups. Springer International Publishing, Birkhäuser Mathematics (2017)
6. Brunton, S.L., Proctor, J.L., Tu, J.H, Kutz, J.N.: Compressive sampling and dynamic mode decomposition. J. Comput. Dyn. **2**(2), 165–191 (2015). arXiv:1312.5186

7. Budišić, M., Mohr, R., Mezić, I.: Applied Koopmanism. Chaos **22**(4), 047,510–047,510 (2012)
8. Carleman, T.: Application de la thorie des quations integrales lineaires aux systmes d'quations diffrentielles nonlinaires. Acta Math. **59**, 63–68 (1932)
9. Dellnitz, M., Junge, O.: Set oriented numerical methods for dynamical systems. In: Handbook of Dynamical Systems, vol. 2, pp. 221–264. Gulf Professional Publishing, Houston (2002)
10. Drmač, Z., Mezić, I., Mohr, R.: Data driven modal decompositions: analysis and enhancements (2017). arXiv:1708.02685
11. Engel, K.J., Nagel, R.: One-Parameter Semigroups for Linear Evolution Equations, vol. 194. Springer Science & Business Media, Berlin (1999)
12. Gaspard, P.: Chaos, Scattering and Statistical Mechanics, vol. 9. Cambridge University Press, Cambridge (2005)
13. Gaspard, P., Nicolis, G., Provata, A., Tasaki, S.: Spectral signature of the pitchfork bifurcation: Liouville equation approach. Phys. Rev. E **51**(1), 74 (1995)
14. Glaz, B., Mezić, I., Fonoberova, M., Loire, S.: Quasi-periodic intermittency in oscillating cylinder flow. J. Fluid Mech. **828**, 680–707 (2017)
15. Goswami, D., Paley, D.A.: Global bilinearization and controllability of control-affine nonlinear systems: a Koopman spectral approach. In: Proceedings of the 56th IEEE Conference on Decision and Control (2017)
16. Govindarajan, N., Arbabi, H., van Blargian, L., Matchen, T., Tegling, E., Mezić, I.: An operator-theoretic viewpoint to non-smooth dynamical systems: Koopman analysis of a hybrid pendulum. In: 2016 IEEE 55th Conference on Decision and Control (CDC), pp. 6477–6484. IEEE (2016)
17. Heersink, B., Warren, M.A., Hoffmann, H.: Dynamic mode decomposition for interconnected control systems. arXiv:1709.02883 (2017)
18. Hemati, M.S., Rowley, C.W., Deem, E.A., Cattafesta, L.N.: De-biasing the dynamic mode decomposition for applied Koopman spectral analysis of noisy datasets. Theor. Comput. Fluid Dyn. **31**(4), 349–368 (2017)
19. Huang, B., Ma, X., Vaidya, U.: Feedback stabilization using Koopman operator. In: 2018 IEEE Conference on Decision and Control (CDC), pp. 6434–6439. IEEE (2018)
20. Isidori, A.: Nonlinear Control Systems. Springer Science & Business Media, Berlin (2013)
21. Johnson, C., Yeung, E.: A class of logistic functions for approximating state-inclusive Koopman operators (2017). arXiv:1712.03132
22. Jovanović, M.R., Schmid, P.J., Nichols, J.W.: Sparsity-promoting dynamic mode decomposition. Phys. Fluids **26**(2), 024103 (2014)
23. Kaiser, E., Kutz, J.N., Brunton, S.L.: Data-driven discovery of Koopman eigenfunctions for control (2017). arXiv:1707.01146
24. Kawahara, Y.: Dynamic mode decomposition with reproducing kernels for Koopman spectral analysis. In: Advances in Neural Information Processing Systems, pp. 911–919 (2016)
25. Klus, S., Gelß, P., Peitz, S., Schütte, C.: Tensor-based dynamic mode decomposition. Nonlinearity **31**(7), 3359 (2018)
26. Koopman, B.O.: Hamiltonian systems and transformation in Hilbert space. Proc. Natl. Acad. Sci. U. S. A. **17**(5), 315 (1931)
27. Korda, M., Mezić, I.: Linear predictors for nonlinear dynamical systems: Koopman operator meets model predictive control. Automatica **93**, 149–160 (2018)
28. Korda, M., Mezić, I.: On convergence of extended dynamic mode decomposition to the Koopman operator. J. Nonlinear Sci. **28**(2), 687–710 (2018)
29. Korda, M., Susuki, Y., Mezić, I.: Power grid transient stabilization using Koopman model predictive control (2018). arXiv:1803.10744
30. Krener, A.J.: Linearization and bilinearization of control systems. In: Proceedings 1974 Allerton Conference on Circuit and System Theory, vol. 834. Monticello (1974)
31. Kurdila, A.J., Bobade, P.S.: Koopman theory and linear approximation spaces (2018). arXiv:1811.10809
32. Kutz, J.N., Brunton, S.L., Brunton, B.W., Proctor, J.L.: Dynamic Mode Decomposition: Data-Driven Modeling of Complex Systems, vol. 149. SIAM (2016)

33. Lan, Y., Mezić, I.: Linearization in the large of nonlinear systems and Koopman operator spectrum. Phys. D **242**, 42–53 (2013)
34. LaSalle, J.P.: Dissipative systems. In: Weiss, L. (ed.) Ordinary Differential Equations 1971 NRL-MRC Conference. Academic, New York (1972)
35. Lasota, A., Mackey, M.C.: Chaos, Fractals, and Noise: Stochastic Aspects of Dynamics. Springer, Berlin (1994)
36. Lasserre, J.B., Henrion, D., Prieur, C., Trélat, E.: Nonlinear optimal control via occupation measures and LMI-relaxations. SIAM J. Control. Optim. **47**(4), 1643–1666 (2008)
37. Li, Q., Dietrich, F., Bollt, E.M., Kevrekidis, I.G.: Extended dynamic mode decomposition with dictionary learning: a data-driven adaptive spectral decomposition of the Koopman operator. Chaos Interdiscip. J. Nonlinear Sci. **27**, 103111 (2017)
38. Maćešić, S., Črnjarić-Žic, N., Mezić, I.: Koopman operator family spectrum for nonautonomous systems - Part 1 (2017). arXiv preprint arXiv:1703.07324
39. Mauroy, A., Goncalves, J.: Linear identification of nonlinear systems: a lifting technique based on the Koopman operator. In: Proceedings of the 55th IEEE Conference on Decision and Control, pp. 6500–6505 (2016)
40. Mauroy, A., Mezić, I.: On the use of Fourier averages to compute the global isochrons of (quasi)periodic dynamics. Chaos **22**(3), 033112 (2012)
41. Mauroy, A., Mezić, I.: Global stability analysis using the eigenfunctions of the Koopman operator. IEEE Trans. Autom. Control **61**(3), 3356–3369 (2016)
42. Mauroy, A., Mezić, I., Moehlis, J.: Isostables, isochrons, and Koopman spectrum for the action-angle representation of stable fixed point dynamics. Phy. D Nonlinear Phenom. **261**, 19–30 (2013)
43. Mauroy, A., Susuki, Y.: Introduction to the Koopman operator in systems and control. In: Proceedings of the SICE Annual Conference (2018)
44. Mezić, I.: Spectral properties of dynamical systems, model reduction and decompositions. Nonlinear Dyn. **41**(1–3), 309–325 (2005)
45. Mezić, I.: Analysis of fluid flows via spectral properties of Koopman operator. Annu. Rev. Fluid Mech. **45**, 357–378 (2013)
46. Mezić, I.: On applications of the spectral theory of the Koopman operator in dynamical systems and control theory. In: 2015 IEEE 54th Annual Conference on Decision and Control (CDC), pp. 7034–7041. IEEE (2015)
47. Mezić, I.: Koopman operator spectrum and data analysis (2017). arXiv:1702.07597
48. Mezić, I., Banaszuk, A.: Comparison of systems with complex behavior. Phys. D Nonlinear Phenom. **197**(1–2), 101–133 (2004)
49. Mezić, I., Surana, A.: Koopman mode decomposition for periodic/quasi-periodic time dependence. IFAC-PapersOnLine **49**(18), 690–697 (2016)
50. Mezić, I., Wiggins, S.: A method for visualization of invariant sets of dynamical systems based on the ergodic partition. Chaos **9**(1), 213–218 (1999)
51. Mohr, R., Mezić, I.: Construction of eigenfunctions for scalar-type operators via Laplace averages with connections to the Koopman operator (2014). arXiv:1403.6559
52. Mohr, R., Mezić, I.: Koopman principle eigenfunctions and linearization of diffeomorphisms (2016). arXiv:1611.01209
53. Mousavi, H.K., Somarakis, C., Motee, N.: Koopman performance analysis of a class of nonlinear dynamical networks. In: Proceedings of the 55th IEEE Conference on Decision and Control, pp. 117–122. IEEE (2016)
54. von Neumann, J.: Proof of the quasi-ergodic hypothesis. Proc. Nat. Acad. Sci. USA **18**, 70–82 (1932)
55. Peitz, S., Klus, S.: Koopman operator-based model reduction for switched-system control of PDEs (2017). arXiv:1710.06759
56. Proctor, J.L., Brunton, S.L., Kutz, J.N.: Dynamic mode decomposition with control. SIAM J. Appl. Dyn. Syst. **15**(1), 142–161 (2016)
57. Proctor, J.L., Eckhoff, P.A.: Discovering dynamic patterns from infectious disease data using dynamic mode decomposition. Int. Health **7**(2), 139–145 (2015)

58. Proctor, L.P., Brunton, S.L., Kutz, J.N.: Generalizing Koopman operator theory to allow for inputs and control. SIAM J. Appl. Dyn. Syst. **17**(1), 909–930 (2018)
59. Raghunathan, A., Vaidya, U.: Optimal stabilization using Lyapunov measures. IEEE Trans. Autom. Control **59**(5), 1316–1321 (2014)
60. Riseth, A.N., Taylor-King, J.P.: Operator fitting for parameter estimation of stochastic differential equations (2017). arXiv:1709.05153
61. Rowley, C.W., Mezić, I., Bagheri, S., Schlatter, P., Henningson, D.S.: Spectral analysis of nonlinear flows. J. Fluid Mech. **641**, 115–127 (2009)
62. Schmid, P.J.: Dynamic mode decomposition of numerical and experimental data. J. Fluid Mech. **656**, 5–28 (2010)
63. Slawinska, J., Szekely, E., Giannakis, D.: Data-driven Koopman analysis of tropical climate space-time variability (2017). arXiv:1711.02526
64. Sootla, A., Ernst, D.: Pulse-based control using koopman operator under parametric uncertainty. IEEE Trans. Autom. Control **63**(3), 791–796 (2017)
65. Surana, A.: Koopman operator based observer synthesis for control-affine nonlinear systems. In: Proceedings of the 55th IEEE Conference on Decision and Control, pp. 6492–6499 (2016)
66. Surana, A.: Koopman operator framework for time series modeling and analysis. J. Nonlinear Sci. pp. 1–34 (2018)
67. Susuki, Y., Mezic, I.: Nonlinear Koopman modes and coherency identification of coupled swing dynamics. IEEE Trans. Power Syst. **26**(4), 1894–1904 (2011)
68. Susuki, Y., Mezić, I.: Nonlinear Koopman modes and power system stability assessment without models. IEEE Trans. Power Syst. **29**(2), 899–907 (2014)
69. Susuki, Y., Mezić, I.: A prony approximation of Koopman mode decomposition. In: Proceedings of the 54th IEEE Conference on Decision and Control (2015)
70. Susuki, Y., Mezić, I.: Uniformly bounded sets in quasiperiodically forced dynamical systems (2018). arXiv:1808.08340
71. Takata, H.: Transformation of a nonlinear system into an augmented linear system. IEEE Trans. Autom. Control **24**(5), 736–741 (1979)
72. Takeishi, N., Kawahara, Y., Tabei, Y., Yairi, T.: Bayesian dynamic mode decomposition. In: Proceedings of the International Joint Conference on Artificial Intelligence (2017)
73. Tu, J.H., Rowley, C.W., Luchtenburg, D.M., Brunton, S.L., Kutz, J.N.: On dynamic mode decomposition: theory and applications. J. Comput. Dyn. **1**(2), 391–421 (2014)
74. Vaidya, U., Mehta, P.G.: Lyapunov measure for almost everywhere stability. IEEE Trans. Autom. Control. **53**(1), 307–323 (2008)
75. Williams, M.O., Kevrekidis, I.G., Rowley, C.W.: A data-driven approximation of the Koopman operator: extending dynamic mode decomposition. J. Nonlinear Sci. **25**(6), 1307–1346 (2015)
76. Williams, M.O., Rowley, C.W., Kevrekidis, I.G.: A kernel-based approach to data-driven Koopman spectral analysis. J. Comput. Dyn. **2**(2), 247–265 (2015)
77. Yeung, E., Kundu, S., Hodas, N.: Learning deep neural network representations for Koopman operators of nonlinear dynamical systems (2017). arXiv:1708.06850
78. Yeung, E., Liu, Z., Hodas, N.O.: A Koopman operator approach for computing and balancing gramians for discrete time nonlinear systems (2017). arXiv:1709.08712
79. Zeng, S.: On systems theoretic aspects of Koopman operator theoretic frameworks. In: Proceeding of the 57th IEEE Conference on Decision and Control (2018)

# Chapter 2
# Koopman Framework for Global Stability Analysis

**Alexandre Mauroy, Aivar Sootla and Igor Mezić**

**Abstract** In this chapter, we present a new framework to study global stability of nonlinear systems. The proposed approach is based on the stability properties of the Koopman operator and can be seen as an extension of classic stability analysis of linear systems. In the case of (hyperbolic) equilibria, we show that the existence of specific eigenfunctions of the operator is a necessary and sufficient condition for global stability of the attractor. Moreover, using the realization of the operator in a finite-dimensional basis, we provide a systematic method to compute candidate Lyapunov functions of stable systems.

## 2.1 Introduction

The most common technique in nonlinear stability analysis is undeniably Lyapunov's second method. Interestingly this technique is reminiscent of an operator-theoretic approach to stability: a Lyapunov function $V$ is nothing but an observable function that enjoys a specific (decreasing) evolution through the Koopman operator, i.e., $d/dt(U^t V) \leq 0$. It is also quite noticeable that the duality between the Koopman operator and the so-called Perron–Frobenius operator acting on densities [12] directly reflects the duality property of Lyapunov techniques. But surprisingly, while the two dual operators are known for decades, dual density-based stability methods were

A. Mauroy (✉)
Department of Mathematics, University of Namur, Namur, Belgium
e-mail: alexandre.mauroy@unamur.be

A. Sootla
Department of Engineering Science, University of Oxford,
Parks Road, Oxford OX1 3PJ, UK
e-mail: aivar.sootla@eng.ox.ac.uk

I. Mezić
Department of Mechanical Engineering, University of California Santa Barbara,
Santa Barbara, CA, USA
e-mail: mezic@engineering.ucsb.edu



A. Mauroy et al. (eds.), *The Koopman Operator in Systems and Control*,
Lecture Notes in Control and Information Sciences 484,
https://doi.org/10.1007/978-3-030-35713-9_2

discovered only recently [23, 27], an observation which clearly suggests that the operator-theoretic approach to nonlinear stability theory has been overlooked. We also refer to Chap. 8 for a detailed description of the duality between Lyapunov and density-based approaches.

As illustrated above, the Koopman operator is implicitly used in stability theory and naturally appears as a relevant tool to deal with stability analysis of nonlinear systems. It provides a global picture of the system, as the action of the Koopman operator on one observable function is equivalent to the action of the flow generated by the system over *all* initial conditions in the state space. In addition, the Koopman operator framework can provide a data-driven approach to stability, through the Koopman mode analysis (see, e.g., [26] in the context of power grids).

In this chapter, we mostly report on the results presented in [16] to demonstrate how the Koopman operator can be used to reveal the stability properties of the underlying dynamical system. In particular, we show that the spectral properties of the Koopman operator capture the stability properties of the system, a result which mirrors classic spectral results in linear stability analysis. Although the framework is general and may encompass a broad variety of dynamics and attractors, we mainly focus on the case of (possibly non-hyperbolic) equilibria. In this case, specific Koopman eigenfunctions are directly related to the region of attraction of stable equilibria (see also [24] in the case of monotone systems) and the Koopman operator can also be used to design Lyapunov functions with a convergence rate determined by the spectrum of the operator. Finally, the framework yields systematic numerical methods for stability analysis. These methods are based on a finite-dimensional approximation of the operator and hinge upon simple linear algebraic techniques.

The chapter is organized as follows. General properties of the Koopman operator are discussed in Sect. 2.2 and spectral properties are investigated in Sect. 2.3. In Sect. 2.4, the Koopman operator framework is used to develop systematic numerical methods for nonlinear stability analysis. Section 2.5 investigates the Lyapunov function design problem. Concluding remarks and perspectives are given in Sect. 2.6.

## 2.2 Stability and Koopman Operator

In this section, we recall some preliminaries on stability and present general properties of the Koopman operator related to stability. Note that the results provided in this chapter focus on continuous-time systems, but most of them can be extended to discrete-time systems in a straightforward way.

### 2.2.1 Preliminaries

We consider continuous-time systems of the form

$$\dot{\mathbf{x}} = \mathbf{F}(\mathbf{x}) \qquad \mathbf{x} \in \mathbb{R}^n \tag{2.1}$$

and we assume that the vector field $\mathbf{F}$ is Lipschitz continuous, so that the solution

$$\mathbf{x}(t) = \mathbf{S}^t(\mathbf{x}_0)$$

associated with the initial condition $\mathbf{x}_0$ exists and is unique. The family of maps $\mathbf{S}^t : \mathbb{R}^n \to \mathbb{R}^n, t \in \mathbb{R}^+$, is the flow generated by (2.1).

Our goal is to characterize the stability properties of the system (2.1), which we recall in the following definition [9, 14].

**Definition 2.1** (*Global stability*) Consider a forward-invariant set $A$ (i.e., $\mathbf{S}^t(A) \subseteq A$ for all $t > 0$) and the distance

$$d(\mathbf{x}, A) = \min_{\mathbf{y} \in A} \|\mathbf{x} - \mathbf{y}\|,$$

where $\|\cdot\|$ denotes the Euclidean norm. The set $A$ is

- *stable* if, for all $\varepsilon > 0$, there exists $\delta > 0$ such that

$$d(\mathbf{x}, A) < \delta \Rightarrow d(\mathbf{S}^t(\mathbf{x}), A) < \varepsilon \quad \forall t \geq 0;$$

- *globally pointwise attractive* in $X \subset \mathbb{R}^n$ if, for all $\varepsilon > 0$ and $\mathbf{x} \in X$, there exists $T > 0$ such that

$$t > T \Rightarrow d(\mathbf{S}^t(\mathbf{x}), A) < \varepsilon;$$

- *globally uniformly attractive* in $X \subset \mathbb{R}^n$ if, for all $\varepsilon > 0$, there exists $T > 0$ such that

$$t > T \Rightarrow d(\mathbf{S}^t(\mathbf{x}), A) < \varepsilon \quad \forall \mathbf{x} \in X;$$

- *globally (uniformly) asymptotically stable* in $X \subset \mathbb{R}^n$ if it is stable and globally (uniformly) attractive in $X$.

△

Uniform attractivity of $A$ implies that $A$ is stable (and therefore globally uniformly asymptotically stable). If the set $A$ is uniformly attractive and does not contain a strictly smaller set that is stable, it is an *attractor* of the system. Unless otherwise stated, we will consider that $A$ is an attractor.

In general, for most systems (e.g., with Lipschitz vector fields), the property of uniform attractivity is not satisfied for $X = \mathbb{R}^n$ but is typically satisfied if $X$ is a compact set containing a unique attractor. Moreover, we will consider that $X$ is forward-invariant under the flow, so that the Koopman semigroup of operators is well defined for observables $f : X \to \mathbb{C}$. These conditions are summarized in the following standing assumption.

**Assumption 2.1** *The set $X \subset \mathbb{R}^n$ is compact and forward-invariant, i.e.,* $\mathbf{S}^t(X) \subseteq X$ $\forall t \geq 0$. △

### 2.2.2 A Stability Property of the Koopman Operator

As a preliminary step, we study the interplay between global stability of the system (2.1) on a set $X$ and stability of the associated Koopman semigroup of operators $U^t f = f \circ \mathbf{S}^t$, $t \geq 0$, acting on observables $f : X \to \mathbb{C}$. The operator is defined on a (Banach) space of functions $\mathscr{F}$, equipped with the norm $\| \cdot \|$. For basic definitions and concepts on the Koopman operator, we refer to Chap. 1.

**Definition 2.2** ([4]) A semigroup of operators $(U^t)_{t\geq 0}$ on a Banach space $\mathscr{F}$ is *strongly stable* if

$$\lim_{t\to\infty} \|U^t f\| = 0 \qquad \forall f \in \mathscr{F}.$$

△

It is clear that the Koopman semigroup of operators is not strongly stable, since the function $\mathbf{1}(\mathbf{x}) = 1\ \forall \mathbf{x} \in \mathbb{R}^n$ is a fixed point of the operator, i.e., $U^t \mathbf{1} = \mathbf{1}$ for all $t \geq 0$. However the restriction of the operator to a proper subspace can be strongly stable. Consider the subspace $\mathscr{F}_{A_c} \subseteq \mathscr{F}$ of functions with support on $A_c = X \setminus A$, i.e.,

$$\mathscr{F}_{A_c} = \{f \in \mathscr{F} | f(\mathbf{x}) = 0\ \ \forall \mathbf{x} \in A\}.$$

Since $A$ is forward-invariant under the flow, we have $U^t f \in \mathscr{F}_{A_c}$ for all $t \geq 0$ and all $f \in \mathscr{F}_{A_c}$, so that we can define the restriction of $U^t$ to the subspace $\mathscr{F}_{A_c}$. This restriction is denoted $U^t_{A_c} : \mathscr{F}_{A_c} \to \mathscr{F}_{A_c}$. We note that a Lyapunov function is a specific observable $f \in \mathscr{F}_{A_c}$ that satisfies $U^t_{A_c} f(\mathbf{x}) < U^s_{A_c} f(\mathbf{x})$ for all $s < t < 0$ and $\mathbf{x} \in X$.

We are now in position to show that the strong stability property of $U^t_{A_c}$ is equivalent to the attractivity property of the attractor.

**Proposition 2.1** *The attractor $A$ of* (2.1) *is globally uniformly attractive in $X$ if and only if $U^t_{A_c}$ is strongly stable in the space of continuous functions $\mathscr{F} = C(X)$ (endowed with the norm $\|f\|_\infty = \sup_{\mathbf{x}\in X} |f(\mathbf{x})|$.*

*Proof Sufficiency.* The result is proved by considering the continuous observable $f(\mathbf{x}) = d(A, \mathbf{x})$.
*Necessity.* Consider $f \in \mathscr{F}_{A_c}$ and $\varepsilon > 0$. Since $f$ is continuous and has zero value on the attractor, there exists $\delta > 0$ such that $\sup_{\mathbf{x}\in A} |f(\mathbf{S}^t(\mathbf{x}))| < \varepsilon$ if $d(\mathbf{S}^t(\mathbf{x}), A) < \delta$. Moreover, since the system is globally uniformly attractive, there exists $T > 0$ such that $d(\mathbf{S}^t(\mathbf{x}), A) < \delta$ for all $\mathbf{x} \in X$ if $t > T$. Thus, for all $\varepsilon > 0$ and all $f \in \mathscr{F}_{A_c}$, there exists $T > 0$ such that $\|U^t f\|_\infty < \varepsilon$ for all $t > T$. □

A weaker result related to pointwise attractivity can also be found in [16]. Moreover, we refer to the recent work [10] for a thorough study of the connections between various stability properties of the Koopman semigroup of operators and stability properties of the attractor.

*Remark 2.1* (*Uniform stability*) A semigroup of operators $(U^t)_{t\geq 0}$ is uniformly (exponentially) stable if $\lim_{t\to\infty} \|U^t\| = 0$ (or equivalently $\|U^t\| \leq Ce^{-\alpha t}$, with $C, \alpha > 0$) [3, 4]. A result similar to Proposition 2.1 but related to uniform stability is not obvious. In fact, the semigroup $U^t$ (and its restriction $U^t_{A_c}$) is not uniformly stable in $\mathscr{F} = C(X)$, but the restriction $U^t_{A_c}$ might be uniformly stable in specific cases, in other spaces. We leave this question for future research. △

Proposition 2.1 is very general, but difficult to use in practice since it provides a stability condition that should be satisfied for all continuous functions with support on $A_c$. As shown in the next section, this limitation is overcome by considering the spectral properties of the Koopman operator.

## 2.3 Spectral Analysis

Since the Koopman operator is linear, it is natural to study its spectral properties, which are related to the stability properties of the operator. This leads to spectral stability analysis of nonlinear systems which mirrors classic linear stability analysis.

### 2.3.1 Spectrum of the Koopman Operator

We will consider the spectrum $\sigma(L)$ of the infinitesimal generator $L = \mathbf{F} \cdot \nabla$ of the Koopman semigroup, i.e., the set of values $\lambda$ such that $L - \lambda I$ is not invertible. The point spectrum $\sigma_p(L)$ is the set of *(Koopman) eigenvalues* $\lambda \in \sigma(L)$ such that

$$L\phi_\lambda = (\nabla\phi_\lambda)^T \mathbf{F} = \lambda\phi_\lambda \tag{2.2}$$

for some *(Koopman) eigenfunction* $\phi_\lambda \in \mathscr{F} \setminus \{0\}$.

*Remark 2.2* Although the spectrum of $U^t$ is more directly connected to the stability properties, it is easier to compute the spectrum of $L$, which is related to the (known) vector field $\mathbf{F}$. Since we will focus on the point spectrum of the operator, considering $U^t$ or $L$ is equivalent: the spectral mapping theorem [5] implies

$$\sigma_p(U^t) \setminus \{0\} = e^{\sigma_p(L)t} \qquad t > 0$$

and the eigenvalues and eigenfunctions satisfy

$$U^t\phi_\lambda = e^{\lambda t}\phi_\lambda.$$

△

In the case $\mathscr{F} = C(X)$, the spectral bound of $L$ is smaller than or equal to 0, i.e., $\text{Re}\{\lambda\} \leq 0$ for all $\lambda \in \sigma(L)$. This follows from the fact that the spectral radius of

$U^t$ is smaller or equal to 1 [5, Chap. IV, Proposition 2.2] since $\|U^t\|_\infty \leq 1$. We can further consider the decomposition of the spectrum

$$\sigma(L) = \sigma^-(L) \cup \sigma^0(L)$$

with

$$\sigma^-(L) = \{\lambda \in \sigma(L) : \text{Re}\{\lambda\} < 0\}.$$
$$\sigma^0(L) = \{\lambda \in \sigma(L) : \text{Re}\{\lambda\} = 0\}.$$

Similarly, for the point spectrum, we note $\sigma_p(L) = \sigma_p^-(L) \cup \sigma_p^0(L)$ with $\sigma_p^-(L) \subseteq \sigma^-(L)$ and $\sigma_p^0(L) \subseteq \sigma^0(L)$.

### 2.3.2 General Results

We can now focus on the relationship between the stability properties of $U^t$ and the spectral properties of $L$. It can be shown that $U^t_{A_c}$ is strongly stable if its infinitesimal generator $L_{A_c}$ satisfies [4, Corollary 3.14]

$$\sigma(L_{A_c}) \cap i\mathbb{R} = \emptyset. \tag{2.3}$$

Hence, according to Proposition 2.1, the spectrum $\sigma(L_{A_c})$ (in the case $\mathscr{F} = C(X)$) can be used to prove that the attractor is globally uniformly attractive. In particular, if $\sigma(L)$ has two disconnected parts $\sigma^-(L)$ and $\sigma^0(L)$ and if the spectral projection $P$ on $\sigma^-$ satisfies $P\mathscr{F} = \mathscr{F}_{A_c}$, then $\sigma(L_{A_c}) = \sigma^-(L)$ [3, Lemma 2.5.7], so that (2.3) is satisfied. In this case, $U^t_{A_c}$ is strongly stable and the attractor is globally uniformly attractive.

In the sequel, we will rather focus on the point spectrum, whose decomposition satisfies the following property.

**Proposition 2.2** *Suppose that $\mathscr{F} = C(X)$. If the attractor $A$ is globally (pointwise or uniformly) attractive in $X$, the Koopman eigenfunctions satisfy*

$$\phi_\lambda \in \mathscr{F}_{A_c} \Leftrightarrow \lambda \in \sigma_p^-(L). \tag{2.4}$$
$$\phi_\lambda \notin \mathscr{F}_{A_c} \Leftrightarrow \lambda \in \sigma_p^0(L). \tag{2.5}$$

*Proof* If the attractor is uniformly attractive, the implication $\phi_\lambda \in \mathscr{F}_{A_c} \Rightarrow \text{Re}\{\lambda\} < 0$ follows from Proposition 2.1. The proof of the implication $\text{Re}\{\lambda\} < 0 \Rightarrow \phi_\lambda \in \mathscr{F}_{A_c}$ directly follows from the continuity of $\phi_\lambda$ and from the fact that $A$ is globally attractive (see [16] for the details of the proof in the case of pointwise attractivity). □

The eigenfunctions $\phi_\lambda \notin \mathscr{F}_{A_c}$ are associated with eigenvalues $\lambda \in \sigma_p^0(L)$ that do not capture the stability properties of the system. Instead these eigenfunctions are related to asymptotic dynamics on the attractor (e.g., periodic invariant sets of ergodic dynamics [18], isochrons of limit cycles [15]), see also Chap. 1. In contrast, the eigenfunctions $\phi_\lambda \in \mathscr{F}_{A_c}$ are associated with eigenvalues $\lambda \in \sigma_p^-(L)$ and capture the stability properties of the system. This is made more precise with the following result.

**Theorem 2.2** *Consider a finite set of eigenvalues $\Lambda \subseteq \sigma_p^-(L)$ with $\phi_\lambda \in C(X)$ for all $\lambda \in \Lambda$. Then the intersection of the zero level sets*

$$M = \bigcap_{\lambda \in \Lambda} \{\mathbf{x} \in X | \phi_\lambda(\mathbf{x}) = 0\}$$

*is invariant under $\mathbf{S}^t$ and globally uniformly asymptotically stable.*

*Proof* Consider the function $\Phi = \sum_{\lambda \in \Lambda} |\phi_\lambda|$ and the family of sets

$$M_\alpha = \{\mathbf{x} \in X | \Phi(\mathbf{x}) \leq \alpha\} \qquad \alpha > 0.$$

*Invariance.* We first observe that

$$\Phi(\mathbf{S}^t(\mathbf{x})) = \sum_{\lambda \in \Lambda} e^{\mathrm{Re}\{\lambda\}t} |\phi_\lambda(\mathbf{x})| \leq e^{st} \Phi(\mathbf{x}) \tag{2.6}$$

with $s = \max_{\lambda \in \Lambda}(\mathrm{Re}\{\lambda\}) < 0$. For all $\mathbf{x} \in M_\alpha$ with $\alpha > 0$ and $t > 0$, it follows that we have $\Phi(\mathbf{S}^t(\mathbf{x})) < \Phi(\mathbf{x}) = \alpha$, so that $M_\alpha$ is forward-invariant. For $\alpha = 0$ and for all $t \in \mathbb{R}$, $\Phi(\mathbf{S}^t(\mathbf{x})) = \Phi(\mathbf{x}) = 0$ so that the set $M = M_0$ is invariant.
*Global uniform attractivity.* For some $\varepsilon > 0$, there exists $\alpha > 0$ such that

$$\mathbf{S}^t(\mathbf{x}) \in M_\alpha \Rightarrow d(\mathbf{S}^t(\mathbf{x}), M_0) < \varepsilon$$

by continuity of $\Phi$ and since $\Phi(\mathbf{x}) = 0$ for all $\mathbf{x} \in M_0$. Moreover, (2.6) implies that $\mathbf{S}^t(\mathbf{x}) \in M_\alpha$ for all $t > T$ with

$$T = \frac{1}{|s|} \log \frac{\max_{\mathbf{x} \in X} \Phi(\mathbf{x})}{\alpha}.$$

*Stability.* Stability follows from uniform attractivity (see, e.g., [10]). Alternatively, we also verify that, for some $\varepsilon > 0$, there exist $\alpha, \delta > 0$ such that

$$d(\mathbf{x}, M_0) < \delta \Rightarrow \mathbf{x} \in M_\alpha \Rightarrow \mathbf{S}^t(\mathbf{x}) \in M_\alpha \Rightarrow d(\mathbf{S}^t(\mathbf{x}), M_0) < \varepsilon$$

for all $t > 0$ since $M_\alpha$ is forward-invariant. □

As illustrated in the next section, this result can be used to show that the attractor is globally asymptotically stable. Since the eigenfunctions $\phi_\lambda$ with $\lambda \in \sigma_p^-(L)$ are zero on the attractor (Proposition 2.2), we have that $A \subseteq M$. Therefore a sufficiently large (but finite) number of eigenfunctions must be considered to ensure that the intersection of their zero level sets satisfy $M = A$.

### 2.3.3 *Specific Cases*

We can now consider specific attractors, with a focus on equilibria.

#### 2.3.3.1 Hyperbolic Equilibria

Suppose that the system (2.1) admits an equilibrium $\mathbf{x}^* \in X$ such that $\mathbf{F}(\mathbf{x}^*) = \mathbf{0}$. Let $\mathbf{J} = \frac{\partial \mathbf{F}}{\partial \mathbf{x}}(\mathbf{x}^*)$ be the Jacobian matrix evaluated at $\mathbf{x}^*$ and denote by $\lambda_j$, $j = 1, \ldots, n$, the eigenvalues of $\mathbf{J}$. For continuously differentiable eigenfunctions, a first-order Taylor expansion of the eigenvalue Eq. (2.2) yields

$$\mathbf{J}^T \nabla \phi_\lambda(\mathbf{x}^*) = \lambda \nabla \phi_\lambda(\mathbf{x}^*) . \tag{2.7}$$

This implies that Koopman eigenvalues with nonzero gradient at the equilibrium are associated with the eigenvalues $\lambda_j$ of the Jacobian matrix $\mathbf{J}$. If the equilibrium is stable and hyperbolic (e.g., $\mathrm{Re}\{\lambda_j\} < 0$ for all $j$), it turns out that these eigenfunctions can be used to prove that the equilibrium is globally asymptotically stable in $X$.

**Proposition 2.3** *Let $X \subset \mathbb{R}^n$ be a connected set that satisfies Assumption* 2.1. *Assume that* (2.1) *with* $\mathbf{F} \in C^2(X)$ *admits an equilibrium* $\mathbf{x}^* \in X$ *and that the Jacobian matrix* $\mathbf{J}$ *is diagonalizable. Then, the equilibrium is hyperbolic and globally uniformly asymptotically stable in X if and only if the Koopman operator associated with* (2.1) *admits n (distinct) eigenfunctions* $\phi_{\lambda_j} \in C^1(X)$, *with* $\mathrm{Re}\{\lambda_j\} < 0$ *and such that* $\nabla \phi_{\lambda_j}(\mathbf{x}^*) \neq \mathbf{0}$. *Moreover, the eigenvalues* $\lambda_j$ *are the eigenvalues of* $\mathbf{J}$.

*Proof Sufficiency.* It follows from (2.7) that the Koopman eigenvalue $\lambda_j$ is an eigenvalue of the Jacobian matrix $J$. Moreover, $\nabla \phi_{\lambda_j}(\mathbf{x}^*)$ is the left eigenvector $\mathbf{w}_j$ of $\mathbf{J}$. Since the $n$ eigenvectors are independent, we have

$$\bigcap_{j=1}^{n} \left\{ \mathbf{x} \in \mathscr{V} \mid \phi_{\lambda_j}(\mathbf{x}) = 0 \right\} = \left\{ \mathbf{x}^* \right\} , \tag{2.8}$$

where $\mathscr{V}$ is a small neighborhood of $\mathbf{x}^*$. This is a simple consequence of intersection theory as the rank of the gradient matrix is full, so that the intersection is a 0-dimensional manifold, i.e., a finite set of points [7]. Note that the argument needs to be extended to complex-valued functions (see [16] for more details).

It remains to show that the zero level sets cannot have another intersection in $X$. Suppose that there is another intersection lying in an invariant set that is not connected to the fixed point. This implies that the boundary $\partial \Omega$ of the basin of attraction $\Omega$ of $\mathbf{x}^*$

has a non empty intersection with $X$, since $X$ is a connected set. Since $X$ is forward-invariant, $\partial\Omega \cap X$ is forward-invariant and contains the limit sets of its trajectories. Consider $\mathbf{x}_\omega \in \partial\Omega \cap X$ that belongs to a limit set. By definition and continuity of the eigenfunctions, we have $\phi_{\lambda_i}(\mathbf{x}_\omega) = 0\,\forall i$. Consider also a neighborhood $\mathscr{V}_\varepsilon$ of $\mathbf{x}_\omega$ such that $\mathscr{V}_\varepsilon \cap \Omega \neq \emptyset$. For all $\mathbf{x}_\varepsilon \in \mathscr{V}_\varepsilon \cap \Omega$, there exist $\mathbf{x}_0 \in \mathscr{V} \setminus \{\mathbf{x}^*\}$ and $T > 0$ such that $\varphi^{-T}(\mathbf{x}_0) = \mathbf{x}_\varepsilon$. Moreover, (2.8) implies that there is at least one eigenfunction that satisfies $|\phi_{\lambda_i}(\mathbf{x}_0)| = C > 0$ and we have $|\phi_{\lambda_i}(\mathbf{x}_\varepsilon)| = C\exp(-\mathrm{Re}\{\lambda_i\}T) > 0$. Therefore, $\phi_{\lambda_i}$ is not continuous in $\mathscr{V}_\varepsilon \subset X$, which is a contradiction.

Finally, since $\mathrm{Re}\{\lambda_i\} < 0$, the result follows from Theorem 2.2 with $M = \{\mathbf{x}^*\}$.
*Necessity.* (The proof is inspired from [11].) Since the equilibrium is globally asymptotically stable in $X$, it follows from Theorem 2.3 in [11] that there exists a $C^1$ diffeomorphism $\mathbf{y} = \mathbf{h}(\mathbf{x})$ such that $\dot{\mathbf{y}} = \mathbf{J}\,\mathbf{y}$, $\mathbf{h}(\mathbf{x}^*) = \mathbf{0}$, and the Jacobian matrix of $\mathbf{h}$ at $\mathbf{0}$ satisfies $\mathbf{J_h} = \mathbf{I}$. For this linear system, there exist $n$ distinct Koopman eigenfunctions $\tilde{\phi}_{\lambda_j}(\mathbf{y}) = \mathbf{y}^T\mathbf{w}_j$ that are associated with the eigenvalues of $\mathbf{J}$. It follows that the Koopman operator of (2.1) has $n$ continuously differentiable eigenfunctions of the form $\phi_{\lambda_j} = \tilde{\phi}_{\lambda_j} \circ \mathbf{h}$. Moreover, we have $\nabla\phi_{\lambda_j}(\mathbf{x}^*) = \mathbf{J_h}^T\,\mathbf{w}_j = \mathbf{w}_j \neq \mathbf{0}$. This concludes the proof. □

The result could potentially be extended to the case of non-diagonalizable Jacobian matrices, using generalized eigenfunctions [20]. We note, however, that the choice of the space $C^1(X)$ is important here, as it allows to select principal eigenfunctions that capture the stability properties of the equilibrium.

For a linear system $\dot{\mathbf{x}} = A x$, we have $\phi_{\lambda_j}(\mathbf{x}) = \mathbf{x}^T\mathbf{w}_j$ where $\mathbf{w}_j$ is a left eigenvector of $\mathbf{A}$. It is clear that $\phi_{\lambda_j} \in C^1(\mathbb{R}^n)$ and $\nabla\phi_{\lambda_j}(\mathbf{x}^*) = \mathbf{w}_j \neq \mathbf{0}$. In this case, Proposition 2.3 implies that the system is globally stable if $\mathrm{Re}\{\lambda_j\} < 0$ for all $j$, so that we recover the classic stability condition for linear systems.

In fact, the result of Proposition 2.3 mirrors classic local stability analysis of nonlinear systems, but is valid globally. While local stability relies on $n$ eigenvalues of the Jacobian matrix, global stability relies on $n$ (continuously differentiable) Koopman eigenfunctions. The support of these specific eigenfunctions corresponds to the basin of attraction of the equilibrium. In fact, the eigenfunctions blow up at the basin boundary as they grow exponentially in backwards time, but they are typically defined everywhere else.

*Example 2.1* The dynamics $\dot{x} = -\mu x + x^2$, with $x \in \mathbb{R}$, are associated with the Koopman eigenfunction of the form $\phi_\lambda(x) = \frac{x}{1-x/\mu}$, with $\lambda = -\mu$. We observe that the eigenfunction is well defined on the basin of attraction $(-\infty, \mu)$ of the equilibrium $x^* = 0$ and blows up at the boundary $x = \mu$. △

The eigenfunctions provide more information than mere stability. In particular, the eigenfunction associated with the dominant eigenvalue $\lambda_1$, with $\mathrm{Re}\{\lambda_1\} > \mathrm{Re}\{\lambda_j\}$ for all $j \neq 1$, captures the dominant asymptotic behavior toward the equilibrium and is related to the notion of isostables [17].

*Remark 2.3* (*Stable, unstable, and center manifolds*) If the equilibrium is not stable, the point spectrum may contain a part $\sigma_p^+(L)$ with $\mathrm{Re}\{\lambda\} > 0$ for all $\lambda \in \sigma_p^+(L)$. The intersection of zero level sets of eigenfunctions $\phi_\lambda$ with $\lambda \in \sigma_p^-(L)$ does not coincide with the equilibrium, but with its *unstable manifold*. Similarly, the intersection of zero level sets of eigenfunctions $\phi_\lambda$ with $\lambda \in \sigma_p^+(L)$ corresponds to the *stable manifold*. If the equilibrium is not a hyperbolic attractor (see below), the intersection of level sets of eigenfunctions $\phi_\lambda$ with $\lambda \in \sigma_p^-(L) \cup \sigma_p^+(L)$ corresponds to a *center manifold* [20]. △

#### 2.3.3.2 Non-hyperbolic Equilibria

If the equilibrium is not hyperbolic, that is, at least one eigenvalue $\lambda_j$ of the Jacobian matrix $\mathbf{J}$ is such that $\mathrm{Re}\{\lambda_j\} = 0$, the result of Proposition 2.3 does not hold. Recall that, according to (2.7), eigenfunctions with nonzero gradient at the equilibrium are associated with the eigenvalues $\lambda_j$. It follows that one cannot find $n$ eigenfunctions $\phi_\lambda$ such that $\nabla\phi_\lambda \neq 0$ and $\mathrm{Re}\{\lambda\} < 0$. Moreover, the necessity part of Proposition 2.3 relies on the Hartman–Grobman theorem, which cannot be applied in this case.

In the case of a non-hyperbolic attractive equilibrium, the Koopman operator defined on the space of continuous functions admits a large set of eigenvalues with strictly negative real parts. Then, provided that one can find a set of associated eigenfunctions such that the intersection of their zero level sets is the equilibrium, Theorem 2.2 can be used to prove global stability.

*Example 2.2* The dynamics $\dot{x} = -x^3$, with $x \in \mathbb{R}$, are associated with Koopman eigenfunctions of the form $\phi_\lambda(x) = \exp(\lambda/(2x^2))$, where $\lambda < 0$. One of those eigenvalues can be used with Theorem 2.2 to prove that the origin is globally asymptotically stable. △

*Remark 2.4* (*Lyapunov stability*) When a non-hyperbolic equilibrium is stable (in Lyapunov sense, see Definition 2.1), Koopman eigenfunctions might be used to claim stability. For instance, this can be useful for systems that have only one eigenfunction $\phi_{\lambda_0}$ with 0 real part. In this case, all the initial conditions starting off the joint zero level set $M_0$ of the stable eigenfunctions will remain on $M_0$. If the level sets of $\phi_{\lambda_0}$ on $M_0$ are closed and bounded, then the system is stable. △

#### 2.3.3.3 Limit Cycles

A result similar to Proposition 2.3 can be obtained in the case of normally hyperbolic limit cycles. In this case, the global stability property is related to the Koopman eigenfunctions associated with the nonzero Floquet exponents of the limit cycle. We do not develop the results here and refer to [16] for the details.

## 2.4 Numerical Methods

In this section, we describe two methods to compute the Koopman eigenfunctions, and therefore infer global stability of the system using the results of Sect. 2.3. The first method relies on the finite-dimensional approximation of the Koopman operator in a basis of functions. The second method is based on polynomial approximations computed through sum-of-squares methods.

### 2.4.1 *Finite-Dimensional Approximation of the Eigenfunctions*

Consider a basis of functions $\{\psi_k\}_{k=1}^N$ that spans a subspace $\mathscr{F}_N \subset \mathscr{F}_{A_c}$ and a projection operator $\Pi : \mathscr{F} \to \mathscr{F}_{A_c}$. With these two ingredients, one can compute a finite-dimensional approximation

$$L_N = \Pi L|_{\mathscr{F}_N} : \mathscr{F}_N \to \mathscr{F}_N$$

of the infinitesimal generator $L$ of the Koopman operator. The operator $L_N$ can be represented by a matrix $\mathbf{L} \in \mathbb{R}^{N \times N}$ whose $k$th column contains the coordinates of $L_N \psi_k$ in the basis of functions, i.e., $L_N \psi_k = \sum_{j=1}^N \mathbf{L}_{jk} \psi_j$. The eigenvalues $\tilde{\lambda}$ of $\mathbf{L}$ approximate the Koopman eigenvalues $\lambda$, and the corresponding eigenvectors $\mathbf{v} = (v_1 \cdots v_N)$ are the coordinates of approximate projections of the Koopman eigenfunctions in the basis of functions, that is $\phi_\lambda \approx \sum_{j=1}^N v_j \psi_j$. More details can be found in Chap. 1.

When the equilibrium is hyperbolic and the vector field is polynomial, a basis of monomials is a natural choice to compute the Koopman eigenfunctions. Moreover, in this case, the eigenvalue problem can be solved iteratively [16]. This is equivalent to Carleman linearization [2].

*Example 2.3 (Hyperbolic equilibrium (Example* 2.1 *continued))* Consider the dynamics $\dot{x} = -\mu x + x^2$, with $x \in \mathbb{R}$, which admits a stable equilibrium at the origin. We compute the Koopman eigenfunction $\phi_{-\mu}$ with a basis of monomials and with the projection operator $\Pi$ such that $\Pi(x^k) = x^k$ if $k \leq n$ and $\Pi(x^k) = 0$ if $k > n$. The matrix $\mathbf{L}$ is given by

$$\mathbf{L} = \begin{bmatrix} -\mu & 0 & 0 & \dots & & 0 \\ 1 & -2\mu & 0 & & & 0 \\ 0 & 2 & -3\mu & \ddots & & \\ & & \ddots & \ddots & & \\ & & & & -(n-1)\mu & 0 \\ & & & & n-1 & -n\mu \end{bmatrix}$$

and its eigenvector associated with the eigenvalue $-\mu$ is equal to $(1, 1/\mu, 1/\mu^2, \dots, 1/\mu^n)^T$, which yields the approximate eigenfunction

$$\tilde{\phi}_{-\mu}(x) = x\left(1 + \frac{x}{\mu} + \frac{x^2}{\mu^2} + \cdots + \frac{x^n}{\mu^n}\right) .$$

When $n \to \infty$, the series converges on $(-\infty, \mu)$ to the eigenfunction $\phi_{-\mu}$ (see Example 2.1). Note that even for finite values $n$, the function $\tilde{\phi}_{-\mu}$ diverges at $x = \mu$ and therefore accurately captures the boundary of the basin of attraction. △

*Example 2.4* (*Hyperbolic equilibrium*) Consider the system

$$\dot{x}_1 = x_2 \tag{2.9}$$

$$\dot{x}_2 = -2x_1 + \frac{1}{3}x_1^3 - 3x_2 \tag{2.10}$$

which admits a stable equilibrium at the origin and two saddle nodes at $(\pm\sqrt{6}, 0)$. The eigenvalues of the Jacobian matrix at the origin are $\lambda_1 = -1$ and $\lambda_2 = -2$. We compute the Koopman eigenfunctions $\phi_{\lambda_1}$ and $\phi_{\lambda_2}$ with a basis of monomials and with the projection operator

$$\Pi(x_1^{k_1} x_2^{k_2}) = \begin{cases} x_1^{k_1} x_2^{k_2} & \text{if } k_1 + k_2 \leq 20 \\ 0 & \text{if } k_1 + k_2 > 20 \end{cases} . \tag{2.11}$$

The level sets of the Koopman eigenfunctions provide a good estimation of the basin of attraction (Fig. 2.1). △

The choice of a basis of monomials is motivated by the following property. If the vector field is analytic then, under mild additional conditions, the Koopman eigenfunctions are analytic in some neighborhood of a stable hyperbolic equilibrium. This follows from Poincaré linearization theorem (see, e.g., [6]). However, the radius

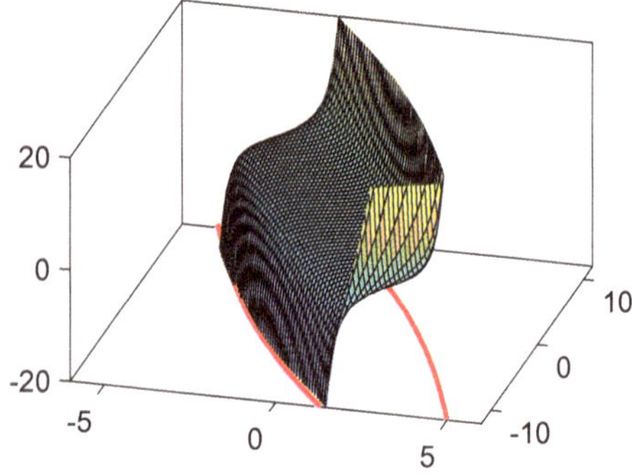


**Fig. 2.1** The Koopman eigenfunction $\phi_{\lambda_1}$ can be used to obtain an estimation of the geometry of the basin of attraction (Example 2.4). The red curves are the boundary of the basin of attraction

of the analyticity disk is constrained by other (unstable) equilibria, so that it might be difficult to obtain tight bounds for large regions of attraction.

The case of non-hyperbolic equilibria is more delicate. Indeed, Koopman eigenfunctions relevant to stability (that is, associated with eigenvalues $\lambda$ such that $\mathrm{Re}\{\lambda\} < 0$) are not all analytic at a non-hyperbolic equilibrium. For instance, in Example 2.2, we have $\phi_\lambda(x) = \exp(\lambda/(2x^2))$ and $d^k\phi_\lambda/dx^k(0) = 0$ for all $k \in \mathbb{N}$, so that the eigenfunctions $\phi_\lambda$, with $\lambda < 0$, are not analytic at the origin. It follows that numerical computations using a basis of monomials can only capture analytic eigenfunctions $\phi_\lambda$ with $\mathrm{Re}\{\lambda\} = 0$, so that the method is not appropriate to infer stability. However, other bases of functions can still provide good approximations of the eigenfunctions relevant to stability (e.g., basis of polynomials, Fourier basis). This is illustrated in the following example.

*Example 2.5 (Non-hyperbolic equilibrium)* Consider the system

$$
\begin{aligned}
\dot{x}_1 &= -x_2 - x_1(x_1^2 + x_2^2) \\
\dot{x}_2 &= x_1 - x_2(x_1^2 + x_2^2)
\end{aligned}
$$

which admits a stable non-hyperbolic equilibrium at the origin. The Koopman eigenfunctions are computed with a basis of Legendre polynomials $p_{k_1}(x_1)p_{k_2}(x_2)$ and with the projection operator

$$
\Pi(p_{k_1}(x_1)p_{k_2}(x_2)) = \begin{cases} p_{k_1}(x_1)p_{k_2}(x_2) & \text{if } k_1 + k_2 \leq 10 \\ 0 & \text{if } k_1 + k_2 > 10 \end{cases}.
$$

For comparison purposes, we also consider a Fourier basis of functions $e^{i(k_1\omega_1 x_1 + k_2\omega_2 x_2)}$ with the projection operator

$$
\Pi f(x_1, x_2) = \sum_{k_1,k_2=-10}^{10} c_{k_1,k_2} e^{\mathrm{i}(k_1\omega_1 x_1 + k_2\omega_2 x_2)},
$$

where the Fourier coefficients $c_{k_1,k_2}$ are computed over the interval $[-5, 5] \times [-5, 5]$ ($\omega_1 = \omega_2 = 2\pi/10$). In both cases, a Koopman eigenfunction $\phi_\lambda$ with $\mathrm{Re}\{\lambda\} < 0$ is obtained and therefore implies global asymptotic stability (Fig. 2.2). △

### 2.4.2 Polynomial Approximations Using Sum-of-Squares Programming

Computation of Koopman eigenfunctions using finite-dimensional representations of the Koopman operator is associated with approximation errors which are hard to quantify. In order to obtain certain approximation guarantees, we can optimize over

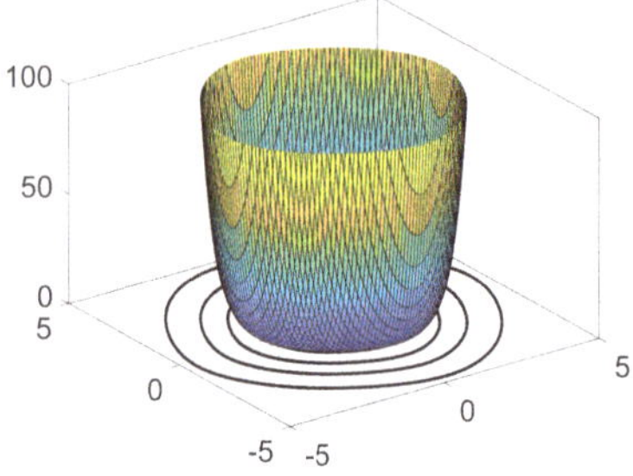


(a) Eigenfunction $|\phi_\lambda|$ with $\lambda = -0.87$ (basis of Legendre polynomials)

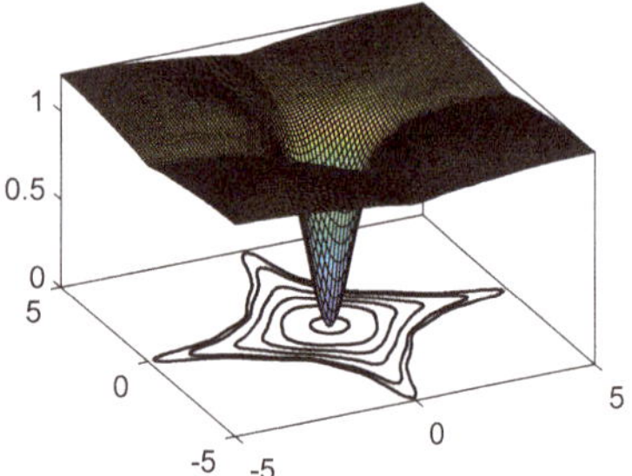


(b) Eigenfunction $|\phi_\lambda|$ with $\lambda = -0.11 + i$ (Fourier basis)

**Fig. 2.2** A Koopman eigenfunction $\phi_\lambda$ can be computed and used to prove that a non-hyperbolic equilibrium is globally stable (Example 2.5). The black curves are the level sets of $|\phi_\lambda|$ are in black. **a** The Koopman eigenfunction $\phi_\lambda$, with $\lambda = -0.11 + \mathrm{i}$, is computed with a basis of Legendre polynomials. **b** The Koopman eigenfunction $\phi_\lambda$, with $\lambda = -0.87$, is computed with a Fourier basis

the tail of the finite-dimensional representations using polynomial optimization tools such as sum-of-squares relaxations.[1]

We consider the system $\dot{\mathbf{x}} = \mathbf{F}(\mathbf{x})$ on the connected and compact set $X$ (but not necessarily forward-invariant). Without loss of generality, we assume that the equilibrium satisfies $\mathbf{x}^* = \mathbf{0}$ and $\mathbf{x}^* \in X$. Furthermore, we require that

(i) the set $X$ can be represented by polynomial constraints, i.e.,

$$X = \{\mathbf{x} \in \mathbb{R}^n | g_j(\mathbf{x}) \leq 0,\ j = 1, \ldots, K\},$$

where $g_j(\mathbf{x})$ are finite degree polynomials;

(ii) the eigenfunctions $\phi_\lambda$ are analytic on $X$.

These assumptions are quite restrictive, but we can always choose a set $X$ (for example, a Euclidean ball around the equilibrium) inside the region of analyticity of $\phi_\lambda(\mathbf{x})$. In the further analysis we focus on the case $X = \{\mathbf{x} \in \mathbb{R}^n | g(\mathbf{x}) \leq 0\}$ for simplicity. We represent the eigenfunctions in the basis of monomials for which sum-of-squares techniques are well developed. Consider the polynomial approximation obtained above (see Sect. 2.4.1), which we denote by $\phi_\lambda^{N_0}(\mathbf{x})$ and represent as follows:

$$\phi_\lambda^{N_0}(\mathbf{x}) = \sum_{(i_1,\ldots,i_n) \in \mathscr{I}^{N_0}} \beta_{i_1,\ldots,i_n} x_1^{i_1} \cdot \ldots \cdot x_n^{i_n},$$

where $N_0$ is the predefined (total) degree of the polynomial, the set of indices is given by

[1] A sum-of-squares relaxation for polynomial optimization problem is technically a restriction since the space of possible solutions is smaller. However, this nomenclature became standard and the relaxation should be understood as making the problem tractable.

$$\mathscr{I}^{N_0} = \left\{ (i_1, \dots, i_n) \in \mathbb{N}^n | i_j \geq 0, \sum_{j=1}^{n} i_j \leq N_0 \right\}$$

and the coefficients $\beta_{i_1,\dots,i_n}$ are real when $\lambda$ is real and complex when $\lambda$ is complex. We will focus on the case where $\lambda$ is real. A similar technique can be used in the complex case by considering the real and complex parts separately. We extend the polynomial approximation $\phi_\lambda^{N_0}(\mathbf{x})$ by adding the tail expression as follows:

$$\tilde{\phi}_\lambda^N(\mathbf{x}) = \phi_\lambda^{N_0}(\mathbf{x}) + \sum_{(i_1,\dots,i_n)\in\mathscr{I}^N\setminus\mathscr{I}^{N_0}} \alpha_{i_1,\dots,i_n} x_1^{i_1} \cdot \dots \cdot x_n^{i_n},$$

where $N$ is the predefined (total) degree of the polynomial, the coefficients $\alpha_{i_1,\dots,i_n}$ are the decision variables, and the coefficients $\beta_{i_1,\dots,i_n}$ describing the polynomial $\phi_\lambda^{N_0}(\mathbf{x})$ are given. Now we can cast the polynomial eigenfunction matching by using the following optimization program:

$$\min_{\tilde{\phi}_\lambda^N(\mathbf{x}),\gamma} \quad \gamma \tag{2.12}$$

$$\text{s.t.} \quad -\gamma \leq (\mathbf{F}(\mathbf{x}))^T \nabla\tilde{\phi}_\lambda^N(\mathbf{x}) - \lambda\tilde{\phi}_\lambda^N(\mathbf{x}) \leq \gamma, \ \forall \mathbf{x} \in X, \tag{2.13}$$

where with a slight abuse of notation we indicate that $\tilde{\phi}_\lambda^N(\mathbf{x})$ is a decision variable, while in reality the decision variables are the coefficients $\alpha_{i_1,\dots,i_n}$. With $N \to \infty$ the solution of (2.12)–(2.13) approximates the solution of the eigenvalue Eq. (2.2), and hence the eigenfunction $\phi_\lambda(\mathbf{x})$. With a finite $N$ we are trying to find the polynomial that matches the eigenvalue equation the best while keeping the values of the first $N_0$ derivatives in the Taylor expansion.

The optimization problem (2.12) belongs to the class of *polynomial optimization problems*. The inequality constraints (2.13) can be represented as polynomial positivity constraints, which are convex but generally not tractable. Indeed, while the number of decision variables in this program is finite, the program itself is infinite dimensional since $\mathbf{x}$ varies over the set $X$. This type of programs can be addressed using a sum-of-squares relaxation [22], where a constraint "$g(\mathbf{x}) \geq 0$ for all $\mathbf{x} \in \mathbb{R}^n$" is replaced by "$g(\mathbf{x})$ is a sum-of-squares of polynomials", i.e., $g(\mathbf{x}) = (\Psi(\mathbf{x}))^T \mathbf{Q}\Psi(\mathbf{x})$ for all $\mathbf{x} \in \mathbb{R}^n$, where $\mathbf{Q}$ is a positive semidefinite matrix and $\Psi(\mathbf{x})$ is a vector of monomials with fixed maximum total degrees. In short, we say that $g(\mathbf{x})$ is an SOS polynomial. Note that if $\mathbf{Q}$ is positive semidefinite then $g(\mathbf{x}) = (\Psi(\mathbf{x}))^T \mathbf{Q}\Psi(\mathbf{x})$ is nonnegative for all $\mathbf{x}$ by definition. Hence, this parametrization allows to eliminate the variable $\mathbf{x}$ from the constraint on $g(\mathbf{x})$. Similarly we can eliminate the variable $\mathbf{x}$ from other constraints, which results in a semidefinite optimization program. Formulating such programs is straightforward using off-the-shelf SOS problem parsers [13, 21], which also implement substantial pre- and post-processing of the SOS programs, while the solutions can be obtained using off-the-shelf free and commercial semidefinite optimization solvers [1, 25]. The SOS programming can also efficiently handle restrictions of $\mathbf{x}$ to a set represented using polynomial constraints [21]. In particular, the constraint of the type $c(\mathbf{x}) \geq 0$ for all $\mathbf{x} \in \{\mathbf{x} \in \mathbb{R}^n | g(\mathbf{x}) \leq 0\}$ for polynomials $c$ and $g$ can be relaxed as "$c(\mathbf{x}) + a(\mathbf{x})g(\mathbf{x})$ and $a(\mathbf{x})$ are SOS polynomials for all $\mathbf{x} \in \mathbb{R}^n$". The complexity of the program is polynomial in the size of the matrix $\mathbf{Q}$, which, however, grows in a combinatorial fashion with the number of states and degree of monomials. We will not further discuss SOS programming but refer the reader to the references above. In the case of $K = 1$, i.e., $X = \{\mathbf{x} \in \mathbb{R}^n | g(\mathbf{x}) \leq 0\}$, we cast the problem as follows:

$$\begin{aligned}
\min_{a,b,\tilde{\phi}_\lambda^N,\gamma} \quad & \gamma \\
\text{s. t.} \quad & \gamma - (\mathbf{F}(\mathbf{x}))^T \nabla\tilde{\phi}_\lambda^N(\mathbf{x}) + \lambda\tilde{\phi}_\lambda^N(\mathbf{x}) + a(\mathbf{x})g(\mathbf{x}) \text{ is SOS} \\
& \gamma + (\mathbf{F}(\mathbf{x}))^T \nabla\tilde{\phi}_\lambda^N(\mathbf{x}) - \lambda\tilde{\phi}_\lambda^N(\mathbf{x}) + b(\mathbf{x})g(\mathbf{x}) \text{ is SOS} \\
& a(\mathbf{x}), b(\mathbf{x}) \text{ are SOS} \\
& \gamma \geq 0,
\end{aligned}$$

where $a$, $b$ are polynomials of fixed degrees. For a large $K$ we need to introduce further SOS polynomials in the optimization.

*Example 2.6* Consider the FitzHugh–Nagumo system described by the following model

$$\begin{aligned}
\dot{\tilde{x}}_1 &= -\tilde{x}_2 - \tilde{x}_1(\tilde{x}_1 - 1)(\tilde{x}_1 - a) + I \\
\dot{\tilde{x}}_2 &= \varepsilon(\tilde{x}_1 - \gamma\tilde{x}_2)
\end{aligned}$$

with $a = 1$, $I = 0.05$, $\varepsilon = 0.08$, $\gamma = 1$. With this set of parameters the system has a unique equilibrium $\tilde{x}_1^* = 0.0256$, $\tilde{x}_2^* = 0.0256$. For exposition purposes, we move the state space so that the equilibrium lies in the origin, i.e., we define $x_1 = \tilde{x}_1 - \tilde{x}_2^*$, $x_2 = \tilde{x}_2 - \tilde{x}_2^*$. We compute the eigenfunction $\phi_{\lambda_1}(x)$ on the set $X = \{(x_1, x_2) \in \mathbb{R}^2 | x_1^2 + x_2^2 - 0.3^2 \leq 0\}$. We choose $N_0 = 1$, that is, we compute the right eigenvector $\mathbf{w} = (w_1, w_2)$ of the Jacobian at the equilibrium corresponding to the dominant eigenvalue $\lambda_1$ and set $\phi_\lambda^M(\mathbf{x}) = w_1 x_1 + w_2 x_2$. We set the total degree $N = 10$ and the total degrees of $a(\mathbf{x})$ and $b(\mathbf{x})$ to be equal to 20. We also evaluate the eigenfunction $\phi_{\lambda_1}(\mathbf{x})$ using the so-called Laplace averages[2] on the equidistant mesh grid $[-0.3, 0.3] \times [-0.3, 0.3]$ with 40 points each in each dimension. The overall error

[2] The Laplace average computed along a trajectory of the system is given by

$$f_\lambda^*(\mathbf{x}) = \lim_{t\to\infty} \frac{1}{T} \int_0^T (f \circ \mathbf{S}(t, \mathbf{x}))\, e^{-\lambda t} dt\ . \tag{2.14}$$

If $f(\mathbf{x}^*) = 0$, $f^*$ is equal to the eigenfunction $\phi_{\lambda_1}$ if it is finite. See [15, 19] and Chaps. 1, 11, and 15 for more details.

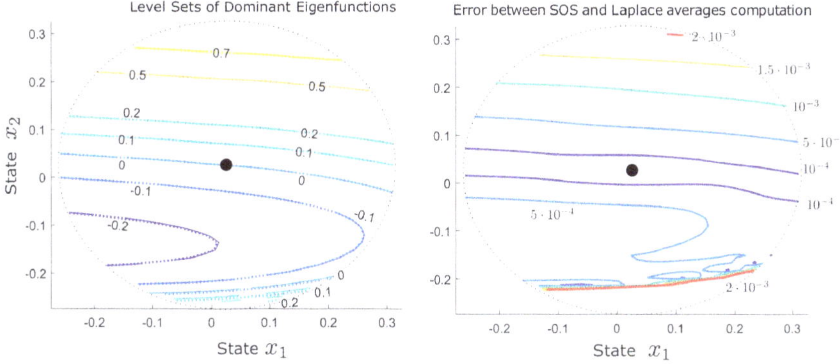


**Fig. 2.3** Computation of $\phi_{\lambda_1}$ using SOS (solid lines in the left panel), Laplace Averages (dashed lines in the left panel) and the error between them (right panel). Note that the solid and dashed lines significantly overlap. The circle in the left and center panel depicts the boundary of the set $X_2$, while the dot stands for the exponentially stable equilibrium point

between the two results is $2.5 \cdot 10^{-3}$. Detailed numerical results are depicted in Fig. 2.3. △

## 2.5 Lyapunov Function Design

The finite-dimensional computation of the eigenfunctions yields approximation errors, so that it does not provide rigorous stability garantees. Focusing on the case of equilibria, we can use the eigenfunctions, and more generally the Koopman operator framework, to construct in a systematic way Lyapunov functions that provide strict stability guarantees.

### *2.5.1 Lyapunov Functions Based on Koopman Eigenfunctions*

In the case of systems with an equilibrium, the eigenfunctions yield candidate Lyapunov functions of the form

$$V(\mathbf{x}) = \left( \sum_{j=1}^{n} c_j \, |\phi_{\lambda_j}(\mathbf{x})|^p \right)^{1/p} \tag{2.15}$$

for some $p \in \mathbb{N}$ and with $c_j > 0$. We verify that

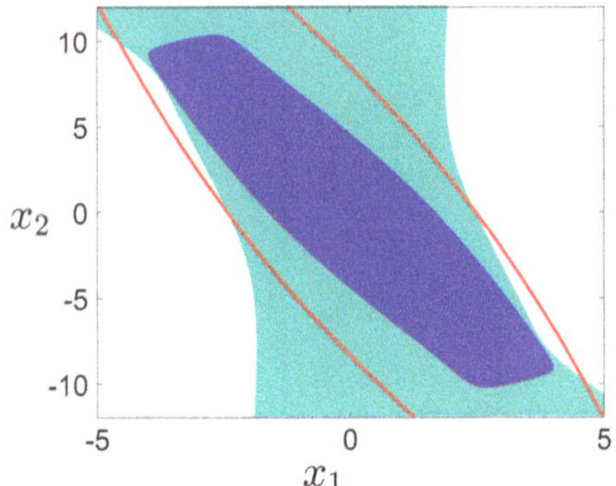


**Fig. 2.4** Koopman eigenfunctions can be used to construct a candidate Lyapunov function (Example 2.7). The largest sublevel set of the Lyapunov function $V = 10\phi_{\lambda_1}^2 + \phi_{\lambda_2}^2$ lying in the region $\overline{X}$ (green region) provides an inner approximation (blue region) of the basin of attraction. The red curves are the boundaries of the basin of attraction

$$U^t V(\mathbf{x}) = \left(\sum_{j=1}^{n} c_j \, |U^t \phi_{\lambda_j}(\mathbf{x})|^p\right)^{1/p} = \left(\sum_{j=1}^{n} e^{p\mathrm{Re}\{\lambda_j\}t} \, c_j \, |\phi_{\lambda_j}(\mathbf{x})|^p\right)^{1/p} < e^{\mathrm{Re}\{\lambda_1\}t} \, V(\mathbf{x})$$

or equivalently $\dot{V}(\mathbf{x}) = LV(\mathbf{x}) < \mathrm{Re}\{\lambda_1\}V(\mathbf{x})$ for all $\mathbf{x} \in \mathrm{Dom}(V) \setminus \{\mathbf{x}^*\}$.

Note that in the case of linear systems $\dot{\mathbf{x}} = \boldsymbol{Ax}$, the Lyapunov function (2.15) with $p = 2$ and $c_j = 1$ takes the form $V(\mathbf{x}) = (\mathbf{x}^T \mathbf{W}\mathbf{W}^T \mathbf{x})^{1/2}$, where the columns of $\mathbf{W}$ are the left eigenvectors of $\mathbf{A}$.

Since the eigenvalues are computed approximately, we can only obtain an approximation of the Lyapunov function (2.15). To overcome this limitation, one needs to evaluate the region $\overline{X}$ where the Lyapunov function is strictly decreasing, that is

$$\overline{X} = \{\mathbf{x} \in X \,|\, \dot{V}(\mathbf{x}) < 0\} \cup \{\mathbf{x}^*\},$$

and consider the largest sublevel set

$$\mathcal{E}_\delta = \{\mathbf{x} \in \mathbb{R}^n \,|\, V(\mathbf{x}) \leq \delta\}$$

such that $\mathcal{E}_\delta \subseteq \overline{X}$. Then $\mathcal{E}_\delta$ provides an inner approximation of the basin of attraction.

This method is illustrated with the following example.

*Example 2.7* Consider the system (2.9)–(2.10). Using the Koopman eigenfunctions computed in Example 2.4, a strict inner approximation of the basin of attraction is obtained with the candidate Lyapunov function $V = 10\phi_{\lambda_1}^2 + \phi_{\lambda_2}^2$, by considering the largest sublevel set in the region $\overline{X}$ where $\dot{V} < 0$ (Fig. 2.4). △

### 2.5.2 General Lyapunov Functions and Approximation Errors

In the previous section, candidate Lyapunov functions have been constructed with the Koopman eigenfunctions. Instead, it is possible to design general Lyapunov functions directly from the Koopman operator. In the finite-dimensional approximation, these Lyapunov functions appear to be related to Lyapunov functions of the Koopman matrix. Just as the Koopman eigenfunctions are approximated as linear combinations of basis functions, Lyapunov functions can also be expressed in terms of the basis functions.

Consider the projection $L_N$ of the infinitesimal generator into the space $\mathscr{F}_N \subset \mathscr{F}_{A_c}$ spanned over the basis functions $\{\psi_k\}_{k=1}^N$, with $\Psi(\mathbf{x}) = \mathbf{0}$ only if $\mathbf{x} = \mathbf{0}$. The dynamics of the lifted states $\mathbf{z} = \Psi(\mathbf{x})$ are approximated by (see Chap. 1)

$$\dot{\mathbf{z}} \approx \mathbf{L}^T \mathbf{z},$$

where $\mathbf{L}$ is the matrix representation of $L_N$. We can take into account the error as follows:

$$\dot{\Psi}(\mathbf{x}) = \mathbf{L}^T \Psi(\mathbf{x}) + \varepsilon(\mathbf{x}), \tag{2.16}$$

where the components of the error vector $\varepsilon(\mathbf{x})$ are given by

$$\varepsilon_k(\mathbf{x}) = ((L - L_N)\psi_k)(\mathbf{x}) = (\nabla\psi_k(\mathbf{x}))^T \mathbf{F}(\mathbf{x}) - \mathbf{L}^T \Psi(\mathbf{x}).$$

We will consider a sum-of-squares function $V(\mathbf{x}) = \Psi^T(\mathbf{x})\mathbf{Q}\Psi(\mathbf{x})$ as a candidate Lyapunov function and will provide stability guarantees by combining linear control theory with Koopman operator. We will use in particular the $\mathscr{H}_\infty$ norm of a transfer function $\mathbf{G}$, which is defined by

$$\|\mathbf{G}(s)\|_{\mathscr{H}_\infty} = \sup_{\omega \in \mathbb{R}} s(\mathbf{G}(i\omega))$$

where $s(\mathbf{G}(i\omega))$ is the largest singular value of the matrix $\mathbf{G}(i\omega)$. We refer the reader to [28] for the background on the $\mathscr{H}_\infty$ norm of linear systems.

**Proposition 2.4** *Let* $\mathbf{L}$ *be the matrix representation of the operator* $L_N$ *in the subspace spanned over* $\{\psi_k\}_{k=1}^N$, *with* $\Psi(\mathbf{x}) = \mathbf{0} \Rightarrow \mathbf{x} = \mathbf{0}$. *If there exists a value* $\gamma > 0$ *such that*

$$\|(sI - \mathbf{L}^T)^{-1}\|_{\mathscr{H}_\infty} < 1/\gamma \tag{2.17}$$

*and*

$$\sup_{\mathbf{x} \in X} \frac{\|\varepsilon(\mathbf{x})\|_2}{\|\Psi(\mathbf{x})\|_2} < \gamma, \tag{2.18}$$

*where $X$ is forward-invariant, then the system admits on $X$ a Lyapunov function of the form $V(\mathbf{x}) = (\Psi(\mathbf{x}))^T \mathbf{Q} \Psi(\mathbf{x})$, with $\mathbf{Q} \succ \mathbf{0}$.*

*Proof* The function $V(\mathbf{x})$ is naturally positive on $X \setminus \{\mathbf{0}\}$ since $\mathbf{Q} \succ \mathbf{0}$ and $\Psi(\mathbf{x})$ is equal to zero only in the origin. Consider the time derivative of $V(\mathbf{x})$:

$$\begin{aligned}\dot{V}(\mathbf{x}) = 2(\dot{\Psi}(\mathbf{x}))^T \mathbf{Q} \Psi(\mathbf{x}) &= 2(\mathbf{L}^T \Psi(\mathbf{x}) + \varepsilon(\mathbf{x}))^T \mathbf{Q} \Psi(\mathbf{x}) \\ &= (\Psi(\mathbf{x}))^T (\mathbf{L}\mathbf{Q} + \mathbf{Q}\mathbf{L}^T) \Psi(\mathbf{x}) + 2(\varepsilon(\mathbf{x}))^T \mathbf{Q} \Psi(\mathbf{x}).\end{aligned}$$

Using the inequality $(\varepsilon(\mathbf{x}) - \mathbf{Q}\Psi(\mathbf{x}))^T (\varepsilon(\mathbf{x}) - \mathbf{Q}\Psi(\mathbf{x})) \geq 0$, we obtain that

$$2(\varepsilon(\mathbf{x}))^T \mathbf{Q} \Psi(\mathbf{x}) \leq (\varepsilon(\mathbf{x}))^T \varepsilon(\mathbf{x}) + (\Psi(\mathbf{x}))^T \mathbf{Q}^2 \Psi(\mathbf{x}).$$

Therefore, we have

$$\begin{aligned}\dot{V}(\mathbf{x}) &\leq (\Psi(\mathbf{x}))^T (\mathbf{L}\mathbf{Q} + \mathbf{Q}\mathbf{L}^T) \Psi(\mathbf{x}) + (\varepsilon(\mathbf{x}))^T \varepsilon(\mathbf{x}) + (\Psi(\mathbf{x}))^T \mathbf{Q}^2 \Psi(\mathbf{x}) \\ &= (\Psi(\mathbf{x}))^T (\mathbf{L}\mathbf{Q} + \mathbf{Q}\mathbf{L}^T + \mathbf{Q}^2 + \gamma^2 I) \Psi(\mathbf{x}) + \|\varepsilon(\mathbf{x})\|_2^2 - \gamma^2 \|\Psi(\mathbf{x})\|_2^2.\end{aligned}$$

It is known that there exists a matrix $\mathbf{Q} \succ \mathbf{0}$ such that

$$\mathbf{L}\mathbf{Q} + \mathbf{Q}\mathbf{L}^T + \mathbf{Q}^2 + \gamma^2 I = 0, \tag{2.19}$$

if and only if $\|(sI - \mathbf{L}^T)^{-1}\|_{\mathscr{H}_\infty} < 1/\gamma$. Therefore we have that $\dot{V}(\mathbf{x}) < 0$ if $\|\varepsilon(\mathbf{x})\|_2^2 - \gamma^2 \|\Psi(\mathbf{x})\|_2^2 < 0$, which completes the proof. □

In the linear case, the error term $\varepsilon(\boldsymbol{x})$ is equal to zero and the result becomes necessary and sufficient. In general, one needs to solve the Riccati equation (2.19) for some value

$$\gamma = \frac{1}{\|(sI - \mathbf{L})^{-1}\|_{\mathscr{H}_\infty} + \eta}$$

with $\eta > 0$ (in the following examples, we set $\eta = 10^{-4}$). The stability region is given by a forward-invariant set lying in the set

$$X_\gamma = \{\mathbf{x} \in \mathbb{R}^n \Big| \|\varepsilon(\mathbf{x})\|_2 < \gamma \|\Psi(\mathbf{x})\|_2\}.$$

For instance, this can be the largest sublevel set of $V$ lying in $X_\gamma$.

The result of Proposition 2.4 relies on two conditions. The inequality (2.17) is an algebraic condition that measures the robustness of the system against (modeling) error while the inequality (2.18) is a functional condition related to the error due to finite-dimensional approximation. Considering (2.16), we can also interpret the result as the application of the small-gain theorem [9] to the feedback interconnection of the linear system $\dot{\mathbf{z}} = \mathbf{L}^T \mathbf{z} + \mathbf{u}$ with the static nonlinearity $T$ such that $\mathbf{u} = T(\mathbf{z}) = \varepsilon(\mathbf{x})$. This interpretation suggests that the conservatism of the result can potentially be improved by exploiting the rich literature in feedback control theory. We also note

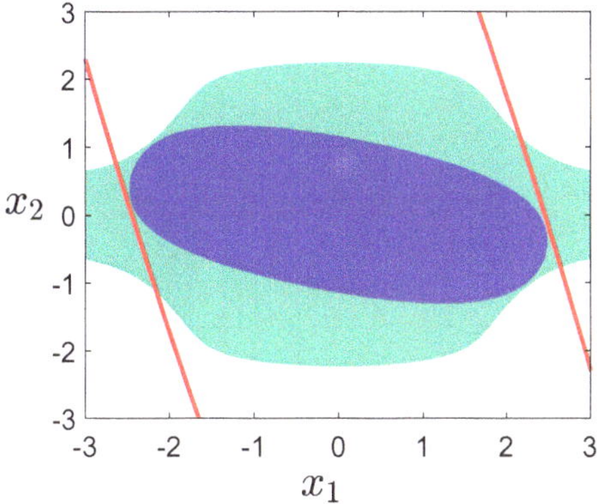


**Fig. 2.5** A general Lyapunov function is constructed with the finite-dimensional approximation of the infinitesimal generator (Example 2.8). The largest sublevel set of the Lyapunov function $V(\mathbf{x}) = \mathbf{x}^T \mathbf{Q} \mathbf{x}$ lying in the region $X_\gamma$ (green region) provides an inner approximation (blue region) of the basin of attraction. The red curves are the boundaries of the basin of attraction

that (2.18) is related to a finite approximate closure property defined in [8], where it is also shown that specific basis functions (i.e., state inclusive logistic functions) can be characterized by global bounds of the error $\varepsilon(x)$. This property could be exploited in future work.

We first illustrate the method with the system of Example 2.4.

*Example 2.8* We consider again the system (2.9)–(2.10), but we rescale the state space with the new coordinates $(x_1', y_2') = (x_1/3, x_2/3)$. The matrix representation of the infinitesimal generator is constructed with the basis of monomials and the projection operator (2.11), but with a maximum total degree equal to 10. We compute the $\mathscr{H}_\infty$ norm, which is equal to 7.476 and select the value $\gamma = 0.134$. The largest sublevel set of the Lyapunov function lying in $X_\gamma$ provides an inner approximation of the basin of attraction (Fig. 2.5). This approximation is smaller but closer to the boundary than the one obtained with Koopman eigenfunctions (Example 2.7). △

The technique derived from Proposition 2.4 is very general and allows to consider non-polynomial vector fields. Indeed, it can take all approximation errors into account, not only the error due to finite-dimensional approximation, but also the error due to the approximation of the projection of the vector field in some basis (e.g., monomials). This is illustrated with the following example.

*Example 2.9* We consider the dynamics

$$\begin{aligned}\dot{\theta}_1 &= K \sin(\theta_1 - \theta_2) - \sin(\theta_1)\\ \dot{\theta}_2 &= K \sin(\theta_2 - \theta_1) - \sin(\theta_2)\end{aligned}$$

with $K = 0.4$, which admits a stable equilibrium at the origin. The dynamics are approximated by a 10th order Taylor approximation

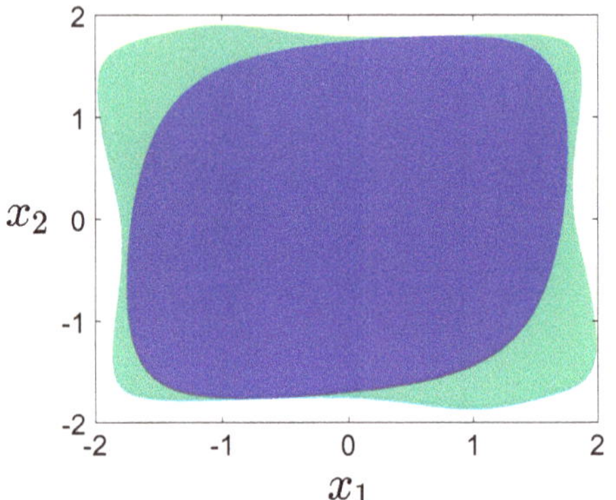


**Fig. 2.6** A general Lyapunov function can also be constructed in the case of non-polynomial vector fields (Example 2.9). The largest sublevel set of the Lyapunov function $V(\mathbf{x}) = \mathbf{x}^T \mathbf{Q} \mathbf{x}$ lying in the region $X_\gamma$ (green region) provides an inner approximation (blue region) of the basin of attraction

$$
\begin{aligned}
\dot{\theta}_1 &= \sum_{\substack{k_1,k_2=0 \\ k_1+k_2 \text{ odd}}}^{10} \frac{K}{k_1!\,k_2!}(-1)^{(k_1+3k_2-1)/2}\,\theta_1^{k_1}\theta_2^{k_2} - \sum_{\substack{k=1 \\ k \text{ odd}}}^{9} \frac{\theta_1^k}{k!} \\
\dot{\theta}_2 &= \sum_{\substack{k_1,k_2=0 \\ k_1+k_2 \text{ odd}}}^{10} \frac{K}{k_1!\,k_2!}(-1)^{(3k_1+k_2-1)/2}\,\theta_1^{k_1}\theta_2^{k_2} - \sum_{\substack{k=1 \\ k \text{ odd}}}^{9} \frac{\theta_2^k}{k!}
\end{aligned}
$$

and rescaled with the new coordinates $(\theta_1', \theta_2') = (\theta_1/2.5, \theta_2/2.5)$. This approximation is used to construct a matrix representation of the infinitesimal generator with the basis of monomials (up to a total degree equal to 10) and with the projection operator (2.11). The $\mathscr{H}_\infty$ norm is equal to 1.519 ($\gamma = 0.658$). The largest sublevel set of the Lyapunov function lying in $X_\gamma$ is shown in Fig. 2.6. △

## 2.6 Conclusion

In this chapter, we have shown that the Koopman operator framework can be used to study the global stability properties of nonlinear systems. This framework mirrors classic tools and results from linear stability analysis, yielding systematic numerical techniques for nonlinear stability analysis. In particular, the spectral properties of the Koopman operator capture the stability properties of the underlying dynamical system and, in this context, it was shown that the support of specific Koopman eigenfunctions reveals the basin of attraction. Our main results were illustrated in the case of (hyperbolic and non-hyperbolic) equilibria, where the Koopman operator can be used to design Lyapunov functions.

The proposed framework is new and opens up to many perspectives and challenges. A main limitation is admittedly the infinite dimension of the operator, yielding inherent approximation errors of (finite-dimensional) numerical methods. To obtain rigorous stability guarantees, a thorough study of these approximation errors related to different finite-dimensional subspaces and projection methods is needed. On a related note, the method presented in Sect. 2.5.2 could be improved by using projections that minimize the finite-dimensional approximations error (e.g., minimax approximations, see also [8]) or small-gain results that exploit the structure of the error term. The methods could also be enhanced to deal with high-dimensional state spaces, possibly by taking benefit of data-driven techniques. Finally, generalizing the present framework to input–output stability does not seem trivial, but would be highly relevant to nonlinear control theory.

**Acknowledgements** Dr. Sootla was supported by the EPSRC grant EP/M002454/1. I. Mezić acknowledges support from ARO-MURI grant W911NF-17-1-0306.

## References

1. Andersen, E.D., Andersen, K.D.: The mosek interior point optimizer for linear programming: an implementation of the homogeneous algorithm. In: High Performance Optimization, pp. 197–232. Springer, Berlin (2000)
2. Carleman, T.: Application de la thorie des quations integrales lineaires aux systmes d'quations diffrentielles nonlinaires. Acta Math. **59**, 63–68 (1932)
3. Curtain, R.F., Zwart, H.: An Introduction to Infinite-Dimensional Linear Systems Theory, vol. 21. Springer Science & Business Media, Berlin (2012)
4. Eisner, T.: Stability of Operators and Operator Semigroups. Birkhäuser, Basel (2010)
5. Engel, K.J., Nagel, R.: One-Parameter Semigroups for Linear Evolution Equations, vol. 194. Springer Science & Business Media, Berlin (1999)
6. Gaspard, P., Nicolis, G., Provata, A., Tasaki, S.: Spectral signature of the pitchfork bifurcation: Liouville equation approach. Phys. Rev. E **51**(1), 74 (1995)
7. Guillemin, V., Pollack, A.: Differential Topology, vol. 370. American Mathematical Soc, Providence (2010)
8. Johnson, C., Yeung, E.: A class of logistic functions for approximating state-inclusive Koopman operators. arXiv:1712.03132 (2017)
9. Khalil, H.: Nonlinear Systems. Prentice Hall, Upper Saddle River (2002)
10. Kühner, V.: What can Koopmanism do for attractors in dynamical systems? arXiv:1902.07487 (2019)
11. Lan, Y., Mezić, I.: Linearization in the large of nonlinear systems and Koopman operator spectrum. Phys. D **242**, 42–53 (2013)
12. Lasota, A., Mackey, M.C.: Chaos, Fractals, and Noise: Stochastic Aspects of Dynamics. Springer, Berlin (1994)
13. Löfberg, J.: YALMIP: A toolbox for modeling and optimization in MATLAB. In: Proceedings of the IEEE International Symposium on Computer-Aided Control System Design, pp. 284–289, Taipei, Taiwan. http://users.isy.liu.se/johanl/yalmip (2004)
14. Lyapunov, A.M.: The general problem of the stability of motion. Int. J. Control **55**(3), 531–534 (1992)
15. Mauroy, A., Mezić, I.: On the use of Fourier averages to compute the global isochrons of (quasi)periodic dynamics. Chaos **22**(3), 033112 (2012)
16. Mauroy, A., Mezić, I.: Global stability analysis using the eigenfunctions of the Koopman operator. IEEE Trans. Autom. Control **61**(11), 3356–3369 (2016)
17. Mauroy, A., Mezić, I., Moehlis, J.: Isostables, isochrons, and Koopman spectrum for the action-angle representation of stable fixed point dynamics. Phys. D **261**, 19–30 (2013)

18. Mezić, I.: Spectral properties of dynamical systems, model reduction and decompositions. Nonlinear Dyn. **41**(1–3), 309–325 (2005)
19. Mezić, I.: Analysis of fluid flows via spectral properties of the Koopman operator. Ann. Rev. Fluid Mech. **45**, 357–378 (2013)
20. Mezić, I.: On applications of the spectral theory of the Koopman operator in dynamical systems and control theory. In: IEEE Conference on Decision and Control, pp. 7034–7041 (2015)
21. Papachristodoulou, A., Anderson, J., Valmorbida, G., Prajna, S., Seiler, P., Parrilo, P.: SOSTOOLS version 3.00 sum of squares optimization toolbox for MATLAB. arXiv:1310.4716 (2013)
22. Parrilo, P.A.: Semidefinite programming relaxations for semialgebraic problems. Math. Program. **96**(2), 293–320 (2003)
23. Rantzer, A.: A dual to Lyapunov's stability theorem. Syst. Control Lett. **42**(3), 161–168 (2001)
24. Sootla, A., Mauroy, A.: Geometric properties of isostables and basins of attraction of monotone systems. IEEE Trans. Autom. Control **62**(12), 6183–6194 (2017)
25. Sturm, J.F.: Using SeDuMi 1.02, a MATLAB toolbox for optimization over symmetric cones. Optim. Method. Softw. **11–12**, 625–653 (1999)
26. Susuki, Y., Mezić, I.: Nonlinear Koopman modes and power system stability assessment without models. IEEE Trans. Power Syst. **29**(2), 899–907 (2014)
27. Vaidya, U., Mehta, P.G.: Lyapunov measure for almost everywhere stability. IEEE Trans. Autom. Control **53**(1), 307–323 (2008)
28. Zhou, K., Doyle, J.C., Glover, K.: Robust and Optimal Control. Prentice Hall, New Jersey (1996)

# Chapter 3
# Koopman Framework for Nonlinear Estimation

**Amit Surana**

**Abstract** In this chapter, we overview a new approach for nonlinear estimation based on Koopman operator-theoretic framework. We exploit Koopman eigenfunctions to create a nonlinear embedding/lifting of underlying nonlinear dynamics to synthesize observer forms (which we call Koopman observer form (KOF)) which enables the use of well-known estimation techniques developed for linear/bilinear systems in context of more general nonlinear systems. Furthermore, we present an extension of this framework for nonlinear constrained state estimation (CSE) with non-convex state constraints. Exploiting the KOF-based representation, we show that under certain conditions the CSE problem can be transformed into a higher dimensional but convex problem. We present a receding horizon estimation formulation based on this transformation, which could provide computational benefit in real-time applications. We also analyze system theoretic properties of KOF in relation to the original nonlinear system, and establish relationship between the original nonlinear estimation problem and the Koopman transformed problem. Finally, we illustrate our approach on a few examples.

## 3.1 Introduction

Nonlinear state observer design has been an area of constant research for several decades and, despite important progress, many outstanding practical problems still remain unsolved [4, 29, 32]. The Extended Kalman filter (EKF) continues to be one of the most widely used practical approaches, but unfortunately has no general guarantees on convergence due to its reliance on linearization. Geometric observer techniques [17, 18, 29, 40] seek a coordinate transformation so that the state estimation error dynamics are linear in the new coordinates. Necessary and sufficient conditions for existence of such transformation have been established but in practice

A. Surana (✉)
United Technologies Research Center, 411 Silver Lane,
East Hartford, CT 06118, USA
e-mail: suranaa@utrc.utc.com



A. Mauroy et al. (eds.), *The Koopman Operator in Systems and Control*,
Lecture Notes in Control and Information Sciences 484,
https://doi.org/10.1007/978-3-030-35713-9_3

are extremely difficult to satisfy. Exact/approximate bilinearization has also been extensively studied [12, 22] but suffers from similar limitations. There has also been work to design observers for more specialized classes of nonlinear systems, such as Lipschitz nonlinear systems [1, 33, 35–37], and bilinear systems [12]. Techniques including Carleman linearization [8, 21] have been proposed (see [12, 41] and references therein) which provides a framework to transform a nonlinear system with control-affine terms approximately into a higher dimensional bilinear form; however, requires analyticity of the underlying system.

Unlike unconstrained state estimation problems, which enjoy recursive closed-form solutions using Kalman filter, (CSE) is typically formulated as an optimization problem and solved in a receding horizon fashion [23, 38, 39, 45], analogous to model predictive control (MPC) [31]. For real time CSE for nonlinear input/ouptut systems and non-convex state constraints, general purpose nonlinear optimization techniques [9] are undesirable as such methods have few known bounds on the computational effort needed, and can become intractable in the sense that a bad initial guess could result in divergence of the numerical algorithm. Convex optimization, on the other hand, can be reliably solved in polynomial time to find the global optimum, see, for example, [5, 11]. In order to take advantage of the powerful convex programming solvers, different techniques for convexification of the originally non-convex problem have been proposed, most common being based on linearization of dynamics and constraints. Such an approach, however, does not offer any convergence guarantees and can introduce artificial infeasibility [26]. Recently, it has been shown that exact convexification of certain types of non-convex constraints related to control inputs can be achieved [2, 24]. Convexification of nonlinear dynamics and non-convex state constraints largely remain a challenge, though some recent work has emerged [25, 26], which offers theoretical guarantees under certain conditions.

The purpose of this chapter is to explore a Koopman operator theoretic framework for nonlinear estimation to alleviate some of the technical challenges stated above, and provide a rigourous yet practical approach. The key idea is to use Koopman eigenfunctions to create a nonlinear embedding/lifting of underlying nonlinear dynamics into a bilinear form, and transform the nonlinearities in the outputs and state constraints into a linear form. Such a transformation then facilitates nonlinear estimation using linear system concepts. This chapter is largely based on the publications [42–44], though some new results establishing relationship between the original nonlinear estimation problem and the Koopman transformed problem are also provided.

## 3.2 Setup and Problem Definition

In this chapter, we consider input/output nonlinear systems of following form:

$$\dot{\mathbf{x}} = \mathbf{f}(\mathbf{x}) + \sum_{i=1}^{n_u} \mathbf{r}_i(\mathbf{x})u_i + \sum_{j=1}^{n_w} \mathbf{s}_j(\mathbf{x})w_j, \tag{3.1}$$

$$\mathbf{y} = \mathbf{h}(\mathbf{x}) + \mathbf{v}, \tag{3.2}$$

where $\mathbf{x} = (x_1, \ldots, x_{n_x})^* \in \mathbb{X} \subset \mathbb{R}^{n_x}$ is state vector, and $\mathbf{f} : \mathbb{X} \to \mathbb{R}^{n_x}$ is the autonomous vector field with $\mathbb{R}$ being the space of reals, and $*$ denotes the standard vector/matrix transpose. Furthermore, $\mathbf{u} = (u_1, \ldots, u_{n_u})^* \in \mathbb{U} \subset \mathbb{R}^{n_u}$ is the control/input vector, $\mathbf{r}_i : \mathbb{X} \to \mathbb{R}^{n_x}, i = 1, \ldots, n_u$ are state-dependent control coupling terms; $\mathbf{w} = (w_1, \ldots, w_{n_w})^* \in \mathbb{R}^{n_w}$ is the unknown/unmeasurable disturbance vector, $\mathbf{s}_i : \mathbb{X} \to \mathbb{R}^{n_x}, i = 1, \ldots, n_w$ are state-dependent disturbance coupling terms; $\mathbf{h} : \mathbb{X} \to \mathbb{R}^{n_y}$ are the outputs; and $\mathbf{v} \in \mathbb{R}^{n_y}$ is the measurement noise. The disturbance is typically modeled as a random process and may account also for modeling uncertainty. Additionally, we allow non-convex state constraints of the form

$$\mathbf{g}(\mathbf{x}) \leq 0, \tag{3.3}$$

where $\mathbf{g}(\mathbf{x}) : \mathbb{X} \to \mathbb{R}^{n_g}$. Such state-dependent equality constraints naturally arise when dealing with differential algebraic systems (DAEs) which are commonly used to model a variety of multiscale engineering and physical systems, such as constrained mechanical systems, electrical systems, and chemical reaction kinetics [6]. The non-convex state inequality constraints, for example, arise in estimation/control problems related to robot navigation in a non-convex obstacle rich environment [25].

Let $C^1(\mathbb{X})$ be the vector space of continuously differentiable scalar complex-valued functions $r : \mathbb{X} \to \mathbb{C}$, where $\mathbb{C}$ is the complex plane. The gradient of a function $r \in C^1(\mathbb{X})$ will be represented as a row vector $\nabla r = (\frac{\partial r}{\partial x_1}, \ldots, \frac{\partial r}{\partial x_{n_x}})$, while Jacobian of any vector-valued function $\mathbf{r} : \mathbb{X} \to \mathbb{C}^p$ with components $\mathbf{r}(\mathbf{x}) = (r_1(\mathbf{x}), \ldots, r_p(\mathbf{x}))^*$, $r_i \in C^1(\mathbb{X})$ will be denoted by

$$\frac{\partial \mathbf{r}}{\partial \mathbf{x}} = \begin{pmatrix} \nabla r_1 \\ \vdots \\ \nabla r_p \end{pmatrix} = \begin{pmatrix} \frac{\partial r_1}{\partial x_1} & \cdots & \frac{\partial r_1}{\partial x_{n_x}} \\ \vdots & \vdots & \vdots \\ \frac{\partial r_p}{\partial x_1} & \cdots & \frac{\partial r_p}{\partial x_{n_x}} \end{pmatrix}. \tag{3.4}$$

Let $\chi(\mathbb{X})$ be a set of all real-valued vector fields $\mathbf{v} : \mathbb{X} \to \mathbb{R}^{n_x}$ on $\mathbb{X}$. Elements of $\chi(\mathbb{X})$ act as linear operators on $C^1(\mathbb{X})$ via Lie differentiation, s.t. for any $r \in C^1(\mathbb{X})$ and $\mathbf{v} \in \chi(\mathbb{X})$,

$$L_{\mathbf{v}} r = < \nabla r, \mathbf{v} >, \tag{3.5}$$

where $< \cdot, \cdot >$ is the standard inner product on the Euclidean space. This definition extends naturally to vector-valued functions $\mathbf{r}(\mathbf{x})$, as follows:

$$L_{\mathbf{v}} \mathbf{r} = \begin{pmatrix} L_{\mathbf{v}} r_1 \\ \vdots \\ L_{\mathbf{v}} r_p \end{pmatrix} = \begin{pmatrix} < \nabla r_1, \mathbf{v} > \\ \vdots \\ < \nabla r_p, \mathbf{v} > \end{pmatrix}. \tag{3.6}$$

We shall denote by: $\mathrm{Re}(c)$, $\mathrm{Im}(c)$, $\overline{c}$, $||c||$ and $\arg(c)$ as the real part, imaginary part, complex conjugate, modulus and argument, respectively, of any complex number $c \in \mathbb{C}$.

### 3.2.1 Estimation Problem

In this chapter our main focus is on constrained state estimation (CSE). Given the measurements $\{\mathbf{y}(t), \forall t \in [0, T]\}$ and the controls $\{\mathbf{u}(t), \forall t \in [0, T]\}$ over the horizon $[0, T]$, the CSE problem comprises of determining the state sequence $\{\mathbf{x}(t), \forall t \in [0, T]\}$ and disturbance $\{\mathbf{w}(t), \forall t \in [0, T]\}$ satisfying the dynamic evolution (3.1), where the states, disturbances, and noise additionally satisfy the following constraints

$$\mathbf{g}(\mathbf{x}(t)) \leq 0, \quad \mathbf{w}(t) \in \mathbb{W}(t), \quad \mathbf{v}(t) \in \mathbb{V}(t), \quad \forall t \in [0, T]. \tag{3.7}$$

Here the state constraints arise from the original requirements (3.3), and $\mathbb{W}(t)$ and $\mathbb{V}(t)$ are additional constraints on disturbances and noise vectors, respectively, which can be added if relevant to the CSE problem at hand. In whatever follows, we will assume that $\mathbb{W}(t)$ and $\mathbb{V}(t)$ are convex sets for all $t \in [0, T]$.

CSE problem is often formulated as a finite horizon optimization problem [23, 38, 39, 45]

$$\mathscr{C}_{\mathbf{x},T}: \begin{cases} \min_{\{\mathbf{x}(t),\mathbf{w}(t):t\in[0,T]\}} \chi(\mathbf{x}(0)) + \int_{t=0}^{T} \xi(\mathbf{w},\mathbf{v})dt \\ \text{subject to:} \\ \dot{\mathbf{x}} = \mathbf{f}(\mathbf{x}) + \sum_{i=1}^{n_u} \mathbf{r}_i(\mathbf{x})u_i + \sum_{j=1}^{n_w} \mathbf{s}_j(\mathbf{x})w_j, \quad \forall t \in [0, T] \\ \mathbf{g}(\mathbf{x}(t)) \leq 0, \quad \forall t \in [0, T] \\ \mathbf{v}(t) = \mathbf{y}(t) - \mathbf{h}(\mathbf{x}(t)) \in \mathbb{V}(t), \quad \forall t \in [0, T] \\ \mathbf{w}(t) \in \mathbb{W}(t), \quad \forall t \in [0, T]. \end{cases} \tag{3.8}$$

The initial penalty term $\chi$ in the objective function is chosen to be

$$\chi(\mathbf{x}) = (\mathbf{x} - \hat{\mathbf{x}}_0)^* \mathbf{P}_0{}^{-1}(\mathbf{x} - \hat{\mathbf{x}}_0), \tag{3.9}$$

and measures deviation of initial condition $\mathbf{x}_0$ from $\hat{\mathbf{x}}_0$ which is a given a priori most likely value of $\mathbf{x}_0$, and $\mathbf{P}_0 > 0$ is a given symmetric positive definite matrix representing confidence that $\hat{\mathbf{x}}_0$ is the most likely value. The second term is taken to be of the form

$$\xi(\mathbf{w}, \mathbf{v}) = \mathbf{w}^*\mathbf{Q}^{-1}\mathbf{w} + \mathbf{v}^*\mathbf{R}^{-1}\mathbf{v}, \tag{3.10}$$

$\mathbf{Q} > 0$, $\mathbf{R} > 0$ are given positive definite matrices. The optimization problem $\mathscr{C}_{\mathbf{x},T}$ (3.8) is in general non-convex due to possibly non-convex state constraints and nonlinear dynamics constraints. We will refer to $\mathscr{C}_{\mathbf{x},T}$ as the nonlinear constrained state estimation (NCSE) problem.

## 3.3 Bilinearization Using Koopman Operator

### 3.3.1 Koopman-Based Transformation

For the input–output system (3.1)–(3.2), consider the flow map $\mathbf{S}(t, \mathbf{x}_0)$ induced by the unactuated part, i.e.,

$$\dot{\mathbf{x}} = \mathbf{f}(\mathbf{x}), \tag{3.11}$$

which is obtained by setting $\mathbf{u} \equiv \mathbf{0}$ and $\mathbf{w} \equiv \mathbf{0}$ in (3.1). The flow map $\mathbf{S}(t, \mathbf{x}_0)$ prescribes the solution of the above ODE starting at the initial condition $\mathbf{x}_0$. Let $\mathbb{F}$ be a space of complex-valued scalar functions $\psi : \mathbb{X} \to \mathbb{C}$, then the Koopman (semi)group of operators $U^t : \mathbb{F} \to \mathbb{F}$ associated with the flow $\mathbf{S}$ is defined by

$$(U^t \psi)(\mathbf{x}) = \psi \circ \mathbf{S}(t, \mathbf{x}). \tag{3.12}$$

We assume $\mathbb{F} \subseteq C^1(\mathbb{X})$, see [27, 30] for discussion on appropriate choices of $\mathbb{F}$.

An eigenfunction of the Koopman operator (or in short Koopman eigenfunction (KEF)) is an observable $\phi \in \mathbb{F}$ that satisfies

$$U^t \phi = e^{\lambda t} \phi, \tag{3.13}$$

where $\lambda \in \mathbb{C}$ is referred to as the Koopman eigenvalue (KE) corresponding to KEF $\phi$, which satisfy the eigenvalue equation

$$L_{\mathbf{f}} \phi = \lambda \phi. \tag{3.14}$$

Let $\phi_i$ be an eigenfunction for the Koopman operator corresponding to the eigenvalue $\lambda_i$. Given a vector-valued observable $\mathbf{r}(\mathbf{x})$, the Koopman mode (KM) $\mathbf{v}_i$, corresponding to $\phi_i$ is the vector of the coefficients of the projection of $\mathbf{r}(\mathbf{x})$ onto the span$\{\phi_i\}$ [7, 28]. $\mathbf{v}_i$ can be thought of as mapping from the observable space into a vector space $V \subset \mathbb{C}^p$; the map $\mathbf{r} \to \phi_i \mathbf{v}_i$ is then a vector-valued projection operator onto the subspace span $\{\phi_i\}$.

Let $\mathbb{S}^{n_z} = \{\phi_1, \phi_2, \ldots, \phi_{n_z}\}$ be a *finite* subset of KEFs for system (3.11), and $\mathbb{F}(\mathbb{S}^{n_z}) = \text{span}(\mathbb{S}^{n_z})$ be the span of $\mathbb{S}^{n_z}$. Associated with $\mathbb{S}^{n_z}$, define a nonlinear change of coordinates $\mathbf{T} : \mathbb{R}^{n_x} \to \mathbb{R}^{n_z}$

$$\mathbf{z}(t) = \mathbf{T}(\mathbf{x}(t)) = \begin{pmatrix} \hat{\phi}_1(\mathbf{x}(t)) \\ \hat{\phi}_2(\mathbf{x}(t)) \\ \vdots \\ \hat{\phi}_{n_z}(\mathbf{x}(t)) \end{pmatrix}, \tag{3.15}$$

where,

- $\hat{\phi}_i = \phi_i$ if $i$th KEF is real, and
- $\hat{\phi}_i = 2\text{Re}(\phi_i)$ and $\hat{\phi}_{i+1} = -2\text{Im}(\phi_i)$, if $i$ and $i+1$th KEFs are complex conjugate pairs.

Following [44], we refer to this transformation as the Koopman canonical transform (KCT) , and the coordinates $\mathbf{z}(t) = (z_1(t), \ldots, z_{n_z}(t))^* \in \mathbb{R}^{n_z}$ as the Koopman canonical coordinates (KCC). It follows then,

$$L_{\mathbf{f}}\mathbf{T}(\mathbf{x}) = \Lambda\mathbf{T}(\mathbf{x}), \tag{3.16}$$

where $\Lambda$ is a $n_z \times n_z$ real block diagonal matrix such that

- $\Lambda$ has a diagonal entry $\Lambda_{i,i} = \lambda_i$, if $i$th KEF is real,
- $\Lambda$ has a block diagonal entry $\begin{bmatrix} \Lambda_{i,i} & \Lambda_{i,i+1} \\ \Lambda_{i+1,i} & \Lambda_{i+1,i+1} \end{bmatrix} = \mathbf{Q}_{\lambda_i}$, if $i$ and $i+1$th KEFs are complex conjugate pairs, where

$$\mathbf{Q}_\lambda = ||\lambda|| \begin{pmatrix} \cos(\arg\lambda) & \sin(\arg\lambda) \\ -\sin(\arg\lambda) & \cos(\arg\lambda) \end{pmatrix}. \tag{3.17}$$

### 3.3.2 Bilinear Input–Output Koopman Observer Form

Let $\mathbb{S}^{n_z}$ be a *finite* subset of KEFs for system (3.11) such that

- *Assumption I*: $\mathbf{x} \in \mathbb{F}(\mathbb{S}^{n_z})$, and so

$$\mathbf{x} = \sum_{i=1}^{n_z} \phi_i(\mathbf{x})\mathbf{v}_i^{\mathbf{x}}, \tag{3.18}$$

  where $\mathbf{v}_i^{\mathbf{x}} \in \mathbb{C}^{n_x}, i = 1, \ldots, n_z$ are KMs for the state vector.
- *Assumption II*: $\mathbf{h}(\mathbf{x}) \in \mathbb{F}(\mathbb{S}^{n_z})$, and so

$$\mathbf{h}(\mathbf{x}) = \sum_{i=1}^{n_z} \phi_i(\mathbf{x})\mathbf{v}_i^{\mathbf{h}}, \tag{3.19}$$

  where $\mathbf{v}_i^{\mathbf{h}} \in \mathbb{C}^{n_y}, i = 1, \ldots, n_z$ are KMs for the output functions.
- *Assumption III*: $\mathbf{g}(\mathbf{x}) \in \mathbb{F}(\mathbb{S}^{n_z})$, and so

$$\mathbf{g}(\mathbf{x}) = \sum_{i=1}^{n_g} \phi_i(\mathbf{x})\mathbf{v}_i^{\mathbf{g}}, \tag{3.20}$$

  where $\mathbf{v}_i^{\mathbf{g}} \in \mathbb{C}^{n_g}, i = 1, \ldots, n_z$ are KMs for the state constraint functions.

- *Assumption IV*: Closure of Lie derivative of KCT, such that

$$L_{\mathbf{r}_i}\mathbf{T}(\mathbf{x}) - \mathbf{b}_i = \sum_{j=1}^{n_z} \phi_j(\mathbf{x})\mathbf{v}_j^{\mathbf{r}_i}, \tag{3.21}$$

$$L_{\mathbf{s}_i}\mathbf{T}(\mathbf{x}) - \mathbf{d}_i = \sum_{j=1}^{n_z} \phi_j(\mathbf{x})\mathbf{v}_j^{\mathbf{s}_i}, \tag{3.22}$$

where $\mathbf{b}_i \in \mathbb{R}^{n_z}$ are constant real-valued vectors $i = 1, \ldots, n_u$, and $\mathbf{v}_j^{\mathbf{r}_i}$, $j = 1, \ldots, n_z$ are the KMs for $L_{\mathbf{r}_i}\mathbf{T}(\mathbf{x}) - \mathbf{b}_i$, $i = 1, \ldots, n_u$; and similarly $\mathbf{d}_i \in \mathbb{R}^{n_z}$ are constant real-valued vectors $i = 1, \ldots, n_w$, and $\mathbf{v}_j^{\mathbf{s}_i}$, $j = 1, \ldots, n_z$ are the KMs for $L_{\mathbf{s}_i}\mathbf{T}(\mathbf{x}) - \mathbf{d}_i$, $i = 1, \ldots, n_w$.

For discussion on implications of *Assumptions I–IV*, we refer the reader to [43, 44].

Using the KCC (3.15), *Assumption I–III* can be expressed as

$$\mathbf{x} = \mathbf{C}^{\mathbf{x}}\mathbf{z}, \tag{3.23}$$

$$\mathbf{h}(\mathbf{x}) = \mathbf{C}^{\mathbf{h}}\mathbf{z}, \tag{3.24}$$

$$\mathbf{g}(\mathbf{x}) = \mathbf{C}^{\mathbf{g}}\mathbf{z}, \tag{3.25}$$

where $\mathbf{C}^{\mathbf{x}} \in \mathbb{R}^{n_x \times n_z}$, $\mathbf{C}^{\mathbf{h}} \in \mathbb{R}^{n_y \times n_z}$ and $\mathbf{C}^{\mathbf{g}} \in \mathbb{R}^{n_g \times n_z}$ are real-valued matrices formed as follows: $i$th column of $\mathbf{C}^{\mathbf{x}}$ is $\mathbf{v}_i^{\mathbf{x}}$ if $i$th KEF is real, and $i, i+1$th columns are $\mathrm{Re}(\mathbf{v}_i^{\mathbf{x}})$ and $\mathrm{Im}(\mathbf{v}_i^{\mathbf{x}})$, respectively, if $i$ and $i+1$th KEFs are complex conjugate pairs. A similar construction applies for $\mathbf{C}^{\mathbf{h}}$ and $\mathbf{C}^{\mathbf{g}}$.

Under *Assumptions II, IV* and change of coordinates (3.15), the system (3.1)–(3.2) can be transformed into a bilinear form

$$\dot{\mathbf{z}} = \Lambda\mathbf{z} + \mathbf{B}^0\mathbf{u} + \mathbf{D}^0\mathbf{w} + \sum_{i=1}^{n_u} \mathbf{B}^i\mathbf{z}u_i + \sum_{i=1}^{n_w} \mathbf{D}^i\mathbf{z}w_i, \tag{3.26}$$

$$\mathbf{y} = \mathbf{C}^{\mathbf{h}}\mathbf{z}, \tag{3.27}$$

where

- $\mathbf{B}^0 = [\mathbf{b}_1, \ldots, \mathbf{b}_{n_u}]$ is a $n_z \times n_u$ matrix, and $\mathbf{B}^i$ is a $n_z \times n_z$ matrix constructed using Koopman modes $\{\mathbf{v}_j^{\mathbf{r}_i}\}$ as discussed above.
- $\mathbf{D}^0 = [\mathbf{d}_1, \ldots, \mathbf{d}_{n_w}]$ is a $n_z \times n_w$ matrix, and $\mathbf{D}^i$ is a $n_z \times n_z$ matrix constructed using Koopman modes $\{\mathbf{v}_j^{\mathbf{s}_i}\}$ as discussed above.

Following [43, 44], we refer to the system (3.26)–(3.27) as bilinear input–output Koopman observer form (KOF).

Note that if $\mathbf{B}^i \equiv 0$ and $L_{\mathbf{r}_i}\mathbf{T}(\mathbf{x}) = \mathbf{b}_i \neq 0$, $i = 1, \ldots, n_u$, the bilinear KOF above reduces to a linear input–output KOF.

The process of obtaining bilinear KOF (or linear KOF when no control inputs are present) is similar to the Carleman approach to linearization [8, 21]: both use an immersion to transform the nonlinear system to a bilinear (or linear) form. The key distinction is that in the bilinear KOF we use a specific set of KEFs (as prescribed by *Assumptions I–IV*) to construct the transformation, while in the Carleman approach the transformation is constructed using tensor products of the state vector and relies on analyticity of the system (3.1)–(3.2).

### 3.3.3 Properties of KOF

Let $\mathbb{Z}$ be the image of statespace $\mathbb{X}$ under the KCT, i.e., $\mathbb{Z} = \mathbf{T}(\mathbb{X})$.

**Theorem 3.1** *Under* Assumption I *KCT* $\mathbf{T}$ *is an injective mapping onto its range space* $\mathbb{Z}$.

*Proof* Let $\mathbf{x} \neq \mathbf{x}'$ s.t. $\mathbf{z} = \mathbf{T}(\mathbf{x}) = \mathbf{T}(\mathbf{x}') = \mathbf{z}'$. Then $\mathbf{x} = \mathbf{C}^{\mathbf{x}}\mathbf{z} = \mathbf{C}^{\mathbf{x}}\mathbf{z}' = \mathbf{x}'$ which is contradiction.

By construction $\mathbf{z} = \mathbf{T}(\mathbf{x}) \longrightarrow \mathbf{C}^{\mathbf{x}}\mathbf{z} = \mathbf{x}$. However, note that $\text{Null}(\mathbf{C}^{\mathbf{x}}) \neq \emptyset$ as typically $n_z > n_x$. Hence, $\mathbf{C}^{\mathbf{x}}\mathbf{z} = \mathbf{x}$ does not necessarily imply that $\mathbf{z} = \mathbf{T}(\mathbf{x})$ as $\mathbf{C}^{\mathbf{x}}(\mathbf{z} + \mathbf{z}_n) = \mathbf{x}$ for all $\mathbf{z}_n \in \text{Null}(\mathbf{C}^{\mathbf{x}})$.

**Theorem 3.2** *For initial condition related by* $\mathbf{z}_0 = \mathbf{T}(\mathbf{x}_0)$, *solution of (3.26) be* $\mathbf{z}(t)$, *then* $\mathbf{x}(t) = \mathbf{C}^{\mathbf{x}}\mathbf{z}(t)$ *is solution of (3.1) for any given* $\mathbf{u}(t)$ *and* $\mathbf{w}(t)$ *with an initial condition* $\mathbf{x}_0$.

*Proof* The proof follows immediately from the construction of the bilinear KOF. □

Note that for any arbitrary $\mathbf{z}_0 \neq \mathbf{T}(\mathbf{x}_0)$, solution of (3.26) may have no relation to the solution of (3.1).

## 3.4 KOF-Based Observability Criterion

In this section, we establish a nonlinear observability criterion for system (3.1)–(3.2) by exploiting the bilinear KOF. We will use the standard notions of nonlinear observability from [15], which we briefly recall first.

**Definition 3.1** (*NO*) A pair of points $\mathbf{x}_0$ and $\mathbf{x}_0'$ are *indistinguishable* (denoted $\mathbf{x}_0 \mathbb{I} \mathbf{x}_0'$) if the system (3.1)–(3.2) with these two initial conditions realizes same input–output map for every admissible control input $\mathbf{u}(t)$, $t \in [t_0, t_1]$. Note that indistinguishability $\mathbb{I}$ is an equivalence relation on $\mathbb{X}$. System (3.1)–(3.2) is said to be *nonlinearly observable at* $\mathbf{x}_0$ if $\mathbb{I}(\mathbf{x}_0) = \{\mathbf{x}_0\}$ and is *nonlinearly observable* if $\mathbb{I}(\mathbf{x}) = \{\mathbf{x}\}$ for every $\mathbf{x} \in \mathbb{X}$.

**Theorem 3.3** *[KOF Observability $\Longrightarrow$ NO] If* Assumptions I–III *hold, and*

$$\begin{aligned}&rank([\mathbf{C^h}, \mathbf{C^h}\Lambda, \mathbf{C^h}B^1, \ldots \mathbf{C^h}B^{n_u}, \mathbf{C^h}\Lambda^2, \mathbf{C^h}\Lambda B^1,\\ &\ldots, \mathbf{C^h}\Lambda B^{n_u}, \mathbf{C^h}B^1\Lambda, \ldots, \mathbf{C^h}(B^{n_u})^{n_z-1}]) = n_z,\end{aligned} \tag{3.28}$$

*then the nonlinear system (3.1)–(3.2) is nonlinearly observable for the space of control inputs $\mathbb{U}$ containing piecewise continuous input signals.*

*Proof* The proof is by contradiction. The condition (3.28) implies that there exists piecewise continuous input signal for which the bilinear KOF (3.26)–(3.27) is observable [14]. Assume (3.1)–(3.2) is not observable, and so there exists two distinct initial conditions $\mathbf{x}_0 \neq \mathbf{x}_0'$ such that they result in same output $\mathbf{y}(t)$ over any interval of time $[0, T]$. Let $\mathbf{z}_0 = \mathbf{T}(\mathbf{x}_0)$ and $\mathbf{z}_0' = \mathbf{T}(\mathbf{x}_0')$, then $\mathbf{z}_0 \neq \mathbf{z}_0'$ by injectivity of $\mathbf{T}$. By construction, the KOF will also produce same outputs when initialized at $\mathbf{z}_0$ or $\mathbf{z}_0'$. This is a contradiction, since under condition (3.28), the KOF (3.26)–(3.27) is observable, and so $\mathbf{z}_0 = \mathbf{z}_0'$. □

The above proposition provides sufficient conditions for nonlinear observability in terms of KOF. However, the observability condition (3.28) required in above theorem is stringent. In fact, as long as the bilinear KOF is observable (as discussed above) restricted to $\mathbb{Z}$ (i.e., any two states $\mathbf{z}, \mathbf{z}' \in \mathbb{Z}$ are indistinguishable), Theorem 3.3 will hold true.

## 3.5 KOF-Based Estimation

In this section, we explore the role of KOF in facilitating solution of the nonlinear estimation problem discussed in Sect. 3.2.1. We consider two cases, estimation with and without constraints.

### *3.5.1 State Estimation Without Constraints*

In absence of state/noise constraints and disturbances, the bilinear KOF (3.26)–(3.27) reduces to

$$\begin{aligned}\dot{\mathbf{z}} &= \Lambda\mathbf{z} + \mathbf{B}^0\mathbf{u} + \sum_{i=1}^{n_u}\mathbf{B}^i\mathbf{z}u_i,\\ \mathbf{y} &= \mathbf{C^h}\mathbf{z}.\end{aligned} \tag{3.29}$$

For bilinear systems of the form (3.29), multiple formulations can be posed for observer design [12]. When the $\mathbf{u}$ is known, one can view (3.29) as a linear time-varying system

$$\dot{\mathbf{z}} = \mathbf{A}(t)\mathbf{z} + \mathbf{B}^0\mathbf{u}, \tag{3.30}$$
$$\mathbf{y} = \mathbf{C}^{\mathbf{h}}\mathbf{z}, \tag{3.31}$$

where $\mathbf{A}(t) = \Lambda + \sum_{i=1}^{n_u} \mathbf{B}^i u_i(t)$, and use well-known Kalman type methods of designing observers of time-varying linear system. For instance, if the system (3.30)–(3.31) is uniformly completely observable, and $\mathbf{A}(t)$ is uniformly bounded in time, one can seek an observer of the form [4, 13],

$$\dot{\hat{\mathbf{z}}} = \mathbf{A}(t)\hat{\mathbf{z}} + \mathbf{B}^0\mathbf{u} + K(t)(\mathbf{y} - \mathbf{C}^{\mathbf{h}}\hat{\mathbf{z}}), \tag{3.32}$$

where the gain $K(t) = \mathbf{M}(t)(\mathbf{C}^{\mathbf{h}})^*\mathbf{W}^{-1}$ is computed based on solution of a matrix Riccati equation

$$\dot{M} = \mathbf{A}(t)\mathbf{M}(t) + \mathbf{M}(t)\mathbf{A}^*(t) - \mathbf{M}(t)(\mathbf{C}^{\mathbf{h}})^*\mathbf{W}^{-1}\mathbf{C}\mathbf{M}(t) + \mathbf{V} + \delta\mathbf{M}(t),$$

where $\mathbf{M}(0) = \mathbf{M}_0 = \mathbf{M}_0^* > 0$, $\mathbf{W} = \mathbf{W}^* > 0$, and with either $\delta \geq 2||\mathbf{A}(t)||$ for all $t$, or $\mathbf{V} = \mathbf{V}^* > 0$. The rate of convergence can be tuned by an appropriate choice of $\delta$ or $\mathbf{V}$.

### 3.5.2 Constrained State Estimation

Under the *Assumption I–IV*, the KCT can be used to transform the NCSE problem $\mathscr{C}_{\mathbf{x},T}$ (3.8) into

$$\mathscr{C}_{\mathbf{z},T}: \begin{cases} \min_{\{\mathbf{z}(t),\mathbf{w}(t):t\in[0,T]\}} \chi(\mathbf{C}^{\mathbf{x}}\mathbf{z}(0)) + \int_{t=0}^{T} \xi(\mathbf{w},\mathbf{v})dt, \\ \text{subject to:} \\ \dot{\mathbf{z}} = \Lambda\mathbf{z} + \mathbf{B}^0\mathbf{u} + \mathbf{D}^0\mathbf{w} + \sum_{i=1}^{n_u} \mathbf{B}^i\mathbf{z}u_i + \sum_{i=1}^{n_w} \mathbf{D}^i\mathbf{z}w_i, \quad \forall t \in [0,T] \\ \mathbf{C}^{\mathbf{g}}\mathbf{z}(t) \leq 0, \quad \forall t \in [0,T] \\ \mathbf{v}(t) = \mathbf{y}(t) - \mathbf{C}^{\mathbf{x}}\mathbf{z}(t) \in \mathbb{V}(t), \quad \forall t \in [0,T] \\ \mathbf{w}(t) \in \mathbb{W}(t), \quad \forall t \in [0,T]. \end{cases} \tag{3.33}$$

We refer to $\mathscr{C}_{\mathbf{z},T}$ as the Koopman-based CSE (KCSE) problem. Note that in KCSE the dynamic constraints are bilinear (and non-convex in general), while all other constraints are linear/convex. In absence of unknown disturbances, the bilinear dynamic constraints also become linear, and hence the non-convex NCSE problem $\mathscr{C}_{\mathbf{x},T}$ is transformed into a convex optimization problem. Theorems below establishes relationship between the feasibility/optimality of the NCSE and KCSE problems.

**Theorem 3.4** *If NCSE problem is feasible $\longrightarrow$ the KCSE is feasible.*

*Proof* Let $\{\mathbf{x}(t), \mathbf{w}(t) : t \in [0, T]\}$ be feasible solution of NCSE, then by Theorem 3.2 solution of (3.26) starting at $\mathbf{z}(0) = \mathbf{T}(\mathbf{x}(0))$ is a feasible solution of KCSE. □

However, note that feasibility of KCSE does not imply feasibility of NCSE. However, NCSE solution constructed using the KCSE solution can be easily checked for feasibility. If feasible than it is also optimal under certain conditions as stated below.

**Theorem 3.5** *Let there be no measurement noise, i.e.,* $\mathbf{v} \equiv 0$ *and no disturbances, i.e.,* $\mathbf{w} \equiv 0$*, and the objective function in the NCSE formulation* $\mathscr{C}_{\mathbf{x},T}$ *has no initial penalty term, i.e.,*

$$\overline{\mathscr{C}}_{\mathbf{x},T} : \begin{cases} \min_{\{\mathbf{x}(t):t\in[0,T]\}} \mathscr{X}(\{\mathbf{x}(t) : t \in [0,T]\}) \\ \qquad\qquad \textit{subject to:} \\ \dot{\mathbf{x}} = \mathbf{f}(\mathbf{x}) + \sum_{i=1}^{n_u} \mathbf{r}_i(\mathbf{x})u_i, \quad \forall t \in [0,T] \\ \qquad \mathbf{g}(\mathbf{x}(t)) \le 0, \quad \forall t \in [0,T] \end{cases} \tag{3.34}$$

*where*

$$\mathscr{X}(\{\mathbf{x}(t) : t \in [0,T]\}) = \int_0^T (\mathbf{y}(t) - \mathbf{h}(\mathbf{x}(t)))^* \mathbf{R}^{-1} (\mathbf{y}(t) - \mathbf{h}(\mathbf{x}(t)))dt. \tag{3.35}$$

*Then the optimal solution of KCSE* $\overline{\mathscr{C}}_{\mathbf{z},T}$ *corresponding to the NCSE problem* $\overline{\mathscr{C}}_{\mathbf{x},T}$ *is an optimal solution of* $\overline{\mathscr{C}}_{\mathbf{x},T}$*.*

*Proof* Let $\{\mathbf{x}^o(t) : t \in [0,T]\}$ be optimal solution of $\overline{\mathscr{C}}_{\mathbf{x},T}$, and $\{\mathbf{z}^o(t) : t \in [0,T]\}$ be corresponding solution of (3.26). Then by Theorem 3.2, $\mathbf{y}(t) = \mathbf{h}(\mathbf{x}^o(t)) = \mathbf{C}^{\mathbf{x}}\mathbf{z}^o(t)$, and hence

$$\mathscr{X}(\{\mathbf{z}^o(t) : t \in [0,T]\}) = 0. \tag{3.36}$$

Let $\{\mathbf{z}'(t) : t \in [0,T]\}$ be any other feasible solution of $\overline{\mathscr{C}}_{\mathbf{z},T}$, then $\mathbf{z}'(t) = \mathbf{z}^o(t) + \Psi^{\mathbf{u}}(t,0)(\mathbf{z}'(0) - \mathbf{z}^o(0))$, where $\Psi^{\mathbf{u}}(t,0)$ is the fundamental matrix solution of the linear system (3.30) for given $\{\mathbf{u}(t) : t \in [0,T]\}$. Hence,

$$\mathscr{X}(\{\mathbf{z}'(t)\}) = \int_0^T (\mathbf{y} - \mathbf{C}^{\mathbf{x}}\mathbf{z}'(t))^* \mathbf{R}^{-1} (\mathbf{y} - \mathbf{C}^{\mathbf{x}}\mathbf{z}'(t))dt > 0, \tag{3.37}$$

and thus $\{\mathbf{z}^o(t) : t \in [0,T]\}$ is an unique global optimal of KCSE. □

## 3.6 Numerical Considerations

### *3.6.1 Discretized NCSE*

For estimation purposes it is commonplace to discretize the continuous model (3.1)–(3.2) with zero order hold on the control and disturbances leading to a nonlinear difference equation:

$$\mathbf{x}_{k+1} = \mathbf{F}(\mathbf{x}_k, \mathbf{u}_k, \mathbf{w}_k), \tag{3.38}$$

$$\mathbf{y}_k = \mathbf{h}(\mathbf{x}_k) + \mathbf{v}_k, \tag{3.39}$$

where the integer $k \in \mathbb{N}$ denotes the discrete-time index and $\mathbb{N}$ is set of whole numbers. A typical choice is $t = k\Delta T_s$ where $\Delta T_s$ denotes the sampling period. The subscripts on the vectors $\mathbf{x}$, $\mathbf{u}$, $\mathbf{w}$, $\mathbf{y}$, $\mathbf{v}$ denote the value at the points of discretization (e.g., $\mathbf{x}_k = \mathbf{x}(k\Delta T_s)$). We also assume that the points of discretization coincide with the measurement times. The discrete form of NCSE problem (3.8) becomes

$$\mathscr{D}_{\mathbf{x},n_T}: \begin{cases} \min_{\{\mathbf{z}_k\}_{k=0}^{n_T}, \{\mathbf{w}_k\}_{k=0}^{n_T-1}} \chi(\mathbf{x}_0) + \sum_{k=0}^{n_T-1} \xi(\mathbf{w}_k, \mathbf{v}_k)\Delta T_s, \\ \text{subject to} \\ \mathbf{F}_k^{\mathbf{x}}(\mathbf{x}_{k+1}, \mathbf{x}_k, \mathbf{w}_k) = 0, k = 0, \dots, n_T - 1, \\ \mathbf{g}(\mathbf{x}_k) \le 0, k = 0, \dots, n_T, \\ \mathbf{v}_k = \mathbf{y}_k - \mathbf{h}(\mathbf{x}_k) \in \mathbb{V}_k, k = 0, \dots, n_T - 1, \\ \mathbf{w}_k \in \mathbb{W}_k, k = 0, \dots, n_T - 1. \end{cases} \tag{3.40}$$

with $\mathbf{F}_k^{\mathbf{x}}(\mathbf{x}_{k+1}, \mathbf{x}_k, \mathbf{w}_k) = \mathbf{x}_{k+1} - \mathbf{F}(\mathbf{x}_k, \mathbf{u}_k, \mathbf{w}_k)$.

The solution $\{\hat{\mathbf{x}}_k\}_{k=0}^{n_T}$ to $\mathscr{D}_{\mathbf{x},n_T}$ at the current time index $n_T$ yields an estimate of the actual sequence $\{\mathbf{x}_k\}_{k=0}^{n_T}$. We refer to $\mathscr{D}_{\mathbf{x},n_T}$ as the *full information problem* and $\{\hat{\mathbf{x}}_k\}_{k=0}^{n_T}$ as the *full information estimate* of $\{\mathbf{x}_k\}_{k=0}^{n_T}$ because all the available information $\{\mathbf{y}_k\}_{k=0}^{n_T-1}$ is used in solving $\mathscr{D}_{\mathbf{x},n_T}$. $\mathscr{D}_{\mathbf{x},n_T}$ has $n_T$ stages, so the computational complexity scales at least linearly with $n_T$, and consequently the online solution of $\mathscr{D}_{\mathbf{x},n_T}$ becomes impractical. To make the estimation problem tractable, one needs to bound the problem size for which a common approach is moving horizon formulation. The basic idea of moving horizon estimation (MHE) is to consider explicitly a fixed amount of data, while approximately summarizing the old data not explicitly accounted for in the estimation by using arrival cost [39].

### 3.6.2 *Discretized KCSE*

A number of different techniques motivated from bilinear system identification can be used to discretize a system of bilinear ODEs [42]. We are interested in techniques that keep the discrete-time models in first-order form, and preserve the stability and the simple bilinear structure of the original continuous-time model. The simpler Euler methods (e.g., one-step Euler) while preserving the bilinearity/first-order form during discretization, could lead to unstable models (when the discretization step is large). Other methods such as discussed in [16] preserve stability, but do not produce models in bilinear form. Adams–Bashforth integration methods result in models in first-order form while maintaining the bilinear structure [34]. Along similar lines, it was noted in [19] that the trapezoidal rule with zero order hold for control and disturbance also preserves bilinearity, and is a good compromise between simplicity and stability of the resulting discrete-time model.

Assuming $\mathbf{u}$ and $\mathbf{w}$ are zero-order held at sample rate $\Delta T_s$, we discretize (3.26) using the trapezoidal method to obtain

$$\frac{\mathbf{z}_{k+1}-\mathbf{z}_k}{\Delta T_s} = \Lambda\left(\frac{\mathbf{z}_{k+1}+\mathbf{z}_k}{2}\right) + \mathbf{B}^0\mathbf{u}_k + \mathbf{D}^0\mathbf{w}_k + \sum_{i=1}^{n_u}\mathbf{B}^i\left(\frac{\mathbf{z}_{k+1}+\mathbf{z}_k}{2}\right)u_{k,i} + \sum_{i=1}^{n_w}\mathbf{D}^i\left(\frac{\mathbf{z}_{k+1}+\mathbf{z}_k}{2}\right)w_{k,i}. \quad (3.41)$$

Rearranging the terms one can express above equation as

$$\mathbf{F}_k^{\mathbf{z}}(\mathbf{z}_{k+1}, \mathbf{z}_k, \mathbf{w}_k) = 0, \quad (3.42)$$

where

$$\mathbf{F}_k^{\mathbf{z}}(\mathbf{z}_{k+1}, \mathbf{z}_k, \mathbf{w}_k) = \mathbf{F}^{\mathbf{z}1}(\mathbf{z}_{k+1}, \mathbf{z}_k) - \mathbf{F}^{\mathbf{z}2}(\mathbf{z}_{k+1}, \mathbf{z}_k, \mathbf{u}_k) - \mathbf{F}^{\mathbf{z}3}(\mathbf{z}_{k+1}, \mathbf{z}_k, \mathbf{w}_k),$$

with,

$$\mathbf{F}^{\mathbf{z}1}(\mathbf{z}_{k+1}, \mathbf{z}_k) = \frac{\mathbf{z}_{k+1}-\mathbf{z}_k}{\Delta T_s} - \Lambda\left(\frac{\mathbf{z}_{k+1}+\mathbf{z}_k}{2}\right),$$

$$\mathbf{F}^{\mathbf{z}2}(\mathbf{z}_{k+1}, \mathbf{z}_k, \mathbf{u}_k) = \mathbf{B}^0\mathbf{u}_k + \sum_{i=1}^{n_u}\mathbf{B}^i\left(\frac{\mathbf{z}_{k+1}+\mathbf{z}_k}{2}\right)u_{k,i},$$

$$\mathbf{F}^{\mathbf{z}3}(\mathbf{z}_{k+1}, \mathbf{z}_k, \mathbf{w}_k) = \mathbf{D}^0\mathbf{w}_k + \sum_{i=1}^{n_w}\mathbf{D}^i\left(\frac{\mathbf{z}_{k+1}+\mathbf{z}_k}{2}\right)w_{k,i}.$$

*Remark 3.1* Constraint (3.42) is bilinear in $\{\mathbf{z}_k\}$ and $\{\mathbf{w}_k\}$ for all $k$.

In absence of disturbances, relation (3.42) simplifies to

$$\overline{\mathbf{F}}_k^{\mathbf{z}}(\mathbf{z}_{k+1}, \mathbf{z}_k) = 0, \quad (3.43)$$

where

$$\overline{\mathbf{F}}_k^{\mathbf{z}}(\mathbf{z}_{k+1}, \mathbf{z}_k) = \mathbf{F}^{\mathbf{z}1}(\mathbf{z}_{k+1}, \mathbf{z}_k) - \mathbf{F}^{\mathbf{z}2}(\mathbf{z}_{k+1}, \mathbf{z}_k, \mathbf{u}_k). \quad (3.44)$$

*Remark 3.2* Note that constraint (3.43) is linear in $\{\mathbf{z}_k\}$ for all $k$.

Using the discretization schemes discussed above, the continuous KCSE problem (3.33) can be expressed as

$$\mathscr{D}_{\mathbf{z},n_T}: \begin{cases} \min_{\{\mathbf{z}_k\}_{k=0}^{n_T}, \{\mathbf{w}_k\}_{k=0}^{n_T-1}} \chi(\mathbf{C}^{\mathbf{x}}\mathbf{z}_0) + \sum_{k=0}^{n_T-1} \xi(\mathbf{w}_k, \mathbf{v}_k)\Delta T_s, \\ \text{subject to:} \\ \mathbf{F}_k^{\mathbf{z}}(\mathbf{z}_{k+1}, \mathbf{z}_k, \mathbf{w}_k) = 0, k = 0, \ldots, n_T - 1, \\ \mathbf{C}^{\mathbf{g}}\mathbf{z}_k \leq 0, k = 0, \ldots, n_T, \\ \mathbf{v}_k = \mathbf{y}_k - \mathbf{C}^{\mathbf{h}}\mathbf{z}_k \in \mathbb{V}_k, k = 0, \ldots, n_T - 1, \\ \mathbf{w}_k \in \mathbb{W}_k, k = 0, \ldots, n_T - 1. \end{cases} \tag{3.45}$$

The initial penalty term can be further expressed as

$$\chi(\mathbf{C}^{\mathbf{x}}\mathbf{z}_0) = (\mathbf{z} - \hat{\mathbf{z}}_0)^*(\mathbf{C}^{\mathbf{x}})^*\mathbf{P}_0{}^{-1}\mathbf{C}^{\mathbf{x}}(\mathbf{z} - \hat{\mathbf{z}}_0), \tag{3.46}$$

where $\hat{\mathbf{z}}_0 = \mathbf{T}(\hat{\mathbf{x}}_0)$. Note that $\hat{\mathbf{z}}_0$ can be approximated as $\hat{\mathbf{z}}_0 \approx (\mathbf{C}^{\mathbf{x}})^{\dagger}\hat{\mathbf{x}}_0$, where $\dagger$ is the pseudoinverse.

In absence of disturbances, the problem $\mathscr{D}_{\mathbf{z},n_T}$ (3.45) simplifies as

$$\overline{\mathscr{D}}_{\mathbf{z},n_T}: \begin{cases} \min_{\{\mathbf{z}_k\}_{k=0}^{n_T}, \{\mathbf{w}_k\}_{k=0}^{n_T-1}} \chi(\mathbf{C}^{\mathbf{x}}\mathbf{z}_0) + \sum_{k=0}^{n_T-1} \xi(\mathbf{v}_k)\Delta T_s, \\ \text{subject to:} \\ \overline{\mathbf{F}}_k^{\mathbf{z}}(\mathbf{z}_{k+1}, \mathbf{z}_k) = 0, k = 0, \ldots, n_T - 1, \\ \mathbf{C}^{\mathbf{g}}\mathbf{z}_k \leq 0, k = 0, \ldots, n_T, \\ \mathbf{v}_k = \mathbf{y}_k - \mathbf{C}^{\mathbf{h}}\mathbf{z}_k \in \mathbb{V}_k, k = 0, \ldots, n_T - 1. \end{cases} \tag{3.47}$$

Once the solution of $\mathscr{D}_{\mathbf{z},n_T}$ or $\overline{\mathscr{D}}_{\mathbf{z},n_T}$ is obtained, one can recover original state sequence estimate simply as $\hat{\mathbf{x}}_k = \mathbf{C}^{\mathbf{x}}\mathbf{z}_k, k = 0, \ldots, n_T$. For MHE Koopman formulation we refer the reader to [42].

### 3.6.3 Solution Approaches

The optimization problem $\mathscr{D}_{\mathbf{x},n_T}$ (3.40) is in general non-convex due to possibly non-convex state constraints and nonlinear dynamics constraints. Such problems have been solved via a variety of approaches, most common being via the use of general purpose nonlinear programming solvers [9]. Such an approach has two main limitations: first, there are few known bounds on the computational effort needed; and second, the solution procedure can become intractable in the sense that a bad initial guess could result in divergence of the numerical algorithm. These limitations make the use of general purpose nonlinear solvers undesirable for automated solutions and real-time applications, where guaranteed convergence and computational speed are key.

Convex optimization, on the other hand, can be reliably solved in polynomial time to the global optimum, see [5]. In order to take advantage of these powerful convex programming solvers, different techniques for convexification of the originally non-convex problem have been proposed. The most common approach is via successive linearization, i.e., at the $i$th succession, the dynamics and constraints are linearized by using the first-order Taylor approximation about the trajectory and the corresponding controls computed in the $(i-1)$th succession. This procedure is repeated until convergence. Thus, while a convex problem (i.e., quadratic program) is solved at each succession, the solution so obtained will not necessarily be same as its non-convex counterpart. Moreover, in addition to the approximation error one can introduce artificial infeasibility [26].

The KCSE optimization problem $\mathscr{D}_{\mathbf{z},n_T}$ (3.45) is an *indefinite* QCQP which is still in general non-convex. One can resort to sequential quadratic programming (SQP) approach to solve the QCQP as proposed in [10, 19], or via semidefinite programming (SDP) based convex relaxations [3]. This could still be advantageous in real time applications, for instance, over using SQP to solve the original nonlinear problem (3.40) as KCSE does not require the access to full nonlinear equations which can be expensive to run in real time, and the associated Jacobians for KCSE can be analytically computed (for which one may have to resort to numerical computations in original nonlinear formulation). In absence of disturbances the benefit is more apparent as the $\mathscr{D}_{\mathbf{z},n_T}$ reduces to a *convex* quadratic program (QP) $\overline{\mathscr{D}}_{\mathbf{z},n_T}$ (3.47) which can be efficiently solved to global optimality using a variety of techniques, see [5].

Note that above benefits offered by KCSE come at the price of increased number of decision variables (denoted by #d in Table 3.1) and number of constraints (denoted by #c in Table 3.1). Table 3.1 compares KCSE with the solution of CSE based on nonlinear optimization (NCSE) and successive linearization (LCSE) as discussed above. We restrict this discussion to the case of no disturbance, and only list state-related decision variables, and constraints arising due to the dynamics and the state constraints (as #d and #c due to disturbance/noise do not change across the three approaches). Note that compared to LCSE which will require multiple successive QP solves (of smaller #d and #c), the KCSE requires only one QP solve but involving higher #d and #c. Moreover, the solution quality of KCSE is expected to be better as the KOF provides a more accurate approximation over larger portion of state space compared to local linearization used in LCSE [43, 44].

**Table 3.1** Comparison of different solution approaches for CSE with no disturbances. Only state-related decision variables/constraints are listed

| | NCSE | LCSE | KCSE |
|---|---|---|---|
| Type | Non-convex | Multiple QP | Single QP |
| #d | $n_T \times n_x$ | $n_T \times n_x$ | $n_T \times n_z$ |
| #c | $n_T \times (n_x + n_g)$ | $n_T \times (n_x + n_g)$ | $n_T \times (n_z + n_g)$ |

## 3.7 Examples

### 3.7.1 Example I

We first illustrate a class of systems for which *Assumption I* and *Assumption II* hold true. Consider a linear system

$$\dot{\mathbf{x}} = \mathbf{A}\mathbf{x} + \sum_{i=1}^{n_u} \overline{\mathbf{b}}_i u_i + \sum_{j=1}^{n_w} \overline{\mathbf{d}}_j \mathbf{w}_j, \tag{3.48}$$

$$\mathbf{y} = \mathbf{h}(\mathbf{x}), \tag{3.49}$$

with non-convex state constraints

$$\mathbf{g}(\mathbf{x}) \leq 0, \tag{3.50}$$

and $\overline{\mathbf{b}}_i$ and $\overline{\mathbf{d}}_i$ are constant real-valued vector of appropriate dimensions. Without loss of generality we assume $\mathbf{A}$ has full set of eigenvalues and has been transformed into eigenbasis so that $\mathbf{A} \leftarrow \text{diag}(\lambda_1, \ldots, \lambda_{n_x})$. Then $\phi_i = x_i$ is a KEF with KE $\lambda_i$. Furthermore, we assume $\lambda_i < 0, i = 1\ldots, n_x$ in which case $\phi_i = x_i$ are the principal KEFs [30], so that any other KEF $\phi_{\mathbf{i}}$ can be expressed as

$$\phi_{\mathbf{i}} = \phi_1^{i_1} \phi_2^{i_2} \cdots \phi_{n_x}^{i_{nx}}, \tag{3.51}$$

with corresponding eigenvalue $\lambda_1^{i_1} \cdots \lambda_{n_x}^{i_{nx}}$. Here, $\mathbf{i} = (i_1, \ldots, i_{n_x})$ with $i_j \in \mathbb{N}$. Let

$$\mathbb{S}^{n_z} = \{\phi_{\mathbf{i}} : i_1 + \cdots i_{n_x} \leq p, i_j \in \mathbb{N}\}, \tag{3.52}$$

be subset of KEFs up to order $p$ such that

$$\mathbf{g}(\mathbf{x}), \mathbf{h}(\mathbf{x}) \in \mathbb{F}(\mathbb{S}^{n_z}). \tag{3.53}$$

Note that by construction $\mathbf{x} \in \mathbb{F}(\mathbb{S}^{n_z})$ and also

$$L_{\overline{\mathbf{b}}_i} \mathbf{T}(\mathbf{x}) - \mathbf{b}_i \in \mathbb{F}(\mathbb{S}^{n_z}), \quad L_{\overline{\mathbf{d}}_i} \mathbf{T}(\mathbf{x}) - \mathbf{d}_i \in \mathbb{F}(\mathbb{S}^{n_z}),$$

for some constant vectors $\mathbf{b}_i$ and $\mathbf{d}_i$. Hence, *Assumptions I–IV* are satisfied, and system (3.48)–(3.49) can be transformed into the KOF (3.26)–(3.27), and the relations (3.23)–(3.25) hold.

### 3.7.2 Example II

Consider the following nonlinear system:

$$\dot{\mathbf{x}} = \begin{pmatrix} \rho x_1 \\ \mu(x_2 - x_1^2) \end{pmatrix} + \mathbf{r}(\mathbf{x})u, \tag{3.54}$$

$$y = h(\mathbf{x}) = x_1^2 + x_2, \tag{3.55}$$

where $\mathbf{r}(\mathbf{x}) = (1, 0)^*$. It can be shown that for autonomous part of (3.54), i.e.,

$$\dot{\mathbf{x}} = \begin{pmatrix} \rho x_1 \\ \mu(x_2 - x_1^2) \end{pmatrix}, \tag{3.56}$$

$\rho, \mu$ are Koopman eigenvalues with eigenfunctions $\phi_\rho(\mathbf{x}) = x_1$, and $\phi_\mu(\mathbf{x}) = x_2 - \alpha x_1^2$, respectively, where $\alpha = \frac{\mu}{\mu - 2\rho}$. Also note that $2\rho, \rho + \mu$ etc. are Koopman eigenvalues with eigenfunctions $\phi_\rho^2, \phi_\rho \phi_\mu$ etc. Let $\phi_1 = \phi_\rho, \quad \phi_2 = \phi_\mu, \quad \phi_3 = \phi_\rho^2$, then it follows that $\mathbf{x} = \sum_{i=1}^3 \phi_i(\mathbf{x})\mathbf{v}_i^{\mathbf{x}}$ where, $\mathbf{v}_1^{\mathbf{x}} = (1, 0)^*$, $\mathbf{v}_2^{\mathbf{x}} = (0, 1)^*$, and $\mathbf{v}_3^{\mathbf{x}} = (0, \alpha)^*$. Similarly, $\mathbf{h}(\mathbf{x}) = \sum_{i=1}^3 \phi_i(\mathbf{x})\mathbf{v}_i^{\mathbf{h}}$, where $\mathbf{v}_1^{\mathbf{h}} = 0$, $\mathbf{v}_2^{\mathbf{h}} = 1$, and $\mathbf{v}_3^{\mathbf{h}} = 1 + \alpha$. Furthermore,

$$L_{\mathbf{r}}\mathbf{T}(\mathbf{x}) = \mathbf{b} + \sum_{j=1}^{n_z} \mathbf{v}_j^{\mathbf{r}} \phi_i(\mathbf{x}), \tag{3.57}$$

where $\mathbf{b} = (1, 0, 0)^*$, and $\mathbf{v}_1^{\mathbf{r}} = (0, -2\alpha, 2)^*$, and $\mathbf{v}_2^{\mathbf{r}} = \mathbf{v}_3^{\mathbf{r}} = (0, 0, 0)^*$.

Using the KCT,

$$\mathbf{z} = \mathbf{T}(\mathbf{x}) = \begin{pmatrix} \phi_1 \\ \phi_2 \\ \phi_3 \end{pmatrix} = \begin{pmatrix} x_1 \\ x_2 - \alpha x_1^2 \\ x_1^2 \end{pmatrix},$$

we get the bilinear KOF (3.26)–(3.27) with

$$\Lambda = \mathrm{diag}(\rho, \mu, 2\rho), \quad \mathbf{C}^{\mathbf{x}} = \begin{pmatrix} 1 & 0 & 0 \\ 0 & 1 & \alpha \end{pmatrix}, \quad \mathbf{C}^{\mathbf{h}} = \begin{pmatrix} 0 & 1 & 1+\alpha \end{pmatrix}.$$

Thus, one can employ recursive observer design techniques as discussed in Sect. 3.5.1.

Next consider estimation with a non-convex state constraint

$$\mathbf{g}(\mathbf{x}) = x_1 + x_2 - \alpha x_1^2 \leq 0 \tag{3.58}$$

for which $\mathbf{g}(\mathbf{x}) = \sum_{i=1}^3 \phi_i(\mathbf{x})\mathbf{v}_i^{\mathbf{g}}$, where $\mathbf{v}_1^{\mathbf{g}} = 1$, $\mathbf{v}_2^{\mathbf{g}} = 1$, and $\mathbf{v}_3^{\mathbf{g}} = 0$, and so $\mathbf{C}^{\mathbf{g}} = \begin{pmatrix} 1 & 1 & 0 \end{pmatrix}$. We compare three solution approaches based on NCSE, KCSE, and LCSE for the full information CSE problem formulation. For KCSE we use both Euler/trapezoidal rule (denoted by KCSE-Euler and KCSE-Trap) for time discretiza-

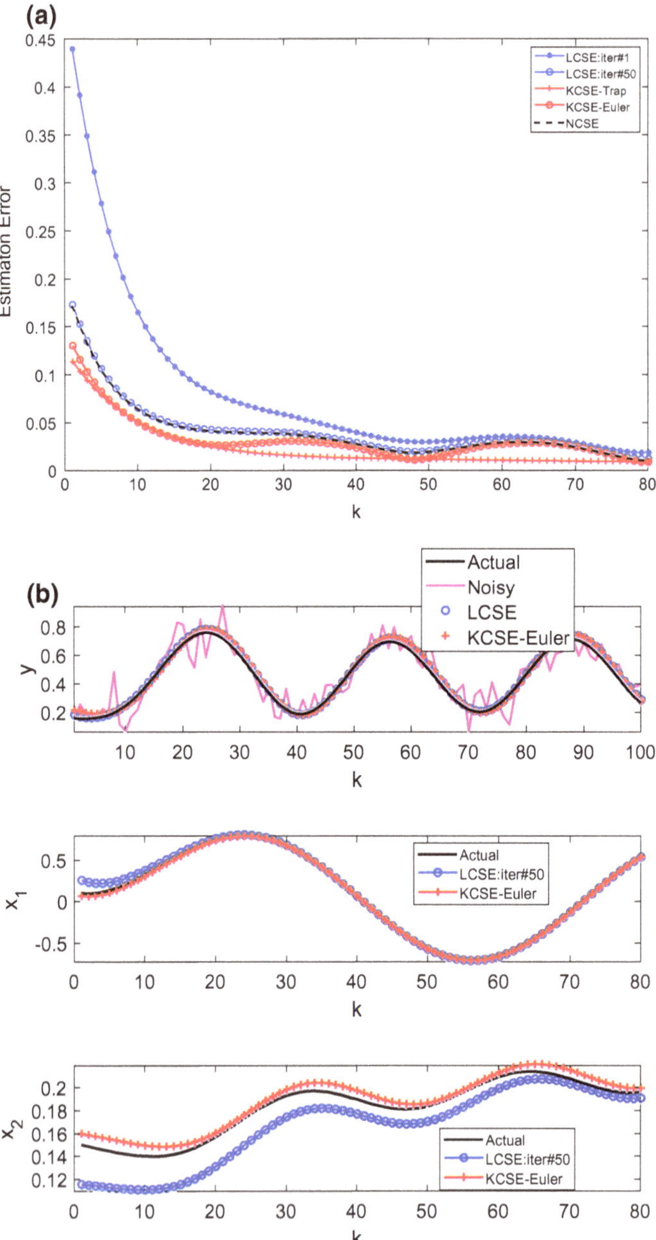


**Fig. 3.1** **a** Comparison of KCSE with LCSE/NCSE in terms of averaged estimation error. **b** Single realization of noisy measurement and corresponding state estimates

tion of KOF as discussed in Sect. 3.6.2. For LCSE we use linearized version of the discrete-time form (3.38)–(3.39) obtained from Euler discretization of (3.54)–(3.55) with the same time step as used in KCSE. To initialize NCSE/LCSE, the initial trajectory is obtained by propagating the discrete-time nonlinear model with initial condition $\hat{\mathbf{x}}_0$ for the given control sequence. The Jacobians used in LCSE are obtained analytically. KCSE/LCSE are solved using QP solver, and NCSE via nonlinear optimization using SQP, respectively, in the MATLAB.

Figure 3.1a shows the comparison of KCSE with LCSE/NCSE over horizon $n_T = 100$ in terms of averaged estimation error. For this $i = 1, \ldots, N = 100$ different realizations (see top subplot in Fig. 3.1b for an instance) of noisy output $\mathbf{y}_t^i = \mathbf{h}(\mathbf{x}_t) + v_t$ (where, $v_t$ is zero mean Gaussian measurement noise with variance $\sigma = 0.1$) is generated using Monte Carlo simulations for a given state sequence $\{\mathbf{x}_t\}_{t=0}^{n_T-1}$. The mean estimation error $\frac{1}{N}\sum_{i=1}^{N} ||\mathbf{x}_t - \hat{\mathbf{x}}_t^i||$ at each time is computed, where $\{\hat{\mathbf{x}}_t^i\}_{t=0}^{n_T-1}$ is the estimated state sequence for the $i-$the realization (see bottom subplots in Fig. 3.1b). Note that KCSE requires one QP solution, while LCSE requires a succession of QP solutions. In Fig. 3.1a we show LCSE solution at the 1st and the 50th succession. The KCSE-Euler/Trap provides more accurate estimates than LCSE/NCSE. As expected KCSE-Trap provides better solution quality than the KCSE-Euler.

## 3.8 Conclusions

In this chapter, we provided a systematic approach for estimation with and without nonlinear state constraints for control-affine input–output nonlinear systems. The bilinear structure of the control system in Koopman eigenfunction coordinate is exploited to provide a linear system-based approach for estimation without state constraints, and a convex optimization-based approach with state constraints. We also established theoretical results regarding the feasibility/optimality of the estimation problem under Koopman eigenfunction coordinate-based transformation. The proposed framework extends naturally for estimation under unknown disturbances, though the advantage of a convex formulation is lost. Simulation results are presented to verify the applicability of the developed framework.

We are currently investigating approaches for data-driven identification of bilinear KOF which will be necessary for applications to more complex problems. Further theoretical and numerical work is required for assessing the tradeoff between computational effort and estimation accuracy resulting from the truncation of the bilinear KOF. The use of bilinear KOF for control synthesis along the lines of [20], and fault detection and isolation in dynamic systems is another avenue of future research.

**Acknowledgements** The funding provided by UTRC is greatly appreciated.

## References


1. Abbaszadeh, M., Marquez, H.J.: A robust observer design method for continuous-time Lipschitz nonlinear systems. In: IEEE CDC (2006)
2. Acıkmese, B., Carson, J.M., Blackmore, L.: Lossless convexification of nonconvex control bound and pointing constraints of the soft landing optimal control problem. IEEE Trans. Control Syst. Technol. **21**(6), 2104–2113 (2013)
3. Bao, X., Sahinidis, N., Tawarmalani, M.: Semidefinite relaxations for quadratically constrained quadratic programming: a review and comparisons. J. Math. Program. Ser. B **129**(1), 129–157 (2011)
4. Besancon, G.: Nonlinear Observers and Applications. Lecture Notes in Control and Information Sciences, vol. 363. Springer, Berlin (2007)
5. Boyd, S., Vandenberghe, L.: Convex optimization. Cambridge University Press, Cambridge (2004)
6. Brenan, K.E., Campbell, S.L., Petzold, L.R.: Numerical solution of initial-value problems in differential-algebraic equations. SIAM, New York (1995)
7. Budisic, M., Mohr, R.M., Mezic, I.: Applied koopmanism. Chaos **22**(4), 047510 (2012)
8. Carleman, T.: Application de la théories des équations intégrales linéaires aux systémes déquations différentielles non linéaires. Acta Math. **59**, 63–87 (1932)
9. Diehl, M., Ferreau, H.J., Haverbeke, N.: Efficient numerical methods for nonlinear MPC and moving horizon estimation. In: Nonlinear Model Predictive Control, pp. 391–417. Springer, Berlin (2009)
10. Diehl, M., Bock, H.G., Schlöder, J.P., Findeisen, R., Nagy, Z., Allgöwer, F.: Real-time optimization and nonlinear model predictive control of processes governed by differential-algebraic equations. J. Process Control **12**(4), 577–585 (2002)
11. Domahidi, A.: Forces: fast optimization for real-time control on embedded systems. http://forces.ethz.ch (2012)
12. Elliott, D.: Bilinear Control Systems: Matrices in Action. Springer, Berlin (2009)
13. Gauthier, J., Kupka, A.: Deterministic Observation Theory and Applications. Cambridge University Press, Cambridge (1997)
14. Grasselli, O.M., Isidori, A.: Deterministic state reconstruction and reachability of bilinear processes. In: IEEE Joint Automatic Control Conference. IEEE, New York (1977)
15. Hermann, R., Krener, A.J.: Nonlinear controllability and observability. IEEE Trans. Autom. Control **22**(5) (1977)
16. Juang, J.N., Lee, C.H.: Continuous-time bilinear system identification using single experiment with multiple pulses. Nonlinear Dyn. **69**(3), 1009–1021 (2012)
17. Kang, W., Krener, A.J., Xiao, M., Xu, L.: A Survey of Observers for Nonlinear Dynamical Systems. Data Assimilation for Atmospheric, Oceanic and Hydrologic Applications, vol. II. Springer, Berlin (2013)
18. Keller, H.: Nonlinear observer design by transformation into a generalized observer canonical form. Int. J. Control **46**(6), 1915–1930 (1987)
19. Kelman, A., Borrelli, F.: Bilinear model predictive control of a hvac system using sequential quadratic programming. IFAC Proc. **44**(1), 9869–9874 (2011)
20. Korda, M., Mezic, I.: Linear predictors for nonlinear dynamical systems: Koopman operator meets model predictive control. Automatica **93**, 149–160
21. Kowalski, K., Steeb, W.H.: Nonlinear Dynamical Systems and Carleman Linearization. World Scientific Publishing Co. Pte. Ltd., Singapore (1991)
22. Krener, A.J.: Bilinear and nonlinear realizations of input output maps. SIAM J. Control Optim. **13**, 827–834 (1975)
23. Kühl, P., Diehl, M., Kraus, T., Schlöder, J.P., Bock, H.G.: A real-time algorithm for moving horizon state and parameter estimation. Comput. Chem. Eng. **35**(1), 71–83 (2011)
24. Liu, X., Lu, P.: Solving nonconvex optimal control problems by convex optimization. J. Guid. Control Dyn. **37**(3), 750–765 (2014)

25. Mao, Y., Dueri, D., Szmuk, M., Acıkmese, B.: Successive convexification of non-convex optimal control problems with state constraints (2017). arXiv:1701.00558
26. Mao, Y., Szmuk, M., Acıkmese, B.: Successive convexification of non-convex optimal control problems and its convergence properties. In: IEEE 55th Conference on Decision and Control, pp. 3636–3641. IEEE, Las Vegas (2016)
27. Mauroy, A., Mezic, I.: Global stability analysis using the eigenfunctions of the Koopman operator. IEEE Trans. Autom. Control **61**(11), 3356–3369 (2016)
28. Mezic, I.: Analysis of fluid flows via spectral properties of the Koopman operator. Annu. Rev. Fluid Mech. **45**, 357–378 (2012)
29. Misawa, E.A., Hedrick, J.K.: Nonlinear observers: a state-of-the-art survey. J. Dyn. Syst. Meas. Control **111**(3), 344–352 (1989)
30. Mohr, R., Mezic, I.: Construction of eigenfunctions for scalar-type operators via Laplace averages with connections to the Koopman operator (2014). arXiv:1403.6559
31. Morari, M., Lee, J.H.: Model predictive control: past, present and future. Comput. Chem. Eng. **23**(4), 667–682 (1999)
32. Nijmeijer, H., Fossen, T.I.: New Directions in Nonlinear Observer Design, vol. 244. Springer, Berlin (1999)
33. Pertew, A.M., Marquez, H.J., Zhao, Q.: H infinity observer design for Lipschitz nonlinear systems. IEEE Trans. Autom. Control **51**(7), 1211–1216 (2006)
34. Phan, M.Q., Shi, Y., Betti, R., Longman, R.W.: Discrete-time bilinear representation of continuous-time bilinear state-space models. Adv. Astronaut. Sci. **143**, 571–589 (2012)
35. Phanomchoeng, G., Rajamani, R.: Observer design for Lipschitz nonlinear systems using Riccati equations. In: Proceedings of the 2010 American Control Conference, pp. 6060–6065 (2010)
36. Raghavan, S., Hedrick, J.K.: Observer design for a class of nonlinear systems. Int. J. Control **59**(2), 515–528 (1994)
37. Rajamani, R.: Observers for Lipschitz nonlinear systems. IEEE Trans. Autom. Control **43**(3), 397–401 (1998)
38. Rao, C.V., Rawlings, J.B., Lee, J.H.: Constrained linear state estimation-a moving horizon approach. Automatica **37**(10), 1619–1628 (2001)
39. Rao, C.V., Rawlings, J.B., Mayne, D.Q.: Constrained state estimation for nonlinear discrete-time systems: stability and moving horizon approximations. IEEE Trans. Autom. Control **48**(2), 246–258 (2003)
40. Respondek, W.: Introduction to Geometric Nonlinear Control; Linearization, Observability, Decoupling. Lectures Given at the Summer School on Mathematical Control Theory. ICTP, Trieste (2001)
41. Rugh, W.J.: Nonlinear System Theory: The Volterra/Wiener Approach. Johns Hopkins University Press, Baltimore (1981)
42. Surana, A., Williams, M.O., Morari, M., Banaszuk, A.: Koopman operator framework for constrained state estimation. In: IEEE 56th Conference on Decision and Control, Melbourne, pp. 94–101 (2017)
43. Surana, A.: Koopman operator framework for observer synthesis for input-output nonlinear systems with control-affine inputs. In: IEEE 55th Conference on Decision and Control (2016)
44. Surana, A., Banaszuk, A.: Linear observer synthesis for nonlinear systems using Koopman Operator framework. IFAC-PapersOnLine **49**(18), 716–723 (2016)
45. Tenny, M.J., Rawlings, J.B.: Efficient moving horizon estimation and nonlinear model predictive control. In: American Control Conference, pp. 4475–4480. Anchorage (2002)

# Chapter 4
# Global Bilinearization and Reachability Analysis of Control-Affine Nonlinear Systems

**Debdipta Goswami and Derek A. Paley**

**Abstract** Nonlinear systems are ubiquitous in real-world applications, but the control design for them is not an easy task. Hence, methods are sought to transform a nonlinear system into linear or bilinear forms to alleviate the problem of nonlinear controllability and control design. While there are linearization techniques like Carleman linearization for embedding a finite-dimensional nonlinear system into an infinite-dimensional space, they depend on the analytic property of the vector fields and work only on polynomial space. The Koopman-based approach described here utilizes the Koopman canonical transform (KCT) to transform the dynamics and ensures bilinearity from the projection of the Koopman operator associated with the control vector fields on the eigenspace of the drift Koopman operator. The resulting bilinear system is then subjected to controllability analysis using the Myhill semigroup method and Lie algebraic structures.

## 4.1 Toward Bilinear Form: Infinitesimal Koopman Operator on Function Space

Operator-theoretic methods essentially work by embedding finite-dimensional dynamics in an infinite-dimensional function space in which functions evolve under a linear operator. The spectral property of the Koopman operator is well analyzed, see, e.g., [12]. The Koopman operator offers effective methods to characterize a nonlinear system in terms of stability [11] and linearization [9]. Reference [16] proposed a framework for designing an observer for a discrete-time unactuated nonlinear system. Reference [15] extends the same framework into continuous time with control-



D. Goswami (✉) · D. A. Paley
University of Maryland, College Park, MD 20740, USA
e-mail: goswamid@umd.edu

D. A. Paley
e-mail: dpaley@umd.edu

 
A. Mauroy et al. (eds.), *The Koopman Operator in Systems and Control*,
Lecture Notes in Control and Information Sciences 484,
https://doi.org/10.1007/978-3-030-35713-9_4

affine dynamics. Reference [15] introduces the Koopman canonical transform (KCT) using Koopman eigenfunctions, which transforms the (nonlinear) dynamics into an observer form. Here, the KCT is utilized to transform a control-affine system into a bilinear one. The bilinearization thus obtained is global and does not rely on the neighborhood of the operating point or trajectory.

### *4.1.1 Koopman Overview and Motivation*

Consider a dynamical system

$$\dot{\mathbf{x}} = \mathbf{f}(\mathbf{x}), \tag{4.1}$$

where $\mathbf{x} \in \mathbb{X} \subseteq \mathbb{R}^d$ and $\mathbf{f} : \mathbb{X} \to \mathbb{X}$. Let $\mathbf{S}(t, \mathbf{x})$ be the flow map of the system (4.1). Let $\mathscr{F}$ be the space of all complex-valued observables $\varphi : \mathbb{X} \to \mathbb{C}$. The continuous-time Koopman operator is defined as $U^t : \mathscr{F} \to \mathscr{F}$ such that

$$(U^t \varphi)(\cdot) = \varphi \circ \mathbf{S}(t, \cdot). \tag{4.2}$$

Unlike the original system, the Koopman operator is linear over its arguments, i.e., observable functions, and therefore can be characterized by its eigenvalues and eigenfunctions. A function $\phi : \mathbb{X} \to \mathbb{C}$ is an eigenfunction of $U^t$ if

$$(U^t \phi)(\cdot) = e^{\lambda t} \phi(\cdot), \tag{4.3}$$

with eigenvalue $\lambda \in \mathbb{C}$. It can be shown [11] that the infinitesimal generator of $U^t$, i.e., $\lim_{t \to 0} \dfrac{U^t - I}{t}$, is $\mathbf{f} \cdot \nabla = L_{\mathbf{f}}$, where $L_{\mathbf{f}}$ is the Lie derivative with respect to $\mathbf{f}$. The infinitesimal generator satisfies the eigenvalue equation

$$L_{\mathbf{f}} \phi = \lambda \phi. \tag{4.4}$$

Hence, the time-varying observable $\psi(t, \mathbf{x}) \triangleq U^t \varphi(\mathbf{x})$ is the solution of the PDE [11]

$$\begin{aligned} \frac{\partial \psi}{\partial t} &= L_{\mathbf{f}} \psi, \\ \psi(0, \mathbf{x}) &= \varphi(\mathbf{x}_0). \end{aligned} \tag{4.5}$$

In spite of its linearity, the Koopman operator is infinite dimensional, and has an infinite number of eigenfunctions. In fact, if $\phi_1$ and $\phi_2$ are eigenfunctions of $U^t$ with eigenvalues $\lambda_1$ and $\lambda_2$, respectively, then $\phi_1^k \phi_2^l$ is also an eigenfunction with eigenvalue $k\lambda_1 + l\lambda_2$ for any $k, l \in \mathbb{N}$. Moreover, the Koopman operator, being an infinite-dimensional operator, may contain continuous and residual spectra with a generalized eigendistribution [13]. The discussions in this chapter are restricted to the point spectra of the Koopman operator.

Let $\mathbf{g}(\cdot) \in \mathscr{F}^p$, $p \in \mathbb{N}$, be a vector-valued observable expressed in terms of Koopman eigenfunctions $\phi_i(\cdot)$ as follows:

$$\mathbf{g}(\cdot) = \sum_{i=1}^{\infty} \phi_i(\cdot)\mathbf{v}_i^{\mathbf{g}}, \tag{4.6}$$

where $\mathbf{v}_i^{\mathbf{g}} \in \mathbb{R}^p$, $i = 1, 2, \ldots$, are called the *Koopman modes* of the observable $\mathbf{g}(\cdot)$. Koopman modes represent the projection of the observable onto the span of Koopman eigenfunctions [4]. The Koopman eigenvalues and the eigenfunctions are properties of the dynamics only, whereas the Koopman modes depend on the observable.

A control-affine system is given by

$$\begin{aligned} \dot{\mathbf{x}} &= \mathbf{f}_0(\mathbf{x}) + \sum_{i=1}^{m} \mathbf{f}_i(\mathbf{x})u_i, \\ \mathbf{y} &= \mathbf{h}(\mathbf{x}), \end{aligned} \tag{4.7}$$

where $\mathbf{x} \in \mathbb{X} \subseteq \mathbb{R}^d$, $u_i \in \mathbb{R}$, for $i = 1, \ldots, m$ and $\mathbf{y} \in \mathbb{R}^p$. Here the control input enters only through the control vector fields $\mathbf{f}_i$. The objective is to transform the system into an appropriate basis in higher or possibly infinite dimensions, so that the drift and control vector fields become linear. Let $\psi(t, \mathbf{x})$ be defined as in (4.5). Note that $\psi(t, \mathbf{x})$ gives the time evolution of the observable quantity along the trajectory. Applying (4.5) to the system (4.7), the evolution PDE is

$$\begin{aligned} \frac{\partial \psi}{\partial t} &= L_{\mathbf{f}_0}\psi + \sum_{i=1}^{m} u_i L_{\mathbf{f}_i}\psi, \\ \psi(0, \mathbf{x}) &= \varphi(\mathbf{x}_0), \end{aligned} \tag{4.8}$$

where $L_{\mathbf{f}_i} \triangleq \mathbf{f}_i \cdot \nabla$, $i = 0, \ldots, m$, are the corresponding Lie derivatives and hence are linear operators on the space of $\psi$. The system of PDE (4.8) looks quite similar to $\dot{\mathbf{x}} = A\mathbf{x} + \sum_{i=1}^{m} B_i\mathbf{x}u_i$, i.e., the usual bilinear system. The system (4.8) differs only in the fact that $L_{\mathbf{f}_i}$ are infinite-dimensional operators operating over function space. The idea here is to choose a suitable collection of scalar observables to transform the original system into a new state space to get a bilinear form (possibly in infinite dimension) and then to project these Lie derivative operators on a finite-dimensional subspace to get a finite-dimensional bilinear system that approximates the original system (4.7).

### 4.1.2 Koopman Canonical Transform: A Review

The natural choice of the basis functions is to use the Koopman eigenfunctions, since these functions, when operated on by the Koopman infinitesimal generator, are multiplied by a scalar only. For this transformation, we use the Koopman Canonical

Transform (KCT) defined in [15] (see also Chap. 3). The KCT relies on the point spectra of the Koopman operator related to the drift vector field and it suffices for most systems because the continuous spectrum is typically empty near an attractor [11]. For the system (4.7), we investigate the Koopman eigenvalues and eigenfunctions of the unactuated dynamics, i.e.,

$$\dot{\mathbf{x}} = \mathbf{f}_0(\mathbf{x}), \tag{4.9}$$

and the flow associated with it. Let $\lambda_i, \phi_i(\cdot)$ for $i = 1, 2, \dots$ be the eigenvalue–eigenfunction pairs of the Koopman operator associated with the system (4.9). KCT [15] transforms the dynamics (4.7) using the eigenfunctions $\phi_i$ in a possibly higher dimensional space. To enable use of KCT, [15] mentions the following assumption.

**Assumption 1** $\exists\ \{\phi_i, i = 1, 2, \dots, n\}$, such that

$$\mathbf{x} = \sum_{i=1}^{n} \phi_i(\mathbf{x})\mathbf{v}_i^{\mathbf{x}}, \quad \mathbf{h}(\mathbf{x}) = \sum_{i=1}^{n} \phi_i(\mathbf{x})\mathbf{v}_i^{\mathbf{h}},$$

where $\mathbf{v}_i^{\mathbf{x}} \in \mathbb{C}^d$ and $\mathbf{v}_i^{\mathbf{h}} \in \mathbb{C}^p$.

This assumption implies that the state vector and the output function can be described in terms of a finite number of Koopman eigenfunctions. Assumption 1 is likely to be satisfied with sufficiently large $n$. If it is not, then $\mathbf{x}$ and $\mathbf{h}(\mathbf{x})$ may be well approximated by $n$ eigenfunctions as in the case of a Fourier series.

KCT consists of the transformation $T(\mathbf{x})$ defined as follows [15]:

$$\begin{aligned} T(\mathbf{x}) &= [\tilde{\phi}_1(\mathbf{x}), \dots, \tilde{\phi}_n(\mathbf{x})]^T, \\ \tilde{\phi}_i(\mathbf{x}) &= \phi_i(\mathbf{x}), \ \text{if } \phi_i : \mathbb{X} \to \mathbb{R}, \\ (\tilde{\phi}_i(\mathbf{x}), \tilde{\phi}_{i+1}(\mathbf{x}))^T &= (2Re(\phi_i(\mathbf{x})), -2Im(\phi_i(\mathbf{x})))^T, \\ &\quad \text{if } \phi_i : \mathbb{X} \to \mathbb{C} \text{ and assuming } \phi_{i+1} = \overline{\phi}_i. \end{aligned} \tag{4.10}$$

Following the transformation $\mathbf{z} = T(\mathbf{x})$, the system (4.7) in the new coordinates is [15]

$$\begin{aligned} \mathbf{x} &= C^{\mathbf{x}}\mathbf{z}, \\ \dot{\mathbf{z}} &= D\mathbf{z} + \sum_{i=1}^{m} L_{\mathbf{f}_i} T(\mathbf{x}) u_i |_{\mathbf{x}=C^{\mathbf{x}}\mathbf{z}}, \\ \mathbf{y} &= C^{\mathbf{h}}\mathbf{z}, \end{aligned} \tag{4.11}$$

where $C^{\mathbf{x}} = [\tilde{\mathbf{v}}_1^{\mathbf{x}}| \dots |\tilde{\mathbf{v}}_n^{\mathbf{x}}]$ and $C^{\mathbf{h}} = [\tilde{\mathbf{v}}_1^{\mathbf{h}}| \dots |\tilde{\mathbf{v}}_n^{\mathbf{h}}]$ with $\tilde{\mathbf{v}}_i^{\mathbf{x}} = \mathbf{v}_i^{\mathbf{x}}$ if $\phi_i$ is real valued, and $[\tilde{\mathbf{v}}_i^{\mathbf{x}}, \tilde{\mathbf{v}}_{i+1}^{\mathbf{x}}]$ is $[\mathrm{Re}\,\mathbf{v}_i^{\mathbf{x}}, \mathrm{Im}\,\mathbf{v}_i^{\mathbf{x}}]$ if $\phi_i$ is complex valued. $\tilde{\mathbf{v}}_i^{\mathbf{h}}$ are defined similarly. $D \in \mathbb{R}^{n\times n}$ is a block diagonal matrix with diagonal entry $D_{i,i} = \lambda_i$ if $\phi_i$ is a real-valued eigenfunction, or $\begin{bmatrix} D_{i,i} & D_{i,i+1} \\ D_{i+1,i} & D_{i+1,i+1} \end{bmatrix} = |\lambda_i| \begin{bmatrix} \cos(\angle\lambda_i) & \sin(\angle\lambda_i) \\ -\sin(\angle\lambda_i) & \cos(\angle\lambda_i) \end{bmatrix}$ if $\phi_i$ is complex.

The transformed system (4.11) is bilinearizable with certain conditions on the control vector fields so that their Lie derivative operators may be represented in terms of the Koopman eigenfunctions of the drift vector field.

## 4.2 Bilinearizability of the Koopman Canonical Transform: Sufficient Conditions

This section derives the sufficient conditions for the bilinearization of the system (4.11) that deals with the Koopman eigenspaces of the drift and control vector fields. This section also describes how we may obtain a bilinear approximation if the conditions are not satisfied. To establish bilinearizability of the system (4.11), we need to analyze the control vector fields of the original system. In the transformed system, the control enters through the transformed vector field $L_{\mathbf{f}_i} T(\mathbf{x})|_{\mathbf{x}=C^{\mathbf{x}}\mathbf{z}}$. Note that $L_{\mathbf{f}_i}$ is the infinitesimal Koopman operator with respect to control vector field $\mathbf{f}_i$.

**Definition 4.1** An *invariant subspace* of a linear mapping $T : V \to V$ from a vector space $V$ to itself is a subspace $W \subseteq V$ such that $T(W) \subseteq W$.

**Theorem 4.1** [6] *The system (4.11) (hence the system (4.7) as well) is bilinearizable in a countable (possibly infinite) basis if the eigenspace of $L_{\mathbf{f}_0}$, i.e., the Koopman operator corresponding to the drift vector field, is an invariant subspace of $L_{\mathbf{f}_i}$, $i = 1, \ldots, m$, i.e., the Koopman operators related to the control vector fields.*

*Proof* If the hypothesis is true, then we can choose eigenfunctions of $L_{\mathbf{f}_0}$, $\{\phi_j : j = 1, 2, \ldots\}$, such that $L_{\mathbf{f}_i}\phi_k \in \text{span}\{\phi_j : j = 1, 2, \ldots\}$, $\forall i = 1, \ldots, m$; $k = 1, 2, \ldots$. This outcome is guaranteed because $\text{span}\{\phi_j : j = 1, 2, \ldots\}$, i.e., the eigenspace of $L_{\mathbf{f}_0}$, is invariant under $L_{\mathbf{f}_i}$, $i = 1, \ldots, m$. So, $\forall k = 1, 2, \ldots$ we have $L_{\mathbf{f}_i}\phi_k = \sum_{j=1}^{\infty} v_j^{\mathbf{f}_i}\phi_j$, where $v_j^{\mathbf{f}_i} \in \mathbb{R}$. Now taking $T(\mathbf{x})$ as in (4.10) but without imposing the finite $n$ condition, we get

$$L_{\mathbf{f}_i} T(\mathbf{x}) = \sum_{j=1}^{\infty} \mathbf{v}_j^{\mathbf{f}_i} \phi_j(\mathbf{x}) = \sum_{j=1}^{\infty} \tilde{\mathbf{v}}_j^{\mathbf{f}_i} \tilde{\phi}_j(\mathbf{x}),$$

where $\tilde{\mathbf{v}}_j^{\mathbf{f}_i} \in \mathbb{R}^d$ and $\tilde{\phi}_j$ are defined as in (4.10). Define $B_i = [\tilde{\mathbf{v}}_1^{\mathbf{f}_i} | \tilde{\mathbf{v}}_2^{\mathbf{f}_i} | \ldots]$. Then, with $\mathbf{z} = T(\mathbf{x})$, the system (4.11) can be expressed as follows:

$$\dot{\mathbf{z}} = D\mathbf{z} + \sum_{i=1}^{m} B_i \mathbf{z} u_i. \tag{4.12}$$

Since the system (4.11) is a transformation of (4.7), bilinearization of the former implies the same for the latter. □

Although Theorem 4.1 gives the condition for the bilinearizability of the control-affine system using KCT with a countable number of eigenfunctions, it still does not solve the problem with infinitely many eigenfunctions. However, for an approximate result, we can truncate the number of eigenfunctions to only the dominant ones. This linear approximation, unlike the Jacobian approach, is global, i.e., it is valid over the entire manifold $\mathbb{X}$ on which the dynamics (4.7) is defined.

**Corollary 4.1** [6] *The systems (4.7) and (4.11) are bilinearizable if the drift vector field* $\mathbf{f}_0 \equiv 0$, *i.e., it is a pure control-affine system.*

*Proof* The proof follows from the fact that every function $\phi(\cdot) \in \mathscr{F}$ is an eigenfunction of $L_{\mathbf{f}_0}$ with $\mathbf{f}_0 \equiv 0$ corresponding to the zero eigenvalue. Hence the whole space $\mathscr{F}$ is the eigenspace of $L_{\mathbf{f}_0}$, which is of course invariant under $L_{\mathbf{f}_i}$, $\forall i = 1, \ldots, m$. Therefore, from Theorem 4.1, the system is bilinearizable. □

Theorem 4.1 and Corollary 4.1 essentially embed the finite-dimensional nonlinear dynamics (4.7) in a higher, possibly infinite-dimensional linear system (4.12). There are other embedding techniques that deal with Hermite polynomials, e.g., Carleman embedding [8], but that technique works only on analytic nonlinearities. The method with Koopman eigenfunctions works on a wide variety of systems and can be characterized in terms of the range and eigenspace of the corresponding Koopman operator.

For a finite-dimensional bilinearization of the system (4.7), we need a stronger assumption than invariance of the eigenspace of $L_{\mathbf{f}_0}$. The invariant subspace must be spanned by a finite number of Koopman eigenfunctions, which is the statement of Theorem 4.2.

**Theorem 4.2** [6] *Suppose* $\exists\, \{\phi_j : j = 1, \ldots, n\}$, $n \in \mathbb{N}$, $n < \infty$, *such that* $\phi_j$, $j = 1, \ldots, n$, *are the Koopman eigenfunctions of the unactuated system (4.9) and* $span\{\phi_1, \ldots, \phi_n\}$ *forms an invariant subspace of* $L_{\mathbf{f}_i}$, $i = 1, \ldots, m$. *Then the system (4.7) and, in turn system (4.11), are bilinearizable with an n-dimensional state space.*

*Proof* The hypothesis dictates that $L_{\mathbf{f}_i}\phi_k \in \text{span}\{\phi_j : j = 1, \ldots, n\}\, \forall i = 1, \ldots, m$; $k = 1, \ldots, n$. Therefore, we conclude

$$L_{\mathbf{f}_i}\phi_k = \sum_{j=1}^{n} v_j^{\mathbf{f}_i}\phi_j, \;\; k = 1, \ldots, n,$$

where $v_j^{\mathbf{f}_i} \in \mathbb{R}$. Now consider $T(\mathbf{x})$ as defined in (4.10). Its Lie derivatives with respect to the control vector fields are

$$L_{\mathbf{f}_i}T(\mathbf{x}) = \sum_{j=1}^{n} \mathbf{v}_j^{\mathbf{f}_i}\phi_j(\mathbf{x}) = \sum_{j=1}^{n} \tilde{\mathbf{v}}_j^{\mathbf{f}_i}\tilde{\phi}_j(\mathbf{x}),$$

where $\tilde{\mathbf{v}}_j^{\mathbf{f}_i} \in \mathbb{R}^d$ and $\tilde{\phi}_j$ are defined as in (4.10). Now, as in the proof of Theorem 4.1, let us define $B_i \triangleq [\tilde{\mathbf{v}}_1^{\mathbf{f}_i} | \tilde{\mathbf{v}}_2^{\mathbf{f}_i} | \ldots | \tilde{\mathbf{v}}_n^{\mathbf{f}_i}]$. Unlike in the proof of Theorem 4.1, this $B_i$ is not only countable, but is also a finite-dimensional operator. Now with transformed coordinate $\mathbf{z} = T(\mathbf{x})$, the system becomes

$$\dot{\mathbf{z}} = D\mathbf{z} + \sum_{i=1}^{m} B_i \mathbf{z} u_i, \tag{4.13}$$

with $\mathbf{z} \in \mathbb{R}^n$, $n < \infty$. □

Though the hypothesis of Theorem 4.2 is difficult to satisfy, we can always include more eigenfunctions $\phi_j$ in the span so that $\|L_{\mathbf{f}_i}\phi_j - \sum_{j=1}^{n} v_j^{\mathbf{f}_i}\phi_j\|$ becomes sufficiently small. Note that usually $n \gg d$, i.e., this method of bilinearization lifts the original dynamics (4.7) to a much higher dimensional state space. The resulting bilinear system is relatively easier to work with in terms of controllability analysis and designing a stabilizing control, as illustrated in the next section. The bilinear system defined by (4.13) will be referred as the Koopman bilinear form (KBF) in the sequel.

## 4.3 Reachability Analysis of the Koopman Bilinear Form

This section is devoted to the reachability analysis of the KBF and requires the concatenation-semigroup structure of the control signals [3]. First, the concepts of reachability and reachable sets are briefly described.

**Definition 4.2** Given the KBF

$$\dot{\mathbf{z}} = D\mathbf{z} + \sum_{i=1}^{m} B_i \mathbf{z} u_i, \ \mathbf{z}(0) = \mathbf{z}_0,$$

a point $\mathbf{z}_d$ is said to be *reachable in time* $T$ if $\exists$ an input $\mathbf{u} : t \in [0, T] \mapsto \mathbf{u}(t) \in \mathbb{R}^m$ such that $\mathbf{z}(T) = \mathbf{z}_d$. If $\mathbf{z}_d$ is reachable in time $T$ for all $T > 0$, then $\mathbf{z}_d$ is said to be *reachable*. Moreover, the set of all reachable point of the system, i.e., $\{\mathbf{z}_d : \mathbf{z}_d \text{ is reachable}\}$ is called the *reachable set*.

Next the semigroup is defined and explained in terms of piecewise-continuous control signal.

**Definition 4.3** A *semigroup* is an algebraic structure consisting of a set $S$ and a binary operation "$\circ$" defined as $\circ : S \times S \to S$, i.e., $x \circ y \in S$, $\forall x, y \in S$ such that it satisfies the associative property

$$\forall x, y, z \in S, \ (x \circ y) \circ z = x \circ (y \circ z).$$

Let $\mathbf{u}^i$ be an $m$-dimensional piecewise-continuous control signal. We can form a semigroup from the set $\{\mathbf{u}^i(\cdot)|\mathbf{u}^i : \mathbb{R}^+ \rightarrow \mathbb{R}^m, \mathbf{u}^i \text{ piecewise continuous}\}$ with the concatenation operation. The concatenation operation looks like

$$\mathbf{u}^1 \circ \mathbf{u}^2 = \begin{cases} \mathbf{u}^1(t), & t \in [0, t_1), \\ \mathbf{u}^2(t - t_1), & t \in [t_1, t_2). \end{cases} \tag{4.14}$$

Denote this semigroup as $U^m$. Each $\mathbf{u} \in U^m$, when applied to the dynamics (4.13), generates a one-to-one continuous map from $\mathbb{R}^n$ into $\mathbb{R}^n$ in terms of a flow map. Let $T^n$ be the semigroup of all such maps with the composition operation. The system (4.13) defines a homomorphism $H$ from $U^m$ into $T^n$ [3]. The image of $U^m$ under $H$ is called the Myhill semigroup [3] of the system. The maps of the Myhill semigroup are, in fact, the flow maps of the system with a particular piecewise-continuous control signal $\mathbf{u} \in U^m$ and, therefore, provide all the information about the dynamics.

In general, for an arbitrary nonlinear system, these maps are difficult to obtain analytically and yield no practical use. But for the bilinear system (4.13), the Myhill semigroup maps are the matrices $Z \in \mathbb{R}^{n\times n}$, satisfying the matrix differential equation

$$\begin{aligned} \dot{Z}(t) &= DZ(t) + \sum_{i=1}^{m} B_i Z(t) u_i, \\ Z(0) &= I, \end{aligned} \tag{4.15}$$

with $\mathbf{z}(t) = Z(t)\mathbf{z}(0)$ for any $\mathbf{z}(0) \in \mathbb{R}^n$. Therefore, given any initial state $\mathbf{z}_0$, the states reachable from $\mathbf{z}_0$ are given by all the points in $\mathbb{R}^n$ that can be generated by $Z(t)\mathbf{z}_0$ for some $t \geq 0$, where $Z(t)$ satisfies (4.15). Consequently, the controllability of the system (4.13) can be characterized by the controllability of the matrix differential equation (4.15). The controllability of a bilinear matrix system has been studied widely [2, 3, 5, 7], exploiting the characteristic of matrices as operators and the corresponding Lie algebraic structures.

The Lie bracket of $\mathbb{R}^{n\times n}$ matrices is defined as

$$[\cdot, \cdot] : \mathbb{R}^{n\times n} \times \mathbb{R}^{n\times n} \rightarrow \mathbb{R}^{n\times n},$$

$$[X, Y] \mapsto XY - YX.$$

Any space of matrices over a field $F$ (usually $\mathbb{R}$ or $\mathbb{C}$) closed under the Lie bracket operation forms a Lie algebra. The dimension $l$ of a Lie algebra is its dimension as a vector space over $F$. The matrix exponentials of all elements of a matrix Lie algebra along with usual matrix multiplication form a matrix Lie group associated with the algebra. For example, all $\mathbb{R}^{n\times n}$ matrices form a Lie algebra with dimension $l = n^2$. The corresponding group is known as the *General Linear group* and denoted as $\mathrm{GL}(n, \mathbb{R})$, which corresponds to the multiplicative group of invertible $n \times n$ real matrices. Denote $\{X_i : i = 1, \ldots, n\}_A$ as the smallest Lie algebra containing $\{X_i : i = 1, \ldots, n\}$ and $\{\exp\{X_i\} : i = 1, \ldots, n\}_G$ as the smallest Lie group containing $\{\exp\{X_i\} : i = 1, \ldots, n\}$. Also $\forall\, A, B \in \mathbb{R}^{n\times n}$ and $k = 0, 1, \ldots$, define $\mathrm{ad}_A^{k+1} B \triangleq$

$[A, \mathrm{ad}_A^k B]$ with $\mathrm{ad}_A^0 B \triangleq B$. The controllability results are stated below using the notation of Lie groups and algebras.

The necessary and sufficient conditions for reachability in the KBF (4.13) are described next. First it is derived for the drift-free matrix bilinear system. Then it is extended to the matrix systems with drift. Finally, the relationship between the reachability in matrix bilinear system (4.15) and KBF (4.13) is explained.

### 4.3.1 Conditions for Reachability

To describe the reachability of a drift-free matrix bilinear system, we need a factorization lemma by Wei and Norman [17], which gives the solution of a bilinear system locally in terms of matrix exponentials.

**Lemma 4.1** *Consider the drift-free matrix differential equation in* $\mathbb{R}^{n\times n}$,

$$\dot{Z}(t) = \sum_{i=1}^{m} B_i Z(t) u_i(t),\ \ Z(0) = I, \tag{4.16}$$

*where* $Z \in \mathbb{R}^{n\times n}$. *Let* $l$ *be the dimension of the Lie algebra* $\{B_i : i = 1, \ldots, m\}_A$. *Then there exists a neighborhood of* $t = 0$ *in which the solution of (4.16) may be expressed in the form*

$$Z(t) = \prod_{i=1}^{l} \exp(h_i(t) B_i),$$

*where* $\{B_i : i = 1, \ldots, l\}$ *is the extension of* $\{B_i : i, \ldots, m\}$ *to a basis of* $\{B_i : i = 1, \ldots, m\}_A$.

**Theorem 4.3** [2] *Consider the drift-free matrix differential equation (4.16) from Lemma 4.1, which corresponds to the system (4.13) with* $\mathbf{f}_0 \equiv 0$ *and* $B_i$ *as defined in Theorem 4.2.* $Z_1 \in \mathbb{R}^{n\times n}$ *is in the reachable set of (4.16) if and only if* $Z_1 \in \{\exp\{\{B_i : i = 1, \ldots, m\}_A\}\}_G$, *i.e., it lies within the smallest group generated by the matrix exponential of the elements of the smallest algebra generated by the control matrices.*

*Proof Sufficiency*: Let $Z_1 \in \{\exp\{\{B_i : i = 1, \ldots, m\}_A\}\}_G$. Then from [2], $Z_1$ can be written as a finite product

$$Z_1 = \prod_{k=1}^{m} \exp(B_{i_k} a_k),$$

where $B_{i_k} \in \{B_i : i = 1, \ldots, m\}$ and $a_k \in \mathbb{R}\, \forall\, k = 1, \ldots, m$. For a $T > 0$, partition $[0, T]$ into $m$ equal intervals $[t_{k-1}, t_k]$, $k = 1, \ldots, m$. Define $\tau \triangleq t_k - t_{k-1}$.

Now for $t \in [t_{k-1}, t_k)$, choose $u_{i_{m-k+1}} = \frac{a_{m-k+1}}{\tau}$ and $u_i \equiv 0 \,\forall\, i \neq i_{m-k+1}$. Hence, for each interval $[t_{k-1}, t_k)$, the system becomes

$$\dot{Z}(t) = \frac{a_{m-k+1}}{\tau} B_{i_{m-k+1}} Z(t),$$

with state transition matrix $\exp\left(t\frac{a_{m-k+1}}{\tau} B_{i_{m-k+1}}\right)$. Therefore

$$\begin{aligned} Z(T) &= \exp\left(a_1 B_{i_1}\right) \cdot \ldots \cdot \exp\left(a_m B_{i_m}\right) Z(0) \\ &= \prod_{k=1}^{m} exp\left(a_k B_{i_k}\right), \end{aligned}$$

since $Z(0) = I$. Hence $Z_1$ is reachable using piecewise-continuous control $\mathbf{u}(t)$.

*Necessity*: Let $Z_1$ be reachable in time $T$, i.e., $Z_1 = Z(T)$. From Lemma 4.1, we have a partition $0 = t_0 < t_1 < \cdots < t_p = T$ such that for $t \in [t_k - 1, t_k)$, $k = 1, \ldots, p$,

$$Z(t) = \prod_{i=1}^{l} \exp(h_i(t) B_i),$$

where $\{B_i : i = 1, \ldots, l\}$ is the extension of $\{B_i : i, \ldots, m\}$ to a basis of $\{B_i : i = 1, \ldots, m\}_A$.
Therefore,

$$\begin{aligned} Z_1 &= Z(T) \\ &= \prod_{k=1}^{p} \prod_{i=1}^{l} \exp(h_i(t_k - t_{k-1}) B_i), \end{aligned}$$

i.e., a product of the matrix exponentials on the Lie algebra $\{B_i : i = 1, \ldots, m\}_A$. Hence $Z_1 \in \{\exp\{\{B_i : i = 1, \ldots, m\}_A\}\}_G$. □

The result of the Theorem 4.3 is somewhat incomplete in that the drift term is absent. The following theorem describes one way in which this constraint can be relaxed.

**Theorem 4.4** [2] *Consider the matrix differential equation in* $\mathbb{R}^{n \times n}$,

$$\dot{Z}(t) = DZ(t) + \sum_{i=1}^{m} B_i Z(t) u_i(t), \;\; Z(0) = I, \tag{4.17}$$

*where* $D$ *and* $B_i$ *are defined as in the proof of Theorem 4.2. Assume* $[ad_D^j k B_i, B_j] = 0$ *for* $i, j = 1, \ldots, m$ *and* $k = 0, 1, \ldots, n^2 - 1$. *Let* $\mathscr{L} = span\{ad_D^k B_i : i = 1, \ldots, m, k = 0, 1, \ldots, n^2 - 1\}$. *Then* $Z_1$ *is reachable at time* $t_1$ *through continuous controls if and only if* $\exists\, L \in \mathscr{L}$ *such that*

$$Z_1 = \exp(t_1 D)\exp(L).$$

*Proof Necessity*: Let $Z_1$ is reachable at time $t_1$, i.e., $Z(t_1) = Z_1$. Using the Baker–Hausdorff formula [2],

$$\begin{aligned}
[\exp(tD)B_i\exp(-tD)\,,\ B_j] &= \left[B_i + [D, B_i]t + \frac{1}{2}[D, [D, B_i]]t^2 + \dots,\ B_j\right] \\
&= [B_i, B_j] + [\mathrm{ad}_D B_i, B_j] + \frac{1}{2}[\mathrm{ad}_D^2 B_i, B_j] + \frac{1}{6}[\mathrm{ad}_D^3 B_i, B_j] + \dots \\
&= \sum_{k=0}^{\infty}\frac{1}{k!}[\mathrm{ad}_D^k B_i, B_j]. \qquad (4.18)
\end{aligned}$$

Now we know $[\mathrm{ad}_D^k B_i, B_j] = 0$ for $k = 0, \dots, n^2 - 1$. However, $\mathrm{ad}_D(\cdot)$ is a linear operator from an $n^2$-dimensional space to itself. Hence, by the Cayley–Hamilton Theorem [2], $\mathrm{ad}_D^k(\cdot)$ for $k \geq n^2 - 1$ are linear combinations of the first $n^2 - 1$ powers. Hence, from (4.18), $[\exp(tD)B_i\exp(-tD)\,,\ B_j]$ vanishes identically.

Also we get

$$\begin{aligned}
0 &= \exp(tD)B_i\exp(-tD)B_j - B_j\exp(tD)B_i\exp(-tD) \\
&= \exp(\sigma D)(\exp(tD)B_i\exp(-tD))B_j\exp(-\sigma D) \\
&\quad - \exp(\sigma D)B_j(\exp(tD)B_i\exp(-tD))\exp(-\sigma D),
\end{aligned}$$

for some arbitrary $\sigma \in \mathbb{R}$. Let $\beta \triangleq t + \sigma$ and $\gamma = \sigma$. Therefore, for all $\beta$ and $\gamma \in \mathbb{R}$,

$$[\exp(\beta D)B_i\exp(-\beta D), \exp(\gamma D)B_i\exp(-\gamma D)] = 0.$$

Now in order to solve the matrix differential equation (4.17), let $Y(t) \triangleq \exp(-tD)$ $Z(t)$. Then (4.17) becomes

$$\dot{Y}(t) = \left(\sum_{i=1}^{m} u_i(t)\exp(-tD)B_i\exp(tD)\right)Y(0), \qquad (4.19)$$

which is of the form $\dot{Y}(t) = B(t)Y(t)$ with $B(t) \triangleq \sum_{i=1}^{m} u_i(t)\exp(tD)B_i\exp(-tD)$ and $[B(t), B(\sigma)] = 0$ for all $t$ and $\sigma$. The solution of this system can be written [10] as $\exp\int_0^t B(\sigma)d\sigma$, i.e.,

$$Y(t) = \exp\left(\sum_{i=1}^{m}\int_0^t u_i(\sigma)\exp(-\sigma D)B_i\exp(\sigma D)d\sigma\right)Y(0) = \exp(L(t))Y_0, \qquad (4.20)$$

where $L(t) \triangleq \left( \sum_{i=1}^{m} \int_0^t u_i(\sigma) \exp(-\sigma D) B_i \exp(\sigma D) d\sigma \right)$. In (4.20), each term in the summation can be written as

$$\begin{aligned} &\int_0^t u_i(\sigma) \exp(-\sigma D) B_i \exp(\sigma D) d\sigma \\ &= \int_0^t (B_i - [D, B_i]\sigma + \frac{1}{2}[D, [D, B_i]]\sigma^2 - \ldots) u_i(\sigma) d\sigma \\ &= \sum_{k=0}^{\infty} \frac{(-1)^k}{k!} \mathrm{ad}_D^k B_i \left( \int_0^t \sigma^k u_i(\sigma) d\sigma \right), \end{aligned}$$

i.e., as a linear combination of $\mathrm{ad}_D^k B_i$ for $k = 0, 1, \ldots$. But, again by the Cayley–Hamilton Theorem, $\mathrm{ad}_D^k B_i$ for $k > n^2 - 1$ is the linear combination of the previous $n - 1$ powers. Therefore, the term within the exponential in (4.20), $L(t) \in \mathscr{L} = \mathrm{span}\{\mathrm{ad}_D^k B_i : i = 1, \ldots, m,\ k = 0, 1, \ldots, n^2 - 1\}$ for all $t > 0$. Choose $t = t_1$ and $L \triangleq L(t_1)$. Now, $Z(t_1) = \exp(tD)Y(t_1) = \exp(tD)\exp(L)Y(0) = \exp(tD)\exp(L)$ since $Z(0) = Y(0) = I$ with $L \in \mathscr{L}$.
*Sufficiency:* Notice that

$$\left. \frac{d^k}{dt^k} \exp(-tD) B_i \exp(tD) \right|_{t=0} = (-1)^k \mathrm{ad}_D^k B_i. \tag{4.21}$$

Reference [1] shows that the image space of the operator taking continuous functions $u_i$ to $\mathbb{R}^{n\times n}$ according to the rule $x = L(u_i) = \int_0^{t_1} u_i(\sigma) \exp(-\sigma D) B_i \exp(\sigma D) d\sigma$ is spanned by the first $n^2$ derivatives of $\exp(-tD) B_i \exp(tD)$ (including the zeroth derivative) evaluated at zero. Using this fact with (4.20) and (4.21), we see that for each $L = \mathscr{L} = \mathrm{span}\{\mathrm{ad}_D^k B_i : i = 1, \ldots, m,\ k = 0, 1, \ldots, n^2 - 1\}$ and $t_1 > 0$, there exists a continuous $\mathbf{u} : [0, t_1] \to \mathbb{R}^m$ such that $Y(t_1) = \exp(L)Y(0) = \exp(L)$ since $Y(0) = Z(0) = I$. Hence $Z_1 \triangleq Z(t_1) = \exp(tD)Y(t_1) = \exp(tD)\exp(L)$ is reachable at time $t_1$. □

Theorems 4.3 and 4.4 describe the reachable set for the bilinear matrix differential equations (4.16) and (4.15). The reachability of KBF (4.13) can be expressed with the help of Theorems 4.3 and 4.4, and is derived in Theorem 4.5.

**Theorem 4.5** *(1) Given a transformed state $\mathbf{z}_1$, if $\exists\, Z_1 \in \mathbb{R}^{n\times n}$ such that $\mathbf{z}_1 = Z_1\mathbf{z}_0$, then $\mathbf{z}_1$ is reachable from $\mathbf{z}_0$ in KBF (4.13) if $Z_1$ is reachable from $Z(0) = I$ in the matrix differential equation (4.15).*

*(2) Conversely if $\mathbf{z}_1$ is reachable from $\mathbf{z}_0$ in KBF (4.13), then $\exists\, Z_1$ in the reachable set of the matrix differential equation (4.15) from $Z(0) = I$ such that $\mathbf{z}_1 = Z_1\mathbf{z}_0$.*

*Proof* (1) We have $Z_1 \in \mathbb{R}^{n\times n}$ such that $\mathbf{z}_1 = Z_1\mathbf{z}_0$, and $Z_1$ is reachable at time $T$ in the matrix differential equation (4.15) from $Z(0) = I$. Hence we have $Z(T) = Z_1$ in (4.15). Now define

$$\mathbf{z}(t) = Z(t)\mathbf{z}_0,\ t \geq 0,$$

where $Z(t)$ is the trajectory of the matrix differential equation (4.15). By differentiation, we obtain

$$\begin{aligned}\dot{\mathbf{z}}(t) &= \dot{Z}(t)\mathbf{z}_0\\ &= DZ(t)\mathbf{z}_0 + \sum_{i=1}^{m} B_i Z(t)\mathbf{z}_0 u_i(t)\\ &= D\mathbf{z}(t) + \sum_{i=1}^{m} B_i\mathbf{z}(t)u_i(t),\end{aligned}$$

which is of the same form as of KBF (4.13) with $\mathbf{z}(0) = \mathbf{z}_0$ and $\mathbf{z}(T) = Z(T)\mathbf{z}_0 = Z_1\mathbf{z}_0 = \mathbf{z}_1$. Hence $\mathbf{z}_1$ is reachable from $\mathbf{z}_0$ in KBF (4.13).

(2) Now we have $\mathbf{z}_1$ reachable from $\mathbf{z}_0$ in KBF (4.13). By contradiction, assume $\nexists\ Z_1 \in \mathbb{R}^{n\times n}$ reachable from $Z(0) = I$ in the matrix differential equation (4.15) such that $\mathbf{z}_1 = Z_1\mathbf{z}_0$. Then at any $t > 0$, $Z(t)\mathbf{z}_0 \neq \mathbf{z}_1$. Like the previous part, define

$$\mathbf{z}(t) = Z(t)\mathbf{z}_0,\ t \geq 0,$$

where $Z(t)$ is the trajectory of the matrix differential equation (4.15). By differentiation, we obtain

$$\begin{aligned}\dot{\mathbf{z}}(t) &= \dot{Z}(t)\mathbf{z}_0\\ &= DZ(t)\mathbf{z}_0 + \sum_{i=1}^{m} B_i Z(t)\mathbf{z}_0 u_i(t)\\ &= D\mathbf{z}(t) + \sum_{i=1}^{m} B_i\mathbf{z}(t)u_i(t),\end{aligned} \tag{4.22}$$

which is of the same form as of KBF (4.13) with $\mathbf{z}(0) = \mathbf{z}_0$. But we know $\mathbf{z}_1$ is reachable in the system (4.22) from $\mathbf{z}_0$, i.e., $\exists\ T$ such that $\mathbf{z}(T) = \mathbf{z}_1$, and again from (4.22) $\mathbf{z}(T) = Z(T)\mathbf{z}_0$. Hence we have $Z_1 \triangleq Z(T)$, such that $\mathbf{z}_1 = Z_1\mathbf{z}_0$, which establishes the contradiction. □

The controllability and reachable sets of the system (4.7) may be characterized by the transformed bilinearized system with $D$, $B_i, i = 1, \ldots, m$, from the Koopman bilinear form in Sect. 4.2. However, because the transformed KBF usually has more dimensions, it may not achieve complete controllability even when the original system does.

## 4.4 Numerical Examples

This section is devoted to the illustration of bilinearization and successive controllability analysis of control-affine system. Here we have used systems with analytically derivable Koopman eigenfunction for the purpose of illustration. The technique is applicable to systems where Koopman eigenfunctions and modes have to be approximated using extended dynamic mode decomposition (EDMD) [18].

To demonstrate the effectiveness of the bilinearization technique described, we choose the system

$$\dot{\mathbf{x}} = \mathbf{f}_0(\mathbf{x}) + \mathbf{f}_1(\mathbf{x})u_1 + \mathbf{f}_2(\mathbf{x})u_2, \tag{4.23}$$

where the drift $\mathbf{f}_0$ is

$$\mathbf{f}_0(\mathbf{x}) = \begin{pmatrix} \lambda x_1 \\ \mu x_2 + (2\lambda - \mu)cx_1^2 \end{pmatrix}.$$

This choice of $\mathbf{f}_0$ is inspired from [14] so that the eigenfunctions may be obtained by inspection. For the demonstration, we choose different $\mathbf{f}_1$ and $\mathbf{f}_2$.

It can be verified that the Koopman eigenvalue–eigenfunction pairs for $L_{\mathbf{f}_0}$ are as follows [6]:

- $\phi_1(\mathbf{x}) = x_1$ with eigenvalue $\lambda$,
- $\phi_2(\mathbf{x}) = x_2 - cx_1^2$ with eigenvalue $\mu$,
- $\phi_3(\mathbf{x}) = x_1^2$ with eigenvalue $2\lambda$,
- $\phi_4(\mathbf{x}) = 1$ with eigenvalue 0.

Any multiplicative combination of these eigenfunctions will yield another eigenfunction with a suitable eigenvalue. However, for our discussion, it is sufficient to consider only these four. $\phi_4$ is the trivial constant eigenfunction with zero eigenvalue, introduced to deal with constant control vector fields.

The Koopman canonical transformation is

$$\mathbf{z} = T(\mathbf{x}) = \begin{pmatrix} \phi_1(\mathbf{x}) \\ \phi_2(\mathbf{x}) \\ \phi_3(\mathbf{x}) \\ \phi_4(\mathbf{x}) \end{pmatrix} = \begin{pmatrix} x_1 \\ x_2 - cx_1^2 \\ x_1^2 \\ 1 \end{pmatrix}$$

and matrix $D$ is given by $D = \mathrm{diag}(\lambda, \mu, 2\lambda, 0)$.

### 4.4.1 Completely Bilinearizable System

Now let us choose $\mathbf{f}_1$ and $\mathbf{f}_2$ such that the system becomes completely bilinearizable in four dimensions according to Theorem 4.2. Let

$$\mathbf{f}_1(\mathbf{x}) = \begin{pmatrix} 1 \\ x_1^2 \end{pmatrix} \text{ and } \mathbf{f}_2(\mathbf{x}) = \begin{pmatrix} 0 \\ 1 \end{pmatrix}.$$

Then

$$L_{\mathbf{f}_1} T(\mathbf{x}) = \begin{pmatrix} 1 \\ -2cx_1 + x_1^2 \\ 2x_1 \\ 0 \end{pmatrix} = B_1 \mathbf{z},$$

where

$$B_1 = \begin{bmatrix} 0 & 0 & 0 & 1 \\ -2c & 0 & 1 & 0 \\ 2 & 0 & 0 & 0 \\ 0 & 0 & 0 & 0 \end{bmatrix}.$$

Similarly $L_{\mathbf{f}_2} T(\mathbf{x}) = B_2 \mathbf{z}$ with

$$B_2 = \begin{bmatrix} 0 & 0 & 0 & 0 \\ 0 & 0 & 0 & 1 \\ 0 & 0 & 0 & 0 \\ 0 & 0 & 0 & 0 \end{bmatrix}.$$

For this simulation, we set $\lambda = 0.3$, $\mu = 0.2$, and $c = -0.5$. We applied $u_1 = \cos(2\pi t)$, a sinusoidal excitation, and $u_2 = -x_2 = -(z_2 + cz_1^2)$, a state feedback. Figure 4.1 shows that the original system response is identical to the response from the bilinearized system after transforming back to the original coordinates.

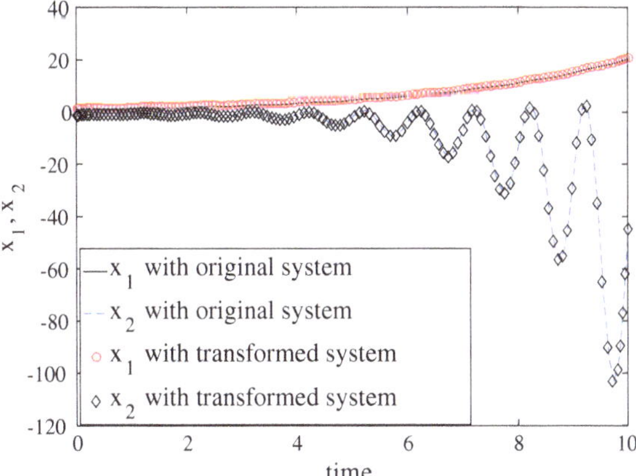


**Fig. 4.1** System response in original and transformed coordinates. © [2017] IEEE. Reprinted, with permission, from [6]

### 4.4.2 Approximately Bilinearized System

Now choose $\mathbf{f}_1(\mathbf{x}) = \begin{pmatrix} 1 \\ \cos x_1 \end{pmatrix}$ and keep everything else the same as Sect. 4.4.1. But now $L_{\mathbf{f}_1} T(\mathbf{x})$ does not lie in the span of $\phi_i, i = 1, \ldots, 4$. So we can only approximately bilinearize the system (4.23) by taking the projection of $L_{\mathbf{g}_1} T(\mathbf{x})$ into the span of these four eigenfunctions. Here

$$L_{\mathbf{f}_1} T(\mathbf{x}) = \begin{pmatrix} 1 \\ -2cx_1 + \cos x_1 \\ 2x_1 \\ 0 \end{pmatrix}.$$

Using a cosine series expansion $\cos x_1 \approx 1 - \frac{x_1^2}{2} = \phi_4(\mathbf{x}) - \frac{1}{2}\phi_3(\mathbf{x})$. With this approximation,

$$B_1 = \begin{bmatrix} 0 & 0 & 0 & 1 \\ -2c & 0 & -\frac{1}{2} & 1 \\ 2 & 0 & 0 & 0 \\ 0 & 0 & 0 & 0 \end{bmatrix}.$$

The transformed bilinearized system and the original system do not give exactly the same response, but they closely follow each other. The resultant responses are shown in Fig. 4.2. The accuracy can be increased by including higher order eigenfunctions, thereby increasing the number of terms in the cosine series.

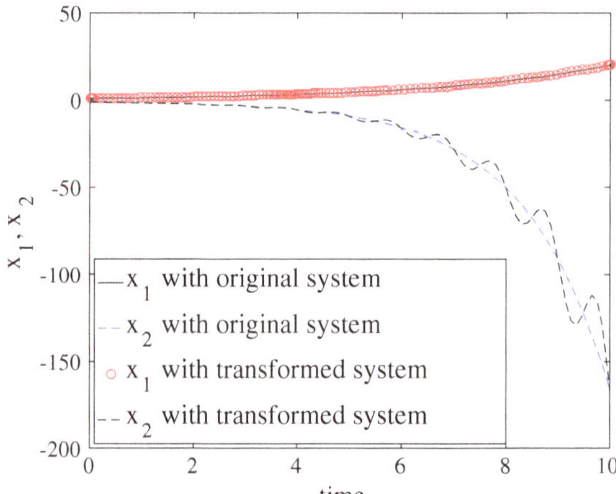


**Fig. 4.2** System response in original and transformed coordinates for approximate bilinearization. © [2017] IEEE. Reprinted, with permission, from [6]

### 4.4.3 Controllability of the System

It can be shown that the bilinearized system $\dot{\mathbf{z}}(t) = D\mathbf{z}(t) + B_1\mathbf{z}(t)u_1(t) + B_2\mathbf{z}(t)$ $u_2(t)$ satisfies the hypothesis of Theorem 4.4. So we resort to finding $Z(t) \in \mathbb{R}^{n\times n}$ where $Z(t)$ satisfies the matrix differential equation (4.15) with $m = 2$ and $\mathbf{z}(t) = Z(t)\mathbf{z}(0)$. According to Theorem 4.4, $Z(t)$ must take the form

$$Z(t) = \exp(tD)\exp(L),$$

where $L \in \mathscr{L} = \text{span}\{\text{ad}_D^k B_i : i = 1, \dots, m,\ k = 0, 1, \dots, n^2 - 1\}$. By explicitly calculating $\exp(tD)\exp(L_1)$, where $L_1 = c_1B_1 + c_2B_2 \in \text{span}\{\text{ad}_D^k B_i : i = 1, \dots, m,\ k = 0, 1, \dots, n^2 - 1\}$, we see that the resultant matrix is

$$Z(t) = \begin{bmatrix} e^{\lambda t} & 0 & 0 & c_1e^{\lambda t} \\ -c_1e^{\mu t}\left(2c - \dfrac{c_1}{2}\right)e^{\mu t} & c_1e^{\mu t} & e^{\mu t} & \left(c_2 - cc_1^2 + \dfrac{c_1^3}{3}\right) \\ 2c_1e^{2\lambda t} & 0 & e^{2\lambda t} & c_1^2e^{2\lambda t} \\ 0 & 0 & 0 & 1 \end{bmatrix}. \tag{4.24}$$

From any $\mathbf{z}_0$, we can achieve $\mathbf{z}(t) = Z(t)\mathbf{z}(0)$ and, therefore, any $z_1(t)$ and $z_2(t)$ can be achieved by varying the scalars $c_1$ and $c_2$ as $z_4 \equiv 1$. So we have global controllability for the original system (4.23). However, the transformed system is not globally controllable because we have no control authority over $z_4(t) \equiv 1$. To stabilize the system, we choose $u_1(t) = -(\lambda + 0.5)\,z_1(t) = -(\lambda + 0.5)\,x_1(t)$ and $u_2(t) = -(2cz_1(t) - z_3(t))\,u_1(t) - (\mu + 0.5)\,z_2(t) = (2cx_1(t) - x_1^2(t))\,u_1(t) - (\mu + 0.5)(x_2(t) - cx_1^2(t))$. This control effectively reduces the transformed system into $\dot{\mathbf{z}} = A\mathbf{z}$ where $A = \text{diag}(-0.5, -0.5, -0.5, 1)$. This input in turn feedback linearizes the original system (4.23) hinting at a strong connection between KBF and feedback linearizability of the system. The system response under this feedback is shown in Fig. 4.3.

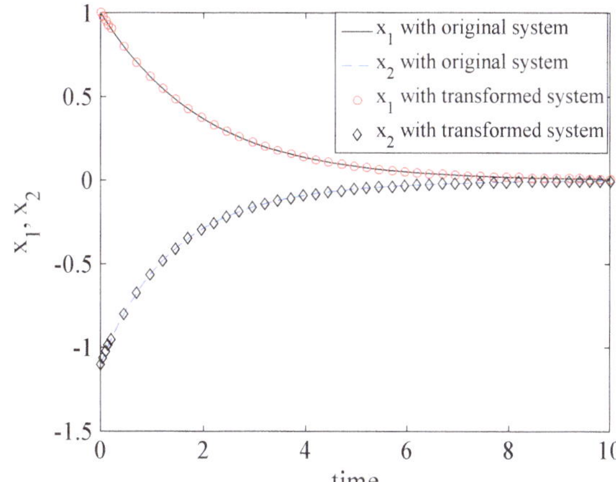


**Fig. 4.3** System response in original and transformed coordinates for approximate bilinearization. © [2017] IEEE. Reprinted, with permission, from [6]

## 4.5 Conclusion

This chapter describes an effective method to globally bilinearize control-affine nonlinear systems using the Koopman canonical transform. The motivation for the bilinearization came from the infinitesimal generator of the Koopman operator where a suitable set of basis functions are sought to project the infinite-dimensional operators. The KCT provides a great framework to accomplish this projection and paves the way to derive the sufficient conditions for the bilinearizability. The bilinear system is subjected to reachability analysis using classical Lie algebraic methods and necessary and sufficient conditions are established. The numerical examples illustrate the power of bilinearization and the corresponding controllability analysis.

## References

1. Brockett, R.W.: Finite Dimensional Linear Systems, pp. 67–122. Wiley, New York (1970)
2. Brockett, R.W.: System theory on group manifolds and coset spaces. SIAM J. Control **10**(2), 265–284 (1972)
3. Brockett, R.W.: Lie Algebras and Lie Groups in Control Theory, pp. 43–82. Springer, Dordrecht (1973)
4. Budisić, M., Mohr, R., Mezić, I.: Applied Koopmanism. Chaos: Interdiscip. J. Nonlinear Sci. **22**(4) (2012)
5. Gantmacher, R.F.: Theory of Matrices. Chelsea, New York (1959)
6. Goswami, D., Paley, D.A.: Global bilinearization and controllability of control-affine nonlinear systems: a Koopman spectral approach. In: 2017 IEEE 56th Annual Conference on Decision and Control, pp. 6107–6112 (2017)
7. Jurdjevic, V., Sussmann, H.J.: Control systems on Lie groups. J. Differ. Equ. **12**(2), 313–329 (1972)
8. Kowalski, K., Steeb, W.H.: Nonlinear Dynamical Systems and Carleman Linearization, pp. 73–102. World Scientific, Singapore (2011)
9. Lan, Y., Mezić, I.: Linearization in the large of nonlinear systems and Koopman operator spectrum. Phys. D: Nonlinear Phenom. **242**(1), 42–53 (2013)
10. Martin, J.: Some results on matrices which commute with their derivatives. SIAM J. Appl. Math. **15**(5), 1171–1183 (1967)
11. Mauroy, A., Mezić, I.: Global stability analysis using the eigenfunctions of the Koopman operator. IEEE Trans. Autom. Control **61**(11), 3356–3369 (2016)
12. Mezić, I.: Spectral properties of dynamical systems, model reduction and decompositions. Nonlinear Dyn. **41**(1), 309–325 (2005)
13. Mohr, R., Mezić, I.: Construction of eigenfunctions for scalar-type operators via Laplace averages with connections to the Koopman operator (2014). arxiv.org/abs/1403.6559
14. Rowley, C.W., Mezić, I., Bagheri, S., Schlatter, P., Henningson, D.S.: Spectral analysis of nonlinear flows. J. Fluid Mech. **641**, 115–127 (2009)
15. Surana, A.: Koopman operator based observer synthesis for control-affine nonlinear systems. In: 55th IEEE Conference on Decision and Control, pp. 6492–6499 (2016)
16. Surana, A., Banaszuk, A.: Linear observer synthesis for nonlinear systems using Koopman operator framework. IFAC-PapersOnLine **49**(18), 716–723 (2016)
17. Wei, J., Norman, E.: On global representations of the solutions of linear differential equations as a product of exponentials. Proc. Am. Math. Soc. **15**(2), 327–334 (1964)
18. Williams, M.O., Kevrekidis, I.G., Rowley, C.W.: A data-driven approximation of the Koopman operator: extending dynamic mode decomposition. J. Nonlinear Sci. **25**(6), 1307–1346 (2015)

# Chapter 5
# Koopman Spectrum and Stability of Cascaded Dynamical Systems

**Ryan Mohr and Igor Mezić**

**Abstract** This chapter investigates the behavior of cascaded dynamical systems through the lens of the Koopman operator and, in particular, its so-called principal eigenfunctions. It is shown that there exist perturbation functions for the initial conditions of each component system that make the orbits for the cascaded system and the decoupled component systems have zero asymptotic relative error. This in turn implies that the evolutions are asymptotically equivalent. By analyzing the exact form of the initial condition perturbation functions, the maximum error between the trajectories of the decoupled systems and the cascaded system can be bounded. More colloquially, these results say that cascaded compositions of stable systems are stable. It is also shown that the process of wiring the component systems together in a cascade structure preserves the principal eigenvalues and these principal eigenvalues are preserved between topologically conjugate systems. Thus, the analysis of cascaded systems is reduced to the determination of the principal eigenvalues and eigenfunctions of each component's Koopman operator and the form of the perturbation functions.

## 5.1 Introduction

Engineered systems are becoming larger and more complex. Additionally, new tools allow us to measure the detailed behavior of increasingly complicated systems, such as biological–chemical networks. This growing complexity presents challenges to traditional tools of modeling and analysis, significantly slowing insight into the behavior of these systems. Predicting the behavior of these systems is hard or

---

R. Mohr (✉)
AIMdyn Inc., Santa Barbara, CA, USA
e-mail: mohrr@amidyn.com

I. Mezić
Department of Mechanical Engineering, University of California Santa Barbara, Santa Barbara, CA, USA
e-mail: mezic@engineering.ucsb.edu



A. Mauroy et al. (eds.), *The Koopman Operator in Systems and Control*,
Lecture Notes in Control and Information Sciences 484,
https://doi.org/10.1007/978-3-030-35713-9_5

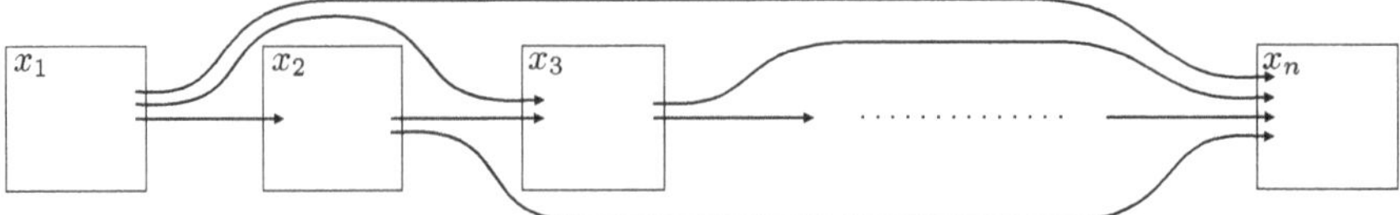


**Fig. 5.1** $n$-level cascade system

insoluble with these tools. For example, any chaotic behavior enforces limits on the possible prediction of the behavior of individual trajectories. This often requires the analysis of statistical properties of ensembles of trajectories [6]. Unfortunately, the numerical computation of trajectories from initial conditions can be intractable due to the memory requirements needed to represent the system. New tools for the modeling, analysis, and prediction of complex systems must be developed.

Many of these systems exhibit a modular-type structure where simple functional units are composed into more complex systems [3, 5, 8, 9, 13, 16, 18]. These systems can often be organized into a forward production unit with slower feedback loops modulating the forward unit's behavior. Modules farther downstream (modulo the slow feedback loops) do not influence the behavior of modules upstream to themselves. Additionally, many system models can be decomposed into these modular components using various techniques [1, 8, 9, 16].

A cascaded dynamical system is constructed via wiring component dynamical systems (submodules) together in such a way that "information" only flows downstream—subsystem $i$ is unaffected by the dynamics of any subsystem $j$ if $j > i$ (Fig. 5.1 gives a graphical depiction). In linear systems, this would be equivalent to a lower block diagonal structure of the matrix. These types of systems can be quite complex, existing in high-dimensional state spaces, for which traditional tools can fail or become intractable. Often though, the behavior of the individual submodules is known in isolation. This can either be by design; they are simple enough to be amenable to analysis by traditional analytical tools; or through physical experimentation. One would then like to characterize the behavior of the full cascaded system from knowledge of the individual components.

This chapter investigates these systems through the lens of the Koopman operator [2, 10–12] and, in particular, its so-called principal eigenfunctions [14]. We study the case where each component system is asymptotically stable, powers of the linear part of each component system approach the zero operator asymptotically faster for upstream systems, and the linearized component systems have disjoint spectrums. We show that there exist perturbation functions for the initial conditions of each component system that make the orbits for the cascaded system and the decoupled component systems have zero asymptotic relative error. This in turn implies that the evolutions are asymptotically equivalent. By analyzing the exact form of the initial condition perturbation functions, the maximum error between the trajectories of the decoupled systems and the cascaded systems can be bounded. More colloquially, these results say that cascaded compositions of stable systems are stable.

In [14, 15] Koopman principal eigenfunctions were defined as the eigenvalues of the linearized dynamics. Here, we show that the principal eigenvalues of the decoupled, component subsystems are also Koopman eigenvalues of the cascaded system; i.e., principal eigenvalues of the component systems, when the systems are considered in isolation, are embedded in the spectrum of the Koopman operator of the cascaded system. Furthermore, the associated Koopman eigenfunctions of the cascaded system are given by composing the principal eigenfunctions of the component subsystem with a perturbation function (that we give explicitly) that maps initial conditions to initial conditions. These results follow from a stronger one that we prove—the dynamics of the full system and the dynamics of the decoupled systems converge exponentially faster to each other than the decoupled component systems converges to their fixed point—in this case, we say that the cascaded system has zero asymptotic relative error. These results allow one to analyze in some detail the behavior of a large cascaded system, without actually having to simulate it.

The work presented here relates the stability of the system to the spectrum of the associated Koopman operator. In connection to this, we must mention the results presented in [7]. That paper deals with the interplay between the stability of a system and the spectral properties of the Koopman operator (see also Chap. 2). It refers to the spectral properties of the Koopman operator for the full system. This differs from our work in that we show we can analyze *pieces* of the system in isolation and still guarantee stability when the systems are wired together.

The rest of this chapter is structured as follows. The next section is devoted to precisely defining the cascaded systems and their associated nominal system (Sect. 5.2); solution operators for these systems (Sect. 5.2.1); and the concepts of asymptotic proportionality, asymptotic equivalence, and zero asymptotic relative error (Sect. 5.2.2). In the following section (Sect. 5.3), we define the Koopman operators and principal eigenfunctions of the component subsystems and extend their definitions to the full cascaded system using a tensor product construction. Here is where we show that the component systems' principal eigenvalues are embedded in the spectrum of the Koopman operator for the cascaded system. Section 5.4 collects the main results of the paper and their corollaries along with a number of remarks. Analytic and numerical examples demonstrating the main theorems for linear cascaded systems are in Sect. 5.4.2. Technical proofs are collected in the Appendix which follows the Conclusions section.

## 5.2 Cascaded Systems

We now define (discrete-time) cascaded systems. In what follows, $t \in \mathbb{Z}$ or $\mathbb{N}_0 = \{0, 1, 2, \ldots\}$.

**Definition 5.1** *(Cascaded system)* Given $\mathbf{x}_i(t) \in \mathbb{C}^{d_i}$ for $i = 1, \ldots, n$, an $n$-level cascaded system is defined, for $i = 2, \ldots, n$, as

$$\begin{aligned}\mathbf{x}_1(t+1) &= \mathbf{L}_1\mathbf{x}_1(t) + \mathbf{N}_1(\mathbf{x}_1(t)) = \mathbf{G}_1(\mathbf{x}_1(t)),\\ \mathbf{x}_i(t+1) &= \mathbf{L}_i\mathbf{x}_i(t) + \sum_{j=1}^{i-1}\mathbf{C}_{i,j}\mathbf{x}_j(t) + \mathbf{N}_i(\mathbf{x}_1(t),\ldots,\mathbf{x}_i(t)) = \mathbf{G}_i(\mathbf{x}_1(t),\ldots,\mathbf{x}_i(t)),\end{aligned} \tag{5.1}$$

where $\mathbf{L}_i : \mathbb{C}^{d_i} \to \mathbb{C}^{d_i}$ and $\mathbf{C}_{i,j} : \mathbb{C}^{d_j} \to \mathbb{C}^{d_i}$ are linear operators and $\mathbf{N}_i : \mathbb{C}^{d_1} \times \cdots \times \mathbb{C}^{d_i} \to \mathbb{C}^{d_i}$ are nonlinear operators having no linear terms so that $\mathbf{N}_i(\mathbf{0}) = \mathbf{0}$ (see Fig. 5.1).

We say that $\mathbf{G}_i$ is called the $i$th component system of the nonlinear cascade. For any fixed $j \in \{1,\ldots,n\}$, if $i < j$, then $\mathbf{G}_i$ is called an *upstream system* to $\mathbf{G}_j$ and if $k > j$, we call $\mathbf{G}_k$ a *downstream system* to $\mathbf{G}_j$. Each component system $\mathbf{G}_i$ depends only upon itself and its upstream systems.

In proving our results, it will be beneficial to analyze the systems linearized around $\mathbf{0}$,

$$\begin{aligned}\mathbf{x}_1(t+1) &= \mathbf{L}_1\mathbf{x}_1(t),\\ \mathbf{x}_i(t+1) &= \mathbf{L}_i\mathbf{x}_i(t) + \sum_{j=1}^{i-1}\mathbf{C}_{i,j}\mathbf{x}_j(t), \quad (i = 2,\ldots,n),\end{aligned} \tag{5.2}$$

and the associated nominal, linear system which is obtained by setting the linear coupling terms $\mathbf{C}_{i,j}$ to $\mathbf{0}$:

$$\mathbf{x}_i(t+1) = \mathbf{L}_i\mathbf{x}_i(t), \qquad (i = 1,\ldots,n). \tag{5.3}$$

### 5.2.1 Solution Operators

There will be three systems for which we need to define solution operators. These systems are the fully connected nonlinear cascaded system (denoted by $\mathsf{NonLin}$) defined via (5.1); the fully connected linear cascade system ($\mathsf{Lin}$) obtained by setting all the nonlinear operators $\mathbf{N}_i = \mathbf{0}$ (see Eq. (5.2)); and the nominal system ($\mathsf{Nom}$) obtained from $\mathsf{Lin}$ by setting all the coupling terms $\mathbf{C}_{i,j} = \mathbf{0}$ (see Eq. (5.3)). The analysis of these operators and their relationship with each other will ultimately give our main results.

For $i = 1,\ldots,n$, let $\Pi_i : \mathbb{C}^{d_1} \times \cdots \times \mathbb{C}^{d_n} \to \mathbb{C}^{d_i}$ denote the canonical projection onto the $i$th system as

$$\Pi_i(\mathbf{x}_1,\ldots,\mathbf{x}_n) = \mathbf{x}_i. \tag{5.4}$$

**Orbits for the Nonlinear Cascade.** The orbit[1] for the nonlinear cascaded system is denoted by the family of nonlinear operators $\mathsf{NonLin}^{\circ t} : \mathbb{C}^{d_1} \times \cdots \times \mathbb{C}^{d_n} \to$

[1] We use $\circ t$ instead of $t$ to remind ourselves that these are compositions of nonlinear operators.

$\mathbb{C}^{d_1} \times \cdots \times \mathbb{C}^{d_n}$. We define $\mathsf{NonLin}^{\circ 0} = (\mathbf{I}_{d_1}, \ldots, \mathbf{I}_{d_n})$, where $\mathbf{I}_{d_i}$ is the identity operator on $\mathbb{C}^{d_i}$. Then, for all $i \geq 1$ and $t \in \mathbb{N}$, these operators are defined recursively as[2]

$$\begin{aligned}&\Pi_i\left(\mathsf{NonLin}^{\circ t}(\mathbf{x}_1, \ldots, \mathbf{x}_n)\right)\\&\quad = \left(\mathbf{L}_i \Pi_i + \sum_{j=1}^{i-1} \mathbf{C}_{i,j} \Pi_j + \mathbf{N}_i \circ (\Pi_1, \ldots, \Pi_i)\right)(\mathsf{NonLin}^{\circ(t-1)}(\mathbf{x}_1, \ldots, \mathbf{x}_n)).\end{aligned} \tag{5.5}$$

For $t = 1$, we denote $\mathsf{NonLin}^{\circ 1} \equiv \mathsf{NonLin}$.

**Orbits for the Linear Cascade.** The orbit for the linear cascade (5.2) is denoted by the family of linear operators $\mathsf{Lin}^{\circ t} : \mathbb{C}^{d_1} \times \cdots \times \mathbb{C}^{d_n} \to \mathbb{C}^{d_1} \times \cdots \times \mathbb{C}^{d_n}$, and these are defined recursively as

$$\Pi_i\left(\mathsf{Lin}^{\circ t}(\mathbf{x}_1, \ldots, \mathbf{x}_n)\right) = \left(\mathbf{L}_i \Pi_i + \sum_{j=1}^{i-1} \mathbf{C}_{i,j} \Pi_j\right)\left(\mathsf{Lin}^{\circ(t-1)}(\mathbf{x}_1, \ldots, \mathbf{x}_n)\right). \tag{5.6}$$

**Orbit for the Nominal (Linear) System.** The orbit for the nominal system is

$$\begin{aligned}\mathsf{Nom}^{\circ t}(\mathbf{x}_1, \ldots, \mathbf{x}_n) &= (\mathbf{L}_1^t \Pi_1, \mathbf{L}_2^t \Pi_2, \ldots, \mathbf{L}_n^t \Pi_n)(\mathbf{x}_1, \ldots, \mathbf{x}_n)\\&= (\mathbf{L}_1^t(\mathbf{x}_1), \mathbf{L}_2^t(\mathbf{x}_2), \ldots, \mathbf{L}_n^t(\mathbf{x}_n)).\end{aligned} \tag{5.7}$$

### 5.2.2 *Asymptotic Equivalence*

The main results of this paper rely on the concept of asymptotic proportionality and asymptotic equivalence between the component systems of the linear cascade and the corresponding nominal component systems. First we define the norm for a cascaded system

**Definition 5.2** Let $\|\cdot\|$ be a norm for $\mathbb{C}^{d_i}$. Define a norm on $\mathbb{C}^{d_1} \times \cdots \times \mathbb{C}^{d_n}$ by

$$\|(\mathbf{x}_1, \ldots, \mathbf{x}_n)\|_\times = \sum_{i=1}^{n} \|\mathbf{x}_i\|. \tag{5.8}$$

For linear maps and operators from one vector space to another, we will also denote their induced norm as $\|\cdot\|$.

**Definition 5.3** *(Asymptotic equivalence)* We say that $\mathsf{Lin}$ is asymptotically equivalent to $\mathsf{Nom}$ if there exists a perturbation function $\mathsf{pert} : \mathbb{C}^{d_1} \times \cdots \times \mathbb{C}^{d_n} \to \mathbb{C}^{d_1} \times \cdots \times \mathbb{C}^{d_n}$ such that for all $i = 1, \ldots, n$ and all $\mathbf{X} = (\mathbf{x}_1, \ldots, \mathbf{x}_n) \in \mathbb{C}^{d_1} \times \cdots \times \mathbb{C}^{d_n}$,

[2]In (5.5) and later, we are using the notation that if we have a collection of maps $f_i : X \to X_i$, $(i = 1, \ldots, n)$, the vector-valued map $\mathbf{f} : X \to X_1 \times \cdots \times X_n$ defined by $\mathbf{f}(\mathbf{x}) := (f_1(\mathbf{x}), \ldots, f_n(\mathbf{x}))$ can be written as $\mathbf{f}(\mathbf{x}) \equiv (f_1, \ldots, f_n)(\mathbf{x})$.

$$\lim_{t\to\infty} \|\Pi_i(\mathsf{Lin}^{\circ t}(\mathbf{X})) - \Pi_i(\mathsf{Nom}^{\circ t}(\mathsf{pert}(\mathbf{X})))\| = 0. \tag{5.9}$$

We say that system $i$ has 0 asymptotic relative error if

$$\lim_{t\to\infty} \frac{\|\Pi_i(\mathsf{Lin}^{\circ t}(\mathbf{X})) - \Pi_i(\mathsf{Nom}^{\circ t}(\mathsf{pert}(\mathbf{X})))\|}{\|\mathbf{L}_i^t\|} = 0. \tag{5.10}$$

A related concept is the asymptotic proportionality of two solution operators. It is easier to satisfy than 0 asymptotic relative error.

**Definition 5.4** *(Asymptotic proportionality)* We say that $\mathsf{Lin}$ is asymptotically proportional to $\mathsf{Nom}$ if there exist a perturbation function $\mathsf{pert} : \mathbb{C}^{d_1} \times \cdots \times \mathbb{C}^{d_n} \to \mathbb{C}^{d_1} \times \cdots \times \mathbb{C}^{d_n}$ and functions $\zeta_i : \mathbb{C}^{d_1} \times \cdots \times \mathbb{C}^{d_{i-1}} \to \mathbb{R}^+$ such that for all $\mathbf{X} = (\mathbf{x}_1, \ldots, \mathbf{x}_n) \in \mathbb{C}^{d_1} \times \cdots \times \mathbb{C}^{d_n}$ and all $i \in \{2, \ldots, n\}$,

$$\lim_{t\to\infty} \frac{\|\Pi_i(\mathsf{Lin}^{\circ t}(\mathbf{X})) - \Pi_i(\mathsf{Nom}^{\circ t}(\mathsf{pert}(\mathbf{X})))\|}{\|\mathbf{L}_i^t\|} \le \zeta_i(\mathbf{x}_1, \ldots, \mathbf{x}_{i-1}). \tag{5.11}$$

Note that the bounding function $\zeta_i$ for system $i$ is only a function of the upstream state vectors $\mathbf{x}_1, \ldots, \mathbf{x}_{i-1}$.

For the linear cascade, a perturbation function $\mathsf{pert} : \mathbb{C}^{d_1} \times \cdots \times \mathbb{C}^{d_n} \to \mathbb{C}^{d_1} \times \cdots \times \mathbb{C}^{d_n}$ will itself have a cascade structure

$$\mathsf{pert}(\mathbf{x}_1, \ldots, \mathbf{x}_n) = \big(\mathsf{pert}_1(\mathbf{x}_1), \mathsf{pert}_2(\mathbf{x}_1, \mathbf{x}_2), \ldots, \mathsf{pert}_n(\mathbf{x}_1, \ldots, \mathbf{x}_n)\big), \tag{5.12}$$

with $\mathsf{pert}_i : \mathbb{C}^{d_1} \times \cdots \times \mathbb{C}^{d_i} \to \mathbb{C}^{d_i}$.

*Remark 5.1* By (5.6) and (5.7), Eqs. (5.9), (5.10), and (5.11) can be rewritten as

$$\lim_{t\to\infty} \|\Pi_i(\mathsf{Lin}^{\circ t}(\mathbf{x}_1, \ldots, \mathbf{x}_n)) - \mathbf{L}_i^t(\mathsf{pert}_i(\mathbf{x}_1, \ldots, \mathbf{x}_i))\| = 0, \tag{5.13}$$

$$\lim_{t\to\infty} \frac{\|\Pi_i(\mathsf{Lin}^{\circ t}(\mathbf{x}_1, \ldots, \mathbf{x}_n)) - \mathbf{L}_i^t(\mathsf{pert}_i(\mathbf{x}_1, \ldots, \mathbf{x}_i))\|}{\|\mathbf{L}_i^t\|} = 0, \tag{5.14}$$

and

$$\lim_{t\to\infty} \frac{\|\Pi_i(\mathsf{Lin}^{\circ t}(\mathbf{x}_1, \ldots, \mathbf{x}_n)) - \mathbf{L}_i^t(\mathsf{pert}_i(\mathbf{x}_1, \ldots, \mathbf{x}_i))\|}{\|\mathbf{L}_i^t\|} \le \zeta_i(\mathbf{x}_1, \ldots, \mathbf{x}_{i-1}). \tag{5.15}$$

Clearly, for $\|\mathbf{L}_i\| < 1$, zero asymptotic relative error implies the other two.

## 5.3 The Koopman Operator and Principal Eigenfunctions for Cascaded Systems

Here, we precisely define the Koopman operator for cascaded systems and show that principal eigenfunctions of component system can be combined to define the principal eigenfunctions of the cascaded system. This extends the original definition of principal eigenfunctions from [14].

Let $\mathscr{A}_i$ be a subalgebra of $C(\mathbb{C}^{d_i}, \mathbb{C})$, the set of continuous complex-valued functions on $\mathbb{C}^{d_i}$, where algebra addition is given by normal function addition, $(f + g)(\mathbf{x}_i) = f(\mathbf{x}_i) + g(\mathbf{x}_i)$, and algebra multiplication is given by pointwise multiplication of functions, $(f \cdot g)(\mathbf{x}_i) = f(\mathbf{x}_i)g(\mathbf{x}_i)$.

Denote the family of Koopman operators associated with the $i$th nominal component system (5.3) as

$$\begin{aligned} U^t_{\mathsf{Nom}_i} &: \mathscr{A}_i \to \mathscr{A}_i \\ (U^t_{\mathsf{Nom}_i} f)(\mathbf{x}_i) &= f(\mathsf{Nom}_i^{\circ t}(\mathbf{x}_i)) = f(\mathbf{L}_i^t \mathbf{x}_i). \end{aligned} \tag{5.16}$$

### 5.3.1 *Principal Eigenfunctions of Component Systems*

In [14, 15], the *principal eigenfunctions* of the Koopman operator were defined with respect to a linearized system. Assume that $\mathbf{L}_i$ is diagonalizable

$$\mathbf{L}_i = \mathbf{V}_i \Lambda_i \mathbf{V}_i^{-1}, \tag{5.17}$$

with the columns of $\mathbf{V}_i$ being the eigenvectors of $\mathbf{L}_i$ and $\Lambda_i = \text{diag}(\lambda_{i,1}, \ldots, \lambda_{i,d_i})$ being a diagonal matrix containing the eigenvalues of $\mathbf{L}_i$. Define the $s$th principal eigenfunction of the $i$th system as

$$\phi_{i,s}(\mathbf{x}_i) = (\mathbf{e}^*_{d_i,s} \mathbf{V}_i^{-1})\mathbf{x}_i, \qquad (s = 1, \ldots, d_i) \tag{5.18}$$

where $\mathbf{e}_{d_i,s}$ is the $s$th canonical basis vector of $\mathbb{C}^{d_i}$ and $\mathbf{e}^*_{d_i,s}$ is its conjugate transpose.[3] The function $\phi_{i,s}$ is the $s$th coordinate functional corresponding to the eigenbasis of the $i$th system. These are indeed eigenfunctions of $U_{\mathsf{Nom}_i}$ at eigenvalue $\lambda_{i,s}$ as shown by the following calculation: for any $s \in \{1, \ldots, d_i\}$

[3]Note that $\mathbf{w}_{i,s} := (\mathbf{e}^*_{i,s}\mathbf{V}_i^{-1})^* = (\mathbf{V}_i^*)^{-1}\mathbf{e}_{i,s}$ is the $s$th dual basis vector in system $i$; that is $\langle \mathbf{v}_{i,t}, \mathbf{w}_{i,s} \rangle_{\mathbb{C}^{d_i}} = \mathbf{w}^*_{i,s}\mathbf{v}_{i,t} = \delta_{s,t}$, where $\mathbf{v}_{i,t}$ is the $t$th eigenvector of $\mathbf{L}_i$.

$$(U_{\mathsf{Nom}_i}\phi_{i,s})(\mathbf{x}_i) = \phi_{i,s}(\mathbf{L}_i\mathbf{x}_i) = (\mathbf{e}^*_{d_i,s}\mathbf{V}_i^{-1})(\mathbf{L}_i\mathbf{x}_i) = \mathbf{e}^*_{d_i,s}\Lambda_i\mathbf{V}_i^{-1}\mathbf{x}_i = \lambda_{i,s}(\mathbf{e}^*_{d_i,s}\mathbf{V}_i^{-1})\mathbf{x}_i$$
$$= \lambda_{i,s}\phi_{i,s}(\mathbf{x}_i). \tag{5.19}$$

For all $i \in \{1, \ldots, n\}$, we define $\phi_{i,0}(\mathbf{x}_i) = 1$ so that $\phi_{i,0}$ is an eigenfunction at 1:

$$(U_{\mathsf{Nom}_i}\phi_{i,0})(\mathbf{x}_i) = 1 = \phi_{i,0}(\mathbf{x}_i). \tag{5.20}$$

## 5.3.2 Principal Eigenfunctions for Cascaded Systems

We use the algebra of observables $\mathscr{A}_i$ on $\mathbb{C}^{d_i}$ to build a space of observables $\mathscr{A}$ for the cascade system. Let $\mathscr{A} = \mathscr{A}_1 \otimes \cdots \otimes \mathscr{A}_n$ and denote $U_{\mathsf{Nom}} : \mathscr{A} \to \mathscr{A}$ and $U_{\mathsf{Lin}} : \mathscr{A} \to \mathscr{A}$ as the Koopman operators associated with the solution operators $\mathsf{Nom}$ and $\mathsf{Lin}$, respectively. Recall that a tensor product $f_1 \otimes \cdots \otimes f_n \in \mathscr{A}_1 \otimes \cdots \otimes \mathscr{A}_n$ acts on multilinear functionals on $\mathscr{A}_1 \times \cdots \times \mathscr{A}_n$ (see [17]). Let $\delta_{(\mathbf{x}_1,\ldots,\mathbf{x}_n)} : \mathscr{A}_1 \times \cdots \times \mathscr{A}_n \to \mathbb{C}$ be the evaluation functional defined as

$$\delta_{(\mathbf{x}_1,\ldots,\mathbf{x}_n)}(f_1, \ldots, f_n) = f_1(\mathbf{x}_1) \cdots f_n(\mathbf{x}_n), \tag{5.21}$$

where the dots on the right side indicate multiplication in $\mathbb{C}$. The tensor product is defined as

$$(f_1 \otimes \cdots \otimes f_n)(\delta_{(\mathbf{x}_1,\ldots,\mathbf{x}_n)}) = \delta_{(\mathbf{x}_1,\ldots,\mathbf{x}_n)}(f_1, \ldots, f_n). \tag{5.22}$$

Due to this, there is no confusion in writing

$$(f_1 \otimes \cdots \otimes f_n)(\mathbf{x}_1, \ldots, \mathbf{x}_n) \equiv f_1(\mathbf{x}_1) \cdots f_n(\mathbf{x}_n) \tag{5.23}$$

and we can consider $f_1 \otimes \cdots \otimes f_n$ as a function on the cascaded system's state space $\mathbb{C}^{d_1} \times \cdots \times \mathbb{C}^{d_n}$.

Principal eigenfunctions for $U_{\mathsf{Nom}} : \mathscr{A} \to \mathscr{A}$ can be defined from the component systems' principal eigenfunctions (5.18) by trivially extending their domain from the component system's state space $\mathbb{C}^{d_i}$ to the cascaded system's state space $\mathbb{C}^{d_1} \times \cdots \times \mathbb{C}^{d_n}$. To this end, for each $i \in \{1, \ldots, n\}$ and $s_i \in \{1, \ldots, d_i\}$, define a principal eigenfunction for $U_{\mathsf{Nom}}$ as

$$\phi_{(0,\ldots,0,s_i,0,\ldots,0)}(\mathbf{x}_1, \ldots, \mathbf{x}_n) = (\phi_{i,s_i} \circ \Pi_i)(\mathbf{x}_1, \ldots, \mathbf{x}_n) \equiv \phi_{i,s_i}(\mathbf{x}_i). \tag{5.24}$$

We can do this since the subsystems are decoupled in the nominal system. Often we will write $\phi_{s_i\mathbf{e}_{n,i}} \equiv \phi_{(0,\ldots,0,s_i,0,\ldots,0)}$. Since by convention $\phi_{i,0} \equiv 1$ for all $i$, this principal eigenfunction for $U_{\mathsf{Nom}}$ can be written as a tensor product of principal eigenfunctions from the component systems,

$$\begin{aligned}\phi_{(0,\ldots,0,s_i,0,\ldots,0)}(\mathbf{x}_1,\ldots,\mathbf{x}_n) &= \phi_{i,s_i}(\mathbf{x}_i)\\ &\equiv (1\otimes\cdots\otimes 1\otimes\phi_{i,s_i}\otimes 1\otimes\cdots\otimes 1)(\mathbf{x}_1,\ldots,\mathbf{x}_n)\\ &= (\phi_{1,0}\otimes\cdots\otimes\phi_{i-1,0}\otimes\phi_{i,s_i}\otimes\phi_{i+1,0}\otimes\cdots\otimes\phi_{n,0})(\mathbf{x}_1,\ldots,\mathbf{x}_n).\end{aligned} \tag{5.25}$$

The multiplication operation in each algebra $\mathscr{A}_i$ is given by pointwise products of functions. We can easily define a multiplication operation, $\bullet : (\mathscr{A}_1\otimes\cdots\otimes\mathscr{A}_n)\times(\mathscr{A}_1\otimes\cdots\otimes\mathscr{A}_n)\to(\mathscr{A}_1\otimes\cdots\otimes\mathscr{A}_n)$, for the tensor product as

$$(f_1\otimes\cdots\otimes f_n)\bullet(g_1\otimes\cdots\otimes g_n) = (f_1\cdot g_1)\otimes\cdots\otimes(f_n\cdot g_n). \tag{5.26}$$

Products of principal eigenfunctions of the form (5.25) give eigenfunctions of $U_{\mathsf{Nom}}$. Denote $\phi_{(s_1,\ldots,s_n)}\in\mathscr{A}$ as

$$\phi_{(s_1,\ldots,s_n)}(\mathbf{x}_1,\ldots,\mathbf{x}_n) = (\phi_{1,s_1}\otimes\cdots\otimes\phi_{n,s_n})(\mathbf{x}_1,\ldots,\mathbf{x}_n), \tag{5.27}$$

where $(s_1,\ldots,s_n)\in\{0,\ldots d_1\}\times\cdots\times\{0,\ldots,d_n\}$. It is clear that this observable can be constructed as a $\bullet$-product of principal eigenfunctions of the form (5.25); namely

$$\phi_{(s_1,\ldots,s_n)} = \phi_{(s_1,0,\ldots,0)}\bullet\phi_{(0,s_2,0,\ldots,0)}\bullet\cdots\bullet\phi_{(0,\ldots,0,s_n)}. \tag{5.28}$$

The function defined in (5.27) is an eigenfunction at $\lambda_{1,s_1}\cdots\lambda_{n,s_n}$ as shown by the following computation:

$$\begin{aligned}(U_{\mathsf{Nom}}\phi_{(s_1,\ldots,s_n)})(\mathbf{x}_1,\ldots,\mathbf{x}_n) &= \phi_{(s_1,\ldots,s_n)}(\mathsf{Nom}(\mathbf{x}_1,\ldots,\mathbf{x}_n))\\ &= \phi_{(s_1,\ldots,s_n)}(\mathbf{L}_1\mathbf{x}_1,\ldots,\mathbf{L}_n\mathbf{x}_n)\\ &= (\phi_{1,s_1}\otimes\cdots\otimes\phi_{n,s_n})(\mathbf{L}_1\mathbf{x}_1,\ldots,\mathbf{L}_n\mathbf{x}_n)\\ &= \phi_{1,s_1}(\mathbf{L}_1\mathbf{x}_1)\cdots\phi_{n,s_n}(\mathbf{L}_n\mathbf{x}_n)\\ &= (U_{\mathsf{Nom}_1}\phi_{1,s_1})(\mathbf{x}_1)\cdots(U_{\mathsf{Nom}_n}\phi_{n,s_n})(\mathbf{x}_n)\\ &= (\lambda_{1,s_1}\phi_{1,s_1})(\mathbf{x}_1)\cdots(\lambda_{n,s_n}\phi_{n,s_n})(\mathbf{x}_n)\\ &= (\lambda_{1,s_1}\cdots\lambda_{n,s_n})(\phi_{1,s_1}(\mathbf{x}_1)\cdots\phi_{n,s_n}(\mathbf{x}_n))\\ &= (\lambda_{1,s_1}\cdots\lambda_{n,s_n})(\phi_{1,s_1}\otimes\cdots\otimes\phi_{n,s_n})(\mathbf{x}_1,\ldots,\mathbf{x}_n)\\ &= (\lambda_{1,s_1}\cdots\lambda_{n,s_n})\phi_{(s_1,\ldots,s_n)}(\mathbf{x}_1,\ldots,\mathbf{x}_n).\end{aligned}$$

The preceding derivation is an example of the more general result that eigenfunctions of the Koopman operator form a semigroup under pointwise multiplication of functions, as pointed out in [2].

*Remark 5.2* Each subalgebra $\mathscr{A}_i$ for the component system $i$ is embedded in $\mathscr{A}$. The embedding is given by the map

$$f\mapsto\underbrace{1\otimes\cdots\otimes 1}_{i-1\text{ times}}\otimes f\otimes\underbrace{1\otimes\cdots\otimes 1}_{n-i\text{ times}},$$

where the $f$ on the right-hand side is in the $i$th position. Furthermore, if each $\mathscr{A}_i$ is generated by the principal eigenfunctions $\phi_{i,1}, \ldots, \phi_{i,d_i}$, then $\mathscr{A}$ is generated by the set of principal eigenfunctions

$$\Phi = \bigcup_{i=1}^{n} \left\{ \phi_{(0,\ldots,0,s_i,0,\ldots,0)} : s_i \in \{1, \ldots, d_i\} \right\}. \tag{5.29}$$

## 5.4 Main Results

We consider the special case of the nonlinear and linear cascades (Eqs. (5.1) and (5.2), respectively) where system $i$ is only affected by system $i-1$ (see Fig. 5.2). This corresponds to the situation where $\mathbf{C}_{i,j}$ can only be nonzero if $j = i - 1$ and $\mathbf{N}_i(\mathbf{x}_1, \ldots, \mathbf{x}_i) = \mathbf{N}_i(\mathbf{x}_{i-1}, \mathbf{x}_i)$:

$$\begin{aligned} \mathbf{x}_1(t+1) &= \mathbf{L}_1\mathbf{x}_1(t) + \mathbf{N}_1(\mathbf{x}_1(t)), \\ \mathbf{x}_i(t+1) &= \mathbf{L}_i\mathbf{x}_i(t) + \mathbf{C}_{i,i-1}\mathbf{x}_{i-1}(t) + \mathbf{N}_i(\mathbf{x}_{i-1}(t), \mathbf{x}_i(t)) \quad (i = 2, \ldots, n), \end{aligned} \tag{5.30}$$

and

$$\begin{aligned} \mathbf{x}_1(t+1) &= \mathbf{L}_1\mathbf{x}_1(t), \\ \mathbf{x}_i(t+1) &= \mathbf{L}_i\mathbf{x}_i(t) + \mathbf{C}_{i,i-1}\mathbf{x}_{i-1}(t) \quad (i = 2, \ldots, n). \end{aligned} \tag{5.31}$$

We will call cascades having the form (5.30) and (5.31) *chained cascades*.

**Condition 5.1** *The following conditions will be in force for all following results:*

1. $\mathbf{L}_i$ *is invertible and diagonalizable for all* $i = 1, \ldots, n$,

$$\mathbf{L}_i\mathbf{V}_i = \mathbf{V}_i\Lambda_i, \tag{5.32}$$

*where* $\mathbf{V}_i$ *is the matrix of eigenvectors and* $\Lambda_i$ *is the diagonal matrix of eigenvalues.*

2. *(Disjoint spectrums) The spectrums of each layer are pairwise disjoint. That is for* $i, j \in \{1, \ldots, n\}$ *satisfying* $i \neq j$

$$\sigma(\mathbf{L}_i) \cap \sigma(\mathbf{L}_j) = \emptyset. \tag{5.33}$$

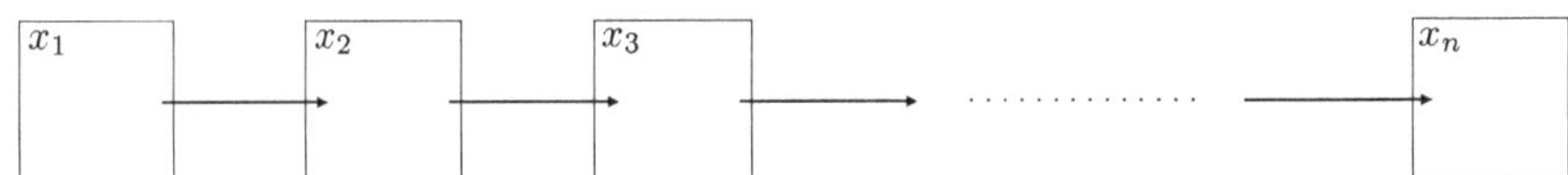


**Fig. 5.2** Chained cascade system

*3. For $i = 1, \ldots, n$, each $j < i$, the linearized systems satisfy*

$$\lim_{t\to\infty} \frac{\|\mathbf{L}_j^t\|}{\|\mathbf{L}_i^t\|} = 0. \tag{5.34}$$

*Remark 5.3* We require condition 2 due to the form of the entries of the coupling matrix $\tilde{\mathbf{C}}_{i,j}$ (Eq. (5.39)) in Theorem 5.2. If any pair of eigenvalues from $\mathbf{L}_i$ and $\mathbf{L}_j$ were equal, $\tilde{\mathbf{C}}_{i,j}$ would not be well defined since its matrix elements have a term of the form $(1 - \lambda_{j,m}/\lambda_{i,\ell})^{-1}$. The requirement of disjointness can be thought of as a non-resonance condition.

*Remark 5.4* Clearly, condition 3 is satisfied if $\|\mathbf{L}_i\| < \|\mathbf{L}_{i+1}\|$, however, as stated, it is more general in that an upstream system's may decay slower than a downstream system's solution for some time as long as eventually the upstream solution approaches 0 faster than the downstream solution. For example, condition 3 would hold true if an upstream system, having a smaller spectral radius than a downstream system's spectral radius, exhibited transient growth (due to non-normality of the upstream matrix) when the downstream system's solution decayed monotonically to 0. In such a case, the solution of the upstream system would grow for a time and we would have that $\|\mathbf{L}_j^t(\mathbf{x})\|/\|\mathbf{L}_i^t\|$ (where $j$ is the upstream system and $i$ the downstream) grows for some finite interval of time $t \in [0, T]$.

*Remark 5.5* Condition 3 is used to prove the results on 0 asymptotic relative error between the nominal and linearized system. While, it is true that the linear cascaded system is still exponentially stable to 0, the condition guarantees that asymptotically the solutions of the linear cascades and the perturbed nominal system's solutions approach each other faster than they approach the fixed point.

## 5.4.1 Theorems for Linear, Chained Cascades

**Theorem 5.2** (0 asymptotic relative error for chained, linear cascades) *Assume Condition 5.1 is in effect. Then* (5.31) *has 0 asymptotically relative error in the sense of* (5.10). *In particular, for all $i \geq 1$*[4] *and all $t \geq 0$,*

$$\begin{aligned}
&\|\Pi_i \circ \mathsf{Lin}^{\circ t}(\mathbf{x}_1, \ldots, \mathbf{x}_n) - \Pi_i \circ \mathsf{Nom}^{\circ t}(\mathsf{pert}(\mathbf{x}_1, \ldots, \mathbf{x}_n))\| \\
&\quad \leq \sum_{j=1}^{i-1} \|\mathbf{D}_{i,j}\| \|\mathbf{L}_j^t \mathsf{pert}_j(\mathbf{x}_1, \ldots, \mathbf{x}_j)\| \qquad (5.35)\\
&\quad \leq \left( \sum_{j=1}^{i-1} \|\mathbf{D}_{i,j}\| \|\mathsf{pert}_j(\mathbf{x}_1, \ldots, \mathbf{x}_j)\| \right) \|\mathbf{L}_i\|^t, \qquad (5.36)
\end{aligned}$$

[4]We take the empty sum $\sum_{j=1}^{0}$ to be 0.

*where*

$$\mathbf{D}_{i,i} = \mathbf{I}_{d_i} \qquad \forall i \in \{1, \ldots, n\}, \tag{5.37}$$

$$\mathbf{D}_{i,j} = \mathbf{L}_i^{-1}\mathbf{V}_i\tilde{\mathbf{C}}_{i,j}\mathbf{V}_j^{-1}, \qquad i \in \{2, \ldots, n\},\ j \in \{1, \ldots, i-1\}, \tag{5.38}$$

*and the matrix* $\tilde{\mathbf{C}}_{i,j} \in \mathbb{C}^{d_i \times d_j}$ *has elements*

$$[\tilde{\mathbf{C}}_{i,j}]_{\ell,m} = \left[\mathbf{V}_i^{-1}\mathbf{C}_{i,i-1}\mathbf{D}_{i-1,j}\mathbf{V}_j\right]_{\ell,m}\left(1 - \frac{\lambda_{j,m}}{\lambda_{i,\ell}}\right)^{-1}, \tag{5.39}$$

*for* $i \in \{2, \ldots, n\}$, $j \in \{1, \ldots, i-1\}$. *The perturbation function for the initial condition,* $\mathsf{pert} : \mathbb{C}^{d_1} \times \cdots \times \mathbb{C}^{d_n} \to \mathbb{C}^{d_1} \times \cdots \times \mathbb{C}^{d_n}$, *is defined as*

$$\begin{aligned}&\mathsf{pert}(\mathbf{x}_1, \ldots, \mathbf{x}_n)\\&\quad = \left(\mathsf{pert}_1 \circ \Pi_1, \mathsf{pert}_2 \circ (\Pi_1, \Pi_2), \ldots, \mathsf{pert}_{n-1} \circ (\Pi_1, \ldots, \Pi_{n-1}), \mathsf{pert}_n\right)(\mathbf{x}_1, \ldots, \mathbf{x}_n)\end{aligned} \tag{5.40}$$

*and the perturbations for each system* $i$, $\mathsf{pert}_i : \mathbb{C}^{d_1} \times \cdots \times \mathbb{C}^{d_i} \to \mathbb{C}^{d_i}$, *are defined recursively by*

$$\mathsf{pert}_1(\mathbf{x}_1) = \mathbf{x}_1 \tag{5.41}$$

$$\mathsf{pert}_i(\mathbf{x}_1, \ldots, \mathbf{x}_i) = \mathbf{x}_i + \sum_{j=1}^{i-1}(-1)^{i-1-j}\mathbf{D}_{i,j}\mathsf{pert}_j(\mathbf{x}_1, \ldots, \mathbf{x}_j), \quad i \in \{2, \ldots, n\}. \tag{5.42}$$

*Furthermore, for all* $i \geq 1$

$$\lim_{t\to\infty} \frac{\|\Pi_i \circ \mathsf{Lin}^{\circ t}(\mathbf{x}_1, \ldots, \mathbf{x}_n) - \Pi_i \circ \mathsf{Nom}^{\circ t}(\mathsf{pert}(\mathbf{x}_1, \ldots, \mathbf{x}_n))\|}{\|\mathbf{L}_i^t\|} = 0. \tag{5.43}$$

*Proof* (*Proof of Theorem* 5.2) Equations (5.36) and (5.43) follow from Corollary 5.3 in the Appendix (Sect. 5.6.1). The expressions for the perturbation terms $\mathsf{pert}_j$ and the coupling matrices $\mathbf{D}_{i,j}$ and $\tilde{\mathbf{C}}_{i,j}$ are derived in the proof of Lemma 5.3 (Sect. 5.6.1). □

Combining the asymptotic equivalence of each of the components gives that the full system is also asymptotically equivalent. This is recorded in the following Corollary.

**Corollary 5.1** (Asymptotic equivalence for chained, linear cascades)

$$\lim_{t\to\infty} \|\mathsf{Lin}^{\circ t}(\mathbf{x}_1, \ldots, \mathbf{x}_n) - \mathsf{Nom}^{\circ t}(\mathsf{pert}(\mathbf{x}_1, \ldots, \mathbf{x}_n))\|_\times = 0. \tag{5.44}$$

The following result is a reinterpretation of Theorem 5.2 in terms of eigenfunctions of the Koopman operator.

**Theorem 5.3** (Perturbation of principal eigenfunctions) *Assume Condition 5.1 is in effect. For any $i \geq 1$, $s_i \in \{1, \ldots, d_i\}$, and $t \in \mathbb{N}$*

$$\left|\left(U^t_{\mathsf{Lin}}\phi_{(0,\ldots,0,s_i,0,\ldots,0)}\right)(\mathbf{x}_1, \ldots, \mathbf{x}_n) - \left(U^t_{\mathsf{Nom}}\phi_{(0,\ldots,0,s_i,0,\ldots,0)}\right) \circ \mathsf{pert}(\mathbf{x}_1, \ldots, \mathbf{x}_n)\right| \leq \|\phi_{i,s_i}\| \sum_{j=1}^{i-1} \|\mathbf{D}_{i,j}\| \|\mathbf{L}^t_j \mathsf{pert}_j(\mathbf{x}_1, \ldots, \mathbf{x}_j)\|. \tag{5.45}$$

*Furthermore, for any $i \in \{1, \ldots, n\}$*

$$\lim_{t\to\infty} \frac{\left|\left(U^t_{\mathsf{Lin}}\phi_{(0,\ldots,0,s_i,0,\ldots,0)}\right)(\mathbf{x}_1, \ldots, \mathbf{x}_n) - \left(U^t_{\mathsf{Nom}}\phi_{(0,\ldots,0,s_i,0,\ldots,0)}\right) \circ \mathsf{pert}(\mathbf{x}_1, \ldots, \mathbf{x}_n)\right|}{\|\mathbf{L}^t_i\|} = 0. \tag{5.46}$$

*Proof* See the Appendix, Sect. 5.6.2 below. □

*Remark 5.6* Recall that $\phi_{(0,\ldots,0,s_i,0,\ldots,0)}(\mathbf{x}_1, \ldots, \mathbf{x}_n) = (\phi_{i,s_i} \circ \Pi_i)(\mathbf{x}_1, \ldots, \mathbf{x}_n)$. Due to the definitions of $\mathsf{Nom}$ and $\mathsf{pert}$,

$$\begin{aligned}\left(U^t_{\mathsf{Nom}}\phi_{(0,\ldots,0,s_i,0,\ldots,0)}\right) \circ \mathsf{pert}(\mathbf{x}_1, \ldots, \mathbf{x}_n) &= (\phi_{i,s_i} \circ \Pi_i \circ \mathsf{Nom}^{\circ t}) \circ \mathsf{pert}(\mathbf{x}_1, \ldots, \mathbf{x}_n)\\ &= (\phi_{i,s_i} \circ \mathsf{Nom}_i^{\circ t}) \circ \mathsf{pert}_i(\mathbf{x}_1, \ldots, \mathbf{x}_i)\\ &= (U^t_{\mathsf{Nom}_i}\phi_{i,s_i}) \circ \mathsf{pert}_i(\mathbf{x}_1, \ldots, \mathbf{x}_i)\\ &= (\lambda^t_{i,s_i}\phi_{i,s_i}) \circ \mathsf{pert}_i(\mathbf{x}_1, \ldots, \mathbf{x}_i)\\ &= \lambda^t_{i,s_i}\phi_{(0,\ldots,0,s_i,0,\ldots,0)} \circ \mathsf{pert}(\mathbf{x}_1, \ldots, \mathbf{x}_n).\end{aligned}$$

The next result shows that eigenfunctions for the linear cascade can be obtained by composing the principal eigenfunctions of the nominal system with the perturbation function.

**Corollary 5.2** *Fix $i \in 1, \ldots, n$, and let $\lambda_{i,s_i} \in \sigma(\mathbf{L}_i)$. Then $\phi_{(0,\ldots,0,s_i,0,\ldots,0)} \circ \mathsf{pert}$ is an eigenfunction of $U_{\mathsf{Lin}}$ at eigenvalue $\lambda_{i,s_i}$.*

*Proof* This is a straightforward application of the GLA Theorem [14] (see the Appendix, Sect. 5.6). We only show it for a peripheral eigenvalue ($|\lambda_{i,s_i}| = \|L_i\|$). This is accomplished by showing that a Laplace average of $\phi_{(0,\ldots,0,s_i,0,\ldots,0)}$ converges to $\phi_{(0,\ldots,0,s_i,0,\ldots,0)} \circ \mathsf{pert}$.

Let $\mathbf{e}_{n,i}$ be the $i$th canonical basis vector of length $n$ and write $\phi_{s_i\mathbf{e}_{n,i}} = \phi_{(0,\ldots,0,s_i,0,\ldots,0)}$. Form the Laplace average,

$$
\begin{aligned}
&\frac{1}{N}\sum_{t=0}^{N-1}\lambda_{i,s_i}^{-t}U_{\mathsf{Lin}}^{t}\phi_{s_i\mathbf{e}_{n,i}}\\
&=\frac{1}{N}\sum_{t=0}^{N-1}\lambda_{i,s_i}^{-t}\left[(U_{\mathsf{Nom}}^{t}\phi_{s_i\mathbf{e}_{n,i}})\circ\mathsf{pert}+U_{\mathsf{Lin}}^{t}\phi_{s_i\mathbf{e}_{n,i}}-(U_{\mathsf{Nom}}^{t}\phi_{s_i\mathbf{e}_{n,i}})\circ\mathsf{pert}\right]\\
&=\frac{1}{N}\sum_{t=0}^{N-1}(\phi_{s_i\mathbf{e}_{n,i}}\circ\mathsf{pert})+\lambda_{i,s_i}^{-t}\left(U_{\mathsf{Lin}}^{t}\phi_{s_i\mathbf{e}_{n,i}}-(U_{\mathsf{Nom}}^{t}\phi_{s_i\mathbf{e}_{n,i}})\circ\mathsf{pert}\right),
\end{aligned}
$$

where we have used $U_{\mathsf{Nom}}^{t}\phi_{s_i\mathbf{e}_{n,i}}=\lambda_{i,s_i}^{t}\phi_{s_i\mathbf{e}_{n,i}}$. Then

$$
\begin{aligned}
&\left\|\frac{1}{N}\sum_{t=0}^{N-1}\lambda_{i,s_i}^{-t}U_{\mathsf{Lin}}^{t}\phi_{s_i\mathbf{e}_{n,i}}-\phi_{s_i\mathbf{e}_{n,i}}\circ\mathsf{pert}\right\|\\
&\leq\frac{1}{N}\sum_{t=0}^{N-1}\left\|\lambda_{i,s_i}^{-t}\left(U_{\mathsf{Lin}}^{t}\phi_{s_i\mathbf{e}_{n,i}}-(U_{\mathsf{Nom}}^{t}\phi_{s_i\mathbf{e}_{n,i}})\circ\mathsf{pert}\right)\right\|\\
&=\frac{1}{N}\sum_{t=0}^{N-1}\frac{\left\|\left(U_{\mathsf{Lin}}^{t}\phi_{s_i\mathbf{e}_{n,i}}-(U_{\mathsf{Nom}}^{t}\phi_{s_i\mathbf{e}_{n,i}})\circ\mathsf{pert}\right)\right\|}{\left\|\mathbf{L}_i^t\right\|}.
\end{aligned}
$$

It is clear from (5.46) that the right-hand side converges to 0 as $N\to\infty$.

In the general case, to project onto the $\lambda$ eigenspace, $U_{\mathsf{Lin}}$ in the above average is replaced with $U_{\mathsf{Lin}}(I-P)$ where $P$ is the projection onto the direct sum of $\mu$-eigenspace, for $\mu$ satisfying $|\mu|>|\lambda|$. □

*Remark 5.7* It can be shown that $\phi_{s_i\mathbf{e}_{n,i}}\circ\mathsf{pert}$ is an eigenfunction of $U_{\mathsf{Lin}}$, without appeal to the GLA Theorem, by direct computation, but it is more involved since $\mathsf{pert}$ for an $n$-layer cascade consists of a product of $n-1$ lower block triangular matrices. Section 5.4.2.1 below shows the computation for just a two-layer system.

### 5.4.2 Examples

To give a feel for the above results for linear, chained cascades, we give an analytically worked out example of a two-layer, chained cascade. We then demonstrate the results numerically for a seven-layer chained cascade.[5]

[5]We choose seven layers here merely for display purposes—graphs for the behavior of the six downstream component systems are easily displayed in a 2-by-3 table.

#### 5.4.2.1 Analytic Example

We demonstrate the result of Corollary 5.2 explicitly for a two-layer chained cascade,

$$\begin{aligned}\mathbf{x}_1(t+1) &= \mathbf{L}_1\mathbf{x}_1(t)\\ \mathbf{x}_2(t+1) &= \mathbf{L}_2\mathbf{x}_2(t) + \mathbf{C}_{2,1}\mathbf{x}_1(t).\end{aligned}$$

Using Theorem 5.2,

$$\mathsf{pert}_1(\mathbf{x}_1,\mathbf{x}_2) = \begin{bmatrix}\mathbf{I}_{d_1} & \mathbf{0}\end{bmatrix}\begin{bmatrix}\mathbf{x}_1\\ \mathbf{x}_2\end{bmatrix} \text{ and}$$

$$\mathsf{pert}_2(\mathbf{x}_1,\mathbf{x}_2) = \begin{bmatrix}\mathbf{D}_{2,1} & \mathbf{I}_{d_2}\end{bmatrix}\begin{bmatrix}\mathbf{x}_1\\ \mathbf{x}_2\end{bmatrix}.$$

Therefore,

$$\mathsf{pert}(\mathbf{x}_1,\mathbf{x}_2) = \begin{bmatrix}\mathsf{pert}_1(\mathbf{x}_1)\\ \mathsf{pert}_2(\mathbf{x}_1,\mathbf{x}_2)\end{bmatrix} = \begin{bmatrix}\mathbf{I}_{d_1} & \mathbf{0}\\ \mathbf{D}_{2,1} & \mathbf{I}_{d_2}\end{bmatrix}\begin{bmatrix}\mathbf{x}_1\\ \mathbf{x}_2\end{bmatrix}. \tag{5.47}$$

Furthermore, we have the principal eigenfunction for the second system

$$\phi_{(0,s_2)}(\mathbf{x}_1,\mathbf{x}_2) = \begin{bmatrix}\mathbf{0} & \mathbf{e}^*_{d_2,s_2}\mathbf{V}_2^{-1}\end{bmatrix}\begin{bmatrix}\mathbf{x}_1\\ \mathbf{x}_2\end{bmatrix}. \tag{5.48}$$

By Lemma 5.3,

$$\mathsf{Lin}^{\circ t}(\mathbf{x}_1,\mathbf{x}_2) = \begin{bmatrix}\mathbf{L}_1^t\mathsf{pert}_1(\mathbf{x}_1)\\ \mathbf{L}_2^t\mathsf{pert}_2(\mathbf{x}_1,\mathbf{x}_2) - \mathbf{D}_{2,1}\mathbf{L}_1^t\mathsf{pert}_1(\mathbf{x}_1)\end{bmatrix}. \tag{5.49}$$

We can write $\mathsf{Lin}^{\circ t}(\mathbf{x}_1,\mathbf{x}_2)$ as

$$\mathsf{Lin}^{\circ t}(\mathbf{x}_1,\mathbf{x}_2) = \begin{bmatrix}\mathbf{I}_{d_1} & \mathbf{0}\\ -\mathbf{D}_{2,1} & \mathbf{I}_{d_2}\end{bmatrix}\begin{bmatrix}\mathbf{L}_1^t & \mathbf{0}\\ \mathbf{0} & \mathbf{L}_2^t\end{bmatrix}\begin{bmatrix}\mathsf{pert}_1(\mathbf{x}_1)\\ \mathsf{pert}_2(\mathbf{x}_2)\end{bmatrix}.$$

Our goal is to show that $\phi_{(0,s_2)}\circ\mathsf{pert}$ is an eigenfunction at eigenvalue $\lambda_{2,s_2}$ for $U^{\circ t}_{\mathsf{Lin}}$. To this end, we compute

$$\begin{aligned}&U^{\circ t}_{\mathsf{Lin}}(\phi_{(0,s_2)}\circ\mathsf{pert})(\mathbf{x}_1,\mathbf{x}_2)\\ &= \phi_{(0,s_2)}\circ\mathsf{pert}\circ\mathsf{Lin}^{\circ t}(\mathbf{x}_1,\mathbf{x}_2)\\ &= \underbrace{\begin{bmatrix}\mathbf{0} & \mathbf{e}^*_{d_2,s_2}\mathbf{V}_2^{-1}\end{bmatrix}}_{\phi_{(0,s_2)}}\underbrace{\begin{bmatrix}\mathbf{I}_{d_1} & \mathbf{0}\\ \mathbf{D}_{2,1} & \mathbf{I}_{d_2}\end{bmatrix}}_{\mathsf{pert}}\underbrace{\begin{bmatrix}\mathbf{I}_{d_1} & \mathbf{0}\\ -\mathbf{D}_{2,1} & \mathbf{I}_{d_2}\end{bmatrix}\begin{bmatrix}\mathbf{L}_1^t & \mathbf{0}\\ \mathbf{0} & \mathbf{L}_2^t\end{bmatrix}\begin{bmatrix}\mathsf{pert}_1(\mathbf{x}_1)\\ \mathsf{pert}_2(\mathbf{x}_2)\end{bmatrix}}_{\mathsf{Lin}^{\circ t}(\mathbf{x}_1,\mathbf{x}_2)}\\ &= \begin{bmatrix}\mathbf{0} & \mathbf{e}^*_{d_2,s_2}\mathbf{V}_2^{-1}\end{bmatrix}\begin{bmatrix}\mathbf{L}_1^t & \mathbf{0}\\ \mathbf{0} & \mathbf{L}_2^t\end{bmatrix}\begin{bmatrix}\mathsf{pert}_1(\mathbf{x}_1)\\ \mathsf{pert}_2(\mathbf{x}_2)\end{bmatrix}\end{aligned}$$

$$
\begin{aligned}
&= \mathbf{e}^*_{d_2,s_2}\mathbf{V}_2^{-1}L_2^t \mathsf{pert}_2(\mathbf{x}_2) \\
&= \mathbf{e}^*_{d_2,s_2}\Lambda_2^t\mathbf{V}_2^{-1}\mathsf{pert}_2(\mathbf{x}_2) \\
&= \lambda^t_{2,s_2}\mathbf{e}^*_{d_2,s_2}\mathbf{V}_2^{-1}\mathsf{pert}_2(\mathbf{x}_2) \\
&= \lambda^t_{2,s_2}\begin{bmatrix}\mathbf{0} & \mathbf{e}^*_{d_2,s_2}\mathbf{V}_2^{-1}\end{bmatrix}\begin{bmatrix}\mathsf{pert}_1(\mathbf{x}_1) \\ \mathsf{pert}_2(\mathbf{x}_2)\end{bmatrix} \\
&= \lambda^t_{2,s_2}(\phi_{(0,s_2)}\circ \mathsf{pert})(\mathbf{x}_1,\mathbf{x}_2).
\end{aligned}
$$

This completes the example.

### *5.4.3 Numerical Example*

The results of Theorem 5.2 were confirmed with simulation. The simulation consisted of seven-layer linear chained, cascaded system with randomly generated dimensions $d_i$ for each system $i$. System matrices $\mathbf{L}_i$ were randomly generated with entries uniformly in the interval $[-1, 1]$ and then scaled to have $\|\mathbf{L}_i\| = (0.9)^{8-i}$ for $i = 1,\dots,7$. The coupling matrices $\mathbf{C}_{i,i-1}$ were also randomly generated with entries uniformly in $[-1, 1]$. Initial conditions for each system $i$ were randomly generated and scaled to have $\|\mathbf{x}_i\| = 1$.

Figures 5.3 and 5.4 show the log absolute error and log relative errors of a typical run of the simulation. The black asterisks in Fig. 5.3 are the predicted upper bound (5.36). The colored lines in each plot correspond to the log of the absolute error,

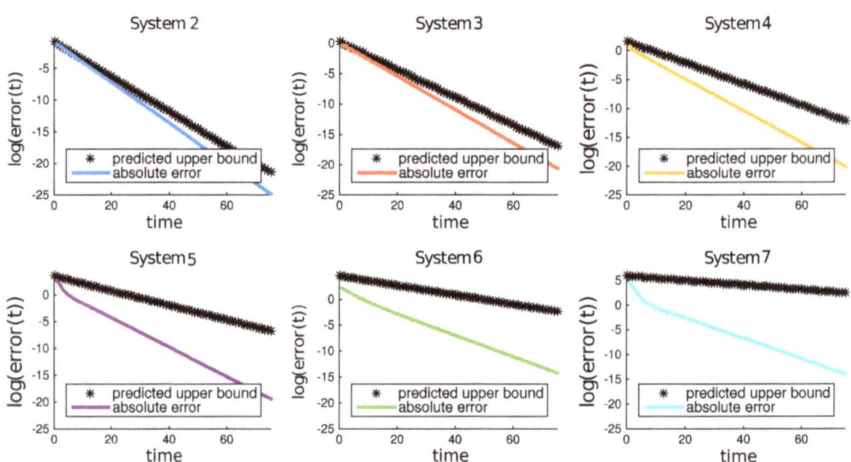


**Fig. 5.3** Log absolute error between $\Pi_i \circ \mathsf{Lin}^{\circ t}$ and $\Pi_i \circ \mathsf{Nom}^{\circ t}$, Eq. (5.50). The black asterisks $(*)$ correspond to the upper bound given by (5.36). The colored lines correspond to the absolute error given by the right-hand side of (5.35). Traces for system 1 are not plotted since by construction $\Pi_1 \circ \mathsf{Lin}^{\circ t} = \Pi_1 \circ \mathsf{Nom}^{\circ t}$ for all $t$

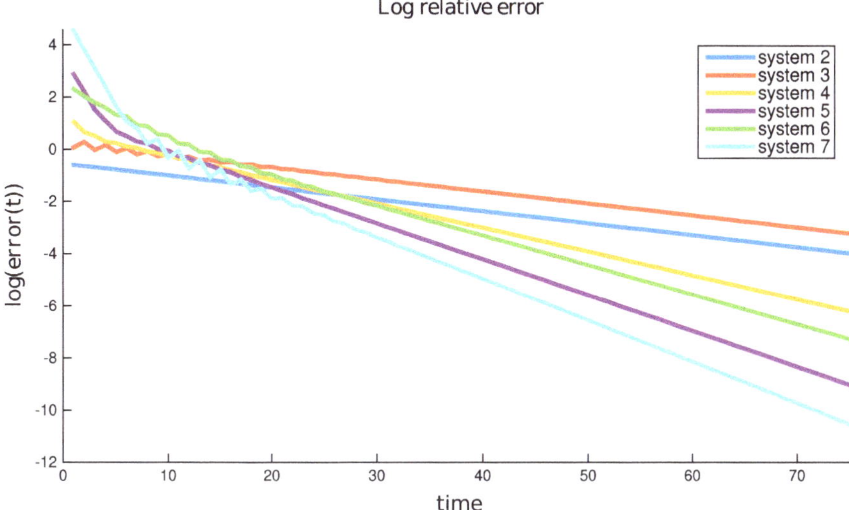


**Fig. 5.4** Log relative error between $\Pi_i \circ \mathsf{Lin}^{\circ t}(\mathbf{x}_1, \dots, \mathbf{x}_n)$ and $\Pi_i \circ \mathsf{Nom}^{\circ t}(\mathsf{pert}(\mathbf{x}_1, \dots, \mathbf{x}_n))$. The colored lines correspond to the log relative error (5.51)

$$\log\left(\|\Pi_i(\mathsf{Lin}^{\circ t}(\mathbf{x}_1, \dots, \mathbf{x}_n)) - \Pi_i(\mathsf{Nom}^{\circ t}(\mathsf{pert}(\mathbf{x}_1, \dots, \mathbf{x}_n)))\|\right), \qquad (5.50)$$

between the linear and the nominal systems. Figure 5.4 shows the log of the relative error

$$\log\left(\frac{\|\Pi_i(\mathsf{Lin}^{\circ t}(\mathbf{x}_1, \dots, \mathbf{x}_n)) - \Pi_i(\mathsf{Nom}^{\circ t}(\mathsf{pert}(\mathbf{x}_1, \dots, \mathbf{x}_n)))\|}{\|\mathbf{L}_i\|^t}\right). \qquad (5.51)$$

As can be seen, the log absolute error and log relative error decrease linearly, confirming that the absolute error and relative error decrease exponentially fast to zero.

### *5.4.4 Theorems for Nonlinear, Chained Cascades*

The asymptotic equivalence for the chained, linear cascades can be pushed to chained, nonlinear cascades with asymptotically stable fixed points through the use of a topological conjugacy. Let $\tau = (\tau_1, \dots, \tau_n) : \mathbb{C}^{d_1} \times \cdots \times \mathbb{C}^{d_n} \to \mathbb{C}^{d_1} \times \cdots \times \mathbb{C}^{d_n}$ be a topological conjugacy from the linear system to the nonlinear system and $\tau^{-1} = (\rho_1, \dots, \rho_n) : \mathbb{C}^{d_1} \times \cdots \times \mathbb{C}^{d_n} \to \mathbb{C}^{d_1} \times \cdots \times \mathbb{C}^{d_n}$ its inverse. In general, both $\tau_i$ and $\rho_i$ are maps from $\mathbb{C}^{d_1} \times \cdots \times \mathbb{C}^{d_n} \to \mathbb{C}^{d_i}$. The topological conjugacy makes the following diagram commute:

$$\begin{array}{ccc} \mathbb{C}^{d_1} \times \cdots \times \mathbb{C}^{d_n} & \xrightarrow{\mathsf{Lin}^{\circ t}} & \mathbb{C}^{d_1} \times \cdots \times \mathbb{C}^{d_n} \\ \downarrow \tau & & \downarrow \tau \\ \mathbb{C}^{d_1} \times \cdots \times \mathbb{C}^{d_n} & \xrightarrow{\mathsf{NonLin}^{\circ t}} & \mathbb{C}^{d_1} \times \cdots \times \mathbb{C}^{d_n} \end{array} \tag{5.52}$$

**Theorem 5.4** (Asymptotic equivalence for nonlinear cascaded systems) *Let the conditions of Theorem 5.2 be satisfied and let* $\tau = (\tau_1, \ldots, \tau_n) : \mathbb{C}^{d_1} \times \cdots \times \mathbb{C}^{d_n} \to \mathbb{C}^{d_1} \times \cdots \times \mathbb{C}^{d_n}$ *be a topological conjugacy satisfying* $\mathsf{Lin} = \tau^{-1} \circ \mathsf{NonLin} \circ \tau$. *Then, for each initial condition* $\mathbf{Y} = (\mathbf{y}_1, \ldots, \mathbf{y}_n)$ *for the nonlinear system,* $\mathsf{NonLin}$ *is asymptotically equivalent to* $\tau \circ \mathsf{Nom} \circ \tau^{-1}$ *with the perturbation function* $\tau \circ \mathsf{pert} \circ \tau^{-1}$*:*

$$\lim_{t\to\infty} \left\| \mathsf{NonLin}^{\circ t}(\mathbf{Y}) - \left(\tau \circ \mathsf{Nom} \circ \tau^{-1}\right)^{\circ t} (\tau \circ \mathsf{pert} \circ \tau^{-1})(\mathbf{Y}) \right\| = 0, \tag{5.53}$$

*where* $\mathsf{pert}$ *is given by* (5.40).

*Proof* See the Appendix, Sect. 5.6.3 for the proof. □

*Remark 5.8* Recall that $\mathsf{Nom}$ is the nominal decoupled linear system and $\tau$ is a map from the linear to the nonlinear system. Then $\tau \circ \mathsf{Nom} \circ \tau^{-1}$ is a map on the same state space as the nonlinear system $\mathsf{NonLin}$ and can be thought of as the nominal nonlinear system. Furthermore, since $\mathsf{pert}$ is the perturbation function for the initial conditions of the linear system, then $\tau \circ \mathsf{pert} \circ \tau^{-1}$ is the perturbation function for the nonlinear system's initial conditions.

**Theorem 5.5** (Perturbation of eigenfunctions for nonlinear cascades) *Let the conditions of Theorem 5.4 be satisfied. For any* $\mathbf{Y} = (\mathbf{y}_1, \ldots, \mathbf{y}_n) \in \mathbb{C}^{d_1} \times \cdots \times \mathbb{C}^{d_n}$,

$$\lim_{t\to\infty} \frac{\left| U^t_{\mathsf{NonLin}}(\phi_{s_i \mathbf{e}_{n,i}} \circ \tau^{-1})(\mathbf{Y}) - U^t_{\tau\circ\mathsf{Nom}\circ\tau^{-1}}(\phi_{s_i \mathbf{e}_{n,i}} \circ \tau^{-1})((\tau \circ \mathsf{pert} \circ \tau^{-1})(\mathbf{Y})) \right|}{\|\mathbf{L}_i\|^t} = 0. \tag{5.54}$$

*Proof* See Appendix, Sect. 5.6.4.

*Remark 5.9* It was shown in [2] that if $\phi$ was an eigenfunction corresponding to the Koopman operator associated with the linearized system and $\tau$ was a topological conjugacy from the linear to the nonlinear system, then $\phi \circ \tau^{-1}$ was an eigenfunction of the Koopman operator associated with the nonlinear system. Recall that $\phi_{s_i \mathbf{e}_{n,i}} = \phi_{(0,\ldots,0,s_i,0,\ldots,0)}$ is a principal eigenfunction for the Koopman operator $U_{\mathsf{Nom}}$ associated with the nominal linear system. Then $\phi_{s_i \mathbf{e}_{n,i}} \circ \tau^{-1}$ is a principal eigenfunction for the Koopman operator $U_{\tau\circ\mathsf{Nom}\circ\tau^{-1}}$ associated with the *nonlinear nominal system* $\tau \circ \mathsf{Nom} \circ \tau^{-1}$.

*Remark 5.10* Theorem 5.5 says that the action of the Koopman operator associated with the nonlinear cascade $\mathsf{NonLin}$ on the observable $\phi_{s_i \mathbf{e}_{n,i}} \circ \tau^{-1} = \phi_{(0,\ldots,0,s_i,0,\ldots,0)} \circ \tau^{-1}$ is asymptotically equivalent to the action of the Koopman operator associated with $\tau \circ \mathsf{Nom} \circ \tau^{-1}$ (the nominal nonlinear system) on $\phi_{(0,\ldots,0,s_i,0,\ldots,0)} \circ \tau^{-1}$ but at a perturbed initial condition, $(\tau \circ \mathsf{pert} \circ \tau^{-1})(\mathbf{y})$.

*Remark 5.11* While the preceding results are proved for chained cascades, they should be easily extensible to the general cascade systems. The difference should only be in the exact form of the perturbation functions and the bounds. For example, consider a three-layer cascade. In a chained cascade, only layer 2 feeds into layer 3, whereas for a general cascade, layer 1 can feed directly into layer 3. However, the perturbation function for layer 3 in the chained cascade case is actually a function of the state variables from layer 2 *and* layer 1. Thus layer 1 already affects the perturbation of the initial condition of layer 3 in the chained cascade case. If we then wire layer 1 directly to layer 3 to get a general cascade, only the effect of layer 1 on the perturbation of layer 3 is changed. The asymptotic results should remain the same.

## 5.5 Conclusions

We have analyzed the Koopman spectrum of cascaded dynamical systems; systems formed by wiring component subsystems together in a lower block triangular form. We show the existence of a perturbation function that maps initial conditions to initial conditions such that the evolution of the cascaded system from any initial condition is asymptotically equivalent to the evolution due to each component subsystem (decoupled from the others) if the initial conditions for the component subsystems are given by the perturbation function applied to original initial condition. The rate of convergence of the orbits in each subsystem is faster than the decay rate of the decoupled subsystem to its fixed point. This is captured in our results by saying that each subsystem has zero asymptotic relative error. We also show a bound on the distance between the trajectories that hold for all time. From these results, it follows that the Koopman principal eigenvalues of each subsystem are also Koopman eigenvalues of the cascaded system. The principal eigenfunctions for a subsystem become eigenfunctions for the cascaded system when composed with the perturbation function. These results also hold for a nonlinear cascade if it is topologically conjugate to a linear cascade for which the results hold.

These results are useful in the analysis of large interconnected systems. Often, in order to analyze these systems efficiently, a decomposition into a lower block diagonal form (cascade structure) or block diagonal form must be performed. Various techniques have been proposed to do this, with more research currently being done. Once in this structure, the results in this chapter allow a further analysis of the system without having to simulate it. The results on the principal Koopman eigenfunctions tell, a priori, how any observable of interest on the system would behave under the dynamics. Such observables would only need to be expanded into the principal eigenfunctions and their products, whose eigenvalues are given by products of the principal eigenvalues. In connection to data-driven analysis of dynamical systems, once some decomposition technique reveals the component systems, they can be studied individually with dynamic mode decomposition (DMD) methods in order to determine their spectrum. See the recent work [4] for a variant of DMD tailored to interconnected systems. As long as the discovered spectrum satisfies the conditions of this chapter, the results presented here guarantee stability of the full, cascaded system.

**Acknowledgements** This research was partially funded under a subcontract from HRL Laboratories, LLC under DARPA contract N66001-16-C-4053 and additionally funded by the DARPA Contract HR0011-16-C-0116.

The views expressed are those of the authors and do not reflect the official policy or position of the Department of Defense or the U.S. Government. Distribution Statement "A": Approved for Public Release, Distribution Unlimited.

## 5.6 Appendix

Here, we collect all the technical proofs of the above results. It can be skipped on a first reading.

### 5.6.1 *Proof of Theorem 5.2: 0 Asymptotic Relative Error—Linear, Chained Cascades*

The first lemma gives the general solution for the $i$th level of the chained linear cascade system.

**Lemma 5.1** *For all $i = 1, \ldots, n$ and $t \geq 0$, denote by $\mathbf{x}_i(t)$ the solution $\Pi_i \circ \mathsf{Lin}^{\circ t}(\mathbf{x}_1, \ldots, \mathbf{x}_n)$ of the $i$th level of* (5.31). *For $i \geq 2$, the general solution satisfies*

$$\mathbf{x}_i(t) \equiv \Pi_i \circ \mathsf{Lin}^{\circ t}(\mathbf{x}_1, \ldots, \mathbf{x}_n) = \mathbf{L}_i^t \mathbf{x}_i + \mathbf{L}_i^{t-1} \mathbf{V}_i \sum_{k=0}^{t-1} \Lambda_i^{-k} \mathbf{V}_i^{-1} \mathbf{C}_{i,i-1} \mathbf{x}_{i-1}(k). \tag{5.55}$$

*Proof* Repeatedly using (5.31), we have

$$\begin{aligned}
\mathbf{x}_i(t) &= \mathbf{L}_i \mathbf{x}_i(t-1) + \mathbf{C}_{i,i-1} \mathbf{x}_{i-1}(t-1) \\
&= \mathbf{L}_i \left[ \mathbf{L}_i \mathbf{x}_i(t-2) + \mathbf{C}_{i,i-1} \mathbf{x}_{i-1}(t-2) \right] + \mathbf{C}_{i,i-1} \mathbf{x}_{i-1}(t-1) \\
&= \mathbf{L}_i^2 \mathbf{x}_i(t-2) + \left[ \mathbf{L}_i \mathbf{C}_{i,i-1} \mathbf{x}_{i-1}(t-2) + \mathbf{C}_{i,i-1} \mathbf{x}_{i-1}(t-1) \right] \\
&\;\;\vdots
\end{aligned}$$

$$
\begin{aligned}
&= \mathbf{L}_i^t \mathbf{x}_i(0) + \big[\mathbf{L}_i^{t-1}\mathbf{C}_{i,i-1}\mathbf{x}_{i-1}(0) + \mathbf{L}_i^{t-2}\mathbf{C}_{i,i-1}\mathbf{x}_{i-1}(1) + \cdots \\
&\qquad + \mathbf{L}_i^{1}\mathbf{C}_{i,i-1}\mathbf{x}_{t-2}(1) + \mathbf{C}_{i,i+1}\mathbf{x}_{i-1}(t-1)\big] \\
&= \mathbf{L}_i^t \mathbf{x}_i(0) + \sum_{k=0}^{t-1} \mathbf{L}_i^{t-1-k}\mathbf{C}_{i,i-1}\mathbf{x}_{i-1}(k) \\
&= \mathbf{L}_i^t \mathbf{x}_i(0) + \mathbf{L}_i^{t-1}\sum_{k=0}^{t-1} \mathbf{L}_i^{-k}\mathbf{C}_{i,i-1}\mathbf{x}_{i-1}(k).
\end{aligned}
$$

Replacing $\mathbf{L}_i^{-k}$ with $\mathbf{V}_i \Lambda_i^{-k} \mathbf{V}_i^{-1}$ in this final expression gives (5.55). □

**Lemma 5.2** *Assume Condition 5.1 holds for* (5.31) *and each* $\mathbf{L}_i$ *is diagonalized by* $\mathbf{L}_i = \mathbf{V}_i \Lambda_i \mathbf{V}_i^{-1}$. *For any matrix* $\mathbf{B} \in \mathbb{C}^{d_i \times d_j}$, *the following equality holds for any* $i, j \in \{1, \ldots, n\}$ *with* $i \neq j$:

$$
\sum_{k=0}^{t-1} \Lambda_i^{-k} \mathbf{B} \Lambda_j^k = \tilde{\mathbf{B}} - \Lambda_i^{-t} \tilde{\mathbf{B}} \Lambda_j^t, \tag{5.56}
$$

*where* $\tilde{\mathbf{B}} \in \mathbb{C}^{d_i \times d_j}$ *is the matrix whose* $(\ell, m)$*th entry is given by*

$$
[\tilde{\mathbf{B}}]_{\ell,m} = [\mathbf{B}]_{\ell,m} \left(1 - \frac{\lambda_{j,m}}{\lambda_{i,\ell}}\right)^{-1}. \tag{5.57}
$$

*Proof* For any matrix $\mathbf{M}$, we denote the $(\ell, m)$th entry as $[\mathbf{M}]_{\ell,m}$. The $(\ell, m)$th entry of (5.56) is given by

$$
\begin{aligned}
[\Lambda_i^{-k} \mathbf{B} \Lambda_j^k]_{\ell,m} &= \sum_{s=1}^{d_i} [\Lambda_i^{-k}]_{\ell,s} [\mathbf{B}\Lambda_j^k]_{s,m} \\
&= \sum_{s=1}^{d_i} [\Lambda_i^{-k}]_{\ell,s} \sum_{u=1}^{d_j} [\mathbf{B}]_{s,u} [\Lambda_j^k]_{u,m}.
\end{aligned}
$$

Since $\Lambda_i$ is diagonal, $[\Lambda_j^k]_{u,m} = 0$ for $u \neq m$ and $[\Lambda_j^k]_{m,m} = \lambda_{j,m}^k$. This gives

$$
[\Lambda_i^{-k} \mathbf{B} \Lambda_j^k]_{\ell,m} = \sum_{s=1}^{d_i} [\Lambda_i^{-k}]_{\ell,s} [\mathbf{B}]_{s,m} \lambda_{j,m}^k.
$$

Since $\Lambda_i^{-k}$ is diagonal, we have

$$
[\Lambda_i^{-k} \mathbf{B} \Lambda_j^k]_{\ell,m} = \lambda_{i,\ell}^{-k} [\mathbf{B}]_{\ell,m} \lambda_{j,m}^k = [\mathbf{B}]_{\ell,m} \left(\frac{\lambda_{j,m}}{\lambda_{i,\ell}}\right)^k. \tag{5.58}
$$

Summing from $k = 0, \ldots, t-1$, gives

$$\begin{aligned}
\sum_{k=0}^{t-1} [\Lambda_i^{-k} \mathbf{B} \Lambda_j^k]_{\ell,m} &= \sum_{k=0}^{t-1} [\mathbf{B}]_{\ell,m} \left( \frac{\lambda_{j,m}}{\lambda_{i,\ell}} \right)^k \\
&= [\mathbf{B}]_{\ell,m} \frac{1 - \left( \frac{\lambda_{j,m}}{\lambda_{i,\ell}} \right)^t}{1 - \left( \frac{\lambda_{j,m}}{\lambda_{i,\ell}} \right)} \\
&= [\tilde{\mathbf{B}}]_{\ell,m} - [\tilde{\mathbf{B}}]_{\ell,m} \left( \frac{\lambda_{j,m}}{\lambda_{i,\ell}} \right)^t .
\end{aligned}$$

Using (5.58), but with $\mathbf{B}$ and $k$ replaced by $\tilde{\mathbf{B}}$ and $t$, respectively, we get

$$[\tilde{\mathbf{B}}]_{\ell,m} \left( \frac{\lambda_{j,m}}{\lambda_{i,\ell}} \right)^t = [\Lambda_i^{-t} \tilde{B} \Lambda_j^t]_{\ell,m}.$$

Therefore,

$$\begin{aligned}
\left[ \left( \sum_{k=0}^{t-1} \Lambda_i^{-k} \mathbf{B} \Lambda_j^k \right) \right]_{\ell,m} &= \sum_{k=0}^{t-1} [\Lambda_i^{-k} \mathbf{B} \Lambda_j^k]_{\ell,m} \\
&= [\tilde{\mathbf{B}}]_{\ell,m} - [\tilde{\mathbf{B}}]_{\ell,m} \left( \frac{\lambda_{j,m}}{\lambda_{i,\ell}} \right)^t \\
&= [\tilde{\mathbf{B}}]_{\ell,m} - [\Lambda_i^{-t} \tilde{\mathbf{B}} \Lambda_j^t]_{\ell,m} \\
&= [\tilde{\mathbf{B}} - \Lambda_i^{-t} \tilde{\mathbf{B}} \Lambda_j^t]_{\ell,m}.
\end{aligned}$$

This is equivalent to (5.56). □

**Lemma 5.3** *For each* $i = 2, \ldots, n$, *the solution of* (5.31) *is*

$$\Pi_i \circ \mathsf{Lin}^{\circ t}(\mathbf{x}_1, \ldots, \mathbf{x}_n) = \sum_{j=1}^{i} (-1)^{i-j} \mathbf{D}_{i,j} \mathbf{L}_j^t \mathsf{pert}_j(\mathbf{x}_1, \ldots, \mathbf{x}_j), \tag{5.59}$$

*where*

$$\mathbf{D}_{i,i} = \mathbf{I}_{d_i} \qquad i \in \{1, \ldots, n\}, \tag{5.60}$$

$$\mathbf{D}_{i,j} = \mathbf{L}_i^{-1} \mathbf{V}_i \tilde{\mathbf{C}}_{i,j} \mathbf{V}_j^{-1} \qquad i \in \{2, \ldots, n\},\ j \in \{1, \ldots, i-1\}, \tag{5.61}$$

*and the matrix* $\tilde{\mathbf{C}}_{i,j} \in \mathbb{C}^{d_i \times d_j}$ *has elements*

$$[\tilde{\mathbf{C}}_{i,j}]_{\ell,m} = \left[\mathbf{V}_i^{-1}\mathbf{C}_{i,i-1}\mathbf{D}_{i-1,j}\mathbf{V}_j\right]_{\ell,m}\left(1-\frac{\lambda_{j,m}}{\lambda_{i,\ell}}\right)^{-1}, \quad i \in \{2,\dots,n\}, \quad j \in \{1,\dots,i-1\}. \tag{5.62}$$

*The perturbation functions* $\mathsf{pert}_i : \mathbb{C}^{d_1}\times\cdots\times\mathbb{C}^{d_i} \to \mathbb{C}^{d_i}$ *are multilinear maps defined inductively by*

$$\mathsf{pert}_1(\mathbf{x}_1) = \mathbf{x}_1 \tag{5.63}$$

$$\mathsf{pert}_i(\mathbf{x}_1,\dots,\mathbf{x}_i) = \mathbf{x}_i + \sum_{j=1}^{i-1}(-1)^{i-1-j}\mathbf{D}_{i,j}\mathsf{pert}_j(\mathbf{x}_1,\dots,\mathbf{x}_j) \quad i \in \{2,\dots,n\}. \tag{5.64}$$

*Proof* We prove the result using induction. First note that the solution for $\Pi_1 \circ \mathsf{Lin}^{\circ t}(\mathbf{x}_1,\dots,\mathbf{x}_n)$ can be written as

$$\mathbf{x}_1(t) = \Pi_1 \circ \mathsf{Lin}^{\circ t}(\mathbf{x}_1,\dots,\mathbf{x}_n) = \mathbf{L}_1^t\mathbf{x}_1 \equiv \mathbf{D}_{1,1}\mathbf{L}_1^t\mathsf{pert}_1(\mathbf{x}_1). \tag{5.65}$$

**Seed step:** Consider $\mathbf{x}_2(t) = \Pi_2 \circ \mathsf{Lin}^{\circ t}(\mathbf{x}_1,\dots,\mathbf{x}_n)$. By Lemma 5.1, Eq. (5.55), this is

$$\begin{aligned}\mathbf{x}_2(t) &= \mathbf{L}_2^t\mathbf{x}_2 + \mathbf{L}_2^{t-1}\mathbf{V}_2\sum_{k=0}^{t-1}\Lambda_2^{-k}\mathbf{V}_2^{-1}\mathbf{C}_{2,1}\mathbf{x}_1(k)\\ &= \mathbf{L}_2^t\mathbf{x}_2 + \mathbf{L}_2^{t-1}\mathbf{V}_2\sum_{k=0}^{t-1}\Lambda_2^{-k}\mathbf{V}_2^{-1}\mathbf{C}_{2,1}\mathbf{D}_{1,1}\mathbf{L}_1^k\mathsf{pert}_1(\mathbf{x}_1),\end{aligned} \tag{5.66}$$

where in the second line we have replaced $\mathbf{x}_1(k)$ with (5.65) for $t = k$. Using $\mathbf{L}_1^k = \mathbf{V}_1\Lambda_1^k\mathbf{V}_1^{-1}$ in the second line gives

$$\mathbf{x}_2(t) = \mathbf{L}_2^t\mathbf{x}_2 + \mathbf{L}_2^{t-1}\mathbf{V}_2\left(\sum_{k=0}^{t-1}\Lambda_2^{-k}\mathbf{V}_2^{-1}\mathbf{C}_{2,1}\mathbf{D}_{1,1}\mathbf{V}_1\Lambda_1^k\right)\mathbf{V}_1^{-1}\mathsf{pert}_1(\mathbf{x}_1). \tag{5.67}$$

Lemma 5.2, (5.56), with $\mathbf{B} \equiv \mathbf{V}_2^{-1}\mathbf{C}_{2,1}\mathbf{D}_{1,1}\mathbf{V}_1$ gives that

$$\left(\sum_{k=0}^{t-1}\Lambda_2^{-k}\mathbf{V}_2^{-1}\mathbf{C}_{2,1}\mathbf{D}_{1,1}\mathbf{V}_1\Lambda_1^k\right) = \tilde{\mathbf{C}}_{2,1} - \Lambda_2^{-t}\tilde{\mathbf{C}}_{2,1}\Lambda_1^t, \tag{5.68}$$

where the elements of $\tilde{\mathbf{C}}_{2,1}$ are given as

$$[\tilde{\mathbf{C}}_{2,1}]_{\ell,m} = [\mathbf{V}_2^{-1}\mathbf{C}_{2,1}\mathbf{D}_{1,1}\mathbf{V}_1]_{\ell,m}\left(1-\frac{\lambda_{1,m}}{\lambda_{2,\ell}}\right)^{-1}, \quad \ell \in \{1,\dots,d_2\}, m \in \{1,\dots,d_1\}. \tag{5.69}$$

Equation (5.69) is the same as (5.62) for $i = 2$. Using (5.68) in (5.67) gives

$$\begin{aligned} \mathbf{x}_2(t) &= \mathbf{L}_2^t \mathbf{x}_2 + \mathbf{L}_2^{t-1} \mathbf{V}_2 \left( \tilde{\mathbf{C}}_{2,1} - \Lambda_2^{-t} \tilde{\mathbf{C}}_{2,1} \Lambda_1^t \right) V_1^{-1} \mathsf{pert}_1(\mathbf{x}_1) \\ &= \mathbf{L}_2^t \left( \mathbf{x}_2 + \mathbf{L}_2^{-1} \mathbf{V}_2 \tilde{\mathbf{C}}_{2,1} \mathbf{V}_1^{-1} \mathsf{pert}_1(\mathbf{x}_1) \right) - \mathbf{L}_2^{t-1} \mathbf{V}_2 \Lambda_2^{-t} \tilde{\mathbf{C}}_{2,1} \Lambda_1^t \mathbf{V}_1^{-1} \mathsf{pert}_1(\mathbf{x}_1). \end{aligned}$$

Since $\mathbf{L}_2^{t-1} \mathbf{V}_2 \Lambda_2^{-t} = \mathbf{L}_2^{-1} \mathbf{V}_2$ and $\Lambda_1^t \mathbf{V}_1^{-1} = \mathbf{V}_1^{-1} \mathbf{L}_1^t$, we get

$$\mathbf{x}_2(t) = \mathbf{L}_2^t \left( \mathbf{x}_2 + \mathbf{L}_2^{-1} \mathbf{V}_2 \tilde{\mathbf{C}}_{2,1} \mathbf{V}_1^{-1} \mathsf{pert}_1(\mathbf{x}_1) \right) - \mathbf{L}_2^{-1} \mathbf{V}_2 \tilde{\mathbf{C}}_{2,1} \mathbf{V}_1^{-1} \mathbf{L}_1^t \mathsf{pert}_1(\mathbf{x}_1). \tag{5.70}$$

Defining $\mathbf{D}_{2,1}$ as

$$\mathbf{D}_{2,1} = \mathbf{L}_2^{-1} \mathbf{V}_2 \tilde{\mathbf{C}}_{2,1} \mathbf{V}_1^{-1} \tag{5.71}$$

and $\mathsf{pert}_2 : \mathbb{C}^{d_1} \times \mathbb{C}^{d_2} \to \mathbb{C}^{d_2}$ as

$$\mathsf{pert}_2(\mathbf{x}_1, \mathbf{x}_2) = \mathbf{x}_2 + \mathbf{D}_{2,1} \mathsf{pert}_1(\mathbf{x}_1) \tag{5.72}$$

gives

$$\begin{aligned} \mathbf{x}_2(t) &= \mathbf{L}_2^t \mathsf{pert}_2(\mathbf{x}_1, \mathbf{x}_2) - \mathbf{D}_{2,1} \mathbf{L}_1^t \mathsf{pert}_1(\mathbf{x}_1) \\ &= (-1)^0 \mathbf{D}_{2,2} \mathbf{L}_2^t \mathsf{pert}_2(\mathbf{x}_1, \mathbf{x}_2) - \mathbf{D}_{2,1} \mathbf{L}_1^t \mathsf{pert}_1(\mathbf{x}_1), \end{aligned}$$

since $\mathbf{D}_{2,2} = \mathbf{I}_{d_2}$ by definition. Finally,

$$\mathbf{x}_2(t) = \sum_{s=0}^{1} (-1)^s \mathbf{D}_{2,2-s} \mathbf{L}_{2-s}^t \mathsf{pert}_{2-s}(\mathbf{x}_1, \ldots, \mathbf{x}_{2-s}). \tag{5.73}$$

Using the change of variables $j = 2 - s$, we have that

$$\mathbf{x}_2(t) = \sum_{j=1}^{2} (-1)^{2-j} \mathbf{D}_{2,j} \mathbf{L}_j^t \mathsf{pert}_j(\mathbf{x}_1, \ldots, \mathbf{x}_j). \tag{5.74}$$

Equations (5.71)–(5.74) are equivalent to Eqs. (5.59), (5.61), and (5.64), for $j = 2$.

**Induction step:** Assume (5.59)–(5.64) hold for for all $j \leq i$ where $i \in \{2, \ldots, n-1\}$. We show they hold for $i + 1$ as well.

Write $\mathbf{x}_{i+1}(t) = \Pi_{i+1} \mathsf{Lin}^{\circ t}(\mathbf{x}_1, \ldots, \mathbf{x}_n)$. By Lemma 5.1, Eq. (5.55), the solution is

$$\mathbf{x}_{i+1}(t) = \mathbf{L}_{i+1}^t \mathbf{x}_{i+1} + \mathbf{L}_{i+1}^{t-1} \mathbf{V}_{i+1} \sum_{k=0}^{t-1} \Lambda_{i+1}^{-k} \mathbf{V}_{i+1}^{-1} \mathbf{C}_{i+1,i} \mathbf{x}_i(k).$$

By the induction hypothesis,

$$\mathbf{x}_i(k) = \sum_{j=1}^{i} (-1)^{i-j} \mathbf{D}_{i,j} \mathbf{L}_j^k \mathsf{pert}_j(\mathbf{x}_1, \ldots, \mathbf{x}_j). \tag{5.75}$$

which gives that $\mathbf{x}_{i+1}(t)$ is (after interchanging the finite sums)

$$\mathbf{x}_{i+1}(t) = \mathbf{L}_{i+1}^t \mathbf{x}_{i+1} + \sum_{j=1}^{i} (-1)^{i-j} \mathbf{L}_{i+1}^{t-1} \mathbf{V}_{i+1} \sum_{k=0}^{t-1} \Lambda_{i+1}^{-k} \mathbf{V}_{i+1}^{-1} \mathbf{C}_{i+1,i} \mathbf{D}_{i,j} \mathbf{L}_j^k \mathsf{pert}_j(\mathbf{x}_1, \ldots, \mathbf{x}_j).$$

Since $\mathbf{L}_j$ is diagonalizable, we substitute $\mathbf{V}_j \Lambda_j^k \mathbf{V}_j^{-1}$ for $\mathbf{L}_j^k$ in the above equation to get

$$\begin{aligned} &\mathbf{x}_{i+1}(t) \\ &\quad = \mathbf{L}_{i+1}^t \mathbf{x}_{i+1} \\ &\quad\quad + \sum_{j=1}^{i} (-1)^{i-j} \mathbf{L}_{i+1}^{t-1} \mathbf{V}_{i+1} \left( \sum_{k=0}^{t-1} \Lambda_{i+1}^{-k} \mathbf{V}_{i+1}^{-1} \mathbf{C}_{i+1,i} \mathbf{D}_{i,j} \mathbf{V}_j \Lambda_j^k \right) \mathbf{V}_j^{-1} \mathsf{pert}_j(\mathbf{x}_1, \ldots, \mathbf{x}_j). \end{aligned} \tag{5.76}$$

By Lemma 5.2, Eq. (5.56), with $\mathbf{B} \equiv \mathbf{V}_{i+1}^{-1} \mathbf{C}_{i+1,i} \mathbf{D}_{i,j} \mathbf{V}_j$, we have

$$\sum_{k=0}^{t-1} \Lambda_{i+1}^{-k} \mathbf{V}_{i+1}^{-1} \mathbf{C}_{i+1,i} \mathbf{D}_{i,j} \mathbf{V}_j \Lambda_j^k = \tilde{\mathbf{C}}_{i+1,j} - \Lambda_{i+1}^{-t} \tilde{\mathbf{C}}_{i+1,j} \Lambda_j^t, \tag{5.77}$$

where for $j \in \{1, \ldots, i\}$ the matrix $\tilde{\mathbf{C}}_{i+1,j} \in \mathbb{C}^{d_{i+1} \times d_j}$ has elements

$$\left[ \tilde{\mathbf{C}}_{i+1,j} \right]_{\ell,m} = \left[ \mathbf{V}_{i+1}^{-1} \mathbf{C}_{i+1,i} \mathbf{D}_{i,j} \mathbf{V}_j \right]_{\ell,m} \left( 1 - \frac{\lambda_{j,m}}{\lambda_{i+1,\ell}} \right)^{-1}. \tag{5.78}$$

Equation (5.78) is (5.62) for $i + 1$. Plugging (5.77) into (5.76) gives

$$\begin{aligned} &\mathbf{x}_{i+1}(t) \\ &\quad = \mathbf{L}_{i+1}^t \mathbf{x}_{i+1} \\ &\quad\quad + \sum_{j=1}^{i} (-1)^{i-j} \mathbf{L}_{i+1}^{t-1} \mathbf{V}_{i+1} \left( \tilde{\mathbf{C}}_{i+1,j} - \Lambda_{i+1}^{-t} \tilde{\mathbf{C}}_{i+1,j} \Lambda_j^t \right) \mathbf{V}_j^{-1} \mathsf{pert}_j(\mathbf{x}_1, \ldots, \mathbf{x}_j) \\ &\quad = \mathbf{L}_{i+1}^t \mathbf{x}_{i+1} + \sum_{j=1}^{i} (-1)^{i-j} \mathbf{L}_{i+1}^{t-1} \mathbf{V}_{i+1} \tilde{\mathbf{C}}_{i+1,j} \mathbf{V}_j^{-1} \mathsf{pert}_j(\mathbf{x}_1, \ldots, \mathbf{x}_j) \\ &\quad\quad + \sum_{j=1}^{i} (-1)^{i+1-j} \mathbf{L}_{i+1}^{t-1} \mathbf{V}_{i+1} \Lambda_{i+1}^{-t} \tilde{\mathbf{C}}_{i+1,j} \Lambda_j^t \mathbf{V}_j^{-1} \mathsf{pert}_j(\mathbf{x}_1, \ldots, \mathbf{x}_j) \end{aligned}$$

$$= \mathbf{L}_{i+1}^{t} \left[ \mathbf{x}_{i+1} + \sum_{j=1}^{i} (-1)^{i-j} \mathbf{L}_{i+1}^{-1} \mathbf{V}_{i+1} \tilde{\mathbf{C}}_{i+1,j} \mathbf{V}_j^{-1} \mathsf{pert}_j(\mathbf{x}_1, \ldots, \mathbf{x}_j) \right]$$
$$+ \sum_{j=1}^{i} (-1)^{i+1-j} \mathbf{L}_{i+1}^{t-1} \mathbf{V}_{i+1} \Lambda_{i+1}^{-t} \tilde{\mathbf{C}}_{i+1,j} \Lambda_j^{t} \mathbf{V}_j^{-1} \mathsf{pert}_j(\mathbf{x}_1, \ldots, \mathbf{x}_j).$$

Since $\mathbf{L}_{i+1}^{t-1} \mathbf{V}_{i+1} \Lambda_{i+1}^{-t} = \mathbf{L}_{i+1}^{-1} \mathbf{V}_{i+1}$ and $\Lambda_j^t \mathbf{V}_j^{-1} = \mathbf{V}_j^{-1} \mathbf{L}_j^t$, then

$$\mathbf{x}_{i+1}(t) = \mathbf{L}_{i+1}^{t} \left[ \mathbf{x}_{i+1} + \sum_{j=1}^{i} (-1)^{i-j} \left( \mathbf{L}_{i+1}^{-1} \mathbf{V}_{i+1} \tilde{\mathbf{C}}_{i+1,j} \mathbf{V}_j^{-1} \right) \mathsf{pert}_j(\mathbf{x}_1, \ldots, \mathbf{x}_j) \right]$$
$$+ \sum_{j=1}^{i} (-1)^{i+1-j} \left( \mathbf{L}_{i+1}^{-1} \mathbf{V}_{i+1} \tilde{\mathbf{C}}_{i+1,j} \mathbf{V}_j^{-1} \right) \mathbf{L}_j^t \mathsf{pert}_j(\mathbf{x}_1, \ldots, \mathbf{x}_j).$$

For $j = 1, \ldots, i$, define

$$\mathbf{D}_{i+1,j} = \mathbf{L}_{i+1}^{-1} \mathbf{V}_{i+1} \tilde{\mathbf{C}}_{i+1,j} \mathbf{V}_j^{-1} \tag{5.79}$$

as in Eq. (5.61) and $\mathsf{pert}_{i+1} : \mathbb{C}^{d_1} \times \cdots \times \mathbb{C}^{d_{i+1}} \to \mathbb{C}^{d_{i+1}}$ as

$$\mathsf{pert}_{i+1}(\mathbf{x}_1, \ldots, \mathbf{x}_n) = \mathbf{x}_{i+1} + \sum_{j=1}^{i} (-1)^{i-j} \mathbf{L}_{i+1}^{-1} \mathbf{V}_{i+1} \tilde{\mathbf{C}}_{i+1,j} \mathbf{V}_j^{-1} \mathsf{pert}_j(\mathbf{x}_1, \ldots, \mathbf{x}_j)$$
$$= \mathbf{x}_{i+1} + \sum_{j=1}^{i} (-1)^{i-j} \mathbf{D}_{i+1,j} \mathsf{pert}_j(\mathbf{x}_1, \ldots, \mathbf{x}_j)$$

as in Eq. (5.64) with the substitution $i \mapsto i + 1$. Substituting these definitions into the expression for the solution $\mathbf{x}_{i+1}(t)$ and defining $\mathbf{D}_{i+1,i+1} = \mathbf{I}_{d_{i+1}}$, we have

$$\mathbf{x}_{i+1}(t) = \mathbf{L}_{i+1}^{t} \mathsf{pert}_{i+1}(\mathbf{x}_1, \ldots, \mathbf{x}_{i+1}) + \sum_{j=1}^{i} (-1)^{i+1-j} \mathbf{D}_{i+1,j} \mathbf{L}_j^t \mathsf{pert}_j(\mathbf{x}_1, \ldots, \mathbf{x}_j)$$
$$= (-1)^0 \mathbf{D}_{i+1,i+1} \mathbf{L}_{i+1}^t \mathsf{pert}_{i+1}(\mathbf{x}_1, \ldots, \mathbf{x}_{i+1})$$
$$+ \sum_{j=1}^{i} (-1)^{i+1-j} \mathbf{D}_{i+1,j} \mathbf{L}_j^t \mathsf{pert}_j(\mathbf{x}_1, \ldots, \mathbf{x}_j)$$
$$= \sum_{j=1}^{i+1} (-1)^{i+1-j} \mathbf{D}_{i+1,j} \mathbf{L}_j^t \mathsf{pert}_j(\mathbf{x}_1, \ldots, \mathbf{x}_j).$$

Comparing with (5.59) with the substitution $i \mapsto i+1$, we see that the induction is complete. This completes the proof. □

**Corollary 5.3** *Assume that Condition 5.1 holds for* (5.31). *Then for all* $i \in \{2, \ldots, n\}$ *and* $t \in \mathbb{N}$,

$$\left\| \Pi_i \circ \mathsf{Lin}^{\circ t}(\mathbf{x}_1, \ldots, x_n) - \mathbf{L}_i^t(\mathsf{pert}_i(\mathbf{x}_1, \ldots, \mathbf{x}_i)) \right\| \leq \sum_{j=1}^{i-1} \|\mathbf{D}_{i,j}\| \|\mathbf{L}_j^t \mathsf{pert}_j(\mathbf{x}_1, \ldots, \mathbf{x}_j)\|, \quad (5.80)$$

*where* $\mathbf{D}_{i,j}$ *and* $\mathsf{pert}_j$ *are given by* (5.61) *and* (5.64). *Furthermore,*

$$\lim_{t\to\infty} \frac{\left\| \Pi_i \circ \mathsf{Lin}^{\circ t}(\mathbf{x}_1, \ldots, \mathbf{x}_n) - \mathbf{L}_i^t(\mathsf{pert}_i(\mathbf{x}_1, \ldots, \mathbf{x}_i)) \right\|}{\|\mathbf{L}_i\|^t} = 0. \quad (5.81)$$

*Proof* Inequality (5.80) follows directly from Lemma 5.3, Eq. (5.59) and the fact that $\mathbf{D}_{i,i} = \mathbf{I}_{d_i}$. Equation (5.81) follows from the condition 5.1, equation (5.34).

### 5.6.2 *Proof of Theorem 5.3: Perturbation of Principal Eigenfunctions—Nominal, Linear System*

We now prove Theorem 5.3. It is a straightforward application of Theorem 5.2.

*Proof* (*Proof of Theorem* 5.3) We first show that for $i \geq 1$ and $t \geq 0$ that (5.45) holds. By definition,

$$\begin{aligned}
&U_{\mathsf{Lin}}^t \phi_{(0,\ldots,0,s_i,0,\ldots,0)}(\mathbf{x}_1, \ldots, \mathbf{x}_n) \\
&\quad - \phi_{(0,\ldots,0,s_i,0,\ldots,0)}(\mathsf{Lin}^{\circ t}(\mathbf{x}_1, \ldots, \mathbf{x}_n)) \\
&\quad = \phi_{i,s_i}(\Pi_i \circ \mathsf{Lin}^{\circ t}(\mathbf{x}_1, \ldots, \mathbf{x}_n)) \\
&\quad = \phi_{i,s_i}(\Pi_i \circ \mathsf{Nom}^{\circ t}(\mathsf{pert}(\mathbf{x}_1, \ldots, \mathbf{x}_n))) \\
&\qquad + \phi_{i,s_i}(\Pi_i \circ \mathsf{Lin}^{\circ t}(\mathbf{x}_1, \ldots, \mathbf{x}_n) - \Pi_i \circ \mathsf{Nom}^{\circ t}(\mathsf{pert}(\mathbf{x}_1, \ldots, \mathbf{x}_n))) \\
&\quad = (U_{\mathsf{Nom}}^t(\phi_{i,s_i} \circ \Pi_i)) \circ \mathsf{pert}(\mathbf{x}_1, \ldots, \mathbf{x}_n) \\
&\qquad + \phi_{i,s_i}(\Pi_i \circ \mathsf{Lin}^{\circ t}(\mathbf{x}_1, \ldots, \mathbf{x}_n) - \Pi_i \circ \mathsf{Nom}^{\circ t}(\mathsf{pert}(\mathbf{x}_1, \ldots, \mathbf{x}_n))) \\
&\quad = (U_{\mathsf{Nom}}^t \phi_{(0,\ldots,0,s_i,0,\ldots,0)}) \circ \mathsf{pert}(\mathbf{x}_1, \ldots, \mathbf{x}_n) \\
&\qquad + \phi_{i,s_i}(\Pi_i \circ \mathsf{Lin}^{\circ t}(\mathbf{x}_1, \ldots, \mathbf{x}_n) - \Pi_i \circ \mathsf{Nom}^{\circ t}(\mathsf{pert}(\mathbf{x}_1, \ldots, \mathbf{x}_n))).
\end{aligned}$$

Therefore,

$$\left|U^t_{\mathsf{Lin}}\phi_{(0,\dots,0,s_i,0,\dots,0)}(\mathbf{x}_1,\dots,\mathbf{x}_n) - U^t_{\mathsf{Nom}}\phi_{(0,\dots,0,s_i,0,\dots,0)}(\mathsf{pert}(\mathbf{x}_1,\dots,\mathbf{x}_n))\right| \\ \leq \|\phi_{i,s_i}\| \left\|\Pi_i \circ \mathsf{Lin}^{\circ t}(\mathbf{x}_1,\dots,\mathbf{x}_n) - \Pi_i \circ \mathsf{Nom}^{\circ t}(\mathsf{pert}(\mathbf{x}_1,\dots,\mathbf{x}_n))\right\|. \quad (5.82)$$

By Theorem 5.2, Eq. (5.36), for all $t \geq 0$ and $i \geq 1$,

$$\left\|\Pi_i \circ \mathsf{Lin}^{\circ t}(\mathbf{x}_1,\dots,\mathbf{x}_n) - \Pi_i \circ \mathsf{Nom}^{\circ t}(\mathsf{pert}(\mathbf{x}_1,\dots,\mathbf{x}_n))\right\| \\ \leq \sum_{j=1}^{i-1} \|\mathbf{D}_{i,j}\| \|\mathbf{L}_j^t \mathsf{pert}_j(\mathbf{x}_1,\dots,\mathbf{x}_j)\|. \quad (5.83)$$

This estimate along with (5.82) gives (5.45).

By Theorem 5.2, Eq. (5.43), for all $i \in \{1,\dots,n\}$ and any $\varepsilon > 0$,

$$\frac{\left\|\Pi_i \circ \mathsf{Lin}^{\circ t}(\mathbf{x}_1,\dots,\mathbf{x}_n) - \Pi_i \circ \mathsf{Nom}^{\circ t}(\mathsf{pert}(\mathbf{x}_1,\dots,\mathbf{x}_n))\right\|}{\|\mathbf{L}_i\|^t} \leq \varepsilon, \quad (5.84)$$

for all $t$ large enough. This is equivalent to (5.46). □

The Generalized Laplace Analysis theorem uses Laplace averages of the Koopman operator to project a function onto an eigenspace. This is is the proof of Corollary 5.2. The following definition and theorem are taken from [14].

**Definition 5.5** *(Dominating point spectrum)* For $r > 0$, let $\mathbb{D}_r$ be the open disc of radius $r$ centered at 0 in the complex plane and let $\sigma(U;\mathbb{D}_r) = \sigma(U) \cap \mathbb{D}_r$. If there exists an $R > 0$ such that $\sigma(U) \setminus \mathbb{D}_R$ is not empty and for every $r > R$, we have

1. if $\sigma(U;\mathbb{D}_r) \cap \sigma_p(U) \neq \emptyset$, then the peripheral spectrum of $\sigma(U;\mathbb{D}_r)$ is not empty, and
2. the set $\sigma(U) \setminus \mathbb{D}_r$ consists only of eigenvalues (i.e., $\sigma(U) \setminus \mathbb{D}_r \subset \sigma_p(U)$).

**Theorem 5.6** *Let $\sigma(U)$ have a dominating point spectrum and assume that the point spectrum is concentrated on isolated circles in the complex plane. Let $\lambda$ be an eigenvalue of $U$. The projection $P_\lambda$ onto the $N(\lambda I - U)$, the $\lambda$-eigenspace of $U$, can be computed as*

$$P_\lambda = \lim_{n\to\infty} \frac{1}{n} \sum_{k=0}^{n-1} \lambda^{-k} U^k \left( I - \sum_{\mu\in\Omega} P_\mu \right), \quad (5.85)$$

*where the limit exists in the strong operator topology and where $\Omega = \{\mu \in \sigma_p(U) : |\mu| > |\lambda|\}$.*

### 5.6.3 *Proof of Theorem 5.4: Asymptotic Equivalence for Nonlinear, Chained Cascades*

We now prove Theorem 5.4. It is a straightforward application of Theorem 5.2 and the fact that the topological conjugacy is a homeomorphism.

*Proof* (*Proof of Theorem* 5.4) Fix $\varepsilon > 0$ and $\mathbf{Y} = (\mathbf{y}_1, \ldots, \mathbf{y}_n) \in \mathbb{C}^{d_1} \times \cdots \times \mathbb{C}^{d_n}$. Define $\mathbf{X} = \tau^{-1}(\mathbf{Y})$. Denote by $\overline{B_i}$ the closed unit ball of radius centered at the origin in $\mathbb{C}^{d_i}$. Condition 5.1 and Theorem 5.2, Eq. (5.43) guarantee that $\mathsf{Lin}^{\circ t}(\mathbf{X})$ and $\mathsf{Nom}^{\circ t}(\mathsf{pert}(\mathbf{X}))$ are in the compact set $\overline{B_1} \times \cdots \times \overline{B_n}$ for all $t$ large enough.

Since $\tau$ is continuous, it is uniformly continuous on $\overline{B_1} \times \cdots \times \overline{B_r}$. Let $\delta > 0$ be such that if $\mathbf{X}, \mathbf{X}' \in \overline{B_1} \times \cdots \times \overline{B_n}$ and $\|\mathbf{X} - \mathbf{X}'\|_\times < \delta$, then $\|\tau(\mathbf{X}) - \tau(\mathbf{X}')\|_\times < \varepsilon$.

By Corollary 5.1, there is a $T \in \mathbb{N}$ such that $t \geq T$ implies

$$\|\mathsf{Lin}^{\circ t}(\mathbf{X}) - \mathsf{Nom}^{\circ t}(\mathsf{pert}(\mathbf{X}))\|_\times < \delta. \tag{5.86}$$

The uniform continuity of $\tau$ implies that

$$\|\tau \circ \mathsf{Lin}^{\circ t}(\mathbf{X}) - \tau \circ \mathsf{Nom}^{\circ t}(\mathsf{pert}(\mathbf{X}))\|_\times < \varepsilon, \qquad (t \geq T). \tag{5.87}$$

Now, since $\tau$ is a topological conjugacy, $\mathsf{Lin}^{\circ t} = \tau^{-1} \circ \mathsf{NonLin}^{\circ t} \circ \tau$. Plugging this into (5.87) gives, for all $t \geq T$,

$$\begin{aligned}
\varepsilon &> \|\tau \circ (\tau^{-1} \circ \mathsf{NonLin}^{\circ t} \circ \tau)(\mathbf{X}) - \tau \circ \mathsf{Nom}^{\circ t}(\mathsf{pert}(\mathbf{X}))\|_\times \\
&= \|(\mathsf{NonLin}^{\circ t}(\tau(\mathbf{X})) - \tau \circ \mathsf{Nom}^{\circ t}(\mathsf{pert}(\mathbf{X}))\|_\times \\
&= \|(\mathsf{NonLin}^{\circ t}(\tau(\mathbf{X})) - (\tau \circ \mathsf{Nom}^{\circ t} \circ \tau^{-1}) \circ \tau \circ (\mathsf{pert}(\mathbf{X}))\|_\times \\
&= \|(\mathsf{NonLin}^{\circ t})(\tau(\tau^{-1}(\mathbf{Y}))) - (\tau \circ \mathsf{Nom}^{\circ t} \circ \tau^{-1}) \circ \tau \circ (\mathsf{pert}(\tau^{-1}(\mathbf{Y}))\|_\times \\
&= \|\mathsf{NonLin}^{\circ t}(\mathbf{Y}) - (\tau \circ \mathsf{Nom}^{\circ t} \circ \tau^{-1}) \circ (\tau \circ \mathsf{pert} \circ \tau^{-1})(\mathbf{Y})\|_\times .
\end{aligned}$$

Therefore,

$$\lim_{t \to \infty} \|\mathsf{NonLin}^{\circ t}(\mathbf{Y}) - (\tau \circ \mathsf{Nom}^{\circ t} \circ \tau^{-1}) \circ (\tau \circ \mathsf{pert} \circ \tau^{-1})(\mathbf{Y})\|_\times = 0.$$

This completes the proof. □

### 5.6.4 *Proof of Theorem 5.5: Perturbation of Principal Eigenfunctions—Nominal, Nonlinear Cascades*

To save space in the following proof, we will write $\phi_{(0,\ldots,0,s_i,0,\ldots,0)}$ as $\phi_{s_i \mathbf{e}_{n,i}}$, where $\mathbf{e}_{n,i}$ is the $i$th canonical basis vector of length $n$.

*Proof* (*Proof of Theorem* 5.5) Fix $\mathbf{Y} = (\mathbf{y}_1, \ldots, \mathbf{y}_n) \in \mathbb{C}^{d_1} \times \cdots \times \mathbb{C}^{d_n}$ and let $\mathbf{X} = \tau^{-1}(\mathbf{Y})$. The topological conjugacy satisfies

$$\mathsf{Lin}^{\circ t}(\mathbf{X}) = (\tau^{-1} \circ \mathsf{NonLin}^{\circ t} \circ \tau)(\mathbf{X}). \tag{5.88}$$

Using this relation, we get

$$\begin{aligned} U^t_{\mathsf{Lin}} \phi_{s_i \mathbf{e}_{n,i}}(\mathbf{X}) &= \phi_{s_i \mathbf{e}_{n,i}}(\mathsf{Lin}^{\circ t}(\mathbf{X})) \\ &= \phi_{s_i \mathbf{e}_{n,i}}((\tau^{-1} \circ \mathsf{NonLin}^{\circ t} \circ \tau)(\mathbf{X})) \\ &= (\phi_{s_i \mathbf{e}_{n,i}} \circ \tau^{-1})(\mathsf{NonLin}^{\circ t}(\tau(\mathbf{X}))) \\ &= U^t_{\mathsf{NonLin}}(\phi_{s_i \mathbf{e}_{n,i}} \circ \tau^{-1})(\mathbf{Y}). \end{aligned}$$

On the other hand,

$$\begin{aligned} U^t_{\mathsf{Nom}} \phi_{s_i \mathbf{e}_{n,i}}(\mathsf{pert}(\mathbf{X})) &= \phi_{s_i \mathbf{e}_{n,i}}(\mathsf{Nom}^{\circ t}(\mathsf{pert}(\mathbf{X}))) \\ &= \phi_{s_i \mathbf{e}_{n,i}}(\tau^{-1} \circ \tau \circ \mathsf{Nom}^{\circ t} \circ \tau^{-1} \circ \tau(\mathsf{pert}(\mathbf{X}))) \\ &= (\phi_{s_i \mathbf{e}_{n,i}} \circ \tau^{-1})((\tau \circ \mathsf{Nom}^{\circ t} \circ \tau^{-1}) \circ \tau(\mathsf{pert}(\mathbf{X}))) \\ &= (\phi_{s_i \mathbf{e}_{n,i}} \circ \tau^{-1})((\tau \circ \mathsf{Nom}^{\circ t} \circ \tau^{-1})(\tau \circ \mathsf{pert} \circ \tau^{-1})(\mathbf{X})) \\ &= U^{\circ t}_{\tau \circ \mathsf{Nom}^t \circ \tau^{-1}}(\phi_{s_i \mathbf{e}_{n,i}} \circ \tau^{-1})((\tau \circ \mathsf{pert} \circ \tau^{-1})(\mathbf{Y})). \end{aligned}$$

Combining these two expression, we have

$$\begin{aligned} &\left| U^t_{\mathsf{NonLin}}(\phi_{s_i \mathbf{e}_{n,i}} \circ \tau^{-1})(\mathbf{Y}) - U^{\circ t}_{\tau \circ \mathsf{Nom}^{\circ t} \circ \tau^{-1}}(\phi_{s_i \mathbf{e}_{n,i}} \circ \tau^{-1})((\tau \circ \mathsf{pert} \circ \tau^{-1})(\mathbf{Y})) \right| \\ &\quad = \left| U^t_{\mathsf{Lin}} \phi_{s_i \mathbf{e}_{n,i}}(\mathbf{X}) - U^t_{\mathsf{Nom}} \phi_{s_i \mathbf{e}_{n,i}}(\mathsf{pert}(\mathbf{X})) \right|. \end{aligned}$$

Theorem 5.3, Eq. (5.46), implies

$$\begin{aligned} 0 &= \lim_{t \to \infty} \frac{\left| U^t_{\mathsf{Lin}} \phi_{s_i \mathbf{e}_{n,i}}(\mathbf{X}) - U^t_{\mathsf{Nom}} \phi_{s_i \mathbf{e}_{n,i}}(\mathsf{pert}(\mathbf{X})) \right|}{\|\mathbf{L}_i\|^t} \\ &= \lim_{t \to \infty} \frac{\left| U^t_{\mathsf{NonLin}}(\phi_{s_i \mathbf{e}_{n,i}} \circ \tau^{-1})(\mathbf{y}) - U^{\circ t}_{\tau \circ \mathsf{Nom}^t \circ \tau^{-1}}(\phi_{s_i \mathbf{e}_{n,i}} \circ \tau^{-1})((\tau \circ \mathsf{pert} \circ \tau^{-1})(\mathbf{Y})) \right|}{\|\mathbf{L}_i\|^t}. \end{aligned}$$

□

## References

1. Banaszuk, A., Fonoberov, V.A., Frewen, T.A., Kobilarov, M., Mathew, G., Mezić, I., Pinto, A., Sahai, T., Sane, H., Speranzon, A., Surana, A.: Scalable approach to uncertainty quantification and robust design of interconnected dynamical systems. Annu. Rev. Control **35**(1), 77–98 (2011)
2. Budisic, M., Mohr, R., Mezić, I.: Applied koopmanism. Chaos **22**(4), 047,510 (2012)

3. Callier, F., Chan, W., Desoer, C.: Input-output stability theory of interconnected systems using decomposition techniques. IEEE Trans. Circuits Syst. **23**(12), 714–729 (1976)
4. Heersink, B., Warren, M.A., Hoffmann, H.: Dynamic mode decomposition for interconnected control systems (2017). arXiv.org
5. Lan, Y., Mezić, I.: On the architecture of cell regulation networks. BMC Syst. Biol. **5**(1), 37 (2011)
6. Lasota, A., Mackey, M.C.: Chaos, Fractals, and Noise: Stochastic Aspects of Dynamics. Applied Mathematical Sciences, vol. 97, 2nd edn. Springer, Berlin (1994)
7. Mauroy, A., Mezić, I.: Global stability analysis using the eigenfunctions of the Koopman operator. IEEE Trans. Autom. Control **61**(11), 3356–3369 (2016)
8. Mesbahi, A., Haeri, M.: Conditions on decomposing linear systems with more than one matrix to block triangular or diagonal form. IEEE Trans. Autom. Control **60**(1), 233–239 (2015)
9. Mezić, I.: Coupled nonlinear dynamical systems: asymptotic behavior and uncertainty propagation. In: 43rd IEEE Conference on Decision and Control, pp. 1778–1783. Atlantis, Paradise Island, Bahamas (2004)
10. Mezić, I.: Spectral properties of dynamical systems, model reduction and decompositions. Nonlinear Dyn. **41**, 309–325 (2005)
11. Mezić, I., Banaszuk, A.: Comparison of systems with complex behavior: spectral methods. In: 39th IEEE Conference on Decision and Control, pp. 1224–1231. UCSB, Sydney, Australia (2000)
12. Mezić, I., Banaszuk, A.: Comparison of systems with complex behavior. Phys. D: Nonlinear Phenom. **197**(1), 101–133 (2004)
13. Michel, A.N.: On the status of stability of interconnected systems. IEEE Trans. Autom. Control **28**(6), 639–653 (1983)
14. Mohr, R., Mezić, I.: Construction of eigenfunctions for scalar-type operators via laplace averages with connections to the Koopman operator, 1–25 (2014). arXiv.org
15. Mohr, R., Mezić, I.: Koopman principal eigenfunctions and linearization of diffeomorphisms (2016). arXiv.org
16. Pichai, V., Sezer, M.E., Siljak, D.D.: A graph-theoretic algorithm for hierarchical decomposition of dynamic-systems with applications to estimation and control. IEEE Trans. Syst. Man Cybern. **13**(2), 197–207 (1983)
17. Ryan, R.A.: Introduction to Tensor Products of Banach Spaces. Springer Monographs in Mathematics. Springer, London (2002)
18. Shen-Orr, S.S., Milo, R., Mangan, S., Alon, U.: Network motifs in the transcriptional regulation network of Escherichia coli. Nature Genet. **31**(1), 64–68 (2002)

# Chapter 6
# Koopman Operator Theory for Nonautonomous and Stochastic Systems

**Senka Maćešić and Nelida Črnjarić-Žic**

**Abstract** In practice, the dynamics of open systems subject to time-dependent or random forcing is much more present than the dynamics of autonomous systems. Therefore, extension of the Koopman operator theory and applications to such systems is of great importance. At the same time, it brings a new viewpoint to the existing theory of nonautonomous as well as random dynamical systems, particularly, with application of Koopman-based data-driven algorithms. In this chapter, we first review the nonautonomous Koopman operator family based on the two standard nonautonomous dynamical system definitions: skew product and process. Then, we state basic properties of the operator and compare performance of the DMD and Arnoldi-type algorithms in the context of both definitions. In the case of the random dynamical systems (RDS), we introduce the associated stochastic Koopman operator family. We show that when RDS is Markovian, this family satisfies semigroup property and we present some properties for RDS generated by the stochastic differential equations. Finally, we discuss data-driven algorithms in the stochastic framework and illustrate their performance on numerical examples.

## 6.1 Introduction

The Koopman operator framework was first extended to the nonautonomous dynamical systems in [25], where a definition of the nonautonomous Koopman eigenvalues, eigenfunctions, and modes as the building blocks of the nonautonomous Koopman mode decompositions was introduced for the case of linear, periodic, and quasiperiodic nonautonomous systems. Also, in [25], numerical approximations of the associated time-dependent Koopman spectral objects were performed by EDMD and kernel-based DMD methods. Recently, different extensions of the DMD and EDMD

S. Maćešić (✉) · N. Črnjarić-Žic
Faculty of Engineering, University of Rijeka, Vukovarska 58, Rijeka, Croatia
e-mail: senka.macesic@riteh.hr

N. Črnjarić-Žic
e-mail: nelida@riteh.hr



A. Mauroy et al. (eds.), *The Koopman Operator in Systems and Control*,
Lecture Notes in Control and Information Sciences 484,
https://doi.org/10.1007/978-3-030-35713-9_6

algorithm have been proposed for application to particular types of nonautonomous dynamical systems: in [18], the multi resolution DMD for decomposing data with multiple timescales has been proposed with a successful application to the nonstationary data; in [12], a strategy inspired by time-changed dynamical systems that involves rescaling the generator was developed; etc.

The generalization of the Koopman operator to nonautonomous dynamical systems and the extension of the numerical methods, such as DMD and EDMD, was studied in recent papers [17, 26–28], however, with emphasis on input and control. The modification of EDMD capable to recover the Koopman spectral objects for the unforced systems in the presence of actuation was developed in [32].

The time-varying linear systems were studied in [34], where the online DMD and weighted DMD were developed to recover the approximations of time-dependent Koopman eigenvalues and eigenfunctions. Although the weighted approach used there has lead to better approximations of Koopman eigenvalues than the standard approach, it follows from Theorem 6.1, proved here, that there exists an intrinsic error when the DMD method with the state observables is applied.

Another possible generalization of the Koopman operator framework is its extension to the random dynamical systems (RDS). In practical applications, the discrete-time RDS generated by random difference equations, or the continuous-time RDS generated by random differential equations or stochastic differential equations could be of particular importance. In [19, 23, 24], the stochastic analogue of the Koopman operator was associated with the discrete RDS, which is the product of independent identically distributed random mappings. Since the one-step map is the generator of such discrete RDS, the stochastic Koopman operator was defined as the expected value of the one-step backward evolution of an observable function. The characterization of the dynamics of the systems using the eigenvalues and the eigenfunctions of the related generators was studied in recent papers [12, 29, 31]. Giannakis in [12], develops a framework for the KMD, based on the representation of the Koopman operator in a smooth orthonormal basis, determined from the time-ordered noisy data through the diffusion map algorithm. Using this representation, the Koopman eigenfunctions are determined as the eigenfunctions of the related advection–diffusion operator. A similar approach, by using the manifold learning technique via diffusion maps, was used in [29], to capture the inherent coordinates for building an intrinsic representation of the dynamics driven by the Langevin stochastic differential equation. The obtained coordinates are actually the approximations of the eigenfunctions of the stochastic Koopman generator.

A review and application of different numerical techniques for approximating the Perron–Frobenious and Koopman operators in nondeterministic settings, such as the Ulam's method, the generalized Galerkin method, and the EDMD method, is given in [16]. Williams et al. in [33], applied EDMD algorithm to a nondeterministic system generated by a Markov process for approximating the eigenfunctions of the backward Kolmogorov equation. The application of the DMD algorithm to noisy data is studied, for example, in recent papers [13, 30]. In order to remove the bias errors produced by using the standard DMD algorithms on data with the observation noise that can arise, for example, as a consequence of imprecise measurements, Hemati et al. developed in

[13], the total least-squares DMD, which for the described type of data outperforms the standard DMD. Takeishi et al. considered in [30], the numerical approximations of spectral objects of the stochastic Koopman operator for the RDS with observation noise by using the DMD algorithm. Due to the systematic error produced by the standard DMD algorithm, they developed the version of the algorithm that takes into account the observation noise and refer to it as the subspace DMD algorithm.

## 6.2 Nonautonomous Koopman Operator

The following definitions of the skew product flow formulation and of the process formulation for the nonautonomous dynamical system follow the terminology and results in [15]. We use both formulations for two alternative definitions of the nonautonomous Koopman operator.

### 6.2.1 *Skew Product Flow Formulation*

Let $\mathbb{T}$ be an additive semigroup (or group) with a metric space structure. It actually stands for $\mathbb{Z}$, $\mathbb{Z}_0^+$, $\mathbb{R}$, or $\mathbb{R}_0^+$. We call $\mathbb{T}$ the time set.

Let $(P, d_P)$ be a metric state space and $\mathbb{T}$ a time set. Let $\theta : \mathbb{T} \times P \to P$ be a continuous mapping such that family $\theta^t = \theta(t, \cdot)$, $t \in \mathbb{T}$ forms a group of bi-continuous mappings and satisfies the cocycle property

$$\theta^0 = id_P \quad \text{and} \quad \theta^{t+s} = \theta^s \circ \theta^t \quad \text{for } t, s \in \mathbb{T}. \tag{6.1}$$

Let $(X, d_X)$ be a metric state space and let $\mathbf{S} : \mathbb{T}_0^+ \times P \times X \to X$ be a continuous mapping such that the two-parameter family $\mathbf{S}^{t,\mathbf{p}} = \mathbf{S}(t, \mathbf{p}, \cdot)$, $t \in \mathbb{T}$, $\mathbf{p} \in P$ satisfies the cocycle property over $\theta^t$

$$\mathbf{S}^{0,\mathbf{p}} = id_X \quad \text{and} \quad \mathbf{S}^{t+s,\mathbf{p}} = \mathbf{S}^{s,\theta^t(\mathbf{p})} \circ \mathbf{S}^{t,\mathbf{p}}, \quad \text{for } t, s \in \mathbb{T}_0^+, \ \mathbf{p} \in P. \tag{6.2}$$

The mapping pair $(\theta, \mathbf{S})$ is called the nonautonomous dynamical system and $\theta$ its driving dynamical system. Furthermore, the mapping $\mathbf{S}_s : \mathbb{T}_0^+ \times P \times X \to P \times X$ defined by

$$\mathbf{S}_s(t, (\mathbf{p}, \mathbf{x})) = (\theta(t, \mathbf{p}), \mathbf{S}(t, \mathbf{p}, \mathbf{x})) \tag{6.3}$$

forms an autonomous semi-dynamical system on $P \times X$. The family $\mathbf{S}_s^t = \mathbf{S}_s(t, \cdot)$, $t \in \mathbb{T}$ is called the skew product flow associated with the nonautonomous dynamical system $(\theta, \mathbf{S})$.

Consider the case $X = \mathbb{R}^d$, $P = \mathbb{R}^m$, and differential equations

$$\dot{\mathbf{p}} = \mathbf{F}_d(\mathbf{p}), \tag{6.4}$$

$$\dot{\mathbf{x}} = \mathbf{F}(\mathbf{p}, \mathbf{x}). \tag{6.5}$$

Assume $\mathbf{p} = \mathbf{p}(t, \mathbf{p}_0)$ is the solution of (6.4) satisfying the condition $\mathbf{p}(0, \mathbf{p}_0) = \mathbf{p}_0$ and $\mathbf{x} = \mathbf{x}(t, \mathbf{p}_0, \mathbf{x}_0)$ is the solution of (6.5) satisfying the condition $\mathbf{x}(0, \mathbf{p}_0, \mathbf{x}_0) = \mathbf{x}_0$. The conditions for the existence and uniqueness of such solutions can be found in [5] (Chap. 1). Then, Eq. (6.4) generates the driving flow $\theta^t(\mathbf{p}_0) = \mathbf{p}(t, \mathbf{p}_0)$ and Eq. (6.5) generates the nonautonomous flow $\mathbf{S}^{t,\mathbf{p}_0}(\mathbf{x}) = \mathbf{x}(t, \mathbf{p}_0, \mathbf{x}_0)$.

**Definition 6.1** Let $\mathbf{S}^{t,\mathbf{p}_0}$ be a nonautonomous flow and $U^{t,\mathbf{p}_0}$ an operator family defined on the space of observables, i.e., scalar-valued functions $f : X \to \mathbb{C}$ by

$$U^{t,\mathbf{p}_0} f = f \circ \mathbf{S}^{t,\mathbf{p}_0}. \tag{6.6}$$

Then $U^{t,\mathbf{p}_0}$ is called the nonautonomous Koopman operator family. If $\lambda^{t,\mathbf{p}_0} \in \mathbb{C}$ and observable $\phi_{\lambda^{t,\mathbf{p}_0}} : X \to \mathbb{C}$ are such that

$$U^{t,\mathbf{p}_0} \phi_{\lambda^{t,\mathbf{p}_0}} = e^{\lambda^{t,\mathbf{p}_0}} \phi_{\lambda^{t,\mathbf{p}_0}}, \tag{6.7}$$

they are called the nonautonomous Koopman eigenvalue and eigenfunction.

This skew product formulation of the nonautonomous Koopman operator was, for example, used in [17].

From the cocycle property of the nonautonomous flow over $\theta^t$, we get

$$U^{0,\mathbf{p}_0} f = f \circ \mathbf{S}^{0,\mathbf{p}_0} = f \circ id_X = f \tag{6.8}$$

and also

$$\begin{aligned} U^{t+s,\mathbf{p}_0} f = f \circ \mathbf{S}^{t+s,\mathbf{p}_0} &= f \circ \mathbf{S}^{s,\theta^t(\mathbf{p}_0)} \circ \mathbf{S}^{t,\mathbf{p}_0} = U^{s,\theta^t(\mathbf{p}_0)} f \circ \mathbf{S}^{t,\mathbf{p}_0} \\ &= U^{s,\theta^t(\mathbf{p}_0)} (U^{t,\mathbf{p}_0} f) = (U^{s,\theta^t(\mathbf{p}_0)} \circ U^{t,\mathbf{p}_0}) f. \end{aligned} \tag{6.9}$$

Therefore, the nonautonomous Koopman operator family forms a cocycle over $\theta^t$.

Also, we can define an operator family $L^{\mathbf{p}_0}$, $\mathbf{p}_0 \in P$ with the action on any smooth observable $f$

$$L^{\mathbf{p}_0} f(\mathbf{x}_0) = \lim_{t \to 0} \frac{U^{t,\mathbf{p}_0} f(\mathbf{x}_0) - f(\mathbf{x}_0)}{t}. \tag{6.10}$$

We call this operator family the nonautonomous Koopman generator family.

In the case of the nonautonomous dynamical system generated by (6.4)–(6.5), we compute

$$L^{\mathbf{p}_0} f(\mathbf{x}_0) = \lim_{t \to 0} \frac{f(\mathbf{x}(t, \mathbf{p}_0, \mathbf{x}_0)) - f(\mathbf{x}(0, \mathbf{p}_0, \mathbf{x}_0))}{t}$$
$$= \frac{d}{dt} f(\mathbf{x}(t, \mathbf{p}_0, \mathbf{x}_0))\Big|_{t=0} = \mathbf{F}(0, \mathbf{p}_0, \mathbf{x}_0) \cdot \nabla f(\mathbf{x}_0), \tag{6.11}$$

i.e., the following relation is valid

$$L^{\mathbf{p}_0} f(\cdot) = \mathbf{F}(0, \mathbf{p}_0, \cdot) \cdot \nabla f(\cdot). \tag{6.12}$$

**Proposition 6.1** *If $\mathbf{A} : \mathbb{R}^m \to \mathbb{R}^{d \times d}$ is continuous, and $\theta : \mathbb{R} \times \mathbb{R}^m \to \mathbb{R}^m$ satisfies the conditions for the driving dynamical system, then linear nonautonomous differential equation*

$$\dot{\mathbf{x}} = \mathbf{A}(\theta(t, \mathbf{p}_0))\mathbf{x} \tag{6.13}$$

*generates a linear nonautonomous flow $\mathbf{S}^{t,\mathbf{p}_0} : \mathbb{R}^d \to \mathbb{R}^d$ satisfying*

$$\mathbf{S}^{t,\mathbf{p}_0}\mathbf{x} = \mathbf{x} + \int_0^t \mathbf{A}(\theta(\tau, \mathbf{p}_0))\mathbf{S}^{\tau,\mathbf{p}_0}\mathbf{x}\, d\tau. \tag{6.14}$$

*If $\mathbf{S}^{t,\mathbf{p}_0}$ is diagonalizable, with simple eigenvalues $\mu_j^{t,\mathbf{p}_0} = e^{\lambda_j^{t,\mathbf{p}_0}}$ and left and right eigenvectors $\mathbf{w}_j^{t,\mathbf{p}_0}$, $\mathbf{v}_j^{t,\mathbf{p}_0}$, $j = 1, \ldots, d$, then*

$$\phi_j^{t,\mathbf{p}_0}(\mathbf{x}) = \langle \mathbf{x}, \mathbf{w}_j^{t,\mathbf{p}_0} \rangle, \quad j = 1, \ldots, d, \tag{6.15}$$

*are*[1] *the eigenfunctions of the nonautonomous Koopman operator $U^{t,\mathbf{p}_0}$ with the corresponding eigenvalues $\lambda_j^{t,\mathbf{p}_0}$, $j = 1, \ldots, d$. Furthermore, $\mathbf{v}_j^{t,\mathbf{p}_0}$, $j = 1, \ldots, d$ are the Koopman modes of the full-state observable and the following expansion is valid*

$$U^{t,\mathbf{p}_0}\mathbf{x}_0 = \sum_{j=1}^{d} \langle \mathbf{x}_0, \mathbf{w}_j^{t,\mathbf{p}_0} \rangle e^{\lambda_j^{t,\mathbf{p}_0}} \mathbf{v}_j^{t,\mathbf{p}_0}. \tag{6.16}$$

*Proof* It is straightforward to verify that if $\mathbf{S}^{t,\mathbf{p}_0}$ is defined with (6.14), then $\mathbf{x}(t, \mathbf{p}_0, \mathbf{x}_0) = \mathbf{S}^{t,\mathbf{p}_0}\mathbf{x}_0$ is the solution for (6.13).

Let additionally $\mathbf{S}^{t,\mathbf{p}_0}$ be diagonalizable, with simple eigenvalues $\mu_j^{t,\mathbf{p}_0} = e^{\lambda_j^{t,\mathbf{p}_0}}$ and left and right eigenvectors $\mathbf{w}_j^{t,\mathbf{p}_0}$, $\mathbf{v}_j^{t,\mathbf{p}_0}$, $j = 1, \ldots, d$. For the observable defined with (6.15), we compute

$$U^{t,\mathbf{p}_0}\phi_j^{t,\mathbf{p}_0}(\mathbf{x}_0) = \phi_j^{t,\mathbf{p}_0}(\mathbf{x}(t, \mathbf{p}_0, \mathbf{x}_0)) = \langle \mathbf{S}^{t,\mathbf{p}_0}\mathbf{x}_0, \mathbf{w}_j^{t,\mathbf{p}_0} \rangle$$
$$= \langle \mathbf{x}_0, \left(\mathbf{S}^{t,\mathbf{p}_0}\right)^* \mathbf{w}_j^{t,\mathbf{p}_0} \rangle = \mu_j^{t,\mathbf{p}_0} \langle \mathbf{x}_0, \mathbf{w}_j^{t,\mathbf{p}_0} \rangle = e^{\lambda_j^{t,\mathbf{p}_0}} \phi_j^{t,\mathbf{p}_0}(\mathbf{x}_0). \tag{6.17}$$

[1] We use $\langle \cdot, \cdot \rangle$, to denote the standard scalar product of vectors in $\mathbb{R}^m$.

Therefore $\phi_j^{t,\mathbf{p}_0}$ is the eigenfunction related to the eigenvalue $\lambda_j^{t,\mathbf{p}_0}$.

Furthermore, we can decompose the state $\mathbf{x}_0$ in the $\mathbf{v}_j^{t,\mathbf{p}_0}$, $j = 1, \ldots, d$ basis,

$$\mathbf{x}_0 = \sum_{j=1}^{d} \langle \mathbf{x}_0, \mathbf{w}_j^{t,\mathbf{p}} \rangle \mathbf{v}_j^{t,\mathbf{p}}. \tag{6.18}$$

Then, action of the Koopman operator on the full-state observable becomes

$$\begin{aligned} U^{t,\mathbf{p}_0}\mathbf{x}_0 = \mathbf{S}^{t,\mathbf{p}_0}\mathbf{x}_0 &= \sum_{j=1}^{d} \langle \mathbf{x}_0, \mathbf{w}_j^{t,\mathbf{p}_0} \rangle \mathbf{S}^{t,\mathbf{p}_0} \mathbf{v}_j^{t,\mathbf{p}_0} \\ &= \sum_{i=1}^{d} \langle \mathbf{x}_0, \mathbf{w}_j^{t,\mathbf{p}_0} \rangle e^{\lambda_j^{t,\mathbf{p}_0}} \mathbf{v}_j^{t,\mathbf{p}_0}. \end{aligned} \tag{6.19}$$

□

### 6.2.2 Process Formulation

Let $(X, d_X)$ be a metric state space and $\mathbb{T}$ a time set. Let $\mathbf{S} : \mathbb{T} \times \mathbb{T} \times X \to X$ be a continuous mapping such that the two-parameter family $\mathbf{S}^{t,t_0} = \mathbf{S}(t, t_0, \cdot)$, $t, t_0 \in \mathbb{T}$ satisfies the cocycle property

$$\mathbf{S}^{t_0,t_0} = id_X \quad \text{and} \quad \mathbf{S}^{t+s,t_0} = \mathbf{S}^{t+s,t} \circ \mathbf{S}^{t,t_0} \quad \text{for} \quad t_0 \leq t \leq t+s, \; t_0, t, s \in \mathbb{T}. \tag{6.20}$$

The mapping $\mathbf{S}$ is called the process and the two-parameter family $\mathbf{S}^{t,t_0}$ is called the nonautonomous flow.

Consider the case of a nonautonomous differential equation

$$\dot{\mathbf{x}} = \mathbf{F}(t, \mathbf{x}) \tag{6.21}$$

on $X = \mathbb{R}^d$. Assume $\mathbf{x} = \mathbf{x}(t, t_0, \mathbf{x}_0)$ is the solution of (6.21) satisfying the condition $\mathbf{x}(t_0, t_0, \mathbf{x}_0) = \mathbf{x}_0$. The conditions for the existence and uniqueness of such a solution can be found in [5] (Chap. 1). Then, Eq. (6.21) generates the nonautonomous flow $\mathbf{S}^{t,t_0}(\mathbf{x}_0) = \mathbf{x}(t, t_0, \mathbf{x}_0)$.

**Definition 6.2** Let $\mathbf{S}^{t,t_0}$ be a nonautonomous flow and $U^{t,t_0}$ an operator family defined on the space of observables $f : X \to \mathbb{C}$ by

$$U^{t,t_0} f = f \circ \mathbf{S}^{t,t_0}. \tag{6.22}$$

Then $U^{t,t_0}$ is called the nonautonomous Koopman operator family. If $\lambda^{t,t_0} \in \mathbb{C}$ and observable $\phi_{\lambda^{t,t_0}} : X \to \mathbb{C}$ are such that

$$U^{t,t_0}\phi_{\lambda^{t,t_0}} = e^{\lambda^{t,t_0}}\phi_{\lambda^{t,t_0}}, \tag{6.23}$$

they are called the nonautonomous Koopman operator eigenvalue and eigenfunction.

This process formulation was originally introduced in [22, 25].

The cocycle property for the nonautonomous Koopman operator

$$U^{t_0,t_0} = id \quad \text{and} \quad U^{t+s,t_0} = U^{t+s,t} \circ U^{t,t_0}, \quad \text{for } t_0 \le t \le t+s,\ t_0, t, s \in \mathbb{T} \tag{6.24}$$

follows from the cocycle property of the two-parameter nonautonomous flow (6.20), as it can be easily verified. Also, we can define an operator family $L^{t_0}$, $t_0 \in \mathbb{T}$ by its action on any smooth observable $f$

$$L^{t_0} f(\mathbf{x}_0) = \lim_{t\to t_0} \frac{U^{t,t_0} f(\mathbf{x}_0) - f(\mathbf{x}_0)}{t - t_0}. \tag{6.25}$$

We call this operator family the nonautonomous Koopman generator family. As proven in [20], in the case of (6.21), the following relation for this generator family is valid

$$L^{t_0} f(\cdot) = \mathbf{F}(t_0, \cdot) \cdot \nabla f(\cdot). \tag{6.26}$$

Finally, let us rephrase the result for the linear nonautonomous systems [20].

**Proposition 6.2** *If $\mathbf{A} : \mathbb{R} \to \mathbb{R}^{d\times d}$ is continuous, then the linear nonautonomous differential equation*

$$\dot{\mathbf{x}} = \mathbf{A}(t)\mathbf{x} \tag{6.27}$$

*generates a linear nonautonomous flow $\mathbf{S}^{t,t_0} : \mathbb{R}^d \to \mathbb{R}^d$ satisfying*

$$\mathbf{S}^{t,t_0}\mathbf{x} = \mathbf{x} + \int_{t_0}^{t} \mathbf{A}(\tau)\mathbf{S}^{\tau,t_0}\mathbf{x}\, d\tau. \tag{6.28}$$

*If $\mathbf{S}^{t,t_0}$ is diagonalizable, with simple eigenvalues $\mu_j^{t,t_0} = e^{\lambda_j^{t,t_0}}$ and left and right eigenvectors $\mathbf{w}_j^{t,t_0}$, $\mathbf{v}_j^{t,t_0}$, $j = 1, \ldots, d$, then*

$$\phi_j^{t,t_0}(\mathbf{x}) = \langle \mathbf{x}, \mathbf{w}_j^{t,t_0} \rangle, \quad j = 1, \ldots, d, \tag{6.29}$$

*are the eigenfunctions of the nonautonomous Koopman operator $U^{t,t_0}$ with the corresponding eigenvalues $\lambda_j^{t,t_0}$, $j = 1, \ldots, d$. Furthermore, $\mathbf{v}_j^{t,t_0}$, $j = 1, \ldots, d$ are the Koopman modes of the full-state observable and the following expansion is valid*

$$U^{t,t_0}\mathbf{x}_0 = \sum_{j=1}^{d} \langle \mathbf{x}_0, \mathbf{w}_j^{t,t_0} \rangle e^{\lambda_j^{t,t_0}} \mathbf{v}_j^{t,t_0}. \tag{6.30}$$

## 6.3 Stochastic Koopman Operator

Suppose that instead of a flow $\theta^t$ defined over a metric space $(P, d_P)$ and bi-continuous, which is the driving system in the skew product definition of the nonautonomous system, a flow $\theta^t$ is defined over a probability space $(\Omega, \mathscr{F}, \mathbb{P})$ and is measure preserving. Using such type of flow, one can define the random dynamical system in which the driving flow represents the noise of the system.

Let $(\Omega, \mathscr{F}, \mathbb{P})$ be a probability space and $\mathbb{T}$ a time set. Let $\theta : \mathbb{T} \times \Omega \to \Omega$ be a measurable mapping such that the family $\theta^t = \theta(t, \cdot) : \Omega \to \Omega$ forms a group or semigroup of measure preserving mappings, i.e.,

$$\theta^t \mathbb{P} = \mathbb{P} \text{ for all } t \in \mathbb{T}, \tag{6.31}$$

and satisfies the cocycle property (6.1). The mapping $\theta$ is called the measure preserving dynamical system.

Let $(X, \mathscr{B})$ be a measurable space and $\mathbf{S} : \mathbb{T} \times \Omega \times X \to X$ be a measurable mapping such that the two-parameter family $\mathbf{S}^{t,\omega} = \mathbf{S}(t, \omega, \cdot), t \in \mathbb{T}, \omega \in \Omega$ satisfies the cocycle property over $\theta^t$

$$\mathbf{S}^{0,\omega} = id_X \quad \text{and} \quad \mathbf{S}^{t+s,\omega} = \mathbf{S}^{t,\theta^s(\omega)} \circ \mathbf{S}^{s,\omega}, \text{ for all } s, t \in \mathbb{T}, \omega \in \Omega. \tag{6.32}$$

The mapping pair $(\theta, \mathbf{S})$ is called the random dynamical system and $\theta$ is its driving dynamical system.

In what follows, we will limit our considerations to the situation where $(X, \mathscr{B})$ is a Polish space, i.e., $\mathscr{B}$ is the Borel $\sigma$-algebra generated by a complete and separable metric topology on $X$. For fixed $\mathbf{x} \in X$, $(\mathbf{S}^{t,\omega}(\mathbf{x}))_{t\in\mathbb{T},\omega\in\Omega}$ is the family of random variables, therefore, in accordance with the Kolmogorov theorem, it can be seen as a realization of a stochastic (random) process on the space $(X^{\mathbb{T}}, \mathscr{B}^{\mathbb{T}}, \mathbb{P})$. The measure $\mathbb{P}$ on $X^{\mathbb{T}}$ is induced by the probability measure on the probability space $\Omega$, thus we will use the same notation for both. Also, we replace the probability space $(\Omega, \mathscr{F}, \mathbb{P})$ with $(X^{\mathbb{T}}, \mathscr{B}^{\mathbb{T}}, \mathbb{P})$, where the element $\omega : \mathbb{T} \to X$ in it is given as $\omega(t) = \mathbf{S}^{t,\omega}(\mathbf{x})$. The probability measure $\mathbb{P}$ on $X^{\mathbb{T}}$ associated with this process is induced by its finite-dimensional probability measures defined on the cylinder sets. It is clear that the initial distribution of the stochastic process at $t = 0$ is engaged in the induced measure $\mathbb{P}$.

**Definition 6.3** Let $\mathbf{S}^{t,\omega}$ be a random dynamical system flow and $U_S^t$ an operator family defined on the space of observables, i.e., scalar-valued measurable functions $f : X \to \mathbb{C}$ by

$$U_S{}^t f(\mathbf{x}) = \mathbb{E}_{\mathbb{P}}[f(\mathbf{S}^{t,\omega}(\mathbf{x}))], \tag{6.33}$$

where $\mathbb{E}_{\mathbb{P}}$ denotes the expectation with respect to the probability measure $\mathbb{P}$. Then $U_S^t$ is called the stochastic Koopman operator family.

If $\lambda_S(t) \in \mathbb{C}$ and observable $\phi_{\lambda_S}^t : X \to \mathbb{C}$ are such that

$$U_S^t \phi_{\lambda_S}^t(\mathbf{x}) = e^{\lambda_S(t)} \phi_{\lambda_S}^t(\mathbf{x}), \tag{6.34}$$

they are called the stochastic Koopman operator eigenvalue and eigenfunction.

Let us consider the case when $\mathbb{T} = \mathbb{R}$, $X = \mathbb{R}^d$, and $\Omega = \mathscr{C}(\mathbb{R}, \mathbb{R}^m)$ (or $\Omega = \mathscr{D}(\mathbb{R}, \mathbb{R}^m)$ denoting the space of "càdlàg" functions which are bounded on bounded intervals, right continuous with left limits and with at most countably many discontinuities). One type of RDS is the continuous-type RDS that is generated by the random differential equation (RDE) of the following form

$$\dot{\mathbf{x}} = F(\theta^t(\omega), \mathbf{x}), \tag{6.35}$$

defined on $X$, where $\omega \in \Omega$ is an element in the probability space $\Omega$ associated with the random dynamics. Such RDE generates an RDS $\mathbf{S}$ over $\theta$, whose action is defined by

$$\mathbf{S}^{t,\omega}(\mathbf{x}) = \mathbf{x} + \int_0^t F(\theta^s(\omega), \mathbf{S}^{s,\omega}(\mathbf{x}))ds. \tag{6.36}$$

The local and global existence of the solution (6.36), as well as the properties of the generated RDS, depend on different regularity properties of the function $F$ and $\theta^t(\omega)$ and can be found in [4] (Sect. 2.2).

A set of trajectories starting at $\mathbf{x}$ that are generated by (6.35) is given by $\mathbf{S}^{t,\omega}(\mathbf{x})$ and it defines the family of random variables, which is the solution of RDE with the initial condition $\mathbf{S}^{0,\omega}(\mathbf{x}) = \mathbf{x}$. Since the solutions of RDE are defined pathwise, for each fixed $\omega$, the trajectory can be determined as a solution of a deterministic ordinary differential equation, so that the RDE (6.35) can be seen as a family of ordinary differential equations. In this type of equations, the randomness refers just to the random parameters, which do not depend on the state of the system. Despite the fact that under some assumptions for the function $F$ and for the driving system $\theta$, the Eq. (6.35) generates the global RDS satisfying certain regularity properties, in a general case, it is usually impossible to determine the distribution of the solution at time $t$ explicitly and to analyze the solutions from the statistical point of view. Therefore, the evaluation of the stochastic Koopman operator, which is defined as the expectation of the solution at time $t$, in such a general case, becomes a challenge.

At least for some special cases, such as for the case of RDS generated by linear RDE, some properties of the eigenvalues and the eigenvectors of the stochastic Koopman operators are known. These are described in the next proposition, stated in [8].

**Proposition 6.3** *If* $\mathbf{A} : \Omega \to \mathbb{R}^{d \times d}$ *and* $\mathbf{A} \in L^1(\Omega, \mathscr{F}, P)$, *then RDE*

$$\dot{\mathbf{x}} = \mathbf{A}(\theta^t(\omega))\mathbf{x}, \tag{6.37}$$

*generates a linear RDS $\mathbf{S}^{t,\omega}$ satisfying*

$$\mathbf{S}^{t,\omega}\mathbf{x} = \mathbf{x} + \int_0^t \mathbf{A}(\theta^s(\omega))\mathbf{S}^{s,\omega}\mathbf{x}\, ds. \tag{6.38}$$

*Assume that $\hat{\mathbf{S}}^t = \mathbb{E}_{\mathbb{P}}[\mathbf{S}^{t,\omega}]$ is diagonalizable, with simple eigenvalues $\hat{\mu}^t_j = \mathrm{e}^{\hat{\lambda}_j(t)}$ and left and right eigenvectors $\hat{\mathbf{w}}^t_j$, $\hat{\mathbf{v}}^t_j$ $j = 1, \ldots, d$. Then*

$$\phi^t_{S,j}(\mathbf{x}) = \langle \mathbf{x}, \hat{\mathbf{w}}^t_j \rangle, \quad j = 1, \ldots, d, \tag{6.39}$$

*are the eigenfunctions of the stochastic Koopman operator $U^t_S$ with the corresponding eigenvalues $\lambda_{S,j}(t) = \hat{\lambda}_j(t)$, $j = 1, \ldots, d$.*

*Moreover, if matrices $\mathbf{A}(\omega)$ commute and are diagonalizable, with the simple eigenvalues $\lambda_j(\omega)$ and the corresponding left eigenvectors $\mathbf{w}_j$, $j = 1, \ldots, d$, then*

$$\hat{\mathbf{w}}^t_j = \mathbf{w}_j \quad \textit{and} \quad \mathrm{e}^{\lambda_{S,j}(t)} = \mathbb{E}_{\mathbb{P}}\left[\mathrm{e}^{\int_0^t \lambda_j(\theta(s)\omega)ds}\right].$$

*Furthermore, $\hat{\mathbf{v}}^t_j$, $j = 1, \ldots, d$ are the Koopman modes of the full-state observable and the following expansion is valid*

$$U^t_S\mathbf{x} = \sum_{j=1}^{d} \langle \mathbf{x}, \hat{\mathbf{w}}^t_j \rangle \mathrm{e}^{\lambda_{S,j}(t)} \hat{\mathbf{v}}^t_j. \tag{6.40}$$

*Proof* The first part of the proposition follows from [4], Example 2.2.8.

Furthermore, the action of the stochastic Koopman operator on the functions defined by (6.39) is equal to

$$\begin{aligned} U^t_S\phi^{S,t}_j(\mathbf{x}) &= \mathbb{E}_{\mathbb{P}}[\langle \mathbf{S}^{t,\omega}\mathbf{x}, \hat{\mathbf{w}}^t_j \rangle] = \mathbb{E}_{\mathbb{P}}[\langle \mathbf{x}, (\mathbf{S}^{t,\omega})^*\hat{\mathbf{w}}^t_j \rangle] = \langle \mathbf{x}, \mathbb{E}_{\mathbb{P}}[(\mathbf{S}^{t,\omega})^*]\hat{\mathbf{w}}^t_j \rangle \\ &= \langle \mathbf{x}, (\hat{\mathbf{S}}^t)^* \hat{\mathbf{w}}^t_j \rangle = \hat{\mu}^t_j \langle \mathbf{x}, \hat{\mathbf{w}}^t_j \rangle = \mathrm{e}^{\hat{\lambda}_j(t)} \phi^{S,t}_j(\mathbf{x}). \end{aligned} \tag{6.41}$$

In the case when matrices $\mathbf{A}(\omega)$ commute, are diagonalizable, and have simple eigenvalues, they are simultaneously diagonalizable, i.e., there exists a joint set of left and right eigenvectors $(\mathbf{w}_j, \mathbf{v}_j)$, $j = 1, \ldots, d$ for each $\mathbf{A}(\omega)$, $\omega \in \Omega$. If $\mathbf{W}$ and $\mathbf{V}$ are matrices of left and right eigenvectors, respectively, and $\boldsymbol{\Lambda}(\omega) = \mathrm{diag}(\lambda_1(\omega), \ldots, \lambda_d(\omega))$, then $\mathbf{A}(\omega) = \mathbf{V}\boldsymbol{\Lambda}(\omega)\mathbf{W}^*$ and

$$\mathbf{S}^{t,\omega} = \mathrm{e}^{\int_0^t \mathbf{A}(\theta^s(\omega))ds} = \mathbf{V}\mathrm{e}^{\int_0^t \boldsymbol{\Lambda}(\theta^s(\omega))ds}\mathbf{W}^*.$$

Therefore,

$$\hat{\mathbf{S}}^t = \mathbf{V}\mathbb{E}_{\mathbb{P}}\left[\mathrm{e}^{\int_0^t \boldsymbol{\Lambda}(\theta^s(\omega))ds}\right]\mathbf{W}^*,$$

and we easily conclude that $\hat{\mathbf{w}}_j^t = \mathbf{w}_j$ and $e^{\lambda_{S,j}(t)} = \mathbb{E}_{\mathbb{P}}\left[e^{\int_0^t \lambda_j(\theta^s(\omega))ds}\right]$. The validity of (6.40) follows directly from the decomposition of the state $\mathbf{x}$ in the base $\hat{\mathbf{w}}_j^t$, $\hat{\mathbf{v}}_j^t$ $j = 1, \ldots, d$. □

A special case of RDE are the "memoryless" RDE, which are defined by the equation of the form

$$\dot{\mathbf{x}} = G(\xi_t(\omega), \mathbf{x}), \tag{6.42}$$

where $\xi_t(\omega)$ denotes the stochastic process. Here the driving dynamical system $\theta^t$ is not given directly, but in terms of the coordinate process $\omega(t) = \xi_t(\omega)$ on the space $\Omega = \mathscr{C}(\mathbb{R}, \mathbb{R}^m)$, as a shift transformation defined by

$$\theta^t(\omega)(\cdot) = \omega(t + \cdot), \quad t \in \mathbb{T}. \tag{6.43}$$

The shift operator $\theta^t$ on the considered probability space is measure preserving and continuous. The coordinate process and the dynamical system $\theta^t$ are connected through the following relation

$$\omega(t) = \xi_t(\omega) = \pi(\theta^t(\omega)), \quad \text{where} \quad \pi(\omega) = \omega(0). \tag{6.44}$$

Thus, (6.42) is the equation of the form (6.35), where the function $F$ in (6.35) is related to $G$ in (6.42) by $F(\theta^t(\omega), \mathbf{x}) = G(\pi(\theta^t(\omega)), \mathbf{x})$.

Another type of differential equations that generate RDS are the stochastic differential equations (SDE). Consider the case where $\mathbb{T} = \mathbb{R}^+$, $X = \mathbb{R}^d$, and the autonomous SDE is of the form

$$dX_t = F(X_t)dt + \sigma(X_t)dW_t, \tag{6.45}$$

where $F : X \to X$ and $\sigma : X \to \mathbb{R}^{d\times r}$ are $L^2$ measurable. Here, $W_t = (W_t^1, \ldots, W_t^r)^T$ denotes the $r$-dimensional Wiener process with independent components and standard properties, i.e., $\mathbb{E}(W_t^i) = 0, i = 1, \ldots, r, \mathbb{E}(W_t^i W_s^j) = \min\{t, s\}\delta_{ij}, i, j = 1, \ldots, r$ ($\delta_{ij}$ is the Kronecker delta symbol). Since the Wiener process is a continuous process with the stationary increments, the probability space can be identified with $\Omega = \mathscr{C}_0(\mathbb{R}^+, \mathbb{R}^r)$ and $\omega \in \Omega$ with the canonical realization of the Wiener process such that $\omega(t) = W_t(\omega)$. Furthermore, the driving flow $\theta$ is defined by the "Wiener shift"

$$\theta^t(\omega)(\cdot) = \omega(t + \cdot) - \omega(t). \tag{6.46}$$

SDE (6.45) generates the one-parameter family of RDS $\mathbf{S}^{t,\omega} := \mathbf{S}(t, \omega, \cdot) : X \to X$ given by $\mathbf{S}^{t,\omega}(\mathbf{x}) = X_t(\mathbf{x})$, where for the initial condition $X_0(\mathbf{x}) = \mathbf{x}$, $X_t$ denotes the solution of (6.45) in terms of Itô integral

$$X_t(\mathbf{x}) = \mathbf{x} + \int_0^t F(X_s)ds + \int_0^t \sigma(X_s)dW_s. \tag{6.47}$$

Under certain continuity and boundedness conditions for the coefficients $f$ and $\sigma$, such type of differential equations generate RDS for which the Markov property holds [2, 3, 6, 7].

Strong results for the properties of the stochastic Koopman operator family can be obtained in more specific settings such as linear setting we analyzed before and Markov setting. According to [2, 7], the RDS will be Markovian if it is generated by SDE of the form (6.45), or if the process $\xi_t$ entering the Eq. (6.42) has independent components. In stated cases,, the generated RDS is Markovian with respect to its accompanying family of $\sigma$-algebras corresponding to the solution and the driving system $\mathscr{F}_t^{\mathbf{x},\omega} = \sigma(\mathbf{S}^{s,\omega}(\mathbf{x}), \theta^s(\omega), 0 \leq s \leq t)$. The Markov property implies that for every $s \leq t$ and every random variable $Y$, measurable with respect to $\mathscr{F}_t^{\mathbf{x},\omega}$,

$$\mathbb{E}[Y|\mathscr{F}_s^{\mathbf{x},\omega}] = \mathbb{E}[Y|\mathbf{S}^{s,\omega}(\mathbf{x})]. \tag{6.48}$$

Moreover, for such $Y$ the following equality, known as the Chapman–Kolmogorov equation, is valid

$$\mathbb{E}\left[Y|\mathbf{S}^{0,\omega}(\mathbf{x})\right] = \mathbb{E}\left[\mathbb{E}\left[Y|\mathbf{S}^{s,\omega}(\mathbf{x})\right]|\mathbf{S}^{0,\omega}(\mathbf{x})\right]. \tag{6.49}$$

Next proposition about the semigroup property of the stochastic Koopman operators is actually a well-known fact for the transition semigroup of the time-homogeneous Markov processes [6].

**Proposition 6.4** *If RDS is time-homogeneous Markovian, the stochastic Koopman operator family satisfies the semigroup property, i.e.,* $U_S^{t+s} = U_S^s \circ U_S^t$.

*Proof* Observe that

$$U_S^0 f(\mathbf{x}) = \mathbb{E}_{\mathbb{P}}[f(\mathbf{S}^{0,\omega}(\mathbf{x}))] = \mathbb{E}_{\mathbb{P}}[f(\mathbf{x})] = f(\mathbf{x}), \tag{6.50}$$

where the second equality follows from the first equation in (6.32). This means $U_S^0 = id$ on the set of real-valued, bounded, and measurable observables. Further, we have

$$\begin{aligned} U_S^{t+s} f(\mathbf{x}) &= \mathbb{E}_{\mathbb{P}}\left[f\left(\mathbf{S}^{t+s,\omega}(\mathbf{x})\right)\right] = \mathbb{E}_{\mathbb{P}}\left[f\left(\mathbf{S}^{t,\theta^s(\omega)} \circ \mathbf{S}^{s,\omega}(\mathbf{x})\right)|\mathbf{S}^{0,\omega}(\mathbf{x}) = \mathbf{x}\right] \\ &= \mathbb{E}_{\mathbb{P}}\left[\mathbb{E}_{\mathbb{P}}\left[f\left(\mathbf{S}^{t,\theta^s(\omega)} \circ \mathbf{S}^{s,\omega}(\mathbf{x})\right)|\mathbf{S}^{s,\omega}(\mathbf{x})\right]|\mathbf{S}^{0,\omega}(\mathbf{x}) = \mathbf{x}\right] \\ &= \mathbb{E}_{\mathbb{P}}\left[U_S^t f(\mathbf{S}^{s,\omega}(\mathbf{x}))|\mathbf{S}^{0,\omega}(\mathbf{x}) = \mathbf{x}\right] = \mathbb{E}_{\mathbb{P}}\left[U_S^t f(\mathbf{S}^{s,\omega}(\mathbf{x}))\right] \\ &= U_S^s\left(U_S^t f(\mathbf{x})\right), \end{aligned} \tag{6.51}$$

where the equality in the second line follows from (6.49). Also, the fact that the conditional probabilities are homogeneous and that $\mathbb{P}$ is invariant with respect to $\theta^s$ is used. □

If RDS is generated by SDE (6.45), with $F$ and $\sigma$ bounded and continuous, the family of stochastic Koopman operators is a strongly continuous semigroup [4]

(Sect. 2.3), [6] (Sect. 17.3). Define the generator of the stochastic Koopman semigroup family $U_S^t$ acting on the observable functions $f \in C_b^2(\mathbb{R}^d)$ ($C_b^2(\mathbb{R}^d)$ is the space of bounded, real, continuous, and twice differentiable functions on $\mathbb{R}^d$ with bounded and continuous first and second derivatives) by the limit

$$L_S f(\mathbf{x}) = \lim_{t \to 0+} \frac{U_S^t f(\mathbf{x}) - f(\mathbf{x})}{t}, \tag{6.52}$$

if it exists.

Consider the generator of the stochastic Koopman semigroup associated with RDS generated by (6.45). For simplicity, let us focus first on the scalar case ($d = 1$). We derive the action of the generator $L_S$ on $f \in C_b^2(\mathbb{R})$ by using (6.52), as in [14],

$$\begin{aligned}
\lim_{t \to 0+} \frac{U_S^t f(x) - f(x)}{t} &= \lim_{t \to 0+} \frac{\mathbb{E}_{\mathbb{P}}[f(X_t(x))] - f(x)}{t} \\
&= \lim_{t \to 0+} \frac{\mathbb{E}_{\mathbb{P}}\left[f(x) + \int_0^t f'(X_s) dX_s + \int_0^t \frac{1}{2} f''(X_s)\sigma(X_s)^2 ds\right] - f(x)}{t} \\
&= \lim_{t \to 0+} \frac{1}{t} \mathbb{E}_{\mathbb{P}}\left[\int_0^t f'(X_s)F(X_s)ds + \int_0^t f'(X_s)\sigma(X_s)dW_s + \int_0^t \frac{1}{2} f''(X_s)\sigma(X_s)^2 ds\right] \\
&= F(X_0)f'(X_0) + \frac{1}{2} f''(X_0)\sigma(X_0)^2 = F(x)f'(x) + \frac{1}{2}\sigma(x)^2 f''(x).
\end{aligned} \tag{6.53}$$

Here, we applied the Itô's lemma and used the fact that $\mathbb{E}_{\mathbb{P}}\left[\int_0^t f'(X_s)\sigma(X_s)dW_s\right] = 0$. In multidimensional case, it can be determined in a similar way. Thus, we have the following proposition.

**Proposition 6.5** *The action of the generator of the stochastic Koopman family $L_S$ on $f \in C_b^2(\mathbb{R}^d)$ is given by*

$$L_S f(\mathbf{x}) = F(\mathbf{x})\nabla f(\mathbf{x}) + \frac{1}{2} Tr\left(\sigma(\mathbf{x})(\nabla^2 f(\mathbf{x}))\sigma(\mathbf{x})^T\right), \tag{6.54}$$

*where Tr denotes the trace of the matrix.*

In the stochastic literature, the operator (6.54) is known under the name of the Kolmogorov operator. The next proposition follows from the proof of the Feynman–Kac formula [6] (Sect. 17.4).

**Proposition 6.6** *Let $\phi_\lambda \in C_b^2(\mathbb{R}^d)$ be a differentiable eigenfunction of the stochastic Koopman generator $L_S$ associated with RDS generated by SDE (6.45) with the corresponding eigenvalue $\lambda$. Then*

$$d\phi_\lambda(X_t) = \lambda\phi_\lambda(X_t)dt + \nabla\phi_\lambda(X_t)\sigma(X_t)dW_t. \tag{6.55}$$

*Furthermore, $\phi_\lambda$ is the eigenfunction of the stochastic Koopman operators $U_S^t$, $t \in \mathbb{T}$, i.e.,*

$$U_S^t \phi_\lambda(\mathbf{x}) = e^{\lambda t} \phi_\lambda(\mathbf{x}). \tag{6.56}$$

*Proof* Suppose $d = 1$. Based on Ito's lemma, the eigenfunction $\phi(X_t)$ evolves in accordance with

$$\begin{aligned} d\phi_\lambda(X_t) &= \phi_\lambda'(X_t)F(X_t)dt + \frac{1}{2}\phi_\lambda''(X_t)\sigma(X_t)^2 dt + \phi_\lambda'(X_t)\sigma(X_t)dW_t \\ &= L_S\phi_\lambda(X_t)dt + \phi_\lambda'(X_t)\sigma(X_t)dW_t \\ &= \lambda\phi_\lambda(X_t)dt + \phi_\lambda'(X_t)\sigma(X_t)dW_t, \end{aligned} \tag{6.57}$$

where in the last equality we used relation $L_S\phi_\lambda(x) = \lambda\phi_\lambda(x)$. The fact that $\phi_\lambda$ is an eigenfunction of each Koopman semigroup member $U_S^t$ follows from the spectral mapping theorem [11] (Chap. 4.3), but here we give a direct proof as well. For the initial condition $X_0 = x$, from (6.57) by using Itô's formula, we get

$$\phi_\lambda(X_t) = \mathrm{e}^{\lambda t}\phi_\lambda(x) + \int_0^t \mathrm{e}^{\lambda(t-s)}\phi_\lambda'(X_s)\sigma(X_s)dW_s. \tag{6.58}$$

Then, by using the martingale property $\mathbb{E}_\mathbb{P}\left[\int_0^t \mathrm{e}^{\lambda(t-s)}\phi_\lambda'(X_s)\sigma(X_s)dW_s\right] = 0$, we obtain

$$\mathbb{E}_\mathbb{P}\left[\phi(X_t)\right] = \mathrm{e}^{\lambda t}\mathbb{E}_\mathbb{P}\left[\phi(x)\right] = \mathrm{e}^{\lambda t}\phi(x). \tag{6.59}$$

The left-hand side of the obtained equation is equal to $U_S^t\phi_\lambda(x)$, thus we have (6.56). Using definition Eq. (6.34), we conclude that $\phi_\lambda(x)$ is the eigenfunction of the stochastic Koopman operator $U_S^t$, with the corresponding eigenvalue $\lambda^S(t) = \lambda t$.

Using a similar procedure, validity of Eqs. (6.55) and (6.56) in multidimensional case is easily derived. □

## 6.4 Data-Driven Algorithms for the Nonautonomous Koopman Operator

An observable $f : X \to \mathbb{C}$ of the nonautonomous dynamical system can be reinterpreted as an observable of the skew product flow, i.e., $g : P \times X \to \mathbb{C}$ with the definition

$$g(\mathbf{y}(t, \mathbf{y}_0)) = f(\mathbf{x}(t, \mathbf{p}_0, \mathbf{x}_0)), \tag{6.60}$$

where $\mathbf{y} = (\mathbf{p}, \mathbf{x}) \in P \times X$. In that sense, propagation in time of such observable can be covered by the nonautonomous Koopman operator (6.22), but it can also be covered by the autonomous Koopman operator of the skew product flow

$$U^t g = g \circ \mathbf{S}_s^t. \tag{6.61}$$

Now, let us consider what the data-driven algorithms can give us on the nonautonomous Koopman operator family. If we apply the cocycle property to the time $t = k\Delta t$, then for the autonomous Koopman operator (of the skew product flow), we get

$$U_S^{k\Delta t} = (U_S^{\Delta t})^k. \tag{6.62}$$

However, if we apply the same in the case of the nonautonomous Koopman operator family in the process formulation, we obtain

$$U^{t_0+k\Delta t, t_0} = U^{t_0+k\Delta t, t_0+(k-1)\Delta t} \circ U^{t_0+(k-1)\Delta t, t_0(k-2)k\Delta t} \circ \cdots \circ U^{t_0+\Delta t, t_0}. \tag{6.63}$$

Similarly, if we apply the same in the case of the nonautonomous Koopman operator family in the skew product formulation, we obtain

$$U^{k\Delta t, \mathbf{p}_0} = U^{\Delta t, \theta^{(k-1)\Delta t}(\mathbf{p}_0)} \circ U^{\Delta t, \theta^{(k-2)\Delta t}(\mathbf{p}_0)} \circ \cdots \circ U^{\Delta t, \mathbf{p}_0}. \tag{6.64}$$

For a chosen vector observable $\mathbf{f} = (f_1, \ldots, f_N)^T : X \to \mathbb{C}^N$, let us consider the snapshots $\mathbf{f}_k = \mathbf{f}(t_k)$, at the times $t_k = t_0 + k\Delta t$, $k = 0, 1, 2, \ldots$. We can form the Hankel matrix $\mathbf{H}$ with elements

$$\mathbf{H} = \begin{pmatrix} \mathbf{f}_0 & \mathbf{f}_1 & \cdots & \mathbf{f}_{m_H-1} & \mathbf{f}_{m_H} \\ \mathbf{f}_1 & \mathbf{f}_2 & \cdots & \mathbf{f}_{m_H} & \mathbf{f}_{m_H+1} \\ \vdots & \vdots & \ddots & \vdots & \\ \mathbf{f}_{n_H-1} & \mathbf{f}_{n_H} & \cdots & \mathbf{f}_{n_H+m_H-2} & \mathbf{f}_{n_H+m_H-1} \end{pmatrix}. \tag{6.65}$$

As we already commented, the Hankel matrix can be understood as created by the action of the autonomous Koopman operator (of the skew product flow) on the observable (6.60), i.e.,

$$H_{N\cdot i+r, j+1} - U_S^{(i+j)\Delta t} g_r(\mathbf{y}_0) = (U_S^{\Delta t})^{(i+j)} g_r(\mathbf{y}_0), \tag{6.66}$$

for $r = 1, \ldots, N, i = 0, \ldots, n_H - 1,\ j = 0, \ldots, m_H$. In this interpretation, the result from [1] might be applicable. In particular, if the autonomous skew product flow is ergodic and if observables are in an invariant subspace of the Koopman operator, the eigenvalues and eigenvectors obtained by the DMD or Arnoldi-type algorithms used on the Hankel matrix, converge to the true Koopman eigenvalues and eigenfunctions of the considered system.

On the other hand, if we have the ambition to compute the eigenvalues and eigenfunctions of the nonautonomous Koopman operator family, we can interpret the same Hankel matrix in the process formulation

$$H_{N\cdot i+r, j+1} = U^{t_0+(i+j)\Delta t, t_0} f_r(\mathbf{x}_0), \tag{6.67}$$

or in the skew product flow formulation

$$H_{N\cdot i+r,j+1} = U^{(i+j)\Delta t,\mathbf{p}_0} f_r(\mathbf{x}_0), \tag{6.68}$$

for $r = 1, \ldots, N, i = 0, \ldots, n_H - 1,\ j = 0, \ldots, m_H$.

In both cases, we do not have the same operator applied multiple times, therefore application of the DMD or Arnoldi-type algorithms on some small snapshot span $(n_H + m_H)$ in order to obtain the local approximations of eigenvalues and eigenfunctions will give us just an approximation of $U^{t_0+\Delta t,t_0}$, i.e., of $U^{\Delta t,\mathbf{p}_0}$. The following theorem exposes the error of such approximation.

**Theorem 6.1** *If $\tilde{\lambda}$ is the eigenvalue obtained with the Arnoldi-type algorithm applied to a vector observable $\mathbf{f} : X \to \mathbb{C}^N$ such that $\mathbf{f}(\mathbf{x}(t))$ is $C^3$, then*

$$\tilde{\lambda} = \frac{\langle \dot{\mathbf{f}}_0, \mathbf{f}_0 \rangle}{\langle \mathbf{f}_0, \mathbf{f}_0 \rangle} + \mathcal{O}(\Delta t) \tag{6.69}$$

*in the two-snapshots case, and*

$$\tilde{\lambda} = \frac{\langle \ddot{\mathbf{f}}_0, \mathbf{f}_0 \rangle}{\langle \dot{\mathbf{f}}_0, \mathbf{f}_0 \rangle} + \mathcal{O}(\Delta t) \tag{6.70}$$

*in the three-snapshots case.*

*Proof* Let us consider the snapshots $\mathbf{f}_k = \mathbf{f}(t_k)$, at the times $t_k = t_0 + k\Delta t$, $k = 0, 1, 2, \ldots, s$.

Since the observable $\mathbf{f}$ is continuous in time at least up to its third derivative, we can expand it in Taylor series at $t = t_0$

$$\mathbf{f}_k = \mathbf{f}_0 + \dot{\mathbf{f}}_0 \Delta t + \ddot{\mathbf{f}}_0 \frac{\Delta t^2}{2!} + \mathcal{O}(\Delta t^3), \quad k = 0, 1, 2, \ldots, s. \tag{6.71}$$

Here, subscript 0 denotes evaluation at $t = t_0$. Now, we want to project $\mathbf{f}_s$ to the Krylov subspace spanned by $\mathbf{f}_0, \ldots, \mathbf{f}_{s-1}$, i.e., we want to find $c_k$, $k = 0, 1, \ldots, s-1$, such that

$$\sum_{k=0}^{s-1} c_k \mathbf{f}_k = \mathbf{f}_s + \mathbf{r} \text{ and } \mathbf{r} \perp \mathbf{f}_l, \quad l = 0, 1, s-1. \tag{6.72}$$

By using (6.72), we get

$$\sum_{k=0}^{s-1} c_k \langle \mathbf{f}_k, \mathbf{f}_l \rangle = \langle \mathbf{f}_s, \mathbf{f}_l \rangle, \quad l = 0, 1, \ldots, s-1, \tag{6.73}$$

while by using (6.71), we obtain

$$\langle \mathbf{f}_k, \mathbf{f}_l \rangle = \langle \mathbf{f}_0, \mathbf{f}_0 \rangle + \left( k\langle \dot{\mathbf{f}}_0, \mathbf{f}_0 \rangle + l\langle \mathbf{f}_0, \dot{\mathbf{f}}_0 \rangle \right) \Delta t$$
$$+ \left( k^2\langle \ddot{\mathbf{f}}_0, \mathbf{f}_0 \rangle + 2kl\langle \dot{\mathbf{f}}_0, \dot{\mathbf{f}}_0 \rangle + l^2\langle \mathbf{x}_0, \ddot{\mathbf{f}}_0 \rangle \right) \frac{\Delta t^2}{2} + \mathcal{O}(\Delta t^3) \,, \tag{6.74}$$

for all $k, l = 0, 1, 2, \ldots, s$.

Observe that $c_k = c_k(\Delta t)$ so we can also expand

$$c_k(\Delta t) = c_k^{(0)} + c_k^{(1)} \Delta t + c_k^{(2)} \frac{\Delta t^2}{2} + \mathcal{O}(\Delta t^3), \quad k = 0, 1, \ldots, s-1, \tag{6.75}$$

and then by using (6.74) and (6.75), we get

$$\sum_{k=0}^{s-1} \Big\{ c_k^{(0)} \langle \mathbf{f}_0, \mathbf{f}_0 \rangle + \left[ c_k^{(1)} \langle \mathbf{f}_0, \mathbf{f}_0 \rangle + c_k^{(0)} \left( k\langle \dot{\mathbf{f}}_0, \mathbf{f}_0 \rangle + l\langle \mathbf{f}_0, \dot{\mathbf{f}}_0 \rangle \right) \right] \Delta t$$
$$+ \Big[ c_k^{(2)} \langle \mathbf{f}_0, \mathbf{f}_0 \rangle + 2c_k^{(1)} \left( k\langle \dot{\mathbf{f}}_0, \mathbf{f}_0 \rangle + l\langle \mathbf{f}_0, \dot{\mathbf{f}}_0 \rangle \right)$$
$$+ c_k^{(0)} \left( k^2\langle \ddot{\mathbf{f}}_0, \mathbf{f}_0 \rangle + 2kl\langle \dot{\mathbf{f}}_0, \dot{\mathbf{f}}_0 \rangle + l^2\langle \mathbf{f}_0, \ddot{\mathbf{f}}_0 \rangle \right) \Big] \frac{\Delta t^2}{2} + \mathcal{O}(\Delta t^3) \Big\}$$
$$= \langle \mathbf{f}_0, \mathbf{f}_0 \rangle + \left[ s\langle \dot{\mathbf{f}}_0, \mathbf{f}_0 \rangle + l\langle \mathbf{f}_0, \dot{\mathbf{f}}_0 \rangle \right] \Delta t$$
$$+ \left[ s^2\langle \ddot{\mathbf{f}}_0, \mathbf{f}_0 \rangle + 2sl\langle \dot{\mathbf{f}}_0, \dot{\mathbf{f}}_0 \rangle + l^2\langle \mathbf{f}_0, \ddot{\mathbf{f}}_0 \rangle \right] \frac{\Delta t^2}{2} + \mathcal{O}(\Delta t^3), \quad l = 0, 1, \ldots, s-1. \tag{6.76}$$

If we separate the terms that multiply $\Delta t^0$, we obtain

$$\left( \sum_{k=0}^{s-1} c_k^{(0)} \right) \langle \mathbf{f}_0, \mathbf{f}_0 \rangle = \langle \mathbf{f}_0, \mathbf{f}_0 \rangle. \tag{6.77}$$

Then, we separate the terms that multiply $\Delta t^1$

$$\left( \sum_{k=0}^{s-1} c_k^{(1)} \right) \langle \mathbf{f}_0, \mathbf{f}_0 \rangle + \left( \sum_{k=0}^{s-1} k c_k^{(0)} \right) \langle \dot{\mathbf{f}}_0, \mathbf{f}_0 \rangle + \left( l \sum_{k=0}^{s-1} c_k^{(0)} \right) \langle \mathbf{f}_0, \dot{\mathbf{f}}_0 \rangle$$
$$= s\langle \dot{\mathbf{f}}_0, \mathbf{f}_0 \rangle + l\langle \mathbf{f}_0, \dot{\mathbf{f}}_0 \rangle, \quad l = 0, 1, \ldots, s-1, \tag{6.78}$$

and finally terms that multiply $\Delta t^2$

$$\left( \sum_{k=0}^{s-1} c_k^{(2)} \right) \langle \mathbf{f}_0, \mathbf{f}_0 \rangle + 2\left( \sum_{k=0}^{s-1} k c_k^{(1)} \right) \langle \dot{\mathbf{f}}_0, \mathbf{f}_0 \rangle + 2\left( l \sum_{k=0}^{s-1} c_k^{(1)} \right) \langle \mathbf{f}_0, \dot{\mathbf{f}}_0 \rangle$$
$$+ \left( \sum_{k=0}^{s-1} k^2 c_k^{(0)} \right) \langle \ddot{\mathbf{f}}_0, \mathbf{f}_0 \rangle + 2\left( l \sum_{k=0}^{s-1} k c_k^{(0)} \right) \langle \dot{\mathbf{f}}_0, \dot{\mathbf{f}}_0 \rangle + l^2 \left( \sum_{k=0}^{s-1} c_k^{(0)} \right) \langle \mathbf{f}_0, \ddot{\mathbf{f}}_0 \rangle$$
$$= s^2\langle \ddot{\mathbf{f}}_0, \mathbf{f}_0 \rangle + 2sl\langle \dot{\mathbf{f}}_0, \dot{\mathbf{f}}_0 \rangle + l^2\langle \mathbf{f}_0, \ddot{\mathbf{f}}_0 \rangle, \quad l = 0, 1, \ldots, s-1., \tag{6.79}$$

and we obtain a system of equations that can be solved for $c_k$, $k = 0, 1, \ldots, s-1$.

Therefore, from (6.77), we get $\sum_{k=0}^{s-1} c_k^{(0)} = 1$, which applied to (6.78) implies

$$\left(\sum_{k=0}^{s-1} c_k^{(1)}\right) \langle \mathbf{f}_0, \mathbf{f}_0 \rangle + \left(\sum_{k=0}^{s-1} k c_k^{(0)}\right) \langle \dot{\mathbf{f}}_0, \mathbf{f}_0 \rangle = s \langle \dot{\mathbf{f}}_0, \mathbf{f}_0 \rangle, \tag{6.80}$$

and applied to (6.79), it implies

$$\left(\sum_{k=0}^{s-1} c_k^{(2)}\right) \langle \mathbf{f}_0, \mathbf{f}_0 \rangle + 2\left(\sum_{k=0}^{s-1} k c_k^{(1)}\right) \langle \dot{\mathbf{f}}_0, \mathbf{f}_0 \rangle + 2\left(l \sum_{k=0}^{s-1} c_k^{(1)}\right) \langle \mathbf{f}_0, \dot{\mathbf{f}}_0 \rangle l + \left(\sum_{k=0}^{s-1} k^2 c_k^{(0)}\right) \langle \ddot{\mathbf{f}}_0, \mathbf{f}_0 \rangle$$
$$+2\left(l \sum_{k=0}^{s-1} k c_k^{(0)}\right) \langle \dot{\mathbf{f}}_0, \dot{\mathbf{f}}_0 \rangle = s^2 \langle \ddot{\mathbf{f}}_0, \mathbf{f}_0 \rangle + sl \langle \dot{\mathbf{f}}_0, \dot{\mathbf{f}}_0 \rangle, \quad l = 0, 1, \ldots, s-1. \tag{6.81}$$

**Two-snapshots case.** If $s = 1$, then from (6.77)–(6.79), we obtain

$$c_0^{(0)} = 1 \tag{6.82}$$

and

$$c_0^{(1)} \langle \mathbf{f}_0, \mathbf{f}_0 \rangle = \langle \dot{\mathbf{f}}_0, \mathbf{f}_0 \rangle, \tag{6.83}$$

and therefore

$$c_0(\Delta t) = 1 + \frac{\langle \dot{\mathbf{f}}_0, \mathbf{f}_0 \rangle}{\langle \mathbf{f}_0, \mathbf{f}_0 \rangle} \Delta t + \mathscr{O}(\Delta t^2). \tag{6.84}$$

In this simple case, we solve the equation $\mu - c_0 = 0$ and if we write the solution in the form $\mu = e^{\tilde{\lambda} \Delta t}$, we get

$$\tilde{\lambda} = \frac{\langle \dot{\mathbf{f}}_0, \mathbf{f}_0 \rangle}{\langle \mathbf{f}_0, \mathbf{f}_0 \rangle} + \mathscr{O}(\Delta t). \tag{6.85}$$

**Three-snapshots case.** If $s = 2$, Eqs. (6.77)–(6.79) become

$$c_0^{(0)} + c_1^{(0)} = 1, \tag{6.86}$$

$$\left(c_0^{(1)} + c_1^{(1)}\right) \langle \mathbf{f}_0, \mathbf{f}_0 \rangle + c_1^{(0)} \langle \dot{\mathbf{f}}_0, \mathbf{f}_0 \rangle = 2 \langle \dot{\mathbf{f}}_0, \mathbf{f}_0 \rangle, \tag{6.87}$$

$$2 c_1^{(1)} \langle \dot{\mathbf{f}}_0, \mathbf{f}_0 \rangle + c_1^{(0)} \langle \ddot{\mathbf{f}}_0, \mathbf{f}_0 \rangle = 4 \langle \ddot{\mathbf{f}}_0, \mathbf{f}_0 \rangle, \tag{6.88}$$

$$2c_1^{(1)}\langle \dot{\mathbf{f}}_0, \mathbf{f}_0\rangle + 2\left(c_0^{(1)} + c_1^{(1)}\right)\langle \mathbf{f}_0, \dot{\mathbf{f}}_0\rangle + c_1^{(0)}\langle \ddot{\mathbf{f}}_0, \mathbf{f}_0\rangle + 2c_1^{(0)}\langle \dot{\mathbf{f}}_0, \dot{\mathbf{f}}_0\rangle = \\ = 4\langle \ddot{\mathbf{f}}_0, \mathbf{f}_0\rangle + 4\langle \dot{\mathbf{f}}_0, \dot{\mathbf{f}}_0\rangle. \tag{6.89}$$

By taking into account (6.88), the last equation becomes

$$\left(c_0^{(1)} + c_1^{(1)}\right)\langle \mathbf{f}_0, \dot{\mathbf{f}}_0\rangle + c_1^{(0)}\langle \dot{\mathbf{f}}_0, \dot{\mathbf{f}}_0\rangle = 2\langle \dot{\mathbf{f}}_0, \dot{\mathbf{f}}_0\rangle. \tag{6.90}$$

If we solve that system, we obtain

$$c_0^{(0)} = -1, c_1^{(0)} = 2, \tag{6.91}$$

$$c_0^{(1)} = -\frac{\langle \ddot{\mathbf{f}}_0, \mathbf{f}_0\rangle}{\langle \dot{\mathbf{f}}_0, \mathbf{f}_0\rangle}, c_1^{(1)} = -c_0^{(1)}. \tag{6.92}$$

Finally, if we compute the roots of the equation $\mu^2 - c_1(\Delta t)\mu - c_0(\Delta t) = 0$ and write them in the form $\mu = e^{\tilde{\lambda}\Delta t}$, we get

$$\tilde{\lambda} = \frac{\langle \ddot{\mathbf{f}}_0, \mathbf{f}_0\rangle}{\langle \dot{\mathbf{f}}_0, \mathbf{f}_0\rangle} + \mathcal{O}(\Delta t). \tag{6.93}$$

□

Consider the simplest case, when **f** is the full-state observable of a nonautonomous dynamical system that is linearized and diagonalized, i.e., $\mathbf{f}(t) = \mathbf{x}(t)$ such that

$$\dot{\mathbf{x}} = \Lambda(t)\mathbf{x}. \tag{6.94}$$

Then

$$\ddot{\mathbf{x}} = \left(\dot{\Lambda}(t) + \Lambda(t)^2\right)\mathbf{x}, \tag{6.95}$$

and Theorem 6.1 means that for the three-snapshots case, Arnoldi-type algorithms will give us the following approximation of eigenvalues

$$\tilde{\lambda} = \frac{\langle \left(\dot{\Lambda}(t_0) + \Lambda(t_0)^2\right)\mathbf{x}_0, \mathbf{x}_0\rangle}{\langle \Lambda(t_0)\mathbf{x}_0, \mathbf{x}_0\rangle} + \mathcal{O}(\Delta t). \tag{6.96}$$

Therefore, even with the right choice of the initial state $\mathbf{x}_0$ instead of obtaining one of the exact eigenvalues $\lambda(t_0)$ or at least an approximation that decreases with the decrease of the time step, we obtain

$$\tilde{\lambda} = \lambda(t_0) + \frac{\dot{\lambda}(t_0)}{\lambda(t_0)} + \mathcal{O}(\Delta t), \tag{6.97}$$

i.e., the error in the obtained approximation is proportional to the time derivative of that eigenvalue.

### 6.4.1 Numerical Examples

In examples that follow, we apply the Vandermonde–Cauchy algorithm [9] to the Hankel matrix (6.65). The Vandermonde–Cauchy provides us with a set of Ritz values and vectors

$$(\mu_j, \mathbf{R}_j),\ j = 1, 2, \ldots, m_H. \tag{6.98}$$

Observe that the Ritz vectors are of dimension $n_H \cdot N$, which is the number of rows of the Hankel matrix, and that each Ritz vector is an approximation of the corresponding eigenfunction evaluation at $\mathbf{f}_0$ multiplied with the related Koopman mode of the vector observable. From Ritz values, we approximate the Koopman operator eigenvalues $\lambda_j,\ j = 1, 2, \ldots, m_H$

$$e^{\lambda_j \Delta t} = \mu_j,\ j = 1, 2, \ldots, m_H. \tag{6.99}$$

Finally, we use the expression

$$\tilde{\mathbf{f}}(t) = \sum_{j=1}^{m_H} \mu_j^{\frac{t-t_0}{\Delta t}} \mathbf{R}_j \tag{6.100}$$

as an algorithm-given approximation of the observable. This expression corresponds with the decomposition in Propositions 6.1 and 6.2.

*Example 6.1* We consider a nonautonomous differential equation of the form

$$\begin{pmatrix} \dot{x}_1 \\ \dot{x}_2 \end{pmatrix} = \begin{pmatrix} (\sigma_0 + A_d \cos(\omega_d t) + B_d \sin(\omega_d t))x_1 + \omega_0 x_2 \\ -\omega_0 x_1 + (\sigma_0 + A_d \cos(\omega_d t) + B_d \sin(\omega_d t))x_2 \end{pmatrix}. \tag{6.101}$$

Alternatively, we can set an autonomous differential equation that corresponds to the driving parameter

$$\begin{pmatrix} \dot{p}_1 \\ \dot{p}_2 \end{pmatrix} = \begin{pmatrix} \omega_d p_2 \\ -\omega_d p_1 \end{pmatrix}, \tag{6.102}$$

and then reformulate (6.101) into

$$\begin{pmatrix} \dot{x}_1 \\ \dot{x}_2 \end{pmatrix} = \begin{pmatrix} -\omega_0 x_1 + (\sigma_0 + p_1)x_2 \\ (\sigma_0 + p_1)x_1 + \omega_0 x_2 \end{pmatrix}. \tag{6.103}$$

*Example 6.2* Similarly, we can reformulate the nonautonomous differential equation

$$\begin{pmatrix} \dot{x}_1 \\ \dot{x}_2 \end{pmatrix} = \begin{pmatrix} \sigma_0 x_1 + (\omega_0 + A_d \cos(\omega_d t) + B_d \sin(\omega_d t))x_2 \\ -(\omega_0 + A_d \cos(\omega_d t) + B_d \sin(\omega_d t))x_1 + \sigma_0 x_2 \end{pmatrix} \tag{6.104}$$

into

$$\begin{pmatrix} \dot{x}_1 \\ \dot{x}_2 \end{pmatrix} = \begin{pmatrix} \sigma_0 x_1 + (\omega_0 + p_1) x_2 \\ -(\omega_0 + p_1) x_1 + \sigma_0 x_2 \end{pmatrix} \tag{6.105}$$

driven by (6.102).

In both examples, we have a nonautonomous oscillator; in the first one, the amplitude is driven, in the second one, the frequency is driven, and in both, the driving parameter (6.102) is an oscillator too. Observe that both process formulations (6.101) and (6.104) satisfy conditions of the Proposition 6.2, while both skew product formulations (6.102)–(6.103) and (6.102)–(6.105) satisfy conditions of the Proposition 6.1. We concentrate on just one observable—the first component of the full-state observable, i.e., $f = x_1$. For that observable, we form a Hankel matrix ((6.65), $N = 1$). The time we cover with the snapshot span we present as the Hankel matrix time span $T_H = (n_H + m_H + 1)\Delta t$. In all computations, we apply the Vandermonde–Cauchy algorithm (see [9]).

To investigate different settings of the skew product flow, we perform the computations for different periods of the driving parameter ($T_d = 2\pi/\omega_d$) and of the non-driven oscillator ($T_0 = 2\pi/\omega_0$). In both Figs. 6.1 and 6.2, subfigures (a) and (b) correspond to the intrinsic period larger than the driven $T_0 > T_d$, subfigures (c) and (d) correspond to equal intrinsic and driven period $T_0 = T_d$, while subfigures (e) and (f) correspond to the intrinsic period smaller than the driven period $T_0 < T_d$. Also, all presented computations are performed with $A_d = B_d = 1$, i.e., the initial state of the parameter $\mathbf{p}_0 = (1, 1)$.

Our first goal is to compute the nonautonomous Koopman operator eigenvalues and eigenfunctions. In hope to catch these local eigenvalues and eigenfunctions, we use Hankel matrix with a relatively small snapshot span. The corresponding results are presented in Figs. 6.1 and 6.2, subfigures (a), (c), and (e).

Our second goal is to see if we can, through that one observable, discover the eigenvalues and eigenfunctions of the related skew product flow which is an autonomous dynamical system. In that case, we use Hankel matrix with a large snapshot span (see [1]). The corresponding results are presented in Figs. 6.1 and 6.2, subfigures (b), (d), and (f).

In every subfigure, in the upper row, we present the real and the imaginary part of the eigenvalues as they evolve in time. The results are obtained with the Vandermonde–Cauchy algorithm on the Hankel matrix. The colors represent the magnitude of the corresponding mode.

In the second row of every subfigure, we compare the data with the results obtained with expression (6.100). When the expression (6.100) is used for the time moments within the Hankel matrix time span, i.e., $t - t_0 \leq T_H$, then it gives us the reconstruction of the data used to obtain the eigenvalues and eigenvectors. When (6.100) is used for the time moments after that span, i.e., $t - t_0 > T_H$, it gives us an extrapolation of the learned model, i.e., it is an attempt to predict the evolution of the dynamical system.

Results in Figs. 6.1 and 6.2, subfigures (a), (c), and (e), upper row, show what is stated in Theorem 6.1. We are under the conditions of the Proposition 6.1 in the skew product flow formulation, i.e., Proposition 6.2 in the process formulation, so the exact nonautonomous principal eigenvalues are well known. However the approximate eigenvalues given by the data-driven algorithm have the error which is of the same nature as the one calculated in Theorem 6.1. Actually, the error is even more complicated since here we use larger snapshot span. If we look at the lower row of the subfigures (a), (c), and (e) (both Figs. 6.1 and 6.2), we see that the reconstruction is very close to the data, and even the prediction holds for some period of time. The algorithm is set in such a way that it accurately uses given data, so the fact that reconstruction works is not a surprise. We should also take into account the fact that the nonautonomous eigenvalues and eigenfunctions are local, so even if the approximate eigenvalues and eigenfunctions were computed without an error, their extrapolation would still have a limited span.

On the other side, results in the upper rows of both Figs. 6.1 and 6.2, subfigures (b), (d), and (f) show that the Vandermonde–Cauchy algorithm applied on the Hankel matrix with the right time span, even when built on just one state component, reveals the correct eigenvalues and eigenfunctions of the autonomous Koopman operator of the skew product flow. Since now we have all the relevant information about the dynamical system, and since these eigenvalues are constant in time, the prediction of the future behavior is, of course, very precise (lower rows of subfigures (b), (d), and (f), both Figs. 6.1 and 6.2).

Regardless of the example, chosen parameters, or different combinations of intrinsic and driven period, all obtained results show the same behavior of the data-driven algorithms.

## 6.5 Data-Driven Algorithms for the Stochastic Koopman Operator

Let $\mathbf{f} = (f_1, \dots, f_N)^T : X \to \mathbb{C}^N$ be a vector observable. We apply the data-driven algorithm in the stochastic setting in the following way. For the chosen time moments $t_k = k\Delta t$ and chosen $\mathbf{x} \in X$, we have

$$U_S^{k\Delta t}\mathbf{f}(\mathbf{x}) = \mathbb{E}[\mathbf{f}(\mathbf{S}^{k\Delta t,\omega}(\mathbf{x}))],\ k = 1, 2, \dots.$$

As in the nonautonomous case, we apply the data-driven algorithm to the Hankel matrix $\mathbf{H}$ with elements

$$H_{N\cdot i+r,j+1} = U_S^{(i+j)\Delta t} f_r(\mathbf{x}), \tag{6.106}$$

for $r = 1, \dots, N, i = 0, \dots, n_H - 1,\ j = 0, \dots, m_H$. Then we apply the Vandermonde–Cauchy algorithm so that by using (6.100), we expect to obtain the algorithm approximation of the function $\overline{\mathbf{f}}(t) = U_S^t\mathbf{f}(\mathbf{x})$, which corresponds to the average value of the observable at time $t$.

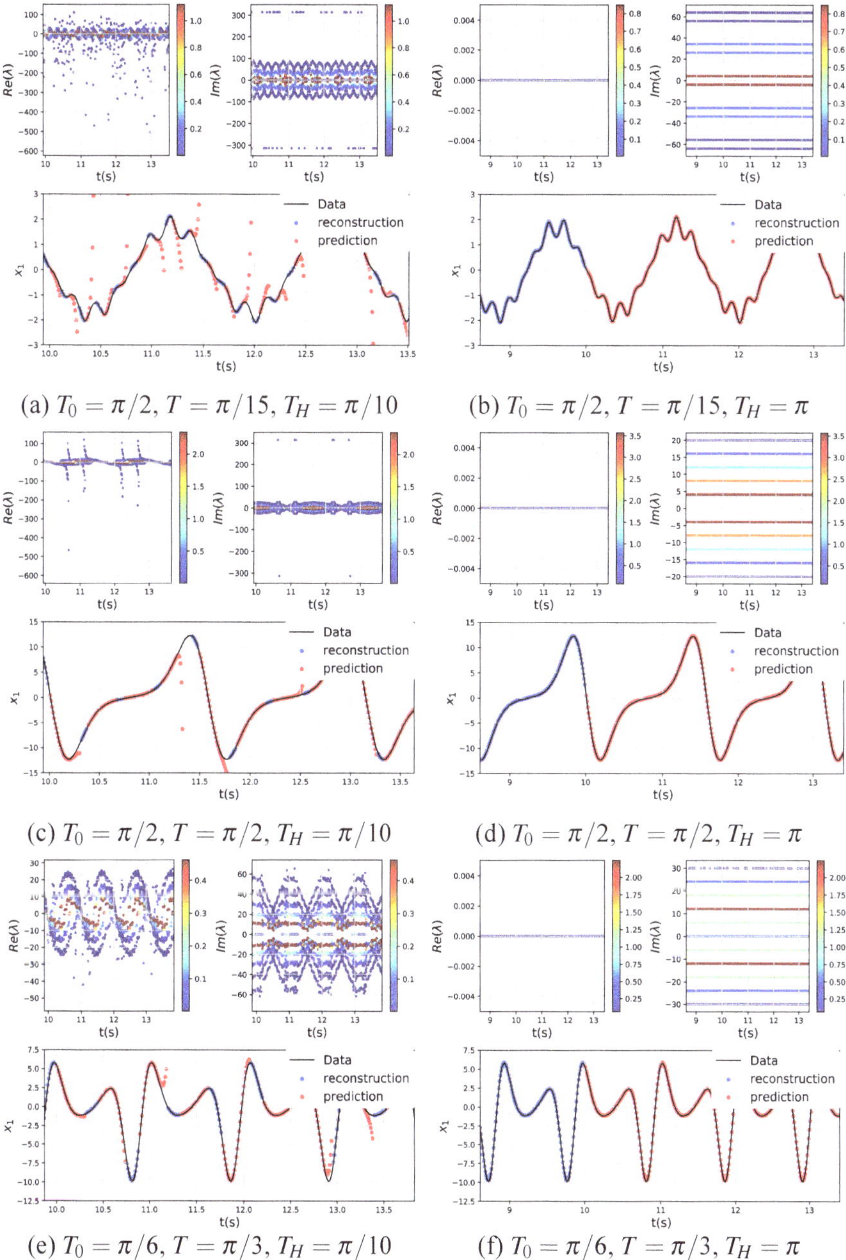


(a) $T_0 = \pi/2, T = \pi/15, T_H = \pi/10$ (b) $T_0 = \pi/2, T = \pi/15, T_H = \pi$

(c) $T_0 = \pi/2, T = \pi/2, T_H = \pi/10$ (d) $T_0 = \pi/2, T = \pi/2, T_H = \pi$

(e) $T_0 = \pi/6, T = \pi/3, T_H = \pi/10$ (f) $T_0 = \pi/6, T = \pi/3, T_H = \pi$

**Fig. 6.1** Oscillator with the driven amplitude: computations obtained with the Vandermonde–Cauchy algorithm on the Hankel matrix for single observable $f = x_1$

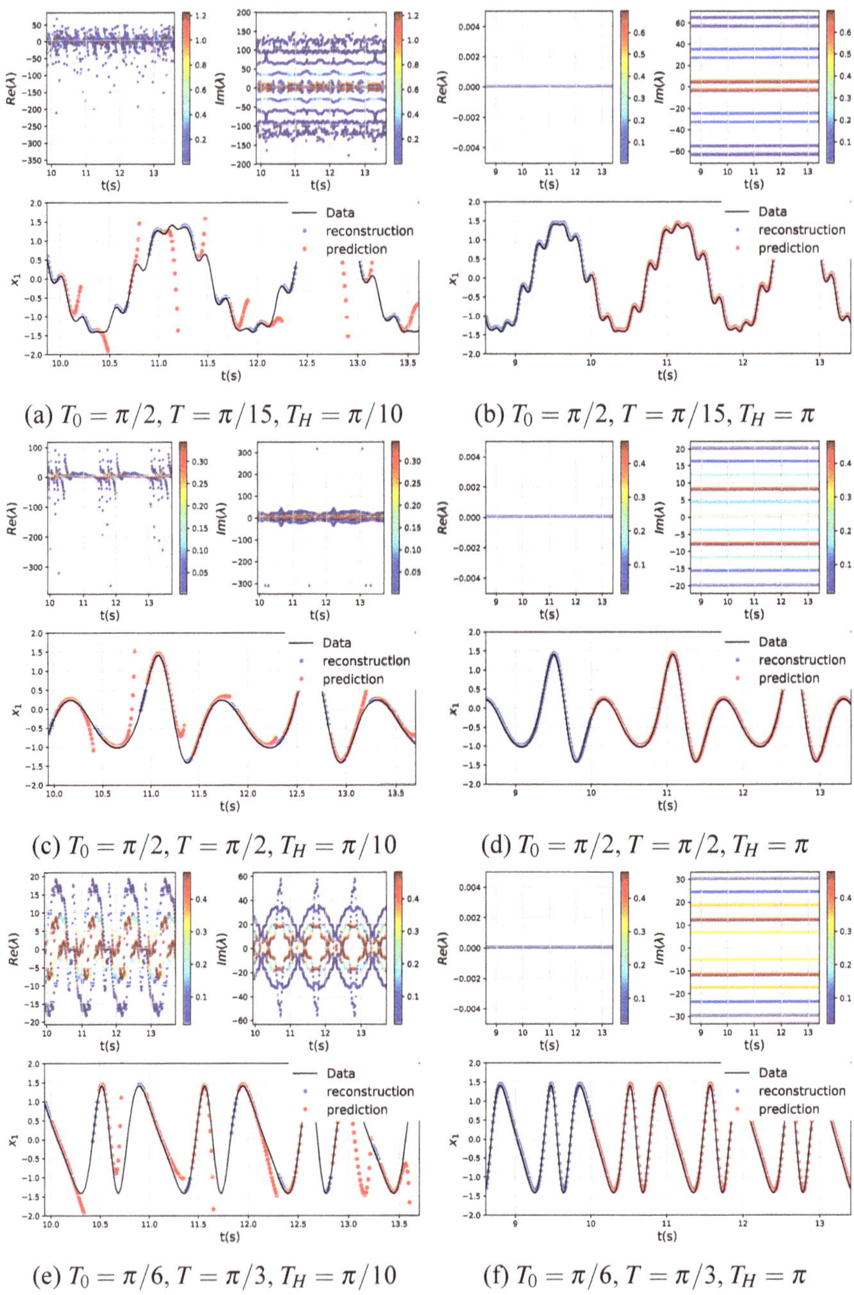


(a) $T_0 = \pi/2, T = \pi/15, T_H = \pi/10$ (b) $T_0 = \pi/2, T = \pi/15, T_H = \pi$

(c) $T_0 = \pi/2, T = \pi/2, T_H = \pi/10$ (d) $T_0 = \pi/2, T = \pi/2, T_H = \pi$

(e) $T_0 = \pi/6, T = \pi/3, T_H = \pi/10$ (f) $T_0 = \pi/6, T = \pi/3, T_H = \pi$

**Fig. 6.2** Oscillator with the driven frequency: computations are obtained with the Vandermonde–Cauchy algorithm on the Hankel matrix for single observable $f = x_1$

If RDS $\mathbf{S}$ is Markovian, it follows from the Proposition 6.4 that

$$U_S^{k\Delta t} = (U_S^{\Delta t})^k,$$

which means that the elements of the Hankel matrix are of the form

$$H_{N\cdot i+r, j+1} = (U_S^{\Delta t})^{i+j} f_r(\mathbf{x}). \tag{6.107}$$

Thus, we expect that $\lambda_j$ in (6.99) are approximations of the eigenvalues of the Koopman generator, if it exists. In such a case, we also expect that (6.100) gives good approximation of $\bar{\mathbf{f}}(t)$.

However, if RDS $\mathbf{S}$ is not Markovian, the probability measure associated with the stochastic process $(\mathbf{S}^{k\Delta t,\omega}(\mathbf{x}))_{k\in\mathbb{N},\omega\in\Omega}$ is not a product measure and the elements of the Hankel matrix do not belong to the same operator applied multiple times. Similarly as in the nonautonomous case, the application of the DMD or Arnoldi-type algorithm should approximate the average change of the eigenvalue $\lambda_{S,j}(t_k)$ in a time interval $[t_k, t_k + \Delta t]$. Due to the fact that these changes are not constant for different values of $t_k$, the application of the algorithm to a snapshot span larger than two could result in a significant error in the approximation of the eigenvalues.

In the second approach, we apply the data-driven DMD algorithm [10] to the matrices

$$\mathbf{X} = [\mathbf{f}(\mathbf{x}_1), \ldots, \mathbf{f}(\mathbf{x}_M)] \text{ and } \mathbf{Y} = [U_S^{\Delta t}\mathbf{f}(\mathbf{x}_1), \ldots, U_S^{\Delta t}\mathbf{f}(\mathbf{x}_M)], \tag{6.108}$$

where $\mathbf{x}_i, i = 1, \ldots, M$ are chosen initial states, in order to obtain the approximations of spectral objects of the stochastic Koopman operator $U_S^{\Delta t}$.

### 6.5.1 Numerical Examples

*Example 6.3* We consider the linear RDE of the form

$$\begin{aligned}\begin{pmatrix}\dot{x}_1\\ \dot{x}_2\end{pmatrix} &= \mathbf{A}(\pi(\theta^t(\omega)))\begin{pmatrix}x_1\\ x_2\end{pmatrix}\\ &= \begin{pmatrix}\mu_0 + a\sin(W_t(\omega)) & \omega_0\\ -\omega_0 & \mu_0 + a\sin(W_t(\omega))\end{pmatrix}\begin{pmatrix}x_1\\ x_2\end{pmatrix},\end{aligned} \tag{6.109}$$

where $W_t(\omega)$ denotes the standard Wiener process, and $\pi$ is defined by $\pi(\omega) = \omega(0)$. Observe that (6.109) is the equation of type (6.42), where the bounded noise is modeled as a function of the Wiener process. As described previously for the "memory-less" RDE, the coordinate process and the shift transformations $\theta^t$ defined by (6.43) are connected through the following relation $\omega(t) = W_t(\omega) = \pi(\theta^t(\omega))$. Since the noise is modeled by a bounded and continuous Markov process, the generated RDS

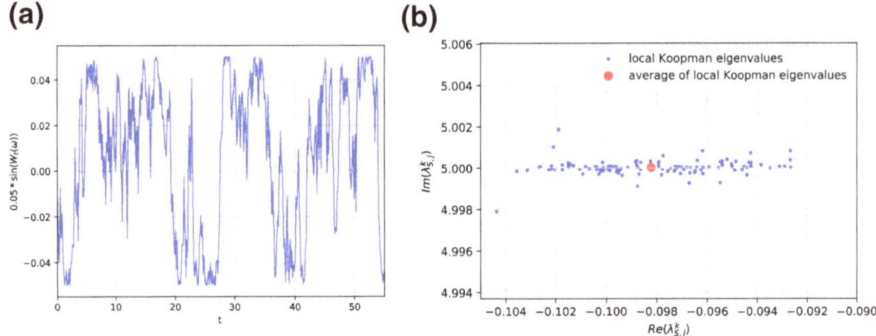


**Fig. 6.3** Example 6.3: **a** the noise for $a = 0.05$; **b** the eigenvalues obtained with the Vandermonde–Cauchy algorithm on the Hankel matrix for single observable $f(\mathbf{x}) = x_1$

is globally defined and differentiable. The matrices $\mathbf{A}(\pi(\theta^t(\omega)))$ are commutative and simultaneously diagonalizable, thus the generated RDS is of the form

$$\mathbf{S}^{t,\omega}(\mathbf{x}) = \mathrm{e}^{a\int_0^t \sin(W_s(\omega))ds}\mathrm{e}^{\hat{\mathbf{A}}t}\mathbf{x},$$

where

$$\hat{\mathbf{A}} = \begin{pmatrix} \mu_0 & \omega_0 \\ -\omega_0 & \mu_0 \end{pmatrix}.$$

It follows from the Proposition 6.3 that the eigenfunctions of the stochastic Koopman operators $U_S^t$ are equal to $\phi_{S,j}^t(\mathbf{x}) = \langle \mathbf{x}, \mathbf{w}_j \rangle$, $j = 1, 2$, where $\mathbf{w}_j$ are left eigenvectors of $\hat{\mathbf{A}}$. Furthermore, the eigenvalues are equal to

$$\lambda_{S,j}(t) = (\mu_0 \pm \omega_0\, i)t + \log \mathbb{E}\left[\mathrm{e}^{a\int_0^t \sin W_s(\omega)ds}\right].$$

We hope to reveal the approximations of the averages of eigenvalues of the operator $U_S^{\Delta t}$ by applying the data-driven algorithm (the Vandermonde–Cauchy algorithm) to the Hankel matrix. For different values of $t_k$, these averages for the operator acting on $[t_k, t_k + \Delta t]$ are equal to

$$\overline{\lambda}_{S,j}^k = (\mu_0 \pm \omega_0\, i) + \frac{1}{\Delta t}\log \mathbb{E}\left[\mathrm{e}^{a\int_{t_k}^{t_k+\Delta t} \sin W_s(\omega)ds}\right].$$

The approximate values of $\overline{\lambda}_{S,j}^k$ obtained by the numerical computations with parameters $\mu_0 = -0.1$, $\omega_0 = 5$, and $a = 0.05$ are presented in Fig. 6.3.

*Example 6.4* We consider the two-dimensional linear SDE of the form

$$dX_t = \mathbf{A}X_t + \sigma dW_t, \tag{6.110}$$

where

$$\mathbf{A} = \begin{pmatrix} \mu_0 & \omega_0 \\ -\omega_0 & \mu_0 \end{pmatrix} \quad \text{and} \quad \sigma = \begin{pmatrix} \sigma_1 & 0 \\ 0 & \sigma_2 \end{pmatrix}. \tag{6.111}$$

SDE (6.110)–(6.111) generates a Markovian RDS. For $Re(\lambda_j) < 0$, where $\lambda_j$, $j = 1, 2$ are the eigenvalues of the matrix $\mathbf{A}$, there exists a generator of the stochastic Koopman operator in $L^2_\mu(\mathbb{R}^2)$ ($\mu$ is an invariant measure associated with the semigroup of Koopman operators) [21]. It has a discrete spectrum of the form

$$\{n_1\lambda_1 + n_2\lambda_2, n_1, n_2 \in \mathbb{N} \cup \{0\}\},$$

while the associated eigenfunctions are polynomials (see [21]). The approximations of the eigenvalues of the stochastic Koopman generator are obtained by applying the data-driven algorithm to the matrices $\mathbf{X}$ and $\mathbf{Y}$ defined by (6.108). For the chosen $K \in \mathbb{N}$, the observable vector $\mathbf{f}$ is formed as $\mathbf{f} = \left(x_1^l x_2^{k-l}\right)_{l=0}^k$, $k = 0, \ldots, K$. The components of the defined observable vector span an invariant subspace of the stochastic Koopman generator, thus we expect that the data-driven algorithm provides us with the approximations of the eigenvalues $e^{\overline{\lambda}\Delta t}$ of the stochastic Koopman operator $U_S^{\Delta t}$ associated to this subspace. The approximating eigenvalues of the associated generator are then equal to $\overline{\lambda}$. In the numerical computations, two different approaches are used to define the set of initial states. In the first one, $M = 400$ initial states are randomly chosen from $[-2, 2] \times [-2, 2]$, while in the second approach, the states $\mathbf{x}_1, \ldots, \mathbf{x}_M$, $M = 400$ are chosen by averaging the states over the set of $N = 1000$ trajectories for one chosen initial condition. We perform the numerical computations

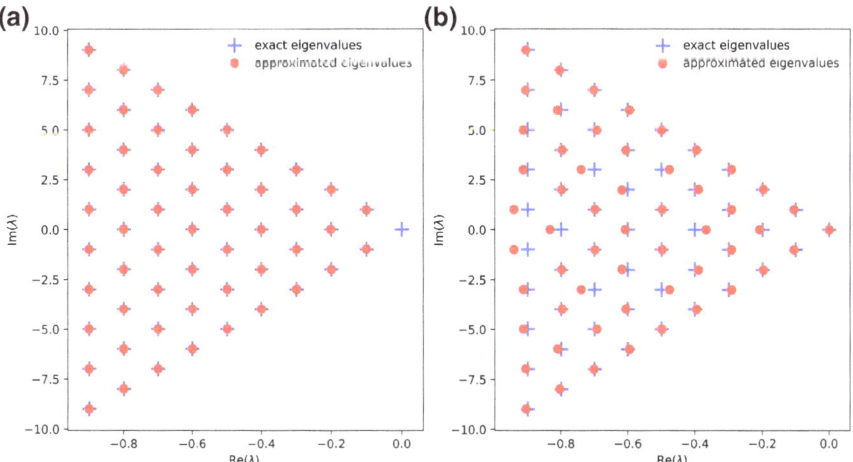


**Fig. 6.4** Example 6.4: the exact eigenvalues of the stochastic Koopman generator and the eigenvalues obtained by the DMD algorithm for **a** randomly chosen states, **b** states chosen on the averaged trajectories

with the following parameters: $\mu_0 = -0.1$, $\omega_0 = 1$, and $\sigma_1 = \sigma_2 = 0.05$. To evaluate the elements of the matrix $\mathbf{Y}$, we average the observable values over $N = 400$ trajectories, approximating in this way the values of the stochastic Koopman operators. The eigenvalues provided by the data-driven algorithm using both described approaches are compared with the exact eigenvalues and presented in Fig. 6.4.

## 6.6 Concluding Remarks

In this chapter, we presented foundations of the Koopman operator theory for nonautonomous dynamical systems, as well as for stochastic dynamical systems. In reality, data collected on complex dynamical systems will exhibit both nonautonomous and stochastic nature, intertwined, not separate. Therefore, future research should be directed to the Koopman framework formulations where nonautonomous and stochastic effects will be simultaneously treated.

Furthermore, we discussed the performance of data-driven algorithms for both types of dynamical systems. We considered applications only on synthetical examples, for the obvious reason: only for such examples we know the exact result so we can judge the quality of the algorithm. The performance actually proved to be excellent. However, there are many additional challenges when Koopman-based data-driven algorithms are applied to real-life data. One of these challenges relates to the hybrid dynamical system, i.e., to the systems that switch discontinuously between different sets of governing laws. If snapshots from which we build the Hankel matrix belong to time intervals with different governing laws, the Koopman eigenvalues and eigenfunctions will not be correctly identified. We presented a possible approach to hybrid dynamical systems in [20], and we plan further research and testing on real-life data.

**Acknowledgements** This research has been supported by the DARPA Contract HR0011-16-C-0116 "On A Data-Driven, Operator-Theoretic Framework for Space–Time Analysis of Process Dynamics". S.M. and N.C-Z. are grateful to Prof. Igor Mezić for helpful mathematical discussions and comments on the manuscript.

## References

1. Arbabi, H., Mezić, I.: Ergodic theory, dynamic mode decomposition and computation of spectral properties of the Koopman operator. SIAM J. Appl. Dyn. Syst. **16**, 2096–2126 (2017)
2. Arnold, L., Kliemann, W.: Qualitative theory of stochastic systems, in Bharucha-Reid, A.T. (ed.) Probabilistic Analysis and Related Topics, vol. 3. Academic Press, New York (1983)
3. Arnold, L.: Stochastic Differential Equations: Theory and Applications. John Wiley Sons, Inc., Hoboken (1974)
4. Arnold, L.: Random Dynamical Systems. Springer, Berlin (1998)
5. Caraballo, T., Han, X.: Applied Nonautonomous and Random Dynamical Systems. BCAM SpringerBriefs. Springer, Cham (2016)

6. Cohen, S.N., Elliot, R.J.: Stochastic Calcukus and Applications. Springer, New York (2015)
7. Crauel, H.: Markov measures for random dynamical systems. Stoch. Stoch. Rep. **37**(3), 153–173 (1991)
8. Črnjarić-Žic, N., Maćešić, S., Mezić, I.: Koopman operator spectrum for random dynamical systems (2017). https://arxiv.org/abs/1711.03146
9. Drmač, Z., Mezić, I., Mohr, R.: Data driven Koopman spectral analysis in Vandermonde–Cauchy form via the DFT: numerical method and theoretical insights (2018). https://arxiv.org/abs/1808.09557
10. Drmač, Z., Mezić, I., Mohr, R.: Data driven modal decompositions: analysis and enhancements. SIAM J. Sci. Comput. **40**(4), A2253–A2285 (2018)
11. Engel, K.J., Nagel, R.: One-parameter semigroups for linear evolution operators. Springer, New York (2001)
12. Giannakis, D.: Data-driven spectral decomposition and forecasting of ergodic dynamical systems. Appl. Comput. Harmon. Anal. (2017). https://doi.org/10.1016/j.acha.2017.09.001
13. Hemati, M.S., Rowley, C.W., Deem, E.A., Cattafesta, L.N.: De-biasing the dynamic mode decomposition for applied Koopman spectral analysis. Theor. Comp. Fluid. Dyn. **31**, 349–368 (2017)
14. Hollingsworth, B.J.: Stochastic Differential Equations: A Dynamical Systems Approach, Dissertation thesis (2008). Auburn University, Auburn
15. Kloeden, P.E., Rasmussen, M.: Nonautonomous Dynamical Systems. Mathematical Surveys and Monographs, vol. 176. AMS, Providence (2011)
16. Klus, S., Koltai, P., Schütte, C.: On the numerical approximation of the Perron–Frobenius and Koopman operator. J. Comp. Dyn. **3**(1), 51–79 (2016). https://doi.org/10.3934/jcd.2016003
17. Korda, M., Mezić, I.: Linear predictors for nonlinear dynamical systems: Koopman operator meets model predictive control. Automatica **93**, 149–160 (2018)
18. Kutz, J.N., Fu, X., Brunton, S.L.: Multiresolution dynamic mode decomposition. SIAM J. Appl. Dyn. Syst. **15**, 713–735 (2016)
19. Lasota, A., Mackey, M.C.: Chaos, Fractals, and Noise. Springer, Berlin (1994)
20. Maćešić, S., Črnjarić-Žic, N., Mezić, I.: Koopman operator family spectrum for nonauonotomus systems. SIAM J. Appl. Dyn. Syst. **17**(4), 2478–2515 (2018)
21. Metafune, G., Pallara, D., Priola, E.: Spectrum of Ornstein–Uhlenbeck operators in $L^p$ spaces with respect to invariant measures. J. Funct. Anal. **196**, 40–60 (2002)
22. Mezić, I.: Analysis of fluid flows via spectral properties of the Koopman operator. Annu. Rev. Fluid Mech. **45**, 357–378 (2013)
23. Mezić, I.: Spectral properties of dynamical systems, model reduction and decompositions. Nonlinear Dynam. **41**, 309–325 (2005)
24. Mezić, I., Banaszuk, A.: Comparison of systems with complex behavior. Physica D **197**, 101–133 (2004)
25. Mezić, I., Surana, A.: Koopman mode decomposition for periodic/quasi-periodic time dependence. IFAC-PapersOnLine **49**, 690–697 (2016). https://doi.org/10.1016/j.ifacol.2016.10.246
26. Proctor, J.L., Brunton, S.L., Kutz, J.N.: Dynamic mode decomposition with control. SIAM J. Appl. Dyn. Syst. **15**, 142–161 (2016)
27. Proctor, J.L., Brunton, S.L., Kutz, J.N.: Including inputs and control within equation-free architectures for complex systems. Eur. Phys. J. Special Topics **225**, 2413–2434 (2016)
28. Proctor, J.L., Brunton, S.L., Kutz, J.N.: Generalizing Koopman theory to allow for inputs and control. SIAM J. Appl. Dyn. Syst. **17**(1), 909–930 (2018)
29. Shnitzer, T., Talmon, R., Slotine, J.J.: Manifold learning with contracting observers for data-driven time-series analysis. IEEE T. Signal Process. **65**, 904–918 (2017)
30. Takeishi, N., Kawahara, Y., Yairi, T.: Subspace dynamic mode decomposition for stochastic Koopman analysis. Phys. Rev. E **96**, 033–310 (2017)
31. Tantet, A., Chekroun, M.D., Dijkstra, H.A., Neelin, J.D.: Mixing spectrum in reduced phase spaces of stochastic differential equations. Part II: Stochastic Hopf bifurcation (2017). https://arxiv.org/abs/1705.07573

32. Williams, M.O., Hemati, M.S., Dawson T.M., Kevrekidis, I.G., Rowley, C.W.: Extending data-driven Koopman analysis to actuated systems. In: Proceedings of the 10th IFAC Symposium on Nonlinear Control Systems, Monterey (2016)
33. Williams, M.O., Kevrekidis, I.G., Rowley, C.W.: A data-driven approximation of the Koopman operator: extending dynamic mode decomposition. J. Nonlinear Sci. **25**(6), 1307–1346 (2015)
34. Zhang, H., Rowley, C.W., Deem, E.A., Cattafesta, L.N.: Online dynamic mode decomposition for time-varying systems (2017). https://arxiv.org/abs/1707.02876

# Chapter 7
# Dynamic Mode Decomposition—A Numerical Linear Algebra Perspective

**Zlatko Drmač**

**Abstract** Data-driven scenarios in analysis and modeling of complex dynamical systems pose formidable challenges to computational science and motivate development of new numerical methods with ever increasing level of sophistication and complexity. An excellent example is computational fluid dynamics (CFD), where the dynamic mode decomposition (DMD) and its enhancement, the sparsity promoting DMD (DMDSP) have emerged as tools of trade for analysis of flow field data, with a host of applications. The problems of high dimension, noisy data, theoretical foundation in connection with the Koopman operator and ergodic theory, and potential applicability in broad spectrum of engineering problems have triggered extensive research and resulted in many theoretical and computational advancements of the DMD framework. This chapter revisits the DMD from the core numerical linear algebra perspective. Recent results on improving numerical robustness and functionality of DMD are reviewed and supplemented with new insights. Further, a new variation of the algorithm is proposed and used as a case study for development of a numerical algorithm in the DMD framework.

## 7.1 Introduction

The basis assumption in DMD analysis [49, 62, 63] is that we are given a sequence of snapshots (vectors of observables) $\mathbf{x}_i \in \mathbb{C}^n$ of an underlying dynamics, that are driven by an unaccessible *black-box* linear operator $\mathbb{A}$;

$$\mathbf{x}_{i+1} = \mathbb{A}\mathbf{x}_i, \quad i = 1, \ldots, m, \quad m < n, \tag{7.1}$$

with some initial state $\mathbf{x}_1$ and with a time lag $\delta t$. No other information is available.

The first task of the DMD is to obtain spectral information, i.e., to approximate some of the eigenvalues and eigenvectors of $\mathbb{A}$. Once the approximate eigenpairs

Z. Drmač (✉)
Department of Mathematics, Faculty of Science, University of Zagreb, Zagreb, Croatia
e-mail: drmac@math.hr



A. Mauroy et al. (eds.), *The Koopman Operator in Systems and Control*,
Lecture Notes in Control and Information Sciences 484,
https://doi.org/10.1007/978-3-030-35713-9_7

$$(\lambda_j, z_j) \text{ such that } \mathbb{A}z_j \approx \lambda_j z_j, \quad j = 1, \ldots, k; \quad k \leq m, \tag{7.2}$$

have been identified, the second task is to derive a spectral spatiotemporal representation of the snapshots $\mathbf{x}_i$:

$$\mathbf{x}_i \approx \sum_{j=1}^{\ell} z_{\varsigma_j} \alpha_j \lambda_{\varsigma_j}^{i-1} \equiv \sum_{j=1}^{\ell} z_{\varsigma_j} \alpha_j |\lambda_{\varsigma_j}|^{i-1} e^{\mathrm{i}\omega_{\varsigma_j}(i-1)\delta t}, \quad i = 1, \ldots, m. \tag{7.3}$$

Here, $\lambda_{\varsigma_j} = |\lambda_{\varsigma_j}| e^{\mathrm{i}\omega_{\varsigma_j}\delta t}$ is the polar representation, $\|z_{\varsigma_j}\|_2 \equiv \sqrt{z_{\varsigma_j}^* z_{\varsigma_j}} = 1$, and the coefficients $\alpha_j$ are determined to minimize the least-square error over all snapshots. It is desirable to have sufficiently small error with small number $\ell$ of the most important modes $z_{\varsigma_1}, \ldots, z_{\varsigma_\ell}$, $\varsigma_j \in \{1, \ldots, k\}$. The decomposition of the snapshots (7.3) can assist in revealing dynamically relevant spatial structures, the $z_{\varsigma_j}$'s, that evolve with amplitudes and frequencies encoded in the corresponding $\lambda_{\varsigma_j}$'s.

For more details and references the reader is referred to Chaps. 1 and 8 in this volume.

### 7.1.1 DMD Framework and Applications

In a carefully designed framework with reach enough set of properly selected observables, the DMD can be considered as a finite-dimensional spectral approximation of the Koopman operator associated with the dynamics under study [4]. This deep theoretical connection gives the DMD a pivotal role in computational study of complex phenomena in fluid dynamics, see, e.g., [61, 70]. Other applications of DMD include e.g., aeroacoustics [50], affective computing (analysis of videos for human emotion recognition [11]), robotics (filtering external perturbation using DMD-based prediction [6]), algorithmic trading on financial markets [52], analysis of infectious disease spread [58], neuroscience [9]—just to name a few. For more detailed analysis of selected applications, we refer the reader to Part III of this volume. For a detailed study of the DMD, see [49].

In general, the data snapshots are acquired under different conditions. If $\mathbb{A}$ represents, e.g., high-performance numerical software for solving systems of ordinary differential equations in high resolution (e.g., in solving semi-discretized partial differential equations), then $n$ can be in hundreds of thousands or even in millions, and an application of $\mathbb{A}$ to a particular vector can be performed with restarting the solver with an appropriate initial condition. In an analysis of combustion instabilities in flame dynamics [32], the snapshots of flames are obtained by vectorizing (reshaping) frames taken by a hi-speed camera with spatial resolution of $1024 \times 1024$ pixels, at 3 KHz rate; after some preprocessing, the number of frames used in DMD is in hundreds. Similarly, in using DMD to detect hidden spontaneous emotions on human faces [11], $\mathbf{x}_i$ is vectorized digital image—with a resolution of $1000 \times 1000$ we already have $n = 10^6$. On the other hand, for a DMD analysis of flu spread, [58] uses the Google Flu Trend tool that generates a spatial–temporal data based on relevant Google search queries and historical records [33].

The tacit assumption in a DMD analysis is that $m \ll n$, i.e., that the number of snapshots $m$ is much smaller than the ambient space dimension $n$. In other words, the action of $\mathbb{A}$ is empirically known only in a small-dimensional subspace of its domain, $\mathbb{C}^n$, as provided by the sequence (7.1). No other information on the action of $\mathbb{A}$ is assumed available. In some applications, it may be known a priori that the underlying operator is (nearly) unitary, and that the eigenvalues will be close to the unit circle. But, in general, having in mind the versatility and adaptivity of the Koopman/DMD framework, we do not make any assumption about the expected locations of the eigenvalues.

#### 7.1.1.1 Computational Aspects of the DMD

A successful application of the DMD requires high level of theoretical and practical expertise for each particular problem, to acquire, process, and interpret the data properly; see, e.g., [32]. The computational part of the DMD, on the other hand, is a black-box containing robust and efficient numerical software based on two fundamental operations in matrix computations, the singular value decomposition (SVD) and computation of eigenvalues and eigenvectors of matrices. Since for both of these operations, good software solutions are available in the state-of-the-art packages such as LAPACK [2], MATLAB [53], and[1] NumPy [56], the numerical aspects and software implementation of DMD seem to be fully satisfactory. Similarly, variations of the DMD, such as the Forward–Backward DMD [13] (designed to curb sensor bias) simply deploy more matrix computational routines, such as a matrix square root, by plugging in off-the-shelf software solutions.[2]

Understanding the numerical aspects of a method is the key for its development (both mathematically and in software implementation) and successful deployment in applications. For example, recent results [22–24] show that considerable improvements of the DMD framework are possible by including state-of-the-art tools of the numerical linear algebra. Our goal in this chapter is to review some of those results and to provide necessary background from numerical analysis. Further, as a case study, we propose a modification of the DMD and discuss numerical issues and implementation details.

[1]Both MATLAB and NumPy actually use LAPACK as a computing engine for most matrix operations, in particular for the SVD and computing eigenvalues and eigenvectors of general matrices.

[2]In MATLAB, the function `sqrtm(.)` computes the matrix principal square root.

### *7.1.2 Overview*

The rest of this chapter is organized as follows. In Sect. 7.2, we briefly review the Krylov decomposition, and numerical difficulties of the algebraically elegant solution of (7.2), (7.3) using spectral decomposition of the companion matrix and the Rayleigh–Ritz extraction of approximate eigenpairs. The DMD is reviewed in Sect. 7.3, where we also discuss the fine details of computing the SVD and truncating it to determine the numerical rank of the data matrix $\mathbf{X}_m \equiv (\mathbf{x}_1 \ \ldots \ \mathbf{x}_m)$. The recently proposed enhancement of the DMD [23] is presented in Sect. 7.4. It includes computable residual bounds, refinement of the Ritz vectors, compression to improve run time efficiency, and extension of the DMD to the general weighted inner product induced by a positive definite matrix of weights. In Sect. 7.5, we present a variation of the DMD that replaces the POD basis with a selection of snapshots that is determined by a rank revealing QR factorization. A brief review of our most recent work is given in Sect. 7.6.

## 7.2 Preliminaries

Before we go into the details of the DMD, we briefly review its main ingredients, the Krylov subspaces and the Rayleigh–Ritz approximation. Also, we discuss in Sect. 7.2.2, the general concepts of numerical analysis of algorithms: backward stability and condition number. In particular, we discuss the numerical difficulties of the straightforward use of the snapshots and the spectral decomposition of the associated companion matrix—these are the main motivation for the SVD-based DMD algorithm.

### *7.2.1 Krylov Decomposition*

The sequence of snapshots $(\mathbf{x}_i)_{i=1}^{m+1}$, with $\mathbf{x}_{i+1} = \mathbb{A}\mathbf{x}_i$, naturally generates a flag of Krylov subspaces $\mathscr{X}_i = \mathrm{range}(\mathbf{X}_i) \subset \mathbb{C}^n$, where

$$\mathbf{X}_i = \left(\mathbf{x}_1 \ \mathbf{x}_2 \ \ldots \ \mathbf{x}_{i-1} \ \mathbf{x}_i\right), \quad i = 1, \ldots, m+1. \tag{7.4}$$

Note that $\mathscr{X}_i \subseteq \mathscr{X}_{i+1}$ and $\mathbb{A}\mathscr{X}_i \subseteq \mathscr{X}_{i+1}$. Further, for each $i = 1, \ldots, m$, the action of $\mathbb{A}$ on $\mathscr{X}_i$ is completely known because $\mathbb{A}\mathbf{X}_i v = \mathbf{X}_{i+1}(:, 2 : i+1)v$ for each $v \in \mathbb{C}^i$. Hence, we have access to the operator $\mathbf{P}_{\mathscr{X}_i}\mathbb{A}\big|_{\mathscr{X}_i}$ via the computable Rayleigh quotient $(\mathbf{X}_i^*\mathbf{X}_i)^{-1}\mathbf{X}_i^*\mathbb{A}\mathbf{X}_i$. If $\mathscr{X}_i$ is $\mathbb{A}$-invariant, then the Rayleigh–Ritz procedure will extract $i$ eigenpairs of $\mathbb{A}$ from $\mathscr{X}_i$.

The dimension argument ensures that at some index $i$ the $\mathbb{A}$-invariance must occur, i.e., at some index $s$, the new snapshot $f_{s+1}$ will not bring any new information not already contained in $\mathscr{X}_s$ and we will have $\mathscr{X}_s = \mathscr{X}_{s+1}$ and $\mathbb{A}\mathscr{X}_s \subseteq \mathscr{X}_s$. If $s$ is the first index with this property, then $\text{rank}(\mathbf{X}_s) = s = \text{rank}(\mathbf{X}_i)$, $i > s$. In matrix notation, we have $\mathbb{A}\mathbf{X}_s = \mathbf{X}_s M_s$, where $M_s = (\mathbf{X}_s^*\mathbf{X}_s)^{-1}\mathbf{X}_s^*\mathbb{A}\mathbf{X}_s$ is the Rayleigh quotient.

We assume the most probable scenario in an application: $\mathbf{X}_m$ is of full column rank and $m \ll n$. Then

$$\mathscr{X}_1 \subsetneq \mathscr{X}_2 \subsetneq \cdots \subsetneq \mathscr{X}_i \subsetneq \mathscr{X}_{i+1} \subsetneq \cdots \subsetneq \mathscr{X}_m \subsetneq \cdots \subsetneq \mathscr{X}_s = \mathscr{X}_{s+1}, \quad \mathbb{A}\mathscr{X}_s \subseteq \mathscr{X}_s, \tag{7.5}$$

i.e., $\dim(\mathscr{X}_i) = i$ for $i = 1, \ldots, m$. If $s = m$, then $\mathbf{x}_{m+1} = \mathbf{X}_m c$, where $c = (c_i)_{i=1}^m$ is the vector of coefficients of the representation of $\mathbf{x}_{m+1}$. Otherwise, let the vector $c$ be computed from the least-squares approximation,

$$c = \underset{v \in \mathbb{C}^m}{\text{argmin}} \|\mathbf{x}_{m+1} - \mathbf{X}_m v\|_2, \tag{7.6}$$

and let $r_{m+1} = \mathbf{x}_{m+1} - \mathbf{X}_m c$ be the corresponding residual. Then, putting together the relations $\mathbf{x}_{i+1} = \mathbb{A}\mathbf{x}_i$, $i = 1, \ldots, m$, and $\mathbf{x}_{m+1} = \mathbf{X}_m c + r_{m+1}$, we obtain the Krylov decomposition

$$\mathbb{A}\mathbf{X}_m = \mathbf{X}_m\mathbf{C}_m + \mathbf{E}_{m+1}, \quad \mathbf{C}_m = \begin{pmatrix} 0 & 0 & \ldots & 0 & c_1 \\ 1 & 0 & \ldots & 0 & c_2 \\ 0 & 1 & \ldots & 0 & c_3 \\ \vdots & \ddots & \ddots & \vdots & \vdots \\ 0 & 0 & \ldots & 1 & c_m \end{pmatrix}, \quad \mathbf{E}_{m+1} = r_{m+1}e_m^T, \quad e_m = \begin{pmatrix} 0 \\ 0 \\ \vdots \\ 0 \\ 1 \end{pmatrix}. \tag{7.7}$$

The least-squares error $r_{m+1}$ is orthogonal to the range of $\mathbf{X}_m$, $\mathbf{X}_m^* r_{m+1} = \mathbf{0}$, i.e., $\mathbf{E}_{m+1}$ is orthogonal to $\mathbf{X}_m$ in the Frobenius inner product on $\mathbb{C}^{n\times m}$: $\text{Trace}(\mathbf{X}_m^*\mathbf{E}_{m+1}) = 0$.

Further, if $\mathbf{X}_m^\dagger = (\mathbf{X}_m^*\mathbf{X}_m)^{-1}\mathbf{X}_m^*$ is the Moore–Penrose pseudoinverse, then

$$\mathbf{X}_m^\dagger \mathbb{A}\mathbf{X}_m = \mathbf{C}_m,$$

i.e., $\mathbf{C}_m$ is the Rayleigh quotient of $\mathbb{A}$ with respect to $\mathscr{X}_m$, represented in the Krylov basis of the columns of $\mathbf{X}_m$. Hence its eigenvalues and eigenvectors provide useful spectral information on $\mathbb{A}$. Namely, if we seek an approximate eigenpair $(\lambda, \mathbf{X}_m v)$, then the Galerkin optimality condition requires the residual to be orthogonal to $\mathbf{X}_m$,

$$\mathbf{X}_m^*(\mathbb{A}\mathbf{X}_m v - \lambda\mathbf{X}_m v) = \mathbf{0}, \quad \text{i.e.,} \quad (\mathbf{X}_m^*\mathbb{A}\mathbf{X}_m)v = \lambda(\mathbf{X}_m^*\mathbf{X}_m)v, \quad \text{i.e.,} \quad (\mathbf{X}_m^\dagger\mathbb{A}\mathbf{X}_m)v = \lambda v.$$

#### 7.2.1.1 Ritz Pairs and Explicit Vandermonde-Based Modal Decomposition

Now, if $\mathbf{C}_m v = \lambda v$, $v \neq \mathbf{0}$, then

$$\mathbb{A}(\mathbf{X}_m v) = \mathbf{X}_m\mathbf{C}_m v + r_{m+1}e_m^T v = \lambda(\mathbf{X}_m v) + r_{m+1}e_m^T v,$$

i.e., $(\lambda, \mathbf{X}_m v)$ represents an approximate eigenpair of $\mathbb{A}$ with the residual norm $\|r_{m+1}\|_2|v_m|$. Hence, from the spectral decomposition of $\mathbf{C}_m$, we can derive $m$ approximations of certain (not necessarily $m$ of them) eigenpairs of $\mathbb{A}$.

The characteristic polynomial of the companion matrix $\mathbf{C}_m$ is $\wp_m(z) = z^m - \sum_{j=1}^m c_j z^{j-1}$. Assume for simplicity that the eigenvalues $\lambda_i$, $i = 1, \ldots, m$, are algebraically simple. Hence, if $\lambda_i$ is an eigenvalue, then $\wp_m(\lambda_i) = 0$ and

$$(1\ \lambda_i\ \ldots\ \lambda_i^{m-1}) \begin{pmatrix} 0 & 0 & \ldots & 0 & c_1 \\ 1 & 0 & \ldots & 0 & c_2 \\ 0 & 1 & \ldots & 0 & c_3 \\ \vdots & \ddots & \ddots & \vdots & \vdots \\ 0 & 0 & \ldots & 1 & c_m \end{pmatrix} = (\lambda_i\ \ldots\ \lambda_i^{m-1}, \sum_{j=1}^m c_j \lambda_i^{j-1}) = \lambda_i (1\ \lambda_i\ \ldots\ \lambda_i^{m-1}),$$

i.e., $(1\ \lambda_i\ \ldots\ \lambda_i^{m-1})$ is the corresponding left eigenvector. Hence, the spectral decomposition of $\mathbf{C}_m$ reads

$$\mathbf{C}_m = \mathbb{V}_m^{-1} \boldsymbol{\Lambda}_m \mathbb{V}_m, \quad \text{where } \boldsymbol{\Lambda}_m = \begin{pmatrix} \lambda_1 & & \\ & \ddots & \\ & & \lambda_m \end{pmatrix}, \quad \mathbb{V}_m = \begin{pmatrix} 1 & \lambda_1 & \ldots & \lambda_1^{m-1} \\ 1 & \lambda_2 & \ldots & \lambda_2^{m-1} \\ \vdots & \vdots & \ldots & \vdots \\ 1 & \lambda_m & \ldots & \lambda_m^{m-1} \end{pmatrix}. \tag{7.8}$$

The assumed algebraic simplicity of the Ritz values $\lambda_j$ implies that the Vandermonde matrix $\mathbb{V}_m$ is nonsingular because $\det(\mathbb{V}_m) \equiv \prod_{j>k}(\lambda_j - \lambda_k) \neq 0$. In particular, the columns of $\mathbb{V}_m^{-1}$ are the eigenvectors of $\mathbf{C}_m$, $\mathbf{C}_m \mathbb{V}_m^{-1} = \mathbb{V}_m^{-1} \boldsymbol{\Lambda}_m$.

Now, the solution of the two fundamental tasks of the dynamic mode analysis are straightforward, at least purely algebraically. The Ritz vectors corresponding to the $\lambda_j$'s are then the columns of

$$\mathbf{W}_m \equiv \mathbf{X}_m \mathbb{V}_m^{-1} \equiv (w_1\ \ldots\ w_m). \tag{7.9}$$

If we define $\alpha_j = \|w_j\|_2$, $z_j = w_j/\alpha_j$, $\mathbf{Z}_m = (z_1\ \ldots\ z_m)$, then the decomposition (7.3), with $\ell = m$, follows from

$$\mathbf{X}_m = \mathbf{W}_m \mathbb{V}_m = \mathbf{Z}_m \mathrm{diag}(\alpha_j)_{j=1}^m \mathbb{V}_m = (z_1\ z_2\ \ldots\ z_m) \begin{pmatrix} \alpha_1 & & & \\ & \alpha_2 & & \\ & & \ddots & \\ & & & \alpha_m \end{pmatrix} \begin{pmatrix} 1 & \lambda_1 & \ldots & \lambda_1^{m-1} \\ 1 & \lambda_2 & \ldots & \lambda_2^{m-1} \\ \vdots & \vdots & \ldots & \vdots \\ 1 & \lambda_m & \ldots & \lambda_m^{m-1} \end{pmatrix}. \tag{7.10}$$

Numerically, however, this is not so simple and we discuss it in Sect. 7.2.2.3.

*Remark 7.1* In the literature on DMD, the above procedure is sometimes called Arnoldi algorithm, which, in our opinion, is misplaced. In our opinion, the proper name should be *Krylov–Rayleigh–Ritz procedure* to point out that the procedure described here is just the Rayleigh–Ritz in a subspace defined by the Krylov basis. The Arnoldi algorithm is the method of applying the Gram–Schmidt orthogonalization

(QR factorization) implicitly on the sequence of the $\mathbf{x}_i$'s. A true Arnoldi scheme in the DMD framework is proposed only recently in [1].

*Remark 7.2* Clearly, if the subspace $\mathscr{X}_m$ determined as the span of the given dataset does not contain information on a desired part of the spectrum of $\mathbb{A}$, then we cannot expect any method to provide detailed insight into the spectral properties of $\mathbb{A}$. On the other hand, if it does, then we must deploy many different techniques to extract relevant spectral information. Any so devised method, in order to be used with confidence, must be accompanied with an error estimate; see Sect. 7.4.1.1.

### 7.2.2 The Curse of Ill-Conditioning

Turning linear algebra schemes into the algorithms of numerical linear algebra in finite precision computer arithmetic is not an easy task. The computed solution to a numerical problem is contaminated, e.g., by the errors of floating-point arithmetic, truncation of an infinite sequence, so that it must be used with care. In addition, the input data is often noisy, so that the whole computational process must be correctly interpreted. We briefly review the basic concepts of numerical analysis (backward stability, condition number) and discuss the potential sources of numerical difficulties related to (7.2) and (7.3).

#### 7.2.2.1 Backward Error and Condition Number

In general, assessing the accuracy of the computed solution is difficult—the error is by definition the difference between the computed value at hand and the unknown exact value. Further, we are often interested not only in the accuracy of the output, but also need to understand its sensitivity to variations on input. This depends both on the computational task, the data, and the algorithm.

The key concept in the analysis is the *backward error.* Suppose, we have to compute $\mathscr{Y} = \mathscr{F}(\mathscr{X})$ for some function $\mathscr{F}$ and some data $\mathscr{X}$ from the domain of $\mathscr{F}$. For instance, for square nonsingular $\mathscr{X} \in \mathbb{C}^{n\times n}$, we need $\mathscr{F}(\mathscr{X}) = \mathscr{X}^{-1} \equiv \mathscr{Y}$. Or, e.g., in (7.6), the data is $\mathscr{X} = (\mathbf{X}_m, \mathbf{x}_{m+1})$, and we need $c = \mathbf{X}_m^{\dagger}\mathbf{x}_{m+1} \equiv \mathscr{Y}$. We use an algorithm that computes an approximation $\widetilde{\mathscr{Y}} \approx \mathscr{Y}$ in computer arithmetic with roundoff $\mathbf{u}$. By back-tracing the steps of the algorithm, we can collect all rounding errors and systematically push them back into the input data $\mathscr{X}$, thus proving that the computed $\widetilde{\mathscr{Y}} = \mathscr{Y} + \delta\mathscr{Y}$ corresponds to exact application of the algorithm to some $\mathscr{X} + \delta\mathscr{X}$. If the *backward error* $\delta\mathscr{X}$ is small, in an appropriate sense, relative to $\mathscr{X}$, we say that the algorithm is *backward stable*. The situation is illustrated by a commutative diagram as in Fig. 7.1. If the input data $\mathscr{X}$ itself is an approximation of some ideal $\mathscr{X}_{exact}$, $\mathscr{X} = \mathscr{X}_{exact} + \delta_0\mathscr{X}$, and if $\delta\mathscr{X}$ and $\delta_0\mathscr{X}$ are comparable in size, then we may consider $\widetilde{\mathscr{Y}}$ as good as $\mathscr{Y}$.

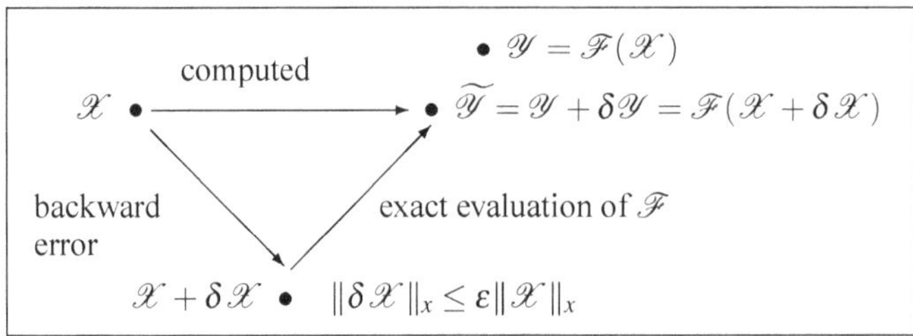


**Fig. 7.1** Commutative diagram for a backward stable algorithm. The numerically computed output $\widetilde{\mathscr{Y}}$ may not be exact, nor even close to the exact value $\mathscr{Y}$, but it can be interpreted as the result of exact evaluation of $\mathscr{F}$ on a slightly perturbed input $\mathscr{X} + \delta\mathscr{X}$

Different algorithms, for the same task, may have differently structured backward errors. Stronger form of backward stability requires, in addition, that $\mathscr{X} + \delta\mathscr{X}$ has same structure as $\mathscr{X}$, e.g., symmetry. For more details, we refer to [38].

Backward stability does not guarantee small *forward error* $\delta\mathscr{Y}$. That will depend on the function $\mathscr{F}$, on $\mathscr{X}$ and on the structure of $\delta\mathscr{X}$. In general, the size of $\delta\mathscr{X}$ on input can be increased by some factor $\mathfrak{C}$, called *condition number*, so that $\|\delta\mathscr{Y}\|_y/\|\mathscr{Y}\|_y \leq \mathfrak{C}\|\delta\mathscr{X}\|_x/\|\mathscr{X}\|_x$. Here we assume that the domain and the codomain of $\mathscr{F}$ are normed spaces with norms $\|\cdot\|_x$ and $\|\cdot\|_y$, respectively. It is the task of *perturbation theory* for $\mathscr{F}$ to identify the optimal $\mathfrak{C}$, i.e.,

$$\liminf_{\delta\to 0}\left\{\mathfrak{C} \geq 0 \;:\; \frac{\|\delta\mathscr{X}\|_x}{\|\mathscr{X}\|_x} < \delta \Longrightarrow \frac{\|\delta\mathscr{Y}\|_y}{\|\mathscr{Y}\|_y} \leq \mathfrak{C}\delta\right\}.$$

As an example, consider $\mathscr{F}(\mathscr{X}) = \mathscr{X}^{-1}$. For sufficiently small $\delta\mathscr{X}$ we have, using the Neumann series, $(\mathscr{X} + \delta\mathscr{X})^{-1} \approx \mathscr{X}^{-1} - \mathscr{X}^{-1}\delta\mathscr{X}\mathscr{X}^{-1} + O(\|\delta\mathscr{X}\|^2)$ and

$$\frac{\|(\mathscr{X} + \delta\mathscr{X})^{-1} - \mathscr{X}^{-1}\|}{\|\mathscr{X}^{-1}\|} \lessapprox (\|\mathscr{X}\|\|\mathscr{X}^{-1}\|)\frac{\|\delta\mathscr{X}\|}{\|\mathscr{X}\|}. \tag{7.11}$$

Here, $\kappa(\mathscr{X}) = \|\mathscr{X}\|\|\mathscr{X}^{-1}\|$ has been identified as the condition number for inverting $\mathscr{X}$, provided that the input error is bounded in terms of $\|\delta\mathscr{X}\|/\|\mathscr{X}\|$—this is in practice the roundoff $\mathbf{u}$ times a moderate factor of the dimension of $\mathscr{X}$. If $\kappa(\mathscr{X})$ is large so that the right-hand side in (7.11) is close to or above one, we say the matrix $\mathscr{X}$ is *ill conditioned* for inverting. Intuitively, ill-conditioned matrix is close to a singular matrix: if $\mathscr{X} + \delta\mathscr{X}$ is singular, then $\|\delta\mathscr{X}\|/\|\mathscr{X}\| \geq 1/\kappa_2(\mathscr{X})$. Here the norm $\|\cdot\|$ is usually the spectral $\|\cdot\|_2$ or the Frobenius matrix norm $\|\cdot\|_F$.

In Sect. 7.3.3, we provide a similar analysis for numerical computation of the SVD. It will be one of the keys for understanding the numerics of the DMD. For theoretical aspects of ill-conditioning, we refer the reader to [8, 14].

#### 7.2.2.2 Near Singularity of the Snapshot Matrix $\mathbf{X}_m$

The matrix $\mathbf{X}_m$ (7.4) of the snapshots (7.1) is, by design, ill conditioned, i.e, nearly rank deficient. Its columns are simply the iterates of the power method for computing the absolutely dominant eigenpairs of the matrix $\mathbb{A}$. For sufficiently large sample, the columns will exhibit near linear dependence.

To illustrate, assume that $\mathbb{A}$ is diagonalizable with eigenpairs $\mathbb{A}\mathbf{a}_i = \alpha_i \mathbf{a}_i$, $\mathbf{a}_i \neq \mathbf{0}$, and that its eigenvalues $\alpha_i$ are enumerated so that $0 \neq |\alpha_1| \geq |\alpha_2| \geq \cdots \geq |\alpha_n|$. Let $\mathbf{x}_1$ be expressed in the eigenvector basis as $\mathbf{x}_1 = \phi_1 \mathbf{a}_1 + \cdots + \phi_n \mathbf{a}_n$. Then

$$\mathbf{x}_{i+1} = \mathbb{A}^i \mathbf{x}_1 = \alpha_1^i \left( \phi_1 \mathbf{a}_1 + (\alpha_2/\alpha_1)^i \phi_2 \mathbf{a}_2 + (\alpha_3/\alpha_1)^i \phi_3 \mathbf{a}_3 + \cdots + (\alpha_n/\alpha_1)^i \phi_n \mathbf{a}_n \right).$$

Hence, if ,e.g., $|\alpha_2| > |\alpha_3|$, then for $j \geq 3$, $\lim_{i\to\infty}(\alpha_j/\alpha_1)^i = 0$, and thus, with big enough $i$ the $\mathbf{x}_i$'s will stay close to the span of $\mathbf{a}_1$ and $\mathbf{a}_2$, provided that $\phi_1 \neq 0, \phi_2 \neq 0$. If $|\alpha_1| > |\alpha_2|$, then the $\mathbf{x}_i$'s will be making smaller and smaller angles with the span of $\mathbf{a}_1$. This means that, depending on the initial $\mathbf{x}_1$ and the index $m$, relatively small changes of $\mathbf{X}_m$ can make it rank deficient; its range may change considerably under tiny perturbations. In the context of spectral approximation by the power method (and subspace iterations), this is desirable and we hope that $\mathbf{X}_i$ will become numerically singular at small index $i$. On the other hand, if we need the singular values of $\mathbf{X}_m$, then we should be aware that the smallest ones may be computed very inaccurately. We discuss this issue in Sect. 7.3.3.

#### 7.2.2.3 Structural Ill-Conditioning of Vandermonde Matrices

The elegant formulation (7.7) with the matrix of eigenvectors of $\mathbf{C}_m$ and the Ritz vectors given explicitly by the inverse of the Vandermonde matrix defined by the eigenvalues (7.8) and (7.9) and the snapshot decomposition (7.10) have not been turned, until only recently,[3] into a numerical software. The main reason is potential ill-conditioning of the Vandermonde matrix, with the consequences discussed in Sect. 7.2.2.1.

Vandermonde matrices are notoriously ill conditioned [57]; for instance an arbitrary real $m \times m$ Vandermonde matrix $\mathbb{V}_m(\lambda_i)$ has condition number $\kappa_2(\mathbb{V}_m) \equiv \|\mathbb{V}_m\|_2 \|\mathbb{V}_m^{-1}\|_2 \geq 2^{m-2}/\sqrt{m}$. In general, the condition number depends on the distribution of the eigenvalues $\lambda_i$ and it can be as small as one and it can grow with the dimension $m$ as fast as $O(m^{m+1})$ for harmonically distributed $\lambda_i$'s, see [30, 31], [38, Chap. 21].

An important remark is in order about computing the condition number. $\kappa_2(\mathbb{V}_m)$ is a function of $\mathbb{V}_m$ and it has its own condition number that determines how accurately $\kappa_2(\mathbb{V}_m)$ can be computed in finite precision arithmetic. It follows from [37] that we cannot accurately estimate $\kappa_2(\mathbb{V}_m)$ if its true value is above $1/\mathbf{u}$. In practice, e.g., using the function `cond()` from MATLAB, this means that extremely high condi-

[3]See Sect. 7.6, where we briefly discuss recently proposed scheme [22] that uses the inverse of $\mathbb{V}_m$.

tion number is usually severely underestimated. (MATLAB computes the condition number as the quotient of the largest and the smallest singular value, and in the ill-conditioned case, the small singular values are computed inaccurately so that those below the round-off level, relative to the matrix norm, usually severely overestimate the true values; see, e.g., [23, Sect. 4.1.3].)

Recall that the classical upper bound on the relative error in the solution of linear systems is $O(n)\mathbf{u}\kappa_2(\mathbb{V}_m)$, so that $\kappa_2(\mathbb{V}_m) > 1/(n\mathbf{u})$ implies no accuracy whatsoever.

## 7.3 Dynamic Mode Decomposition

To avoid numerical ill-conditioning of the Vandermonde matrix, instead of the procedure outlined in Sect. 7.2.1.1, Schmid [62, 63] proposed an approach based on the SVD decomposition of the snapshot matrix $\mathbf{X}_m$. From Sect. 7.2.2.2, we know that $\mathbf{X}_m$ may be nearly rank deficient, which means ill conditioned. The idea is to curb the ill-conditioning by deploying unitary (orthogonal in the case of real data) transformations, and to truncate the smallest singular values.

In this section, we describe the DMD algorithm, and provide necessary material for understanding fine numerical details related to numerical determination of the rank of the snapshot matrix, and, in Sect. 7.3.3, computation of the SVD.

### *7.3.1 Low-Rank Approximation of the Data Matrix*

In the first step of DMD, the data matrix $\mathbf{X}_m$ is approximated by a low-rank matrix whose rank should match the *numerical rank* [34] of $\mathbf{X}_m$. This is achieved in an optimal way using the following optimality property of the singular value decomposition (SVD).

**Theorem 7.1** (Eckart–Young [27], Mirsky [54]) *Let the SVD of* $\mathbf{X}_m \in \mathbb{C}^{n\times m}$ *be*

$$\mathbf{X}_m = \mathbf{U}\boldsymbol{\Sigma}\mathbf{V}^*, \quad \boldsymbol{\Sigma} = \operatorname{diag}(\sigma_i)_{i=1}^{\min(m,n)}, \quad \sigma_1 \geq \cdots \geq \sigma_{\min(m,n)} \geq 0.$$

*For* $k \in \{1, \dots, \min(m,n)\}$*, define* $\mathbf{U}_k = \mathbf{U}(:, 1:k)$, $\boldsymbol{\Sigma}_k = \boldsymbol{\Sigma}(1:k, 1:k)$, $\mathbf{V}_k = \mathbf{V}(:, 1:k)$*, and* $\mathbf{X}_m^{(k)} = \mathbf{U}_k\boldsymbol{\Sigma}_k\mathbf{V}_k^*$*. The optimal rank k approximations of* $\mathbf{X}_m$ *in the spectral* $\|\cdot\|_2$ *and in the Frobenius norm* $\|\cdot\|_F$ *are*

$$\min_{\operatorname{rank}(\mathbf{N})\leq k} \|\mathbf{X}_m - \mathbf{N}\|_2 = \|\mathbf{X}_m - \mathbf{X}_m^{(k)}\|_2 = \sigma_{k+1}, \tag{7.12}$$

$$\min_{\operatorname{rank}(\mathbf{N})\leq k} \|\mathbf{X}_m - \mathbf{N}\|_F = \|\mathbf{X}_m - \mathbf{X}_m^{(k)}\|_F = \sqrt{\sum_{i=k+1}^{\min(n,m)} \sigma_i^2}. \tag{7.13}$$

Suppose that for some $\delta > \varepsilon > 0$, we can determine an index $k < \mathrm{rank}(\mathbf{X}_\mathrm{m})$ such that $\sigma_k \geq \delta > \varepsilon \geq \sigma_{k+1}$. Then $\|\mathbf{X}_m - \mathbf{X}_m^{(k)}\|_2 \leq \varepsilon$, i.e., the $\varepsilon$-ball centered at $\mathbf{X}_m$ contains a matrix of rank $k$. In fact, since $\sigma_k > \varepsilon$, no matrix in that ball can have rank lower than $k$. Moreover, if the radius of the ball is increased to $\delta$, it cannot contain any matrix of rank below $k$ since $\|\mathbf{X}_m - \mathbf{N}\|_2 < \delta \leq \sigma_k$ implies $\mathrm{rank}(\mathbf{N}) \geq k$. For an instructive discussion, we refer to [34], where the numerical rank with respect to $\|\cdot\|_2$ (and analogously with respect to $\|\cdot\|_F$) is defined as the triplet $(\delta, \varepsilon, k)$.

*Remark 7.3* The columns of $\mathbf{U}_k$ span the best $k$-dimensional subspace that captures the snapshot collection in the sense that

$$\sum_{i=1}^{m} \|\mathbf{x}_i - \mathbf{U}_k \mathbf{U}_k^* \mathbf{x}_i\|_2^2 \equiv \|\mathbf{X}_m - \mathbf{U}_k(\mathbf{U}_k^* \mathbf{X}_m)\|_F^2 = \min_{Q_k^* Q_k = \mathbb{I}_k} \|\mathbf{X}_m - Q_k Q_k^* \mathbf{X}_m\|_F^2.$$

The columns of $\mathbf{U}_k$ are the proper orthogonal decomposition (POD) basis; note that this approximation is also known as the principal component analysis (PCA).

### 7.3.2 DMD Algorithm

In the next step of DMD, the Rayleigh–Ritz projection of $\mathbb{A}$ is computed with respect to $\mathbf{U}_k$ using the approximation

$$\mathbf{Y}_m \equiv (\mathbf{x}_2 \ \dots \ \mathbf{x}_{m+1}) = \mathbb{A}\mathbf{X}_m \approx \mathbb{A}\mathbf{X}_m^{(k)} = \mathbb{A}\mathbf{U}_k \boldsymbol{\Sigma}_k \mathbf{V}_k^*. \tag{7.14}$$

Since $(\mathbf{X}_m - \mathbf{X}_m^{(k)})\mathbf{V}_k = \mathbf{0}$ and $\mathbf{Y}_m = \mathbb{A}\mathbf{U}_k \boldsymbol{\Sigma}_k V_k^* + \mathbb{A}(\mathbf{X}_m - \mathbf{X}_m^{(k)})$, the Rayleigh quotient $\mathbf{U}_k^*(\mathbb{A}\mathbf{U}_k)$ can be computed as

$$\mathbf{S}_k \equiv \mathbf{U}_k^*(\mathbb{A}\mathbf{U}_k) = \mathbf{U}_k^*(\mathbf{Y}_m \mathbf{V}_k \boldsymbol{\Sigma}_k^{-1}). \tag{7.15}$$

Now, if $\mathbf{S}_k w_i = \lambda_i w_i$, then DMD uses $(\lambda_i, \mathbf{U}_k w_i)$ as an approximate eigenpair of $\mathbb{A}$.

*Remark 7.4* As noted in [68], in (7.14), (7.15), it is not necessary to work with sequential data as in (7.14); it suffices to have two sets of snapshots $\mathbf{X}_m$, $\mathbf{Y}_m$ such that $\mathbf{Y}_m = \mathbb{A}\mathbf{X}_m$. In the sequel, we will describe all algorithms in terms of the general pair $(\mathbf{X}_m, \mathbf{Y}_m \equiv \mathbb{A}\mathbf{X}_m)$. Of course, in the case of sequential data (7.14), a software implementation will use only $\mathbf{X}_{m+1} \equiv (\mathbf{X}_m \ \mathbf{x}_{m+1}) \in \mathbb{C}^{n \times (m+1)}$, and not two $n \times m$ matrices $\mathbf{X}_m$ and $\mathbf{Y}_m$.

The complete procedure is outlined in Algorithm 7.1.
Although designed to avoid the companion matrix $\mathbf{C}_m$ and its ill-conditioned eigenvector matrix $\mathbb{V}_m$ (see Sect. 7.2.1.1), Algorithm 7.1 uses the Rayleigh quotient $\mathbf{S}_k$ that is closely related to $\mathbf{C}_m$. The following proposition from [23] establishes precise relationship between $\mathbf{S}_k$ and $\mathbf{C}_m$.

**Algorithm 7.1** $[Z_k, \Lambda_k] = \mathrm{DMD}(\mathbf{X}_m, \mathbf{Y}_m)$

**Input:** • $\mathbf{X}_m = (\mathbf{x}_1, \ldots, \mathbf{x}_m)$, $\mathbf{Y}_m = (\mathbf{y}_1, \ldots, \mathbf{y}_m) \in \mathbb{C}^{n\times m}$ that define a sequence of snapshots pairs $(\mathbf{x}_i, \mathbf{y}_i \equiv \mathbb{A}\mathbf{x}_i)$.
(Tacit assumption is that $n$ is large and that $m \ll n$.)
1: $[\mathbf{U}, \boldsymbol{\Sigma}, \mathbf{V}] = svd(\mathbf{X}_m)$ ; {*The thin SVD:* $\mathbf{X}_m = \mathbf{U}\boldsymbol{\Sigma}\mathbf{V}^*$, $\mathbf{U} \in \mathbb{C}^{n\times m}$, $\boldsymbol{\Sigma} = \mathrm{diag}(\sigma_i)_{i=1}^m$, $\mathbf{V} \in \mathbb{C}^{m\times m}$}
2: Determine numerical rank $k$.
3: Set $\mathbf{U}_k = \mathbf{U}(:, 1:k)$, $\mathbf{V}_k = \mathbf{V}(:, 1:k)$, $\boldsymbol{\Sigma}_k = \boldsymbol{\Sigma}(1:k, 1:k)$
4: $\mathbf{S}_k = ((\mathbf{U}_k^*\mathbf{Y}_m)\mathbf{V}_k)\boldsymbol{\Sigma}_k^{-1}$; {*Schmid's formula for the Rayleigh quotient* $\mathbf{U}_k^*\mathbb{A}\mathbf{U}_k$}
5: $[\mathbf{W}_k, \boldsymbol{\Lambda}_k] = \mathrm{eig}(\mathbf{S}_k)$ $\{\boldsymbol{\Lambda}_k = \mathrm{diag}(\lambda_i)_{i=1}^k; \mathbf{S}_k\mathbf{W}_k(:, i) = \lambda_i\mathbf{W}_k(:, i); \|\mathbf{W}_k(:, i)\|_2 = 1\}$
6: $\mathbf{Z}_k = \mathbf{U}_k\mathbf{W}_k$ {*Ritz vectors*}
**Output:** $\mathbf{Z}_k = (z_1 \ \ldots \ z_k)$, $\boldsymbol{\Lambda}_k$

**Proposition 7.1** ([23]) *Let* $\mathbf{S}_k$ *be computed as in Algorithm 7.1. If* $k = m$, *then* $\mathbf{S}_m$ *and* $\mathbf{C}_m$ *are similar matrices, where the similarity is realized by the matrix* $\mathbf{V}\boldsymbol{\Sigma}^{-1}$. *If* $k < m$, *then* $\mathbf{S}_k$ *is the Rayleigh quotient of* $\mathbf{C}_m$, *with the matrix* $\mathbf{V}_k\boldsymbol{\Sigma}_k^{-1}$.

Instead of the detailed proof, we discuss few details, following [23]. Note that

$$\delta\mathbf{X}_m \equiv \mathbf{X}_m - \mathbf{X}_m^{(k)} = \sum_{j=k+1}^{n} \sigma_j \mathbf{U}(:, j)\mathbf{V}(:, j)^*, \quad \|\delta\mathbf{X}_m\|_2 = \sigma_{k+1}, \quad \delta\mathbf{X}_m\mathbf{V}_k = \mathbf{0}.$$

Then we can write relation (7.7) as

$$\mathbb{A}(\mathbf{U}_k\boldsymbol{\Sigma}_k\mathbf{V}_k^* + \delta\mathbf{X}_m) = (\mathbf{U}_k\boldsymbol{\Sigma}_k\mathbf{V}_k^* + \delta\mathbf{X}_m)\mathbf{C}_m + r_{m+1}e_m^T,$$

i.e., $\mathbb{A}(\mathbf{U}_k\boldsymbol{\Sigma}_k\mathbf{V}_k^*) = (\mathbf{U}_k\boldsymbol{\Sigma}_k\mathbf{V}_k^*)\mathbf{C}_m + r_{m+1}e_m^T + \delta\mathbf{X}_m\mathbf{C}_m - \mathbb{A}\delta\mathbf{X}_m$, and

$$\mathbb{A}\mathbf{U}_k = \mathbf{U}_k(\boldsymbol{\Sigma}_k\mathbf{V}_k^*\mathbf{C}_m\mathbf{V}_k\boldsymbol{\Sigma}_k^{-1}) + r_{m+1}e_m^T\mathbf{V}_k\boldsymbol{\Sigma}_k^{-1} + \delta\mathbf{X}_m\mathbf{C}_m\mathbf{V}_k\boldsymbol{\Sigma}_k^{-1}. \tag{7.16}$$

Since $\mathbf{U}_k^*\delta\mathbf{X}_m = \mathbf{0}$ and $\mathbf{U}_k^*r_{m+1} = \mathbf{0}$, $\mathbf{S}_k = \mathbf{U}_k^*\mathbb{A}\mathbf{U}_k = \boldsymbol{\Sigma}_k\mathbf{V}_k^*\mathbf{C}_m\mathbf{V}_k\boldsymbol{\Sigma}_k^{-1}$ is a Rayleigh quotient of $\mathbf{C}_m$, and

$$\mathbb{A}\mathbf{U}_k = \mathbf{U}_k\mathbf{S}_k + r_{m+1}g_k^* + \mathbb{G}_k, \quad g_k^* = e_m^T\mathbf{V}_k\boldsymbol{\Sigma}_k^{-1}, \quad \mathbb{G}_k = \delta\mathbf{X}_m\mathbf{C}_m\mathbf{V}_k\boldsymbol{\Sigma}_k^{-1}. \tag{7.17}$$

Further, if $\mathbf{S}_k w_i = \lambda_i w_i$, then

$$\mathbb{A}(\mathbf{U}_k w_i) = \lambda_i(\mathbf{U}_k w_i) + r_{m+1}g_k^* w_i + \mathbb{G}_k w_i. \tag{7.18}$$

Note that here the second term in the residual is estimated as

$$\|\mathbb{G}_k\|_2 = \|\delta\mathbf{X}_m\mathbf{C}_m\mathbf{V}_k\boldsymbol{\Sigma}_k^{-1}\|_2 \leq \|\mathbf{C}_m\|_2\frac{\sigma_{k+1}}{\sigma_k}. \tag{7.19}$$

Hence, neglecting $G_k$ and using the approximate Krylov decomposition[4]

$$\mathbb{A}\mathbf{U}_k \approx \mathbf{U}_k\mathbf{S}_k + r_{m+1}g_k^* \tag{7.20}$$

is acceptable only if the singular values are distributed so that

$$\sigma_1 \geq \sigma_2 \geq \cdots \geq \sigma_k \gg \sigma_{k+1} \geq \cdots \geq \sigma_m, \tag{7.21}$$

i.e., in the case of sharp drop after the index $k$. This is in accordance with the discussion on the numerical rank in Sect. 7.3.1. This kind of analysis of truncation criterion to determine the numerical rank (Line 2 of Algorithm 7.1) is, to our best knowledge, not discussed in the literature on DMD, until recently in [23].

*Remark 7.5* By using the truncated SVD, the DMD prefers large-energy content and low-energy modes are discarded. This may lead to a loss of important spectral content. If one desires to keep those as well, then the index $k$ should be determined so that $\boldsymbol{\Sigma}_k$ contains small singular values as well. Unfortunately, the small singular values of $\mathbf{X}_m$ may be poorly determined by the data and will most likely be computed with large error. Then, when their inverses are used in (7.15), the matrix $S_k$ may be computed with large errors; see [23, Sect. 4.1.3].

To complete the picture of DMD as a computational tool, in the next section, we briefly review numerical computation of the SVD.

### *7.3.3 Numerical Computation of the SVD*

The SVD is numerically computed using unitary matrices, and it is used with confidence in many applications. The first ground is in the *backward stability* of the state-of-the-art numerical algorithms for computing the SVD. This is usually stated in the literature as follows: If $\mathbf{A} \approx \widetilde{\mathbf{U}}\widetilde{\boldsymbol{\Sigma}}\widetilde{\mathbf{V}}^*$ is the computed SVD of a matrix $\mathbf{A}$, then there exist unitary matrices $\widehat{\mathbf{U}}$, $\widehat{\mathbf{V}}$, and a perturbation $\delta\mathbf{A}$ (*backward error*) such that $\|\widehat{\mathbf{U}} - \widetilde{\mathbf{U}}\|_2 \leq \varepsilon_1$, $\|\widehat{\mathbf{V}} - \widetilde{\mathbf{V}}\|_2 \leq \varepsilon_2$, and

$$\mathbf{A} + \delta\mathbf{A} = \widehat{\mathbf{U}}\widetilde{\boldsymbol{\Sigma}}\widehat{\mathbf{V}}^*, \quad \|\delta\mathbf{A}\|_2 \leq \varepsilon\|\mathbf{A}\|_2. \tag{7.22}$$

(Here $\varepsilon_1$, $\varepsilon_2$, $\varepsilon$ are bounded by modestly growing polynomials in matrix dimensions times the roundoff unit $\mathbf{u}$.) Hence, even if $\widetilde{\mathbf{U}}\widetilde{\boldsymbol{\Sigma}}\widetilde{\mathbf{V}}^*$ is not an exact SVD of $\mathbf{A}$, small changes in $\widetilde{\mathbf{U}}$ and $\widetilde{\mathbf{V}}$ will establish an exact SVD (7.22) of a nearby matrix[5] $\mathbf{A} + \delta\mathbf{A}$. (Note that the computed matrices $\widetilde{\mathbf{U}}$ and $\widetilde{\mathbf{V}}$ are only *numerically unitary*.) Furthermore, the algorithms have been implemented as robust numerical software.

[4]See [66] for a definition and basis properties of Krylov decomposition.

[5]This is sometimes called *mixed stability*: backward error corresponds to slightly changed output.

The second numerically relevant property of the SVD originates in the perturbation theory, shared with the Hermitian spectral decomposition. Namely, the size of the perturbation of any singular value cannot exceed the size of the perturbation of the matrix. The following theorem summarizes the two classical results.

**Theorem 7.2** *Let the singular values of* $\mathbf{A}$ *and* $\mathbf{A} + \delta\mathbf{A}$ *be* $\sigma_1 \geq \cdots \geq \sigma_{\min(m,n)}$ *and* $\widetilde{\sigma}_1 \geq \cdots \geq \widetilde{\sigma}_{\min(m,n)}$, *respectively. Then the distances between the corresponding singular values are estimated by a Weyl-type bound*

$$\max_i |\widetilde{\sigma}_i - \sigma_i| \leq \|\delta\mathbf{A}\|_2.$$

*Further, the Wieland–Hoffman theorem yields* $\sqrt{\sum_{i=1}^{\min(m,n)} |\widetilde{\sigma}_i - \sigma_i|^2} \leq \|\delta\mathbf{A}\|_F$.

Hence, if we combine Theorem 7.2 with the backward stability (7.22), we have that for each computed singular value $\widetilde{\sigma}_i = \sigma_i + \delta\sigma_i$

$$|\delta\sigma_i| \leq \|\delta\mathbf{A}\|_2 \leq \varepsilon\|\mathbf{A}\|_2. \tag{7.23}$$

This kind of accuracy is typical for the bidiagonalization-based methods for computing the SVD, such as the QR algorithm or the divide-and-conquer method. State-of-the-art software implementations of these methods are available in LAPACK [2] as the driver routine `xGESVD(.)` and `xGESDD(.)`, respectively. These two are usually under the hood of functions for computing the SVD in many packages; e.g., the function `svd(.)` in MATLAB.

There is a caveat, however. This may not be satisfactory for the singular values that are much smaller than $\|\mathbf{A}\|_2 = \sigma_1$ because the relative error in a $\sigma_i \neq 0$ satisfies

$$\frac{|\delta\sigma_i|}{\sigma_i} \leq \varepsilon\frac{\sigma_1}{\sigma_i} \leq \varepsilon\|\mathbf{A}\|_2\|\mathbf{A}^\dagger\|_2 \equiv \varepsilon\kappa_2(\mathbf{A}), \quad \kappa_2(\mathbf{A}) = \|\mathbf{A}\|_2\|\mathbf{A}^\dagger\|_2. \tag{7.24}$$

So, any singular value $\sigma_i$ such that $\sigma_i < \mathbf{u}\sigma_1$ may be badly approximated by $\widetilde{\sigma}_i$, with no accurate digit whatsoever. For more details and an illustrative example in the context of DMD, we refer to [23, Sect. 4.1.3].

Furthermore, the backward stability relation (7.22), with a backward error that is bounded as $\|\delta\mathbf{A}\|_2 \leq \varepsilon\|\mathbf{A}\|_2$, may be considered both satisfactory and unsatisfactory, depending on the concrete setting. In the operator norm sense, this is the best one can hope for. On the other hand, it allows that, say, a small column of $\mathbf{A}$ is changed beyond recognition, without affecting the operator norm bound.

If the columns of $\mathbf{A}$ are the snapshots of a dynamical process, then it is desirable to be able to claim that the backward error is small in each snapshot, relative to its own norm, and independent of its size relative to other snapshots or the norm of the matrix. This corresponds to a backward error $\delta\mathbf{A}$ such that

$$\|\delta\mathbf{A}(:,i)\|_2 \leq \varepsilon\|\mathbf{A}(:,i)\|_2, \quad i = 1, \ldots, m. \tag{7.25}$$

In terms of the snapshots (i.e., $\mathbf{A} = \mathbf{X}_m$), this reads $\|\delta \mathbf{x}_i\|_2 \leq \varepsilon \|\mathbf{x}_i\|_2$, for all snapshots. If $\delta\mathbf{A}$ can be bounded only as in (7.22), then we can only claim that $\|\delta \mathbf{x}_i\|_2 \leq \|\delta\mathbf{A}\|_2 \leq \varepsilon\|\mathbf{A}\|_2 \leq \varepsilon\sqrt{m}\max_i \|\mathbf{x}_i\|_2$, which is less satisfactory.

The finer structure (7.25) of the backward error allows for a forward error bound superior to (7.24). To illustrate the main point, assume $\mathbf{A}$ is of full column rank and write the backward perturbation in the multiplicative form

$$\mathbf{A} + \delta\mathbf{A} = (\mathbb{I}_n + \delta\mathbf{A}\mathbf{A}^{\dagger})\mathbf{A}. \tag{7.26}$$

**Theorem 7.3** (Eisenstat and Ipsen, [28]) *Let $\sigma_1 \geq \cdots \geq \sigma_n$ and $\tilde{\sigma}_1 \geq \cdots \geq \tilde{\sigma}_n$ be the singular values of $\mathbf{A}$ and $\mathbf{A} + \delta\mathbf{A}$, respectively. Assume that $\mathbf{A} + \delta\mathbf{A}$ can be written in the form of multiplicative perturbation $\mathbf{A} + \delta\mathbf{A} = \Xi_1\mathbf{A}\Xi_2$ and let $\xi = \max\{\|\Xi_1\Xi_1^T - I\|_2, \|\Xi_2^T\Xi_2 - I\|_2\}$. Then*

$$|\tilde{\sigma}_i - \sigma_i| \leq \xi\sigma_i, \quad i = 1, \ldots, n.$$

Hence, an application of this theorem to (7.26) yields

$$\max_i \frac{|\sigma_i - \tilde{\sigma}_i|}{\sigma_i} \leq 2\|\delta\mathbf{A}\mathbf{A}^{\dagger}\|_2 + \|\delta\mathbf{A}\mathbf{A}^{\dagger}\|_2^2.$$

Now, the key is in bounding $\delta\mathbf{A}\mathbf{A}^{\dagger}$. If we set $\mathbf{D_A} = \mathrm{diag}(\|\mathbf{A}(:,i)\|_2)_{i=1}^m$, then

$$\delta\mathbf{A}\mathbf{D}_{\mathbf{A}}^{-1}\mathbf{D}_{\mathbf{A}}\mathbf{A}^{\dagger} = (\delta\mathbf{A}\mathbf{D}_{\mathbf{A}}^{-1})(\mathbf{A}\mathbf{D}_{\mathbf{A}}^{-1})^{\dagger} \equiv \delta\mathbf{A}_c\mathbf{A}_c^{\dagger}, \quad \text{where } \delta\mathbf{A}_c = \delta\mathbf{A}\mathbf{D}_{\mathbf{A}}^{-1}, \quad \mathbf{A}_c = \mathbf{A}\mathbf{D}_{\mathbf{A}}^{-1},$$

$\|\mathbf{A}_c\|_2 \in [1, \sqrt{m}]$, and

$$\|\delta\mathbf{A}\mathbf{A}^{\dagger}\|_2 \leq \|\delta\mathbf{A}_c\|_2\|\mathbf{A}_c^{\dagger}\|_2 \leq \sqrt{m}\max_{i=1,\ldots,m}\|\delta\mathbf{A}_c(:,i)\|_2\|\mathbf{A}_c^{\dagger}\|_2$$
$$\leq \varepsilon\sqrt{m}\|\mathbf{A}_c^{\dagger}\|_2 \equiv \varepsilon\frac{\sqrt{m}}{\|\mathbf{A}_c\|_2}\kappa_2(\mathbf{A}_c) = \varepsilon_1\kappa_2(\mathbf{A}_c), \quad \varepsilon_1 = \varepsilon\frac{\sqrt{m}}{\|\mathbf{A}_c\|_2} \in [\varepsilon, \sqrt{m}\varepsilon].$$

Hence

$$\max_i \frac{|\sigma_i - \tilde{\sigma}_i|}{\sigma_i} \leq 2\varepsilon_1\kappa_2(\mathbf{A}_c) + (\varepsilon_1\kappa_2(\mathbf{A}_c))^2. \tag{7.27}$$

**Theorem 7.4** (Van der Sluis [65]) *Let $\mathbf{A} \in \mathbb{C}^{n\times m}$ be of full column rank and let $\mathbf{D_A} = \mathrm{diag}(\|\mathbf{A}(:,i)\|_2)_{i=1}^m$ and $\mathbf{A}_c = \mathbf{A}\mathbf{D}_{\mathbf{A}}^{-1}$. Then $\kappa_2(\mathbf{A}_c) \leq \sqrt{m}\min_{\mathbf{D}=\mathrm{diag}}\kappa_2(\mathbf{AD})$.*

Hence, $\kappa_2(\mathbf{A}_c)$ is in the worst case at most $\sqrt{m}$ times larger than $\kappa_2(\mathbf{A})$, but it is usually smaller and often much smaller than $\kappa_2(\mathbf{A})$.

Computational method that has the backward error (7.25) and the accuracy of the singular values as in (7.27) is the Jacobi SVD algorithm [25, 26], that is available in LAPACK in the driver routines `xGESVJ()`, `xGEJSV(.)`. Recently, a preconditioned QR SVD algorithm [19] has been implemented to deliver numerical SVD with the accuracy similar as in the Jacobi SVD. These more accurate methods also compute better approximate singular vectors, but here we omit the details due to limited space, and we refer the reader to [19, 25, 26]. If the columns of $A$ are nearly equilibrated (all of similar Euclidean lengths), then (7.25) and (7.24) imply similar error bounds in the singular values.

There is a trade-off between run time and accuracy—these more accurate methods are somewhat slower than `xGESVD(.)` and `xGESDD(.)`, but in the case of, e.g., tall matrices, the run time difference is marginal because in the case $n \gg m$, all these methods start with the QR factorization $\mathbf{A} = \mathbf{QR}$, where $\mathbf{Q}$ is unitary $n \times m$, and $\mathbf{R}$ is $m \times m$ upper triangular. This most expensive step is then followed by the SVD of $R$. It should be noted that the QR factorization of tall matrices can be computed very efficiently on modern high-performance computing hardware, see [17].

*Remark 7.6* Due to space constrain, we only mention two other important numerical accuracy issues relevant for full analysis of the DMD. Namely, the accuracy of the computed singular vectors of $\mathbf{X}_m$ and the eigenvectors of the Rayleigh quotient $S_k$ in addition depends on the separation (gap) of the corresponding singular values/eigenvalues and their neighbors in the spectrum. See, e.g., [67, Chap. V., Sect. 2], [51].

## 7.4 An Enhanced DMD—Refined Rayleigh–Ritz DDMD with Residual Bound (DDMD_R4)

For further development of the DMD framework, both with respect to numerical robustness and computational efficiency, the following issues emerge as core topics for further study:

1. Among all computed Ritz pairs, how can we distinguish which ones are good approximations of some eigenpairs of the underlying $\mathbb{A}$, and which ones should be discarded?
2. Since it is known that the Ritz vectors are not the best eigenvector approximations from a given subspace, is it possible to extract better information?
3. For high fidelity, numerical simulations are often performed in high resolution resulting in high-dimensional ambient space. What are the possibilities to reduce the computational run time?

In this section, we discuss the above and review results from the recent work[6] [23].

[6]For an earlier, extended version see [21].

## 7.4.1 Data-Driven Residual and Refined Ritz Pairs

The topics 1. and 2. in the above list require computing and minimizing the residual $\mathbb{A}z_i - \lambda_i z_i$ of a Ritz pair $(\lambda_i, z_i)$. This is a well-established technique in the large-scale eigenvalue computations. It only remains to adapt it to the data-driven framework.

### 7.4.1.1 Residual and an Eigenvalue Error Bound

The key observation is that residual can be computed explicitly, using only the available snapshots and without invoking generally unaccessible $\mathbb{A}$. Indeed, as we already saw in Sect. 7.3.2, from $\mathbf{X}_m = \mathbf{U}\boldsymbol{\Sigma}\mathbf{V}^*$ and $\mathbf{V}^*\mathbf{V}_k = (\,\mathbb{I}_k\ \mathbf{0}_{k,m-k}\,)^T$, it follows that $\mathbf{X}_m\mathbf{V}_k = \mathbf{U}_k\boldsymbol{\Sigma}_k$ and, thus, $\mathbb{A}\mathbf{U}_k = \mathbf{Y}_m\mathbf{V}_k\boldsymbol{\Sigma}_k^{-1}$. Since $z_i \in \mathrm{range}(\mathbf{U}_k)$, we can compute $\mathbb{A}z_i$ and the corresponding residual norm as follows:

**Proposition 7.2** ([23]) *For the Ritz pairs $(\lambda_i, \mathbf{Z}_k(:,i) \equiv \mathbf{U}_k\mathbf{W}_k(:,i))$, $i = 1, \ldots, k$, computed in Algorithm 7.1, the residual norms can be computed as follows:*

$$r_k(i) = \|\mathbb{A}(\mathbf{U}_k\mathbf{W}_k(:,i)) - \lambda_i(\mathbf{U}_k\mathbf{W}_k(:,i))\|_2 = \|(\mathbf{Y}_m\mathbf{V}_k\boldsymbol{\Sigma}_k^{-1})\mathbf{W}_k(:,i) - \lambda_i\mathbf{Z}_k(:,i)\|_2. \tag{7.28}$$

*Further, if $\mathbb{A}$ is diagonalizable with spectral decomposition $\mathbb{A} = \boldsymbol{\Psi}\mathrm{diag}(\alpha_i)_{i=1}^n\boldsymbol{\Psi}^{-1}$, then $\lambda_i$ approximates an eigenvalue of $\mathbb{A}$ with the Bauer–Fike type error bound $\min_{\alpha_j}|\lambda_i - \alpha_j| \leq \kappa_2(\boldsymbol{\Psi})r_k(i)$.*

Information provided by the residual allows for selecting the Ritz pairs that provide good approximations to the eigenpairs of the matrix $\mathbb{A}$ and good information on the eigenvalues of the underlying Koopman operator. An example is given in [23, Sect 4.2], where a residual thresholding has selected the Ritz pairs that match the Koopman eigenvalues obtained in a theoretical analysis [5] of a laminar incompressible two-dimensional flow around a cylinder. In that example, the observables (the pressure and the components of the velocity) are obtained from a high-resolution numerical simulation of the incompressible Navier–Stokes equations.

### 7.4.1.2 Refined Ritz Pairs

It is a well-known fact that the Ritz vectors are not optimal in the sense of the residual norm $\|\mathbb{A}z - \lambda z\|_2$; for a given Ritz value $\lambda$, there might be a better vector $\tilde{z}$ in the search space $\mathscr{U}_k = \mathrm{range}(\mathbf{U}_k)$ that yields smaller residual. For a thorough analysis of the refined Ritz vectors, see, e.g., [41–45].

Hence, it makes sense to replace $z$ with the optimal vector that solves the following optimization problem:

$$\min_{z\in\mathbf{U}_k\setminus\{\mathbf{0}\}} \frac{\|\mathbb{A}z-\lambda z\|_2}{\|z\|_2} = \min_{w\neq\mathbf{0}} \frac{\|\mathbb{A}\mathbf{U}_k w-\lambda\mathbf{U}_k w\|_2}{\|\mathbf{U}_k w\|_2} = \min_{\|w\|_2=1} \|(\mathbb{A}\mathbf{U}_k-\lambda\mathbf{U}_k)w\|_2. \tag{7.29}$$

Recall that $\mathbb{A}\mathbf{U}_k = \mathbf{Y}_m\mathbf{V}_k\mathbf{\Sigma}_k^{-1}$, i.e., no access to $\mathbb{A}$ is required. On the other hand, (7.29) is precisely the variational characterization of the smallest singular value of $\mathbb{A}\mathbf{U}_k-\lambda\mathbf{U}_k$ [40, Theorem 3.1.2], i.e., the minimum (7.29) equals $\sigma_{\min}(\mathbb{A}\mathbf{U}_k-\lambda\mathbf{U}_k)$, where $\sigma_{\min}(\cdot)$ denotes the smallest singular value of a matrix, and the minimum is attained at the right singular vector $w_\lambda$ corresponding to $\sigma_\lambda \equiv \sigma_{\min}(\mathbb{A}\mathbf{U}_k-\lambda\mathbf{U}_k)$.

As a result, the refined Ritz vector corresponding to $\lambda_i$ is $\widetilde{z}_i \equiv \mathbf{U}_k w_{\lambda_i}$ and the optimal residual is $\sigma_{\lambda_i} \leq r_k(i)$. We can take this refinement one step further using the optimality property of the Rayleigh quotient:

$$\frac{z^*\mathbb{A}z}{z^*z} = \operatorname*{argmin}_{\zeta\in\mathbb{C}} \|\mathbb{A}z-\zeta z\|_2, \quad z\in\mathbb{C}^n\setminus\{\mathbf{0}\}. \tag{7.30}$$

Hence, after obtaining $\widetilde{z}_i$ that minimizes (7.29), we can replace $\lambda_i$ wit the Rayleigh quotient $\widetilde{\lambda}_i \equiv \widetilde{z}_i^*\mathbb{A}\widetilde{z}_i$, which is better in the sense of the residual (7.30), and easily computed as

$$\widetilde{z}_i^*\mathbb{A}\widetilde{z}_i = w_{\lambda_i}^*\mathbf{U}_k^*\mathbb{A}\mathbf{U}_k w_{\lambda_i} = w_{\lambda_i}^*\mathbf{S}_k w_{\lambda_i}. \text{ (Here } \mathbf{S}_k=\mathbf{U}_k^*\mathbb{A}\mathbf{U}_k \text{ and } \|\widetilde{z}_i\|_2 = \|w_{\lambda_i}\|_2 = 1.) \tag{7.31}$$

In principle, this procedure can be repeated by returning to (7.29) with refined Ritz value $\widetilde{\lambda}_i$, but after the first step, the improvement is usually so marginal that it does not warrant the computational effort.

If $k$ is large, then refining all Ritz pairs is computationally intensive. In practice, however, we may not wish to refine all pairs. We may selected ones with smallest residual, or those with the $\lambda_j$'s in a particular region $\Omega\subset\mathbb{C}$ of interest. But even in such cases, computing the smallest singular values and the corresponding vector, for example, few dozen matrices of the form $\mathbb{A}\mathbf{U}_k-\lambda\mathbf{U}_k$ may considerably increase the run time. In the next section, we review a preprocessing step, proposed in [23], that reduces the cost of the refinement procedure.

#### 7.4.1.3 Technical Detail—Computation of Refined Ritz Vectors

Let us now discuss some implementation details related to computing the refined Ritz vectors, i.e., computing the smallest singular value with the corresponding right singular vector of $\mathbf{B}_k-\lambda\mathbf{U}_k$, where $\mathbf{B}_k=\mathbf{Y}_m\mathbf{V}_k\mathbf{\Sigma}_k^{-1}$ and $k\leq m\ll n$.

The refinement process can be confined to a particular $2k$-dimensional subspace, so that in the case $2k<m\ll n$, the optimization (7.29) becomes more efficient. This is achieved using the QR factorization

$$\begin{pmatrix}\mathbf{U}_k & \mathbf{B}_k\end{pmatrix} = \mathbf{Q}\mathbf{R}, \quad \mathbf{R} = \begin{matrix} & \begin{matrix} k & k \end{matrix} \\ \begin{matrix} k \\ k' \end{matrix} & \begin{pmatrix} \mathbf{R}_{[11]} & \mathbf{R}_{[12]} \\ \mathbf{0} & \mathbf{R}_{[22]} \end{pmatrix} \end{matrix}, \quad k'=\min(n-k,k), \tag{7.32}$$

which can be used to write the pencil $\mathbf{B}_k - \lambda \mathbf{U}_k$ as

$$\mathbf{B}_k - \lambda \mathbf{U}_k = \mathbf{Q}\left(\begin{pmatrix} \mathbf{R}_{[12]} \\ \mathbf{R}_{[22]} \end{pmatrix} - \lambda \begin{pmatrix} \mathbf{R}_{[11]} \\ \mathbf{0} \end{pmatrix}\right) \equiv \mathbf{Q}\mathbf{R}_\lambda, \quad \mathbf{R}_\lambda = \begin{pmatrix} \mathbf{R}_{[12]} - \lambda \mathbf{R}_{[11]} \\ \mathbf{R}_{[22]} \end{pmatrix}. \tag{7.33}$$

Since $\mathbf{Q}$ is unitary, we can proceed with computing the smallest singular value and the corresponding right singular vector of the $2k \times k$ matrix $\mathbf{R}_\lambda$, i.e., solving

$$\min_{\|w\|_2=1} \|\mathbf{R}_\lambda w\|_2. \tag{7.34}$$

Note that the QR factorization (7.32) is computed only once and used for all Ritz pairs that have been selected for refinement. However, its cost is $O(nk^2)$ *flops*, and it could be paid off if its output could be reused to save operations in other part of the computational scheme. This is indeed possible, as the computed triangular factor $\mathbf{R}$ contains information on the Rayleigh quotient matrix $\mathbf{S}_k$.

**Proposition 7.3** ([23]) *In the QR factorization (7.32), define* $\boldsymbol{\Phi} = \mathrm{diag}((\mathbf{R}_{[11]})_{ii})_{i=1}^k$. *Then* $\boldsymbol{\Phi}^* \mathbf{R}_{[12]} = \mathbf{U}_k^* \mathbf{B}_k \equiv \mathbf{S}_k$.

*Remark 7.7* The benefit of Proposition 7.3 is that it saves $2nk^2$ flops needed to compute $\mathbf{S}_k$; instead, at most $k^2$ complex sign changes in $\mathbf{R}_{[12]}$ are needed. Namely, in the numerical software (e.g., MATLAB, LAPACK), the QR factorization is implemented so that the diagonal entries of $\mathbf{R}$ are real even for complex input matrices. Hence, in practical computation, the matrix $\boldsymbol{\Phi}$ is $\mathrm{diag}(\pm 1)_{i=1}^k$.

Finally, we note that, for practical reasons, the full minimization (7.34) can be replaced with few steps of some simple optimization scheme to reduce $\|\mathbf{R}_\lambda w\|_2$. Optimal course of action is to deploy some specialized method, such as [29].

### 7.4.2 Scaling the Data

If $\mathbf{X}_m$ has a column that is small in norm, then it must have small singular value and it will be declared numerically rank deficient. On the other hand, that small column may carry valuable information, despite its low-energy content.

Following the discussion in Sect. 7.3.3, we conclude that the singular values of the column scaled matrix $\mathbf{X}_m^{(cs)} = \mathbf{X}_m \mathbf{D}_\mathbf{X}^\dagger$, $\mathbf{D}_\mathbf{X} = \mathrm{diag}(\|\mathbf{X}_m(:,i)\|_2)_{i=1}^m$ are expected to have smaller ratio of the largest and the smallest one, as compared to the quotient of the two extreme singular values of $\mathbf{X}_m$. Hence, we expect the numerical rank of $\mathbf{X}_m^{(cs)}$ to be larger than of $\mathbf{X}_m$, and the SVD of $\mathbf{X}_m^{(cs)}$ to be computed more accurately than of $\mathbf{X}_m$, even for methods whose accuracy is governed by (7.24). Intuitively, large condition number of $\mathbf{X}_m^{(cs)}$ (i.e., the presence of small singular values) indicates near collinearity of the snapshots in the sense of small angles.

Further, if $\mathbf{Y}_m = \mathbb{A}\mathbf{X}_m$, then $\mathbf{Y}_m \mathbf{D}_\mathbf{X} = \mathbb{A}\mathbf{X}_m \mathbf{D}_\mathbf{X}$, i.e., $\mathbf{Y}_m^{(cs)} = \mathbb{A}\mathbf{X}_m^{(cs)}$, where $\mathbf{Y}_m^{(cs)} = \mathbf{Y}_m \mathbf{D}_\mathbf{X}$. Hence, we can apply the DMD to the input–output pair $(\mathbf{X}_m^{(cs)}, \mathbf{Y}_m^{(cs)})$ as well.

Scaling the rows of the data matrices is also meaningful. Recall that the rows correspond to particular observables, possibly of different physical nature and each given in its own physical units, and possibly with different levels of noise. In such situation, the Euclidean length of a snapshot, $\|\mathbf{x}_i\|_2$, is not easily interpreted; see, e.g., [39, Sect. 3.4.3]. Hence, it may be important to scale the data with a diagonal matrix $\boldsymbol{\Delta}$ and use the scaled data $\mathbf{X}_m^{(rs)} = \boldsymbol{\Delta}\mathbf{X}_m$, $\mathbf{Y}_m^{(rs)} = \boldsymbol{\Delta}\mathbf{Y}_m$; the choice of nonsingular diagonal scaling $\boldsymbol{\Delta}$ depends on a concrete application. Note that this transformation of observables corresponds to a similarity transformation of the operator $\mathbb{A}$, because $\mathbf{Y}_m^{(rs)} = (\boldsymbol{\Delta}\mathbb{A}\boldsymbol{\Delta}^{-1})\mathbf{X}_m^{(rs)}$. Hence, the eigenvalues remain unchanged, and the eigenvectors are scaled accordingly: $\mathbb{A}v = \lambda v$ is equivalent to $(\boldsymbol{\Delta}\mathbb{A}\boldsymbol{\Delta}^{-1})(\boldsymbol{\Delta}v) = \lambda(\boldsymbol{\Delta}v)$.

Proper scaling of the raw data is important. In particular, it influences the numerical computations; e.g., the numerical rank and thus the dimension of the POD basis and the number of Ritz values returned by DMD depend on the scaling, see [23, Sect. 4.1]. This discussion leads to the more general question on proper inner product and normed structure in which we measure the size and truncate parts deemed negligible.

### *7.4.3 DMD in a Weighted Inner Product*

In essence, the DMD Algorithm 7.1 first computes a POD basis $\mathbf{U}_k$ for the snapshots in the data matrix $\mathbf{X}_m$ by truncating its SVD, and proceeds with computing the Rayleigh quotient $\mathbf{S}_k = \mathbf{U}_k^*\mathbb{A}\mathbf{U}_k$.

This Galerkin/POD step in model order reduction may cause loss of important physical characteristics if it is done in an inappropriate Hilbert space structure. Weighted POD [69] assumes that, with the data, we also have an inner product defined by a Hermitian positive definite matrix $\mathbf{M}$ by $(x, y)_\mathbf{M} = y^*\mathbf{M}x$, and the POD basis is computed with respect to $(\cdot,\cdot)_\mathbf{M}$. Here $\mathbf{M}$ can be derived from careful physical consideration (the new inner product should induce norm that is consistent with a definition of energy, see, e.g., [60], [39, Sect. 3.4.3]), statistical analysis (covariance), or it may be the positive definite solution $\mathbf{M}$ of a Lyapunov matrix equation, see, e.g., [48, Sect. 6.1], [59, 64].

Hence, it seems appropriate to perform the DMD analysis in $(\mathbb{C}^n, (\cdot,\cdot)_\mathbf{M})$. For practical work with $(\cdot,\cdot)_\mathbf{M}$, it is useful to have a factorization $\mathbf{M} = \mathbf{L}\mathbf{L}^*$, where $\mathbf{L}$ is a conveniently computed square factor of $\mathbf{M}$; it can be the lower triangular Cholesky factor (practical for computation), or the positive definite square root $\sqrt{\mathbf{M}}$ (elegant for theoretical analysis). Then the induced norm is $\|v\|_\mathbf{M} = \|\mathbf{L}^*v\|_2$, and the corresponding induced operator norm is $\|\mathbb{A}\|_\mathbf{M} = \|\mathbf{L}^*\mathbb{A}\mathbf{L}^{-*}\|_2$.

For the reader's convenience, we recall the weighted POD. The following proposition from [23] gives the necessary ingredients to perform DMD with respect to the inner product $(\cdot,\cdot)_\mathbf{M}$.

**Proposition 7.4** ([23]) *Assume that we are given the data $\mathbf{X}_m$ and $\mathbf{Y}_m = \mathbb{A}\mathbf{X}_m$, and that Algorithm 7.2 is applied to $\mathbf{X}_m$. The $(\cdot,\cdot)_\mathbf{M}$-weighted Rayleigh quotient matrix of $\mathbb{A}$ with respect to the POD basis $\widehat{\mathbf{U}}_k$ can be computed from the available data as*

**Algorithm 7.2** $[\widetilde{\mathbf{U}}_k, \mathbf{L}, [\widehat{\mathbf{U}}_k]] = \text{Weighted\_POD}(\mathbf{X}_m; \mathbf{M}; \varepsilon)$ (See, e.g., [69].)

**Input:**
- The matrix of snapshots $\mathbf{X}_m = (\mathbf{x}_1, \mathbf{x}_2, \ldots, \mathbf{x}_m) \in \mathbb{C}^{n\times m}$. (Assumed $m < n$.)
- Hermitian $n \times n$ positive definite $\mathbf{M}$ that defines the inner product.
- Tolerance level $\varepsilon$ for numerical rank detection.

1: Compute the Cholesky decomposition $\mathbf{M} = \mathbf{L}\mathbf{L}^*$.
2: Compute the thin SVD $\mathbf{L}^*\mathbf{X}_m = \widetilde{\mathbf{U}}\boldsymbol{\Sigma}\mathbf{V}^*$. $\{\boldsymbol{\Sigma} = \text{diag}(\sigma_i)_{i=1}^m; \sigma_1 \geq \cdots \geq \sigma_m.\}$
3: Determine the numerical rank $k$ using the threshold $\varepsilon$.
**Output:** $\widehat{\mathbf{U}}_k \equiv (\mathbf{L}^*)^{-1}\widetilde{\mathbf{U}}_k$, where $\widetilde{\mathbf{U}}_k = \widetilde{\mathbf{U}}(:, 1:k)$. $\widehat{\mathbf{U}}_k$ can be returned implicitly by $\mathbf{L}$ and $\widetilde{\mathbf{U}}_k$.

$\widehat{\mathbf{S}}_k = \widetilde{\mathbf{U}}_k^*\mathbf{L}^*\mathbf{Y}_m\mathbf{V}_k\boldsymbol{\Sigma}_k^{-1}$. *If* $(\lambda, w)$ *is an eigenpair of* $\widehat{\mathbf{S}}_k$ $(\widehat{\mathbf{S}}_k w = \lambda w,\ w \neq 0)$ *then the corresponding Ritz pair* $(\lambda, \widehat{\mathbf{U}}_k w)$ *of* $\mathbb{A}$ *has the residual norm computable as*

$$\|\mathbb{A}(\widehat{\mathbf{U}}_k w) - \lambda(\widehat{\mathbf{U}}_k w)\|_{\mathbf{M}} = \|\mathbf{L}^*\mathbf{Y}_m\mathbf{V}_k\boldsymbol{\Sigma}_k^{-1}w - \lambda\widetilde{\mathbf{U}}_k w\|_2. \tag{7.35}$$

Note that, in (7.35), the weighted norm may be interpreted as the energy of the residual. For more details, we refer the reader to [23, Sect. 5.5].

### 7.4.4 *The Algorithm*

We are now ready to assemble a variation of the DMD, recently proposed in [23], using the elements described in Sect. 7.4.1. For the sake of brevity, we show a generic version and supplement it with a few remarks. Only the unweighted form of the algorithm is presented. For more details, the reader is referred to [23].

*Remark 7.8* In Line 8, only the eigenvalues are computed, which saves $O(k^3)$ flops as compared to Line 5 in Algorithm 7.1. The Ritz vectors are then computed in the refinement procedure. Only this option is shown for the sake of brevity; our software implementation allows first computing the Ritz vectors with the corresponding residuals, and then refining only the selected ones.

*Remark 7.9* The simplest way to implement Line 10 is to compute the full $2k \times k$ SVD, and then to take the smallest singular value $\sigma_{\lambda_i}$ and the corresponding right singular vector $w_{\lambda_i}$. It is an interesting research problem to develop a more efficient method that will target only $\sigma_{\lambda_i}$, $w_{\lambda_i}$ and thus reduce the complexity, from $O(k^3)$ to $O(k^2)$ per refined vector.

*Remark 7.10* Variational characterization identifies the smallest singular value $\sigma_{\lambda_i}$ as the value of the minimal residual (7.29). We should be aware that the smallest singular value may be computed with relatively large error, so in practice, it may not represent the residual norm exactly. Although we are not interested in the number of accurate digits of the residual, but in its order of magnitude, we can also compute it as $\sigma_{\lambda_i} = \|\mathbf{R}_{\lambda_i} w_{\lambda_i}\|_2$, or using the formula (7.28).

**Algorithm 7.3** $[\mathbf{Z}_k, \boldsymbol{\Lambda}_k, r_k, \rho_k] = \mathrm{DDMD\_R4}(\mathbf{X}_m, \mathbf{Y}_m; \varepsilon)$ {*Refined Rayleigh–Ritz Data-Driven Modal Decomposition* [23]}

**Input:**

- $\mathbf{X}_m = (\mathbf{x}_1, \ldots, \mathbf{x}_m), \mathbf{Y}_m = (\mathbf{y}_1, \ldots, \mathbf{y}_m) \in \mathbb{C}^{n\times m}$ that define a sequence of snapshots pairs $(\mathbf{x}_i, \mathbf{y}_i \equiv \mathbb{A}(\mathbf{x}_i))$. (Tacit assumption is that $n$ is large and that $m \ll n$.)
- Tolerance level $\varepsilon$ for numerical rank determination.

1: $\mathbf{D}_x = \mathrm{diag}(\|\mathbf{X}_m(:,i)\|_2)_{i=1}^m$; $\mathbf{X}_m^{(1)} = \mathbf{X}_m\mathbf{D}_x^\dagger$; $\mathbf{Y}_m^{(1)} = \mathbf{Y}_m\mathbf{D}_x^\dagger$
2: $[\mathbf{U}, \boldsymbol{\Sigma}, \mathbf{V}] = svd(\mathbf{X}_m^{(1)})$ ; {*The thin SVD:* $\mathbf{X}_m^{(1)} = \mathbf{U}\boldsymbol{\Sigma}\mathbf{V}^*$, $\mathbf{U} \in \mathbb{C}^{n\times m}$, $\boldsymbol{\Sigma} = \mathrm{diag}(\sigma_i)_{i=1}^m$}
3: Determine numerical rank $k$, with the threshold $\varepsilon$. See Sect. 7.3.1.
4: Set $\mathbf{U}_k = \mathbf{U}(:, 1:k)$, $\mathbf{V}_k = \mathbf{V}(:, 1:k)$, $\boldsymbol{\Sigma}_k = \boldsymbol{\Sigma}(1:k, 1:k)$
5: $\mathbf{B}_k = \mathbf{Y}_m^{(1)}(\mathbf{V}_k\boldsymbol{\Sigma}_k^{-1})$; {*Schmid's data-driven formula for* $\mathbb{A}\mathbf{U}_k$}
6: $[\mathbf{Q}, \mathbf{R}] = qr((\mathbf{U}_k, \mathbf{B}_k))$; {*The thin QR factorization:* $(\mathbf{U}_k, \mathbf{B}_k) = \mathbf{Q}\mathbf{R}$; $\mathbf{Q}$ *not computed*}
7: $\mathbf{S}_k = \mathrm{diag}(\overline{\mathbf{R}_{ii}})_{i=1}^k \mathbf{R}(1:k, k+1:2k)$ {$\mathbf{S}_k = \mathbf{U}_k^*\mathbb{A}\mathbf{U}_k$ *is the Rayleigh quotient*}
8: $\boldsymbol{\Lambda}_k = \mathrm{eig}(\mathbf{S}_k)$ {$\boldsymbol{\Lambda}_k = \mathrm{diag}(\lambda_i)_{i=1}^k$; *Ritz values, i.e., eigenvalues of* $\mathbf{S}_k$}
9: **for** $i = 1, \ldots, k$ **do**
10: $[\sigma_{\lambda_i}, w_{\lambda_i}] = svd_{\min}\left(\begin{pmatrix} \mathbf{R}(1:k,k+1:2k) - \lambda_i\mathbf{R}(1:k,1:k) \\ \mathbf{R}(k+1:2k,k+1:2k) \end{pmatrix}\right)$; {*Minimal singular value of* $\mathbf{R}_{\lambda_i}$ *(7.33) and the corresponding right singular vector*}
11: $\mathbf{W}_k(:,i) = w_{\lambda_i}$ ; $r_k(i) = \sigma_{\lambda_i}$ {*Optimal residual,* $\sigma_{\lambda_i} = \|\mathbf{R}_{\lambda_i}w_{\lambda_i}\|_2$}
12: $\rho_k(i) = w_{\lambda_i}^*\mathbf{S}_k w_{\lambda_i}$ {*Rayleigh quotient,* $\rho_k(i) = (\mathbf{U}_k w_{\lambda_i})^*\mathbb{A}(\mathbf{U}_k w_{\lambda_i})$; *see (7.30), (7.31)*}
13: **end for**
14: $\mathbf{Z}_k = \mathbf{U}_k\mathbf{W}_k$ {*Refined Ritz vectors*}

**Output:** $\mathbf{Z}_k, \boldsymbol{\Lambda}_k, r_k, \rho_k$

### 7.4.5 Compressed DMD

The computation in any variation of the DMD takes place in the linear span of the snapshots. In the case of sequential data, this is the range of $\mathbf{X}_{m+1} = (\mathbf{x}_1, \ldots, \mathbf{x}_{m+1})$—at most $(m+1)$-dimensional subspace of $\mathbb{C}^n$, where $m \ll n$. In the general case of $\mathbf{X}_m \in \mathbb{C}^{n\times m}$ and $\mathbf{Y}_m = \mathbb{A}\mathbf{X}_m$, the ambient space is the at most $2m$-dimensional range of the matrix $(\mathbf{X}_m\ \mathbf{Y}_m)$.

Let us consider the sequential case first; $\mathbf{X}_m = (\mathbf{x}_1, \ldots, \mathbf{x}_m), \mathbf{Y}_m = (\mathbf{x}_2, \ldots, \mathbf{x}_{m+1})$. The most efficient numerically robust way to construct an orthonormal basis for $\mathrm{range}(\mathbf{X}_{m+1})$ is to compute the QR factorization

$$\mathbf{X}_{m+1} = \mathbf{Q}_f \begin{pmatrix} \mathbf{R}_f \\ \mathbf{0} \end{pmatrix} = \widehat{\mathbf{Q}}_f\mathbf{R}_f, \quad \text{where } \mathbf{Q}_f^*\mathbf{Q}_f = \mathbb{I}_n, \quad \widehat{\mathbf{Q}}_f = \mathbf{Q}_f(:, 1:m+1), \tag{7.36}$$

$\mathrm{range}(\widehat{\mathbf{Q}}_f) \supseteq \mathrm{range}(\mathbf{X}_{m+1})$, and $\mathbf{R}_f$ is $(m+1)\times(m+1)$ upper triangular. (If $\mathbf{X}_{m+1}$ is of full column rank, then $\mathrm{range}(\widehat{\mathbf{Q}}_f) = \mathrm{range}(\mathbf{X}_{m+1})$.)

If we set $\mathbf{R}_x = \mathbf{R}_f(:, 1:m)$, $\mathbf{R}_y = \mathbf{R}_f(:, 2:m+1)$, then $\mathbf{X}_m = \widehat{\mathbf{Q}}_f\mathbf{R}_x$, $\mathbf{Y}_m = \widehat{\mathbf{Q}}_f\mathbf{R}_y$, with the structure

$$\mathbf{R}_f = \begin{pmatrix} \times & \circledast & \circledast & \circledast & \div \\ & \circledast & \circledast & \circledast & \div \\ & & \circledast & \circledast & \div \\ & & & \circledast & \div \\ & & & & \div \end{pmatrix}, \quad \mathbf{R}_x = \begin{pmatrix} \times & \circledast & \circledast & \circledast \\ & \circledast & \circledast & \circledast \\ & & \circledast & \circledast \\ & & & \circledast \\ & & & 0 \end{pmatrix}, \quad \mathbf{R}_y = \begin{pmatrix} \circledast & \circledast & \circledast & \div \\ \circledast & \circledast & \circledast & \div \\ & \circledast & \circledast & \div \\ & & \circledast & \div \\ & & & \div \end{pmatrix}, \tag{7.37}$$

and, in the basis of the columns of $\widehat{\mathbf{Q}}_f$, we can identify $\mathbf{X}_m \equiv \mathbf{R}_x$, $\mathbf{Y}_m \equiv \mathbf{R}_y$, i.e., we can think of $\mathbf{X}_m$ and $\mathbf{Y}_m$ as $m$ snapshots in an $(m+1)$-dimensional space.

This change of basis implicitly induces a unitary similarity applied to $\mathbb{A}$, since

$$\mathbf{X}_m = \mathbf{Q}_f \begin{pmatrix} \mathbf{R}_x \\ \mathbf{0} \end{pmatrix}, \quad \mathbf{Y}_m = \mathbb{A}\mathbf{Q}_f \begin{pmatrix} \mathbf{R}_x \\ \mathbf{0} \end{pmatrix} = \mathbf{Q}_f \begin{pmatrix} \mathbf{R}_y \\ \mathbf{0} \end{pmatrix} \Longrightarrow \begin{pmatrix} \mathbf{R}_y \\ \mathbf{0} \end{pmatrix} = (\mathbf{Q}_f^* \mathbb{A} \mathbf{Q}_f) \begin{pmatrix} \mathbf{R}_x \\ \mathbf{0} \end{pmatrix}. \tag{7.38}$$

Note that $\mathbf{R}_y = (\widehat{\mathbf{Q}}_f^* \mathbb{A} \widehat{\mathbf{Q}}_f)\mathbf{R}_x$, and $\mathbf{R}_x = (\widehat{\mathbf{Q}}_f^* \mathbb{A}^{-1} \widehat{\mathbf{Q}}_f)\mathbf{R}_y$, provided that $\mathbb{A}$ is invertible. In the case of the exact DMD with $\widehat{\mathbf{A}} = \mathbf{Y}_m \mathbf{X}_m^\dagger$, the matrix $\mathbf{R}_y \mathbf{R}_x^\dagger$ is a Rayleigh quotient of $\widehat{\mathbf{A}}$ with respect to $\widehat{\mathbf{Q}}_f$, because

$$\mathbf{X}_m^\dagger = (\mathbf{R}_x^\dagger \ 0)\, \mathbf{Q}_f^* = \mathbf{R}_x^\dagger \widehat{\mathbf{Q}}_f^*, \quad \widehat{\mathbf{A}} = \mathbf{Y}_m \mathbf{X}_m^\dagger = \mathbf{Q}_f \begin{pmatrix} \mathbf{R}_y \mathbf{R}_x^\dagger & \mathbf{0} \\ \mathbf{0} & \mathbf{0} \end{pmatrix} \mathbf{Q}_f^* = \widehat{\mathbf{Q}}_f \mathbf{R}_y \mathbf{R}_x^\dagger \widehat{\mathbf{Q}}_f^*. \tag{7.39}$$

In the general case of two sequences of snapshots $\mathbf{X}_m$ and $\mathbf{Y}_m = \mathbb{A}\mathbf{X}_m$, from the QR factorization

$$(\mathbf{X}_m \ \mathbf{Y}_m) = \mathbf{Q}_{xy} \begin{pmatrix} \mathbf{R}_{[11]} & \mathbf{R}_{[12]} \\ \mathbf{0} & \mathbf{R}_{[22]} \\ \mathbf{0} & \mathbf{0} \end{pmatrix} = \widehat{\mathbf{Q}}_{xy} \begin{pmatrix} \mathbf{R}_{[11]} & \mathbf{R}_{[12]} \\ \mathbf{0} & \mathbf{R}_{[22]}, \end{pmatrix}$$

we have the new representations of the snapshots (analogously to (7.37))

$$\mathbf{X}_m = \widehat{\mathbf{Q}}_{xy} \begin{pmatrix} \mathbf{R}_{[11]} \\ \mathbf{0} \end{pmatrix} \equiv \widehat{\mathbf{Q}}_{xy} \mathbf{R}_x, \quad \mathbf{Y}_m = \widehat{\mathbf{Q}}_{xy} \begin{pmatrix} \mathbf{R}_{[12]} \\ \mathbf{R}_{[22]} \end{pmatrix} \equiv \widehat{\mathbf{Q}}_{xy} \mathbf{R}_y.$$

If we apply the DMD algorithm to the data $(\mathbf{R}_x, \mathbf{R}_y)$, it will return the matrix of approximate eigenvalues $\boldsymbol{\Lambda}_k$ with the corresponding eigenvectors as the columns of $\widehat{\mathbf{Z}}_k \in \mathbb{C}^{(m+1)\times k}$ (or $\widehat{\mathbf{Z}}_k \in \mathbb{C}^{2m\times k}$). To transform this output in terms of the original data, it suffices to lift the eigenvectors as $\mathbf{Z}_k = \widehat{\mathbf{Q}}_f \widehat{\mathbf{Z}}_k$ (or $\mathbf{Z}_k = \widehat{\mathbf{Q}}_{xy} \widehat{\mathbf{Z}}_k$).

The benefit of this compressed DMD is that the dimension $n$ is reduced to much smaller dimension $m+1$ (or $2m$) using the QR factorization that can be optimized for tall and skinny matrices [17, 55].

*Remark 7.11* In an efficient software implementation, the matrix $\widehat{\mathbf{Q}}_f$ is not formed explicitly. Instead, the information on the $m+1$ Householder reflectors used in the factorization is stored in the positions of the annihilated entries (see, e.g., `xGEQRF` in LAPACK) and then one can apply such implicitly stored $\mathbf{Q}_f$ and compute $\mathbf{Z}_k = \mathbf{Q}_f \begin{pmatrix} \widehat{\mathbf{Z}}_k \\ \mathbf{0} \end{pmatrix} \equiv \widehat{\mathbf{Q}}_f \widehat{\mathbf{Z}}_k$ (see, e.g., `xORMQR` in LAPACK).

#### 7.4.5.1 Applications of the QR Compression

This compression trick applies to Algorithms 7.1, 7.3 as well as to the Exact DMD [68, Algorithm 2]. It especially greatly improves the efficiency of Algorithm 7.3 because it reduces the overhead of computing the refined Ritz vectors, and of the Forward–Backward DMD which applies the DMD twice (by swapping the roles of $\mathbf{X}_m$ and $\mathbf{Y}_m$). For the readers convenience, in Algorithm 7.4, we show the compressed version of Algorithm 7.3; for the other two algorithms, the corresponding compressed versions are straightforward.

---

**Algorithm 7.4** $[Z_k, \Lambda_k, r_k, \rho_k]$ = DDMD_R4_C($\mathbf{F}_m$; $\varepsilon$) {*QR Compressed Refined DDMD_R4*}

---

**Input:**

- $\mathbf{X}_{m+1} = (\mathbf{x}_1, \ldots, \mathbf{x}_m, \mathbf{x}_{m+1})$ that defines a sequence of snapshots pairs $\mathbf{x}_{i+1} = \mathbb{A}\mathbf{x}_i$. (Tacit assumption is that $n$ is large and that $m \ll n$.)
- Tolerance level $\varepsilon$ for numerical rank determination.

1: $[\widehat{\mathbf{Q}}_f, \mathbf{R}_f] = qr(\mathbf{X}_{m+1}, 0)$ ; {*thin QR factorization*}
2: $\mathbf{R}_x = \mathbf{R}_f(1 : m+1, 1 : m)$, $\mathbf{R}_y = \mathbf{R}_f(1 : m+1, 2 : m+1)$ ; {*New representaitons of* $\mathbf{X}_m$, $\mathbf{Y}_m$.}
3: $[\widehat{\mathbf{Z}}_k, \Lambda_k, r_k, \rho_k]$ = DDMD_R4($\mathbf{R}_x$, $\mathbf{R}_y$; $\varepsilon$); {*Algorithm 7.3 in* $(m+1)$*-dimensional ambient space*}
4: $\mathbf{Z}_k = \widehat{\mathbf{Q}}_f \widehat{\mathbf{Z}}_k$
**Output:** $\mathbf{Z}_k, \Lambda_k, r_k, \rho_k$

---

*Remark 7.12* Using (7.36), (7.37) as in Algorithm 7.4 facilitates efficient updating/downdating if we keep adding new snapshots and/or dropping the ones at the beginning, e.g., if the snapshots are taken from a sliding (in discrete-time steps) window that may even be of variable width. Also, rows (observables) may be added/removed and the decomposition recomputed from the previous one. The key is that only the QR factorization (7.36) needs to be updated, using the well-established algorithms (see, e.g., [36], [7, Sect. 2.4.6]) that are also available for parallel computing (see, e.g., [3]); the rest of the computation takes place in $(m+1)$-dimensional space. We omit the details for the sake of brevity.

## 7.5 DDMD Based on Rank Revealing QR Factorization

One of the keys in the DMD methods is reduction of the dimension of the subspace determined by the supplied snapshots, and computation of the Rayleigh quotient in an orthonormal basis that optimally captures the input data. All this is facilitated by the SVD decomposition.

Although optimal in the sense of Theorem 7.1, the SVD-based low-rank approximation of the range of the snapshots by the POD basis $\mathbf{U}_k$ is *(i)* expensive to compute, *(ii)* difficult to update or downdate, and *(iii)* the range of $\mathbf{U}_k$ may not contain any of the snapshots $\mathbf{x}_i$.

In this section, we explore an alternative to the SVD, to find a low-dimensional subspace that captures all snapshots well in the least-square sense, but that itself is spanned by carefully selected representative snapshots. We use this problem as a case study for a DMD-type algorithm development.

### 7.5.1 Low-Rank Approximation Using QR Factorization

It might be of interest to have a low-rank approximation of the range of $\mathbf{X}_m$ that is spanned by the carefully selected snapshots, and with an approximation error comparable to the one of the POD basis $\mathbf{U}_k$. In that way, the computed approximate eigenvectors will reside in the span of the snapshots that are selected as most important ones—in some cases, this could be a desirable property.

This can be achieved using the column pivoted rank revealing QR factorization

$$\mathbf{X}_m \boldsymbol{\Pi} = \mathbf{Q}\mathbf{R}, \quad \mathbf{Q} \in \mathbb{C}^{n\times m}, \quad \mathbf{Q}^*\mathbf{Q} = \mathbb{I}_m, \quad \mathbf{R} \in \mathbb{C}^{m\times m} \text{ upper triangular}, \tag{7.40}$$

where the permutation matrix $\boldsymbol{\Pi}$ ensures that the numerical rank of $\mathbf{R}$ can be easily read-off from its structure. More precisely, if the numerical rank of $\mathbf{X}_m$ is $k$ (as determined from the singular values of $\mathbf{X}_m$), then the leading $k \times k$ submatrix $\mathbf{R}_k \equiv \mathbf{R}(1:k, 1:k)$ of $\mathbf{R}$ is in certain sense dominant part of $\mathbf{R}$. The most commonly used pivot strategy was introduced by Businger and Golub [10], which ensures

$$|\mathbf{R}_{ii}| \geq \sqrt{\sum_{k=i}^{j} |\mathbf{R}_{kj}|^2}, \quad 1 \leq i \leq j \leq m. \tag{7.41}$$

This pivoting is used in the MATLAB's function `qr(.)`, which is based on a robust implementation [20] of (7.40), (7.41) in the subroutine `xGEQP3(.)` in LAPACK [2].

Then, if $\mathbf{X}_m$ is close to a rank deficient matrix, this will be revealed in $\mathbf{R}$ by a block structure, for some threshold $\eta > 0$ and an index $1 \leq k < m$,

$$\mathbf{R} = \begin{pmatrix} \mathbf{R}_{[11]} & \mathbf{R}_{[12]} \\ \mathbf{0} & \mathbf{R}_{[22]} \end{pmatrix}, \quad \mathbf{R}_{[11]} \in \mathbb{C}^{k\times k}, \quad |\mathbf{R}_{kk}| \geq \eta |\mathbf{R}_{k+1,k+1}| \geq \frac{\eta}{\sqrt{m-k}} \|\mathbf{R}_{[22]}\|_F. \tag{7.42}$$

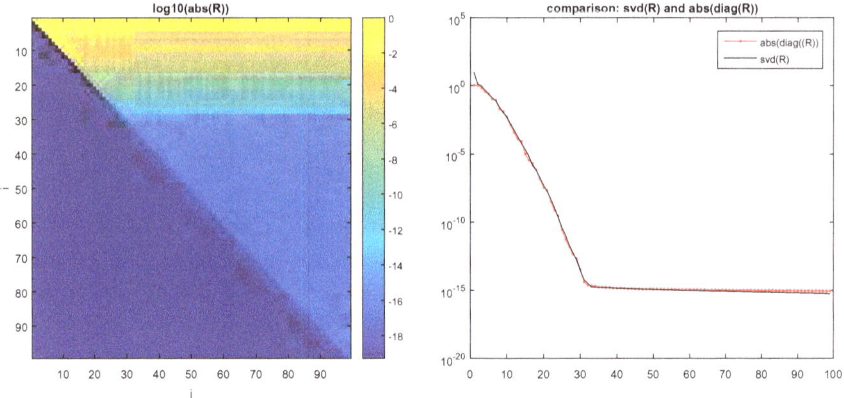


**Fig. 7.2** (Example 7.1.) *Left panel*: $\log_{10}|R|$. Note how the diagonal entries and the rows of **R** gradually decrease in norm, with increasing index. *Right panel*: Comparison of the computed singular values of **R** and the absolute values of the diagonal entries of **R**. The QR factorization has revealed the numerical rank around 30, similarly as the SVD

Except in some rare cases,[7] the values $|\mathbf{R}_{ii}|$, $i = 1, \ldots, m$, and $\|\mathbf{R}(i,:)\|_2$, $i = 1, \ldots, m$, faithfully mimic the distribution of the singular values of $\mathbf{X}_m$, and it can be used with confidence. For instance, the least-squares solution in MATLAB (the *backslash* operator) uses this factorization. If one wants theoretical guarantee with sharp bounds, then one can deploy *strong rank revealing* QR factorization [12, 35].

*Example 7.1* We use the synthetic data of [23, Sect. 4.1], where the snapshots are generated in $\mathbb{R}^{1000}$ as $\mathbf{x}_{i+1} = \mathbb{A}\mathbf{x}_i$, $i = 1, \ldots, 99$, with $\mathbf{x}_1$ taken entry-wise uniformly from [0, 1]. The matrix $\mathbb{A}$ is generated as $\mathbb{A} = \exp(-\mathbf{B}^{-1})$, where **B** is pseudorandom with entries drawn uniformly from [0, 1], and scaled as $\mathbb{A} := \mathbb{A}/\|\mathbb{A}\|_2$. We take $\mathbf{X}_m = (\mathbf{x}_1, \ldots, \mathbf{x}_{99})$ and compute the factorization (7.40) using the function `qr(.)` from MATLAB. Then we inspect the structure of **R** and compare its diagonal with its singular values (that is, the singular values of $\mathbf{X}_m$); see Fig. 7.2.

Hence, if the numerical rank of $\mathbf{X}_m$ is determined as $k < m$, then in the block partition (7.42), the block $\mathbf{R}_{[22]}$ will be small compared to $\mathbf{R}_{[11]}$, so that we can write

$$\widetilde{\mathbf{R}} \equiv \begin{pmatrix} \mathbf{R}_{[11]} & \mathbf{R}_{[12]} \\ \mathbf{0} & \mathbf{0} \end{pmatrix} \approx \mathbf{R}, \quad \text{and} \quad \mathbf{X}_m \boldsymbol{\Pi} \approx \mathbf{Q}(:, 1:k) \left( \mathbf{R}_{[11]} \; \mathbf{R}_{[12]} \right).$$

In other words, in the basis of $\mathbf{Q}(:, 1:k)$, $\mathbf{X}_m$ is approximately represented by $\left( \mathbf{R}_{[11]} \; \mathbf{R}_{[12]} \right) \boldsymbol{\Pi}^T$, and the pivotal $k$ columns (selected by $\boldsymbol{\Pi}$) are represented without error.

[7] In an example constructed by Kahan [47], the pivoting fails to reveal one small singular value.

### 7.5.2 Implementation Details

Suppose now that we have computed the factorization (7.40), and that we have the numerical rank $k$, as discussed in Sect. 7.5.1. It remains to apply the Rayleigh–Ritz procedure, similarly as in Sect. 7.3.2, but now in a subspace determined by a rank revealing QR factorization.

Set $\mathbf{Q}_k = \mathbf{Q}(:, 1:k)$, $\mathbf{R}_k = \mathbf{R}(1:k, 1:k)$, and note that

$$(\mathbf{Y}_m\boldsymbol{\Pi})(:, 1:k) = \mathbb{A}\mathbf{Q}_k\mathbf{R}_k, \quad \mathbb{A}\mathbf{Q}_k = (\mathbf{Y}_m\boldsymbol{\Pi})(:, 1:k)\mathbf{R}_k^{-1}. \tag{7.43}$$

Hence, the action of $\mathbb{A}$ in $\mathscr{Q}_k = \text{range}(\mathbf{Q}_k)$ is known. In particular, the Rayleigh quotient reads

$$\mathbf{S}_k = \mathbf{Q}_k^*\mathbb{A}\mathbf{Q}_k = (\mathbf{Q}_k^*((\mathbf{Y}_m\boldsymbol{\Pi})(:, 1:k)))\mathbf{R}_k^{-1}. \tag{7.44}$$

The rest is completely analogous to the development of the DMD as outlined in Sect. 7.3.2, and the enhancements described in Sect. 7.4.1 follow, *mutatis mutandis*; instead of $\mathbf{U}_k$, $\boldsymbol{\Sigma}_k$, $\mathbf{V}_k$, we now have, respectively, $\mathbf{Q}_k$, $\mathbf{R}_k$, and $\boldsymbol{\Pi}_k \equiv \boldsymbol{\Pi}(:, 1:k)$. Note that $\mathbf{Q}_k$ spans the same subspace as the $k$ columns of $\mathbf{X}_m$ that are selected as leading (in $\mathbf{X}_m\boldsymbol{\Pi}$) by the pivoting $\boldsymbol{\Pi}$.

Due to space limitations, we cannot give all the fine details of a software implementation. Instead, we list the most relevant issues that can be used as a guidelines for an implementation.

*(i)* In an implementation of the large-scale column pivoted QR factorization on a multiprocessor hardware, communication required by pivoting is a bottleneck for high-performance optimization. An efficient and numerically robust alternative to the Businger–Golub pivoting [10] is the tournament pivoting introduced in [16].

*(ii)* Another way to lower the cost of pivoting is to first compute non-pivoted factorization $\mathbf{X}_m = \mathbf{Q}_1\mathbf{R}_1$, and then $\mathbf{R}_1\boldsymbol{\Pi} = \mathbf{Q}_2\mathbf{R}$ which yields $\mathbf{X}_m\boldsymbol{\Pi} = \mathbf{Q}_1\mathbf{R}_1\boldsymbol{\Pi} = \mathbf{Q}_1\mathbf{Q}_2\mathbf{R} \equiv \mathbf{Q}\mathbf{R}$, with $\mathbf{Q} = \mathbf{Q}_1\mathbf{Q}_2$. In our case, $\mathbf{X}_m$ is tall and skinny, $n \gg m$, so that the pivoting is used only on the smaller $m \times m$ factor, and the initial QR factorization of $\mathbf{X}_m$ can be computed very efficiently [17, 55].

*(iii)* And, we can also use the compression step from Sect. 7.4.5, so that in that case the compressed snapshot matrix is $(m+1) \times m$ for sequential data, and $2m \times m$ in the general case. These operations are much easier to adapt to updating/downdating if the snapshots represent a time window that slides in discrete steps in time.

*(iv)* If the snapshots are accompanied with a diagonal matrix $\Delta$ of weights that scale the rows of $\mathbf{X}_m$, as discussed in Sect. 7.4.2, then it may be important for numerical robustness to use row pivoting as well (reordering the observables).

For the readers convenience, we outline a simplified version of the method in Algorithm 7.5. All remarks from Sect. 7.4.4 apply here as well.

*Example 7.2* We first test Algorithm 7.5 using the data from Example 7.1. Since in this case the matrix $\mathbb{A}$ is known, we can compute its eigenvalues (using `eig(.)` from MATLAB), compare them with the computed Ritz vectors, and compute the residuals explicitly. The results are shown in Fig. 7.3.

**Algorithm 7.5** $[\mathbf{Z}_k, \boldsymbol{\Lambda}_k, r_k, \rho_k] = \mathrm{DMD\_R4\_QRCP}(\mathbf{X}_m, \mathbf{Y}_m; \varepsilon)$ *{Refined Rayleigh–Ritz DMD using QR with column pivoting}*

**Input:**

- $\mathbf{X}_m = (\mathbf{x}_1, \ldots, \mathbf{x}_m), \mathbf{Y}_m = (\mathbf{y}_1, \ldots, \mathbf{y}_m) \in \mathbb{C}^{n\times m}$ that define a sequence of snapshots pairs $(\mathbf{x}_i, \mathbf{y}_i \equiv \mathbb{A}(\mathbf{x}_i))$. (Tacit assumption is that $n$ is large and that $m \ll n$.)
- Tolerance level $\varepsilon$ for numerical rank determination.

1: $\mathbf{D}_x = \mathrm{diag}(\|\mathbf{X}_m(:,i)\|_2)_{i=1}^m$; $\mathbf{X}_m^{(1)} = \mathbf{X}_m\mathbf{D}_x^\dagger$; $\mathbf{Y}_m^{(1)} = \mathbf{Y}_m\mathbf{D}_x^\dagger$
2: $[\mathbf{Q}_x, \mathbf{R}_x, \boldsymbol{\Pi}_x] = qr(\mathbf{X}_m^{(1)})$ ; *{The thin QR factorization with column pivoting:* $\mathbf{X}_m^{(1)}\boldsymbol{\Pi}_x = \mathbf{Q}_x\mathbf{R}_x$, $\mathbf{Q}_x \in \mathbb{C}^{n\times m}$, $\mathbf{R}_x \in \mathbb{C}^{m\times m}$.*}*
3: Determine numerical rank $k$, with the threshold $\varepsilon$. See Sect. 7.5.1.
4: Set $\mathbf{Q}_k = \mathbf{Q}(:, 1:k)$, $\boldsymbol{\Pi}_k = \boldsymbol{\Pi}_x(:, 1:k)$, $\mathbf{R}_k = \mathbf{R}(1:k, 1:k)$
5: $\mathbf{B}_k = (\mathbf{Y}_m^{(1)}\boldsymbol{\Pi}_k)\mathbf{R}_k^{-1}$; *{See (7.43).}*
6: $[\mathbf{Q}, \mathbf{T}] = qr((\mathbf{Q}_k, \mathbf{B}_k))$; *{The thin QR factorization:* $(\mathbf{Q}_k, \mathbf{B}_k) = \mathbf{Q}\mathbf{R}$; $\mathbf{Q}$ *not computed}*
7: $\mathbf{S}_k = \mathrm{diag}(\overline{\mathbf{T}_{ii}})_{i=1}^k \mathbf{T}(1:k, k+1:2k)$ *{Or,* $\mathbf{S}_k = \mathbf{Q}_k^*\mathbf{B}_k$; see (7.44).*}*
8: $\boldsymbol{\Lambda}_k = \mathrm{eig}(\mathbf{S}_k)$ *{*$\boldsymbol{\Lambda}_k = \mathrm{diag}(\lambda_i)_{i=1}^k$; *Ritz values, i.e., eigenvalues of* $\mathbf{S}_k$*}*
9: **for** $i = 1, \ldots, k$ **do**
10: $[\sigma_{\lambda_i}, w_{\lambda_i}] = svd_{\min}\left(\begin{pmatrix} \mathbf{T}(1:k,k+1:2k) - \lambda_i\mathbf{T}(1:k,1:k) \\ \mathbf{T}(k+1:2k,k+1:2k)\end{pmatrix}\right)$; *{Minimal singular value and the corresponding right singular vector}*
11: $\mathbf{W}_k(:, i) = w_{\lambda_i}$
12: $r_k(i) = \sigma_{\lambda_i}$ *{Optimal residual}*
13: $\rho_k(i) = w_{\lambda_i}^* \mathbf{S}_k w_{\lambda_i}$ *{Rayleigh quotient wit the refined Ritz vector}*
14: **end for**
15: $\mathbf{Z}_k = \mathbf{Q}_k\mathbf{W}_k$ *{Refined Ritz vectors}*

**Output:** $\mathbf{Z}_k, \boldsymbol{\Lambda}_k, r_k, \rho_k$

*Example 7.3* In the second example, we use the data from [23, Sect. 4.2]. The dimension of the search space is set to $k = 50$, and the results are shown in Figs. 7.4 and 7.5.

## 7.6 Further Developments and Concluding Remarks

The recent improvements of the DMD [23], partially outlined in Sect. 7.4, and the scheme proposed in Sect. 7.5, show that further development of the DMD framework greatly benefits from a core numerical linear algebra approach.

Another example that fortifies this claim is given recently in [22] where the numerically ill-conditioned approach from Sect. 7.2.1.1 is implemented using the state-of-the-art methods of matrix computations. In brief, relation (7.7) is post-multiplied by the DFT matrix $\mathbb{F}$, and $\mathbf{X}_m\mathbf{C}_m\mathbb{F}$ is written, using (7.8), as $(\mathbf{X}_m\mathbb{F})\mathbb{F}^*(\mathbb{V}_m^{-1}\boldsymbol{\Lambda}_m\mathbb{V}_m)\mathbb{F}$. The key observation is that $\mathbb{V}_m\mathbb{F} = \mathscr{D}_1\mathscr{C}\mathscr{D}_2$, where $\mathscr{D}_1$ is diagonal, $\mathscr{D}_2$ is diagonal unitary, and $\mathscr{C}$ is Cauchy matrix defined by the $\lambda_i$'s and the $m$th roots of unity $\omega_j$'s; all three matrices are given explicitly in terms of the $\lambda_i$'s and the $\omega_j$'s, without actual

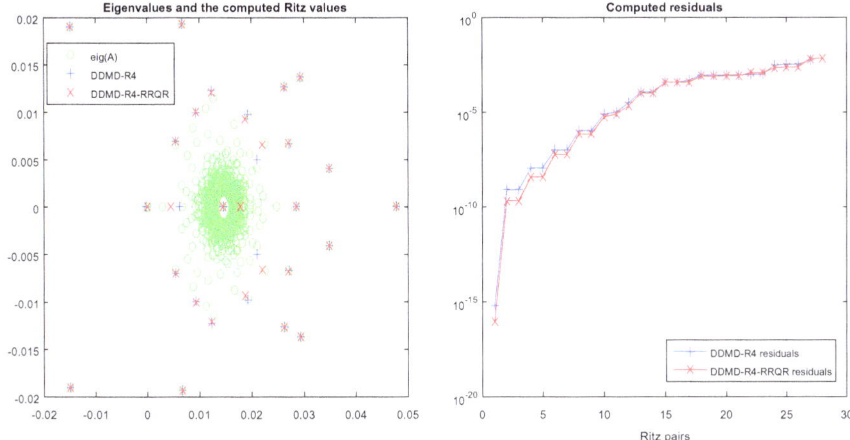


**Fig. 7.3** (Example 7.2.) *Left panel*: The eigenvalues of $\mathbb{A}$ ($\circ$) as computed by `eig(.)` and the Ritz values from Algorithm 7.3 ($+$) and Algorithm 7.5 ($\times$). *Right panel*: The corresponding residuals. These residuals are computed explicitly, using $\mathbb{A}$. It has been checked that the data-driven residuals, as returned by the algorithms (see Proposition 7.2), correspond to these and thus correctly assess the quality of the computed Ritz pairs. Note that, with the same threshold, Algorithm 7.5 determined the numerical rank as $k = 28$, while Algorithm 7.3 has $k = 27$

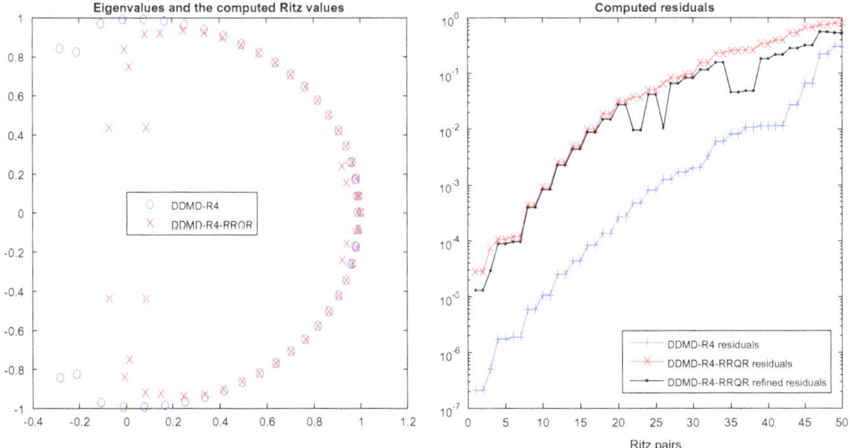


**Fig. 7.4** (Example 7.2.) *Left panel*: The Ritz values from Algorithm 7.3 ($\circ$) and Algorithm 7.5 ($\times$). *Right panel*: The corresponding sorted residuals. The middle graph represents refined residuals from Algorithm 7.5. The $\times$'s can be further improved as discussed in Sect. 7.4.1.2. Recall that this information is extracted under the constraint that it belongs to a subspace spanned by a selection of 50 snapshots

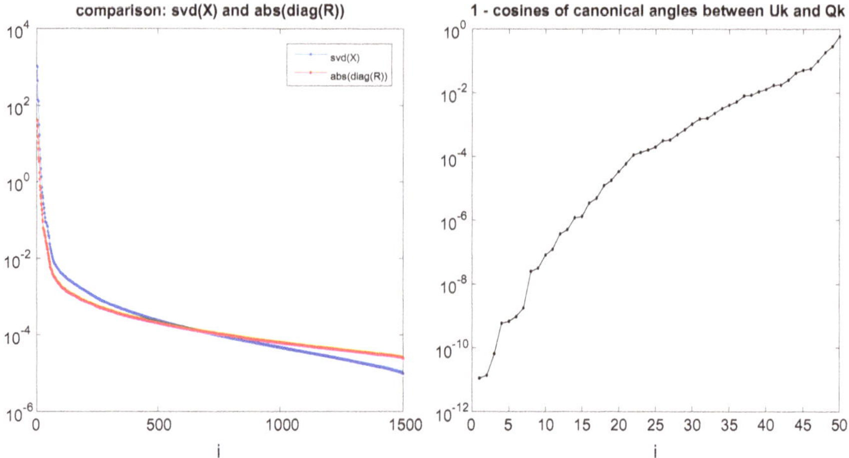


**Fig. 7.5** (Example 7.2.) *Left panel*: The singular values of the data matrix $\mathbf{X}_m$ and the absolute values of the diagonal entries of the matrix $\mathbf{R}_x$ in Line 2. of the Algorithm 7.5. *Right panel*: The values of $1 - \cos\vartheta_i$, where the $\vartheta_i$'s are the cosines of the canonical angles between the ranges of $\mathbf{U}_k$ (Line 4. in Algorithm 7.3) and $\mathbf{Q}_k$ (Line 4. in Algorithm 7.5). Recall that $\cos\vartheta_i$, $i = 1, \ldots, k$, are the singular values of $\mathbf{U}_k^*\mathbf{Q}_k$

matrix multiplication. Moreover, the computation with $\mathscr{D}_1\mathscr{C}\mathscr{D}_2$ (LU decomposition, solving linear systems, SVD) is possible independent of the condition number $\kappa_2(\mathbb{V}_m)$ if special algorithms are used [15, 18, 25, 26]. For instance, in [22], this scheme is successfully applied to a quasi-two-dimensional Kolmogorov-like flow with $\kappa_2(\mathbb{V}_m) > 10^{70}$, using only the standard 64 bit double precision machine arithmetic. Although the full numerical analysis and understanding of the scheme is still challenging problem, it shows that research in this direction is justified.

For the snapshot reconstruction problem (7.3), new algorithms and detailed numerical analysis of the associated structured least-squares problem are offered in [24]. In addition to the normal equation approach and the ADMM solver for a sparsity promoting $\ell_1$ regularized least-squares problem used in [24, 46] proposes a corrected semi-normal approach and a QR factorization based method.

Our current work continues along the principles and techniques presented in [22–24], and briefly reviewed in this chapter, with particular attention to the development of a high-performance software toolbox. We hope that this work will trigger a new line of research of core numerical aspects of the DMD and, more generally, the Koopman operator framework for computational study of dynamical systems.

**Acknowledgements** This research is supported by the DARPA Contract HR0011-16-C-0116 *"On a Data-Driven, Operator-Theoretic Framework for Space–Time Analysis of Process Dynamics"* and the DARPA Contract HR0011-18-9-0033 *"The Physics of Artificial Intelligence"*.

# References


1. Anantharamu, S., Mahesh, K.: A parallel Dynamic Mode Decomposition algorithm using modified Full Orthogonalization Arnoldi for large sequential snapshots. ArXiv e-prints (2018)
2. Anderson, E., Bai, Z., Bischof, C., Blackford, L.S., Demmel, J., Dongarra, J.J., Du Croz, J., Hammarling, S., Greenbaum, A., McKenney, A., Sorensen, D.: LAPACK Users' Guide, 3rd edn. Society for Industrial and Applied Mathematics, Philadelphia (1999)
3. Andrew, R., Dingle, N.: Implementing QR factorization updating algorithms on GPUs. Parallel Comput. **40**(7), 161–172. http://www.sciencedirect.com/science/article/pii/S0167819114000337 (2014). https://doi.org/10.1016/j.parco.2014.03.003. (7th Workshop on Parallel Matrix Algorithms and Applications)
4. Arbabi, H., Mezić, I.: Ergodic theory, dynamic mode decomposition, and computation of spectral properties of the Koopman operator. SIAM J. Appl. Dyn. Syst. **16**(4), 2096–2126 (2017). https://doi.org/10.1137/17M1125236
5. Bagheri, S.: Koopman-mode decomposition of the cylinder wake. J. Fluid Mech. **726**, 596–623 (2013). https://doi.org/10.1017/jfm.2013.249
6. Berger, E., Sastuba, M., Vogt, D., Jung, B., Amor, H.B.: Estimation of perturbations in robotic behavior using dynamic mode decomposition. Adv. Robot. **29**(5), 331–343 (2015)
7. Björck, A.: Numerical Methods in Matrix Computations. Springer, Berlin (2015)
8. Brgisser, P., Cucker, F.: Condition: The Geometry of Numerical Algorithms. Springer Publishing Company, Incorporated, Berlin (2013)
9. Brunton, B.W., Johnson, L.A., Ojemann, J.G., Kutz, J.N.: Extracting spatialtemporal coherent patterns in large-scale neural recordings using dynamic mode decomposition. J. Neurosci. Methods **258**, 1–15. http://www.sciencedirect.com/science/article/pii/S0165027015003829 (2016). https://doi.org/10.1016/j.jneumeth.2015.10.010
10. Businger, P.A., Golub, G.H.: Linear least squares solutions by Householder transformations. Numerische Mathematik **7**, 269–276 (1965)
11. Cat Le Ngo, A., See, J., Chung-Wei Phan, R.: Sparsity in Dynamics of Spontaneous Subtle Emotions: Analysis & Application. ArXiv e-prints (2016)
12. Chandrasekaran, S., Ipsen, I.C.F.: On rank-revealing QR factorisations. SIAM J. Matrix Anal. Appl. **15**(2), 592–622 (1994)
13. Dawson, S.T.M., Hemati, M.S., Williams, M.O., Rowley, C.W.: Characterizing and correcting for the effect of sensor noise in the dynamic mode decomposition. Exp. Fluids **57**(3), 42 (2016). https://doi.org/10.1007/s00348-016-2127-7
14. Demmel, J.: The geometry of iii-conditioning. J. Complex. **3**(2), 201–229. http://www.sciencedirect.com/science/article/pii/0885064X87900276 (1987). https://doi.org/10.1016/0885-064X(87)90027-6
15. Demmel, J.: Accurate singular value decompositions of structured matrices. SIAM J. Matrix Anal. Appl. **21**(2), 562–580 (1999)
16. Demmel, J., Grigori, L., Gu, M., Xiang, H.: Communication avoiding rank revealing qr factorization with column pivoting. SIAM J. Matrix Anal. Appl. **36**(1), 55–89 (2015). https://doi.org/10.1137/13092157X
17. Demmel, J., Grigori, L., Hoemmen, M., Langou, J.: Communication-optimal parallel and sequential QR and LU factorizations. SIAM J. Sci. Comput. **34**(1), A206–A239 (2012). https://doi.org/10.1137/080731992
18. Demmel, J., Gu, M., Eisenstat, S., Slapničar, I., Veselić, K., Drmač, Z.: Computing the singular value decomposition with high relative accuracy. Lin. Alg. Appl. **299**, 21–80 (1999)
19. Drmač, Z.: Algorithm 977: A QR–preconditioned QR SVD method for computing the svd with high accuracy. ACM Trans. Math. Softw. **44**(1), 11:1–11:30 (2017). https://doi.org/10.1145/3061709
20. Drmač, Z., Bujanović, Z.: On the failure of rank revealing QR factorization software - a case study. ACM Trans. Math. Softw. **35**(2), 1–28 (2008)
21. Drmač, Z., Mezić, I., Mohr, R.: Data driven modal decompositions: analysis and enhancements (2017). arXiv:1708.02685

22. Drmač, Z., Mezić, I., Mohr, R.: Data driven Koopman spectral analysis in Vandermonde-Cauchy form via the DFT: numerical method and theoretical insights. SIAM J. Sci. Comput. **41**(5), A3118–A3151 (2018). arXiv:1808.09557
23. Drmač, Z., Mezić, I., Mohr, R.: Data driven modal decompositions: analysis and enhancements. SIAM J. Sci. Comput. **40**(4), A2253–A2285 (2018). https://doi.org/10.1137/17M1144155
24. Drmač, Z., Mezić, I., Mohr, R.: On least squares problems with certain Vandermonde–Khatri–Rao structure with applications to DMD (2018). arXiv:1811.12562
25. Drmač, Z., Veselić, K.: New fast and accurate Jacobi SVD algorithm: I. SIAM J. Matrix Anal. Appl. **29**(4), 1322–1342 (2008)
26. Drmač, Z., Veselić, K.: New fast and accurate Jacobi SVD algorithm: II. SIAM J. Matrix Anal. Appl. **29**(4), 1343–1362 (2008)
27. Eckart, C., Young, G.: The approximation of one matrix by another of lower rank. Psychometrika **1**(3), 211–218 (1936). https://doi.org/10.1007/BF02288367
28. Eisenstat, S., Ipsen, I.: Relative perturbation techniques for singular value problems. SIAM J. Numer. Anal. **32**(6), 1972–1988 (1995)
29. de la Fraga, L.G.: A very fast procedure to calculate the smallest singular value. In: 2015 Eighth International Conference on Advances in Pattern Recognition (ICAPR), pp. 1–4 (2015). https://doi.org/10.1109/ICAPR.2015.7050656
30. Gautschi, W.: Optimally conditioned Vandermonde matrices. Numerische Mathematik **24**(1), 1–12 (1975). https://doi.org/10.1007/BF01437212
31. Gautschi, W.: How (un)stable are Vandermonde systems? In: R. Wong (ed.) Asymptotic and Computational Analysis. Lecture Notes in Pure and Applied Mathematics, vol. 124, pp. 193–210 (1990)
32. Ghosal, S., Ramanan, V., Sarkar, S., Chakravarthy, S., Sarkar, S.: Detection and analysis of combustion instability from hi-speed flame images using dynamic mode decomposition. In: ASME. Dynamic Systems and Control Conference, vol. 1 (2016). https://doi.org/10.1115/DSCC2016-9907
33. Ginsberg, J., Mohebbi, M., Patel, R., Brammer, L., Smolinski, M., Brilliant, L.: Detecting influenza epidemics using search engine query data. Nature **457**, 1012–1014. http://www.nature.com/nature/journal/v457/n7232/full/nature07634.html (2009). https://doi.org/10.1038/nature07634
34. Golub, G.H., Klema, V.C., Stewart, G.W.: Rank degeneracy and least squares problems. Technical Report CS-TR-76-559, Stanford, CA, USA (1976)
35. Gu, M., Eisenstat, S.C.: Efficient algorithms for computing a strong rank-revealing QR factorization. SIAM J. Sci. Comput. **17**(4), 848–869 (1996)
36. Hammarling, S., Lucas, C.: Updating th QR factorization and the least squares problem. Technical Report MIMS EPrint: 2008.111, University of Manchester (2008)
37. Higham, D.J.: Condition numbers and their condition numbers. Linear Algebr. Appl. **214**, 193–213. http://www.sciencedirect.com/science/article/pii/0024379593000669 (1995). https://doi.org/10.1016/0024-3795(93)00066-9
38. Higham, N.J.: Accuracy and Stability of Numerical Algorithms. SIAM, Philadelphia (1996)
39. Holmes, P., Lumley, J.L., Berkooz, G., Rowley, C.W.: Turbulence, Coherent Structures, Dynamical Systems and Symmetry, 2 edn. Cambridge Monographs on Mechanics. Cambridge University Press, Cambridge (2012). https://doi.org/10.1017/CBO9780511919701
40. Horn, R.A., Johnson, C.R.: Topics in Matrix Analysis. Cambridge University Press, Cambridge (1991)
41. Jia, Z.: Refined iterative algorithms based on Arnoldi's process for large unsymmetric eigenproblems. Linear Algebr. Appl. **259**, 1–23. http://www.sciencedirect.com/science/article/pii/S0024379596002388 (1997). https://doi.org/10.1016/S0024-3795(96)00238-8
42. Jia, Z.: Polynomial characterizations of the approximate eigenvectors by the refined Arnoldi method and an implicitly restarted refined Arnoldi algorithm. Linear Algebr. Appl. **287**(1–3), 191–214. http://www.sciencedirect.com/science/article/pii/S0024379598101970 (1999). https://doi.org/10.1016/S0024-3795(98)10197-0

43. Jia, Z.: Residuals of refined projection methods for large matrix eigenproblems. Comput. Math. Appl. **41**(7), 813–820. http://www.sciencedirect.com/science/article/pii/S0898122100003217 (2001). https://doi.org/10.1016/S0898-1221(00)00321-7
44. Jia, Z.: Some theoretical comparisons of refined Ritz vectors and Ritz vectors. Sci. China Ser. Math. **47**, 222–233 (2004)
45. Jia, Z., Stewart, G.W.: An analysis of the Rayleigh-Ritz method for approximating eigenspaces. Math. Comput. **70**(234), 637–647 (2001). https://doi.org/10.1090/S0025-5718-00-01208-4
46. Jovanović, M.R., Schmid, P.J., Nichols, J.W.: Sparsity-promoting dynamic mode decomposition. Phys. Fluids **26**(2), 024103 (2014)
47. Kahan, W.: Numerical linear algebra. Can. Math. Bull. **9**(6), 757–801 (1965)
48. Kalashnikova, I., Arunajatesan, S., Barone, M.F., van Bloemen Waanders, B.G., Fike, J.A.: Reduced order modeling for prediction and control of large–scale systems. Sandia Report SAND2014–4693, Sandia National Laboratories (2014)
49. Kutz, J.N., Brunton, S.L., Brunton, B.W., Proctor, J.L.: Dynamic Mode Decomposition: Data-Driven Modeling of Complex Systems. SIAM-Society for Industrial and Applied Mathematics, USA (2016)
50. Lele, S.K., Nichols, J.W.: A second golden age of aeroacoustics? Philos. Trans. R. Soc. Lond. A Math. Phys. Eng. Sci. **372**(2022). http://rsta.royalsocietypublishing.org/content/372/2022/20130321 (2014). https://doi.org/10.1098/rsta.2013.0321
51. Li, R.: Relative perturbation theory: Ii. eigenspace and singular subspace variations. SIAM J. Matrix Anal. Appl. **20**(2), 471–492 (1998). https://doi.org/10.1137/S0895479896298506
52. Mann, J., Kutz, J.N.: Dynamic Mode Decomposition for Financial Trading Strategies. ArXiv e-prints (2015)
53. The Mathworks, Inc., Natick, Massachusetts: MATLAB version 9.2.0.556344 (R2017a) (2017)
54. Mirsky, L.: Symmetric gauge functions and unitarily invariant norms. Q. J. Math. **11**(1), 50 (1960). https://doi.org/10.1093/qmath/11.1.50
55. Nguyen, H.D., Demmel, J.: Reproducible tall-skinny QR. In: 2015 IEEE 22nd Symposium on Computer Arithmetic, pp. 152–159 (2015). https://doi.org/10.1109/ARITH.2015.28
56. Oliphant, T.: Guide to NumPy. Trelgol Publishing (2006)
57. Pan, V.Y.: How bad are Vandermonde matrices? ArXiv e-prints http://adsabs.harvard.edu/abs/2015arXiv150402118P (2015)
58. Proctor, J.L., Eckhoff, P.A.: Discovering dynamic patterns from infectious disease data using dynamic mode decomposition. Int. Health **7**(2), 139–145 (2015). https://doi.org/10.1093/inthealth/ihv009
59. Rowley, C.W.: Model reduction for fluids, using balanced proper orthogonal decomposition. Int. J. Bifur. Chaos Appl. Sci. Eng. **15**(3), 997–1013 (2005)
60. Rowley, C.W., Colonius, T., Murray, R.M.: Model reduction for compressible flows using pod and galerkin projection. Phys. D: Nonlinear Phenom **189**(1), 115–129. http://www.sciencedirect.com/science/article/pii/S0167278903003841 (2004). https://doi.org/10.1016/j.physd.2003.03.001
61. Rowley, C.W., Mezić, I., Bagheri, S., Schlatter, P., Henningson, D.S.: Spectral analysis of nonlinear flows. J. Fluid Mech. **641**, 115–127 (2009)
62. Schmid, P.J.: Dynamic mode decomposition of numerical and experimental data. J. Fluid Mech. **656**, 5–28. https://www.cambridge.org/core/article/dynamic-mode-decomposition-of-numerical-and-experimental-data/AA4C763B525515AD4521A6CC5E10DBD4 (2010). https://doi.org/10.1017/S0022112010001217
63. Schmid, P.J., Sesterhenn, J.: Dynamic mode decomposition of numerical and experimental data. In: Sixty-First Annual Meeting of the APS Division of Fluid Dynamics. San Antonio, Texas, USA (2008)
64. Serre, G., Lafon, P., Gloerfelt, X., Bailly, C.: Reliable reduced-order models for time-dependent linearized Euler equations. J. Comput. Phys. **231**(15), 5176–5194 (2012)
65. van der Sluis, A.: Condition numbers and equilibration of matrices. Numerische Mathematik **14**, 14–23 (1969)

66. Stewart, G.W.: A Krylov-Schur algorithm for large eigenproblems. SIAM J. Matrix Anal. Appl. **23**(3), 601–614 (2001)
67. Stewart, G.W., Sun, J.G.: Matrix Perturbation Theory. Academic, Cambridge (1990)
68. Tu, J.H., Rowley, C.W., Luchtenburg, D.M., Brunton, S.L., Kutz, J.N.: On Dynamic Mode Decomposition: Theory and Applications. ArXiv e-prints (2013)
69. Volkwein, S.: Model reduction using proper orthogonal decomposition. Lecture Notes, Institute of Mathematics and Scientific Computing, University of Graz. http://www.uni-graz.at/imawww/volkwein/POD.pdf (2011)
70. Williams, M.O., Kevrekidis, I.G., Rowley, C.W.: A data-driven approximation of the Koopman operator: extending dynamic mode decomposition. J. Nonlinear Sci. **25**(6), 1307–1346 (2015). https://doi.org/10.1007/s00332-015-9258-5

# Part II
# Methodologies

# Chapter 8
# Data-Driven Approximations of Dynamical Systems Operators for Control

**Eurika Kaiser, J. Nathan Kutz and Steven L. Brunton**

**Abstract** The Koopman and Perron Frobenius transport operators are fundamentally changing how we approach dynamical systems, providing linear representations for even strongly nonlinear dynamics. Although there is tremendous potential benefit of such a linear representation for estimation and control, transport operators are infinite dimensional, making them difficult to work with numerically. Obtaining low-dimensional matrix approximations of these operators is paramount for applications, and the dynamic mode decomposition has quickly become a standard numerical algorithm to approximate the Koopman operator. Related methods have seen rapid development, due to a combination of an increasing abundance of data and the extensibility of DMD based on its simple framing in terms of linear algebra. In this chapter, we review key innovations in the data-driven characterization of transport operators for control, providing a high-level and unified perspective. We emphasize important recent developments around sparsity and control, and discuss emerging methods in big data and machine learning.

## 8.1 Introduction

Data-driven modeling using linear operators has the potential to transform the estimation and control of strongly nonlinear systems. Linear operators, such as Koopman and Perron–Frobenius operators, provide a principled linear embedding of nonlinear dynamics, extending the application of standard linear methods to nonlinear systems, and thus significantly simplifying the control design and reducing the com-

E. Kaiser (✉) · S. L. Brunton
Department of Mechanical Engineering, University of Washington, Seattle, WA 98195, USA
e-mail: eurika@uw.edu

S. L. Brunton
e-mail: sbrunton@uw.edu

J. N. Kutz
Department of Applied Mathematics, University of Washington, Seattle, WA 98195, USA
e-mail: kutz@uw.edu



A. Mauroy et al. (eds.), *The Koopman Operator in Systems and Control*,
Lecture Notes in Control and Information Sciences 484,
https://doi.org/10.1007/978-3-030-35713-9_8

putational burden. More broadly, data-driven discovery of dynamical systems is undergoing rapid development, driven by the lack of (simple) equations and the increasing abundance of high-fidelity data. There have been recent successes in the discovery of functional representations of nonlinear dynamical systems, e.g., using evolutionary optimization techniques [22, 161] and sparse optimization [31]. However, control based on nonlinear equations is particularly challenging, becoming infeasible for higher dimensional problems, lacking guarantees, and often requiring problem-tailored formulations. In contrast, the emerging field of linear operators in dynamical systems seeks to embed nonlinear dynamics in a globally linear representation, providing a compelling mathematical framework for the linear estimation, prediction, and control of strongly nonlinear systems. The rise of advanced data science and machine learning algorithms, vastly expanded computational resources, and advanced sensor technologies make this a fertile ground for the rapid development of data-driven approximations to these linear operators for control. In this chapter, we review key innovations, discuss major challenges and promising future directions, and provide a high-level and unified perspective on data-driven approximations of transfer operators for control.

System identification has reached a high degree of maturity. There exist a plethora of techniques that identify linear and nonlinear systems [129] based on data, including state-space modeling via the eigensystem realization algorithm (ERA) [76] and other subspace identification methods, Volterra series [25, 108], linear and nonlinear autoregressive models [3] (e.g., ARX, ARMA, NARX, and NARMAX), and neural network models [47, 103, 193], to name only a few. The reader is referred to [104, 105] for a compressed overview of identification techniques for linear and nonlinear systems. In the machine learning community, manifold learning, e.g., locally linear embedding and self-organizing maps, and nonparametric modeling, e.g., Gaussian processes, have been proven to be useful for identifying nonlinear systems [87, 88, 143]. Most of these models are considered data driven [166], as they do not impose a specific model structure based on the governing equations. However, there is an increasing shift from black-box modeling to inferring unknown physics and constraining models with known prior information. For instance, the recent sparse identification of nonlinear dynamics (SINDy) [31], which has been extended to incorporate the effect of control [32, 79], is able to take into account known expert knowledge such as symmetries and conservation laws [106]. Learning accurate nonlinear models is particularly challenging as small deviations in the parameters may produce fundamentally different system behavior. Importantly, it is possible to control many nonlinear systems using linear models. Examples include weakly nonlinear systems for which models are obtained based on a local linearization of the nonlinear dynamics at a reference point. However, the pronounced nonlinearities present in many applications generally require nonlinear control design. The lack of robustness and stability guarantees except for special cases and the increased computational burden during the online phase, which becomes prohibitive for high-dimensional systems, restricts the application of nonlinear control to low-dimensional systems and requires specialized design techniques. Encoding nonlinear dynamics in linear models through operator-theoretic approaches provides a new opportunity for the control of previously intractable systems.

Linear embedding theory for nonlinear systems goes back to seminal works by B.O. Koopman in 1931 [89] and T. Carleman in 1932 [35]. Koopman showed that Hamiltonian dynamics can be described through an infinite-dimensional linear operator acting on the Hilbert space of all possible observables that can be measured from the underlying state. Closely related, Carleman demonstrated that systems of ordinary differential equations with polynomial nonlinearities can be represented as an infinite-dimensional system of linear differential equations. Since its introduction, the so-called Carleman linearization has been applied to a wide range of problems [15, 24, 93] including for the Lyapunov exponent calculation [6] and for finding first integrals [94]. The emerging Koopman operator perspective provides an alternative direction and generalizes beyond polynomial systems. The ultimate goal is to learn a globally linear embedding of the nonlinear dynamics such that powerful linear methods become immediately applicable and are useful in a larger domain. Of particular interest are the spectral properties of these operators that encode global information and can be related to geometrical properties of the underlying dynamical system [120]. The potential to discover global properties of dynamical systems through operator-theoretic methods for diagnostic purposes, i.e., improving understanding of the underlying dynamics, has driven continued efforts to develop improved algorithms. The confluence of big data, advances in machine learning, and new sensor technologies has further fueled the progress in data-driven methods, facilitating equation-free approximations of these operators. These efforts will be discussed below particularly in the context of system identification for control.

We introduce the Koopman and Perron–Frobenius operators by considering the following autonomous nonlinear dynamical system:

$$\frac{d}{dt}\mathbf{x}(t) = \mathbf{F}(\mathbf{x}(t)), \quad \mathbf{x}(0) = \mathbf{x}_0 \tag{8.1}$$

with $\mathbf{x} \in \mathbb{X} \subset \mathbb{R}^n$, initial condition $\mathbf{x}_0 \in \mathbb{X}$, and the flow is denoted by $\mathbf{S}^t$ so that $\mathbf{x}(t) = \mathbf{S}^t(\mathbf{x}_0)$. The nonlinear system (8.1) can be equivalently described by infinite-dimensional, linear operators acting on observable or density functions (Fig. 8.1). The linearity of these operators is appealing; however, their infinite dimensionality poses issues for representation and computation and current research aims to approximate the evolution instead on a finite-dimensional subspace facilitating a finite-dimensional matrix representation [30].

Let $(\mathbb{X}, \mathfrak{B}, \mu)$ be a measure space with state space $\mathbb{X}$, $\sigma$-algebra $\mathfrak{B}$, and measure $\mu$. The Koopman operator [89, 90, 120, 121] is an infinite-dimensional linear operator that advances measurement functions $f \in L^2(\mathbb{X})$:

$$f(t, \mathbf{x}_0) = U^t f(\mathbf{x}_0) = f(\mathbf{S}^t(\mathbf{x}_0)). \tag{8.2}$$

Any set of eigenfunctions of the Koopman operator spans an invariant subspace, where the dynamics evolve linearly along these basis directions. Thus, the spectral

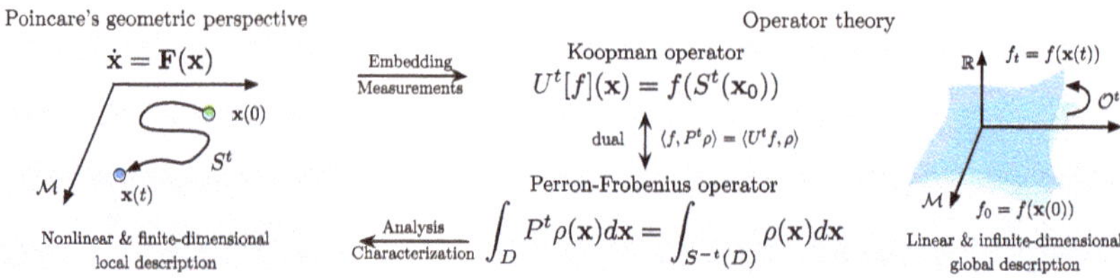


**Fig. 8.1** Poincaré's local phase space characterization versus the global operator-theoretic perspective

analysis of the Koopman operator is of particular interest and spectral properties have been shown to be related to intrinsic time scales, geometrical properties, and long-term behavior of the underlying dynamical system [122]. Koopman eigenfunctions $\phi(\mathbf{x})$ associated with a particular eigenvalue $\lambda$ are special or intrinsic observables, which evolve linearly according to

$$U^t\phi(\mathbf{x}_0) = \phi(\mathbf{S}^t(\mathbf{x}_0)) = e^{\lambda t}\phi(\mathbf{x}_0), \tag{8.3}$$

and which can generally be discontinuous [122]. It can be shown [98] for continuously differentiable functions $f$ with compact support, that the function $f(t, \mathbf{x}) := U^t[f](\mathbf{x})$ satisfies the first-order partial differential equation (PDE):

$$\frac{\partial}{\partial t} f(t, \mathbf{x}) = \mathbf{F}(\mathbf{x}) \cdot \nabla f(t, \mathbf{x}) = L_U f(\mathbf{x}), \quad f_0 := f(0, \mathbf{x}), \tag{8.4}$$

where $L_U$ is the infinitesimal generator of the semigroup of Koopman operators $\{U^t\}_{t\geq 0}$, for which an exponential representation $U^t = e^{L_U t}$ exists. Smooth Koopman eigenfunctions can then be interpreted as the eigenfunctions of the generator and satisfy

$$\frac{d}{dt}\phi(\mathbf{x}) = \mathbf{F}(\mathbf{x}) \cdot \nabla\phi(\mathbf{x}) = \lambda\phi(\mathbf{x}). \tag{8.5}$$

The Koopman operator evolves measurement functions, a perspective that is particularly amenable to data-driven approaches. Further details on the Koopman operator and numerical methods for its approximation are provided in Chap. 1. The Koopman operator is dual to the Perron–Frobenius operator [21, 38, 98, 130], i.e., $\langle P^t\rho, f\rangle = \langle \rho, U^t f\rangle$ for any $\rho \in L^1$ and $f \in L^\infty$, and as a consequence these share the same spectral properties. The Perron–Frobenius operator propagates densities $\rho \in L^1(\mathbb{X})$ and is defined as

$$\int_B P^t\rho(\mathbf{x})\mu(d\mathbf{x}) = \int_{\mathbf{S}^{-t}(B)} \rho(\mathbf{x})\,\mu(d\mathbf{x}) \quad \forall B \in \mathfrak{B}. \tag{8.6}$$

Further,

$$\rho(t, \mathbf{x}) = P^t \rho(\mathbf{x}) = \int_{\mathfrak{B}} \delta(\mathbf{x} - \mathbf{S}^t(\mathbf{x}_0))\rho(\mathbf{x}_0)\mu(d\mathbf{x}_0), \tag{8.7}$$

where $\delta(\mathbf{x} - \mathbf{S}^t(\mathbf{x}_0))$ represents the deterministic kernel. For an invertible system, this becomes $P^t\rho(\mathbf{x}) = J^{-t}(\mathbf{x})\rho(\mathbf{S}^{-t}(\mathbf{x}))$, where $J^{-t}(\mathbf{x}) := \det(d\mathbf{S}^{-t}(\mathbf{x})/d\mathbf{x})$ is the determinant of the Jacobian of $\mathbf{S}^{-t}(\mathbf{x})$; thus, the density varies inversely with the infinitesimal volume occupied by the trajectories. For invertible and conservative systems, we have $P^t\rho(\mathbf{x}) = \rho(\mathbf{S}^{-t}(\mathbf{x}))$ with volume preservation $\operatorname{div} \mathbf{F} = 0$. Eigenfunctions of the Perron–Frobenius operator satisfy

$$J^{-1}(\mathbf{x})\nu(\mathbf{S}^{-1}(\mathbf{x})) = \lambda\nu(\mathbf{x}). \tag{8.8}$$

Of particular interest is the *physical* invariant measure $\mu^*(B) = \mu^*(\mathbf{S}^{-t}(B))$ for all sets $B \in \mathfrak{B}$, which is stationary under the evolution of the flow. The associated invariant density $P\rho^*(\mathbf{x}) = \rho^*(\mathbf{x})$ corresponds to an eigenfunction at eigenvalue 1, which describes the asymptotic behavior of the underlying dynamics. Control is often designed to alter the observed invariant measure or density. Spectral properties of the Perron–Frobenius operator are related, e.g., to almost-invariant sets, meta-stable states, mixing properties, decay of correlations [43, 44, 61, 65].

The infinitesimal generator of the semigroup of Perron–Frobenius operators $\{P^t\}_{t\geq 0}$ is given by the Liouville operator $L_P$ [64, 102]:

$$\frac{\partial}{\partial t}\rho(t, \mathbf{x}) = -\nabla \cdot (\mathbf{F}(\mathbf{x})\,\rho(t, \mathbf{x})) = L_P[\rho](\mathbf{x}), \quad \rho_0 := \rho(0, \mathbf{x}), \tag{8.9}$$

for continuously differentiable $\rho$ with compact support and appropriate boundary conditions. The Liouville equation (8.9) describes how the flow transports densities in phase space and has a very intuitive interpretation as the conservation of probability or mass in terms of trajectories in the phase space. Alternatively, the evolution of the density may be interpreted as propagated uncertainty of an initial state. This is a first-order PDE that can be solved with the method of characteristics [50, 202]. The invariant density satisfies $L_P[\rho^*](\mathbf{x}) = 0$ and is an eigenfunction at eigenvalue 0.

The aim of this chapter is to provide an overview of major developments and advances in data-driven control using operator-theoretic methods. The chapter is organized as follows: In Sect. 8.2, the general control problem for a nonlinear system is formulated from an optimal control perspective. Control-dependent Koopman and Perron–Frobenius operators are introduced in Sect. 8.3. The main objective of the operator-theoretic approach is to find a linear representation of the underlying dynamical system. Data-driven system identification methods for the controlled nonlinear system are summarized in Sect. 8.4. Important aspects in control theory such as observability, controllability, state estimation, and control design in the operator-theoretic framework are discussed in Sect. 8.5. The chapter is concluded in Sect. 8.6 with a discussion on major challenges, open problems, and possible future directions of transfer operators approximations and their application for control.

## 8.2 Control Problem Formulation

Reformulating strongly nonlinear dynamics in a linear framework via transfer operators is appealing as it enables the application of powerful optimal and robust estimation and control techniques available for linear systems [49, 167, 171]. The formulation of an optimal control problem [171] appears in many applications, such as trajectory control in robotics or boundary layer stabilization in fluids, and has been considered widely in the context of Koopman operators for control [30, 77, 78, 91]. For that reason, we discuss in this chapter the operator-theoretic framework in the context of optimal control as an illustrative example.

Optimization-based approaches for control provide a highly suitable framework for the control of nonlinear systems, e.g., including constraints and flexible cost formulations. The optimal control problem can be stated as follows:

$$\min_{\mathbf{u}\in\mathbb{U}} \int_{t_0}^{t_f} l[\mathbf{x}(t), \mathbf{u}(t), t]\, dt + l_f[\mathbf{x}(t_f), t_f] \tag{8.10}$$

subject to

$$\frac{d}{dt}\mathbf{x}(t) = \mathbf{F}(\mathbf{x}, \mathbf{u}), \quad \mathbf{x}(0) = \mathbf{x}_0, \tag{8.11a}$$

$$\mathbf{y}(t) = \mathbf{H}(\mathbf{x}, \mathbf{u}), \tag{8.11b}$$

and possibly additional constraints on states, measurements, control inputs, or time. We consider the nonlinear system (8.1) extended to include an external input $\mathbf{u} \in \mathbb{U} \subset \mathbb{R}^q$, where $\mathbb{U}$ is the space of admissible control values, and assume continuously differentiable state dynamics $\mathbf{F} : \mathbb{X} \times \mathbb{U} \to T\mathbb{X}$, where $T\mathbb{X}$ denotes the tangent bundle. The state $\mathbf{x}$ may not be fully accessible and instead a limited set of output measurements $\mathbf{y} \in \mathbb{Y} \subset \mathbb{R}^p$ may be collected, which are prescribed by the measurement function $\mathbf{H} : \mathbb{X} \times \mathbb{U} \to \mathbb{Y}$. The nature of the solution of the optimization problem is determined by the choice of the terminal state cost $l_f[\cdot]$ and the running cost $l[\cdot]$. The objective is to determine a control law or policy that minimizes the cost functional (8.10).

A nonlinear optimal control formulation for the system in (8.11) can be established using dynamic programming [18]. This relies on Bellman's principle of optimality [16] and leads to the Hamilton–Jacobi–Bellman (HJB) equation [14], a nonlinear PDE for the globally optimal solution. Solving this nonlinear PDE is computationally challenging and as a result, a variational argument and Pontryagin's maximum principle [142] is often used instead, leading to a set of coupled ordinary differential equations (ODEs), the Euler–Lagrange equations. While this two-point boundary value problem is solvable for higher dimensional problems in contrast to the HJB equation, it is still too computationally demanding in real-time applications for many nonlinear systems, e.g., in autonomous flight [181]. Moreover, for high-dimensional systems, such as fluid flows, expensive direct and adjoint simulations render this approach infeasible, instead motivating the use of reduced-order models.

Controlling nonlinear, and possibly high-dimensional, systems is generally computationally demanding and performance and robustness guarantees exist only for certain classes of dynamical systems. However, the control problem above simplifies considerably for quadratic cost functions and linear dynamics of the form

$$\frac{\mathrm{d}}{\mathrm{d}t}\mathbf{x} = \mathbf{A}\mathbf{x} + \mathbf{B}\mathbf{u}, \quad \mathbf{x}(0) = \mathbf{x}_0, \tag{8.12}$$

where the matrix $\mathbf{A}$ may be obtained by a suitable linearization of the nonlinear dynamics (8.11) around an equilibrium point or operating condition. Considering linear dynamics, the optimization problem can be simplified tremendously and becomes solvable even for high-dimensional systems. In the simplest case, without any constraints, this reduces to solving an algebraic Riccati equation (ARE) and yields the linear quadratic regulator (LQR) [171].

Many problems may not permit a fully linear representation or the approximation may only be valid close to the linearization point. Moreover, models estimated from data may quickly become invalid, either due to the application of control or changing system parameters and conditions. For these cases, there exist adaptive and nonlinear variants of the formulation above. For instance, control-affine, nonlinear systems, whose governing equations can be factored into a linear-like structure, permit a state-dependent transition matrix $\mathbf{A}(\mathbf{x})$ and actuation matrix $\mathbf{B}(\mathbf{x})$ so that the ARE may be solved pointwise as a state-dependent Riccati equation (SDRE) [137]. The SDRE generalizes LQR for nonlinear systems, retaining a simple implementation and often yielding near-optimal controllers [37]. Alternatively, multiple model systems can be considered, which usually consist of a large set of locally valid models, for which control is then determined either by using a specific model selected based on some metric or by averaging/interpolating the control action from the model set [127].

A widely adopted adaptive variant is model predictive control (MPC) [4, 63, 126], which solves the optimal control problem over a receding horizon subject to the modeled dynamics and system constraints. The receding horizon problem is formulated as an open-loop optimization over a finite-time horizon. At each time step and given the current measurement, a sequence of future control inputs minimizing the cost $J$ over the time horizon is determined. The first control value of this sequence is then applied, and the optimization is re-initialized and repeated at each subsequent time step. This results in an implicit feedback control law $\mathbf{C}(\mathbf{x}_j) := \mathbf{u}_{j+1}(\mathbf{x}_j)$, where $\mathbf{u}_{j+1}$ is the first input in the optimized actuation sequence starting at the initial condition $\mathbf{x}_j := \mathbf{x}(t_j)$. MPC is particularly ubiquitous in industrial applications including process industries [118] and aerospace [56], as it enables more general formulations of control objectives and the control of strongly nonlinear systems with constraints, which are difficult to handle using traditional linear control approaches.

Linear models arising as approximations of the Koopman or Perron–Frobenius operators may be combined with any of these approaches. The potential benefit is the increased accuracy and validity as compared with linear models based on a local linearization and a reduced computational burden compared with schemes based on nonlinear models. Moreover, many data-driven approximation algorithms for these operators are easy to implement and efficient to compute as these are found on standard linear algebra techniques.

## 8.3 Control-Oriented Transfer Operators

Controlled dynamical systems have been increasingly considered in the operator-theoretic framework, particularly for state estimation, to disambiguate dynamics from the effect of control, and for related problems such as optimized sensor and actuator placement. While the Liouville equation has been examined for control purposes for a longer time [23, 26, 155], and interestingly also in the context of artificial intelligence [95], the control-oriented formulations of the Koopman and Perron–Frobenius operators have gained attraction only recently [39, 60, 146, 195].

### *8.3.1 Non-affine Control Systems*

We consider a non-affine control dynamical system and assume access to the full state $\mathbf{y} = \mathbf{x}$:

$$\frac{d}{dt}\mathbf{x}(t) = \mathbf{F}(\mathbf{x}, \mathbf{u}), \quad \mathbf{x}(0) = \mathbf{x}_0. \tag{8.13}$$

The flow associated with (8.13) is referred to as the *control flow* and is denoted by $\mathbf{S}^t(\mathbf{x}, \mathbf{u})$. Skew-product flows, such as $\mathbf{S}^t(\mathbf{x}, \mathbf{u})$, arise in topological dynamics to study nonautonomous systems, e.g., with explicit time dependency. Thus, we consider the action of an operator in the extended state space $\mathbb{X} \times \mathbb{U}$.

The Koopman operator is defined as acting on the extended state:

$$U^t f(\mathbf{x}, \mathbf{u}) = f(\mathbf{S}^t(\mathbf{x}, \mathbf{u}), \mathbf{u}). \tag{8.14}$$

Here the inputs $\mathbf{u}$ vary in time and their dynamics are governed, e.g., by a prescribed exogenous behavior or a given state-dependent feedback control law $\mathbf{u} = \mathbf{K}(\mathbf{x})$. If the inputs themselves are not evolving dynamically, this would reduce to $U^t f(\mathbf{x}, \mathbf{u}) = f(\mathbf{S}^t(\mathbf{x}, \mathbf{u}), \mathbf{0})$, in which case the inputs parameterize the dynamics [146]. The spectral properties of $U$ should then contain information about the unforced system with $\mathbf{u} = \mathbf{0}$. Koopman eigenfunctions associated with (8.14) satisfy

$$U^t \phi(\mathbf{x}, \mathbf{u}) = e^{\lambda t} \phi(\mathbf{x}, \mathbf{u}). \tag{8.15}$$

Assuming smooth dynamics and observables, the generator equation for the Koopman family is

$$\frac{\partial}{\partial t} f(t, \mathbf{x}, \mathbf{u}) = \mathbf{F}(\mathbf{x}, \mathbf{u}) \cdot \nabla_{\mathbf{x}} f(t, \mathbf{x}, \mathbf{u}) + \dot{\mathbf{u}} \cdot \nabla_{\mathbf{u}} f(t, \mathbf{x}, \mathbf{u}). \tag{8.16}$$

As above, smooth Koopman eigenfunctions can be considered eigenfunctions of the infinitesimal generator satisfying

$$\frac{d}{dt}\phi(\mathbf{x}, \mathbf{u}) = \mathbf{F}(\mathbf{x}, \mathbf{u}) \cdot \nabla_{\mathbf{x}}\phi(\mathbf{x}, \mathbf{u}) + \dot{\mathbf{u}} \cdot \nabla_{\mathbf{u}}\phi(\mathbf{x}, \mathbf{u}) = \lambda\phi(\mathbf{x}, \mathbf{u}). \tag{8.17}$$

Representing the system in terms of a finite set of observables or eigenfunctions generally requires a reformulation of the cost functional (8.10) in terms of these eigenfunctions:

$$J = \int_{t_0}^{t_f} l^{Kq}[\mathbf{z}(t), \mathbf{u}(t), t]\, dt + l_f^{Kq}[\mathbf{z}(t_f), t_f], \tag{8.18}$$

where the superscript "$Kq$" refers to the Koopman representation and $\mathbf{z}(t) = \mathbf{f}(\mathbf{x}(t), \mathbf{u}(t))$ describes the nonlinear transformation of the state $\mathbf{x}$ through a vector-valued observable $\mathbf{f}(\mathbf{x}, \mathbf{u}) = [f_1(\mathbf{x}, \mathbf{u}), \ldots, f_d(\mathbf{x}, \mathbf{u})]^T$. It may be of interest to directly control specific observables, e.g., Koopman eigenfunctions, that are associated with a particular physical behavior. If the state $\mathbf{x}$ is included as an observable, it is possible to transform the cost functional (8.18) into (8.10) by modifying $l^{Kq}$ and $l_f^{Kq}$ accordingly. However, there does not exist a finite-dimensional Koopman-invariant subspace explicitly including the state, that is topologically conjugate to a system representing multiple fixed points, limit cycles, or more complicated structures [30]. Nevertheless, it maybe possible to obtain a linearization in the entire basin of attraction for a single fixed point or periodic orbit [97, 196]. Thus, cost functionals (8.18) and (8.10) are generally different. Further, Koopman eigenfunctions maybe considered as observables and the state may be recovered via inversion, e.g., approximated from data using multidimensional scaling [84], to estimate (8.18).

Analogously, we can consider the Perron–Frobenius operator acting on densities

$$P^t\rho(\mathbf{x}, \mathbf{u}) = J^{-1}(\mathbf{x})\rho(S^{-t}(\mathbf{x}, \mathbf{u}), \mathbf{u}), \tag{8.19}$$

for which eigenfunctions satisfy

$$P^t\nu(\mathbf{x}, \mathbf{u}) = e^{\lambda t}\nu(\mathbf{x}, \mathbf{u}). \tag{8.20}$$

The control-oriented Liouville equation describing the transport of densities under the influence of an exogenous input is given by

$$\frac{\partial}{\partial t}\rho(t,\mathbf{x},\mathbf{u}) = -\nabla_{\mathbf{x}} \cdot (\mathbf{F}(\mathbf{x},\mathbf{u})\rho(t,\mathbf{x},\mathbf{u})) - \nabla_{\mathbf{u}} \cdot (\dot{\mathbf{u}}\rho(t,\mathbf{x},\mathbf{u})), \quad \rho_0(\mathbf{x},\mathbf{u}) := \rho(0,\mathbf{x},\mathbf{u}), \tag{8.21}$$

where the initial condition is, e.g., $\rho_0(\mathbf{x},\mathbf{u}) = \delta(\mathbf{x}'-\mathbf{x})\delta(\mathbf{u}'-\mathbf{u})$, a point mass at $\mathbf{x}'$ and $\mathbf{u}'$ for deterministic dynamics. This assumes the conservation of probability in the state–action space. The first term on the right-hand side describes the local change in density due to the flow, while the second term describes the local change due to exogenous inputs.

We are often interested in changing the long-term density $\rho(t,\mathbf{x})$. The objective is then to determine inputs $\mathbf{u}$ so that $\rho(t,\mathbf{x})$ becomes close to a desired target density $\rho^T(\mathbf{x})$ over some time horizon $[0, t_f]$ or in the limit $t_f \rightarrow \infty$, which can be interpreted as steering a density of initial conditions or particles governed by the underlying dynamical system (8.13). Assuming that the probability density function (PDF) is not an explicit function of the control input $\mathbf{u}$ but indirectly affected through the state dynamics $\dot{\mathbf{x}} = \mathbf{F}(\mathbf{x},\mathbf{u})$, we have

$$\frac{\partial}{\partial t}\rho(t,\mathbf{x}) = -\nabla_{\mathbf{x}} \cdot (\mathbf{F}(\mathbf{x},\mathbf{u})\rho(t,\mathbf{x})), \tag{8.22}$$

which is considered in the seminal works of Brockett [23, 26], particularly in the context of ensemble control. The finite-time control objective may be formulated as

$$J = \int_{t_0}^{t_f} \int_{\mathbb{X}} \left[\rho(t,\mathbf{x}) - \rho^T(\mathbf{x})\right] l_x^P(\mathbf{x}) d\mathbf{x} + l_u^P[\mathbf{u}(t), t] dt \tag{8.23}$$

measuring the weighted deviation from the desired density and penalizing control input and time. Ideally, the controlled system has a unique equilibrium density $\rho^T(\mathbf{x})$, corresponding to the sole eigenfunction of the Perron–Frobenius operator with eigenvalue 1, with all others decaying exponentially. In [26], a general cost functional of the following form is proposed:

$$J = \int_{t_0}^{t_f} \int_{\mathbb{X}} \rho(t,\mathbf{x}) l(\mathbf{x},\mathbf{u}) d\mathbf{x} dt + \int_{\mathbb{X}} \left(\frac{\partial \mathbf{u}}{\partial \mathbf{x}}\right)^2 d\mathbf{x} + \int_{t_0}^{t_f} \left(\frac{\partial \mathbf{u}}{\partial t}\right)^2 dt, \tag{8.24}$$

where the first term evaluates the average performance of the system at the current time, the second term penalizes high gain feedback, and the third term may promote linearity of the control law.

In general, different control perspectives can be assumed: (1) actively modifying the observable or density function, or (2) directly controlling the trajectory informed by the evolution dynamics of the observable or density function, e.g., ensemble control in a drift field. An illustrative schematic is provided in Fig. 8.2 using a particle in a field analogy. It may be possible to steer particles by controlling the external field, in which these particles "live", e.g., by modifying the Hamiltonian function, an eigenfunction of the Koopman operator, or the flow or an electromagnetic field, so that the particle trajectories will change. Alternatively, it may be possible to control directly the particles, e.g., steering gliders in the ocean, while the external

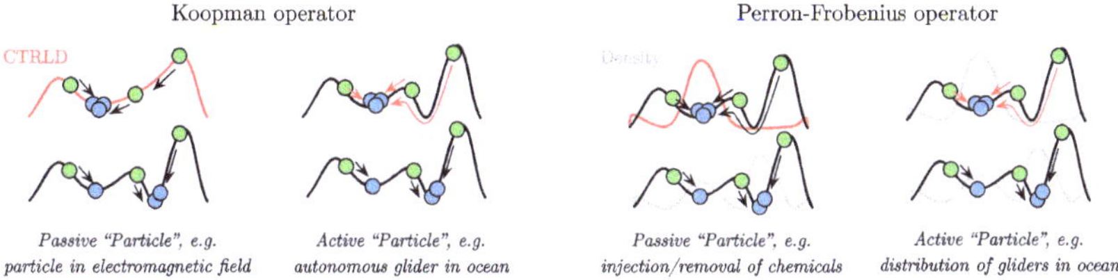


**Fig. 8.2** Controlling actively the observable or density function versus the trajectory in the operator-theoretic framework

field remains unaffected. A similar analogy can be provided for the Perron–Frobenius operator. We may be able to either modify a distribution by adding/removing sources and sinks, e.g., adding sorbents to remove oil spill in the ocean or heating/cooling elements in a room, or control the ensemble giving rise to the distribution directly, such as ocean gliders.

### 8.3.2 *Affine Control Systems*

A control-affine dynamical system is given by

$$\frac{d}{dt}\mathbf{x}(t) = \mathbf{F}(\mathbf{x}) + \mathbf{G}(\mathbf{x})\mathbf{u}, \quad \mathbf{x}(0) = \mathbf{x}_0, \tag{8.25}$$

where $\mathbf{G}(\mathbf{x})\mathbf{u} := \sum_i g_i(\mathbf{x})u_i$ represents mixed terms in $\mathbf{x}$ and $\mathbf{u}$. Above we sought a representation on the extended state $(\mathbf{x}, \mathbf{u})$, in which case the spectral properties of the associated operators will contain knowledge on the unforced system with $\mathbf{u} = \mathbf{0}$. Alternatively, we can consider operators associated with the unforced dynamics $\dot{\mathbf{x}} = \mathbf{F}(\mathbf{x})$ and how these are affected by the external control input, rendering the representation nonautonomous. Then, additional terms will arise and inform how control affects observables, densities, and eigenfunctions:

$$\frac{\partial}{\partial t} f(t, \mathbf{x}) = (\mathbf{F}(\mathbf{x}) + \mathbf{G}(\mathbf{x})\mathbf{u}) \cdot \nabla f(t, \mathbf{x}) = \mathbf{F}(\mathbf{x}) \cdot \nabla f(t, \mathbf{x}) + \mathbf{G}(\mathbf{x})\mathbf{u} \cdot \nabla f(t, \mathbf{x}), \tag{8.26}$$

where $\nabla := \nabla_{\mathbf{x}}$. We assume Koopman eigenfunctions are associated with the unforced dynamics $\mathbf{F}(\mathbf{x})$. Then, the smooth Koopman eigenfunctions satisfy

$$\frac{d}{dt}\phi(\mathbf{x}) = \lambda\phi(\mathbf{x}) + \nabla\phi(\mathbf{x}) \cdot \mathbf{G}(\mathbf{x})\mathbf{u}, \tag{8.27}$$

which renders the dynamics governing the eigenfunctions generally nonlinear. In the special case, where $\mathbf{G}(\mathbf{x}) = \mathbf{B}$ is constant, this remains linear, and the equation becomes bilinear if $\mathbf{G}(\mathbf{x}) = \mathbf{B}\mathbf{x}$ is a linear function of the state, for which a rich control literature exists [55]. A reduced system description for state estimation or model-based control and the control objective (8.18) may then be provided in eigenfunction coordinates (8.27), which is further discussed in Sects. 8.4.1 and 8.5.3.

The Liouville equation for the density becomes

$$\frac{\partial}{\partial t}\rho(t,\mathbf{x}) = -\nabla\cdot((\mathbf{F}(\mathbf{x})+\mathbf{G}(\mathbf{x})\mathbf{u})\rho(t,\mathbf{x})) = -\nabla\cdot((\mathbf{F}(\mathbf{x})\rho(t,\mathbf{x})) - \nabla\cdot(\mathbf{G}(\mathbf{x})\rho(t,\mathbf{x}))\,\mathbf{u}, \tag{8.28}$$

and correspondingly eigenfunctions satisfy

$$\frac{d}{dt}\nu(\mathbf{x}) = \lambda\nu(\mathbf{x}) - \nabla\cdot(\mathbf{G}(\mathbf{x})\nu(\mathbf{x}))\mathbf{u}. \tag{8.29}$$

The controlled Liouville equation has also been examined in other contexts, where a scalar observable is advected with the flow, e.g., concentration of chemical components or temperature distribution, in cases where diffusion is assumed negligible. It may be possible that the controller directly injects or removes content to modify the density, e.g., for carbon removal strategies in the atmosphere or oil spill mitigation in the ocean. Then, (8.28) is modified to

$$\frac{\partial}{\partial t}\rho(t,\mathbf{x}) = -\nabla\cdot((\mathbf{F}(\mathbf{x})\rho(t,\mathbf{x})) + \sum_{i=1}^{p} g_i(\mathbf{x})u_i, \tag{8.30}$$

where $g_i(\mathbf{x})$ is a set of functions prescribing the spatial impact of the actuators and the vector field $\mathbf{F}$ remains constant in time.

## 8.4 System Identification

In this section, we outline data-driven strategies to learn control-oriented representations of dynamical systems in the operator-theoretic framework. Most approaches rely on dynamic mode decomposition for the Koopman operator and the classical Ulam's method for the Perron–Frobenius operator. Emerging techniques such as machine learning and feature engineering aim to provide generalizations for systems of higher complexity, while exploiting and promoting sparsity often targets interpretability and efficient computation. A general overview of algorithms for data-driven approximation of the Koopman and the Perron–Frobenius operators has been provided in [86], which will not be repeated here. Here, we discuss approaches that specifically include the effect of control or external parameters.

### *8.4.1 Koopman Operator*

System identification for the Koopman operator can be broadly divided into three main directions: (1) linear models as approximations to the Koopman operator describing the evolution of a, usually large, set of observable functions, (2) (bi)linear models for Koopman eigenfunctions, which constitute a Koopman-invariant subspace, and (3) linear models parametrized by the control input. We assume a collection of data tuples $(\mathbf{x}_k, \mathbf{x}'_k, \mathbf{u}_k)$, where the state $\mathbf{x}_k := \mathbf{x}(t_k) \in \mathbb{R}^n$, the time-shifted state $\mathbf{x}'_k := \mathbf{x}(t_k + \Delta t) \in \mathbb{R}^n$, and inputs $\mathbf{u}_k := \mathbf{u}(t_k) \in \mathbb{R}^q$ are sampled at discrete times $t_k$, $k = 1, \ldots, m$. We define a vector-valued observable function $\mathbf{f}(\mathbf{x}) = [f_1(\mathbf{x}), f_2(\mathbf{x}), \ldots, f_d(\mathbf{x})]^T$ with $\mathbf{f} : \mathbb{R}^n \rightarrow \mathbb{R}^d$ evaluated on data, which is generally a nonlinear transformation of the state so that $\mathbf{z} = \mathbf{f}(\mathbf{x})$ embeds the state in a feature space of possibly higher dimension, i.e., $d \gg n$.

**Dynamic mode decomposition with control:** Building on DMD, the control-oriented extension *DMD with control* (DMDc) was proposed to disambiguate the effect of control from the natural dynamics in high-dimensional systems [145]. Assuming the availability of full-state measurements, observables are taken as linear measurements of the state, i.e., $\mathbf{f} = \mathbf{x}$. Then, the dynamical behavior of coordinates $\mathbf{z}$ are modeled as $\dot{\mathbf{z}}(t) = \mathbf{A}\mathbf{z}(t) + \mathbf{B}\mathbf{u}(t)$ or

$$\mathbf{z}_{k+1} = \mathbf{A}\mathbf{z}_k + \mathbf{B}\mathbf{u}_k \tag{8.31}$$

in discrete time, which makes standard linear control techniques readily applicable. The dimensionality of the problem is determined by the dimension of the state. However, many systems evolve on a low-dimensional attractor so that the singular value decomposition (SVD) can be used to solve the problem efficiently. Further, DMDc has been combined with compressed DMD [33] for compressive system identification [12]. While DMDc is feasible for high-dimensional systems with $n \ll m$, such as fluids that may have millions or billions of degrees of freedom, linear measurements are known to not be sufficiently rich to capture strongly nonlinear behavior, motivating the use of nonlinear observables. Nevertheless, recent work [79] indicates that even models with relatively little predictive power may still be sufficient for model predictive control schemes (see also Sect. 8.4.3).

**Extended DMD with control:** Extended DMD (eDMD) has been proposed as a generalization of DMD employing a dictionary of nonlinear functions as observables [196]. DMD is then equivalent to eDMD with the special set of observables corresponding to linear monomials, $\mathbf{f}(\mathbf{x}) = [x_1, x_2, \ldots, x_n]^T$. In domains where such a local linearization would be insufficient, more complex basis functions, such as high-order monomials, Hermite polynomials, or radial basis functions, can be employed. The number of observables typically exceeds the state dimension, i.e., $d \gg n$; however, the algorithm can be reformulated to scale with the dimension of samples instead of the dimension of features/states. A similar technique to eDMD has been used to recover the vector field of unforced dynamics [114]. eDMD has also been extended for nonlinear system identification, which we refer to as *eDMD with control* (eDMDc) [195], and been combined with MPC [91, 92] (see also Chap. 9). By transforming the system so that the dynamics evolve linearly, linear MPC methods become readily applicable, which require solving a convex quadratic program instead of a nonconvex program as in nonlinear MPC. In [91], the MPC optimization is further reformulated to scale with the dimension of the state $n$ and not with the number of observables $d$, which is advantageous if $d \gg n$. However, the quality of the resulting model and its prediction is highly sensitive to the choice of function basis.

**Time-delay coordinates and DMDc:** Often we do not have access to the full state, but instead can only measure a single or few variables. Then time-delay coordinates, i.e., $\mathbf{f}(\mathbf{x}(t)) = [x(t), x(t-1\tau), x(t-2\tau), \ldots, x(t-(d-1)\tau)]$ with time delay $\tau$, have been proven to be useful for analyzing and modeling dynamics. The Taken's embedding theorem [180] provides conditions under which the original system can be faithfully reconstructed in time-delay coordinates, providing a diffeomorphism between the two systems. Recently, time-delay coordinates have been shown to provide a Koopman-invariant subspace and used to model unforced systems with bimodal or bursting behavior as linear systems closed by an intrinsic forcing term [29]. This formulation is achieved by performing DMD on the delay coordinates arranged in a Hankel matrix [9, 29] (see also Chap. 19). Hankel matrix based system identification has been used for decades in the eigensystem realization algorithm (ERA) [76]. Assuming sample pairs $(x_k, u_k)$ of a single long trajectory $x(t)$ generated by a continuous system with single control input $u(t)$, the vector-valued observable is given by $\mathbf{z}_k = \mathbf{f}(\mathbf{x}_k) = [x_k, x_{k-1}, x_{k-2}, \ldots, x_{k-d_1+1}]$, where $f_i(x_k) = x_{k-(i-1)}$ and $\mathbf{v}_k = [u_k, u_{k-1}, u_{k-2}, \ldots, u_{k-d_2+1}]$. The embedding dimensions $d_1 = d_2 = d$ are assumed equal for simplicity. The system is then given by

$$\mathbf{z}_{k+1} = \mathbf{A}\mathbf{z}_k + \mathbf{B}\mathbf{v}_k, \tag{8.32a}$$

$$\mathbf{x}_k = \begin{bmatrix} 1 & 0 & \ldots & 0 \end{bmatrix} \mathbf{z}_k, \tag{8.32b}$$

where the state is recovered from the first component of $\mathbf{z}$. The control matrix $\mathbf{B}$ must satisfy a lower triangular structure to not violate causality. If the number of time-delay coordinates is large, an SVD can be applied and the model is built on the first $r$ eigen-time-delay coordinates. It can be advantageous to embed the full state $\mathbf{x}$ if available instead of just a single component, as more information is used to improve prediction accuracy. In Eq. 8.32, the control history appears as a separate vector of control inputs; it can also be useful to augment the state with the past control values so that the current actuation value appears as single control input:

$$\hat{\mathbf{z}}_{k+1} = \begin{bmatrix} \mathbf{z}_k \\ u_{k-1} \\ \vdots \\ u_{k-d+1} \end{bmatrix}_{k+1} = \begin{bmatrix} \mathbf{A} & \mathbf{B}_{[2,d]} \\ \mathbf{0} & \begin{bmatrix} \mathbf{0} & \mathbf{0} \\ \mathbf{0} & \mathbf{I} \end{bmatrix} \end{bmatrix} \begin{bmatrix} \mathbf{z}_k \\ u_{k-1} \\ \vdots \\ u_{k-d+1} \end{bmatrix}_{k} + \begin{bmatrix} \mathbf{B}_{[1,1]} \\ 1 \\ 0 \\ \vdots \\ 0 \end{bmatrix} u_k \tag{8.33a}$$

$$\mathbf{x}_k = \begin{bmatrix} 1 & 0 & \ldots & 0 \end{bmatrix} \hat{\mathbf{z}}_k, \tag{8.33b}$$

where $\mathbf{I}$ denotes the $d-2 \times d-2$ identity matrix and $\mathbf{B}_{[a,b]}$ contain columns $a$ through $b$ of $\mathbf{B}$.

**Control in eigenfunction coordinates:** Eigenfunctions of the Koopman operator provide a natural set of intrinsic observables as these themselves behave linearly in time and correspond to global properties of the system. Particularly, control-affine systems Eq. (8.25) have been studied in this context. The general approach relies on estimating eigenfunctions associated with the autonomous system and then identifying how these are affected by the control input. The observables are then given by a set of eigenfunctions $\mathbf{z} = \mathbf{f}(\mathbf{x}) = [\phi_1(\mathbf{x}), \ldots, \phi_r(\mathbf{x})]$. Their continuous-time dynamics are given by

$$\dot{\mathbf{z}} = (\mathbf{F}(\mathbf{x}) + \mathbf{G}(\mathbf{x})\mathbf{u}) \cdot \nabla_{\mathbf{x}}\mathbf{f}(\mathbf{x}) = \mathbf{A}\mathbf{z} + \mathbf{B}(\mathbf{z})\mathbf{u}, \tag{8.34}$$

where $\mathbf{A} := \mathbf{F}(\mathbf{x}) \cdot \nabla_{\mathbf{x}}\mathbf{f}(\mathbf{x}) = \boldsymbol{\Lambda} := \text{diag}(\lambda_1, \ldots, \lambda_r)$. The term $\mathbf{B}(\mathbf{z}) := \mathbf{G}(\mathbf{x})\mathbf{u} \cdot \nabla_{\mathbf{x}}\mathbf{f}(\mathbf{x})$ is linear in $\mathbf{z}$ since $\mathbf{G}(\mathbf{x})\mathbf{u} \cdot \nabla_{\mathbf{x}}$ is a linear operator. For $\mathbf{u} = \mathbf{0}$, this constitutes a Koopman-invariant subspace for the unforced system; otherwise, the control term $\mathbf{B}(\mathbf{z})$ prescribes how these eigenfunctions are modified by $\mathbf{u}$. The advantage of this formulation is that the dimension of the system scales with the number of eigenfunctions, and we are often interested in a few dominant eigenfunctions associated with persistent dynamics. This formulation has been examined for state estimation and observer design [173, 175] and data-driven control [77]. Spectral properties of the projected infinitesimal generator have also been related to stability properties of the underlying dynamical system [115, 116] (see also Sect. 8.5.1 and Chaps. 2 and 5). In general, the eigenfunctions can be identified using eDMD, kernel-DMD, or other variants, and the model (8.34) may be well represented as long as their span contains $\mathbf{F}(\mathbf{x})$, $\mathbf{G}(\mathbf{x})$, and $\mathbf{x}$ [175]. However, the eigenfunctions obtained from DMD/eDMD may be spurious, i.e., they do not behave linearly. For instance, noise can produce physically irrelevant but energetically important, unstable eigenfunctions. An alternative [77] seeks a functional representation of the eigenfunctions in a dictionary of basis functions $\boldsymbol{\Theta}(\mathbf{x}) = [\theta_1(\mathbf{x}), \theta_2(\mathbf{x}), \ldots, \theta_p(\mathbf{x})]$ so that $\phi(\mathbf{x}) \approx \sum_{i=1}^{p} \theta_i(\mathbf{x})\xi_i = \boldsymbol{\Theta}(\mathbf{x})\xi$. A sparsity constraint on the vector of coefficients $\xi \in \mathbb{R}^d$ can then be used to identify an analytical expression of the eigenfunction. This approach is restricted to smooth eigenfunctions in the point spectrum of the Koopman operator and results in an optimization problem to find a sparse vector in the nullspace of the matrix $(\dot{\mathbf{x}} \cdot \nabla_{\mathbf{x}}\boldsymbol{\Theta}(\mathbf{x}) - \lambda\boldsymbol{\Theta}(\mathbf{x}))\,\xi = 0$ for a given eigenvalue $\lambda$. This requires an accurate estimation of the time derivative of $\mathbf{x}$ and is sensitive to noise. The control term $\mathbf{B}(\mathbf{z})$ can either be determined from the gradient of the identified eigenfunctions for given $\mathbf{G}(\mathbf{x})$ or identified directly [78] for unknown $\mathbf{G}(\mathbf{x})$ by similarly expanding it in terms of a dictionary, $\mathbf{G}(\mathbf{x}) \approx \sum_{i=1}^{q} \psi_i(\mathbf{x})\eta_i$. The approach can be further extended to non-affine control systems, for which control-dependent eigenfunctions may be considered, using a library on the extended state, i.e., $\boldsymbol{\Theta}(\mathbf{x}, \mathbf{u})$. This data-driven characterization has been combined with SDRE [77] and MPC [78] for controlling the underlying dynamical system. The cost functional generally needs to be formulated in eigenfunction coordinates as discussed in the previous section.

**Parameterized linear models:** Another class of linear parameter-varying models have been considered to disambiguate the approximation of the Koopman operator of the unforced system from the effect of exogenous forcing [195] and for model predictive control of switched systems [138]. In particular, a set of linear models parametrized by the control input is considered as an approximation to the nonautonomous Koopman operator:

$$\mathbf{z}_{k+1} = \mathbf{A}(\mathbf{u})\mathbf{z}_k. \tag{8.35}$$

For discrete-valued inputs $\mathbf{u}_i \in \{\mathbf{u}_1, \ldots, \mathbf{u}_q\}$, a finite set of matrices defined as $\{\mathbf{A}_i := \mathbf{A}(\mathbf{u}_i)\}_{i=1}^q$ may be estimated, in the simplest case with DMD or a variant by determining these for each discrete control input separately. This replaces the nonautonomous Koopman operator with a set of autonomous Koopman operators, each element associated with a constant control input, as considered in [146]. A more robust way that generalizes to continuous input data has been proposed in [195], where $\mathbf{A}(\mathbf{u})$ is expanded in terms of basis functions $\psi_i : \mathbb{U} \to \mathbb{R}$ as in eDMD for the state variable, so that $\mathbf{A}(\mathbf{u}) = \sum_{k=1}^d \psi_k(\mathbf{u})\mathbf{A}_k$. Then, for a given value of $\mathbf{u}$, matrix $\mathbf{A}$ is constructed using coefficients $\mathbf{A}_k$ and its spectral properties can be examined. Since eDMD may be prone to overfitting, the estimation problem can be regularized using the group sparsity penalty [163].

Determining an optimal control sequence for Eq. (8.35) is a combinatorial problem, which can be efficiently solved for low-dimensional problems using iterative algorithms from dynamic programming [18]. In [138], the MPC problem for the sequence of control inputs is transformed into a time-switching optimization problem assuming a fixed sequence of consecutive discrete control inputs. The model can further be modified as a bilinear continuous, piecewise-smooth variant with constant system matrices, which does not suffer from the curse of dimensionality and allows for continuous control inputs via linear interpolation [139]. Open- and closed-loop control of switched systems is also considered in Chap. 10.

### *8.4.2 Perron–Frobenius Operator*

A classical method to approximate the Perron–Frobenius operator is the Ulam–Galerkin method, a particular Galerkin method where the test functions are characteristic functions. Given an equipartition of the phase space into disjoint sets $\{B_1, \ldots, B_d\}$ and data, the action of the operator can be approximated as

$$\mathbf{P}_{ij}^{\tau} = \frac{\mathrm{card}(\mathbf{x}_k | \mathbf{x}_k \in B_j \wedge \mathbf{F}^{\tau}(\mathbf{x}_k) \in B_i)}{\mathrm{card}(\mathbf{x}_k \in B_j)}, \tag{8.36}$$

where $\mathbf{F}^{\tau}$ is the flow map and $\tau$ is the time duration. The $\tau-$step stochastic transition matrix acts on a discrete probability vector $\mathbf{p} = [p_1, \ldots, p_d]^T$ and satisfies the properties $\sum_{j=1^d} P_{ij} = 1$ and $0 \leq P_{ij} \leq 1$. A widely used software package based on set-oriented methods is GAIO [42]. Each element $B_i$ can be associated with a distinct symbol. The dynamics are then propagated on the partition and can be analyzed in terms of the sequence of symbols. The approach has been generalized to

PDEs [81, 187], i.e., where the evolution of a probability density function $p(\mathbf{u}, t)$ of a spatiotemporal vector field $\mathbf{u}(\mathbf{x}, t)$ is of interest. After expanding the vector field $\mathbf{u}(\mathbf{x}, t) = \sum_i a_i(t)\mathbf{u}_i(\mathbf{x})$, e.g., using POD, Ulam's method can be applied to the $a_i$'s. Other data-driven methods to approximate the Perron–Frobenius operator include blind source separation, the variational approach for conformation dynamics (VAC) [131] and DMD-based variants exploiting the duality between Koopman and Perron–Frobenius operators, e.g., naturally structured DMD (NSDMD) [74] and constrained Ulam dynamic mode decomposition [67].

Data-driven control based on the Perron–Frobenius operator appears more challenging as it relies on good statistical estimates requiring large amounts of data. In practice, methods often use a Monte Carlo sampling to estimate the transition probabilities, which suffers from the curse of dimensionality. However, it is particularly appealing as it provides a natural framework to incorporate uncertainties and transients, allows the control of ensembles and mixing properties of the underlying dynamical system, and yields a more global controller.

**Parametrized models:** Identifying a probabilistic model disambiguating the unforced dynamics and the effect of actuation is commonly realized by a parametrized representation. Incorporating the effect of control can be achieved via a simple extension of Ulam's method by considering a discretized control input that parametrizes the transition matrix in (8.36):

$$\mathbf{P}_{ij}^{u_l} = \frac{\text{card}(\mathbf{x}_k | \mathbf{x}_k \in B_j \wedge \mathbf{F}^{\tau}(\mathbf{x}_k, \mathbf{u}_l) \in B_i)}{\text{card}(\mathbf{x}_k \in B_j)}, \tag{8.37}$$

where each $\mathbf{P}(\mathbf{u}_l) := \mathbf{P}_{ij}^{u_l}$ satisfies the normalization and positivity properties. The dynamics of the probability vector can then be expressed as

$$\mathbf{p}_{k+1} = \mathbf{P}(\mathbf{u}_l)\mathbf{p}_k. \tag{8.38}$$

Approximations for (8.37) have been obtained employing set-oriented methods [40] and applying NSDMD with Gaussian radial basis functions [39], which are then used to derive optimal feedback control strategies. The model (8.38) can be derived from a discretization of the Liouville equation in space, time, and control [82] and represents a control-dependent Markov chain. The corresponding control problem can be associated with a Markov decision process (MDP) and appears in discrete optimal control [18]. Therein, it is used as a model to predict the expectation of the value function in dynamic programming, without reference to the Liouville equation. Building on these ideas, cluster-dependent feedback control has been developed by examining the spectral properties of the control-dependent transition matrix family [82], where state discretization of a fluid from the high-dimensional snapshot data is achieved using POD and k-means clustering.

**Multiplicative models:** Alternatively, the effect of control may be encoded in a separate stochastic matrix $\mathbf{P}^u$ approximating the action of a stochastic kernel:

$$\mathbf{p}_{k+1} = \mathbf{P}\mathbf{P}^u\mathbf{p}_k, \tag{8.39}$$

which recovers the unforced dynamics when $\mathbf{P}^u = \mathbf{I}$ is the identity matrix. This model appears, e.g., in the context of optimizing mixing properties [60, 62], where $\mathbf{P}$ represents advection via the Perron–Frobenius operator and $\mathbf{P}^u$ represents the discretized diffusion kernel to be optimized. In [119], limitations of nonlinear stabilization are examined when $\mathbf{P}^u = \mathbf{P}^{1+C}$ encodes a state feedback law $u = (1 + C)(x) = x + C(x)$, where $\mathbf{P}^{1+C} = \mathbf{I}$ if control via $C$ is inactive.

**Additive models:** It may also be possible to formulate additive models, which exhibit some similarity to traditional state-space models, of the form

$$\mathbf{p}_{k+1} = (\mathbf{P} + \mathbf{P}^u)\mathbf{p}_k, \tag{8.40}$$

where additional constraints are necessary, so that $\mathbf{P} + \mathbf{P}^u$ satisfies the positivity and conservation constraints from probability. The matrix $\mathbf{P}^u$ may be interpreted as a disturbance to the unforced transition probabilities $\mathbf{P}$. It may also be possible to represent $\mathbf{P}^u$ as $\mathbf{P}^c\mathbf{u}_k$, where $\mathbf{P}^c$ is a constant matrix and $\mathbf{u}_k$ is the control input. A similar formulation to (8.40) incorporating control appears from a state discretization of the controlled Liouville equation as outlined in [81], resulting in a discrete-state, continuous-time Markov model. Variations of this type of model have been used to study optimal response of Markov chains [7] and to control chaos [19].

**Remark:** Until recently, the focus on control in the Perron–Frobenius and Liouville operator framework has been mostly analytic and theoretical. Further, most problems studied have been low dimensional. On the other hand, there exists an extensive literature on probabilistic models for control, which can benefit the operator-theoretic framework but which have not been fully explored yet. Examples include Hidden Markov models, see [5, 199] in the context of the Perron–Frobenius and Fokker–Planck operators, and partial-observation Markov decision processes (POMDP) [11].

### *8.4.3 Numerical Example*

We compare the effectiveness of model predictive control for different DMDc-based models, which differ in their choice of dictionary functions. The van der Pol oscillator is considered as an illustrative example:

$$\frac{d^2x}{dt^2} - \mu(1 - x^2)\frac{dx}{dt} + x = u \tag{8.41}$$

with $\mu = 0.2$ and where $u$ is the external control input. In the following, the state vector is defined as $\mathbf{x} := [x_1, x_2]^T = [x, dx/dt]^T$. The training data consists of 200 trajectories in the box $[-6, 6] \times [-6, 6]$ integrated until $T = 1$ with time step $\Delta t = 0.05$, i.e., 4000 sample points in total. The forcing $u(t) = 5\sin(|\omega_1|t)\sin(|\omega_2|t)$, where $\omega_i \sim \mathcal{N}(0, 10)$, is applied for the identification task. The models under consideration include DMDc on the state $\mathbf{x}$, eDMDc using monomials up to order 5, and delayDMDc with embedding dimension $d = 5$, as compared in Fig. 8.3. The control objective is the stabilization of the unstable fixed point at the origin using the following cost functional

$$J = \sum_{k=1}^{T/\Delta t} \mathbf{x}_k^T \mathbf{Q} \mathbf{x}_k + R_u u_k^2 + R_{\Delta u}(u_{k+1} - u_k) \tag{8.42}$$

with weight matrices $\mathbf{Q} = \left(\begin{smallmatrix} 1 & 0 \\ 0 & 1 \end{smallmatrix}\right)$, $R_u = 0.1$, and $R_{\Delta u} = 0.1$, and input constraints $-5 < u < 5$, $-50 < \Delta u < 50$, $u_0 = 0$. The prediction/control horizon is $T = 0.75 = 15\Delta t$.

The models are evaluated on a test dataset that is different from the training set, visualized for a subset of trajectories in Fig. 8.3a. The smallest error is achieved by eDMDc, followed by delayDMDc and DMDc, with the latter exhibiting a large deviation from the actual evolution. MPC control results are displayed for $\mu = 0.2$ in Fig. 8.3b, c. Three main observations can be made: (1) despite the prediction inaccuracies, all three models are able to successfully control the system; (2) best performance is achieved with eDMDc and delayDMDc, while delayDMDc generally performs slightly better; (3) the farther away the initial condition is from the fixed point to be stabilized, the worse the performance for DMDc (compare results in Fig. 8.3d). It is important to point out that the specific control performance is highly sensitive to the chosen prediction horizon, weights, and the forcing for the training data; however, delayDMDc appears to be most robust. Observed here and in previous work [79], superior prediction performance may be overrated in combination with predictive control schemes. The overall performance lies more in the robustness of MPC than in the specific model choice due to the repeated initialization with the updated measurement vector and the sufficiently short prediction horizon.

## 8.5 Control Theory in the Operator-Theoretic Framework

In this section, we review control-theoretic advances using transfer operators and their connections to classical concepts in control such as stability analysis and observability.

### *8.5.1 Stability Analysis*

Analyzing the stability of dynamical systems is an important aspect of control theory, e.g., in order to develop stabilizing controllers. In particular, the Lyapunov function plays an essential role in the global stability analysis and control design of nonlinear

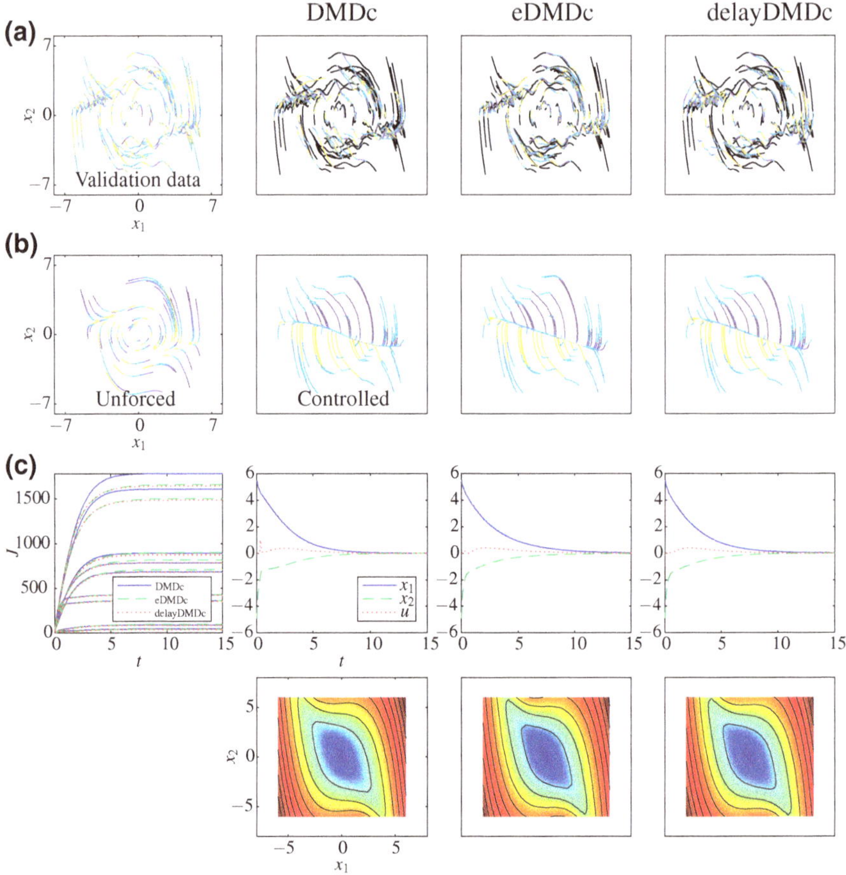


**Fig. 8.3** Van der Pol oscillator for $\mu = 0.2$ comparing DMDc, eDMDc, and delay DMDc (from second to fourth column): **a** validation data and prediction (colored by control input), **b** unforced and forced phase plots with MPC for initial conditions of the validation dataset (colored by control input), **c** cumulative cost of subset of the trajectories and a specific example trajectory, and **d** color-coded initial conditions with respect to converged control cost

systems. For stable, linear systems, the Lyapunov function is the positive solution of a corresponding Lyapunov equation. In contrast, finding and constructing a Lyapunov function poses a severe challenge for general nonlinear systems. There has been important progress within the operator-theoretic context for nonlinear stability analysis as spectral properties of the operators can be associated with geometric properties such as Lyapunov functions and measures, contracting metrics, and isostables (Fig. 8.4).

Some of this work relates back to a weaker notion of stability in terms of a density function as introduced by Rantzer [151]. In particular, the concept of asymptotic stability in an almost everywhere (a.e.) sense is guaranteed through the existence

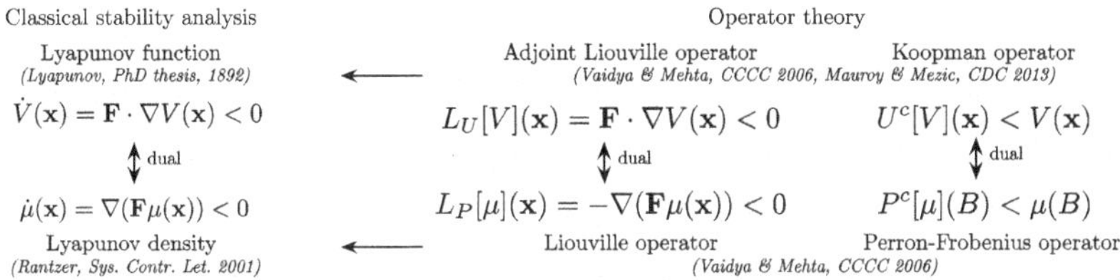


**Fig. 8.4** Classical stability analysis and within the operator-theoretic framework. The duality between the Lyapunov function and Lyapunov density [151] can be derived from the duality of the adjoint transport operators. Operators $U^c$ and $P^c$ are restricted to the corresponding function space with support on $\mathbb{X}\backslash\mathbb{A}$, where $\mathbb{A}$ is the attractor

of a density function. Further, for certain systems usually with polynomial vector fields, Lyapunov functions can be constructed using a suitable choice of polynomials, resulting in a linear problem formulation [136]. These ideas were later combined in a convex, linear formulation based on the discretization of the density function via polynomials, to jointly synthesize the density function and state feedback. The crucial connection of this line of work to the Perron–Frobenius operator was made by Vaidya and Mehta [182] demonstrating that these ideas of duality and linearity can be expressed using methods from ergodic theory, and the Lyapunov measure represents the dual of the Lyapunov function, including its relationship to the Koopman operator. Spectral properties, specifically invariant sets, are then used to determine stability properties. The Lyapunov measure is closely related to the density function introduced in [151] and captures the weaker a.e. notion of stability. Analogously to the Lyapunov equation, the solution to the Lyapunov measure equation provides necessary and sufficient conditions for a.e. stability of nonlinear systems and the Lyapunov measure can be connected to the Perron–Frobenius operator, which generalizes stability analysis to invariant sets of nonlinear systems [183]. Set-oriented methods have been used for solving the Lyapunov measure equation leading to a linear program [188]. A weaker notion of *coarse stability* has been further introduced to characterize the stability properties obtained from the finite-dimensional approximation. A formulation for the infinitesimal generator of the Perron–Frobenius operator, the Liouville operator, has also been proposed and can be solved numerically using the finite-element method [150]. In particular, for a stable fixed point $\mathbf{x}^* \in \mathbb{X}$, a Lyapunov measure (or density) $\mu$ satisfies a.e.

$$L_P[\mu](\mathbf{x}) < 0 \quad \forall \mathbf{x} \neq \mathbf{x}^*, \tag{8.43}$$

which corresponds to $\nabla(\mathbf{F}\mu(\mathbf{x})) > 0$ as originally introduced in [151]. Thus, the Lyapunov measure (density) decays a.e. under the action of the Liouville operator and is related to its spectral properties. These ideas have further been extended to stochastic systems [185, 186], used to compute the domain of attraction [192], and employed for control [148, 189] (see also Chap. 12). We also note that a connection between mixing properties and stability of stochastic (Markov) matrices has been established using perturbation analysis [123, 124].

The Koopman operator has also been used to study nonlinear stability (see also Chaps. 2 and 5). Certain sets of points in phase space that exhibit the same asymptotic behavior are referred to as isostables and play an important role in characterizing transient behavior. Isostables correspond to level sets of a particular Koopman eigenfunction and can be computed in the entire basin of attraction using, e.g., Laplace averages [117]. It has been further shown [115, 117] that special Lyapunov functions from classical stability theory can be determined from Koopman eigenfunctions and a sufficient condition for global stability is the existence of stable eigenfunctions. As with the Lyapunov measure, it can be shown that for a stable fixed point there exists a special, nonnegative observable $V(\mathbf{x})$ that satisfies

$$L_U[V](\mathbf{x}) < 0 \quad \forall \mathbf{x} \neq \mathbf{x}^*. \tag{8.44}$$

Thus, the Lyapunov function decays everywhere under the action of the adjoint Liouville operator and is related to its properties and those of the Koopman operator family. In [115], it has been shown that global stability of a fixed point can be established through the existence of a set of $C^1$ eigenfunctions of the Koopman operator associated with the eigenvalues of the Jacobian of the vector field, and that Koopman eigenfunctions can be used to define a Lyapunov function and contracting metrics. A numerical scheme based on a Taylor expansion is proposed to compute stability properties including the domain of attraction [115]. These ideas have been further extended to examine global stability properties of hyperbolic fixed points and limit cycles and also for non-analytical eigenfunctions [116]. We further point out that there exists an extensive literature on characterization and stability analysis of complicated and real-world systems via the Koopman operator, often facilitated through dynamic mode decomposition, e.g., for fluid dynamics [159, 160], power grids [177], neuroscience [28], traffic [45, 101].

### 8.5.2 *Observability and Controllability*

Characterizing a system's observability and controllability is crucial for designing sensor-based estimation and control strategies. Linear observability and controllability criteria have been extended to characterize nonlinear systems [72, 85], e.g., local observability can be tested by constructing an observability matrix using Lie derivatives. Balanced model reduction has also been extended to nonlinear systems [96, 158, 200]. Operator-theoretic approaches provide new opportunities for determining observability and controllability properties of high-dimensional nonlinear systems from data using linear techniques.

There has been considerable work on nonlinear observability analysis based on the Perron–Frobenius operator formulation for the Lyapunov measure equation. Analogously to the observability Gramian for linear systems, which can be obtained as the solution to a Lyapunov equation, the nonlinear equivalent can be obtained from the Lyapunov measure equation. In [184], set-oriented methods have been used to compute the observability Gramian for nonlinear systems with output measurements, based on partitioning the phase space into slow and fast time regimes using the Lyapunov measure. The linear observability Gramian appears here as a special case of the transfer operator based approach. The degree of observability of a set can further be related to the residence time. These ideas have been extended to determine sensor and actuator placement for the controlled PDE associated with the infinitesimal generator of the transfer operator [190]. In particular, the observability and controllability are characterized in terms of the support of the observability and controllability matrices, respectively, and the finite-time and infinite-time Gramians are formulated in terms of the transfer operators. The infinite-time controllability is only well defined when the vector field $\mathbf{F}(\mathbf{x})$ is a.e. uniformly stable; thus, an alternative approach to compute the infinite-time controllability set is based on an exponential discounting of the density, which does not rely on the stability property [164]. The terms *coarse observability* and *coarse controllability* have been introduced to characterize the weaker notion obtained from the finite-dimensional approximation [165].

The Koopman operator perspective may also be used to assess nonlinear observability and controllability. Generally, two states are indistinguishable if their future output sequences are identical. Thus, a nonlinear system is considered nonlinearly observable if any pair of states are distinguishable [72]. In [174], nonlinear observability is evaluated in the proposed Koopman-based linear observer design framework, which provides a linear representation of the underlying system in terms of the Koopman eigenfunctions (see Sect. 8.5.3). In particular, the nonlinear system is nonlinearly observable if the pair $(\mathbf{A}, \mathbf{C}^H)$ is observable, which can be determined via the rank condition of the corresponding observability matrix. These ideas have been applied to study pedestrian crowd flow [17], extended further to input–output nonlinear systems [173] resulting in bilinear or Lipschitz formulations, and used to compute controllability and reachability [66] (see also Chap. 4). The observability and controllability Gramians, which are used to examine the degree of observability and controllability, can be computed in the lifted observable space given by a dictionary of functions [198]. In this case, the observability/controllability of the underlying system is related to the observability/controllability of the observables. The underlying assumption is that the state, input, and output are representable in terms of a few Koopman eigenfunctions, which relies on a suitable choice of observable functions to retain the relevant information.

### 8.5.3 Observer Synthesis for State Estimation

Observer design is critical for sensor-based estimation of high-dimensional systems. The observer is a dynamical system that estimates the current state $\hat{\mathbf{x}}_k$ from the history of output measurements of the system. The extended Kalman filter (EKF) is perhaps the most widely used nonlinear observer, benefiting from high performance and a simple formulation based on linearization of the nonlinear state equations. However, EKFs are sensitive to noise, do not have convergence guarantees, have limited performance for strongly nonlinear systems, and assume Gaussian noise and disturbances [36, 152].

Observer design in the Koopman operator framework involves constructing a state estimator for the system $\mathbf{x}_{k+1} = \mathbf{F}(\mathbf{x}_k, \mathbf{u}_k), \quad \mathbf{y}_k = \mathbf{H}(\mathbf{x}_k)$ using a nonlinear transformation of states $\mathbf{x}$ and outputs $\mathbf{y}$ (and inputs for systems with actuation). This results in the following estimator dynamical system:

$$\mathbf{z}_{k+1} = \mathbf{A}\mathbf{z}_k, \quad \mathbf{y}_k = \mathbf{C}^y(\mathbf{z}_k), \quad \mathbf{x}_k = \mathbf{C}^x(\mathbf{z}_k), \tag{8.45}$$

where $\mathbf{C}^x$ and $\mathbf{C}^H$ are formed from the Koopman modes and observer Koopman modes. System (8.45) can be obtained from a spectral decomposition of the Koopman operator, where $\mathbf{z}$ represents Koopman eigenfunction coordinates, which has been referred to as Koopman observer form [174] (see also Sect. 8.4.1 and Chap. 3). The resulting system requires that the state and output lie in the span of a subset of eigenfunctions. Luenberger or Kalman observers can then be synthesized based on (8.45), where convergence can be shown under the observability condition based on the Koopman spectral properties, and shown to exhibit increased performance over the extended Kalman filter [174]. This work has been extended to input–output systems yielding Lipschitz or bilinear observers, resulting from the representation of the unforced system in eigenfunction coordinates [173], and for constrained state estimation based on a moving horizon analogous to model predictive control [175]. Building on kernel DMD and the Koopman observer form, a Luenberger Koopman observer has been applied to pedestrian crowd flow to estimate the full state from limited measurements [17].

The probabilistic perspective based on the Perron–Frobenius operator allows one to incorporate uncertainty into observer synthesis. Particle filters based on Monte Carlo sampling are commonly used for forecasting and estimation under uncertainty by approximating the PDF with a finite number of samples. However, this approach may suffer from the curse of dimensionality [168], particle degeneracy [10], and loss of diversity [153]. A particular challenge is the accurate estimation of the filter weights. The Perron–Frobenius and Liouville operator based approaches have been shown to be advantageous over Monte Carlo methods [41, 156]. More recently [50], a Bayesian formalism involving the Perron–Frobenius operator has been proposed, which is shown to achieve superior performance over a generic particle filter and bootstrap filter. This work has been extended to stochastic systems using a Karhunen–Loève decomposition of the process noise for uncertainty quantification and state estimation [51–53].

### 8.5.4 Control Design

Linear models that capture the dynamics of a nonlinear system provide tremendous opportunities for model-based control, making linear techniques readily applicable to nonlinear systems. In particular, Koopman-based frameworks are amenable to the application of standard linear control theory. Koopman models (see Sect. 8.4.1) have been increasingly used in combination with optimal control for trajectory steering, e.g., with LQR [30], SDRE [77], and MPC [8, 78, 91, 138]. Besides theoretical developments, Koopman-based MPC has been increasingly applied in realistic problems, such as power grids [92], high-dimensional fluid flows [70], and experimental robotic systems [1, 2]. Although the predictive power of Koopman models may be sensitive to the particular choice of observables and the training data, MPC provides a robust control framework that systematically compensates for model uncertainty by taking into account new measurements (see also Sect. 8.4.3). Optimal control formulations have also been considered for switching problems [138, 169, 170]. Switching problems are also considered in Chaps. 10 and 11. Based on a global bilinearization [66], the underlying dynamical system can be stabilized using feedback linearization [77]. A different strategy aims to shape Koopman eigenfunctions directly, referred to as eigenstructure assignment, which has been examined for fluid flow problems [71]. In contrast to pole placement, which aims to design the eigenvalues of the closed-loop system, eigenstructure assignment aims to additionally modify the structure of the eigenfunctions in a desirable way, which requires additional inputs. More recently, feedback stabilization for nonlinear systems has been achieved via a control Lyapunov function based approach based on the bilinear Koopman system in eigenfunction coordinates [173]. In particular, a convex optimization problem is formulated to search for the control Lyapunov function for the bilinear Koopman system [73].

In many applications it is of interest to shape the distribution of some quantity, e.g., swarms of UAVs or satellites, spin dynamics in nuclear magnetic resonance (NMR) spectroscopy and imaging [99], dispersions in process industries [191], oil spill in the ocean, search mission design [140], and beam dynamics [135, 147]. Of particular interest is the shaping of the asymptotic PDF [59, 68, 201], so that the stationary PDF of the closed-loop system coincides with a desired distribution. This relies on the idea that temporal averages can be replaced by spatial analogues under Birkoff's ergodic theorem [98], after which the mean or variance of a quantity can be controlled by modifying the stationary PDF of the controlled system. In [113], centralized feedback laws are determined for multi-agent systems to achieve a particular target distribution (here a uniform coverage in a domain with particular geometries) based on an ergodicity metric. These ideas have been later used to formulate a particular sampling strategy to reproduce paintings [157]. Optimal control approaches have been proposed to modify the PDF in the presence of uncertainty in initial conditions or parameters [141, 154]. The process of driving one distribution to another one is further intimately related to Monge–Kantorovich optimal transport theory [54, 69, 83, 125]. In [69], optimal transport theory has been used to solve a finite-horizon control problem to achieve a desired distribution, where the optimal control vector field is estimated by solving the associated Liouville equation over the finite horizon. Set-oriented and graph-based methods have also been used to study controllability and optimal transport [54].

Set-oriented representations of the dynamics (see Sect. 8.4.2) are amenable to optimal control approaches based on dynamic programming [18] for discrete-time, discrete-state, and discrete-action problems. The discretized control problem can be interpreted as a Markov decision problem (MDP). The optimization problem for the MDP can then be posed as a linear program, which has been demonstrated for communication networks [5]. An optimal control formulation in terms of an MDP has been formulated for the discretization Liouville equation and demonstrated for navigation in a maze [95]. Similar ideas have been used for controlling communication networks [5] and optimizing the asymptotic PDF over the velocity field in high-dimensional fluid flows [82]. The inverse Frobenius–Perron problem (IFPP) [19] formulates the control problem for the stationary PDF as a perturbation problem, which aims to find a perturbed transition matrix and the associated perturbed dynamics that give rise to the target density. The Markov transition matrix approximating the Perron–Frobenius operator can also be interpreted as a graph, where the nodes correspond to the symbols associated with the discrete states and the vertex weights are the transition probabilities facilitating the application of graph-theoretic methods for optimal or minimal-path searches, e.g., in the context of IFPP [20]. Stabilization problems have been successfully solved via the Lyapunov measure equation [148, 189], where the problem can be formulated as a linear program, extended for stochastic systems [40], and applied to high-dimensional fluid flows [187]. A different perspective is assumed in [60], where a convex optimization problem is formulated to determine local perturbations in the context of optimal mixing. In particular, the stochastic kernel is designed as a perturbation to the deterministic kernel of the Perron–Frobenius operator, which controls the diffusion and thus the mixing properties of the underlying system. A similar problem has been studied based on the infinitesimal generator [62], but resulting in a convex quadratic program. More recent work [7] considers the perturbation of the stochastic part of the dynamical system and provides a solution in closed form.

The success of MPC has also driven its use for controlling probability distributions. A nonlinear MPC problem using set-oriented representations [132] suffers, however, from the curse of dimensionality. Thus, the work has been extended using Monte Carlo and particle filter methods to estimate the PDF, which is referred to as particle MPC [133]. In particle MPC, the Liouville equation enters the MPC formulation as a constraint to control the mean and variance of the state. It is worth noting that data-driven PDF prediction using neural nets is increasingly explored due to faster evaluation/prediction of the PDF in contrast to numerically integrating sample points. The loss metric, i.e., the error between the true and estimated PDF from the neural net, can be computed using automatic differentiation of neural nets [128], building on physics informed neural net via PDE constraints [149], e.g., the Liouville equation.

### 8.5.5 Sensor and Actuator Placement

Determining suitable locations of sensors and actuators for data collection and decision-making is a crucial task in any real-world application. Generally, sensor and actuator placement may refer to selecting spatial locations or specific variables, which are used synonymously here. Operator-theoretic placement provides a significant opportunity, as they capture global properties, such as meta-stable states, associated with persistent dynamics, and they generalize to nonlinear systems, e.g., for estimating nonlinear observability and controllability. Selecting optimal sensor and actuator locations amounts to a combinatorially hard problem. Compressed sensing [34, 46], sparsity-promoting algorithms employing an $l_1$-penalty term, and sparse sampling techniques such as gappy POD [57] have played an increasingly important role in the context of sensor/actuator selection [109].

In particular, the underlying low-rank structure of the data can be exploited for efficient sensing. Thus, compressed sensing and sparsity-promoting algorithms have also been increasingly combined with POD [13] and DMD [12, 33, 75, 144]. Optimized sensor locations can be tailored to a specific cluster-based reduced-order model (CROM) [80], which provides a coarse approximation of the Perron–Frobenius operator for PDEs [81]. Here, the sparse sensor optimization for classification (SSPOC) [27] algorithm or alternatively QR factorization (see, e.g., discrete empirical interpolation methods (DEIM) [48]) is applied to the dominant POD modes to learn sensor locations that are able to classify the full high-dimensional state into the correct cluster. The probabilistic dynamics given by the model are preserved if the resulting sensing matrix satisfies the restricted isometry property from compressed sensing. Recently [110], sensor placement has been explored for multiscale systems using multi-resolution DMD and DEIM, where sensor locations are tailored to intermittent or transient mode activity. The control-oriented framework DMD with control [146] has been unified with compressive DMD [33] for compressive system identification [12] to identify low-order models from limited input–output data and enable reconstruction of the high-dimensional, interpretable state. This framework, and similarly eDMDc, provides opportunities for optimized placement of sensors and actuators using linear systems techniques. A greedy submodular approach was proposed in [172] based on the Koopman observer form and a mutual information or observability based criterion.

Sensor placement and actuator placement can also be performed by leveraging the generator PDEs of the Perron–Frobenius and Koopman families. Controllability and observability Gramians can be generalized for nonlinear systems based on the (controlled) Liouville and adjoint Liouville equation and subsequently used for sensor and actuator placement by maximizing the support (or the $l_2$ norm) of the finite-time Gramians [164, 165, 190]. The approach utilizes set-oriented methods, where sensors/actuators are placed in certain cells, and the location can be optimized by solving a convex optimization problem [164]. A greedy heuristic approach based on these ideas using set-oriented methods is proposed in [58], which further investigates different criteria such as maximizing the sensing volume (sensor coverage), response time, and accuracy (relative measure transported to the sensor in finite time) and incorporating spatial constraints. The framework has been further extended to incorporate uncertainty [162]. Balanced truncation has recently been used for efficient placement of sensors and actuators to simultaneously maximize the controllability and observability Gramians [111], which may be promising for Koopman-based Gramians.

## 8.6 Conclusions and Future Directions

In this chapter, we have explored a number of applications of the Koopman and Perron Frobenius operators for the control of nonlinear systems. Although operator-theoretic analysis has a long history in dynamical systems and control, there has been considerable renewed interest in the past decade. Much of this recent progress has been driven by the increased availability of data, along with advanced machine learning and optimization algorithms. These data-driven approaches are providing approximations to the Koopman and Perron Frobenius operators that are then used for control. We have introduced a general control formulation and discussed how this may benefit from a coordinate transformation where nonlinear dynamics become approximately linear. Considerations such as stability, controllability and observability, controller and observer design, and sensor/actuator placement were discussed in the context of operator-theoretic approaches, highlighting recent theoretical and applied advances. We also discussed a number of important embedding strategies, such as extended DMD, delay coordinates, and eigenfunctions, which may all be used for these control objectives. Finally, we provided an example that leverages these various embeddings for model predictive control, showing that MPC is remarkably robust to model uncertainty.

Despite the tremendous recent progress in operator-based control, there are a number of limitations that must be addressed in the future. Obtaining accurate representations of the Koopman operator from data is a key enabler of model-based control. However, identifying a coordinate system that is closed under the Koopman operator is notoriously difficult, as the eigenfunctions may have arbitrarily complex representations. Fortunately, emerging techniques in deep learning are providing powerful representations for these linearizing transformations [100, 107, 112, 134, 179, 194, 197]. Another key challenge is the existence of a continuous eigenvalue spectrum in chaotic dynamical systems, which complicates matters considerably. Recent methods in delay coordinates [9, 29, 176, 178] and tailored neural network architectures [107] are making headway, although this is far from resolved. However, at this point, these advanced embedding techniques have not been adapted for control design, which is an exciting avenue of ongoing work.

Although much of the focus in data-driven Koopman and Perron–Frobenius theory has been on obtaining increasingly accurate approximate embeddings, it may be that improved models have marginal benefit in control design. Certainly improved models are useful for prediction and estimation, but advanced control techniques, such as model predictive control, are remarkably robust to lackluster models. In many cases, it may be that a simple DMDc model may work nearly as well as a sophisticated model based on the latest embedding techniques. However, future operator-based controllers will need to be certified, requiring guarantees on the model fidelity and a rigorous quantification of errors and uncertainties. Importantly, the model uncertainty is intimately related to the quality and quantity of the training data, the richness of the dynamics sampled, and the basis used for regression. Future efforts will undoubtedly continue to incorporate known or partially known dynamics, structure, and constraints in the data-driven regressions, improving both the models and the uncertainty bounds.

The current state of the art in operator-theoretic control, with a compelling set of success stories and a long list of future work, makes it likely that this field will continue to grow and develop for decades to come. It is inevitable that these efforts will increasingly intersect with the fields of machine learning, optimization, and control. Advances in machine learning and sparse optimization will continue to drive innovations, both for model discovery and for sensor placement, which are closely related to Koopman and Perron Frobenius frameworks. Likewise, concrete successes in the control of nonlinear systems will continue to motivate these advanced operator-based approaches.

**Acknowledgements** EK gratefully acknowledges support by the "Washington Research Foundation Fund for Innovation in Data-Intensive Discovery" and a Data Science Environments project award from the Gordon and Betty Moore Foundation (Award #2013-10-29) and the Alfred P. Sloan Foundation (Award #3835) to the University of Washington eScience Institute, and funding through the Mistletoe Foundation. SLB and JNK acknowledge support from the Defense Advanced Research Projects Agency (DARPA contract PA-18-01-FP-125). SLB acknowledges support from the Army Research Office (W911NF-17-1-0306 and W911NF-17-1-0422). JNK acknowledges support from the Air Force Office of Scientific Research (FA9550-19-1-0011).

## References

1. Abraham, I., De La Torre, G., Murphey, T.D.: Model-based control using Koopman operators (Neuroscience and Robotics Lab (NxR Lab)). https://vimeo.com/219458009
2. Abraham, I., De La Torre, G., Murphey, T.D.: Model-based control using Koopman operators (2017). arXiv preprint arXiv:1709.01568
3. Akaike, H.: Fitting autoregressive models for prediction. Ann. Inst. Stat. Math. **21**(1), 243–247 (1969)
4. Allgöwer, F., Findeisen, R., Nagy, Z.K.: Nonlinear model predictive control: from theory to application. J. Chin. Inst. Chem. Eng. **35**(3), 299–315 (2004)
5. Alpcan, T., Mehta, P.G., Basar, T., Vaidya, U.: Control of non-equilibrium dynamics in communication networks. In: 2006 45th IEEE Conference on Decision and Control, pp. 5216–5221. IEEE (2006)
6. Andrade, R.F.S.: Carleman embedding and Lyapunov exponents. J. Math. Phys. **23**(12), 2271–2275 (1982)
7. Antown, F., Dragičević, D., Froyland, G.: Optimal linear responses for Markov chains and stochastically perturbed dynamical systems (2018). arXiv preprint arXiv:1801.03234

8. Arbabi, H., Korda, M., Mezić, I.: A data-driven Koopman model predictive control framework for nonlinear flows (2018). arXiv preprint arXiv:1804.05291
9. Arbabi, H., Mezić, I.: Ergodic theory, dynamic mode decomposition and computation of spectral properties of the Koopman operator. SIAM J. Appl. Dyn. Syst. **16**(4), 2096–2126 (2017)
10. Arulampalam, M.S., Maskell, S., Gordon, N., Clapp, T.: A tutorial on particle filters for online nonlinear/non-Gaussian Bayesian tracking. IEEE Trans. Signal Process. **50**(2), 174–188 (2002)
11. Åström, K.J.: Optimal control of Markov processes with incomplete state information. J. Math. Anal. Appl. **10**(1), 174–205 (1965)
12. Bai, Z., Kaiser, E., Proctor, J.L., Kutz, J.N., Brunton, S.L.: Dynamic mode decomposition for compressive system identification (2017). arXiv preprint arXiv:1710.07737
13. Bai, Z., Wimalajeewa, T., Berger, Z., Wang, G., Glauser, M., Varshney, P.K.: Low-dimensional approach for reconstruction of airfoil data via compressive sensing. AIAA J. **53**(4), 920–933 (2014)
14. Bellman, R., Kalaba, R.: Selected Papers on Mathematical Trends in Control Theory. Dover, New York (1964)
15. Bellman, R., Richardson, J.M.: On some questions arising in the approximate solution of nonlinear differential equations. Q. Appl. Math. **20**(4), 333–339 (1963)
16. Bellman, R.E.: Dynamic Programming. Princeton University Press, Princeton (1957)
17. Benosman, M., Mansour, H., Huroyan, V.: Koopman-operator observer-based estimation of pedestrian crowd flows. IFAC-PapersOnLine **50**(1), 14028–14033 (2017)
18. Bertsekas, D.P.: Dynamic Programming and Optimal Control, vol. I. Athena Scientific, Nashua (2005)
19. Bollt, E.M.: Controlling chaos and the inverse Frobenius-Perron problem: global stabilization of arbitrary invariant measures. Int. J. Bifurc. Chaos **10**(05), 1033–1050 (2000)
20. Bollt, E.M.: Combinatorial control of global dynamics in a chaotic differential equation. Int. J. Bifurc. Chaos **11**(08), 2145–2162 (2001)
21. Bollt, E.M., Santitissadeekorn, N.: Applied and Computational Measurable Dynamics. SIAM, Philadelphia (2013)
22. Bongard, J., Lipson, H.: Automated reverse engineering of nonlinear dynamical systems. Proc. Natl. Acad. Sci. **104**(24), 9943–9948 (2007)
23. Brockett, R.: Notes on the control of the Liouville equation. In: Control of partial Differential Equations, pp. 101–129. Springer, Berlin (2012)
24. Brockett, R.W.: Nonlinear systems and differential geometry. Proc. IEEE **64**(1), 61–72 (1976)
25. Brockett, R.W.: Volterra series and geometric control theory. Automatica **12**(2), 167–176 (1976)
26. Brockett, R.W.: Optimal control of the Liouville equation. AMS IP Stud. Adv. Math. **39**, 23 (2007)
27. Brunton, B.W., Brunton, S.L., Proctor, J.L., Kutz, J.N.: Sparse sensor placement optimization for classification. AMS IP Stud. Adv. Math. **76**(5), 2099–2122 (2016)
28. Brunton, B.W., Johnson, L.A., Ojemann, J.G., Kutz, J.N.: Extracting spatial-temporal coherent patterns in large-scale neural recordings using dynamic mode decomposition. J. Neurosci. Methods **258**, 1–15 (2016)
29. Brunton, S.L., Brunton, B.W., Proctor, J.L., Kaiser, E., Kutz, J.N.: Chaos as an intermittently forced linear system. Nat. Commun. **8**(19), 1–9 (2017)
30. Brunton, S.L., Brunton, B.W., Proctor, J.L., Kutz, J.N.: Koopman invariant subspaces and finite linear representations of nonlinear dynamical systems for control. PLoS ONE **11**(2), e0150,171 (2016)
31. Brunton, S.L., Proctor, J.L., Kutz, J.N.: Discovering governing equations from data by sparse identification of nonlinear dynamical systems. Proc. Natl. Acad. Sci **113**(15), 3932–3937 (2016)
32. Brunton, S.L., Proctor, J.L., Kutz, J.N.: Sparse identification of nonlinear dynamics with control (SINDYc). IFAC NOLCOS **49**(18), 710–715 (2016)

33. Brunton, S.L., Proctor, J.L., Tu, J.H., Kutz, J.N.: Compressed sensing and dynamic mode decomposition. J. Comput. Dyn. **2**(2), 165–191 (2015)
34. Candès, E.J., Romberg, J., Tao, T.: Robust uncertainty principles: exact signal reconstruction from highly incomplete frequency information. IEEE Trans. Inf. Theory **52**(2), 489–509 (2006)
35. Carleman, T.: Application de la théorie des équations intégrales linéaires aux systèmes d'équations différentielles non linéaires. Acta Mathematica **59**(1), 63–87 (1932)
36. Chaves, M., Sontag, E.D.: State-estimators for chemical reaction networks of Feinberg-Horn-Jackson zero deficiency type. Eur. J. Control **4**(8), 343–359 (2002)
37. Clautier, J.R., D'Souza, N., Mracek, C.P.: Nonlinear regulation and nonlinear $H^{\infty}$ control via the state-dependent Riccati equation technique: Part 1, theory. In: Proceedings of the International Conference on Nonlinear Problems in Aviation and Aerospace, pp. 117–131 (1996)
38. Cvitanović, P., Artuso, R., Mainieri, R., Tanner, G., Vattay, G.: Chaos: Classical and Quantum. Niels Bohr Institute, Copenhagen (2012). http://ChaosBook.org
39. Das, A.K., Huang, B., Vaidya, U.: Data-driven optimal control using Perron-Frobenius operator (2018). arXiv preprint arXiv:1806.03649
40. Das, A.K., Raghunathan, A.U., Vaidya, U.: Transfer operator-based approach for optimal stabilization of stochastic systems. In: American Control Conference (ACC), 2017, pp. 1759–1764. IEEE (2017)
41. Daum, F.: Non-particle filters. In: Signal and Data Processing of Small Targets 2006, vol. 6236, p. 623614. International Society for Optics and Photonics (2006)
42. Dellnitz, M., Froyland, G., Junge, O.: The algorithms behind GAIO – Set-oriented numerical methods for dynamical systems. Ergodic Theory. Analysis, and Efficient Simulation of Dynamical Systems, pp. 145–174. Springer, Berlin (2001)
43. Dellnitz, M., Junge, O.: On the approximation of complicated dynamical behavior. SIAM J. Numer. Anal. **36**(2), 491–515 (1999)
44. Dellnitz, M., Froyland, G., Sertl, S.: On the isolated spectrum of the Perron-Frobenius operator. Nonlinearity **13**(4), 1171 (2000)
45. Dicle, C., Mansour, H., Tian, D., Benosman, M., Vetro, A.: Robust low rank dynamic mode decomposition for compressed domain crowd and traffic flow analysis. In: IEEE International Conference on Multimedia and Expo (ICME), pp. 1–6. IEEE (2016)
46. Donoho, D.L.: Compressed sensing. IEEE Trans. Inf. Theory **52**(4), 1289–1306 (2006)
47. Draeger, A., Engell, S., Ranke, H.: Model predictive control using neural networks. IEEE Control Syst. Mag. **15**(5), 61–66 (1995)
48. Drmac, Z., Gugercin, S.: A new selection operator for the discrete empirical interpolation method-improved a priori error bound and extensions. SIAM J. Sci. Comput. **38**(2), A631–A648 (2016)
49. Dullerud, G.E., Paganini, F.: A Course in Robust Control Theory: A Convex Approach. Texts in Applied Mathematics. Springer, Berlin (2000)
50. Dutta, P., Bhattacharya, R.: Hypersonic state estimation using the Frobenius-Perron operator. J. Guid., Control., Dyn. **34**(2), 325–344 (2011)
51. Dutta, P., Halder, A., Bhattacharya, R.: Uncertainty quantification for stochastic nonlinear systems using Perron-Frobenius operator and Karhunen-Loève expansion. In: 2012 IEEE International Conference on Control Applications (CCA), pp. 1449–1454. IEEE (2012)
52. Dutta, P., Halder, A., Bhattacharya, R.: Nonlinear filtering with transfer operator. In: American Control Conference (ACC), 2013, pp. 3069–3074. IEEE (2013)
53. Dutta, P., Halder, A., Bhattacharya, R.: Nonlinear estimation with Perron-Frobenius operator and Karhunen-Loève expansion. IEEE Trans. Aerosp. Electron. Syst. **51**(4), 3210–3225 (2015)
54. Elamvazhuthi, K., Grover, P.: Optimal transport over nonlinear systems via infinitesimal generators on graphs (2016). arXiv preprint arXiv:1612.01193
55. Elliott, D.: Bilinear Control Systems: Matrices in Action, vol. 169. Springer Science & Business Media, Berlin (2009)

56. Eren, U., Prach, A., Koçer, B.B., Raković, S.V., Kayacan, E., Açıkmeşe, B.: Model predictive control in aerospace systems: Current state and opportunities. J. Guid., Control., Dyn. (2017)
57. Everson, R., Sirovich, L.: Karhunen-Loève procedure for gappy data. JOSA A **12**(8), 1657–1664 (1995)
58. Fontanini, A.D., Vaidya, U., Ganapathysubramanian, B.: A methodology for optimal placement of sensors in enclosed environments: a dynamical systems approach. Build. Environ. **100**, 145–161 (2016)
59. Forbes, M., Guay, M., Forbes, J.: Control design for first-order processes: shaping the probability density of the process state. J. Process Control **14**(4), 399–410 (2004)
60. Froyland, G., González-Tokman, C., Watson, T.M.: Optimal mixing enhancement by local perturbation. SIAM Rev. **58**(3), 494–513 (2016)
61. Froyland, G., Padberg, K.: Almost-invariant sets and invariant manifolds - connecting probabilistic and geometric descriptions of coherent structures in flows. Phys. D **238**, 1507–1523 (2009)
62. Froyland, G., Santitissadeekorn, N.: Optimal mixing enhancement (2016). arXiv preprint arXiv:1610.01651
63. Garcia, C.E., Prett, D.M., Morari, M.: Model predictive control: theory and practice - a survey. Automatica **25**(3), 335–348 (1989)
64. Gaspard, P.: Chaos, Scattering and Statistical Mechanics, vol. 9. Cambridge University Press, Cambridge (2005)
65. Gaspard, P., Nicolis, G., Provata, A.: Spectral signature of the pitchfork bifurcation: Liouville equation approach. Phys. Rev. E **51**(1), 74–94 (1995). https://doi.org/10.1103/PhysRevE.51.74
66. Goswami, D., Paley, D.A.: Global bilinearization and controllability of control-affine nonlinear systems: a Koopman spectral approach. In: 2017 IEEE 56th Annual Conference on Decision and Control (CDC), pp. 6107–6112. IEEE (2017)
67. Goswami, D., Thackray, E., Paley, D.A.: Constrained Ulam dynamic mode decomposition: approximation of the Perron-Frobenius operator for deterministic and stochastic systems. IEEE Control Syst. Lett. **2**(4), 809–814 (2018)
68. Guo, L., Wang, H.: Generalized discrete-time PI control of output PDFs using square root B-spline expansion. Automatica **41**(1), 159–162 (2005)
69. Halder, A., Bhattacharya, R.: Geodesic density tracking with applications to data driven modeling. In: American Control Conference (ACC), 2014, pp. 616–621. IEEE (2014)
70. Hanke, S., Peitz, S., Wallscheid, O., Klus, S., Böcker, J., Dellnitz, M.: Koopman operator based finite-set model predictive control for electrical drives (2018). arXiv preprint arXiv:1804.00854
71. Hemati, M., Yao, H.: Dynamic mode shaping for fluid flow control: New strategies for transient growth suppression. In: 8th AIAA Theoretical Fluid Mechanics Conference, p. 3160 (2017)
72. Hermann, R., Krener, A.: Nonlinear controllability and observability. IEEE Trans. Autom. Control **22**(5), 728–740 (1977)
73. Huang, B., Ma, X., Vaidya, U.: Data-driven nonlinear stabilization using Koopman operator (2019). arXiv preprint arXiv:1901.07678
74. Huang, B., Vaidya, U.: Data-driven approximation of transfer operators: naturally structured dynamic mode decomposition (2017). arXiv preprint arXiv:1709.06203 (2017)
75. Jovanović, M.R., Schmid, P.J., Nichols, J.W.: Sparsity-promoting dynamic mode decomposition. Phys. Fluids **26**(2), 024,103 (2014)
76. Juang, J.N., Pappa, R.S.: An eigensystem realization algorithm for modal parameter identification and model reduction. J. Guid., Control., Dyn. **8**(5), 620–627 (1985)
77. Kaiser, E., Kutz, J.N., Brunton, S.L.: Data-driven discovery of Koopman eigenfunctions for control (2017). arXiv Preprint arXiv:1707.01146
78. Kaiser, E., Kutz, J.N., Brunton, S.L.: Discovering conservation laws from data for control. In: 2018 IEEE Conference on Decision and Control (CDC), pp. 6415–6421. IEEE (2018)
79. Kaiser, E., Kutz, J.N., Brunton, S.L.: Sparse identification of nonlinear dynamics for model predictive control in the low-data limit. Proc. R. Soc. Lond. A **474**(2219), (2018)

80. Kaiser, E., Morzyński, M., Daviller, G., Kutz, J.N., Brunton, B.W., Brunton, S.L.: Sparsity enabled cluster reduced-order models for control. J. Comput. Phys. **352**, 388–409 (2018)
81. Kaiser, E., Noack, B.R., Cordier, L., Spohn, A., Segond, M., Abel, M., Daviller, G., Osth, J., Krajnovic, S., Niven, R.K.: Cluster-based reduced-order modelling of a mixing layer. J. Fluid Mech. **754**, 365–414 (2014)
82. Kaiser, E., Noack, B.R., Spohn, A., Cattafesta, L.N., Morzyński, M.: Cluster-based control of a separating flow over a smoothly contoured ramp. Theor. Comput. Fluid Dyn. **31**(5–6), 579–593 (2017)
83. Kantorovich, L.V.: On the translocation of masses. Dokl. Akad. Nauk. USSR (NS) **37**, 199–201 (1942)
84. Kawahara, Y.: Dynamic mode decomposition with reproducing kernels for Koopman spectral analysis. In: Advances in Neural Information Processing Systems, pp. 911–919 (2016)
85. Khalil, H.K.: Noninear Systems. Prentice-Hall, New Jersey (1996)
86. Klus, S., Nüske, F., Koltai, P., Wu, H., Kevrekidis, I., Schütte, C., Noé, F.: Data-driven model reduction and transfer operator approximation (2017). arXiv:1703.10112
87. Ko, J., Klein, D.J., Fox, D., Haehnel, D.: Gaussian processes and reinforcement learning for identification and control of an autonomous blimp. In: 2007 IEEE International Conference on Robotics and Automation, pp. 742–747. IEEE (2007)
88. Kocijan, J., Murray-Smith, R., Rasmussen, C.E., Girard, A.: Gaussian process model based predictive control. In: American Control Conference, 2004. Proceedings of the 2004, vol. 3, pp. 2214–2219. IEEE (2004)
89. Koopman, B.O.: Hamiltonian systems and transformation in Hilbert space. Proc. Natl. Acad. Sci. **17**(5), 315–318 (1931)
90. Koopman, B.O., Neumann, J.V.: Dynamical systems of continuous spectra. Proc. Natl. Acad. Sci. **18**(3), 255 (1932)
91. Korda, M., Mezić, I.: Linear predictors for nonlinear dynamical systems: Koopman operator meets model predictive control (2016). arXiv:1611.03537
92. Korda, M., Susuki, Y., Mezić, I.: Power grid transient stabilization using Koopman model predictive control (2018). arXiv preprint arXiv:1803.10744
93. Kowalski, K., Steeb, W.H.: Nonlinear Dynamical Systems and Carleman Linearization. World Scientific, Singapore (1991)
94. Kus, M.: Integrals of motion for the Lorenz system. J. Phys. A: Math. Gen. **16**(18), L689 (1983)
95. Kwee, I., Schmidhuber, J.: Optimal control using the transport equation: the Liouville machine. Adapt. Behav. **9**(2), 105–118 (2001)
96. Lall, S., Marsden, J.E., Glavaški, S.: A subspace approach to balanced truncation for model reduction of nonlinear control systems. Int. J. Robust Nonlinear Control.: IFAC-Affil. J. **12**(6), 519–535 (2002)
97. Lan, Y., Mezić, I.: Linearization in the large of nonlinear systems and Koopman operator spectrum. Phys. D **242**(1), 42–53 (2013)
98. Lasota, A., Mackey, M.C.: Chaos, Fractals, and Noise: Stochastic Aspects of Dynamics, vol. 97. Springer Science & Business Media, Berlin (2013)
99. Li, J.S., Khaneja, N.: Ensemble control of Bloch equations. IEEE Trans. Autom. Control **54**(3), 528–536 (2009)
100. Li, Q., Dietrich, F., Bollt, E.M., Kevrekidis, I.G.: Extended dynamic mode decomposition with dictionary learning: A data-driven adaptive spectral decomposition of the Koopman operator. Chaos **27**, 103,111 (2017)
101. Ling, E., Ratliff, L., Coogan, S.: Koopman operator approach for instability detection and mitigation in signalized traffic. In: 21st International Conference on Intelligent Transportation Systems (ITSC), pp. 1297–1302. IEEE (2018)
102. Liouville, J.: Note sur la théorie de la variation des constantes arbitraires. Journal de mathématiques pures et appliquées, 342–349 (1838)
103. Lippmann, R.: An introduction to computing with neural nets. IEEE Assp Mag. **4**(2), 4–22 (1987)

104. Ljung, L.: Approaches to identification of nonlinear systems. In: Proceedings of the 29th Chinese Control Conference, Beijing. IEEE (2010)
105. Ljung, L.: Perspectives on system identification. Ann. Rev. Control **34**(1), 1–12 (2010)
106. Loiseau, J.C., Noack, B.R., Brunton, S.L.: Sparse reduced-order modelling: sensor-based dynamics to full-state estimation. J. Fluid Mech. **844**, 459–490 (2018)
107. Lusch, B., Kutz, J.N., Brunton, S.L.: Deep learning for universal linear embeddings of nonlinear dynamics (2017). arXiv preprint arXiv:1712.09707
108. Maner, B.R., Doyle, F.J., Ogunnaike, B.A., Pearson, R.K.: A nonlinear model predictive control scheme using second order Volterra models. In: American Control Conference, 1994, vol. 3, pp. 3253–3257 (1994)
109. Manohar, K., Brunton, B.W., Kutz, J.N., Brunton, S.L.: Data-driven sparse sensor placement. IEEE Control Syst. Mag. **38**(3), 63–86 (2018)
110. Manohar, K., Kaiser, E., Brunton, S.L., Kutz, J.N.: Optimized sampling for multiscale dynamics. SIAM Multiscale Model. Simul. **17**(1), 117–136 (2019)
111. Manohar, K., Kutz, J.N., Brunton, S.L.: Optimal sensor and actuator placement using balanced model reduction (2018). arXiv preprint arXiv:1812.01574
112. Mardt, A., Pasquali, L., Wu, H., Noé, F.: Vampnets: Deep learning of molecular kinetics (2017). arXiv preprint arXiv:1710.06012
113. Mathew, G., Mezić, I.: Metrics for ergodicity and design of ergodic dynamics for multi-agent systems. Phys. D: Nonlinear Phenom. **240**(4–5), 432–442 (2011)
114. Mauroy, A., Goncalves, J.: Koopman-based lifting techniques for nonlinear systems identification (2017). arXiv preprint arXiv:1709.02003
115. Mauroy, A., Mezić, I.: A spectral operator-theoretic framework for global stability. In: 2013 IEEE 52nd Annual Conference on Decision and Control (CDC), pp. 5234–5239. IEEE (2013)
116. Mauroy, A., Mezić, I.: Global stability analysis using the eigenfunctions of the Koopman operator. IEEE Trans. Autom. Control **61**, 3356–3369 (2016)
117. Mauroy, A., Mezić, I., Moehlis, J.: Isostables, isochrons, and Koopman spectrum for the action-angle representation of stable fixed point dynamics. Phys. D: Nonlinear Phenom. **261**, 19–30 (2013)
118. Mayne, D.Q.: Model predictive control: recent developments and future promise. Automatica **50**(12), 2967–2986 (2014)
119. Mehta, P.G., Vaidya, U., Banaszuk, A.: Markov chains, entropy, and fundamental limitations in nonlinear stabilization. IEEE Trans. Autom. Control **53**(3), 784–791 (2008)
120. Mezić, I.: Spectral properties of dynamical systems, model reduction and decompositions. Nonlinear Dyn. **41**(1–3), 309–325 (2005)
121. Mezić, I.: Analysis of fluid flows via spectral properties of the Koopman operator. Annu. Rev. Fluid Mech. **45**(1), 357–378 (2013). https://doi.org/10.1146/annurev-fluid-011212-140652
122. Mezić, I.: Spectral operator methods in dynamical systems: Theory and applications. Draft manuscript UCSB (2017)
123. Mitrophanov, A.Y.: Stability and exponential convergence of continuous-time Markov chains. J. Appl. Probab. **40**(4), 970–979 (2003)
124. Mitrophanov, A.Y.: Sensitivity and convergence of uniformly ergodic Markov chains. J. Appl. Probab. **42**(4), 1003–1014 (2005)
125. Monge, G.: Mémoire sur la théorie des déblais et des remblais. Histoire de l'Académie Royale des Sciences de Paris (1781)
126. Morari, M., Lee, J.H.: Model predictive control: past, present and future. Comput. Chem. Eng. **23**(4), 667–682 (1999)
127. Murray-Smith, R., Johansen, T.: Multiple Model Approaches to Nonlinear Modelling and Control. CRC Press, Boca Raton (1997)
128. Nakamura-Zimmerer, T., Venturi, D., Gong, Q.: Data-driven computational optimal control for nonlinear systems under uncertainty. In: SIAM Annual Meeting 2018. Portland, Oregon
129. Nelles, O.: Nonlinear System Identification: From Classical Approaches to Neural Networks and Fuzzy Models. Springer, Berlin (2013)

130. Nicolis, G.: Introduction to Nonlinear Science. Cambridge University Press, Cambridge (1995)
131. Noé, F., Nuske, F.: A variational approach to modeling slow processes in stochastic dynamical systems. Multiscale Model Simul. **11**(2), 635–655 (2013)
132. Ohsumi, K., Ohtsuka, T.: Nonlinear receding horizon control of probability density functions. IFAC Proc. **43**(14), 735–740 (2010)
133. Ohsumi, K., Ohtsuka, T.: Particle model predictive control for probability density functions. IFAC Proc. **44**(1), 7993–7998 (2011)
134. Otto, S.E., Rowley, C.W.: Linearly-recurrent autoencoder networks for learning dynamics (2017). arXiv preprint arXiv:1712.01378
135. Ovsyannikov, D., Ovsyannikov, A., Vorogushin, M., Svistunov, Y.A., Durkin, A.: Beam dynamics optimization: models, methods and applications. Nucl. Instrum. Methods Phys. Res. Sect. A: Accel., Spectrometers, Detect. Assoc. Equip. **558**(1), 11–19 (2006)
136. Parrilo, P.A.: Structured semidefinite programs and semialgebraic geometry methods in robustness and optimization. Ph.D. thesis, California Institute of Technology, Pasadena, CA (2000)
137. Pearson, J.D.: Approximation methods in optimal control i. Sub-optimal control. I. J. Electron. **13**(5), 453–469 (1962)
138. Peitz, S., Klus, S.: Koopman operator-based model reduction for switched-system control of PDEs (2017). arXiv preprint arXiv:1710.06759
139. Peitz, S., Klus, S.: Feedback control of nonlinear PDEs using data-efficient reduced order models based on the Koopman operator (2018). arXiv preprint arXiv:1806.09898
140. Phelps, C., Gong, Q., Royset, J.O., Walton, C., Kaminer, I.: Consistent approximation of a nonlinear optimal control problem with uncertain parameters. Automatica **50**(12), 2987–2997 (2014)
141. Phelps, C., Royset, J.O., Gong, Q.: Optimal control of uncertain systems using sample average approximations. SIAM J. Control Optim. **54**(1), 1–29 (2016)
142. Pontryagin, L., Boltyanskii, V., Gamkrelidze, R., Mishchenko, E.: The Mathematical Theory of Optimal Processes. Interscience, Woburn (1962)
143. Principe, J.C., Wang, L., Motter, M.A.: Local dynamic modeling with self-organizing maps and applications to nonlinear system identification and control. Proc. IEEE **86**(11), 2240–2258 (1998)
144. Proctor, J.L., Brunton, S.L., Brunton, B.W., Kutz, J.N.: Exploiting sparsity and equation-free architectures in complex systems (invited review). Eur. Phys. J. Spec. Top. **223**(13), 2665–2684 (2014)
145. Proctor, J.L., Brunton, S.L., Kutz, J.N.: Dynamic mode decomposition with control. SIAM J. Appl. Dyn. Sys. **15**(1), 142–161 (2016)
146. Proctor, J.L., Brunton, S.L., Kutz, J.N.: Generalizing Koopman theory to allow for inputs and control. SIAM J. Appl. Dyn. Syst. **17**(1), 909–930 (2018)
147. Propoi, A.: Problems of the optimal control of mixed states. Avtomatika i Telemekhanika **3**, 87–98 (1994)
148. Raghunathan, A., Vaidya, U.: Optimal stabilization using Lyapunov measures. IEEE Trans. Autom. Control **59**(5), 1316–1321 (2014)
149. Raissi, M., Perdikaris, P., Karniadakis, G.E.: Physics informed deep learning (part i): data-driven solutions of nonlinear partial differential equations (2017). arXiv preprint arXiv:1711.10561
150. Rajaram, R., Vaidya, U., Fardad, M., Ganapathysubramanian, B.: Stability in the almost everywhere sense: a linear transfer operator approach. J. Math. Anal. Appl. **368**(1), 144–156 (2010)
151. Rantzer, A.: A dual to Lyapunov's stability theorem. Syst. Control. Lett. **42**, 161–168 (2001)
152. Reif, K., Gunther, S., Yaz, E., Unbehauen, R.: Stochastic stability of the continuous-time extended Kalman filter. IEE Proc. Control. Theory Appl. **147**(1), 45–52 (2000)
153. Ristic, B., Arulampalam, S., Gordon, N.: Beyond the Kalman Filter: Particle Filters for Tracking Applications. Artech House (2003)

154. Ross, I.M., Karpenko, M., Proulx, R.J.: Path constraints in tychastic and unscented optimal control: Theory, application and experimental results. In: American Control Conference (ACC), 2016, pp. 2918–2923. IEEE (2016)
155. Roy, S., Borzì, A.: Numerical investigation of a class of Liouville control problems. J. Sci. Comput. **73**(1), 178–202 (2017)
156. Runolfsson, T., Lin, C.: Computation of uncertainty distributions in complex dynamical systems. In: American Control Conference, 2009. ACC'09., pp. 2458–2463. IEEE (2009)
157. Sahai, T., Mathew, G., Surana, A.: A chaotic dynamical system that paints and samples. IFAC-PapersOnLine **50**(1), 10760–10765 (2017)
158. Scherpen, J.: Balancing for nonlinear systems. Syst. Control. Lett. **21**(2), 143–153 (1993)
159. Schmid, P.J.: Application of the dynamic mode decomposition to experimental data. Exper. Fluids **50**(4), 1123–1130 (2011)
160. Schmid, P.J., Li, L., Juniper, M., Pust, O.: Applications of the dynamic mode decomposition. Theor. Comput. Fluid Dyn. **25**(1–4), 249–259 (2011)
161. Schmidt, M., Lipson, H.: Distilling free-form natural laws from experimental data. Science **324**(5923), 81–85 (2009)
162. Sharma, H., Fontanini, A.D., Vaidya, U., Ganapathysubramanian, B.: Transfer operator theoretic framework for monitoring building indoor environment in uncertain operating conditions (2018). arXiv preprint arXiv:1807:04781
163. Simon, N., Tibshirani, R.: Standardization and the group lasso penalty. Statistica Sinica **22**(3), 983 (2012)
164. Sinha, S., Vaidya, U., Rajaram, R.: Optimal placement of actuators and sensors for control of nonequilibrium dynamics. In: 2013 European Control Conference (ECC), July 17–19, 2013, Zrich, Switzerland (2013)
165. Sinha, S., Vaidya, U., Rajaram, R.: Operator theoretic framework for optimal placement of sensors and actuators for control of nonequilibrium dynamics. J. Math. Anal. Appl. **440**(2), 750–772 (2016)
166. Sjöberg, J., Zhang, Q., Ljung, L., Benveniste, A., Delyon, B., Glorennec, P.Y., Hjalmarsson, H., Juditsky, A.: Nonlinear black-box modeling in system identification: a unified overview. Automatica **31**(12), 1691–1724 (1995)
167. Skogestad, S., Postlethwaite, I.: Multivariable Feedback Control: Analysis and Design, 2nd edn. Wiley, Inc., New Jersey (2005)
168. Snyder, C., Bengtsson, T., Bickel, P., Anderson, J.: Obstacles to high-dimensional particle filtering. Mon. Weather Rev. **136**(12), 4629–4640 (2008)
169. Sootla, A., Mauroy, A., Ernst, D.: An optimal control formulation of pulse-based control using Koopman operator. Automatica **91**, 217–224 (2018)
170. Sootla, A., Mauroy, A., Goncalves, J.: Shaping pulses to control bistable monotone systems using Koopman operator. IFAC-PapersOnLine **49**(18), 698–703 (2016)
171. Stengel, R.F.: Optimal Control and Estimation. Courier Corporation (2012)
172. Surana, A.: Koopman Operator Based Nonlinear Systems and Control Framework. MoDyL Kickoff, Santa Barbara (2016)
173. Surana, A.: Koopman operator based observer synthesis for control-affine nonlinear systems. In: 55th IEEE Conference on Decision and Control (CDC), pp. 6492–6499 (2016)
174. Surana, A., Banaszuk, A.: Linear observer synthesis for nonlinear systems using Koopman operator framework. IFAC-PapersOnLine **49**(18), 716–723 (2016)
175. Surana, A., Williams, M.O., Morari, M., Banaszuk, A.: Koopman operator framework for constrained state estimation. In: 2017 IEEE 56th Annual Conference on Decision and Control (CDC), pp. 94–101. IEEE (2017)
176. Susuki, Y., Mezić, I.: A prony approximation of Koopman mode decomposition. IEEE Conference on Decision and Control (CDC), pp. 7022–7027 (2015)
177. Susuki, Y., Mezić, I., Raak, F., Hikihara, T.: Applied Koopman operator theory for power systems technology. Nonlinear Theory Appl. IEICE **7**(4), 430–459 (2016)
178. Susuki, Y., Sako, K., Hikihara, T.: On the spectral equivalence of Koopman operators through delay embedding (2017). arXiv preprint arXiv:1706.01006

179. Takeishi, N., Kawahara, Y., Yairi, T.: Learning Koopman invariant subspaces for dynamic mode decomposition. In: Advances in Neural Information Processing Systems, pp. 1130–1140 (2017)
180. Takens, F.: Detecting strange attractors in turbulence. Lect. Notes Math. **898**, 366–381 (1981)
181. Tang, S., Kumar, V.: Autonomous flying. Annu. Rev. Control., Robot., Auton. Syst. **1**(6), 1–24 (2018)
182. Vaidya, U.: Duality in stability theory: Lyapunov function and Lyapunov measure. In: 44th Allerton Conference on Communication, Control and Computing, pp. 185–190 (2006)
183. Vaidya, U.: Converse theorem for almost everywhere stability using Lyapunov measure. In: American Control Conference, 2007. ACC'07, pp. 4835–4840. IEEE (2007)
184. Vaidya, U.: Observability Gramian for nonlinear systems. In: 2007 46th IEEE Conference on Decision and Control, pp. 3357–3362. IEEE (2007)
185. Vaidya, U.: Stochastic stability analysis of discrete-time system using Lyapunov measure. In: American Control Conference (ACC), 2015, pp. 4646–4651. IEEE (2015)
186. Vaidya, U., Chinde, V.: Computation of the Lyapunov measure for almost everywhere stochastic stability. In: 2015 IEEE 54th Annual Conference on Decision and Control (CDC), pp. 7042–7047. IEEE (2015)
187. Vaidya, U., Ganapathysubramanian, B., Raghunathan, A.: Transfer operator method for control in fluid flows. In: Proceedings of the 48th IEEE Conference on Decision and Control, 2009 held jointly with the 2009 28th Chinese Control Conference. CDC/CCC 2009, pp. 1806–1811. IEEE (2009)
188. Vaidya, U., Mehta, P.G.: Lyapunov measure for almost everywhere stability. IEEE Trans. Autom. Control **53**(1), 307–323 (2008)
189. Vaidya, U., Mehta, P.G., Shanbhag, U.V.: Nonlinear stabilization via control Lyapunov measure. IEEE Trans. Autom. Control **55**(6), 1314–1328 (2010)
190. Vaidya, U., Rajaram, R., Dasgupta, S.: Actuator and sensor placement in linear advection PDE with building system application. J. Math. Anal. Appl. **394**(1), 213–224 (2012)
191. Wang, H., Baki, H., Kabore, P.: Control of bounded dynamic stochastic distributions using square root models: an applicability study in papermaking systems. Trans. Inst. Meas. Control. **23**(1), 51–68 (2001)
192. Wang, K., Vaidya, U.: Transfer operator approach for computing domain of attraction. In: 2010 49th IEEE Conference on Decision and Control (CDC), pp. 5390–5395. IEEE (2010)
193. Wang, T., Gao, H., Qiu, J.: A combined adaptive neural network and nonlinear model predictive control for multirate networked industrial process control. IEEE Trans. Neural Netw. Learn. Syst. **27**(2), 416–425 (2016)
194. Wehmeyer, C., Noé, F.: Time-lagged autoencoders: deep learning of slow collective variables for molecular kinetics. J. Chem. Phys. **148**(24), 241,703 (2018)
195. Williams, M.O., Hemati, M.S., Dawson, S.T.M., Kevrekidis, I.G., Rowley, C.W.: Extending data-driven Koopman analysis to actuated systems. IFAC-PapersOnLine **49**(18), 704–709 (2016)
196. Williams, M.O., Kevrekidis, I.G., Rowley, C.W.: A data-driven approximation of the Koopman operator: extending dynamic mode decomposition. J. Nonlinear Sci. (2015)
197. Yeung, E., Kundu, S., Hodas, N.: Learning deep neural network representations for Koopman operators of nonlinear dynamical systems (2017). arXiv preprint arXiv:1708.06850
198. Yeung, E., Liu, Z., Hodas, N.O.: A Koopman operator approach for computing and balancing gramians for discrete time nonlinear systems. In: 2018 Annual American Control Conference (ACC), pp. 337–344. IEEE (2018)
199. Yu, S., Mehta, P.G.: Fundamental performance limitations via entropy estimates with hidden Markov models. In: 2007 46th IEEE Conference on Decision and Control, pp. 3982–3988. IEEE (2007)
200. Zhang, L., Lam, J.: On H2 model reduction of bilinear systems. Automatica **38**(2), 205–216 (2002)

201. Zhu, C., Zhu, W., Yang, Y.: Design of feedback control of a nonlinear stochastic system for targeting a pre-specified stationary probability distribution. Probab. Eng. Mech. **30**, 20–26 (2012)
202. Zwillinger, D.: Handbook of Differential Equations. Academic Press, San Diego (1989)

# Chapter 9
# Koopman Model Predictive Control of Nonlinear Dynamical Systems

**Milan Korda and Igor Mezić**

**Abstract** This chapter presents a class of linear predictors for nonlinear controlled dynamical systems. The basic idea is to lift (or embed) the nonlinear dynamics into a higher dimensional space where its evolution is approximately linear. This is achieved by extending the Koopman operator framework to controlled dynamical systems and applying the extended dynamic mode decomposition (EDMD) with a particular choice of basis functions leading to a predictor in the form of a finite-dimensional linear controlled dynamical system. In numerical examples, the linear predictors obtained in this way exhibit a performance superior to existing linear predictors such as those based on local linearization or the so-called Carleman linearization. Importantly, the procedure to construct these linear predictors is completely data-driven and extremely simple—it boils down to a nonlinear transformation of the data (the lifting) and a linear least-squares problem in the lifted space that can be readily solved for large datasets. These linear predictors can be readily used to design controllers for the nonlinear dynamical system using linear controller design methodologies. We focus in particular on model predictive control (MPC) and show that MPC controllers designed in this way enjoy computational complexity of the underlying optimization problem comparable to that of MPC for a linear dynamical system with the same number of control inputs and the same dimension of the state space. Importantly, linear inequality constraints on the state and control inputs as well as nonlinear constraints on the state can be imposed in a linear fashion in the proposed MPC scheme. Similarly, cost functions nonlinear in the state variable can be handled in a linear



M. Korda (✉)
LAAS-CNRS, 7 Avenue du Colonel Roche, 31400 Toulouse, France

Faculty of Electrical Engineering, Czech Technical University in Prague, Technická 4, 16206 Prague, Czech Republic
e-mail: korda@laas.fr

I. Mezić
Department of Mechanical Engineering, University of California Santa Barbara, Santa Barbara, CA, USA
e-mail: mezic@engineering.ucsb.edu



A. Mauroy et al. (eds.), *The Koopman Operator in Systems and Control*,
Lecture Notes in Control and Information Sciences 484,
https://doi.org/10.1007/978-3-030-35713-9_9

fashion. We treat the full-state measurement case as well as the input–output case and demonstrate the approach with numerical examples.

## 9.1 Problem Description

This chapter uses the concept of nonlinear *embedding* (or lifting) for control of *nonlinear* dynamical systems using *linear* control techniques. The embedded system in the form of a linear controlled dynamical system is obtained as a finite-dimensional approximation of the Koopman operator generalized to systems with control inputs. This linear controlled dynamical system, acting as a predictors for the original nonlinear system, is then used within a model predictive control (MPC) framework. See Fig. 9.1. The material of this chapter is based on [5].

To begin, consider the nonlinear controlled dynamical system

$$x^+ = f(x, u), \tag{9.1}$$

where $x \in \mathbb{R}^n$ is the state, $x^+ \in \mathbb{R}^n$ the successor state, and $u \in \mathbb{R}^m$ the control input. The aim is to construct a *linear predictor* of the form

$$z^+ = Az + Bu, \tag{9.2a}$$

$$\hat{x} = Cz, \tag{9.2b}$$

$$z_0 = \Psi(x_0), \tag{9.2c}$$

where

$$\Psi = [\psi_1, \ldots, \psi_N]^\top$$

is a nonlinear embedding mapping from the original state space $\mathbb{R}^n$ to a possible much larger dimensional space $\mathbb{R}^N$. In particular, the initial state $z_0$ of the predictor is given by $\Psi(x_0)$. The output of this linear predictor $\hat{x}$ is then a prediction of the true state $x$. The embedding mapping $\Psi$ and the predictor matrices $A$, $B$, $C$ need to be constructed in such a way that the predicted sequence

$$\hat{x}_0, \hat{x}_1, \ldots, \hat{x}_{N_{\text{pred}}}$$

is a good approximation of the true state sequence

$$x_0, x_1, \ldots, x_{N_{\text{pred}}}$$

for any initial condition $x_0$ of practical interest (e.g., satisfying all physical constraints on the state) and any admissible (i.e., satisfying all input constraints) control sequence $(u_0, u_1, \ldots, u_{N_{\text{pred}}-1})$. We emphasize that the predictions $\hat{x}_0, \hat{x}_1, \ldots \hat{x}_{N_{\text{pred}}}$ are obtained in an "open-loop" fashion, i.e., the embedding mapping $\Psi$ is applied only to the initial state $x_0$ and the predictions are obtained as the solution to (9.2), i.e.,

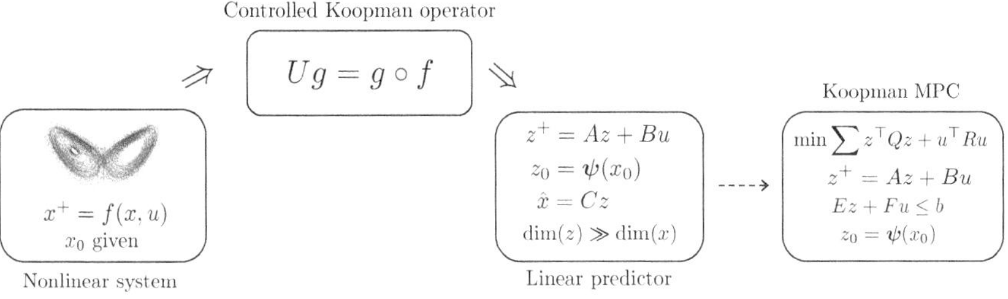


**Fig. 9.1** Linear predictor for a nonlinear controlled dynamical system—$z$ is the lifted state evolving on a higher dimensional state space, $\hat{x}$ is the prediction of the true state $x$, and $\psi$ is a nonlinear embedding mapping. This predictor is then used within a *linear* MPC framework for control. Adapted from Ref. [5], Copyright 2018, with permission from Elsevier

$$\hat{x}_k = CA^k z_0 + \sum_{i=0}^{k-1} CA^{k-i-1} Bu_i$$

for $k \in \{0, \ldots, N_{\text{pred}}\}$.

## 9.2 Koopman Operator for Controlled Systems

In this section, we generalize the Koopman operator to controlled systems and obtain the linear predictor (9.2) as a finite-dimensional approximation of this operator. In order to do so, we consider the extended dynamical system

$$\chi^+ = F(\chi) := \begin{bmatrix} f(x, u_0) \\ \mathscr{S}\big((u_i)_{i=0}^{\infty}\big) \end{bmatrix}, \tag{9.3}$$

where $\mathscr{S}$ is the left shift operator, i.e., $\mathscr{S}\big((u_i)_{i=0}^{\infty}\big) = (u_i)_{i=1}^{\infty}$ and

$$\chi = \begin{bmatrix} x \\ (u_i)_{i=0}^{\infty} \end{bmatrix}$$

is the extended state evolving on $\mathbb{R}^n \times \ell(\mathbb{R}^m)$, where $\ell(\mathbb{R}^m)$ is the space of all control sequences $(u_i)_{i=0}^{\infty}$.

For any observable $\phi : \mathbb{R}^n \times \ell((u_i)_{i=0}^{\infty}) \to \mathbb{R}$, the controlled Koopman operator $U$ is then defined by the composition relation

$$U\phi = \phi \circ F. \tag{9.4}$$

### 9.2.1 Approximation of the Controlled Koopman Operator

In this section, we apply the extended dynamic mode decomposition (EDMD) algorithm of [11] to the controlled Koopman operator (9.4). With a special choice of basis functions in EDMD, the resulting approximation of the Koopman operator will be in the form of the linear predictor (9.2).

Given a dataset

$$(\chi^j, \chi^{j+}), \quad \chi^{j+} = F(\chi^j), \quad j = 1, \ldots, M,$$

the EDMD algorithm constructs a finite-dimensional approximation of the Koopman operator by seeking a matrix $\mathscr{A}$ minimizing the one-step prediction error

$$\sum_{j=1}^{M} \|\Phi(\chi^{j+}) - \mathscr{A}\Phi(\chi^j)\|_2^2, \tag{9.5}$$

where

$$\Phi(\chi) = \left[\phi_1(\chi) \ldots \phi_{N_\phi}(\chi)\right]^\top$$

is a vector of possibly nonlinear basis functions $\phi_i : \mathbb{R}^n \times \ell(\mathbb{R}^m) \to \mathbb{R}, i \in \{1, \ldots, N_\phi\}$.

Note that in general, the optimization problem (9.5) involves infinite-dimensional quantities $(u_i)_{i=0}^\infty$ and therefore cannot be solved easily. In order to be able to solve this optimization problem tractably and in order to obtain a linear predictor in the form of (9.2), we impose a special form on the basis functions:

$$\phi_i(x, (u_i)_{i=0}^\infty) = \psi_i(x) + \mathscr{L}_i\big((u_i)_{i=0}^\infty\big), \tag{9.6}$$

where $\psi_i : \mathbb{R}^n \to \mathbb{R}$ is in general nonlinear but $\mathscr{L}_i : \ell(\mathbb{R}^m) \to \mathbb{R}$ is linear. Once this form of $\phi_i$'s is imposed, we can without loss of generality assume (by linearity of $\mathscr{L}_i$ and causality) $\Phi$ of the form

$$\Phi\big(x, (u_i)_{i=0}^\infty\big) = \begin{bmatrix} \Psi(x) \\ u_0 \end{bmatrix}, \tag{9.7}$$

where

$$\Psi = [\psi_1, \ldots, \psi_N]^\top$$

is a vector of possibly nonlinear basis functions (i.e., the embedding mapping) depending only on $x$. Denoting by $\bar{\mathscr{A}}$ the first $N$ rows of $\mathscr{A}$ and decomposing this matrix such that $\bar{\mathscr{A}} = [A, B]$ with $A \in \mathbb{R}^{N\times N}$, $B \in \mathbb{R}^{N\times m}$ and using the notation $\chi^j = (x^j, (u_i^j)_{i=0}^\infty)$ in (9.5), leads to the minimization problem

$$\min_{A,B} \sum_{j=1}^{M} \|\boldsymbol{\psi}(x^{j+}) - A\boldsymbol{\psi}(x^j) - B\boldsymbol{u}_0^j\|_2^2. \tag{9.8}$$

Minimizing (9.8) over $A$ and $B$ leads to the predictor of the form (9.2).

The matrix $C$ is obtained as the least-squares projection of the state $x$ onto the span of $\{\psi_1, \dots, \psi_N\}$ by solving

$$\min_{C} \sum_{j=1}^{M} \|x^j - C\Psi(x^j)\|_2^2. \tag{9.9}$$

Problems (9.8) and (9.9) are linear least-squares problem that can be readily solved using linear algebra.

*Remark 9.1* Note that the solution to (9.9) is trivial if the set of embedding functions $\{\psi_1, \dots, \psi_N\}$ contains the state observable, i.e., if, after possible reordering, $\psi_i(x) = x_i$ for all $i \in \{1, \dots, n\}$. In this case, the solution to (9.9) is $C = [I, 0]$, where $I$ is the identity matrix of size $n$.

### 9.2.2 Numerical Algorithm—Finding A, B, C

This section summarizes the EDMD algorithm applied to the controlled Koopman operator. Assume that a set of data

$$\mathbf{X} = \left[x^1, \dots, x^M\right], \ \mathbf{Y} = \left[y^1, \dots, y^M\right], \mathbf{U} = \left[u^1, \dots, u^M\right] \tag{9.10}$$

satisfying $y^i = f(x^i, u^i)$ is available. No temporal ordering of the data—in particular, the data is not required to lie on a single trajectory of (16.1).

Given the data $\mathbf{X}, \mathbf{Y}, \mathbf{U}$ in (9.10), the matrices $A \in \mathbb{R}^{N\times N}$ and $B \in \mathbb{R}^{N\times m}$ in (9.2) are obtained as the best linear one-step predictor in the embedding space in a least-squares sense by solving

$$\min_{A,B} \|\mathbf{Y}_{\text{lift}} - A\mathbf{X}_{\text{lift}} - B\mathbf{U}\|_F, \tag{9.11}$$

where

$$\mathbf{X}_{\text{lift}} = \left[\Psi(x^1), \dots, \Psi(x^M)\right], \ \mathbf{Y}_{\text{lift}} = \left[\Psi(y^1), \dots, \Psi(y^M)\right], \tag{9.12}$$

with

$$\Psi(x) = [\psi_1(x), \dots, \psi_N(x)]^\top \tag{9.13}$$

being a given basis (or dictionary) of nonlinear functions (the embedding mapping). The symbol $\|\cdot\|_F$ denotes the Frobenius norm of a matrix. The matrix $C \in \mathbb{R}^{n\times N}$

is obtained by solving the problem of the best least-squares projection of $x$ onto the span of the embedding mapping, which, when expressed using the available data, leads to

$$\min_{C} \|\mathbf{X} - C\mathbf{X}_{\text{lift}}\|_F. \tag{9.14}$$

The analytical solutions to (9.11) and (9.14) are

$$[A, B] = \mathbf{Y}_{\text{lift}}[\mathbf{X}_{\text{lift}}, \mathbf{U}]^{\dagger} \tag{9.15}$$

and

$$C = \mathbf{X}\mathbf{X}_{\text{lift}}^{\dagger}. \tag{9.16}$$

### 9.2.3 *Practical Considerations*

The analytical solution (9.15) is not the preferred method of solving the least-squares problem (9.11) in practice. In particular for larger datasets with $M \gg N$, it is beneficial to solve instead the normal equations associated to (9.11). The normal equations read

$$\mathbf{V} = \mathscr{M}\mathbf{G} \tag{9.17}$$

with variable $\mathscr{M} = [A, B]$ and data

$$\mathbf{G} = \begin{bmatrix} \mathbf{X}_{\text{lift}} \\ \mathbf{U} \end{bmatrix} \begin{bmatrix} \mathbf{X}_{\text{lift}} \\ \mathbf{U} \end{bmatrix}^{\top}, \quad \mathbf{V} = \mathbf{Y}_{\text{lift}} \begin{bmatrix} \mathbf{X}_{\text{lift}} \\ \mathbf{U} \end{bmatrix}^{\top}.$$

Any solution to (9.17) is a solution to (9.11). Importantly, the size of the matrices $\mathbf{G}$ and $\mathbf{V}$ is $(N + m) \times (N + m)$ and $N \times (N + m)$, respectively, and hence independent of the number of samples $M$ in the dataset (9.10). The same considerations hold for the least-squares problem (9.14). Note that, in practice, the embedding functions $\psi_i$ will typically contain the state itself in which case the solution to (9.14) is just the selection of appropriate indices of $\mathbf{X}_{\text{lift}}$, i.e., after possible reordering, $C = [I, 0]$.

If embedding to a very high-dimensional space is required, it may be possible to utilize the so-called kernel methods known from machine learning which do not require an explicit evaluation of the embedding mapping $\Psi$. These methods were successfully applied to the standard EDMD algorithm in [12], leading to substantial computational savings.

## 9.3 Koopman MPC

Model predictive control (MPC) is an optimization-based control framework where a user-specified cost function (expressing the control objective) is minimized over a *finite* prediction horizon subject to constraints on the control inputs and states (expressing, e.g., the physical limitations of the control setup). The decision variables in the optimization problems are the predicted control inputs and states along the prediction horizon. Once this optimization problem is solved, only the first computed control input is applied to the system and the entire procedure is repeated in a recursive fashion at each time step of the closed-loop operation, giving rise to the so-called receding horizon implementation. For nonlinear dynamical systems, the optimization problem solved at each time step is typically a non-convex optimization problem, which makes the solution extremely computationally costly, thereby typically resorting to local optimization methods.

The Koopman MPC, on the other hand, replaces the nonlinear dynamics by the *linear* predictor (9.2) and therefore, provided that all non-convex functions appearing in the objective and constraints are including in the embedding mapping $\Psi$, the resulting optimization problem is a *convex quadratic program* (QP) for which extremely efficient tailored solvers exist (e.g., [3]). This dramatically reduces the computational cost and thereby allows for deployment of Koopman MPC in real-world, real-time applications requiring extremely fast sampling rates.

We first describe the Koopman MPC problem and its computational properties; after that, in Sect. 9.3.2, we describe how a classical nonlinear MPC problem can be transformed to the Koopman MPC. The Koopman MPC solves at each time step of the closed-loop operation the optimization problem

$$\begin{array}{lll} \underset{u_i, z_i}{\text{minimize}} & J\left((u_i)_{i=0}^{N_p-1}, (z_i)_{i=0}^{N_p}\right) & \\ \text{subject to} & z_{i+1} = Az_i + Bu_i, & i = 0, \ldots, N_p - 1 \\ & E_i z_i + F_i u_i \leq b_i, & i = 0, \ldots, N_p - 1 \\ & E_{N_p} z_{N_p} \leq b_{N_p} & \\ \text{parameter} & z_0 = \Psi(x_{\text{current}}), & \end{array} \tag{9.18}$$

where $N_p$ is the prediction horizon and the *convex* quadratic cost function $J$ is given by

$$\begin{aligned} J\left((u_i)_{i=0}^{N_p-1}, (z_i)_{i=0}^{N_p}\right) &= z_{N_p}^\top Q_{N_p} z_{N_p} + q_{N_p}^\top z_{N_p} \\ &+ \sum_{i=0}^{N_p-1} z_i^\top Q_i z_i + u_i^\top R_i u_i + q_i^\top z_i + r_i^\top u_i \end{aligned}$$

with $Q_i \in \mathbb{R}^{N\times N}$ and $R_i \in \mathbb{R}^{m\times m}$ positive semidefinite. The matrices $E_i \in \mathbb{R}^{n_c\times N}$ and $F_i \in \mathbb{R}^{n_c\times m}$ and the vector $b_i \in \mathbb{R}^{n_c}$ define state and input polyhedral constraints.

The optimization problem (MPC) is parametrized by the current state of the nonlinear dynamical system $x_{\text{current}}$. This optimization problem defines a feedback controller

$$\kappa(x) = u_0^\star(x),$$

where $u_0^\star(x)$ denotes an optimal solution to problem (MPC) with $x_{\text{current}} = x$. The optimization problem (MPC) is a convex QP.

### 9.3.1 Eliminating Dependence on the Embedding Dimension

In this section, we show that the computational complexity of solving the MPC problem (MPC) can be rendered independent of the dimension of the lifted state $N$. This is achieved by transforming (MPC) in the so-called *dense form*

$$\begin{array}{ll}\underset{\mathbf{u}\in\mathbb{R}^{mN_p}}{\text{minimize}} & \mathbf{u}^\top H\mathbf{u}^\top + h^\top\mathbf{u} + z_0^\top G\mathbf{u} \\ \text{subject to} & L\mathbf{u} + Mz_0 \le c \\ \text{parameter} & z_0 = \Psi(x_k),\end{array} \tag{9.19}$$

for some positive semidefinite matrix $H \in \mathbb{R}^{mN_p\times mN_p}$ and some matrices and vectors $h \in \mathbb{R}^{mNp}$, $G \in \mathbb{R}^{N\times mNp}$, $L \in \mathbb{R}^{n_cN_p\times mNp}$, $M \in \mathbb{R}^{n_cN_p\times N}$ and $c \in \mathbb{R}^{n_cN_p}$. The optimization is over the vector of predicted control inputs

$$\mathbf{u} = [u_0^\top, u_1^\top, \ldots, u_{N_p-1}^\top]^\top$$

This "dense" formulation can be readily derived from the "sparse" formulation (MPC) by solving explicitly for $z_i$'s and concatenating the point-wise-in-time stage costs and constraints; see the Appendix for explicit expressions for the data matrices of (9.19) in terms of those of (MPC).

Similarly to (MPC), the optimization problem (9.19) is a convex QP. Crucially, however, the size of the Hessian $H$ or the number of the constraints $n_c$ in the dense formulation (9.19) is *independent* of the size of the lift $N$. Hence, once the nonlinear mapping $z_0 = \Psi(x_k)$ is evaluated, the computational cost of solving (9.19) is comparable to solving a standard *linear* MPC on the same prediction horizon, with the same number of control inputs and with the dimension of the state space equal to $n$ rather than $N$. This comes from the fact that the cost of solving an MPC problem in a dense form is independent of the dimension of the state space, once the data matrices in (9.19) are formed. Importantly, these matrices are *fixed* and *precomputed offline* before deploying the controller (with the exception of the inexpensive matrix–vector multiplication $z_0^\top G$). This is in contrast with other nonlinear MPC schemes where these matrices have to be recomputed at each time step of the closed-loop operation, thereby greatly increasing the computational cost.

The closed-loop operation of the Koopman MPC can be summarized by the following algorithm, where

$$\mathbf{u}^\star = [u_0^{\star\top}, u_1^{\star\top}, \ldots, u_{N_p-1}^{\star\top}]^\top$$

denotes the optimal solution to (9.19).

**Algorithm 9.1** Koopman MPC—closed-loop operation. From Ref. [5], Copyright 2018, with permission from Elsevier.

1: **for** $k = 0, 1, \ldots$ **do**
2: Compute $z_0 = \Psi(x_k)$
3: Solve (9.19) to get an optimal solution $\mathbf{u}^\star$
4: Set $u_k = u_0^\star$
5: $x_{k+1} = f(x_k, u_k)$ [apply $u_k$ to the system (16.1)]
6: **end for**

## 9.3.2 *Transforming NMPC to Koopman MPC*

Here we describe in detail how a traditional nonlinear MPC problem translates[1] to the proposed MPC (MPC). We assume a nonlinear MPC problem which at each time step of the closed-loop operation solves the optimization problem

$$\begin{array}{ll} \underset{u_i, \overline{x}_i}{\text{minimize}} & l_{N_p}(x_{N_p}) + \sum_{i=0}^{N_p-1} l_i(\overline{x}_i) + u_i^\top \overline{R}_i u_i + \overline{r}_i^\top u_i \\ \text{subject to} & \overline{x}_{i+1} = f(\overline{x}_i, u_i), \qquad i = 0, \ldots, N_p - 1 \\ & c_{x_i}(\overline{x}_i) + c_{u_i}^\top u_i \le 0, \qquad i = 0, \ldots, N_p - 1 \\ & c_{x_{N_p}}(\overline{x}_{N_p}) \le 0 \\ & \overline{x}_0 = x_{\text{current}}, \end{array} \tag{9.20}$$

where the notation $\overline{x}$ is used to distinguish the predicted state $\overline{x}$, used only within the optimization problem (9.20), from the true measured state $x$ of the dynamical system (16.1). Notice that the true nonlinear dynamics $x^+ = f(x, u)$ appears as a constraint of (9.20); in addition, the functions $l_i$ and $c_{x_i}$ can be nonlinear and hence the optimization problem (9.20) is in general non-convex and extremely hard to solve to global optimality.

In order to translate (9.20) to the proposed form (MPC), we assume that a predictor of the form (9.2) with matrices $A$ and $B$ has been constructed as described in Sect. 9.2.2, using an embedding mapping $\Psi = [\psi_1, \ldots, \psi_N]^\top$. The matrices $A$, $B$ appear in the first constraint of (MPC) and the embedding mapping $\Psi$ is used for initialization $z_0 = \Psi(x_{\text{current}})$. In order to obtain the remaining data matrices of (MPC), we assume without loss of generality that $\psi_i(x) = l_i(x)$, $i = 0, \ldots, N_p$ and $\psi_{N_p+i}(x) = c_{x_i}(x)$, $i = 0, \ldots, N_p$. With this assumption, the remaining data is given by $Q_i = 0$, $R_i = \overline{R}_i$, $r = \overline{r}_i$, $q_i = [0_{1\times i}, 1, 0_{1\times N-1-i}]$, $E_i = [0_{1\times N_p+i}, 1, 0_{1\times N-N_p-1-i}]$, $F_i = c_{u_i}^\top$, $b_i = 0$, where $0_{i\times j}$ denotes the matrix of zeros of size $i \times j$. Note that this derivation assumed that the constraint functions $c_{x_i}$ and $c_{u_i}$ are scalar valued; for vector-valued constraint functions, the approach is analogous, setting the embedding functions $\psi_i$ equal to the individual components of the constraint functions $c_{x_i}$.

[1]By translate, we do not mean to rewrite in an equivalent form. The problem (MPC) is of course only an approximation to (9.20) since linear predictor is not exact unless the input-to-state mapping of the nonlinear dynamical system (16.1) is linear for all fixed initial conditions.

We also note that this canonical approach always leads to a linear cost function in (MPC). However, in special cases, when some of the cost functions $l_i(x_i)$ are convex quadratic, one may want to use the freedom of the formulation (MPC) and instead of setting $\psi_i = l_i$, use the quadratic terms in the cost function of (MPC), thereby reducing the dimension of the lift. See Sect. 9.5.2 for an example.

## 9.4 Extensions

In this section, we describe extensions of the proposed approach to input–output dynamical systems and to systems with disturbances/noise.

### *9.4.1 Input–Output Dynamical Systems*

Here we describe how the approach can be generalized to the case when full-state measurements are not available, but rather only certain output is measured. To this end, consider the dynamical system

$$\begin{aligned} x^+ &= f(x, u), \\ y &= h(x), \end{aligned} \tag{9.21}$$

where $y$ is the measured output and $h : \mathbb{R}^n \to \mathbb{R}^{n_h}$.

#### 9.4.1.1 Dynamics (9.21) Is Known

If the dynamics (9.21) is known, one can construct a predictor of the form (9.2) by applying the algorithm of Sect. 9.2.2 to data obtained from simulation of the dynamical system (9.21). In closed-loop operation, the predictor (9.2) is then used in conjunction with a state estimator for the dynamical system (9.21). Alternatively, one can design, using linear observer design methodologies, a state estimator directly for the linear predictor (9.2). Interestingly, doing so obviates the need to evaluate the embedding mapping $z = \Psi(x)$ in closed loop, as the lifted state is directly estimated. This idea is closely related to the use of the Koopman operator for state estimation proposed for the first time in [10].

#### 9.4.1.2 Dynamics (9.21) Is Not Known

If the dynamics (9.21) is not known and only input–output data is available, one could construct a predictor of the form (9.2) by taking as embedding functions only functions of the output $y$, i.e., $\psi_i(x) = \phi_i(h(x))$. This, however, would be extremely restrictive as this severely restricts the class of embedding functions available. Indeed, if, for example, $h(x) = x_1$, only functions of the first component are available. However, this problem can be circumvented by utilizing the fact that subsequent measurements, $h(x)$, $h(f(x))$, $h(f(f(x)))$, etc., are available and therefore we can define observables depending not only on $h$ but also on repeated composition of $h$ with $f$. In practice, this corresponds to having the embedding functions that depend not only on the current measured output but also on several previous measured outputs (and inputs, in the controlled setting); this is a classical practice in system identification theory (see, e.g., [6]).

Assume therefore that we are given a collection of data

$$\tilde{\mathbf{X}} = [\boldsymbol{\zeta}^1, \ldots, \boldsymbol{\zeta}^M],\ \ \tilde{\mathbf{Y}} = [\boldsymbol{\zeta}^{1+}, \ldots, \boldsymbol{\zeta}^{M+}],\ \ \tilde{\mathbf{U}} = [u^1, \ldots, u^M]$$

where

$$\begin{aligned} \boldsymbol{\zeta}^i &= \left[y_{i,n_d}^\top\ \overline{u}_{i,n_d-1}^\top\ y_{i,n_d-1}^\top \cdots \overline{u}_{i,0}^\top\ y_{i,0}^\top\right]^\top \in \mathbb{R}^{(n_d+1)n_h+n_d m} \\ \boldsymbol{\zeta}^{i+} &= \left[y_{i,n_d+1}^\top\ \overline{u}_{i,n_d}^\top\ y_{i,n_d}^\top \cdots \overline{u}_{i,1}^\top\ y_{i,1}^\top\right]^\top \in \mathbb{R}^{(n_d+1)n_h+n_d m} \\ u_i &= \overline{u}_{i,n_d} \end{aligned} \tag{9.22}$$

with $(y_{i,j})_{j=0}^{n_d+1}$ being a vector of consecutive output measurements generated by $(\overline{u}_{i,j})_{j=0}^{n_d}$ consecutive inputs. We note that there does not need to be any temporal relation between $\boldsymbol{\zeta}_i$ and $\boldsymbol{\zeta}_{i+1}$. If, however, $\boldsymbol{\zeta}_i$ and $\boldsymbol{\zeta}_{i+1}$ are in fact successors, then the matrices $\tilde{\mathbf{X}}$ and $\tilde{\mathbf{Y}}$ have the familiar (quasi-) Hankel structure known from system identification theory.

Computation of a linear predictor then proceeds in the same way as for the full-state measurement: We embed the collected data and look for the best one-step ahead predictor in the embedded space. This leads to the optimization problem

$$\min_{A,B} \left\| \tilde{\mathbf{Y}}_{\text{lift}} - A\tilde{\mathbf{X}}_{\text{lift}} - B\tilde{\mathbf{U}} \right\|_F, \tag{9.23}$$

where

$$\tilde{\mathbf{X}}_{\text{lift}} = \left[\Psi(\boldsymbol{\zeta}^1),\ \ldots, \Psi(\boldsymbol{\zeta}^M)\right],\ \tilde{\mathbf{Y}}_{\text{lift}} = \left[\Psi(\boldsymbol{\zeta}^{1+}), \ldots, \Psi(\boldsymbol{\zeta}^{M+})\right] \tag{9.24}$$

and

$$\Psi(\boldsymbol{\zeta}) = [\psi_1(\boldsymbol{\zeta}), \ldots, \psi_N(\boldsymbol{\zeta})]^\top$$

is a vector of real or complex valued, possibly nonlinear basis functions (i.e., the embedding). This leads to a linear predictor in the embedding space

$$z^+ = Az + Bu, \tag{9.25a}$$

$$\hat{y} = Cz, \tag{9.25b}$$
$$z_0 = \Psi(\boldsymbol{\zeta}_0), \tag{9.25c}$$

where the initial condition is given by the vector of $n_d + 1$ most recent output measurements and $n_d$ input measurements

$$\boldsymbol{\zeta}_0 = \left[y_0^\top \; \overline{u}_{-1}^\top \; y_{-1}^\top \; \ldots \; \overline{u}_{-n_d}^\top \; y_{-n_d}^\top\right]^\top$$

and where $\hat{y}$ is the prediction of $y$ with $C$ being the solution to

$$\min_C \left\| [y_{1,n_d}, \ldots, y_{M,n_d}] - C\tilde{\mathbf{X}}_{\text{lift}} \right\|_F. \tag{9.26}$$

We remark that the solution to (9.26) is trivial provided that the outputs and its delays are included among the embedding functions; see Remark 9.1.

*Remark 9.2* (*Closed-loop operation*) The closed-loop operation of the resulting MPC controller follows the steps of Algorithm 9.1, only the initialization $z_0 = \Psi(x_k)$ in line 2 is replaced by $z_0 = \Psi(\boldsymbol{\zeta}_k)$, where $\boldsymbol{\zeta}_k = [y_k^\top, u_{k-1}^\top, y_{k-1}^\top, \ldots, u_{k-n_d}^\top, y_{k-n_d}^\top]^\top$.

### 9.4.2 Disturbance/Noise Propagation

The approach can be readily extended to systems affected by a disturbance or noise of the form

$$x^+ = f(x, u, w), \tag{9.27}$$

where $w$ is the disturbance or process noise. The goal is to construct a predictor for (9.27) of the form

$$z^+ = Az + Bu + Dw, \tag{9.28a}$$
$$\hat{x} = Cz, \tag{9.28b}$$
$$z_0 = \Psi(x_0), \tag{9.28c}$$

where, as before, $\Psi$ is a nonlinear embedding mapping defined in (9.13). For example, if $w$ is a stochastic disturbance with known distribution, the linear predictor of the form (9.28) can be used to approximately compute the distribution of the state $x$ at a future time instance, given a sequence of control inputs up to that time.

In order to obtain the matrices of the predictor (9.28), we assume that we are given data of the form

$$\mathbf{X} = \left[x^1, \ldots, x^M\right], \; \mathbf{Y} = \left[y^1, \ldots, y^M\right], \tag{9.29a}$$

$$\mathbf{U} = \left[u^1, \ldots, u^M\right], \quad \mathbf{W} = \left[w^1, \ldots, w^M\right] \tag{9.29b}$$

satisfying $y_i = f(x_i, u_i, w_i)$ for all $i = 1, \ldots, M$. As in Sect. 9.2.2, the matrices are then obtained using least-squares regression:

$$\min_{A,B,D} \|\mathbf{Y}_{\text{lift}} - A\mathbf{X}_{\text{lift}} - B\mathbf{U} - D\mathbf{W}\|_F, \quad \min_{C} \|\mathbf{X} - C\mathbf{X}_{\text{lift}}\|_F. \tag{9.30}$$

If measurements of the disturbance $w$ are not available, then these must be best estimated from the available data, either using one of the nonlinear estimation techniques or using the Koopman operator-based estimator proposed in [9]. Alternatively, if the mapping $f$ is known (either analytically or in the form of an algorithm) and an algorithm to draw samples from the distribution of $w$ is available, one can obtain data (9.29) by simulation.

*Remark 9.3* If full-state measurements are not available, the approach can be readily combined with the approach of Sect. 9.4.1.1 or 9.4.1.2.

## 9.5 Numerical Examples

In this section, we compare the prediction accuracy of the linear embedding-based predictor (9.2) with several other predictors and demonstrate the use of the Koopman MPC proposed in Sect. 9.3 for feedback control of a bilinear model of a motor and of the Korteweg–de Vries nonlinear partial differential equation. The source code for the numerical examples is available online.[2]

### 9.5.1 Prediction Comparison

In order to evaluate the proposed predictor, we compare its prediction quality with that of several commonly used predictors. The system to compare the predictors on is the classical forced Van der Pol oscillator with dynamics given by

$$\begin{aligned} \dot{x}_1 &= 2x_2, \\ \dot{x}_2 &= -0.8x_1 + 2x_2 - 10x_1^2 x_2 + u. \end{aligned}$$

The predictors compared are

1. Predictor based on local linearization of the dynamics at the origin,
2. Predictor based on local linearization of the dynamics at a given initial condition $x_0$,
3. Carleman linearization predictor [1],
4. The proposed embedding-based predictor (9.2).

---

[2] https://github.com/MilanKorda/KoopmanMPC/raw/master/KoopmanMPC.zip.

In order to obtain the linear predictor (9.2), we discretize the dynamics using the Runge–Kutta four method with discretization period $T_s = 0.01$ s and simulate 200 trajectories over 1000 sampling periods (i.e., 10 s per trajectory). The control input for each trajectory is a random signal with uniform distribution over the interval $[-1, 1]$. The trajectories start from initial conditions generated randomly with uniform distribution on the unit box $[-1, 1]^2$. This data collection process results in the matrices $\mathbf{X}$ and $\mathbf{Y}$ of size $2 \times 2 \cdot 10^5$ and matrix $\mathbf{U}$ of size $1 \times 10^5$. The embedding functions $\psi_i$ are chosen to be the state itself (i.e., $\psi_1 = x_1$, $\psi_2 = x_2$) and 100 thin plate spline radial basis functions[3] with centers selected randomly with uniform distribution on the unit box. The dimension of the lifted state space is therefore $N = 102$.

The degree of the Carleman linearization is set to 14, resulting in the size of the Carleman linearization predictor of 120 (= the number of monomials of degree less than or equal to 14 in two variables). The $B$ matrix for the Carleman linearization predictor is set to $[0, 1, 0, \dots, 0]^\top$.

Figure 9.2 compares the predictions starting from two initial conditions $x_0^1 = [0.5, 0.5]^\top$, $x_0^2 = [-0.1, -0.5]^\top$ generated by a control signal $u(t)$ being a square wave with unit magnitude and period 0.3 s. Table 9.1 reports the relative root-mean-squared errors (RMSE)

$$\mathrm{RMSE} = 100 \cdot \frac{\sqrt{\sum_k \|x_{\mathrm{pred}}(kT_s) - x_{\mathrm{true}}(kT_s)\|_2^2}}{\sqrt{\sum_k \|x_{\mathrm{true}}(kT_s)\|_2^2}} \tag{9.31}$$

for each predictor averaged over 100 randomly sampled initial conditions with the same square wave forcing. We observe that the Koopman linear predictor is far superior to the remaining predictors. Finally, Table 9.2 reports the prediction accuracy of the Koopman predictor in terms of the average RMSE error as a function of the dimension of the lift $N$; we observe that, as expected, the prediction error decreases with increasing $N$, albeit not monotonously.

## 9.5.2 Feedback Control of a Bilinear Motor

In this section, we apply the proposed approach to the control of a bilinear model of a DC motor [2]. The model reads

$$\begin{aligned}
\dot{x}_1 &= -(R_a/L_a)x_1 - (k_m/L_a)x_2 u + u_a/L_a, \\
\dot{x}_2 &= -(B/J)x_2 + (k_m/J)x_1 u - \tau_l/J, \\
y &= x_2,
\end{aligned}$$

[3] A thin plate spline radial basis function with center at $x_0$ is defined by $\psi(x) = \|x - x_0\|^2 \log(\|x - x_0\|)$.

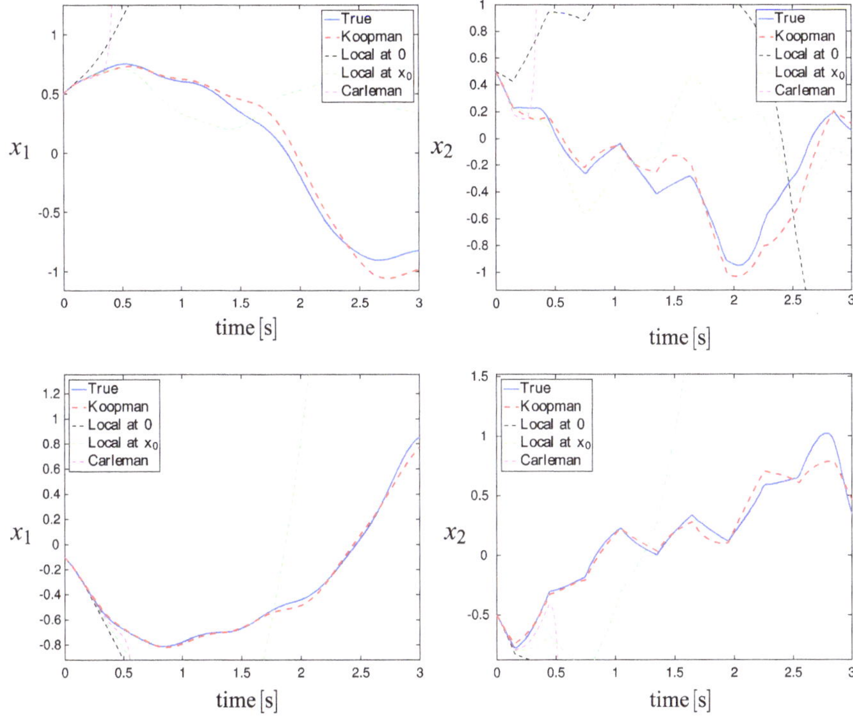


**Fig. 9.2** Prediction comparison for the forced Van der Pol oscillator. Top: initial condition $x_0 = [0.5, 0.5]^\top$. Bottom: initial condition $x_0 = [-0.1, -0.5]^\top$. The forcing $u(t)$ is in both cases a square wave with unit amplitude and period 0.3 s. Reprinted from Ref. [5], Copyright 2018, with permission from Elsevier

**Table 9.1** Prediction comparison—average RMSE (9.31) over 100 randomly sampled initial conditions: comparison among different predictors. From Ref. [5], Copyright 2018, with permission from Elsevier

| $x_0$ | Average RMSE (%) |
|---|---|
| Koopman | 24.4 |
| Local linearization at $x_0$ | $2.83 \cdot 10^3$ |
| Local linearization at 0 | 912.5 |
| Carleman | $5.08 \cdot 10^{22}$ |

where $x_1$ is the rotor current, $x_2$ the angular velocity, and the control input $u$ is the stator current and the output $y$ is the angular velocity. The parameters are $L_a = 0.314$, $R_a = 12.345$, $k_m = 0.253$, $J = 0.00441$, $B = 0.00732$, $\tau_l = 1.47$, $u_a = 60$. Notice in particular the bilinearity between the state and the control input. The physical constraints on the control input are $u \in [-4, 4]$, which we scale to $[-1, 1]$.

**Table 9.2** Prediction comparison—Koopman predictor—average prediction RMSE over 100 randomly sampled initial conditions as a function of the dimension of the lift. From Ref. [5], Copyright 2018, with permission from Elsevier

| $N$ | 5 | 10 | 25 | 50 | 75 | 100 |
|---|---|---|---|---|---|---|
| Average RMSE (%) | 66.5 | 44.9 | 47.0 | 38.7 | 30.6 | 24.4 |

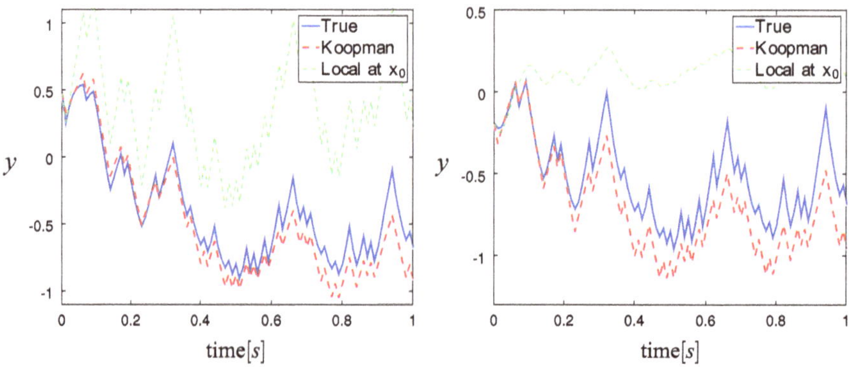


**Fig. 9.3** Feedback control of a bilinear motor—predictor comparison. Left: initial condition $x_0 = [0.887, 0.587]^\top$. Right: initial condition $x_0 = [-0.404, -0.126]^\top$. Reprinted from Ref. [5], Copyright 2018, with permission from Elsevier

The goal is to design an MPC controller based on Sect. 9.4.1.2, i.e., assuming only input–output data available and no explicit knowledge of the model. In order to obtain the Koopman predictor (9.25), we discretize the scaled dynamics using the Runge–Kutta four method with discretization period $T_s = 0.01$ s and simulate 200 trajectories over 1000 sampling periods (i.e., 10 s per trajectory). The control input for each trajectory is a random signal uniformly distributed on $[-1, 1]$. The trajectories start from initial conditions generated randomly with uniform distribution on the unit box $[-1, 1]^2$. We choose the number of delays $n_d = 1$. The embedding functions $\psi_i$ are chosen to be the time-delayed vector $\boldsymbol{\zeta} \in \mathbb{R}^3$, defined in (9.22), and 100 thin plate spline radial basis functions (see Footnote 4) with centers selected randomly with uniform distribution over $[-1, 1]^3$. The dimension of the lifted state space is therefore $N = 103$. First, in Fig. 9.3, we compare the output predictions for two different, randomly chosen, initial conditions against the predictor based on local linearization at a given initial condition. The prediction accuracy of the proposed predictor is superior, especially for longer prediction times. This is documented further in Table 9.3 by the relative root-mean-squared errors (9.31) over a one-second prediction horizon averaged over 100 randomly sampled initial conditions. Both in Fig. 9.3 and Table 9.3, the control signal was a pseudorandom binary signal generated anew for each initial condition.

**Table 9.3** Feedback control of a bilinear motor—prediction RMSE (9.31) for 100 randomly generated initial conditions. From Ref. [5], Copyright 2018, with permission from Elsevier

| | Koopman | Local linearization at $x_0$ |
|---|---|---|
| Average RMSE (%) | 32.3 | 135.5 |

The control objective is to track a given angular velocity reference $y_r$, which translates into the objective function minimized in the MPC problem

$$\begin{aligned} J = &\ (Cz_{N_p} - y_r)^\top Q_{N_p}(Cz_{N_p} - y_r) \\ &+ \sum_{i=0}^{N_p-1} (Cz_i - y_r)^\top Q(Cz_i - y_r) + u_i^\top R u_i \end{aligned} \tag{9.32}$$

with $C = [1, 0, \ldots, 0]$. This tracking objective function readily translates to the canonical form (MPC) by expanding the quadratic forms and neglecting constant terms. The cost function matrices were chosen as $Q = Q_{N_p} = 1$ and $R = 0.01$. The prediction horizon was set to one second, which results in $N_p = 100$. We compare the Koopman operator-based MPC controller (K-MPC) with an MPC controller based on local linearization (L-MPC) in two scenarios. In the first one, we do not impose any constraints on the output and track a piecewise constant reference. In the second one, we impose the constraint $y \in [-0.4, 0.4]$ and track a time-varying reference $y_r(t) = 0.5\cos(2\pi t/3)$, which violates the output constraint for some portion of the simulated period. The simulation results are shown in Fig. 9.4. We observe a virtually identical tracking performance in the first case. In the second case, however, the local linearization controller becomes infeasible and hence cannot complete the entire simulation period.[4] This infeasibility occurs due to the inaccurate predictions of the local linearization predictor over longer prediction horizons. The proposed K-MPC controller, on the other hand, does not run infeasible and completes the simulation period without violating the constraints.

Note that, even in the first scenario where the two controllers perform equally, the K-MPC controller has the benefit of being completely *data-driven* and requiring only *output measurements*, whereas the L-MPC controller requires a model (to compute the local linearization) and full-state measurements. In addition, the average computation time[5] required to evaluate the control input of the K-MPC controller was 6.86 ms (including the evaluation of the embedding mapping $\Psi(\zeta)$), as opposed to 103 ms for the L-MPC controller. This discrepancy is due to the fact that the local linearization and all data defining the underlying optimization problem that depend on it have to be recomputed at every iteration, which is costly on its own and also

[4]Infeasibility of the underlying optimization problem is a common problem encountered in predictive control with various heuristic (e.g., soft constraints) or theoretically substantiated (e.g., set invariance) approaches trying to address them. See, e.g., [4, 8] for more details.

[5]The optimization problems were solved by qpOASES [3] running on MATLAB and 2 GHz Intel Core i7 with 8 GB RAM.

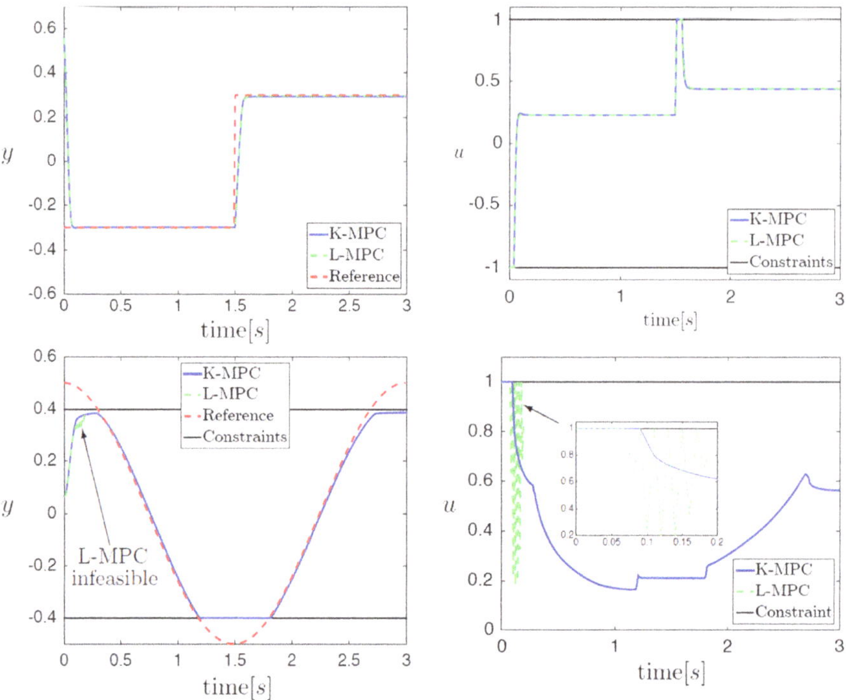


**Fig. 9.4** Feedback control of a bilinear motor—reference tracking. Top: piecewise constant reference, $x_0 = [0, 0.6]^\top$, no state constraints. Bottom: time-varying reference, $x_0 = [-0.1, 0.1]^\top$, constraints on the output imposed. Reprinted from Ref. [5], Copyright 2018, with permission from Elsevier

precludes efficient warm-starting; on the other hand, all data (except for the initial condition) of the underlying optimization problem of K-MPC are precomputed offline. In both cases, the computation times could be significantly reduced with a more efficient implementation. However, we believe, that the proposed approach would still be superior in terms of computational speed.

### *9.5.3 Nonlinear PDE Control*

In order to demonstrate the scalability and versatility of the approach, we use it to control the nonlinear Korteweg–de Vries (KdV) equation modeling the propagation of acoustic waves in a plasma or shallow-water waves [7]. The equation reads

$$\frac{\partial y(t,x)}{\partial t} + y(t,x)\frac{\partial y(t,x)}{\partial x} + \frac{\partial^3 y(t,x)}{\partial x^3} = u(t,x),$$

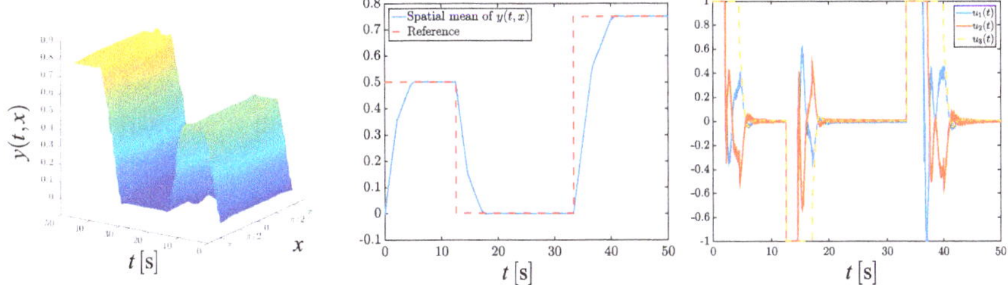


**Fig. 9.5** Nonlinear PDE control—Tracking of a time-varying constant-in-space reference profile for the Korteweg–de Vries equation. Left: closed-loop solution. Middle: spatial mean of the solution. Right: control inputs. Reprinted from Ref. [5], Copyright 2018, with permission from Elsevier

where $y(t, x)$ is the unknown function and $u(t, x)$ the control input. We consider a periodic boundary condition on the spatial variable $x \in [-\pi, \pi]$. The nonlinear PDE is discretized using the split-stepping method with a spatial mesh of 128 points and time discretization of $\Delta t = 0.01$ s, resulting in a computational state space of dimension $n = 128$. The control input $u$ is considered to be of the form $u(t, x) = \sum_{i=1}^{3} u_i(t)v_i(x)$, where the coefficients $u_i(t)$ are to be determined by the controller and $v_i$ are fixed spatial profiles given by $v_i(x) = e^{-25(x-c_i)^2}$ with $c_1 = -\pi/2, c_2 = 0, c_3 = \pi/2$. The control inputs are constrained to $u_i(t) \in [-1, 1]$. The Koopman predictors are constructed from data in the form of 1000 trajectories of length 200 samples. The initial conditions of the trajectories are random convex combinations of three fixed spatial profiles given by $y_0^1 = e^{-(x-\pi/2)^2}$, $y_0^2 = -\sin(x/2)^2$, $y_0^3 = e^{-(x+\pi/2)^2}$; the control inputs $u_i(t)$ are distributed uniformly in $[-1, 1]^3$. The embedding mapping $\Psi$ is composed of the state itself, the elementwise square of the state, the elementwise product of the state with its periodic shift and the constant function, resulting in the dimension of the lifted state $N = 3 \cdot 128 + 1 = 385$. The control goal is to track a constant-in-space reference that varies in time in a piecewise constant manner. In order to do so, we design the Koopman MPC (MPC) with the reference tracking objective (9.32) with $Q = Q_{N_p} = I$, $R = 0$, $C = [I_{128}, 0]$ and prediction horizon $N_p = 10$ (i.e., 0.1 s). The results are depicted in Fig. 9.5; we observe a fast and accurate tracking of the reference profile. The average computation time to evaluate the control input was 0.28 ms (using the dense form (9.19) and the hardware configuration described in Footnote 6), allowing for deployment in real-time applications requiring very fast sampling rates.

## 9.6 Conclusion and Outlook

This chapter described a model predictive control (MPC) framework of nonlinear systems based on the Koopman operator. The underlying idea is to embed the nonlinear dynamics to a higher dimensional space where its evolution is approximately predictable in a linear fashion. The predictors obtained in this way exhibit superior performance compared to other linear predictors (e.g., local linearization) and can be readily used for feedback control using MPC in a purely convex fashion. Computational complexity of the underlying optimization problem is comparable to that of an MPC problem for a linear dynamical system of the same size. This is achieved by using the so-called dense form of an MPC problem whose computational complexity depends only on the number of control inputs and is virtually independent of the number of states. Importantly, the entire control design procedure is data driven, requiring only input–output measurements.

Future work should focus on imposing or proving closed-loop guarantees (e.g., stability or degree of suboptimality) of the controller designed using the presented methodology and on optimal selection of the embedding mapping, informed by the dynamical system at hand.

**Acknowledgements** This research was supported in part by the ARO-MURI grants W911NF-14-1-0359 and W911NF-17-1-0306 and the DARPA grant HR0011-16-C-0116. The research of M. Korda was supported by the Swiss National Science Foundation under grant P2ELP2_165166.

## Appendix

This appendix expresses explicitly the matrices in the "dense-form" MPC problem (9.19) as a function of the data defining the "sparse-form" MPC problem (MPC). The matrices are given by

$$H = \mathbf{R} + \mathbf{B}^\top \mathbf{Q}\mathbf{B}, \quad h = \mathbf{B}^\top \mathbf{q} + \mathbf{r}, \quad G = 2\mathbf{A}^\top \mathbf{Q}\mathbf{B},$$

$$L = \mathbf{F} + \mathbf{E}\mathbf{B}, \quad M = \mathbf{E}\mathbf{A}, \quad c = [b_0^\top, \ldots, b_{N_p}^\top]^\top,$$

where

$$\mathbf{A} = \begin{bmatrix} I \\ A \\ A^2 \\ \vdots \\ A^{N_p} \end{bmatrix}, \quad \mathbf{B} = \begin{bmatrix} 0 & 0 & \ldots & 0 \\ B & 0 & \ldots & 0 \\ AB & B & \ldots & 0 \\ \vdots & \ddots & \ddots & \\ A^{N_p-1}B & \ldots & AB & B \end{bmatrix}, \quad \mathbf{F} = \begin{bmatrix} F_0 & 0 & \ldots & 0 \\ 0 & F_1 & \ldots & 0 \\ \vdots & & \ddots & \vdots \\ 0 & 0 & \ldots & F_{N_p-1} \\ 0 & 0 & \ldots & 0 \end{bmatrix}$$

$$\mathbf{Q} = \mathrm{diag}(Q_0, \ldots, Q_{N_p}), \quad \mathbf{R} = \mathrm{diag}(R_0, \ldots, R_{N_p-1}),$$

$$\mathbf{E} = \mathrm{diag}(E_0, \ldots, E_{N_p}), \quad \mathbf{q} = [q_0^\top, \ldots, q_{N_p}^\top]^\top, \quad \mathbf{r} = [r_0^\top, \ldots, r_{N_p-1}^\top]^\top,$$

with $\mathrm{diag}(\cdot, \ldots, \cdot)$ denoting a block diagonal matrix composed of the arguments.

## References

1. Carleman, T.: Application de la théorie des équations intégrales linéaires aux systémes d'équations différentielles non linéaires. Acta Math. **59**(1), 63–87 (1932)
2. Daniel-Berhe, S., Unbehauen, H.: Experimental physical parameter estimation of a thyristor driven DC-motor using the HMF-method. Control Eng. Pract. **6**(5), 615–626 (1998)
3. Ferreau, H.J., Kirches, C., Potschka, A., Bock, H.G., Diehl, M.: qpoases: a parametric active-set algorithm for quadratic programming. Math. Program. Comput. **6**(4), 327–363 (2014)
4. Grüne, L., Pannek, J.: Nonlinear model predictive control. In: Nonlinear Model Predictive Control. Springer, Berlin (2011)
5. Korda, M., Mezić, I.: Linear predictors for nonlinear dynamical systems: Koopman operator meets model predictive control. Automatica **93**, 149–160 (2018)
6. Ljung, L.: System identification. In: Signal Analysis and Prediction, pp. 163–173. Springer, Berlin (1998)
7. Miura, R.M.: The korteweg-de Vries equation: a survey of results. SIAM Rev. **18**(3), 412–459 (1976)
8. Rawlings, J.B., Mayne, D.Q.: Model Predictive Control: Theory and Design, 1st edn. Nob Hill Publishing (2009)
9. Surana, A.: Koopman operator based observer synthesis for control-affine nonlinear systems. In: Conference on Decision and Control (CDC) (2016)
10. Surana, A., Banaszuk, A.: Linear observer synthesis for nonlinear systems using Koopman operator framework. In: IFAC Symposium on Nonlinear Control Systems (NOLCOS) (2016)
11. Williams, M.O., Kevrekidis, I.G., Rowley, C.W.: A data-driven approximation of the Koopman operator: extending dynamic mode decomposition. J. Nonlinear Sci. **25**(6), 1307–1346 (2015)
12. Williams, M.O., Rowley, C.W., Kevrekidis, I.G.: A kernel-based approach to data-driven Koopman spectral analysis. J. Comput. Dyn. **2**(2), 247–265 (2015)

# Chapter 10
# Feedback Control of Nonlinear PDEs Using Data-Efficient Reduced Order Models Based on the Koopman Operator

**Sebastian Peitz and Stefan Klus**

**Abstract** In the development of model predictive controllers for PDE-constrained problems, the use of reduced order models is essential to enable real-time applicability. Besides local linearization approaches, proper orthogonal decomposition (POD) has been most widely used in the past in order to derive such models. Due to the huge advances concerning both theory as well as the numerical approximation, a very promising alternative based on the Koopman operator has recently emerged. In this chapter, we present two control strategies for model predictive control of nonlinear PDEs using data-efficient approximations of the Koopman operator. In the first one, the dynamic control system is replaced by a small number of autonomous systems with different yet constant inputs. The control problem is consequently transformed into a switching problem. In the second approach, a bilinear surrogate model is obtained via a convex combination of these autonomous systems. Using a recent convergence result for extended dynamic mode decomposition (EDMD), convergence of the reduced objective function can be shown. We study the properties of these two strategies with respect to solution quality, data requirements, and complexity of the resulting optimization problem using the 1-dimensional Burgers equation and the 2-dimensional Navier–Stokes equations as examples. Finally, an extension for online adaptivity is presented.

## 10.1 Introduction

The control of systems governed by nonlinear partial differential equations (PDEs) is very challenging. Since classical control strategies are difficult to develop in this context, advanced techniques such as *model predictive control (MPC)* [14] are very

S. Peitz (✉)
Department of Mathematics, Paderborn University, Paderborn, Germany
e-mail: speitz@math.upb.de

S. Klus
Department of Mathematics and Computer Science, Freie Universität Berlin, Berlin, Germany
e-mail: stefan.klus@fu-berlin.de



A. Mauroy et al. (eds.), *The Koopman Operator in Systems and Control*,
Lecture Notes in Control and Information Sciences 484,
https://doi.org/10.1007/978-3-030-35713-9_10

popular. In MPC, an open-loop optimal control problem is repeatedly solved online over a finite-time horizon using a model of the system dynamics, which then results in a closed-loop controller. The PDE-constrained optimal control problems we are interested in are of the following form:

$$\begin{aligned} \min_u J(\mathbf{y}) &= \min_u \int_{t_0}^{t_e} L(\mathbf{y}(\cdot,t))\,dt \\ \text{s.t.}\quad \dot{\mathbf{y}}(\cdot,t) &= G(\mathbf{y}(\cdot,t),\mathbf{u}(t)), \\ \mathbf{y}(\cdot,0) &= \mathbf{y}^0, \end{aligned} \tag{OCP}$$

where $\mathbf{y}$ is the system state and $\mathbf{y}(\cdot,t)$ is an element of an appropriate function space $\mathscr{Y}$ the system state (depending on the $d$-dimensional space coordinate $\mathbf{x} \in \Omega \subseteq \mathbb{R}^d$ and the time $t \in \mathbb{R}^{\geq 0}$), $u \in L^2([t_0,t_e],\mathfrak{U})$ is the control function, and the partial differential operator $G : \mathscr{Y} \times \mathfrak{U} \to \mathscr{Y}$ describes the system dynamics. For ease of notation, the term in the objective function $L : \mathscr{Y} \to \mathbb{R}$ does not depend on $u$ explicitly.

The downside of MPC is that the open-loop control problem (OCP) has to be solved in a short amount of time, which is generally not possible for PDE-constrained problems when using a standard discretization approach such as finite elements or finite volumes. A remedy to this issue is *reduced order modeling (ROM)*, where the high-fidelity model is replaced by a low-dimensional surrogate model, see [4, 28] for overviews. In the nonlinear case, the method of *proper orthogonal decomposition (POD)* [43] has been successfully applied in a large variety of problems concerning both simulation and control. In the latter case, there exist different approaches to ensure convergence toward the true optimum. The two most popular are to derive error bounds based on the singular values associated with the POD modes [17, 26, 40, 46] and to adapt classical *trust-region* approaches to surrogate modeling [5, 12, 39].

A much more recent approach to develop a ROM is via the linear but infinite-dimensional *Koopman operator* [23], which describes the dynamics of observables. In the past decade, significant advances were obtained concerning theoretical aspects of the Koopman operator [7, 29–31] as well as its numerical approximation via *dynamic mode decomposition (DMD)* [41, 42, 47] or *extended dynamic mode decomposition (EDMD)* [20, 21, 48]. An advantage over POD is that this approach can also be applied in situations where the underlying system dynamics is unknown.

The above-mentioned advancements have led to several approaches for including the Koopman operator in control frameworks. In [6, 37, 38], extensions of DMD and the Koopman operator framework are presented for systems with inputs. In [2, 24], a similar approach is embedded into a feedback framework via MPC, and in [19], eigenfunctions of the Koopman operator are computed from data in order to realize a linearization of nonlinear control systems. In many of these approaches, the Koopman operator is approximated for an augmented state (consisting of the actual state and the control) in order to deal with the nonautonomous control system. For this reason, a large amount of data is necessary to cover a sufficient range of the dynamics. Alternative approaches have recently been presented by the authors in [34, 36]. Since the Koopman operator is only applicable to autonomous systems in its original formulation, we take the following two steps:

(i) replace the control system $G$ by a finite number of autonomous systems $G_{u^j}$ with constant input $u^j$;
(ii) construct reduced order models for low-dimensional observations (instead of the entire state) of $G_{u^j}$ using the corresponding Koopman operator $U_{u^j}$.

In a third step, the PDE constraint in Problem (OCP) is replaced by the reduced model. We perform this step in two different ways (cf. [34, 36], respectively, for details):

(iii-a) transform the optimization problem into a switching problem (which of the autonomous systems has to be applied in each time step?);
(iii-b) construct a bilinear surrogate model via linear interpolation between two Koopman operators $U_{u^0}$ and $U_{u^1}$.

In this way, convergence toward the true optimum can be shown by utilizing a recent convergence result for EDMD [25].

Since reduced order modeling approaches using the Koopman operator are relatively new, people have much less experience in this direction compared to more established methods such as POD. The purpose of this chapter is, therefore, to study the two approaches described above regarding the numerical performance. We address both the quality of the solution compared to the PDE-constrained problem as well as the influence of the training data. Furthermore, the effect of introducing the switching problem transformation is studied.

The remainder of this chapter is structured as follows. In Sect. 10.2, the notation for the Koopman operator is introduced and the reduced order modeling approach for low-dimensional observations is presented. In Sect. 10.3, we give a short introduction to model predictive control which we use to realize feedback behavior. We then introduce the two reduced order modeling strategies in Sect. 10.4 before studying numerical properties and the control performance in Sect. 10.5. Finally, we use the concept from [16] to obtain online updates for the reduced models in Sect. 10.6 before drawing a conclusion in Sect. 10.7.

## 10.2 Reduced Order Modeling Using the Koopman Operator

Let $\Phi : \mathscr{Y} \to \mathscr{Y}$ be a discrete deterministic dynamical system defined on the state space $\mathscr{Y}$ and let $f : \mathscr{Y} \to \mathbb{R}^q$ be a real-valued observable of the system. Then the Koopman operator $U : \mathscr{F} \to \mathscr{F}$ with $\mathscr{F} = L^2(\mathscr{Y})$, see [7, 27, 30, 48], which describes the evolution of the observable $f$, is defined by

$$(Uf)(\mathbf{y}) = f(\Phi(\mathbf{y})).$$

The Koopman operator is linear but infinite dimensional. Its adjoint, the Perron–Frobenius operator, describes the evolution of densities. The definition of the Koopman operator can be naturally extended to continuous-time dynamical systems as described in [7, 27]. Given an autonomous system of the form

$$\dot{\mathbf{y}}(\cdot, t) = G(\mathbf{y}(\cdot, t)),$$

the *Koopman semigroup* of operators $\{U^t\}$ is defined as

$$(U^t f)(\mathbf{y}(\cdot, t)) = f(\Phi^t(\mathbf{y}(\cdot, t))),$$

where $\Phi^t$ is the flow map associated with $G$. In what follows, we will mainly consider discrete dynamical systems, given by the discretization of ODEs or PDEs. That is, $\Phi = \Phi^h$ for a fixed time step $h$.

One method to compute a numerical approximation of the Koopman operator from data is EDMD [21, 48]. The following brief description is based on the review paper [22]. EDMD is a generalization of DMD [42, 47] and can be used to compute a finite-dimensional approximation of the Koopman operator, its eigenvalues, eigenfunctions, and modes. In contrast to DMD, EDMD allows arbitrary basis functions—which could be, for instance, monomials, Hermite polynomials, or trigonometric functions—for the approximation of the dynamics. We do not observe the full (potentially infinite-dimensional) state of the system, but consider only a finite number of measurements, given by $\mathbf{z}_i = f(\mathbf{y}(\cdot, t_i)) \in \mathbb{R}^q$. The special case $f = \mathsf{Id}$ is known as the *full state observable*. For a given set of basis functions $\{\psi_1, \psi_2, \ldots, \psi_k\}$, we then define a vector-valued function $\psi\colon \mathbb{R}^q \to \mathbb{R}^k$ by

$$\psi(\mathbf{z}) = \begin{bmatrix}\psi_1(\mathbf{z})\ \psi_2(\mathbf{z})\ \ldots\ \psi_k(\mathbf{z})\end{bmatrix}^\top.$$

If $\psi(\mathbf{z}) = \mathbf{z}$, we obtain DMD as a special case of EDMD. We assume that we have either measurement or simulation data, written in matrix form as

$$\mathbf{Z} = \begin{bmatrix}\mathbf{z}_1\ \mathbf{z}_2\ \cdots\ \mathbf{z}_m\end{bmatrix} \quad\text{and}\quad \widetilde{\mathbf{Z}} = \begin{bmatrix}\widetilde{\mathbf{z}}_1\ \widetilde{\mathbf{z}}_2\ \cdots\ \widetilde{\mathbf{z}}_m\end{bmatrix},$$

where $\widetilde{\mathbf{z}}_i = f(\Phi(\mathbf{y}_i))$. The data could either be obtained via many short simulations or experiments with different initial conditions or one long-term trajectory or measurement. If the data is extracted from one long trajectory, then $\widetilde{\mathbf{z}}_i = \mathbf{z}_{i+1}$. The data matrices are embedded into the typically higher dimensional feature space by

$$\Psi_{\mathbf{Z}} = \begin{bmatrix}\psi(\mathbf{z}_1)\ \psi(\mathbf{z}_2)\ \ldots\ \psi(\mathbf{z}_m)\end{bmatrix} \quad\text{and}\quad \Psi_{\widetilde{\mathbf{Z}}} = \begin{bmatrix}\psi(\widetilde{\mathbf{z}}_1)\ \psi(\widetilde{\mathbf{z}}_2)\ \ldots\ \psi(\widetilde{\mathbf{z}}_m)\end{bmatrix}.$$

With these data matrices, we then compute the matrix $\mathbf{U} \in \mathbb{R}^{k\times k}$ defined by

$$\mathbf{U}^\top = \Psi_{\widetilde{\mathbf{Z}}}\Psi_{\mathbf{Z}}^+ = \left(\Psi_{\widetilde{\mathbf{Z}}}\Psi_{\mathbf{Z}}^\top\right)\left(\Psi_{\mathbf{Z}}\Psi_{\mathbf{Z}}^\top\right)^+,$$

where $^+$ denotes the pseudoinverse. The matrix $\mathbf{U}$ can be viewed as a finite-dimensional approximation of the Koopman operator.

Convergence of EDMD in the infinite-data limit for a fixed set of basis functions was first analyzed in [21, 48]. Convergence toward the Koopman operator for the case that also the number of basis functions goes to infinity has recently been proven in [25]. Under some assumptions such as independent drawing of data points with respect to some given probability measure $\mu$ and boundedness of the Koopman operator, the EDMD approximation $\mathbf{U}$ converges to the Koopman operator for $k \to \infty$ and $m \to \infty$ provided that $(\psi_i)_{i=1}^{\infty}$ is an orthonormal basis of $\mathscr{F}$. For the convergence results below, we assume that these conditions are satisfied.

The decomposition of the Koopman operator into modes, eigenvalues, and eigenfunctions is commonly used to analyze the system dynamics as well as predict the future state. In the situation we are presenting here, we can pursue an even simpler approach and obtain the update for the observable $\mathbf{z}$ directly using $\mathbf{U}$ which yields the Koopman operator based reduced order model:

$$\boldsymbol{\eta}_{i+1} = \mathbf{U}^\top \boldsymbol{\eta}_i, \quad i = 0, 1, \ldots \tag{10.1}$$

with $\boldsymbol{\eta} = \psi(\mathbf{z})$ and $\boldsymbol{\eta}_0 = \psi(f(\mathbf{y}^0))$. This approach is visualized in Fig. 10.1, and we see that

$$f \circ \Phi = Uf \approx P \circ \mathbf{U}^\top \circ \psi \circ f,$$

where $P$ is the projection from the feature space to the observable space. If we let the number of data points as well as the number of basis functions go to infinity, then the EDMD approximation converges to the Koopman operator as discussed above.

Table 10.1 shows the efficiency of the K-ROM for the two examples which we will consider throughout this chapter, the 1-dimensional Burgers equation and the 2-dimensional Navier–Stokes equations. Here, we have chosen a monomial basis for the dictionary $\Psi$. The maximum order of the monomials and the number of

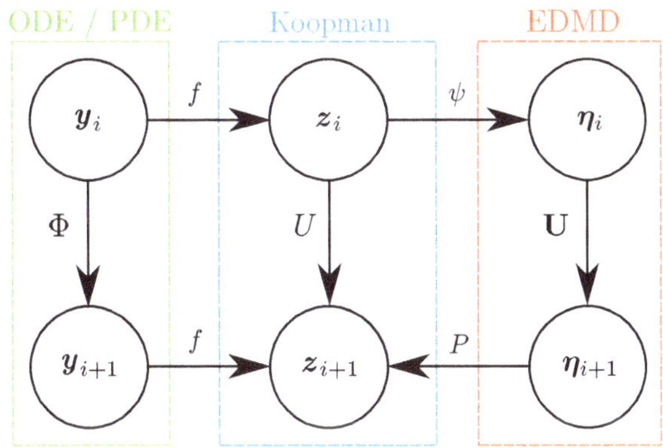


**Fig. 10.1** Relation between the system dynamics $\Phi$, the corresponding Koopman operator $U$ and its finite-dimensional representation $\mathbf{U}$ computed via EDMD

**Table 10.1** Numerical setup and efficiency analysis for two example problems

| Problem | 1-dimensional Burgers | 2-dimensional NSE |
|---|---|---|
| Order of monomials | 3 | 2 |
| $\dim(\mathbf{z})$ | 4 | 8 |
| $\dim(\psi(\mathbf{z})) = \dim(\mathbf{U}_{u^j})$ | 35 | 45 |
| $\dim(\mathbf{y})$ | 49 (FD) | 22,000 (FV) |
| $\dim(\mathbf{y})/\dim(\psi(\mathbf{z}))$ | 1.4 | 488.9 |
| Speedup (10.1) versus PDE | $\approx$100 | $\approx 7.5 \cdot 10^4$ |

observations yield the K-ROM dimension. We see that the dimension of the PDE-constrained problem (obtained by a finite-difference (FD) approximation) is reduced by a factor of 1.4 in the Burgers example. Although this does not seem like a large reduction, we obtain a speedup of approximately 100 since (10.1) is linear on the one hand and we can choose larger step sizes in the reduced model on the other hand. For the Navier–Stokes example, we additionally have a major reduction of the dimension of the state compared to the finite volume (FV) discretization by a factor of almost 500. This way, a speedup of 75,000 is achieved.

## 10.3 Model Predictive Control

Due to uncertainties and noise, open-loop control strategies are insufficient for applications. To overcome this issue, in MPC [14], the infinite-horizon open-loop problem is split into finite-horizon problems with a *prediction horizon* of length $p$:

$$\begin{aligned} &\min_{\mathbf{u}\in\mathfrak{U}^p} \sum_{i=s}^{s+p-1} L(\mathbf{y}_i) \qquad\qquad \text{(MPC)} \\ \text{s.t.}\quad &\mathbf{y}_{i+1} = \Phi(\mathbf{y}_i, \mathbf{u}_{i-s+1}), \quad i = s, \ldots, s+p-1, \\ &\mathbf{y}_s = \mathbf{y}^s, \end{aligned}$$

where $\mathbf{y}^s$ is the initial condition obtained (or approximated) from sensor data. The first part of this solution is then applied to the real system while the optimization is repeated with the prediction horizon moving forward by one sample time. In Problem (MPC), the system dynamics are of discrete form since the control input is constant over each sample time interval. When dealing with continuous-time systems such as Problem (OCP), this formulation can be regarded as the flow map $\Phi^h$ (cf. Sect. 10.2) of the continuous dynamics with time step $h$.

A consequence of the MPC method is that Problem (MPC) has to be solved online, i.e., within the time step $h$. Since this is in general impossible for PDE-constrained problems, we will present two approaches to replace the PDE constraint in (MPC) by a Koopman operator based reduced order model in the next section.

## 10.4 A Data-Efficient Method to Construct Koopman Operator Based Surrogate Models

As already outlined in the introduction, we will construct K-ROMs with significant speedup factors. In order to do this in a data-efficient way, we perform steps (i) and (ii) mentioned there. The first step ensures that our data requirements are moderate since we only have to collect data for a finite set of inputs. The second step allows us to derive linear models with a low dimension. Since the approach is entirely data based and hence equation free, we can use any observation and in particular those which are relevant for the control task at hand.

In a third step, we have to transform the optimal control problem accordingly which we will do in two different ways. In Sect. 10.4.1, the optimization problem is transformed into a switching problem and in Sect. 10.4.2, a bilinear model is constructed via linear interpolation.

### *10.4.1 Transformation to Switched Systems*

A large variety of technical systems is controlled via switching between different inputs. Examples are valves in chemical reactors which are either open or closed or the switching of gears in electrical drives. These so-called *switched systems* can be regarded as a special case of hybrid systems, which possess both continuous and discrete-time control inputs (cf. [49] for a survey). We here make use of the concept of switched systems in order to reduce the data that is required for the training process of the K-ROM. To this end, we replace the right-hand side of the dynamical control system by a finite set of autonomous systems, which is achieved by fixing the input to $n_c$ different constant values $\{u^0, \ldots, u^{n_c-1}\} = \hat{\mathfrak{U}} \subset \mathfrak{U}$ in (OCP). This yields $n_c$ different differential operators, $G_{u^0}, \ldots, G_{u^{n_c-1}}$, and the respective flow maps $\Phi_{u^0}, \ldots, \Phi_{u^{n_c-1}}$.

A consequence of this approach is that the optimization problem is transformed into a combinatorial problem, where we have to select the optimal right-hand side in each time step. Since combinatorial problems are often more challenging to solve, we can fix the sequence in which the different autonomous systems are used, i.e., we switch from $G_{u^0}$ to $G_{u^1}$, from $G_{u^1}$ to $G_{u^2}$, and so on. Having reached the final system, we go back from $G_{u^{n_c-1}}$ to $G_{u^0}$. This way, the optimization variable again becomes real-valued, as we now have to compute the time instants for the switches $\boldsymbol{\tau} \in \mathbb{R}^{p+2}$ (with $\boldsymbol{\tau}_0 = t_0$ and $\boldsymbol{\tau}_{p+1} = t_e$):

$$\begin{aligned} \dot{\mathbf{y}}(\cdot,t) &= G_{u^j}(\mathbf{y}(\cdot,t)) \quad \text{for } t \in [\boldsymbol{\tau}_{l-1}, \boldsymbol{\tau}_l), \\ \mathbf{y}(\cdot,0) &= \mathbf{y}^0, \\ j &= l \bmod n_c. \end{aligned} \tag{10.2}$$

Using (10.2), we can reformulate the open-loop problem (OCP) in terms of the switching instants

$$\begin{aligned} \min_{\boldsymbol{\tau}\in\mathbb{R}^{p+2}} J(\mathbf{y}) &= \min_{\boldsymbol{\tau}\in\mathbb{R}^{p+2}} \int_{t_0}^{t_e} L(\mathbf{y}(\cdot,t))\,dt \\ \text{s.t.} \qquad & (10.2). \end{aligned} \tag{OCP$^s$}$$

Different methods exist for efficiently computing solutions to (OCP$^s$), see, e.g., [9, 10, 44] for continuous-time or [13] for discrete-time problems. A second-order method has recently been proposed in [45]. The following example illustrates the consequences of introducing such a switching control.

*Example 10.1* Assume we want to control the behavior of the Van der Pol oscillator given by

$$\begin{aligned} \dot{\mathbf{y}}(t) &= G(\mathbf{y}(t), u(t)) = \begin{pmatrix} \mathbf{y}_2(t) \\ \left(1 - \mathbf{y}_1(t)^2\right)\mathbf{y}_2(t) - \mathbf{y}_1(t) + u(t) \end{pmatrix}, \\ \mathbf{y}(0) &= \mathbf{y}^0. \end{aligned}$$

By restricting the input $u$ to $n_c$ values, we can transform the control system into $n_c$ autonomous systems of the form

$$\begin{aligned} \dot{\mathbf{y}}(t) &= G_{u^j}(\mathbf{y}(t)) = \begin{pmatrix} \mathbf{y}_2(t) \\ \left(1 - \mathbf{y}_1(t)^2\right)\mathbf{y}_2(t) - \mathbf{y}_1(t) \end{pmatrix} + \begin{pmatrix} 0 \\ u^j \end{pmatrix}, \\ \mathbf{y}(0) &= \mathbf{y}^0. \end{aligned}$$

The system dynamics for different input functions $u$ are shown in Fig. 10.2. △

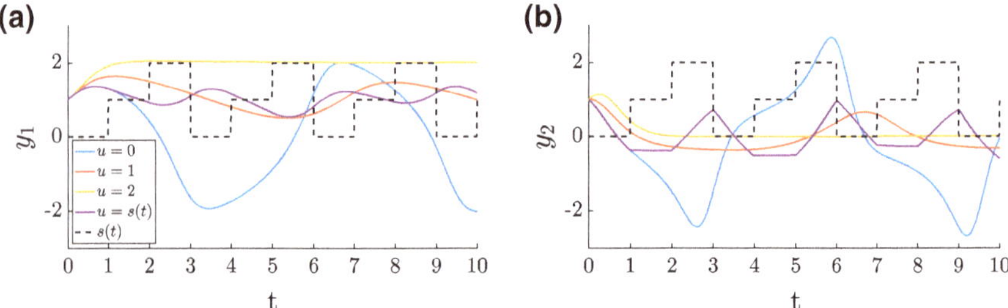


**Fig. 10.2** The trajectories of three different autonomous systems with starting point $\mathbf{y}^0 = (1, 1)^\top$ and the trajectory of the switched system with switching according to the dashed line

If we consider the discrete-time formulation (MPC) for the closed-loop controller, the transformation is achieved in a very similar fashion

$$\begin{aligned} \min_{\mathbf{u}\in\hat{\mathfrak{U}}^p} \ & \sum_{i=s}^{s+p-1} L(\mathbf{y}_i) \\ \text{s.t.} \ & \mathbf{y}_{i+1} = \Phi_{\mathbf{u}_{i-s+1}}(\mathbf{y}_i), \quad i = s, \dots, s+p-1, \\ & \mathbf{y}_s = \mathbf{y}^s. \end{aligned} \tag{MPC$^s$}$$

Since the input is constant over the sample time interval, we here obtain a combinatorial problem without a workaround as in the continuous-time case. Each entry of $\mathbf{u}$ now describes which flow map $\Phi_{\mathbf{u}_i}$ to apply in the $i$th step.

A popular approach to solve discrete-time optimal control problems of such form is via dynamic programming [3]. However, this is only advisable if we are interested in larger prediction horizons $p$. For low values of $p$ (say, 3), the most efficient way is indeed to evaluate all ${n_c}^p$ values of $\mathbf{u}$.

#### 10.4.1.1 Example: The 1-Dimensional Burgers Equation

We now study the difference between a continuous control input and a switched system approximation using the 1-dimensional Burgers equation with periodic boundary conditions and a distributed control:

$$\begin{aligned} \dot{y}(x,t) - \nu \Delta y(x,t) + y(x,t)\nabla y(x,t) &= u(t)\chi(x), \\ y(x,0) &= y^0(x). \end{aligned} \tag{10.3}$$

The viscosity is set to $\nu = 0.01$ and the distributed control is realized by a time-dependent scalar input $u$ and a shape function $\chi$ which is shown in Fig. 10.3a.

In order to enable comparability to our K-ROM approach, the objective function only depends on a few points in space (the black dots in Fig. 10.3a), i.e.,

$$\mathbf{z}(t) = f(y(\cdot,t)) = (y(0,t),\ y(0.5,t),\ y(1,t),\ y(1.5,t))^\top,$$

and we formulate the tracking type objective function in terms of these observations only. This yields the following MPC problem

$$\begin{aligned} \min_{\mathbf{u}\in\mathfrak{U}^p} \ & \sum_{i=s}^{s+p-1} J_i = \sum_{i=s}^{s+p-1} \|\mathbf{z}_i - \mathbf{z}_i^{\text{opt}}\|_2^2 \\ \text{s.t.} \ & y_{i+1} = \Phi(y_i, \mathbf{u}_{i-s+1}), \quad i = s, \dots, s+p-1 \\ & y_s = y^s \end{aligned} \tag{10.4}$$

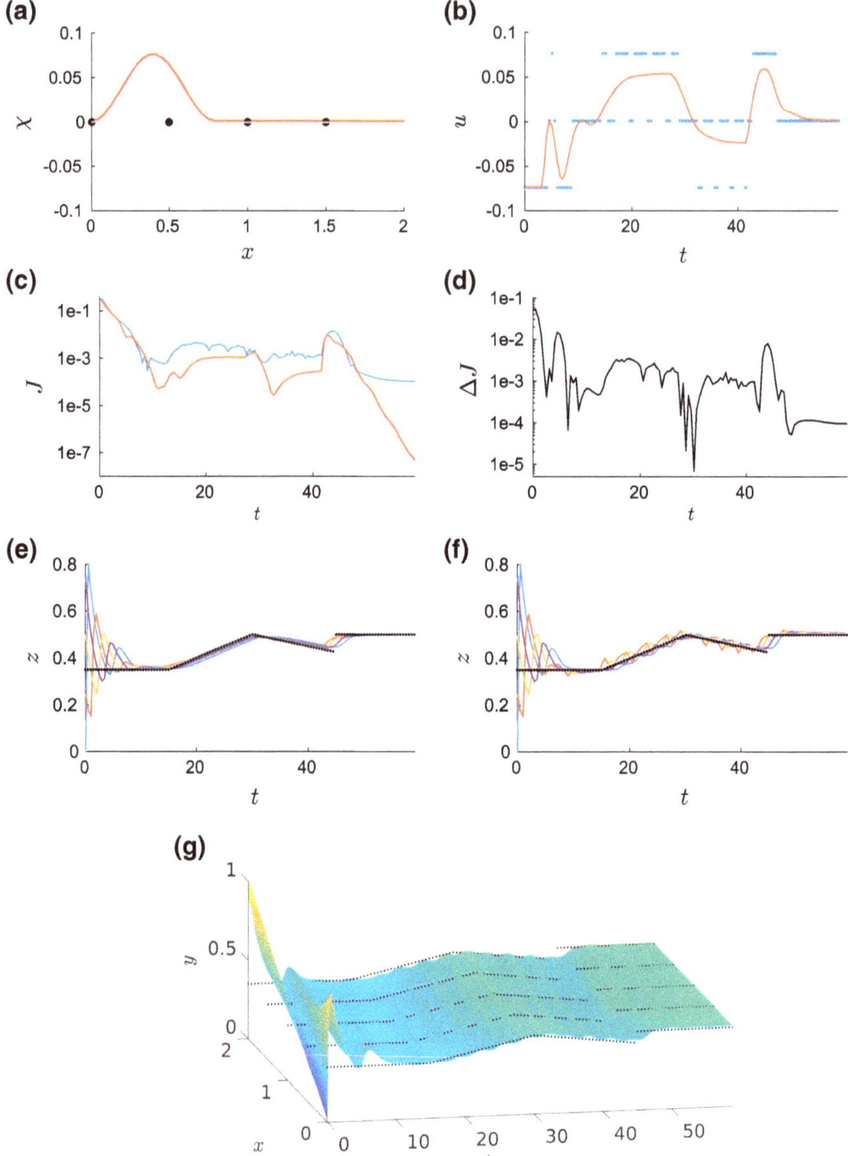


**Fig. 10.3** **a** The shape function $\chi$ and the positions where the state is observed by $f$ for the objective function. **b** Comparison of the optimal control obtained by MPC using the continuous control system (10.4) (orange) and the switched system (10.5) (blue). **c** The corresponding objective function values. **d** Difference between the trajectories in (**c**), $\Delta J = |J_{(10.4)} - J_{(10.5)}|$. **e** Optimal trajectory of the observation **z** and the corresponding $\mathbf{z}^{\text{opt}}$ for the continuous problem (10.4). **f** Like **e** but for the switching problem (10.5). **g** Optimal state trajectory $y$ corresponding to (**f**). The reference points $\mathbf{z}^{\text{opt}}$ are shown in black

with $\Phi(y, \mathbf{u})$ being the flow map of (10.3) and $y^s$ the initial value for the current iteration. Similar to Example 10.1, we now limit the control input to three constant values $u^0 = 0.075$, $u^1 = 0$, and $u^2 = -0.075$. By this, the right-hand side of (10.3) is transformed to $u^j \chi(x)$ which yields $\Phi_{u^j}(y)$ as the flow maps corresponding to the constant inputs $u^0$, $u^1$ and $u^2$. The MPC problem is transformed accordingly:

$$\begin{aligned} \min_{\mathbf{u} \in \{u^0, u^1, u^2\}^p} \sum_{i=s}^{s+p-1} J_i &= \sum_{i=s}^{s+p-1} \|\mathbf{z}_i - \mathbf{z}_i^{\text{opt}}\|_2^2 \\ \text{s.t.} \quad y_{i+1} &= \Phi_{\mathbf{u}_{i-s+1}}(y_i), \quad i = s, \dots, s+p-1, \\ y_s &= y^s. \end{aligned} \tag{10.5}$$

The results for the two MPC problems (10.4) and (10.5) are compared in Fig. 10.3. We see in (b) that the switching approach results in inputs which are close to the continuous solution when averaging over small time frames. In (c) and (d) we see that this results in a solution with slightly reduced quality, i.e., the distance to the reference trajectory is slightly larger. Figure 10.3e, f shows the observation $\mathbf{z}$ for the two problems, i.e., the dynamics the controller acts on. Finally, the state $y$ corresponding to (e) is shown in (g).

#### 10.4.1.2 Switched System MPC Based on K-ROMs

Assuming that we have computed the $n_c$ K-ROMs (i.e., $\mathbf{U}_{u^0}$ to $\mathbf{U}_{u^{n_c}}$) via EDMD from data, we can now replace the expensive PDE evaluation in (MPC$^s$) by the much cheaper K-ROM:

$$\begin{aligned} &\min_{\hat{\mathbf{u}} \in \hat{\mathfrak{U}}^p} \sum_{i=s}^{s+p-1} \hat{L}(\eta) \\ \text{s.t.} \quad &\eta_{i+1} = \mathbf{U}_{\hat{\mathbf{u}}_{i-s+1}}^\top \eta_i \quad \text{for } i = s, \dots, s+p-1, \\ &\eta_s = \psi(f(\mathbf{y}^s)), \end{aligned} \tag{K-MPC$^s$}$$

where $\hat{L}$ is the reduced objective function formulated with respect to the observables. We now compare the two problem formulations (MPC$^s$) and (K-MPC$^s$), i.e., the switching time MPC problem and the corresponding approximation using the K-ROM. To this end, we first assume that the full objective function $L$ can be expressed in terms of the observations $\mathbf{z}$.

**Assumption 10.1** $L(\mathbf{y}(t)) = \hat{L}(\eta_i)$ *for all* $t \in [t_0, t_0 + h, t_0 + 2h, \dots, t_e]$ *and the corresponding* $i = (t - t_0)/h$.

The assumption is not restrictive since this has to be satisfied in every application, where only observations are available (e.g., sensor data). Consequently, the objective $L$ has to be defined accordingly.

This assumption allows us to prove that the objective function of the K-ROM based problem converges to the full objectives as the number of measurements $m$ and the basis size $k$ for the dictionary $\Psi$ tend to infinity (i.e., we have convergence for EDMD, see [25] for details).

**Theorem 10.2** ([36]) *Consider Problem (MPC$^s$) and the corresponding approximation (K-MPC$^s$) using the K-ROM and assume that we have convergence of EDMD toward the Koopman operator as the basis size and number of sampled data points tend to infinity (according to [25]), i.e.,*

$$\lim_{k\to\infty}\lim_{m\to\infty} \mathbf{U}_{u^j}^{\top} = U_{u^j} \quad \textit{for } j = 1,\ldots,n_c.$$

*Then, under Assumption 10.1, the objective function value of Problem (K-MPC$^s$) converges in measure (with respect to the initial condition $z_0 = f(y^0)$) to that of Problem (MPC$^s$) for every $u \in \hat{\mathfrak{U}}^p$, $p < \infty$. That is, for all $\varepsilon > 0$ we obtain*

$$\lim_{k\to\infty}\lim_{m\to\infty} \mu\left(\left\{z_0 \in \mathscr{Z} \,\middle|\, \sum_{i=0}^{p} |\hat{L}(\eta_i) - L(y_i)| \geq \varepsilon\right\}\right) = 0.$$

*Here $\mathscr{Z} \subset \mathbb{R}^q$ is the space of all measurements with an associated measure $\mu$.*

*Remark 10.1* Note that Theorem 10.2 does not necessarily imply convergence for optimization algorithms but only states that the K-ROM and the full system very likely possess the same minimizers. However, this gives us confidence to use K-ROMs within a feedback loop. Finally, the result will benefit from possible future convergence results for EDMD with stronger statements (as are available for POD, for instance). △

As we have seen in Table 10.1 for two example problems, this reduced order modeling approach allows us to reduce the numerical effort by several orders of magnitude. A more detailed discussion regarding the numerical benefits will follow in Sect. 10.5.

### *10.4.2 Bilinear Models via Linear Interpolation*

The switched systems approach allows us to replace the system dynamics by a K-ROM in a straightforward manner. However, this approach comes at a prize, namely that we now have a combinatorial optimization problem which is often more expensive to solve. Furthermore, we have limited the control input to a small number of predefined values. Although this does not necessarily have a major impact on the control performance as we have seen in Sect. 10.4.1, we would nonetheless like to allow for arbitrary control inputs. One possibility to do so is to approximate a Koopman operator for an augmented observation, i.e.,

$$\hat{\mathbf{z}} = \begin{pmatrix} \mathbf{z} \\ \mathbf{u} \end{pmatrix}.$$

This approach is pursued in [37] for open-loop and in [24] for closed-loop control problems. As a consequence, data for different combinations of states and inputs has to be collected to approximate the Koopman operator for $\hat{\mathbf{z}}$. To reduce the amount of required data, we here pursue an alternative approach where we still only approximate Koopman operators for a small number of autonomous systems. In order to allow for continuous controls, we now define the matrices

$$\mathbf{A} = \mathbf{U}_{\mathbf{u}^0}^\top \quad \text{and} \quad \mathbf{B} = \begin{bmatrix} \mathbf{B}_1 & \cdots & \mathbf{B}_{n_c-1} \end{bmatrix} = \begin{bmatrix} \mathbf{U}_{\mathbf{u}^1}^\top - \mathbf{U}_{\mathbf{u}^0}^\top & \cdots & \mathbf{U}_{\mathbf{u}^{n_c-1}}^\top - \mathbf{U}_{\mathbf{u}^0}^\top \end{bmatrix}$$

and introduce the bilinear control system

$$\begin{aligned} \eta_{i+1} &= \mathbf{A}\eta_i + \sum_{j=1}^{n_c-1} \mathbf{B}_j \eta_i \mathbf{u}_{i,j}, \\ \eta_0 &= \psi(f(\mathbf{y}^0)). \end{aligned} \qquad \text{(K-ROM)}$$

The term bilinear refers to the fact that (K-ROM) contains a term $\eta \cdot \mathbf{u}$ but is otherwise linear both in $\eta$ and in $\mathbf{u}$ [11]. By restricting $\mathbf{u}_i$ to the convex hull of the individual control inputs, i.e.,

$$\mathbf{u}_i \in \overline{\mathfrak{U}} = \left\{ \overline{\mathbf{u}} \in \mathbb{R}^{n_c-1} \,\middle|\, \overline{\mathbf{u}}_j \geq 0 \,\forall\, j \in \{1, \dots, n_c - 1\}, \sum_{j=1}^{n_c-1} \overline{\mathbf{u}}_j \leq 1 \right\},$$

the system (K-ROM) performs a linear interpolation between the different systems with inputs $\mathbf{u}^0, \dots, \mathbf{u}^{n_c-1}$. Note that (K-ROM) yields the exact dynamics for the individual systems (up to a set of measure zero) due to the convergence result for EDMD. For intermediate values of $\mathbf{u}$, we can exploit the linearity of the Koopman operator [34]. Assuming that the observation map $f$ is linear and that the system dynamics $\Phi$ are linear in $\mathbf{u}$, a convex combination of operators is—for small time steps $h$—approximately equal to the Koopman operator for the convex combination of the controls $\mathbf{u}^0, \dots, \mathbf{u}^{n_c-1}$, i.e.,

$$\sum_{j=0}^{n_c-1} \alpha_j U_{\mathbf{u}^j} \approx U_{\overline{\mathbf{u}}},$$

where

$$\overline{\mathbf{u}} = \sum_{j=0}^{n_c-1} \alpha_j \mathbf{u}^j, \quad \sum_{j=1}^{n_c-1} \alpha_j \leq 1, \; \alpha_j \geq 0 \,\forall\, j \in \{0, \dots, n_c - 1\}.$$

The above consideration hence justifies the use of this relaxation approach in order to avoid the combinatorial problem. Similar to the switched systems approach, we can now easily reduce the numerical effort of (MPC) (i.e., the MPC problem with continuous inputs) by replacing the system dynamics by the bilinear surrogate model:

$$\begin{aligned} \min_{\hat{\mathbf{u}} \in \mathbb{R}^p} & \sum_{i=s}^{s+p-1} \hat{L}(\boldsymbol{\eta}_i) && \text{(K-MPC)} \\ \text{s.t.} \quad \boldsymbol{\eta}_{i+1} &= \mathbf{A}\boldsymbol{\eta}_i + \sum_{j=1}^{n_c-1} \mathbf{B}_j \boldsymbol{\eta}_i u_{i,j} \quad \text{for } i = s, \ldots, s+p-1, \\ \boldsymbol{\eta}_s &= \psi(f(\mathbf{y}^s)). \end{aligned}$$

*Remark 10.2* In theory, the linear interpolation is valid for the generators corresponding to the respective Koopman operators, i.e.,

$$\mathscr{L} f = \lim_{t \to 0} \frac{U^t f - f}{t}.$$

Linear interpolation of the corresponding Koopman operators can be viewed as an explicit Euler approximation of the differential equation for the observable defined by the generator. Thus, it is valid for small time steps $h$. △

*Remark 10.3* Equation (K-MPC) is a bilinear control problem. Efficient algorithms specifically tailored to this problem class exist, see, e.g., [11, 33]. An alternative to using MPC would be, for instance, to solve a state-dependent Riccati equation [8] in order to obtain a closed-loop controller. In the experiments conducted here, we simply use an SQP solver for nonlinear optimization problems. In this case, the acceleration is entirely due to the faster model evaluation. △

*Remark 10.4* The linear interpolation approach is only valid if the control system $\Phi(\mathbf{y}, \mathbf{u})$ depends linearly on $\mathbf{u}$ which is not always the case. (In [34], the influence of nonlinear control dependencies has been studied.) In these situations, a way to reduce the inaccuracy of the bilinear system (K-ROM) is to introduce multiple bilinear K-ROMs which are valid locally, which is inspired by similar concepts in the reduced-basis community, see, e.g., [1]. Consequently, the linear interpolation is performed between two operators which are less far apart in terms of the control input. (We are here using single input for ease of notation, but the concept can be easily extended to higher dimensional control inputs.) This means that we approximate several Koopman operators $\mathbf{U}_{u^j}$ corresponding to $u^0 < u^1 < \cdots < u^{n_c-1}$. The K-ROM then consists of several locally valid bilinear models:

$$
\left[\mathbf{A},\ \mathbf{B}\right] = \begin{cases} \left[\mathbf{U}_{u^0},\ \mathbf{U}_{u^1} - \mathbf{U}_{u^0}\right] & \text{for } u \in [u^0, u^1) \\ \left[\mathbf{U}_{u^1},\ \mathbf{U}_{u^2} - \mathbf{U}_{u^1}\right] & \text{for } u \in [u^1, u^2) \\ \quad\vdots & \quad\vdots \\ \left[\mathbf{U}_{u^{n_c-2}},\ \mathbf{U}_{u^{n_c-1}} - \mathbf{U}_{u^{n_c-2}}\right] & \text{for } u \in [u^{n_c-2}, u^{n_c-1}]. \end{cases} \tag{10.6}
$$

Note that due to this, the control system is continuous and piecewise smooth with possible kinks at $u^1, u^2, \ldots, u^{n_c-2}$, which has to be taken into account by the optimization routine solving (K-MPC). △

## 10.5 On the Influence of the Amount of Data and the Selection of Basis Functions

In this section, we study the two examples already mentioned in Table 10.1 in more detail. Since the convergence result for EDMD only holds for infinitely large dictionaries $\Psi$ as well as infinitely many data points, the assumptions of the convergence theorems are obviously not satisfied in a practical setting. This means that we need to investigate the influence of the basis functions used in the construction of the dictionary $\Psi$ as well as the impact of the amount of training data on the controller performance. Furthermore, we compare the two K-ROM approaches against each other and against the full solution. The latter is only possible for the Burgers equation as the numerical effort for solving the Navier–Stokes-based MPC problem is prohibitively large.

### 10.5.1 *Test Cases and Reference Setup*

Here, we first introduce the two test cases and validate the K-ROM approach using one particular numerical setup. All algorithms except the Navier–Stokes simulations are implemented in MATLAB. For the switched systems optimization, all possible $n_c{}^p$ inputs $\mathbf{u}$ are evaluated as motivated in Sect. 10.4.1. For the bilinear K-ROM, the MATLAB function *fmincon* is applied which uses a sequential quadratic programming (SQP, see [32]) approach.

#### 10.5.1.1 The 1-Dimensional Burgers Equation

We now compare the solutions obtained by the full control problems and their K-ROM approximations, respectively, using the problem setup introduced in Sect. 10.4.1.1. For the switched system approach, we choose the inputs $u^0 = -0.075$, $u^1 = 0.075$, and $u^2 = 0$. For the bilinear surrogate model, we use $u^0$ and $u^1$ to construct $A$ and $B$.

For the data collection process, we use three different initial conditions and for each of these, we perform one simulation with constant inputs $u^0$, $u^1$ and $u^2$, respectively. Finally, we perform an additional simulation with a constant switching sequence between the inputs. This yields 12 simulations in total, each of which is 60 s long, and we collect a snapshot every 0.005 s. The switched sequences are then split into three matrices according to which input is active during which time step. These data points are then attached to the respective snapshot matrices with constant inputs. The time step $h$ for the flow map $\Phi$ in the control problem as well as for the approximation of the Koopman operator is set to 0.5 s.

The performance of the reduced approaches is visualized in Fig. 10.4 (for a prediction horizon of length $p = 3$) where the switched approaches are compared in the left column and the continuous ones in the right column.

We see that in the switched systems approach, the inputs (Fig. 10.4a) vary significantly. This is due to the MPC framework. As soon as the two $\mathbf{z}$ trajectories differ slightly, the corresponding optimization problems do not necessarily possess the same optimal solution any longer. Consequently, small inaccuracies may result in different control trajectories. However, we see in Fig. 10.4c, e that the value of the objective function is of comparable quality. In some parts (e.g., at $t \approx 10\,s$), the ripples around the reference trajectory (Fig. 10.4g) are slightly larger than in the PDE-constrained case (cf. Fig. 10.3f on p. 264). Nevertheless, it can be concluded that the switched systems K-ROM approach is very well suited for real-time control of the Burgers equation.

Looking at the continuous K-ROM, we see that we have an even better performance (cf. Fig. 10.3d), as can be expected due to the larger freedom in choosing the input to the system. We see that the difference between the PDE based and the K-ROM based solutions is similar to the switched systems case. However, a significant advantage is that we now require only data for two autonomous systems instead of three. This means that the data requirements can be further reduced by 33%. For the same reasons as in the switched systems case, we do not have a very good agreement between the optimal control. We will further study this effect in Sect. 10.5.2.

#### 10.5.1.2 The 2-Dimensional Navier–Stokes Equations

The second example is the flow around a cylinder described by the 2-dimensional incompressible Navier–Stokes equations at a Reynolds number of $Re = 100$:

$$
\begin{aligned}
\dot{\mathbf{y}}(\mathbf{x},t) + \mathbf{y}(\mathbf{x},t)\cdot\nabla\mathbf{y}(\mathbf{x},t) &= \nabla p(\mathbf{x},t) + \frac{1}{Re}\Delta\mathbf{y}(\mathbf{x},t),\\
\nabla\cdot\mathbf{y}(\mathbf{x},t) &= 0,\\
\mathbf{y}(\mathbf{x},0) &= \mathbf{y}^0(\mathbf{x}),
\end{aligned}
$$

see Fig. 10.5a for the problem setup and a snapshot of the solution computed with *OpenFOAM* [18] using a finite volume discretization with 22,000 cells. The system

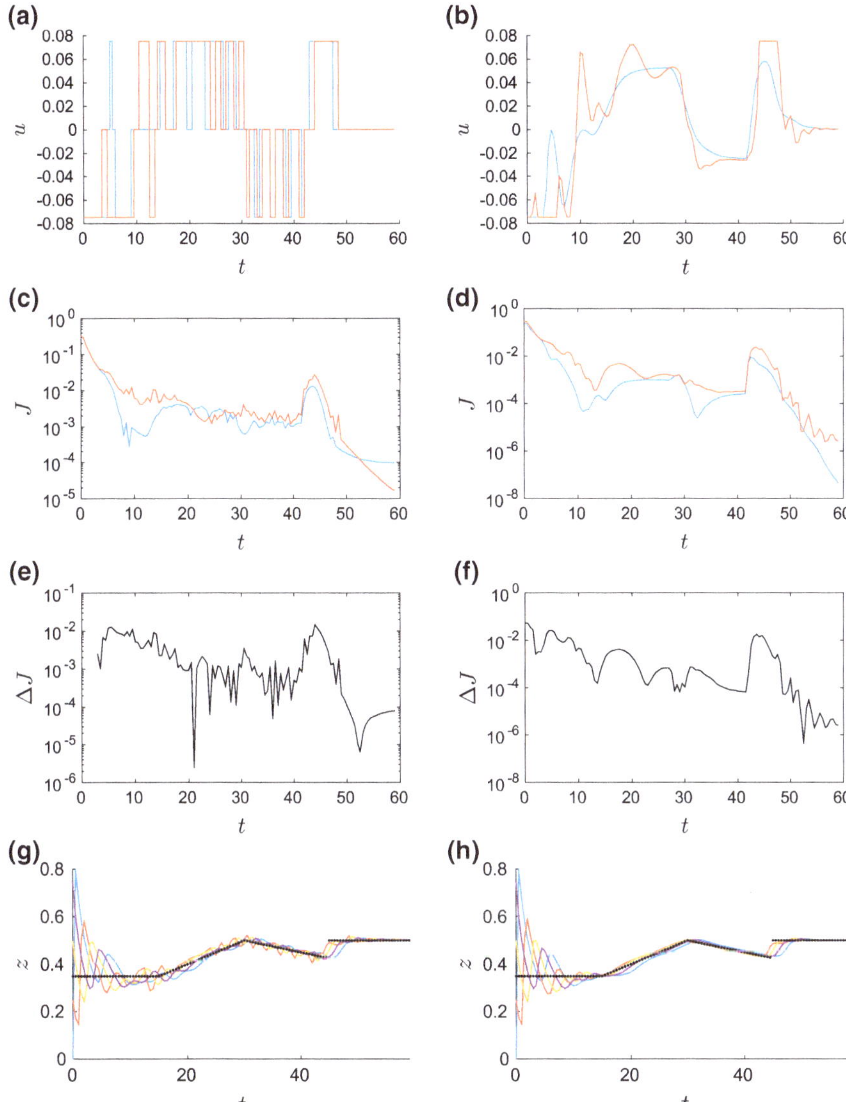


**Fig. 10.4** Left column: (MPC$^s$) versus (K-MPC$^s$). Right column: (MPC) versus (K-MPC). **a**, **b** Optimal control for the full problem (blue) and for the K-ROM based problem (orange). **c**, **d** The corresponding objective function values. **e**, **f** Absolute value of the difference between the full and the reduced solution. **g**, **h** Optimal trajectories of the observation **z**. For a comparison to the full system see Fig. 10.3f, e

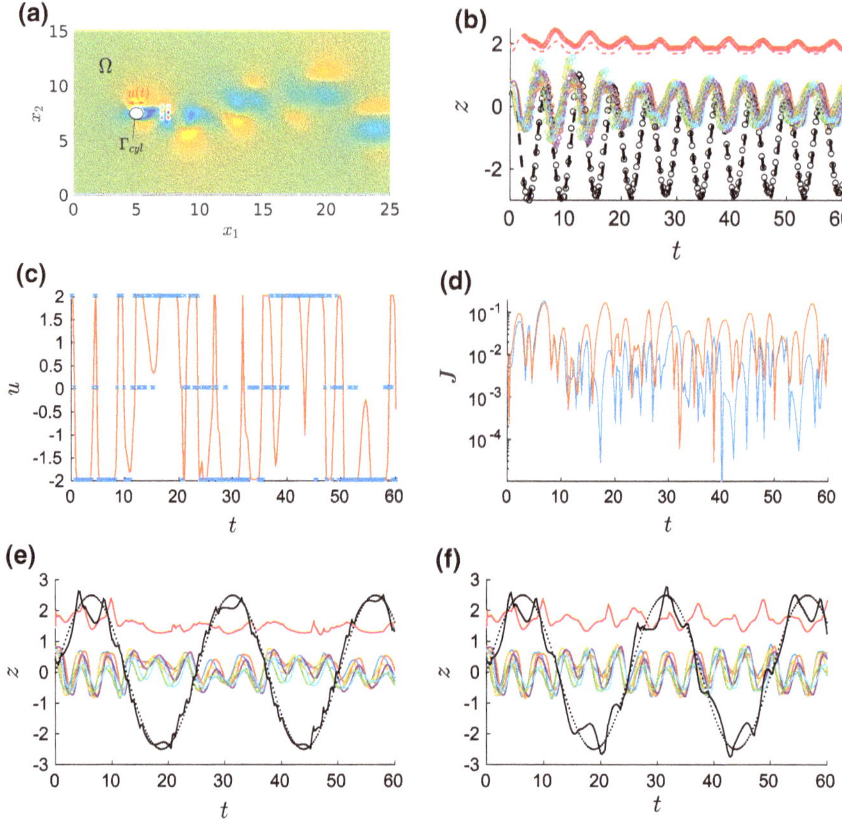


**Fig. 10.5** **a** Snapshot of the PDE simulation. Colored dots: sensor positions $(\mathbf{x}_1, \ldots, \mathbf{x}_6)$. **b** Comparison between the observation of the PDE solution (dots) and the bilinear K-ROM (K-ROM) $(u^0 = 0,\, u^1 = 2)$ for $u(t) = 1 + sin(t)$. **c** Control trajectories obtained by (K-MPC$^s$) (blue) and (K-MPC) (orange). **d** Objective function values corresponding to (**c**). **e** Trajectories of the observations $\mathbf{z} = f(\mathbf{y})$ corresponding to (K-MPC$^s$). Balck dots: reference trajectory for the lift coefficient (black line). Red line: drag coefficient. Horizontal velocities are colored according to the sensor positions in (**a**). **f** As in **e** but for (K-MPC)

is controlled via rotation of the cylinder, i.e., $u(t)$ is the angular velocity. Without control, the well-known *von Kármán vortex street* occurs.

Similar to the Burgers example, we do not observe the full state. Since we want to control the vertical force on the cylinder (i.e., the lift), we directly observe the lift coefficient $C_l$. In addition, we observe the drag coefficient $C_d$ and the vertical velocity at six points $(\mathbf{x}_1, \ldots, \mathbf{x}_6)$ in the cylinder wake (see Fig. 10.5a), which yields the following observation:

$$\mathbf{z}(t) = (C_l(t),\; C_d(t),\; \mathbf{y}_2(\mathbf{x}_1, t), \ldots,\; \mathbf{y}_2(\mathbf{x}_6, t))^\top .$$

As already mentioned, we want to influence the lift by rotating the cylinder. Since the lift coefficient is one of the observables, we simply have to track the corresponding entry of $\mathbf{z}$ in the MPC problem:

$$\min_{\mathbf{u}\in\{u^0,u^1,u^2\}^p} \sum_{i=s}^{s+p-1} \left(\mathbf{z}_{i,1} - \mathbf{z}_i^{\text{opt}}\right)^2.$$

Here, we introduce three autonomous systems with the constant cylinder rotations $u_0 = -2$, $u_1 = 0$, $u_2 = 2$ for both K-ROM approaches. The data is collected from one long-time simulation over 3000 s with a random switching. As the lag time, we choose $h = 0.25$. The fact that we have three autonomous systems means that we use the localized ROM concept (10.6) for the bilinear model. The MPC solutions to both reduced problem formulations (both with a prediction horizon of length $p = 5$) are compared in Fig. 10.5. We see in (b) a comparison between a PDE simulation and the bilinear K-ROM, and very good agreement is observed despite the significant dimension reduction. In Fig. 10.5c, d, the two K-ROM based solutions are compared and we observe almost equal quality of the solution. Interestingly, the switching approach is superior to the bilinear K-ROM. The reason for this is likely that due to the localized K-ROM approach, the objective function possesses many non-smooth kinks for $p = 5$. Consequently, the true optimum is difficult to compute numerically without using algorithms specifically tailored to continuous, piecewise-smooth problems. Furthermore, inaccuracies introduced via linear interpolation become more significant for longer prediction horizons, i.e., for larger $p$.

### 10.5.2 Data Sampling and Basis Selection

We have seen in the previous section that both approaches are capable of controlling complex PDE constrained problems in real time. As already mentioned, the convergence result for EDMD only holds for infinitely large dictionaries $\Psi$ and infinitely many data points. Consequently, we now study the influence of different numerical parameters on the solution quality since these assumptions are not met in a practical setting.

In Fig. 10.6, the influence of the maximal order of the monomial basis and the size of the training dataset is visualized for the Burgers example. In accordance with Sect. 10.5.1.1, we have taken three K-ROMs for the switched systems approach and two for the bilinear model. The order of the polynomials directly influences the dimension of the K-ROM such that the speedup crucially depends on this choice, see Fig. 10.6a, where we have speedup factors of approximately 200 for a maximum order of 1 (i.e., standard DMD). This factor then reduces to roughly 50 for polynomials up to order 5. However, we see that for both K-ROM approaches, it is sufficient to consider monomials of order 2. Another benefit of these lower dimensional K-ROMs is that the amount of required data is smaller. The larger the surrogate model

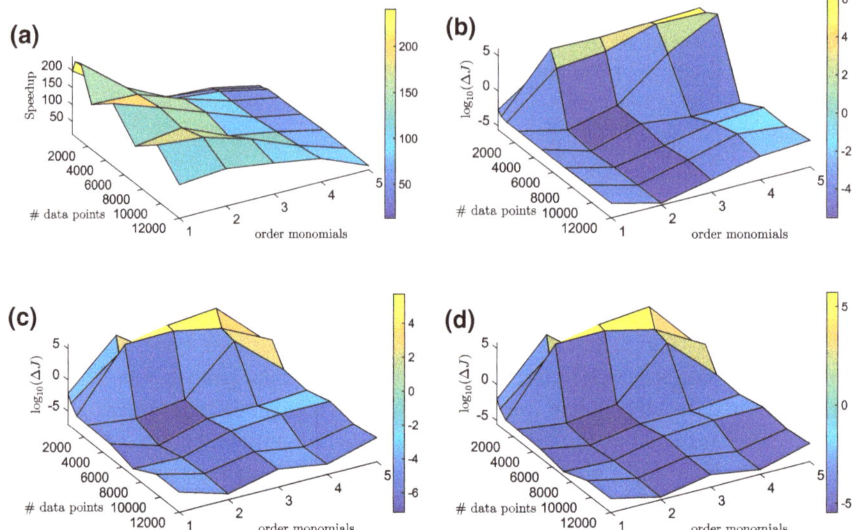


**Fig. 10.6** Analysis of the performance of the two K-ROM approaches depending on the order of monomials as well as the number of data points used for the training for the 1-dimensional Burgers equation. **a** Speedup in comparison to the PDE solver. **b** to **d** Integrated error $\Delta J$ between the K-ROM and the full solution. **b** Equation (K-MPC) versus (MPC). **c** (K-MPC$^s$) versus (MPC$^s$). **d** (K-MPC$^s$) versus (MPC)

is, the more data we need to compute satisfactory approximations of the Koopman operator. In fact, it appears that the standard DMD approach is the most robust concerning the amount of training data. When comparing Fig. 10.6b, d, where the distance between the K-ROM approaches and the continuous PDE-constrained MPC problem are compared, we see that—surprisingly—the switched systems approach yields a feedback behavior of similar quality compared to the bilinear K-ROM. Note, however, that the amount of training data is smaller for the bilinear model.

When studying the Navier–Stokes example, the picture is very similar, cf. Fig. 10.7. Since the PDE-constrained solution is not available, we here plot the integrated objective function value of the PDE model: $\overline{J} = \int_{t_0}^{t_e} J(t)\,dt$. Similar to the previous example, the DMD based version appears to be the most robust concerning the data requirements. Again, both approaches are very similar in performance, and considering monomials of order larger than two is significantly inferior.

*Remark 10.5* (*Influence of the number of K-ROMs*) As a final experiment, we study how the number of localized K-ROMs influences the solution. To this end, we revisit the Burgers example with three different discretizations of $u$. We consider the cases with two, three, and five inputs at which data is available. We see in Fig. 10.8 that no real trend can be identified. It is interesting to see, however, that no improvement can be observed when increasing the number of control inputs from three to five. On the contrary, increasing the number of ROMs may even be disadvantageous,

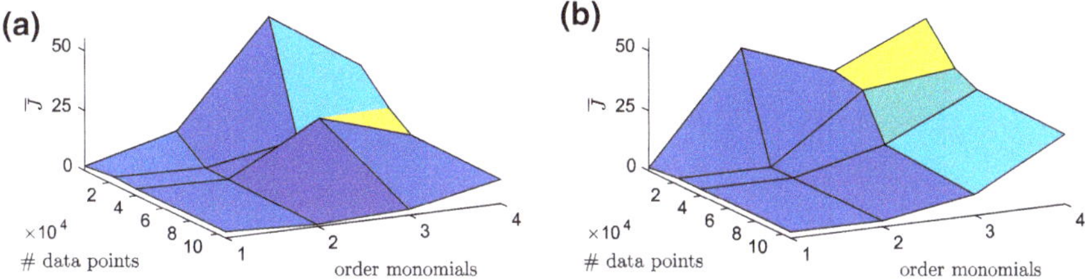


**Fig. 10.7** Analysis of the performance of the two K-ROM approaches depending on the order of monomials as well as the number of data points used for the training for the 2-dimensional Navier–Stokes equations. **a** Integrated objective function value for (K-MPC$^S$). **b** Integrated objective function value for (K-MPC).

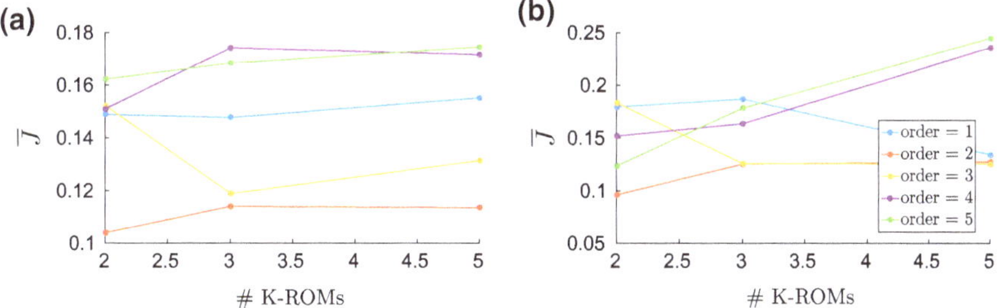


**Fig. 10.8** Integrated objective function value for (K-MPC$^S$) (**a**) and (K-MPC) (**b**), evaluated at a medium data value (i.e., # data points ≈ 4000, cf. Fig. 10.6d)

in particular if the amount of data is insufficient for a good approximation of the Koopman operator. In conclusion, a small to moderate number of control inputs in combination with the classical DMD approximation appears to be the most efficient and robust choice for the problems studied here. △

## 10.6 Online Updates Using Sensor Data

During the MPC algorithm, new sensor data is obtained at every sample time $h$. Furthermore, these data points are collected in regions of state space which are of particular interest since they are close to the desired state. Therefore, it is a natural idea to use this data to further improve the quality of the K-ROM. To this end, we make use of the idea developed in [16], where incremental updates of the DMD approximation are performed in a way that we do not have to store data points that have already been taken into account. We make use of the transformation

$$\mathbf{U}^\top = \mathbf{A}\mathbf{G}^+$$

with

$$\mathbf{A} = \frac{1}{m} \sum_{i=1}^{m} \Psi(\widetilde{\mathbf{z}}_i)\Psi(\mathbf{z}_i)^\top,$$
$$\mathbf{G} = \frac{1}{m} \sum_{i=1}^{m} \Psi(\mathbf{z}_i)\Psi(\mathbf{z}_i)^\top,$$

see [21, 48] for details. We see that in order to update $\mathbf{U}^\top$, we merely have to store the (in our case low-dimensional) matrices $\mathbf{A}$ and $\mathbf{G}$. Each time we obtain a new snapshot pair $(\mathbf{z}_{m+1}, \widetilde{\mathbf{z}}_{m+1})$ we can update the EDMD approximation via

$$\begin{aligned} \widehat{\mathbf{A}} &= \frac{m\mathbf{A} + q\left(\Psi(\widetilde{\mathbf{z}}_{m+1})\Psi(\mathbf{z}_{m+1})^\top\right)}{m+q}, \\ \widehat{\mathbf{G}} &= \frac{m\mathbf{G} + q\left(\Psi(\mathbf{z}_{m+1})\Psi(\mathbf{z}_{m+1})^\top\right)}{m+q}, \\ \widehat{\mathbf{U}}^\top &= \widehat{\mathbf{A}}\widehat{\mathbf{G}}^+, \end{aligned}$$

where $q \in \mathbb{N}^{\geq 1}$ is a weight parameter. Note that we have to set $q = 1$ in order to obtain the standard EDMD procedure. Alternatively, we can determine $q$ in such a way that the update has a higher impact, e.g., by some prescribed percentage $\varepsilon$:

$$q = \left\lfloor m \frac{\varepsilon}{1-\varepsilon} \right\rfloor.$$

The most expensive part for this update is to compute the pseudoinverse of $\widehat{\mathbf{G}}$. However, since the K-ROM dimension is generally low, it can be computed efficiently.

We now apply this procedure to the Navier–Stokes example. As we have seen in Sect. 10.5.2, the standard DMD approximation is most stable in the low data limit. To illustrate the data efficiency of our approach, we consequently choose DMD for the K-ROM computation in this section and start with only 50 data points for each of the three autonomous systems $u^0 = -2$, $u^1 = 0$, and $u^2 = 2$.

We set $\varepsilon = 0.025$ and store the matrices $\mathbf{A}$ and $\mathbf{G}$ and sensor data that we measure during the MPC algorithm. We then update the three autonomous systems every 10 s. Note that this approach is only viable for the switching approach (K-MPC$^s$) since otherwise, we collect data at intermediate control values for which we do not have a reduced order model. For this approach, the setup proposed in [2, 24] would be more appropriate.

Figure 10.9 shows the results for the lift tracking problem, where we compare the online adaptation with the pure offline training considered before. For the comparison, we consider a longer trajectory of 500 s with

$$z^{\mathsf{opt}}(t) = 2.5 \sin\left(\frac{t}{4}\right) \cos\left(\frac{t}{20}\right).$$

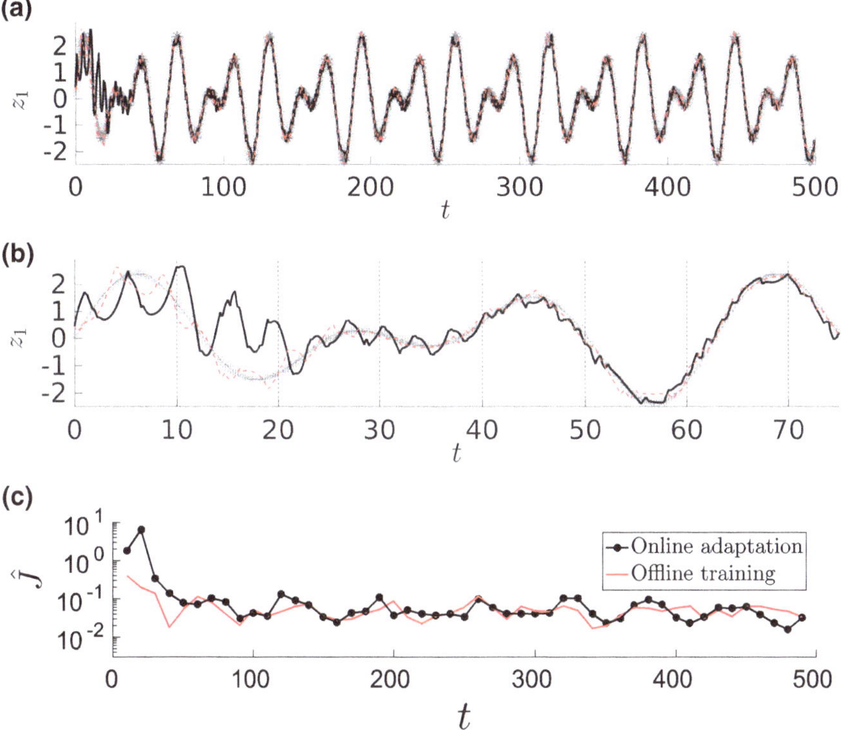


**Fig. 10.9** Online K-ROM adaptation within (K-MPC$^s$). **a** Optimal lift trajectory obtained via online adaptation (black) and the standard approach with offline training (red). The reference trajectory is marked by the gray stars. **b** As in **a**, zoom into the first 75 s. The vertical dashed lines indicate the time instances at which the K-ROMs are updated. **c** Objective function value integrated over the past 10 s. The dots mark the points where the K-ROMs are updated

In order to evaluate the online procedure, we compare the objective function value integrated over the past 10 s, i.e., over the period with the current K-ROM approximation:

$$\hat{J}(t) = \int_{t-10}^{t} \left(\mathbf{z}_1(\tau) - z^{\text{opt}}(\tau)\right)^2 \, d\tau.$$

We see that after the first four to five updates, we already have a very good agreement. Consequently, a K-ROM based controller can be set up with a very small amount of data and then be updated regularly. After a short period of time, the performance is equal to that with an extensive offline phase.

It should be noted that this approach is only applicable if the system cannot stay arbitrarily far from the reference trajectory. Otherwise, it could happen that a strong deterioration occurs from the beginning. Consequently, the resulting training data would not be suited for improving the control performance.

## 10.7 Conclusion

We have presented two methods for solving PDE-constrained optimal control problems using the Koopman operator for significant speedup. In order to increase the data efficiency, we only collect data for a small number of constant inputs. The corresponding control problem then becomes a switching problem. Alternatively, we obtain a bilinear surrogate model via linear interpolation between two K-ROMs. Based on a recent convergence result for EDMD, convergence of the K-ROM based optimization problems can be shown. Extensive numerical studies show the applicability of both methods to nonlinear PDE-constrained problems. Interestingly, a simple DMD approximation of the Koopman operator is beneficial both for the solution quality and robustness with respect to the amount of training data. The reason for the good performance is that predictability only has to be accurate for a small number of time steps. Finally, an extension to online adaptations using sensor data has been presented and validated. This way, we can set up real-time controllers with very limited data.

For future work, it will be interesting to validate the methodology in experiments. First results for ODE-constrained problems (electrical drives) are promising [15]. Furthermore, the high efficiency is a great motivator to employ this model reduction technique in even more computationally challenging tasks such as multi-objective optimal control [35]. Finally, the system dynamics of the examples considered here are fairly well behaved. Consequently, it will be of great interest to study the presented approaches for more complex dynamical systems, e.g., for decreasing viscosity/increasing Reynolds numbers.

## References

1. Albrecht, F., Haasdonk, B., Kaulmann, S., Ohlberger, M.: The localized reduced basis multiscale method. Proceedings of ALGORITHMY **2012**, 393–403 (2012)
2. Arbabi, H., Korda, M., Mezic, I.: A data-driven Koopman model predictive control framework for nonlinear flows (2018). arXiv:1804.05291
3. Bellmann, R.E., Stuart, E.D.: Applied Dynamic Programming. Princeton University Press, Princeton (2015)
4. Benner, P., Gugercin, S., Willcox, K.: A survey of projection-based model reduction methods for parametric dynamical systems. SIAM Rev. **57**(4), 483–531 (2015)
5. Bergmann, M., Cordier, L.: Optimal control of the cylinder wake in the laminar regime by trust-region methods and POD reduced-order models. J. Comput. Phys. **227**(16), 7813–7840 (2008)
6. Brunton, S.L., Brunton, B.W., Proctor, J.L., Kutz, J.N.: Koopman invariant subspaces and finite linear representations of nonlinear dynamical systems for control. PLoS ONE **11**(2), 1–19 (2016)
7. Budišić, M., Mohr, R., Mezić, I.: Applied Koopmanism. Chaos **22** (2012)
8. Cimen, T.: State-dependent Riccati Equation (SDRE) control: a survey. In: IFAC Proceedings Volumes, vol. 41, pp. 3761–3775. IFAC (2008)
9. Egerstedt, M., Wardi, Y., Axelsson, H.: Transition-time optimization for switched-mode dynamical systems. IEEE Trans. Autom. Control **51**(1), 110–115 (2006)

10. Egerstedt, M., Wardi, Y., Delmotte, F.: Optimal control of switching times in switched dynamical systems. In: 42nd IEEE International Conference on Decision and Control (CDC), pp. 2138–2143 (2003)
11. Elliott, D.: Bilinear Control Systems. Springer Science + Business Media, Berlin (2009)
12. Fahl, M.: Trust-region methods for flow control based on reduced order modelling. Ph.D. Thesis, University of Trier (2000)
13. Flaßkamp, K., Murphey, T., Ober-Blöbaum, S.: Discretized switching time optimization problems. In: 12th European Control Conference, pp. 3179–3184 (2013)
14. Grüne, L., Pannek, J.: Nonlinear Model Predictive Control, 2nd edn. Springer International Publishing, Berlin (2017)
15. Hanke, S., Peitz, S., Wallscheid, O., Klus, S., Böcker, J., Dellnitz, M.: Koopman operator based finite-set model predictive control for electrical drives (2018). arXiv:1804.00854
16. Hemati, M.S., Williams, M.O., Rowley, C.W.: Dynamic mode decomposition for large and streaming datasets. Phys. Fluids **26**(111701), 1–6 (2014)
17. Hinze, M., Volkwein, S.: Proper orthogonal decomposition surrogate models for nonlinear dynamical systems: error estimates and suboptimal control. In: Benner, P., Sorensen, D.C., Mehrmann, V. (eds.) Reduction of Large-Scale Systems, vol. 45, pp. 261–306. Springer, Berlin (2005)
18. Jasak, H., Jemcov, A., Tukovic, Z.: OpenFOAM : A C++ Library for Complex Physics Simulations. In: International Workshop on Coupled Methods in Numerical Dynamics, pp. 1–20 (2007)
19. Kaiser, E., Kutz, J.N., Brunton, S.L.: Data-driven discovery of Koopman eigenfunctions for control (2017). arXiv:1707.0114
20. Klus, S., Gelß, P., Peitz, S., Schütte, C.: Tensor-based dynamic mode decomposition. Nonlinearity **31**(7), 3359–3380 (2018)
21. Klus, S., Koltai, P., Schütte, C.: On the numerical approximation of the Perron–Frobenius and Koopman operator. J. Comput. Dyn. **3**(1), 51–79 (2016)
22. Klus, S., Nüske, F., Koltai, P., Wu, H., Kevrekidis, I., Schütte, C., Noé, F.: Data-driven model reduction and transfer operator approximation. J. Nonlinear Sci. **28**(3), 985–1010 (2018)
23. Koopman, B.O.: Hamiltonian systems and transformations in Hilbert space. Proc. Natl. Acad. Sci. **17**(5), 315–318 (1931)
24. Korda, M., Mezić, I.: Linear predictors for nonlinear dynamical systems: Koopman operator meets model predictive control. Automatica **93**, 149–160 (2018)
25. Korda, M., Mezić, I.: On convergence of extended dynamic mode decomposition to the Koopman operator. J. Nonlinear Sci. **28**(2), 687–710 (2018)
26. Kunisch, K., Volkwein, S.: Control of the burgers equation by a reduced-order approach using proper orthogonal decomposition. J. Optim. Theory Appl. **102**(2), 345–371 (1999)
27. Lasota, A., Mackey, M.C.: Chaos, fractals, and noise: stochastic aspects of dynamics. Applied Mathematical Sciences, vol. 97, 2nd edn. Springer, Berlin (1994)
28. Lassila, T., Manzoni, A., Quarteroni, A., Rozza, G.: Model order reduction in fluid dynamics: challenges and perspectives. In: Quarteroni, A., Rozza, G. (eds.) Reduced Order Methods for Modeling and Computational Reduction, pp. 235–273. Springer, Cham (2014)
29. Mezić, I.: Spectral properties of dynamical systems, model reduction and decompositions. Nonlinear Dyn. **41**, 309–325 (2005)
30. Mezić, I.: Analysis of fluid flows via spectral properties of the Koopman operator. Annu. Rev. Fluid Mech. **45**, 357–378 (2013)
31. Mezić, I., Banaszuk, A.: Comparison of systems with complex behavior. Phys. D Nonlinear Phenom. **197**, 101–133 (2004)
32. Nocedal, J., Wright, S.J.: Numerical Optimization, 2nd edn. Springer Series in Operations Research and Financial Engineering. Springer Science & Business Media, Berlin (2006)
33. Pardalos, P.M., Yatsenko, V.: Optimization and Control of Bilinear Systems. Springer, Berlin (2008)
34. Peitz, S.: Controlling nonlinear PDEs using low-dimensional bilinear approximations obtained from data (2018). arXiv:1801.06419

35. Peitz, S., Dellnitz, M.: A survey of recent trends in multiobjective optimal control surrogate models, feedback control and objective reduction. Math. Comput. Appl. **23**(2) (2018)
36. Peitz, S., Klus, S.: Koopman operator-based model reduction for switched-system control of PDEs. Automatica **106**, 184–191 (2019)
37. Proctor, J.L., Brunton, S.L., Kutz, J.N.: Dynamic mode decomposition with control. SIAM J. Appl. Dyn. Syst. **15**(1), 142–161 (2015)
38. Proctor, J.L., Brunton, S.L., Kutz, J.N.: Generalizing Koopman Theory to allow for inputs and control. SIAM J. Appl. Dyn. Syst. **17**(1), 909–930 (2018)
39. Qian, E., Grepl, M., Veroy, K., Willcox, K.: A Certified Trust Region Reduced Basis Approach to PDE-Constrained Optimization. ACDL Technical Report TR16-3 (2016)
40. Rowley, C.W.: Model reduction for fluids, using balanced proper orthogonal decomposition. Int. J. Bifurc. Chaos **15**(3), 997–1013 (2005)
41. Rowley, C.W., Mezić, I., Bagheri, S., Schlatter, P., Henningson, D.S.: Spectral analysis of nonlinear flows. J. Fluid Mech. **641**, 115–127 (2009)
42. Schmid, P.J.: Dynamic mode decomposition of numerical and experimental data. J. Fluid Mech. **656**, 5–28 (2010)
43. Sirovich, L.: Turbulence and the dynamics of coherent structures part I: coherent structures. Q. Appl. Math. **45**(3), 561–571 (1987)
44. Stellato, B., Ober-Blöbaum, S., Goulart, P.J.: Optimal control of switching times in switched linear systems. In: IEEE 55th Conference on Decision and Control, pp. 7228–7233 (2016)
45. Stellato, B., Ober-Blöbaum, S., Goulart, P.J.: Second-order switching time optimization for switched dynamical systems. IEEE Trans. Autom. Control. **62**(10), 5407–5414 (2017)
46. Tröltzsch, F., Volkwein, S.: POD a-posteriori error estimates for linear-quadratic optimal control problems. Comput. Optim. Appl. **44**(1), 83–115 (2009)
47. Tu, J.H., Rowley, C.W., Luchtenburg, D.M., Brunton, S.L., Kutz, J.N.: On dynamic mode decomposition: theory and applications. J. Comput. Dyn. **1**(2), 391–421 (2014)
48. Williams, M.O., Kevrekidis, I.G., Rowley, C.W.: A data-driven approximation of the Koopman operator: extending dynamic mode decomposition. J. Nonlinear Sci. **25**(6), 1307–1346 (2015)
49. Zhu, F., Antsaklis, P.J.: Optimal control of hybrid switched systems: a brief survey. Discret. Event Dyn. Syst. **25**(3), 345–364 (2015)

# Chapter 11
# Solving Optimal Control Problems for Monotone Systems Using the Koopman Operator

**Aivar Sootla, Guy-Bart Stan and Damien Ernst**

**Abstract** Koopman operator theory offers numerous techniques for analysis and control of complex systems. In particular, in this chapter we will argue that for the problem of convergence to an equilibrium, the Koopman operator can be used to take advantage of the geometric properties of controlled systems, thus making the optimal solutions more transparent, and easier to analyze and implement. The motivation for the study of the convergence problem comes from biological applications, where easy-to-implement and easy-to-analyze solutions are of particular value. At the moment, theoretical results have been developed for a class of nonlinear systems called monotone systems. However, the core ideas presented here can be applied heuristically to non-monotone systems. Furthermore, the convergence problem can serve as a building block for solving other control problems such as switching between stable equilibria or inducing oscillations. These applications are illustrated in biologically inspired numerical examples.

## 11.1 Introduction

Research in this chapter was motivated by applications in a rather new field of science called Synthetic Biology, which aims at the design of biological devices for useful purposes, such as cell-based production of valuable chemical compounds (cellular biofactories), detection of toxins or viruses in a sample (whole-cell biosensors),

---

A. Sootla (✉)
Department of Engineering Science, University of Oxford, Parks Road, Oxford OX1 3PJ, United Kingdom
e-mail: aivar.sootla@eng.ox.ac.uk

G.-B. Stan
Department of Bioengineering, Imperial College London, South Kensington Campus, London SW7 2AZ, United Kingdom
e-mail: g.stan@imperial.ac.uk

D. Ernst
Montefiore Institute, University of Liège, Allée de la Découverte 10, 4000 Liège, Belgium
e-mail: dernst@uliege.be



A. Mauroy et al. (eds.), *The Koopman Operator in Systems and Control*,
Lecture Notes in Control and Information Sciences 484,
https://doi.org/10.1007/978-3-030-35713-9_11

etc. [4, 8, 25]. Engineering novel biological devices are of high interest since the cell machinery opens the door to the synthesis of many useful organic compounds, which cannot be easily realized by other means. This opens avenues to novel applications in biotechnology, bioprocess engineering, and cell-based medicine. This introduction contains only the basic information on challenges and advances of control theory applications in the field of Synthetic Biology, and we refer the interested reader to [10] for additional background.

The DNA of living organisms encodes information about their biological functions such as growth, development, replication, etc. The DNA is a collection of genes, which are read in the processes called gene expression into molecules such as RNAs, proteins, etc., which serves a particular purpose within a cell. Linking the information contained in the DNA to the genetic functions it encodes is not straightforward, since proteins, RNAs, and genes interact with each other, creating complex and often poorly characterized networks. For example, some proteins are known to block or activate the expression of particular genes, while other proteins can bind to each other forming inert complexes. The abundance of complex and poorly characterized systems was one of the reasons why synthetic biology attracted interest from the feedback control community. A simple example of a feedback control problem in the context of cell biology is forcing the amount of a particular protein in the cell to reach a particular level using exogenous stimuli (i.e., control signals). In control theory, we typically design these signals based on observations of the system (called measurements) to serve a particular purpose (mathematically formulated as minimizing a *predefined* objective function). External control of protein levels in microorganisms was shown to be possible in [21, 22, 39], where sensing and actuating are implemented using native or engineered cellular mechanisms. A typical way of performing real-time measurements is fusing a fluorescent protein (for example, mCherry, sfGFP, or Venus) [11] to a protein of interest, and using microscopy to measure the fluorescence levels. Actuating can be performed using chemical, light, or temperature induction, where a specific chemical, light wavelength, or temperature profile (heat shock), respectively, are used to activate or shut down gene expression and thereby the production of certain proteins [11]. While the research mentioned above was successful in controlling simple biochemical processes at both single cell and population levels, many problems were identified. Quantitative measurement of biological signals, which is of paramount importance for effective control, is still a problem due to current technological limitations. This is because fluorescence measurements depend on the measurement device. Furthermore, optimal control methods, such as dynamic programming, prohibit temporal constraints on control signals. For instance, it is not straightforward to constrain control signals to be step functions in time, however, such restrictions are often met in biological applications.

Considering the modeling and control challenges delineated above, we looked for other techniques that can be used for control of biological processes. In what follows, we focus on the Koopman operator framework, which is a linear infinite-dimensional representation of nonlinear dynamical systems. Due to its linearity, the Koopman operator allows extending the notions of eigenvalues and eigenvectors (called eigenfunctions in the Koopman framework) to nonlinear systems.

Following [19], we consider the Koopman operator on a basin of attraction of exponentially stable equilibrium. If the slowest Koopman eigenvalue is simple, the corresponding eigenfunction (called the dominant eigenfunction) offers convenient geometric analysis tools [20]. In particular, (under some technical conditions) the trajectories originating on one level set of the dominant eigenfunction reach other level sets simultaneously. Furthermore, the equilibrium lies on the zero level set of the dominant eigenfunction. Therefore, the values of the dominant eigenfunction at a point in the state space can be used to estimate the convergence time from this point to a neighborhood around the equilibrium [20]. We will elaborate on this property in Sect. 11.2.2, but we note that the convergence time (and the values of the dominant eigenfunction) can serve as a qualitative measure of distance to the equilibrium.

As a starting point, we consider the study [35], in which the authors restricted the class of considered systems to be monotone (see [2, 9]) and solved the problem of switching between the equilibria of a bistable system using very simple open-loop control signals. Although monotone systems represent a (theoretically) restrictive class of nonlinear systems, many biological systems can be modeled as monotone systems [28]. While the authors of [35] did not present a mathematical formulation for an optimal control problem, which yields the open-loop control signals computed in [35], the study itself was an indication that such a problem formulation may exist. The Koopman operator approach to the problem was proposed in [29], but an optimal control interpretation was still lacking. In this chapter, we focus on the results from [33, 36], where the problem of convergence to an exponentially stable equilibrium was solved for monotone systems. The main idea in [33, 36] was replacing the Euclidean distance to the equilibrium with the convergence time to a specific level set of the dominant eigenfunction, which can be seen as a qualitative measure of distance to the equilibrium. The new problem formulation also allows casting a dynamic optimization program as a static one, thus significantly simplifying the computations while providing easy-to-implement control signals.

It is worth mentioning that the Koopman operator has been employed in optimal control applications, albeit, in a different context. In [12, 14, 23, 24] a data-based approach was taken, which allowed inferring control laws based on the observed or simulated trajectories. In [18] the convergence problem was formulated using the Koopman operator. While the class of systems was not restricted, full controllability was assumed, i.e., control signals affect all the states in an affine manner. In [40] it was proposed to synchronize cardiac cells by formulating an optimal control problem with the state at the terminal time restricted to a specified level set of the dominant eigenfunction. However, the authors did not parametrize the control signal, which led to complicated optimal control signals.

The rest of the chapter is organized as follows. In Sect. 11.2, we cover basic definitions, pertaining to the Koopman operator and monotone systems theory. We set up the convergence problem in Sect. 11.3 on a specific case study and discuss previous research addressing the problem, its limitations and main outcomes. In Sect. 11.4, we discuss how to formulate the convergence problem in an easier way so as to provide efficient solutions to the convergence problem. Section 11.5 is devoted to possible generalizations, while in Sect. 11.6 we illustrate the results on numerical examples. We conclude in Sect. 11.7.

## 11.2 Theoretical Background

### *11.2.1 Dynamical Systems*

Consider a system of the form

$$\dot{\mathbf{x}} = \mathbf{F}(\mathbf{x}, u), \quad \mathbf{x}(0) = \mathbf{x}^0, \tag{11.1}$$

with $\mathbf{F} : \mathscr{D} \times \mathscr{U} \to \mathbb{R}^n$, $u : \mathbb{R} \to \mathscr{U}$, and where the sets $\mathscr{D} \subset \mathbb{R}^n$, $\mathscr{U} \subset \mathbb{R}$ are convex, compact, and $u$ belongs to the space $\mathscr{U}_\infty$ of Lebesgue measurable functions with values from $\mathscr{U}$. We assume that $\mathbf{F}(\mathbf{x}, u)$ is twice continuously differentiable ($C^2$) in $(\mathbf{x}, u)$ on an open set containing $\mathscr{D} \times \mathscr{U}$. Compactness of $\mathscr{D}, \mathscr{U}$ is required for monotonicity certificates, but convexity can be relaxed [2]. Smoothness of $F$ is required for Koopman operator analysis [19, 20]. The flow map $\mathbf{S} : \mathbb{R} \times \mathscr{D} \times \mathscr{U}_\infty \to \mathbb{R}^n$ induced by the system is such that $\mathbf{S}(t, \mathbf{x}^0, u)$ is a solution of (11.1) with an initial condition $\mathbf{x}^0$ and a control signal $u$. We denote the closure of the set $X$ as $\text{cl}(X)$, and its interior as $\text{int}(X)$. We denote the differential of a function $\mathbf{G}(\mathbf{x}, \mathbf{y}) : \mathbb{R}^n \times \mathbb{R}^m \to \mathbb{R}^k$ with respect to $\mathbf{x}$ as $\partial_{\mathbf{x}}\mathbf{G}(\mathbf{x}, \mathbf{y})$. Let $\mathbf{J}(\mathbf{x})$ denote the Jacobian matrix of $\mathbf{F}(\mathbf{x}, 0)$ (i.e., $\mathbf{J}(\mathbf{x}) = \partial_{\mathbf{x}}\mathbf{F}(\mathbf{x}, 0)$). For an equilibrium $\mathbf{x}^*$, the eigenvalues of $\mathbf{J}(\mathbf{x}^*)$ are $\{\lambda_1, \ldots, \lambda_n\}$ (counted with their algebraic multiplicities) and are ordered by their real part, that is $\text{Re}(\lambda_i) \geq \text{Re}(\lambda_j)$ for all $i < j$. We also assume the eigenvectors of $\mathbf{J}(\mathbf{x}^*)$ are linearly independent (i.e., $\mathbf{J}(\mathbf{x}^*)$ is diagonalizable).

### *11.2.2 Koopman Operator*

We denote a semigroup of Koopman operators as $U^t$, where $U^t f(\mathbf{x}) = f \circ \mathbf{S}(t, \mathbf{x})$ for all $t \geq 0$, and we restrict the observables to be $f : \mathbb{R}^n \to \mathbb{C}$. We consider only systems with $\mathbf{F} \in C^2$ around an exponentially stable equilibrium $\mathbf{x}^*$ (that is, the eigenvalues $\lambda_j$ of $\mathbf{J}(\mathbf{x}^*)$ are such that $\text{Re}(\lambda_j) < 0$ for all $j$) with a diagonalizable $\mathbf{J}(\mathbf{x}^*)$. In this case, the eigenvalues $\lambda_j$ of the Jacobian matrix $\mathbf{J}(\mathbf{x}^*)$ are also the eigenvalues of the Koopman operator and are called Koopman eigenvalues, which are associated with the Koopman eigenfunctions $\phi_{\lambda_j} : \mathscr{C} \to \mathbb{C}$ satisfying

$$U^t \phi_{\lambda_j}(\mathbf{x}) = \phi_{\lambda_j}(\mathbf{S}(t, \mathbf{x})) = \phi_{\lambda_j}(\mathbf{x})\, e^{\lambda_j t}, \quad \mathbf{x} \in \mathscr{C}, \tag{11.2}$$

where $\mathscr{C}$ is a subset of $\mathscr{B}(\mathbf{x}^*) = \{\mathbf{x} \in \mathbb{R}^n | \lim_{t\to\infty} \mathbf{S}(t, \mathbf{x}) = \mathbf{x}^*\}$—the basin of attraction of $\mathbf{x}^*$. Furthermore, Koopman eigenfunctions $\phi_{\lambda_j}$ belong to $C^1$ [19]. If $\mathscr{C}$ is compact, then the infinitesimal generator for the semigroup can be written as $L = \mathbf{F}^T \nabla$, and we have the following relation for Koopman eigenfunctions and eigenvalues:

$$\mathbf{F}^T(\mathbf{x}) \nabla \phi_{\lambda_j}(\mathbf{x}) = \lambda_j \phi_{\lambda_j}(\mathbf{x}). \tag{11.3}$$

We refer to an eigenvalue $\lambda_1$ satisfying $\text{Re}(\lambda_1) > \text{Re}(\lambda_j)$ for all $\lambda_1 \neq \lambda_j$ as *the dominant eigenvalue*. We assume that such an eigenvalue exists (it is the case for monotone systems) and we call the associated eigenfunction $\phi_{\lambda_1}$ *the dominant eigenfunction*. Under the assumptions presented above, the dominant eigenfunction $\phi_{\lambda_1}$ can be computed through the Laplace average

$$f_\lambda^*(\mathbf{x}) = \lim_{t \to \infty} \frac{1}{T} \int_0^T (f \circ \mathbf{S}(t, \mathbf{x})) e^{-\lambda t} dt, \tag{11.4}$$

for some $f \in C^1$ satisfying $f(\mathbf{x}^*) = 0$ and $(\nabla f(\mathbf{x}^*))^T \mathbf{v}_1 \neq 0$, where $\mathbf{v}_1$ is a right eigenvector of $\mathbf{J}(\mathbf{x}^*)$ corresponding to $\lambda_1$. The Laplace average $f_{\lambda_1}^*$ is equal to $\phi_{\lambda_1}(\mathbf{x})$ up to multiplication by a scalar factor. In order to compute $f_{\lambda_1}^*$, we can, for example, choose $f(\mathbf{x}) = \mathbf{w}_1^T(\mathbf{x} - \mathbf{x}^*)$, where $\mathbf{w}_1$ is a left eigenvector of $\mathbf{J}(\mathbf{x}^*)$ corresponding to $\lambda_1$. Other eigenfunctions are typically hard to compute using Laplace averages, but they can be estimated using linear algebraic [19] or data-based methods such as dynamic mode decomposition (DMD) (cf. [26, 38]).

The Koopman eigenfunctions capture important geometric properties of the system. In particular, the dominant Koopman eigenfunction $\phi_{\lambda_1}$ is related to the notion of *isostables* [20]. If the eigenfunction $\phi_{\lambda_1}$ is $C^1$, an isostable $\partial\mathscr{B}_\alpha$ associated with the value $\alpha > 0$ is the boundary of the set $\mathscr{B}_\alpha = \{\mathbf{x} \in \mathbb{R}^n \mid |\phi_{\lambda_1}(\mathbf{x})| \leq \alpha\}$. It can be shown that the equilibrium $\mathbf{x}^*$ lies on $\partial\mathscr{B}_0$ and trajectories with initial conditions on the same isostable $\partial\mathscr{B}_{\alpha_1}$ reach other isostables $\partial\mathscr{B}_{\alpha_2}$, with $\alpha_2 < \alpha_1$, after the same time [20]

$$\mathscr{T} = \frac{1}{|\text{Re}(\lambda_1)|} \ln\left(\frac{\alpha_1}{\alpha_2}\right). \tag{11.5}$$

Hence isostables capture the dominant (or asymptotic) behavior of the unforced system. For instance, in the case of a simple and real $\lambda_1$, for example, it can be shown that the trajectories starting from $\partial\mathscr{B}_\alpha$ share the same asymptotic evolution

$$\mathbf{S}(t, \mathbf{x}) \to \mathbf{x}^* + \mathbf{v}_1 \alpha e^{\lambda_1 t}, \quad t \to \infty,$$

where $\mathbf{x} \in \partial\mathscr{B}_\alpha$, $\mathbf{v}_1$ is a right eigenvector of $\mathbf{J}(\mathbf{x}^*)$ corresponding to $\lambda_1$.

### 11.2.3 Partial Orders and Monotone Systems

In the monotone systems theory *partial orders*, denoted as $\succeq$, are defined using cones $\mathscr{K}$ in $\mathbb{R}^n$ under some conditions on $\mathscr{K}$. We, however, consider only the partial order induced by a nonnegative orthant $\mathbb{R}^n_{\geq 0} = \{\mathbf{x} \in \mathbb{R}^n | x_i \geq 0,\ i = 1, \ldots, n\}$, which is defined as follows: $\mathbf{x} \succeq \mathbf{y}$ if and only if $\mathbf{x} - \mathbf{y} \in \mathbb{R}^n_{\geq 0}$ (we write $\mathbf{x} \not\succeq y$ if the relation $\mathbf{x} \succeq \mathbf{y}$ does not hold). We will also write $\mathbf{x} \succ \mathbf{y}$ if $\mathbf{x} \succeq \mathbf{y}$ and $\mathbf{x} \neq \mathbf{y}$, and $\mathbf{x} \gg \mathbf{y}$ if

$\mathbf{x} - \mathbf{y} \in \mathbb{R}^n_{>0} = \{\mathbf{x} \in \mathbb{R}^n | x_i > 0,\ i = 1, \ldots, n\}$. A partial order on the space of signals $u \in \mathscr{U}_\infty$ is defined in a similar manner: $u \succeq v$ if $u(t) - v(t) \in \mathbb{R}_{\geq 0}$ for all $t \geq 0$. We now introduce concepts that are important for our subsequent discussion. The set $[\mathbf{x},\ \mathbf{y}] = \{\mathbf{z} \in \mathbb{R}^n | \mathbf{x} \preceq \mathbf{z} \preceq \mathbf{y}\}$ is called *an order-interval* in the order $\succeq$. For a function $W : \mathbb{R}^n \to \mathbb{R} \cup \{-\infty, +\infty\}$, we refer to the set $\mathrm{dom}(W) = \{\mathbf{x} \in \mathbb{R}^n || W(\mathbf{x})| < \infty\}$ as *its effective domain*. A function $W : \mathbb{R}^n \to \mathbb{R} \cup \{-\infty, +\infty\}$ is called *monotone increasing* if $W(\mathbf{x}) \geq W(\mathbf{y})$ for all $\mathbf{x} \succeq \mathbf{y}$ on $\mathrm{dom}(W)$. We can now define *control monotone systems*.

**Definition 11.1** The system $\dot{\mathbf{x}} = \mathbf{F}(\mathbf{x}, u)$ is called *monotone* if $\mathbf{S}(t, \mathbf{x}, u) \preceq \mathbf{S}(t, \mathbf{y}, v)$ for all $t \geq 0$, and for all $\mathbf{x} \preceq \mathbf{y}$, $u \preceq v$.

In the case of unforced dynamical systems we will also consider a stricter definition of monotonicity.

**Definition 11.2** The unforced system $\dot{\mathbf{x}} = \mathbf{F}(\mathbf{x}, 0)$ is *strongly monotone* if it is monotone and $\mathbf{x} \prec \mathbf{y}$ implies that $\mathbf{S}(t, \mathbf{x}, 0) \ll \mathbf{S}(t, \mathbf{y}, 0)$ for all $t > 0$.

A certificate for monotonicity is a condition on the system's vector field and can be found in [2]. Monotone systems possess a number of strong stability properties. For example, stable periodic orbits are not possible with $u = 0$ [9], the small gain theorem can be cast as a condition on the static input–output response of the systems [2]. A rather minor property of monotone systems that we will use is the geometric interpretation of basins of attraction, the proof of which can be found in, e.g., [35].

**Proposition 11.1** *Let the system* $\dot{\mathbf{x}} = \mathbf{F}(\mathbf{x})$ *be monotone on the basin of attraction* $\mathscr{B}(\mathbf{x}^*)$ *of an asymptotically stable fixed point* $\mathbf{x}^*$*, then for any* $\mathbf{x}$*,* $\mathbf{y}$ *in* $\mathscr{B}(\mathbf{x}^*)$ *the order-interval* $[\mathbf{x},\ \mathbf{y}]$ *lies in* $\mathscr{B}(\mathbf{x}^*)$.

One of the fundamental results for this chapter is the description of spectral properties of monotone systems using the Koopman operator [34].

**Proposition 11.2** *Assuming that the system* $\dot{\mathbf{x}} = \mathbf{F}(\mathbf{x})$ *with* $\mathbf{F} \in C^2$ *has an exponentially stable equilibrium* $\mathbf{x}^*$ *and that* $\mathbf{J}(\mathbf{x}^*)$ *is diagonalizable, let* $\lambda_j$ *be the eigenvalues of* $\mathbf{J}(\mathbf{x}^*)$ *such that* $\mathrm{Re}(\lambda_i) \geq \mathrm{Re}(\lambda_j)$ *for all* $i \leq j$.
*(i) If the system is monotone with respect to* $\mathscr{K}$ *on a set* $\mathscr{C} \subseteq \mathscr{B}(\mathbf{x}^*)$*, then* $\lambda_1$ *is real and negative, the right eigenvector* $\mathbf{v}_1$ *of* $\mathbf{J}(\mathbf{x}^*)$ *can be chosen such that* $\mathbf{v}_1 \succ 0$*, while the eigenfunction* $\phi_{\lambda_1}$ *can be chosen such that* $\phi_{\lambda_1}(\mathbf{x}) \geq \phi_{\lambda_1}(\mathbf{y})$ *for all* $\mathbf{x}, \mathbf{y} \in \mathscr{C}$ *satisfying* $\mathbf{x} \succeq \mathbf{y}$.
*(ii) Furthermore, if the system is strongly monotone with respect to* $\mathscr{K}$ *on a set* $\mathscr{C} \subseteq \mathscr{B}(\mathbf{x}^*)$ *then* $\lambda_1$ *is simple, real and negative,* $\lambda_1 > \mathrm{Re}(\lambda_j)$ *for all* $j \geq 2$*,* $\mathbf{v}_1$ *and* $\phi_{\lambda_1}$ *can be chosen such that* $\mathbf{v}_1 \gg 0$ *and* $\phi_{\lambda_1}(\mathbf{x}) > \phi_{\lambda_1}(\mathbf{y})$ *for all* $\mathbf{x}, \mathbf{y} \in \mathscr{C}$ *satisfying* $\mathbf{x} \succ \mathbf{y}$.

Without loss of generality, we will assume that a dominant eigenfunction $\phi_{\lambda_1}$ is monotone increasing even if $\lambda_1$ is not simple. In the linear case, this result reduces to the celebrated Perron–Frobenius theorem (cf. [3]). Intuitively, this should not be surprising in light of Koopman operator theory. In [15] it was shown that the Koopman operator provides a global extension of the Hartman–Grobman linearization theorem (cf. [27]) for globally exponentially stable nonlinear systems under sufficient smoothness assumptions.

## 11.3 Problem Motivation and Issues with Naive Formulations

Due to many challenges in the control of biological systems, it is reasonable to start by solving simple control problems. With this in mind, we formulated the following problem.

*Problem 1. Converging to an equilibrium.* Consider the system $\dot{\mathbf{x}} = \mathbf{F}(\mathbf{x}, u)$ and the initial state $\mathbf{x}^0$. Compute a control signal $u(t) \in \mathscr{U}_\infty$ such that the flow $\mathbf{S}(t, \mathbf{x}^0, u(\cdot))$ reaches an $\varepsilon$-ball around $\mathbf{x}^*$ for some small $\varepsilon > 0$ in minimum time units $\mathscr{T}_{\text{conv}}$ subject to the energy budget $\|u\|_{\mathbb{L}_1} \leq \mathscr{E}_{\max}$.

One straightforward application of the convergence problem is the problem of switching between stable equilibria in multistable systems. As an initial case study, consider a simplified model of a bistable genetic toggle switch system:

$$\begin{aligned} \dot{x}_1 &= p_{11} + \frac{p_{12}}{1 + (x_2)^{p_{13}}} - p_{14}x_1 + bu, \\ \dot{x}_2 &= p_{21} + \frac{p_{22}}{1 + (x_1)^{p_{23}}} - p_{24}x_2, \end{aligned} \tag{11.6}$$

where $u$ is assumed to be nonnegative, $p_{ij}$, $b$ are nonnegative, and the parameters $p_{13}$, $p_{23}$ are typically integers larger than one and smaller than six. For a range of parameter values this model is forward invariant on $\mathbb{R}^2_{\geq 0}$ with two exponentially stable equilibria in $\mathbb{R}^2_{\geq 0}$, which we denote as $\mathbf{x}^\bullet$ and $\mathbf{x}^*$. In $\mathbf{x}^*$ the value $x_1^*$ is much larger than $x_2^*$, while in $\mathbf{x}^\bullet$ the value $x_2^\bullet$ is much larger than $x_1^\bullet$.

In [32], $\mathbf{x}^*$ was chosen as the target equilibrium and $\mathbf{x}^\bullet$ as the initial point $\mathbf{x}^0$, and the authors opted for an approximate dynamic programming solution. Specifically, they used a batch-mode reinforcement learning approach [6]. In order to apply this method they discretized the model using $d\mathbf{x}/dt = (\mathbf{x}(k+1) - \mathbf{x}(k))/\Delta$ for some small positive $\Delta$ and formulated the optimal control problem as

$$V(\mathbf{x}) = \inf_{u(\cdot)} \sum_{k=0}^{\infty} \gamma^k (d(\mathbf{x}(k), \mathbf{x}^*) + \beta \|u(k)\|),$$

where $\gamma \in (0, 1)$ is called the "discount factor" and is used to guarantee that the problem is well posed, $d(\cdot, \cdot)$ is a distance function in $\mathbb{R}^n$ and $\|\cdot\|$ is a norm, while $\beta$ is a positive scalar. Optimal solutions for different functions $d(\cdot, \cdot)$, norms $\|\cdot\|$ and scalars $\beta$ were computed. In what follows we will illustrate the results with $\|\cdot\|$ being the absolute value, while $d(\cdot, \cdot)$ is chosen to be the Hilbert metric, which provided more robust results to changes in $\mathbf{x}^*$ than the Euclidean metric. Note that the objective function is only a mathematical representation of the problem we want to solve. In particular, we do not impose a hard constraint on the norm of the control signals and we require the trajectory to converge to $\mathbf{x}^*$. While it is possible to impose the specified constraints theoretically, in practical terms we can pick (by tuning) a large enough $\beta$, and ensure that the constraint is satisfied. Furthermore, the penalty on the control signal will ensure that we stop applying control in a vicinity of $\mathbf{x}^*$. The main idea of the algorithm in [6] is to use the simulated or observed trajectories of the model in order to solve the optimal control problem. Using the dynamic programming principle the problem is cast as an approximation of the objective function $V$. With a growing number of samples in the trajectories, the solution of the approximate dynamic programming converges to a unique solution under some mild conditions [6].

We depict numerical solutions in Fig. 11.1. The optimal control signal forces jittering in the trajectory by constantly switching on and off the control action. This is not an artefact of our approximations: decreasing $\Delta$ and increasing the number of computed trajectories still yields a jittering behavior in the computed solution. The underlying issue is the singularity of the continuous-time optimal control problem. We will explain this behavior using geometrical arguments. We first note that the control signal will mostly take minimal (0) and maximal ($u_{\max}$) values for this kind of problems following the bang–bang control principle. If the trajectory of the system is in the basin of attraction of $\mathbf{x}^*$ (denoted as $\mathscr{B}(\mathbf{x}^*)$), then the trajectories will converge to $\mathbf{x}^*$ even if the control is set to zero. If the trajectories are in $\mathscr{B}(\mathbf{x}^\bullet)$, then trajectories will converge to $\mathbf{x}^\bullet$ in the absence of control signal. If the manifold where the optimal control signal switches between zero and $u_{\max}$ lies in $\mathscr{B}(\mathbf{x}^\bullet)$, then we will observe jittering in the trajectory until it leaves $\mathscr{B}(\mathbf{x}^\bullet)$. In order to avoid jittering, one needs to employ singular optimal control methods, however, using this mathematical apparatus seemed rather excessive. Indeed, one can easily show that applying constant control signals is enough to ensure convergence to $\mathbf{x}^*$ when starting from $\mathbf{x}^\bullet$. More generally, we can parametrize control signals as follows:

$$u = \kappa h(t, \tau), \text{ where } h(t, \tau) = \begin{cases} 1 & t \le \tau, \\ 0 & t > \tau. \end{cases} \tag{11.7}$$

Furthermore, if a pair $(\kappa_1, \tau_1)$ forces the trajectory to move from $\mathbf{x}^\bullet$ to $\mathbf{x}^*$ then for any $\kappa_2 \ge \kappa_1$ and $\tau_2 \ge \tau_1$, the pair $(\kappa_2, \tau_2)$ will also force a switch from $\mathbf{x}^\bullet$ to $\mathbf{x}^*$ (11.6) [35]. This property was used to provide efficient and tight estimates on the set of all $(\kappa, \tau)$ allowing the switch. Finally, a closed-loop control law (control law depending on the state) can be constructed in a straightforward manner provided we can build an outer approximation of $\mathscr{B}(\mathbf{x}^\bullet)$—the basin of attraction of $\mathbf{x}^\bullet$. Indeed, in our case $\mathbb{R}^2_{\ge 0} \subset \mathrm{cl}(\mathscr{B}(\mathbf{x}^\bullet) \cup \mathscr{B}(\mathbf{x}^*))$, therefore if $\mathscr{C}$ contains $\mathscr{B}(\mathbf{x}^\bullet)$, then the set $\mathbb{R}^2_{\ge 0}/\mathscr{C}$ is contained in $\mathscr{B}(\mathbf{x}^*)$ and we can define the closed-loop control law as follows:

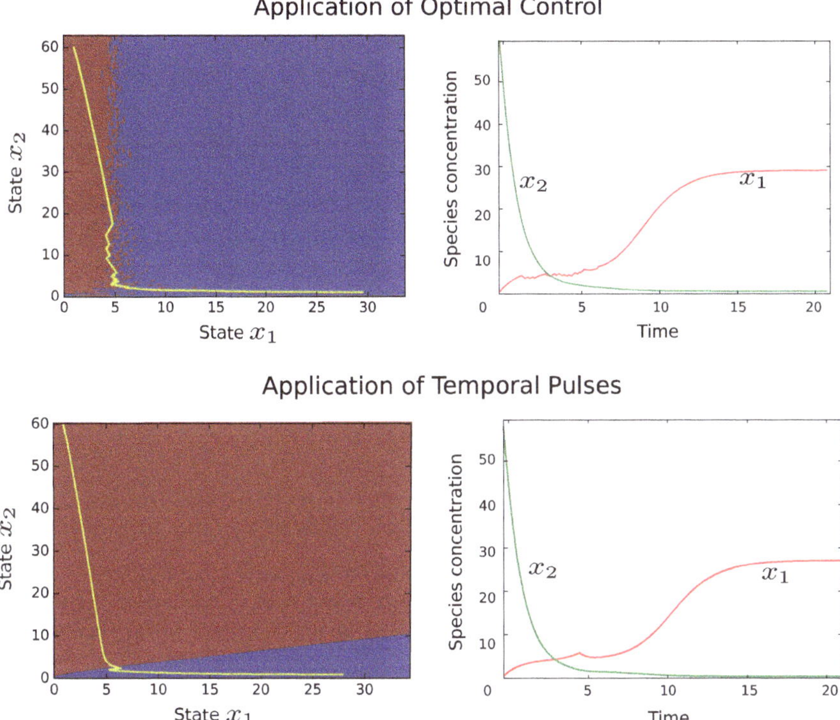


**Fig. 11.1** Switching the toggle. We set $p_{11} = p_{21} = 1$, $p_{12} = 60$, $p_{22} = 30$, $p_{13} = 3$, $p_{23} = 2$, $p_{14} = p_{24} = 1$, $b = 5$, $\Delta = 0.1$. Top panel: Application of approximate dynamic programming. We formulated the cost with $\gamma = 0.7$, $\beta = 0.8$, $\|u\| = |u|$, $u \in [0, 1]$, and $d(\mathbf{x}, \mathbf{y}) = \log(\max(x_1/y_1, \ldots, x_n/y_n)/\min(x_1/y_1, \ldots, x_n/y_n))$ as a measure of distance. We ran the algorithm from [6] with $3 \cdot 10^5$ data points and 70 iterations. In the left panel, the red area indicates the set where $u = 1$; the blue area denotes the set where $u = 0$; and the yellow curve denotes the controlled trajectory. In the right panel, we plot the controlled trajectories versus time. Bottom Panel: Application of temporal pulses. The red and blue areas delineate the basins of attraction of the two equilibria and the yellow curve denotes the trajectory controlled with a temporal pulse with the following characteristics: $\kappa = 5$, $\tau = 4.7$. In the right panel, we plot the controlled trajectories versus time

$$u(\mathbf{x}) = \begin{cases} \kappa & \mathbf{x} \in \mathscr{C}, \\ 0 & \mathbf{x} \in \mathbb{R}^2_{\geq 0}/\mathscr{C}. \end{cases} \tag{11.8}$$

This idea is simple and intuitive and can be applied to the class of *monotone* systems [35], to which the system (11.6) belongs.

The control signal (11.7) falls into the category of *open-loop control* signals. That is, the control signal does not depend on the states $\mathbf{x}$, but only on time, and the control signal is computed in advance and cannot be changed later. Closed-loop solutions, such as solutions arising from approximate dynamic programming, are preferable due to their inherent ability to deal with measurement noise, modeling errors, and disturbances. While we also constructed a closed-loop control law (11.8) and there are efficient methods for estimating basins of attraction for monotone systems [34], this solution does not offer much insight into the "geometry" of the problem. Therefore one of the main outcomes of these studies was not a solution, but a series of questions: considering all the benefits of open-loop controls (11.7), is it possible to formulate an optimal control problem, which gives (11.7) (or (11.8)) as optimal solutions? Can we explicitly use geometric properties of the system to design and tune the control law? Can the control laws be computed efficiently? We will provide answers to these questions for the case of monotone systems with the help of the Koopman operator.

## 11.4 Convergence to an Isostable Problem

### *11.4.1 Problem Formulation*

The eigenfunctions of the Koopman operator can be used to define a metric, which endows some geometric properties of the system. In our case, we need to use only the dominant eigenfunction, therefore, we resort to the use of a (contracting) pseudometric $d_K(\mathbf{x}, \mathbf{y}) = |\phi_{\lambda_1}(\mathbf{x}) - \phi_{\lambda_1}(\mathbf{y})|$ on a basin of attraction $\mathscr{B}(\mathbf{x}^*)$. While this pseudometric does not generally generate compact sublevel sets (especially for monotone systems), we can still use it for approximation of the convergence problem, provided we ensure that $d_K(\mathbf{x}, \mathbf{y})$ is sufficiently small. Note that if $d_K(\mathbf{x}, \mathbf{x}^*) \le \alpha$, then $\mathbf{x}$ lies in the set $\mathscr{B}_\alpha(\mathbf{x}^*)$, whose boundary is an isostable, and hence we can formulate the problem as follows:

*Problem* 1*. *Convergence to an isostable.* Consider the system $\dot{\mathbf{x}} = \mathbf{F}(\mathbf{x}, u)$ at the initial state $\mathbf{x}^0$. Compute a control signal $u(t) = \kappa h(t, \tau)$ such that the flow $\mathbf{S}(t, \mathbf{x}^0, u(\cdot))$ reaches the set $\mathscr{B}_\varepsilon(\mathbf{x}^*)$ for some small $\varepsilon > 0$ in minimum time units $\mathscr{T}_{\mathrm{conv}}$ subject to the energy budget $\|u\|_{\mathbb{L}_1} \le \mathscr{E}_{\max}$.

The main challenge in solving this problem is the parametrization of the control signal. Most of the control methods (such as dynamic programming or Pontryagin's maximum principle) are not tailored to deal with time-parametrized control signals since they rely on the semigroup property of the objective function or the dual variable. Hence it is not entirely clear how to systematically approach this problem using optimal control. Equipped with the Koopman operator framework, we will show how to solve Problem 1* using *a static optimization program* instead of the dynamic optimization program arising from the optimal control formulation. In order to do so we need to restrict the class of considered systems using the following assumptions:

**A1.** The vector field $\mathbf{F}(\mathbf{x}, u)$ in (11.1) is twice continuously differentiable in $(\mathbf{x}, u)$ on an open set containing $\mathscr{D} \times \mathscr{U}$. The unforced system (11.1) has an exponentially stable equilibrium $\mathbf{x}^*$ in $\mathscr{D}$ and $\mathbf{J}(\mathbf{x}^*)$ is diagonalizable.

**A2.** The system is monotone with respect to $\mathbb{R}^n_{\ge 0} \times \mathbb{R}$ and for all $\mathbf{x} \in \mathscr{D}$, $u \in \mathscr{U}_\infty$, the flow $\mathbf{S}(t, \mathbf{x}, u(\cdot))$ belongs to $\mathscr{D}$.

**A3.** There exists an eigenfunction $\phi_{\lambda_1}(\mathbf{x})$ such that $\nabla\phi_{\lambda_1}(\mathbf{x}) \gg 0$ for all $\mathbf{x} \in \mathrm{dom}(\phi_{\lambda_1})$. Furthermore, $\mathbf{F}(\mathbf{x}, \kappa_1) \succ \mathbf{F}(\mathbf{x}, \kappa_2)$ for all $\mathbf{x} \in \mathscr{D}$ and $\kappa_1 > \kappa_2 \geq 0$.

Assumption A1 guarantees existence and uniqueness of solutions, and the existence of continuously differentiable ($C^1$) Koopman eigenfunctions around the exponentially stable equilibrium $\mathbf{x}^*$. Monotonicity in Assumption A2 is the crucial assumption that requires careful checking. Forward-invariance is relatively straightforward to check: in our case $\mathscr{U}$ is a compact interval $[0, u_{\max}]$, and we only need to make sure that $\mathscr{D} = \mathrm{cl}(\mathscr{B}(\mathbf{x}^*) \cup \mathscr{B}(\mathbf{x}^\bullet))$ and $\mathbf{S}(t, \mathbf{x}, u_{\max})$ does not leave $\mathscr{D}$ for all $t > 0$, $\mathbf{x} \in \mathscr{D}$. If Assumptions A1–A2 hold, then we have $\nabla\phi_{\lambda_1}(\mathbf{x}) \succ 0$ and $\mathbf{F}(\mathbf{x}, \kappa_1) \succeq \mathbf{F}(\mathbf{x}, \kappa_2)$ for $\kappa_1 > \kappa_2 \geq 0$. Hence, Assumption A3 serves as a technical assumption that guarantees uniqueness of solutions to the posed control problem and a certain degree of regularity. We will comment throughout the chapter on the case when Assumption A3 does not hold. Parametrization of control signals in *Problem* 1* can potentially be relaxed, since the only fundamental limitation is to set $u(t)$ to zero after some time $\tau$. We discuss this relaxation in Sect. 11.5.2.

### *11.4.2 Existence of Solutions*

Restriction of control signals to temporal pulses will be shown to be beneficial from both a computational and an implementation viewpoint. A possible price to pay for this simplification is the loss of feasibility due to the restriction of the space of control signals. This can potentially occur in multistable systems, however, in the case of bistable monotone systems, a few simple arguments show that the space of temporal pulses is rich enough to solve the convergence problem.

**Proposition 11.3** *Consider a monotone system* $\dot{\mathbf{x}} = \mathbf{F}(\mathbf{x}, u)$ *on* $\mathscr{D} \times \mathscr{U}$ *under standard conditions for uniqueness and existence of solutions. If*
*(i) for* $u = 0$ *there are two asymptotically stable equilibria* $\mathbf{x}^*$ *and* $\mathbf{x}^\bullet$ *in* $\mathscr{D}$ *such that* $\mathbf{x}^* \gg \mathbf{x}^\bullet$.
*(ii) for any* $u \in \mathscr{U}_\infty$ *the flow* $\mathbf{S}(t, \mathbf{x}, u(\cdot))$ *lies in the set* $\mathscr{D} = \mathrm{cl}(\mathscr{B}(\mathbf{x}^*) \cup \mathscr{B}(\mathbf{x}^\bullet))$.
*(iii) for every* $\mathbf{x}$ *in a small* $\varepsilon$*-ball (in the Euclidean metric) around* $\mathbf{x}^\bullet$, *there exists* $u^0 \in \mathscr{U}_\infty$ *such that* $\mathbf{S}(t, \mathbf{x}, u^0)$ *converges to* $\mathbf{x}^*$.
*Then there exist* $\kappa$ *and* $\tau$ *such that* $\mathbf{S}(t, \mathbf{x}, \kappa h(\cdot, \tau))$ *converges to* $\mathbf{x}^*$.

*Proof* If $\mathbf{x}^0 \in \mathscr{B}(\mathbf{x}^*)$, then we can choose $u^0 = 0$, which is a temporal pulse with $\tau = 0$.

Consider now the case when $\mathbf{x}^0$ lies in the $\varepsilon$-ball around $\mathbf{x}^\bullet$, and $\varepsilon$ is small enough such that $\mathbf{x}^0 \in \mathscr{B}(\mathbf{x}^\bullet)$. According to the point (iii) there exists a control signal $u^0 \in \mathscr{U}_\infty$ driving the system from $\mathbf{x}^0$ to $\mathbf{x}^*$. Due to monotonicity, we have $\mathbf{S}(t, \mathbf{x}^0, u^0) \preceq \mathbf{S}(t, \mathbf{x}^0, \kappa)$, where $u^0(t) \leq \kappa$ for (almost) all $t$. Since for all $t > 0$

$$\mathbf{S}(t, \mathbf{x}^0, 0) \preceq \mathbf{S}(t, \mathbf{x}^0, u^0) \preceq \mathbf{S}(t, \mathbf{x}^0, \kappa),$$

we have that $\mathbf{S}(t, \mathbf{x}^0, u^0) \in [\mathbf{S}(t, \mathbf{x}^0, 0),\ \mathbf{S}(t, \mathbf{x}^0, \kappa)]$. According to the premise, at some time $\tau$, the flow $\mathbf{S}(\tau, \mathbf{x}^0, u^0)$ will be in the basin of attraction of $\mathscr{B}(\mathbf{x}^*)$. Now if $\mathbf{S}(\tau, \mathbf{x}^0, \kappa) \in \mathscr{B}(\mathbf{x}^\bullet)$ we have a contradiction, since $\mathbf{S}(\tau, \mathbf{x}^0, u^0)$ belongs to $[\mathbf{S}(\tau, \mathbf{x}^0, 0),\ \mathbf{S}(\tau, \mathbf{x}^0, \kappa)]$, which is a subset of $\mathscr{B}(\mathbf{x}^\bullet)$ according to Proposition 11.1. Hence if we can switch from $\mathbf{x}^0$ to $\mathbf{x}^*$ with a control signal $u(t)$, then we can switch with a temporal pulse $\kappa h(t, \tau)$.

The case when $\mathbf{x}^0$ lies in $\mathscr{B}(\mathbf{x}^\bullet)$, but not in an $\varepsilon$ ball around $\mathbf{x}^\bullet$, is treated in a similar manner by first allowing the trajectory to converge to a neighborhood of $\mathbf{x}^0$ with $u^1 = 0$ and then applying the argument above. □

The condition $\mathbf{x}^* \gg \mathbf{x}^\bullet$ (or $\mathbf{x}^* \ll \mathbf{x}^\bullet$ which can be cast in the form above by changing the order to one induced by $-\mathbb{R}^n_{\geq 0}$) is typically fulfilled for a large class of bistable monotone systems. Condition (iii) is a controllability condition, which ensures the existence of solutions for non-parametrized control signals. Condition (ii) is quite easy to verify as discussed above. Note that in Proposition 11.3 we relax Assumptions A1 and A2 and this result is valid for a more general class of systems.

We can strengthen the argument for using constant control signals by showing that *constant controls are optimal* in the absence of energy constraints. Consider the following optimal control problem over bounded and measurable control signals:

$$\begin{aligned} V(\mathbf{z}, \kappa, \beta) = & \inf_{\tau, u \in \mathscr{U}_\infty([0,\kappa])} \tau, \\ & \text{subject to } (11.1),\ \mathbf{x}(0) = \mathbf{z}, \\ & \qquad \mathbf{x}(\tau) \in \mathscr{C}_\beta = \{\mathbf{y} \in \mathbb{R}^n | \phi_{\lambda_1}(\mathbf{y}) = \beta\}, \end{aligned} \tag{11.9}$$

where $\phi_{\lambda_1}(\mathbf{x})$ is a $C^1$ increasing dominant eigenfunction defined on the basin of attraction of $\mathbf{x}^*$. Under our assumptions, the solution to this problem is surprisingly straightforward.

**Proposition 11.4** *Let the system* (11.1) *satisfy Assumptions A1–A3. Then*
*(i) If $\phi_{\lambda_1}(\mathbf{z}) < \beta$, then the optimal solution to (11.9), if it exists, is $u(t) = \kappa$ for all $t \in [0, \tau]$.*
*(ii) If $\phi_{\lambda_1}(\mathbf{z}) \geq \beta$, then the optimal solution to (11.9) is $u(t) = 0$ for all $t \geq 0$.*

*Proof* (i) Let $u^0(t) = \kappa$ for all $t > 0$, and $u^\delta(t)$ be any admissible control signal. Then $u^0(t) \succeq u^\delta(t)$ for all $t \in [0,\ \tau]$. By monotonicity we then have $\mathbf{S}(t, \mathbf{z}, u^0(\cdot)) \succeq \mathbf{S}(t, \mathbf{z}, u^\delta(\cdot))$, which leads to $\phi_{\lambda_1}(\mathbf{S}(t, \mathbf{z}, u^0(\cdot))) \geq \phi_{\lambda_1}(\mathbf{S}(t, \mathbf{z}, u^\delta(\cdot)))$ for all $t \geq 0$ due to Proposition 11.2. Hence

$$\begin{aligned} \beta &= \phi_{\lambda_1}(\mathbf{S}(\tau, \mathbf{z}, u^0(\cdot))) \geq \phi_{\lambda_1}(\mathbf{S}(\tau, \mathbf{z}, u^\delta(\cdot))), \\ \beta &> \phi_{\lambda_1}(\mathbf{S}(t, \mathbf{z}, u^0(\cdot))) \geq \phi_{\lambda_1}(\mathbf{S}(t, \mathbf{z}, u^\delta(\cdot))) \text{ for } t < \tau, \end{aligned}$$

which implies that the target set $\mathscr{C}_\beta$ is reached with $u^0(\cdot)$ at least as fast as with any other admissible control signal $u^\delta(\cdot)$. Therefore, the control signal $u^0(t) = \kappa$ is an optimal solution of the problem.

(ii) The proof is similar to the point (i). □

### 11.4.3 Pulse Control Function

The computational solution to *Problem* 1* relies on a function of state $\mathbf{x}$, of the pulse magnitude $\kappa$ and of the pulse length $\tau$, which we call *the pulse control function*.

**Definition 11.3** Let the function $r : \mathscr{D} \times \mathbb{R}_{\geq 0} \times \mathbb{R}_{\geq 0} \to \mathbb{C} \bigcup \{\infty\}$ such that

$$r(\mathbf{x}, \kappa, \tau) = \phi_{\lambda_1}(\mathbf{S}(\tau, \mathbf{x}, \kappa)),$$

where $\phi_{\lambda_1}$ is a dominant eigenfunction on the basin of attraction of $\mathbf{x}^*$, be called *the pulse control function*. By convention $r(\mathbf{x}, \kappa, \tau) = \infty$, if $\mathbf{S}(\tau, \mathbf{x}, \kappa) \notin \mathscr{B}(\mathbf{x}^*)$.

If $\phi_{\lambda_1}$ is real valued and monotone increasing on $\text{dom}(\phi_{\lambda_1}) = \mathscr{B}(\mathbf{x}^*)$, then we assume it is extended to $\mathscr{D}$ so that $\phi_{\lambda_1} : \mathscr{D} \to \mathbb{R} \cup \{-\infty, +\infty\}$ is monotone increasing on $\mathscr{D}$. We note that the case when the point $\mathbf{x}$ is another equilibrium $\mathbf{x}^{\bullet}$ was considered in [29]. Estimation of the pulse control function also relies on a Laplace average (11.4), similarly to what was done for the estimation of $\phi_{\lambda_1}$

$$\begin{aligned} r(\mathbf{x}, \kappa, \tau) &= \lim_{\bar{t} \to \infty} \frac{1}{\bar{t}} \int_0^{\bar{t}} f \circ \mathbf{S}(t, \mathbf{S}(\tau, \mathbf{x}, \kappa), 0) e^{-\lambda_1 t} dt \\ &= \lim_{\bar{t} \to \infty} \frac{1}{\bar{t}} \int_{\tau}^{\bar{t}} f \circ \mathbf{S}(t, \mathbf{x}, \kappa h(\cdot, \tau)) e^{-\lambda_1 (t-\tau)} dt, \end{aligned} \quad (11.10)$$

where $\lambda_1$ is the dominant eigenvalue of $\mathbf{J}(\mathbf{x}^*)$ with a corresponding right eigenvector $\mathbf{v}_1$, $f \in C^1$ satisfies $f(\mathbf{x}^*) = 0$, $\mathbf{v}_1^T \nabla f(\mathbf{x}^*) \neq 0$, and $h(t, \tau)$ is the step function defined in (11.7). In practice, we choose $f(\mathbf{x}) = \mathbf{w}_1^T(\mathbf{x} - \mathbf{x}^*)$, where $\mathbf{w}_1$ is the dominant left eigenvector of $\mathbf{J}(\mathbf{x}^*)$. If Assumptions A1 and A2 hold then $\lambda_1$ is real according to Proposition 11.2, and we have

$$r(\mathbf{x}, \kappa, \tau) \approx \mathbf{w}_1^T(\mathbf{S}(\bar{t}, \mathbf{x}, \kappa h(\cdot, \tau)) - \mathbf{x}^*) e^{-\lambda_1(\bar{t}-\tau)},$$

where the time $\bar{t}$ should be chosen large enough. In particular, the value of $\bar{t}$ should be chosen such that $\mathbf{w}_1^T(\mathbf{S}(t, \mathbf{x}, \kappa h(\cdot, \tau)) - \mathbf{x}^*) e^{-\lambda_1(t-\tau)}$ converges to a constant value. Naturally, with $\bar{t}$ tending to infinity the numerical integration errors can lead to divergence of $\mathbf{w}_1^T(\mathbf{S}(t, \mathbf{x}, \kappa h(\cdot, \tau)) - \mathbf{x}^*) e^{-\lambda_1(t-\tau)}$. Therefore, the tolerance of the differential equation solver should be set to $O(e^{\lambda_1(\bar{t}-\tau)})$.

For monotone systems the pulse control function $r$ possesses strong properties, which stem from Proposition 11.2 and are the key element for this approach.

**Lemma 11.1** *Let the system* (11.1) *satisfy Assumptions A1–A3. Then $r$ is a $C^1$ function on its effective domain* $\text{dom}(r)$. *Furthermore, for all* $(\mathbf{x}, \kappa, \tau) \in \text{dom}(r)$
*(i)* $\partial_{\mathbf{x}} r(\mathbf{x}, \kappa, \tau) \gg 0$, $\partial_{\kappa} r(\mathbf{x}, \kappa, \tau) > 0$ *and* $\partial_{\tau} r(\mathbf{x}, \kappa, \tau) > \lambda_1 r(\mathbf{x}, \kappa, \tau)$;
*(ii) If* $r(\mathbf{x}, \kappa, \tau) \leq 0$, *then* $\partial_{\tau} r(\mathbf{x}, \kappa, \tau) > 0$.

*Proof* (o) First, we need to show that under the assumptions above for all $t > 0$ and $\kappa_1 > \kappa_2$, we have

$$\mathbf{S}(t, \mathbf{x}, \kappa_1) \succ \mathbf{S}(t, \mathbf{x}, \kappa_2), \tag{11.11}$$

$$\phi_{\lambda_1}(\mathbf{S}(t, \mathbf{x}, \kappa_1)) > \phi_{\lambda_1}(\mathbf{S}(t, \mathbf{x}, \kappa_2)). \tag{11.12}$$

Since $\mathbf{S}(t, \mathbf{x}, \kappa_1) \succeq \mathbf{S}(t, \mathbf{x}, \kappa_2)$ for all $t > 0$ and $\kappa_1 > \kappa_2$ due to Assumption A2 (monotonicity), we only need to show that $\mathbf{S}(t, \mathbf{x}, \kappa_1) \neq \mathbf{S}(t, \mathbf{x}, \kappa_2)$ for all finite $t > 0$. At $t = 0$, the time derivatives of the flows $\mathbf{S}(t, \mathbf{x}, \kappa_1)$ and $\mathbf{S}(t, \mathbf{x}, \kappa_2)$ are equal to $\mathbf{F}(\mathbf{x}, \kappa_1)$ and $\mathbf{F}(\mathbf{x}, \kappa_2)$, respectively. Due to Assumption A3 we have that $\mathbf{F}(\mathbf{x}, \kappa_1) \succ \mathbf{F}(\mathbf{x}, \kappa_2)$ and consequently there exists a $T > 0$ such that $\mathbf{S}(t, \mathbf{x}, \kappa_1) \succ \mathbf{S}(t, \mathbf{x}, \kappa_2)$ for all $t < T$. If for some $\xi \geq T$ we have that $\mathbf{S}(\xi, \mathbf{x}, \kappa_1) = \mathbf{S}(\xi, \mathbf{x}, \kappa_2)$ and $\mathbf{S}(\xi, \mathbf{x}, \kappa_1) \succ \mathbf{S}(\xi, \mathbf{x}, \kappa_2)$ for all $t < \xi$, then for some index $i$ we have

$$\left.\frac{d\mathbf{S}_i(t, \mathbf{x}, \kappa_1)}{dt}\right|_{t=\xi} < \left.\frac{d\mathbf{S}_i(t, \mathbf{x}, \kappa_2)}{dt}\right|_{t=\xi}.$$

This leads to $F_i(\mathbf{S}(\xi, \mathbf{x}, \kappa_1), \kappa_1) < F_i(\mathbf{S}(\xi, \mathbf{x}, \kappa_2), \kappa_2)$, which together with $\mathbf{S}(\xi, \mathbf{x}, \kappa_1) = \mathbf{S}(\xi, \mathbf{x}, \kappa_2)$ contradicts Assumption A3. Therefore, $\mathbf{S}(t, \mathbf{x}, \kappa_1) \succ \mathbf{S}(t, \mathbf{x}, \kappa_2)$ for all finite $t > 0$. Due to Assumption A3 we also have that $\nabla\phi_{\lambda_1}(\mathbf{x}) \gg 0$, which in particular means that $\phi_{\lambda_1}(\mathbf{x}) > \phi_{\lambda_1}(\mathbf{y})$ for all $\mathbf{x} \succ \mathbf{y}$, and (11.12) follows from (11.11).

(i) Assumption A1 guarantees that $\mathbf{F}(\mathbf{x}, u) \in C^2$ and hence the flow $\mathbf{S}(t, \mathbf{x}, u)$ is continuously differentiable for constant control signals. Combining this with the fact that $\phi_{\lambda_1} \in C^1$ it follows that $r(\mathbf{x}, \kappa, \tau) = \phi_{\lambda_1}(\mathbf{S}(\tau, \mathbf{x}, \kappa))$ is a $C^1$ function.

For $(\mathbf{x}, \kappa, \tau)$ and $(\mathbf{y}, \kappa, \tau) \in \mathrm{dom}(r)$ such that $\mathbf{x} \succ \mathbf{y}$, monotonicity and $\nabla\phi_{\lambda_1}(\mathbf{x}) \gg 0$ (Assumption A3) ensure that $\phi_{\lambda_1}(\mathbf{S}(\tau, \mathbf{x}, \kappa)) > \phi_{\lambda_1}(\mathbf{S}(\tau, \mathbf{y}, \kappa))$, which implies that $\nabla_{\mathbf{x}} r(\mathbf{x}, \kappa, \tau) \gg 0$.

For $(\mathbf{x}, \kappa, \tau)$ and $(\mathbf{x}, \nu, \tau) \in \mathrm{dom}(r)$ such that $\kappa > \nu$, monotonicity and point (o) ensure that (11.12) holds, which implies that $\partial_\kappa r(\mathbf{x}, \kappa, \tau) > 0$.

Finally, the following derivation leads to $\partial_\tau r(\mathbf{x}, \kappa, \tau) > \lambda_1 r(\mathbf{x}, \kappa, \tau)$:

$$\begin{aligned}\partial_\tau r(\mathbf{x}, \kappa, \tau) = \left.\frac{d\phi_{\lambda_1}(\mathbf{S}(t, \mathbf{x}, \kappa))}{dt}\right|_{t=\tau} &= \\ \nabla\phi_{\lambda_1}(\mathbf{S}(\tau, \mathbf{x}, \kappa))^T \mathbf{F}(\mathbf{S}(\tau, \mathbf{x}, \kappa), \kappa) &>^* \\ \nabla\phi_{\lambda_1}(\mathbf{S}(\tau, \mathbf{x}, \kappa))^T \mathbf{F}(\mathbf{S}(\tau, \mathbf{x}, \kappa), 0) &=^\dagger \\ \lambda_1\phi_{\lambda_1}(\mathbf{S}(\tau, \mathbf{x}, \kappa)) &= \lambda_1 r(\mathbf{x}, \kappa, \tau),\end{aligned}$$

where the inequality $*$ is due to Assumption A3, and the equality $\dagger$ is due to (11.3).

(ii) follows directly from point (i). □

If Assumption A3 does not hold, then all the inequalities in Lemma 11.1 are not strict. For instance, we have that $\partial_{\mathbf{x}} r(\mathbf{x}, \kappa, \tau) \succeq 0$, $\partial_\kappa r(\mathbf{x}, \kappa, \tau) \geq 0$ and $\partial_\tau r(\mathbf{x}, \kappa, \tau) \geq \lambda_1 r(\mathbf{x}, \kappa, \tau)$ in point (i).

In light of Lemma 11.1, the pulse control function $r$ has a direct relation to the optimization problem formulated in (11.9). In particular, $\phi_{\lambda_1}(\mathbf{x}) < \beta$ implies that $r(\mathbf{x}, \kappa, V(\mathbf{x}, \kappa, \beta)) = \beta$ provided that the optimization problem in (11.9) has a solution. On the other hand, there might be some values $\tau \neq V(\mathbf{x}, \kappa, \beta)$ such that $r(\mathbf{x}, \kappa, \tau) = \beta$. In general, if $\phi_{\lambda_1}(\mathbf{x}) < \beta$, then

$$V(\mathbf{x}, \kappa, \beta) = \min\{\tau \in \mathbb{R}_{>0} | r(\mathbf{x}, \kappa, \tau) = \beta\}, \tag{11.13}$$

provided that the solution exists. In particular, if $r$ is strictly monotone increasing in $\tau$ on some subset of $(\mathbf{x}, \kappa, \tau)$, then $V(\mathbf{x}, \kappa, \beta) = \tau$ if and only if $r(\mathbf{x}, \kappa, \tau) = \beta$, which again justifies our use of the pulse control function in these cases.

If the premise of the point (ii) in Lemma 11.1 holds, then the level sets of $r$ are the graphs of strictly decreasing functions, a result which simplifies their computations as discussed in [29]. The algorithms developed in [13, 30, 34] can be used to estimate these level sets.

**Corollary 11.1** *Let the system* (11.1) *satisfy Assumptions A1–A3, and* $\alpha > 0$. *If* $\phi_{\lambda_1}(\mathbf{x}) < -\alpha$, *then* $\{(\kappa, \tau) \in \mathbb{R}^2_{>0} | r(\mathbf{x}, \kappa, \tau) = -\alpha\}$ *is the graph of a strictly decreasing function, i.e., this set does not contain pairs* $(\kappa_1, \tau_1) \neq (\kappa_2, \tau_2)$ *such that* $\kappa_1 \leq \kappa_2$ *and* $\tau_1 \leq \tau_2$*;*

*Proof* Due to Lemma 11.1 we have that $\partial_\tau r(\mathbf{x}, \kappa, \tau) > 0$ as long as $r(\mathbf{x}, \kappa, \tau) \leq 0$, and that $\partial_\kappa r(\mathbf{x}, \kappa, \tau) > 0$. Now since $\partial_\tau r(\mathbf{x}, \kappa, \tau) > 0$ (for all $(\kappa, \tau)$ such that $r(\mathbf{x}, \kappa, \tau) = -\alpha < 0$), the implicit function theorem implies that there exists a function $\tau = g(\kappa, \alpha)$ such that $r(\mathbf{x}, \kappa, g(\kappa, \alpha)) = -\alpha$. Furthermore, the function $g$ is $C^1$ in $\kappa$ and $\partial_\kappa g = -\partial_\kappa r(\mathbf{x}, \kappa, \tau)/\partial_\tau r(\mathbf{x}, \kappa, \tau) < 0$ in the neighborhood of the level set $\{(\kappa, \tau) | r(\mathbf{x}, \kappa, \tau) = -\alpha\}$. Therefore, the level set $\{\tau, \kappa | r(\mathbf{x}, \kappa, \tau) = -\alpha\}$ is the graph of a strictly decreasing function in $\kappa$. It also directly follows that the level set $\{(\kappa, \tau) | r(\mathbf{x}, \kappa, \tau) = -\alpha\}$ is the graph of a strictly decreasing function in $\tau$. □

### *11.4.4 Solution to the Problem of Convergence to Isostables*

We can now reformulate *Problem* 1* using the pulse control function. If $\phi_{\lambda_1}(\mathbf{x}^0) \geq \varepsilon$, then according to Proposition 11.4 the optimal convergence to the level set $\phi_{\lambda_1}(\mathbf{x}) = -\varepsilon$ is solved by a zero control signal. Therefore, we can assume that $\phi_{\lambda_1}(\mathbf{x}^0) \leq -\varepsilon$, that is we approach the equilibrium $\mathbf{x}^*$ from below in the partial order. Since $r(\mathbf{x}, \kappa, \tau) = -\varepsilon$, $\mathbf{x}$ belongs to $\{\mathbf{x} \in \mathbb{R}^n | \phi_{\lambda_1}(\mathbf{S}(\tau, \mathbf{x}, \kappa)) = -\varepsilon\}$ and as a consequence $\mathbf{x} \in \mathscr{B}_\varepsilon$. Therefore, the terminal constraint becomes simply $r(\mathbf{x}, \kappa, \tau) \leq -\varepsilon$. The convergence time formula is computed using Koopman operator theory as

$$\mathscr{T}_{\text{conv}} = \frac{1}{|\text{Re}(\lambda_1)|} \ln\left(\frac{|\phi_{\lambda_1}(\mathbf{S}(\tau, \mathbf{x}, \kappa))|}{\varepsilon}\right) + \tau = \frac{1}{|\text{Re}(\lambda_1)|} \ln\left(\frac{|r(\mathbf{x}, \kappa, \tau)|}{\varepsilon}\right) + \tau,$$

where the total convergence time is computed as the sum of the time of free motion of the flow (the first term in the sum) and the time $\tau$ during which the control is applied (the second term in the sum). Finally, the energy budget constraint in $\mathbb{L}_1$ norm is $\kappa \cdot \tau \leq \mathscr{E}_{\max}$ and *Problem* 1* becomes a *static* optimization program

$$\gamma^* = \min_{\kappa \geq 0, \tau \geq 0} \frac{1}{|\text{Re}(\lambda_1)|} \ln\left(|r(\mathbf{x}, \kappa, \tau)|\right) + \tau, \tag{11.14}$$

$$\text{subject to: } r(\mathbf{x}, \kappa, \tau) \leq -\varepsilon, \tag{11.15}$$

$$\kappa \cdot \tau \leq \mathscr{E}_{\max}, \tag{11.16}$$

where in the objective function we removed the constant terms from $\mathscr{T}_{\text{conv}}$, that is $\gamma^* = \mathscr{T}_{\text{conv}} + \frac{1}{|\text{Re}(\lambda_1)|} \ln(\varepsilon)$.

**Theorem 11.1** *Consider the system* (11.1), Problem 1* *under Assumptions A1–A3 and the optimization program* (11.14). *If* $\phi_{\lambda_1}(\mathbf{x}^0) \leq -\varepsilon$, *an optimal solution to* (11.14) *is an optimal solution to* Problem 1*, *if the former is feasible. Furthermore, the objective is nonincreasing in* $\kappa$ *and* $\tau$ *and an optimal solution to* (11.14), *if it exists, is achieved at the boundary of the admissible set to the constraint* (11.15) *and/or the constraint* (11.16).

*Proof* It is straightforward to verify that all the constraints and optimization objectives are the same for *Problem 1* and problem (11.14). Hence, by construction, the first part of the statement is fulfilled.
Since the system is monotone, $\lambda_1$ is real and therefore $\lambda_1 = \text{Re}(\lambda_1)$. Now according to the constraint (11.15), we have that $r(\mathbf{x}^0, \kappa, \tau) < 0$, which implies the following derivation:

$$\partial_\tau (\ln(|r(\mathbf{x}^0, \kappa, \tau)| e^{|\lambda_1|\tau})) = \frac{\partial_\tau (|r(\mathbf{x}^0, \kappa, \tau)| e^{|\lambda_1|\tau})}{|r(\mathbf{x}^0, \kappa, \tau)| e^{|\lambda_1|\tau}} =$$

$$\frac{-\partial_\tau (r(\mathbf{x}^0, \kappa, \tau)) \cdot e^{|\lambda_1|\tau} + |\lambda_1||r(\mathbf{x}^0, \kappa, \tau)| e^{|\lambda_1|\tau}}{|r(\mathbf{x}^0, \kappa, \tau)| e^{|\lambda_1|\tau}} <^*$$

$$\frac{\lambda_1 |r(\mathbf{x}^0, \kappa, \tau)| e^{|(\lambda_1)|\tau} + |\lambda_1||r(\mathbf{x}^0, \kappa, \tau)| e^{|\lambda_1|\tau}}{|r(\mathbf{x}^0, \kappa, \tau)| e^{|\lambda_1|\tau}} = 0,$$

where the inequality $*$ follows from Lemma 11.1. Hence, the derivative of the objective function in (11.14) with respect to $\tau$ is negative. Also, $\partial_\kappa \ln(|r(\mathbf{x}^0, \kappa, \tau)| e^{|\lambda_1|\tau})$ is also negative according to Lemma 11.1. This shows that if there is a feasible point, the constraints (11.15), (11.16) are reached in order to minimize the objective. □

If Assumption A3 does not hold, then the function $r$ is not strictly monotone increasing and multiple minima in (11.14) are possible, including the points which do not activate the constraints, but we can still compute a minimizing solution using the same program. If the premise of Theorem 11.1 holds then the optimization program can be solved by a line search over $\kappa$ (or $\tau$) over the constraints boundaries, since the minimum is attained when one of the constraints is active. The two terms in the sum of the objective function (11.14) establish the trade-off on the choice of the intermediate target isostable (which is to be reached after a time $\tau$) and the convergence time to the target isostable $\mathscr{B}_\varepsilon(\mathbf{x}^*)$. For instance, choosing an isostable close to the equilibrium can lead to a large pulse duration $\tau$ (second term), but a small convergence time of the free motion (first term). Visualization of the level sets of the function $r$ and the boundaries of the constraints also allows to understand the trade-off between the constraints (energy spent) and the objective (convergence time), which is not straightforward using standard optimal control theory. To summarize, we derived a *static optimization problem*, which has the same solution as the *dynamic optimization problem* (Problem 1* under Assumptions A1–A3).

In the case of the switching problem, that is, converging from another equilibrium $\mathbf{x}^\bullet$ to the target equilibrium $\mathbf{x}^*$, we can make a connection to the open-loop solution from [29, 35]. In particular, in [35] the authors compute the effective domain of the function $r(\mathbf{x}^\bullet, \kappa, \tau)$ or the set $\{(\kappa, \tau) \in \mathbb{R}^2 || r(\mathbf{x}^\bullet, \kappa, \tau)| < \infty\}$. The open-loop solutions $(\kappa, \tau)$ are then picked from this set. In [29], the authors compute the level sets of the function $r(\mathbf{x}^\bullet, \kappa, \tau)$ and the convergence time from $\mathbf{x}^\bullet$.

## 11.5 Generalizations

### *11.5.1 Dealing with Parametric Uncertainty*

One of the advantages of monotone systems appears when dealing with parametric uncertainty, provided the system is monotone with respect to parameter changes as well. Consider the system

$$\dot{\mathbf{x}} = \mathbf{F}(\mathbf{x}, \mathbf{p}, u), \ \mathbf{x}(0) = \mathbf{x}^0, \tag{11.17}$$

where $\mathbf{p}$ belongs to a compact set $\mathscr{P} \in \mathbb{R}^l$, the flow of the system is $\mathbf{S} : \mathbb{R}_{>0} \times \mathscr{D} \times \mathscr{P} \times \mathscr{U}_\infty \to \mathscr{D}$, and for every $\mathbf{p}$ the system (11.17) satisfies Assumptions A1–A3. Monotonicity with respect to the parameter $\mathbf{p}$ can be treated as monotonicity with respect to constant input signals. Monotonicity with respect to parameters allows to study a family of systems using upper and lower bounding systems. For instance, if $\mathbf{p} \in [\mathbf{p}_1, \ \mathbf{p}_2] \subseteq \mathscr{P}$, then for all $\mathbf{x}$, $u$, and $t \geq 0$ we have

$$\mathbf{S}(t, \mathbf{x}, \mathbf{p}_1, u(\cdot)) \preceq \mathbf{S}(t, \mathbf{x}, \mathbf{p}, u(\cdot)) \preceq \mathbf{S}(t, \mathbf{x}, \mathbf{p}_2, u(\cdot)).$$

In biological applications, this property is particularly useful since precise values of parameters are hard to estimate, but intervals of realistic parameter values are readily accessible.

In this section, we discuss the results from [33]. First an auxiliary problem is formulated

$$V(\mathbf{z}, \kappa, \beta, \mathbf{p}) = \inf_{\tau, u \in \mathscr{U}_\infty([0, \kappa])} \tau, \tag{11.18}$$

$$\text{subject to: } \dot{\mathbf{x}} = \mathbf{F}(\mathbf{x}, \mathbf{p}, u), \ \mathbf{x}(0) = \mathbf{z}, \ \mathbf{x}(\tau) \in \mathscr{A}_{\text{target}} = \{\tilde{\mathbf{x}} \in \mathbb{R}^n | g(\tilde{\mathbf{x}}) = \beta\},$$

where $g \in C^1$ and $\nabla g \gg 0$ on $\text{dom}(g)$. In comparison to (11.9) there is one major difference: the target set is parametrized by an arbitrary monotone increasing function $g$. Since the objective function depends on parameter $\mathbf{p}$, parameter-dependence is not taken into account in this formulation. The solution to the problem (11.18) can be obtained similarly to Proposition 11.4, that is if $g(\mathbf{z}) < \beta$ (respectively, $g(\mathbf{z}) \geq \beta$), then $u^0(\cdot) = \kappa$ (respectively, $u^0(\cdot) = 0$) is an optimal solution of (11.18). Using this observation, it is possible to derive a bound on $V$ given parametric uncertainties. In particular, if the system (11.17) is monotone with respect to parameter variations then

$$V(\mathbf{z}, \kappa, \beta, \mathbf{p}_1) \geq V(\mathbf{z}, \kappa, \beta, \mathbf{p}) \geq V(\mathbf{z}, \kappa, \beta, \mathbf{p}_2),$$

for all $\mathbf{p} \in [\mathbf{p}_1,\ \mathbf{p}_2]$ under some technical conditions. The objective functions $V$ can be computed using an extended pulse control function. To that effect, we simply introduce $\tilde{r}(\mathbf{x}, \kappa, \tau, \mathbf{p}) = \phi_{\lambda_1,\mathbf{p}}(\mathbf{S}(\tau, \mathbf{x}, \mathbf{p}, \kappa))$, where $\phi_{\lambda_1,\mathbf{p}}$ is the dominant eigenfunction of (11.17) for a fixed value $\mathbf{p}$. Note, however, that properties of $\tilde{r}$ with respect to changes in parameters are harder to ascertain. This is because the Koopman eigenfunctions and eigenvalues change under parameter variations. Nevertheless the order is preserved in terms of the convergence times to a specific set $\mathscr{A}_{\text{target}}$.

Given the variations in the equilibria, eigenfunctions and eigenvalues, it seems almost impossible to estimate the convergence time to an isostable of the system (11.17) for a specific parameter value $\mathbf{p}$ lying in the order-interval $\mathbf{p} \in [\mathbf{p}_1,\ \mathbf{p}_2]$. Interestingly, some estimates can still be provided. The result is quite technical and therefore, in what follows, we only sketch the main idea. Let us first introduce a few notations:

$$\begin{aligned}
&\mathscr{T}(\mathbf{x}, \kappa, \tau, \mathbf{p}, \varepsilon) = \frac{1}{|\lambda_1(\mathbf{p})|} \ln\left(\frac{|\tilde{r}(\mathbf{x}, \kappa, \tau, \mathbf{p})|}{\varepsilon}\right),\\
&\widetilde{\mathscr{T}}_\sigma(\mathbf{x}, \mathbf{p}, \varepsilon) = \{(\kappa, \tau) \in \mathbb{R}^2_{\geq 0} | \mathscr{T}(\mathbf{x}, \kappa, \tau, \mathbf{p}, \varepsilon) = \sigma\},\\
&\partial_- \mathscr{B}_\varepsilon(\mathbf{x}^*(\mathbf{p})) = \{\mathbf{x} \in \mathbb{R}^n | \phi_{\lambda_1,\mathbf{p}} = -\varepsilon\} \text{ for } \varepsilon > 0,\\
&\mathscr{S}_\sigma(\mathbf{x}, \mathbf{p}_1, \mathbf{p}_2, \varepsilon) = \Big\{(\kappa, \tau) \in \mathbb{R}^2 \Big| \nu_1 \leq \mu \leq \nu_2, \xi_1 \leq \tau \leq \xi_2\\
&\qquad (\nu_1, \xi_1) \in \widetilde{\mathscr{T}}_\sigma(\mathbf{x}, \mathbf{p}_1, \varepsilon), (\nu_2, \xi_2) \in \widetilde{\mathscr{T}}_\sigma(\mathbf{x}, \mathbf{p}_2, \varepsilon)\Big\}.
\end{aligned}$$

It was shown in [33], that for any $(\kappa, \tau) \in \mathscr{S}_\sigma(\mathbf{x}, \mathbf{p}_1, \mathbf{p}_2, \varepsilon)$ and $\mathbf{p} \in [\mathbf{p}_1,\ \mathbf{p}_2]$, the flow $\mathbf{S}(t, \mathbf{x}, \mathbf{p}, \mu h(\cdot, \tau))$ at time $\sigma + \tau$ belongs to the set $\mathscr{A}_{\text{target}} = \{\mathbf{x} \in \mathbb{R}^n | \mathbf{z}_1 \preceq \mathbf{x} \preceq \mathbf{z}_2, \mathbf{z}_1 \in \partial_- \mathscr{B}_\varepsilon(\mathbf{x}^*(\mathbf{p}_1)), \mathbf{z}_2 \in \partial_- \mathscr{B}_\varepsilon(\mathbf{x}^*(\mathbf{p}_2))\}$. This means that the set $\widetilde{\mathscr{T}}_\sigma(\mathbf{x}, \mathbf{p}, \varepsilon)$ will intersect with $\mathscr{S}_\sigma(\mathbf{x}, \mathbf{p}_1, \mathbf{p}_2, \varepsilon)$, and for a small enough $\varepsilon$ the set $\mathscr{S}_\sigma(\mathbf{x}, \mathbf{p}_1, \mathbf{p}_2, \varepsilon)$ provides a good estimate on $\widetilde{\mathscr{T}}_\sigma(\mathbf{x}, \mathbf{p}, \varepsilon)$ as demonstrated on numerical examples in [33].

### *11.5.2 Generalizing the Input Space*

As we mentioned above the restrictions to single-input systems and temporal pulses can be relaxed. We will provide a generalization of the function $r$ to control signals dependent on a vector of parameters as follows:

$$\mathbf{u}(t, \mathbf{p}, \tau) = \begin{cases} \mathbf{v}(t, \mathbf{p}) & \text{if } t \le \tau, \\ 0 & \text{otherwise,} \end{cases} \tag{11.19}$$

where $\mathbf{p} \in \mathscr{P} \subset \mathbb{R}^k_{\ge 0}$ is a parameter vector, $\mathscr{P}$ is a convex, compact set, and $\mathbf{v}(t, \mathbf{p}) : \mathbb{R} \times \mathbb{R}^k_{\ge 0} \to \mathbb{R}^m_{\ge 0}$ is an a priori given family of time-dependent measurable functions. The dependence of $\mathbf{v}$ on $\mathbf{p}$ should be such that for any $\mathbf{p}_1$ larger than $\mathbf{p}_2$ in the standard partial order we have that $\mathbf{u}(\cdot, \mathbf{p}_1)$ is larger than $\mathbf{u}(\cdot, \mathbf{p}_2)$ in the standard partial order. For example, we can define a family of basis functions $\{\mathbf{v}^i(t)\}_{i=1}^k : \mathbb{R} \to \mathbb{R}^m_{\ge 0}$ and parametrize control signals as

$$\mathbf{v}(t, \mathbf{p}) = \sum_{i=1}^{k} p_i \mathbf{v}^i(t).$$

Now we can define an extension of a pulse control function as

$$r(\mathbf{x}, \mathbf{p}, \tau) = \phi_{\lambda_1}(\mathbf{S}(\tau, \mathbf{x}, \mathbf{u}(\cdot, \mathbf{p}, \tau))).$$

This function can be computed in a similar manner to a pulsed control function $r$ and it can also be verified that $r(\mathbf{x}, \mathbf{p}, \tau)$ is a real function provided that the dominant eigenvalue $\lambda_1$ is real.

Another parametrization can be useful in the multi-input case. We can define $v_i$, the $i$th entry of $\mathbf{v}$, as $v_i = \kappa_i h_i(t, p_i)$ and $\tau = \max\{p_i\}$. In this case, we can reproduce in a straightforward manner the results of Propositions 11.2, 11.3 and derive an optimization program similar to (11.14). Optimality can be obtained with additional regularity assumptions such as, for example, strong monotonicity for every $\mathbf{p} \in \mathscr{P}$.

### 11.5.3 Possible Relaxations of the Monotonicity Assumption

In order to use our results we need to make sure that the conclusion of Lemma 11.1 holds, that is $\partial_\mu r(\mathbf{x}, \mu, \tau)$, $\partial_\mathbf{x} r(\mathbf{x}, \mu, \tau)$ have nonnegative entries and $\partial_\tau r(\mathbf{x}, \mu, \tau) > \lambda_1 r(\mathbf{x}, \mu, \tau)$. It is fairly straightforward to show that we can relax monotonicity with respect to $\mathbb{R}^n_{\ge 0} \times \mathbb{R}_{\ge 0}$ to monotonicity with respect to $\mathscr{K} \times \mathbb{R}_{\ge 0}$, where $\mathscr{K}$ is a proper cone. Furthermore, the derivation of the optimization program (11.14) does not explicitly rely on the monotonicity assumption, except for the implicit use of Propositions 11.3 and 11.4. The key property of monotone systems allowing a "simple" optimal solution to (11.14) is the fact that the temporal pulses $\kappa h(t, \tau)$ act in the direction of the gradient of $\phi_{\lambda_1}(\mathbf{x})$. Therefore, if $\phi_{\lambda_1}(\mathbf{x})$ is negative then temporal pulses are aligned with the flow of the unforced system, and if $\phi_{\lambda_1}(\mathbf{x})$ is positive then temporal pulses act in the opposite direction. There is no indication that monotonicity is necessary and sufficient for this property. A particularly appealing relaxation for monotonicity is *differential positivity* [7] , which was studied from the Koopman operator point of view in [17]. Differentially positive systems are defined for systems with a sufficiently smooth vector field $F$ using a *prolonged* dynamical system:

$$\begin{aligned} \dot{\mathbf{x}} &= \mathbf{F}(\mathbf{x}, u) \\ \dot{\delta \mathbf{x}} &= \partial_{\mathbf{x}} \mathbf{F}(\mathbf{x}, u) \delta \mathbf{x} + \partial_u \mathbf{F}(\mathbf{x}, u) \delta u \end{aligned} \tag{11.20}$$

with $(\mathbf{x}, \delta\mathbf{x}) \in \mathbb{R}^n \times \mathbb{R}^n$ and where $\partial_{\mathbf{x}}$, $\partial_u$ denote the differentials with respect to $\mathbf{x}$ and $u$, respectively. Following the definitions by [7], we let *a smooth cone field* $\mathscr{K}(\mathbf{x})$ be defined as

$$\mathscr{K}(\mathbf{x}) = \left\{ \delta\mathbf{x} \in \mathbb{R}^n \middle| k_i(\mathbf{x}, \delta\mathbf{x}) \geq 0,\ i = 1, \ldots, m \right\},$$

where $\mathscr{K}(\mathbf{x})$ is a proper cone for every $\mathbf{x} \in \mathbb{R}^n$, and $k_i(\cdot, \cdot)$ are smooth functions. We will also use the dual cone field defined as

$$\mathscr{K}^*(\mathbf{x}) = \left\{ \mathbf{y} \in \mathbb{R}^n \middle| \mathbf{y}^T \delta\mathbf{x} \geq 0, \delta\mathbf{x} \in \mathscr{K}(\mathbf{x}) \right\}.$$

Using the concepts above, the class of differentially positive systems is defined in the following way:

**Definition 11.4** The system $\dot{\mathbf{x}} = \mathbf{F}(\mathbf{x}, u)$ with a sufficiently smooth $\mathbf{F}$ is called differentially positive with respect to the cone field $\mathscr{K}_{\mathbf{x}}(\mathbf{x}, u) \times \mathscr{K}_u(\mathbf{x}, u)$ if the prolonged system leaves the cone field $\mathscr{K}_{\mathbf{x}}(\mathbf{x}, u) \times \mathscr{K}_u(\mathbf{x}, u)$ invariant. Namely,

$$\left.\begin{aligned} \delta\mathbf{x}(t_0) &\in \mathscr{K}_{\mathbf{x}}(\mathbf{x}(t_0), u(t_0)) \\ \delta u(t) &\in \mathscr{K}_u(\mathbf{x}(t), u(t))\ \forall t \geq t_0 \end{aligned}\right\} \Rightarrow \delta\mathbf{x}(t) \in \mathscr{K}_{\mathbf{x}}(\mathbf{x}(t), u(t))\ \forall t \geq t_0\,. \tag{11.21}$$

In the case of dynamical systems $\dot{\mathbf{x}} = \mathbf{F}(\mathbf{x})$ the prolonged system induces a flow $(\mathbf{S}, \partial_{\mathbf{x}}\mathbf{S})$ such that $(t, \mathbf{x}, \delta\mathbf{x}) \mapsto (\mathbf{S}(t, \mathbf{x}), \partial_{\mathbf{x}}\mathbf{S}(t, \mathbf{x})\delta x)$ is a solution of (11.20). Furthermore, if an exponentially stable equilibrium $\mathbf{x}^*$ is such that the dominant eigenvalue $\lambda_1$ of the Jacobian $\mathbf{J}(\mathbf{x}^*)$ is simple and real then the system is differentially positive on the basin of attraction $\mathscr{B}(\mathbf{x}^*)$. Additionally, the cone field $\mathscr{K}(\mathbf{x})$ can be built (under diagonalizability assumption on $\mathbf{J}(\mathbf{x}^*)$) in such a way that the corresponding eigenfunction $\phi_{\lambda_1}(\mathbf{x})$ lies in $\mathscr{K}^*(\mathbf{x})$ for all $\mathbf{x}$. These properties seem to indicate that our results such as Proposition 11.4 and Lemma 11.1 can potentially be extended to differentially positive systems under additional restrictions. For instance, the case, where the cone field $\mathscr{K}_{\mathbf{x}}(\mathbf{x}, u)$ does not depend on $u$ ($\mathscr{K}_{\mathbf{x}}(\mathbf{x}, u) = \mathscr{K}_{\mathbf{x}}(\mathbf{x})$) and $\mathscr{K}_u(\mathbf{x}, u)$ is a positive orthant (i.e., $\mathscr{K}_u(\mathbf{x}, u) = \mathbb{R}^m_{\geq 0}$), appears as the simplest relaxation of monotonicity. However, differentially positive systems depending on control signals (i.e., $\dot{\mathbf{x}} = \mathbf{F}(\mathbf{x}, u)$) are not well studied and many properties relevant to our problems are still not established. Furthermore, certificates for such systems would potentially involve checking conditions on the cone field for all $\mathbf{x}$ in the state space, which appears to be computationally expensive. Therefore, the computational burden will be shifted from computing the optimal control to verifying system properties. On the other hand, if the solutions to (11.14) exist, then we can still compute them while perhaps sacrificing optimality. In this case, the approach can be used as a heuristic with a justification based on the results obtained for monotone systems.

## 11.6 Numerical Examples

### *11.6.1 Controlling the Generalized Repressilator*

The generalized repressilator represents an interesting case study analyzed theoretically in [37]. We consider a generalized repressilator with eight chemical species (states), which interact only with their direct neighbors in a ring topology. The corresponding dynamic equations are as follows:

$$\begin{aligned} \dot{x}_1 &= \frac{p_1}{1+(x_8/p_2)^{p_3}} + p_4 - p_5 x_1 + u_1, \\ \dot{x}_2 &= \frac{p_1}{1+(x_1/p_2)^{p_3}} + p_4 - p_5 x_2 + u_2, \\ \dot{x}_i &= \frac{p_1}{1+(x_{i-1}/p_2)^{p_3}} + p_4 - p_5 x_i, \ \forall i = 3, \dots 8, \end{aligned} \tag{11.22}$$

where $p_1 = 40$, $p_2 = 1$, $p_3 = 2$, $p_4 = 1$, and $p_5 = 1$. This system has two exponentially stable equilibria $\mathbf{x}^*$ and $\mathbf{x}^\bullet$ and is monotone with respect to the cones $\mathscr{K}_\mathbf{x} = \mathbf{P}_\mathbf{x}\mathbb{R}^8$ and $\mathscr{K}_\mathbf{u} = \mathbf{P}_\mathbf{u}\mathbb{R}^2$, where $\mathbf{P}_\mathbf{x} = \text{diag}([1, \ -1, \ 1, \ -1, \ 1, \ -1, \ 1, \ -1])$, $\mathbf{P}_\mathbf{u} = \text{diag}([1, \ -1])$, and $\mathbf{x}^\bullet \preceq_{\mathscr{K}_\mathbf{x}} \mathbf{x}^*$. It can actually be shown that the unforced system is strongly monotone in the interior of $\mathbb{R}^8_{\geq 0}$ for all positive parameter values. The control signal $u_1$ can switch the system from the state $\mathbf{x}^\bullet$ to the state $\mathbf{x}^*$, while the control signal $u_2$ can switch the system from the state $\mathbf{x}^*$ to the state $\mathbf{x}^\bullet$.

#### 11.6.1.1 Switching Problem

We first consider the problem of switching from $\mathbf{x}^\bullet$ to $\mathbf{x}^*$. We present the comparison from [36] between the closed-loop solution, which was discussed in this chapter, and the open-loop solutions from [29]. The open-loop solution computes the level sets of $r(\mathbf{x}^\bullet, \kappa, \tau)$, which allows to pick $(\kappa, \tau)$ with an appropriate convergence time and spent energy. In terms of analysis it is more convenient to plot the level sets of the function $\mathscr{T}_{\text{conv}}(\mathbf{x}^\bullet, \kappa, \tau, 10^{-2}) = \frac{1}{|\lambda_1|} \ln\left(\dfrac{r(\mathbf{x}^\bullet, \kappa, \tau)}{10^{-2}}\right) + \tau$ instead of the function $r(\mathbf{x}^\bullet, \kappa, \tau)$. Note that the function $r$ is computed with the dominant eigenfunction associated with the target equilibrium $\mathbf{x}^*$. In Fig. 11.2, we plot the level set of the

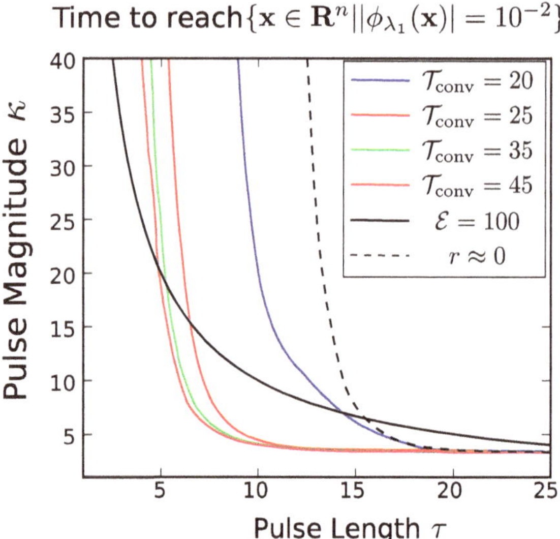


**Fig. 11.2** Level sets of $\mathscr{T}_{\text{conv}}$ with $\varepsilon = 10^{-2}$, level set $r = 0$, and energy budget curve $\kappa\tau = \mathscr{E}_{\max}$. Reprinted from [36], Copyright 2016, with permission from Elsevier

function $\mathscr{T}_{\text{conv}}(\mathbf{x}^\bullet, \kappa, \tau, 10^{-2}) = \frac{1}{|\lambda_1|} \ln\left(\frac{r(\mathbf{x}^\bullet, \kappa, \tau)}{10^{-2}}\right) + \tau$, the level set $\mathscr{E}_{\max} = 100$ of the function $\kappa\tau = \mathscr{E}_{\max}$, and the level set $r(\mathbf{x}^\bullet, \kappa, \tau) = 0$. The last two are related to the constraints of the static optimization program (11.14).

The function $\mathscr{T}_{\text{conv}}$ can escape to $-\infty$ around the level set $r(\mathbf{x}^\bullet, \kappa, \tau) \approx 0$. This is not a conflict with the interpretation of the function $\mathscr{T}_{\text{conv}}$, since it represents the convergence time only if the value of $|r(\mathbf{x}^\bullet, \kappa, \tau)|$ is larger than $10^{-2}$. Otherwise $\frac{1}{|\lambda_1|} \ln\left(\frac{|r(\mathbf{x}^\bullet, \kappa, \tau)|}{10^{-2}}\right)$ is negative, and the computational results are meaningless. This also explains why the level sets of $\mathscr{T}_{\text{conv}}$ appear to have the same asymptotics as the level set $r(\mathbf{x}^\bullet, \kappa, \tau) = 0$ in Fig. 11.2.

The closed-loop approach relies on computing the function $r(\mathbf{x}, \kappa, \tau)$ at a given point $\mathbf{x}$, which is not necessarily an equilibrium. While it is fairly clear that the closed-loop solutions are preferable, we still compare the closed- and open-loop strategies subject to perturbations of parameters $p_{ij}$. In particular we compute the control signals using nominal parameter values, but the simulations are performed using modified parameters

*Setting A*. We set $p_{i1}^A = 50$ for odd $i$.
*Setting B*. We set $p_{i1}^B = 30$ for odd $i$.

The Euclidean distance between the nominal initial point and the actual initial point in Settings A and B is equal to 0.025 and 0.031, respectively, which is negligible.

In order to compute an open-loop optimal control policy based on the nominal model (i.e., with parameter values $p_{ij}$), one can solve the static optimization program (11.14). Figure 11.2 also offers a graphical solution to the problem and a depiction of possible trade-offs in the problem. In our case, the optimal solution lies at the intersection of the constraint curves (i.e., energy budget curve and level

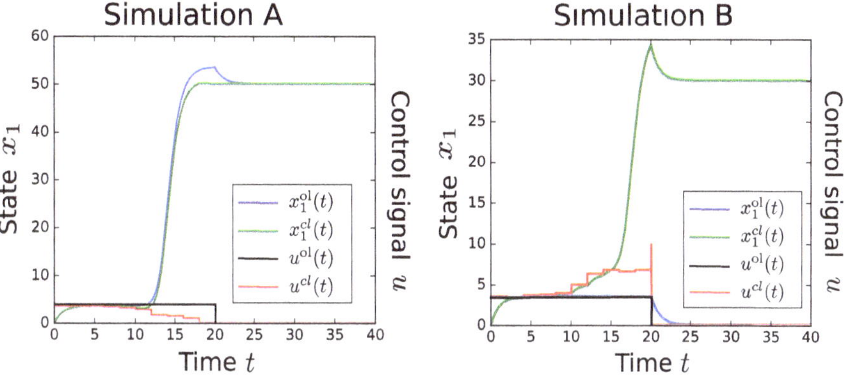


**Fig. 11.3** Closed-loop and open-loop switching. Left and right panels depict simulations for Settings A and B, respectively. In both figures, $x_1^{\mathrm{ol}}$, $x_1^{\mathrm{cl}}$ stand for the trajectories of the state $x_1$ in the open- and closed-loop settings, respectively, and $u^{\mathrm{ol}}$, $u^{\mathrm{cl}}$ stand for the corresponding control signals. Reprinted from [36], Copyright 2016, with permission from Elsevier

set $r(\mathbf{x}^{\bullet}, \kappa, \tau) = 0$). We set $\mathscr{E}_{\max} = 100$ and for both closed- and open-loop strategies we pick a pair $(\kappa^0,\ \tau^0) = (3.53,\ 20)$ lying near the zero level set of $r$, which is not on the energy budget constraint curve. We make this choice, in order to be able to react to possible disturbances/modeling errors. The closed-loop control is updated every $t_{\mathrm{samp}} = 2$ time units. At $i$th iteration the energy budget is updated for the energy spent. Then the values of $r(\mathbf{x}((i-1)t_{\mathrm{samp}}), \kappa, \tau - (i-1))t_{\mathrm{samp}})$ are computed, where $\mathbf{x}((i-1)t_{\mathrm{samp}})$ is the current state of the system, and the values of $\kappa$ are taken on a uniform grid of 100 points in [2, 10]. Thereafter $\kappa^0$ is updated by minimizing the convergence time $\mathscr{T}_{\mathrm{conv}}$ using computed values of $r$ satisfying the terminal set and the energy budget constraint. The simulation results are depicted in Fig. 11.3. In Simulation A, the system converges to the target equilibrium faster than the nominal one (i.e., with parameters $p_{ij}$) and the closed-loop solution saves energy and limits the overshoot in comparison with the open-loop solution. In Simulation B, the system converges to the target equilibrium slower than the nominal one, and all the energy budget is spent. In this case, the closed-loop solution allows the switch, while the open-loop (i.e., [29]) does not.

#### 11.6.1.2 "Stabilizing" an Unstable Periodic Orbit in the Generalized Repressilator

In [37] it was shown that the generalized repressilator admits an unstable periodic orbit, which lies around the saddle point located in the order-interval $[\mathbf{x}^{\bullet},\ \mathbf{x}^{*}]$. Therefore, oscillations can be induced by pushing the trajectories from one stable equilibrium to another repeatedly. Here we also rely on pulse control functions, but in this case we have two: $r^{*}(\mathbf{x}^{\bullet}, \kappa, \tau)$ (for converging to $\mathbf{x}^{*}$) and $r^{\bullet}$ (for converging to $\mathbf{x}^{\bullet}$) defined using the eigenfunctions $\phi^{*}_{\lambda_1}(\mathbf{x})$ and $\phi^{\bullet}_{\lambda_1}(\mathbf{x})$, respectively, which in turn are

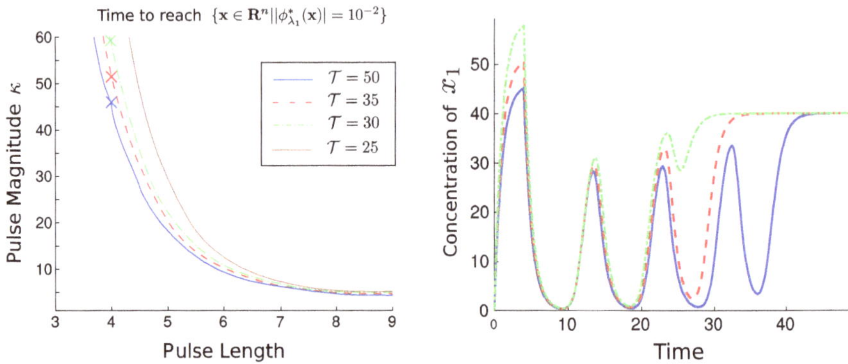


**Fig. 11.4** Switching between equilibria in a generalized repressilator system. In the left panel, the level sets of the curve $1/\text{Re}(\lambda_1)\ln(\phi^*_{\lambda_1}(\mathbf{x})/10^{-2})$ are depicted. The trajectories in the right panel are generated by the pairs $(\kappa, \tau)$ corresponding to the crosses on the left panel. Furthermore, for any pair $(\kappa, \tau)$ lying on green dash-dotted, red dashed, blue solid level set of $1/\text{Re}(\lambda_1)\ln(\phi^*_{\lambda_1}(\mathbf{x})/10^{-2})$ (the left panel) the corresponding trajectory will have a similar convergence time $\mathcal{T}$ as green dash-dotted, red dashed, blue solid trajectories (the right panel), respectively

defined on the basins of attraction $\mathcal{B}(\mathbf{x}^*)$ and $\mathcal{B}(\mathbf{x}^\bullet)$, respectively. Since the parameters in every $i$th equation are the same we can compute only the function $r^*$ and deduce $r^\bullet$ by renumbering the states.

In the left panel Fig. 11.4 we plot the level sets of the function $\mathcal{T} = \ln(\frac{|r^*(\mathbf{x}^\bullet,\kappa,\tau)|}{10^{-2}})$, where crosses stand for representative pairs $(\kappa, \tau)$ and the trajectories corresponding to those crosses are plotted in right panel of Fig. 11.4. As the reader may notice, once the function $r^*$ is computed the choice of the control signal is rather straightforward. As demonstrated in [36], the pulse control function can also be used for closed-loop switching where the pair $(\kappa, \tau)$ is updated at prescribed time intervals. Here, we will use the pulse control functions in order to induce oscillations. Numerical simulations suggest that the trajectories follow a quasi-periodic orbit in Fig. 11.4, while switching between the stable equilibria using a pulse. Furthermore, the oscillations can be sustained for an arbitrary long time, provided that the flow $\mathbf{S}(t, x_0, \kappa h(\cdot, \tau))$ at time $\tau$ lies on the manifold separating the basins of attraction $\mathcal{B}(\mathbf{x}^*)$ and $\mathcal{B}(\mathbf{x}^\bullet)$. This orbit, however, is unstable since an arbitrary small deviation from this manifold will result in convergence of the flow to $\mathbf{x}^*$ or $\mathbf{x}^\bullet$. For this problem we will force the flow to remain in a compact set $\mathcal{M} = [\mathbf{x}^\bullet + \delta\mathbf{P_x 1}, \mathbf{x}^* - \delta\mathbf{P_x 1}]$, where $\delta = 1$ and $\mathbf{P_x}$ is the matrix defining the partial order in the state variables. Note that the set $\mathcal{M}$ contains the unstable periodic orbit for the chosen parameter values. We will update the control signal only when the flow leaves $\mathcal{M}$. For both $u_1$, and $u_2$ we use the pair $(\kappa, \tau) = (44.5, 4)$ to compute the control signals. The control signal $u_1 = \kappa h(\cdot, \tau)$ (respectively, $u_2 = \kappa h(\cdot, \tau)$ ) forces the switch from $\mathbf{x}^\bullet$ to $\mathbf{x}^*$ (respectively, from $\mathbf{x}^*$ to $\mathbf{x}^\bullet$), while resulting in a trajectory with lightly damped oscillations. The control signal is switched from $u_1$ to $u_2$, when the flow leaves $\mathcal{M}$ while approaching $\mathbf{x}^*$. Meaning that we set $u_2 = \kappa h(\cdot, \tau)$ and $u_1 = 0$. The control is switched from $u_2$

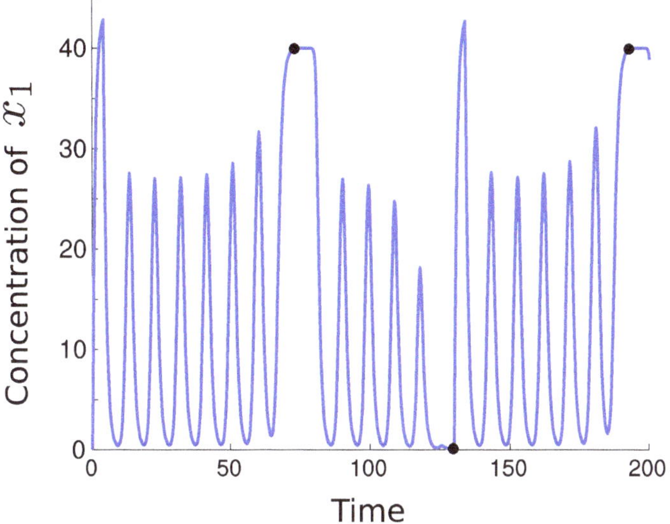


**Fig. 11.5** Inducing oscillatory behavior in the generalized repressilator system with eight states. The pulses for both $u_1$ and $u_2$ are equal, and are generated using a pair $(\kappa, \ \tau) = (44.5, \ 4)$. The control signal $u_1$ is applied at times $t = 0, 111.92$, while the control signal $u_2$ is applied at times $t = 57.97, 165.8$

to $u_1$, when the flow leaves $\mathscr{M}$ while approaching $\mathbf{x}^\bullet$. We depict the simulations in Fig. 11.5.

More uniform oscillations may potentially be obtained if we keep the flow inside the set

$$\widetilde{\mathscr{M}} = \{\mathbf{z} \in \mathbb{R}^8_{\geq 0} | \phi^\bullet_{\lambda_1}(\mathbf{z}) \geq 1/\delta \text{ and } \phi^*_{\lambda_1}(\mathbf{z}) \leq -1/\delta\},$$

for a small positive $\delta$. This can be done by applying the control signal $u_1 = \kappa_1 h(\cdot, t_1 + \tau_1)$ if the flow reaches the set $\{\mathbf{z} \in \mathbb{R}^8_{\geq 0} | \phi^\bullet_{\lambda_1}(\mathbf{z}) = 1/\delta\}$ at the time $t_1$, and the control signal $u_2 = \kappa_2 h(\cdot, t_2 + \tau_2)$ if the flow reaches the set $\{\mathbf{z} \in \mathbb{R}^8_{\geq 0} | \phi^*_{\lambda_1}(\mathbf{z}) = -1/\delta\}$ at the time $t_2$. The values of $(\kappa_1, \tau_1)$, $(\kappa_2, \tau_2)$ are chosen using the functions $r^*(\mathbf{x}(t_1), \cdot, \cdot)$, $r^\bullet(\mathbf{x}(t_2), \cdot, \cdot)$, respectively, where $\mathbf{x}(t)$ denotes the value of the state $\mathbf{x}$ at the time $t$.

We note that a similar control strategy was used in [37]. However, no computational approach to estimate the settling time of the trajectory, the pulse magnitude or its length was provided. In [31], it was proposed to track other periodic trajectories (such as sine functions of time) instead using approximate dynamic programming. However, the solution was very computationally expensive and offered little insight into the problem. We finally note that the existence of an unstable periodic orbit is not required to induce a periodic orbit in a monotone system as was demonstrated in vivo on a toggle switch system implemented in *E. coli* in [16]. Again the theoretical results in [33] justified an intuitive approach in [16].

### 11.6.2 HIV Viral Load Control

Viral load dynamics of the human immunodeficiency virus (HIV) in the bloodstream can be modeled by a bistable system, where one stable equilibrium corresponds to the high viral load (non-healthy state) and the other to the low viral load (healthy state). Even though HIV is still present in "the healthy state", the virus population is suppressed by the immune system and cannot increase to life-threatening levels. Drug therapies can lower the viral load, once a therapy is interrupted, however, the viral load can return to the non-healthy state. Mathematically, this means that the trajectory of the model does not reach the basin of attraction of the healthy state under this therapy protocol. Typically, a combination of therapies is used, which mathematically can be expressed as temporal control signals. The objective is to design the therapies (control signals) so as to drive the viral load to the "healthy state". The therapies typically follow structured treatment interruption (STI) protocols, that is, the therapies are interrupted at specific times. In this example, we will validate this approach based on mathematical modeling. We take the modeling approach developed in [1] and consider the following differential equations:

$$\begin{aligned}
\dot{T}_1 &= s_1 - \gamma_1 T_1 - (1-u_1)\beta_1 V T_1,\\
\dot{T}_2 &= s_2 - \gamma_2 T_2 - (1-fu_1)\beta_2 V T_2,\\
\dot{T}_1^* &= (1-u_1)\beta_1 V T_1 - c_{T^*} T_1^* - \kappa_1 E T_1^*,\\
\dot{T}_2^* &= (1-fu_1)\beta_2 V T_2 - c_{T^*} T_2^* - \kappa_2 E T_2^*,\\
\dot{V} &= (1-u_2) N_T k (T_1^* + T_2^*) - ((1-u_1)\rho_1\beta_1 T_1 + (1-fu_1)\rho_2\beta_2 T_2) V - c_V V,\\
\dot{E} &= \lambda_E + \frac{b_E (T_1^* + T_2^*)}{(T_1^* + T_2^*) + K_b} E - \frac{d_E (T_1^* + T_2^*)}{(T_1^* + T_2^*) + K_d} E - c_E E,
\end{aligned}$$

where $T_1$ is the number of healthy $CD4^+$ T-lymphocytes, $T_2$ is the number of healthy macrophages, $T_1^*$ is the number of infected $CD4^+$ T-lymphocytes, $T_2^*$ is the number of infected macrophages, $V$ is the number of free virus particles, and $E$ is the number of HIV-specific cytotoxic T-cells. Given the parameters

$$\begin{gathered}
s_1 = 10^4,\ \gamma_1 = \gamma_2 = 0.01,\ \beta_1 = 8 \cdot 10^{-7},\ s_2 = 31.98,\\
f = 0.34, \beta_2 = 10^{-4}, c_{T^*} = 0.7,\ \kappa_1 = \kappa_2 = 10^{-5},\ c_V = 13,\\
N_T = K_b = 100,\ \rho_1 = \rho_2 = \lambda_E = 1,\ k = 0.7,\ b_E = 0.3,\\
d_E = 0.25,\ K_d = 500, c_E = 0.1,
\end{gathered}$$

the system is not monotone and it has three equilibria, which will denote $\mathbf{x}^u$, $\mathbf{x}^*$, and $\mathbf{x}^\bullet$. Computation with the MATLAB symbolic toolbox gives the following values:

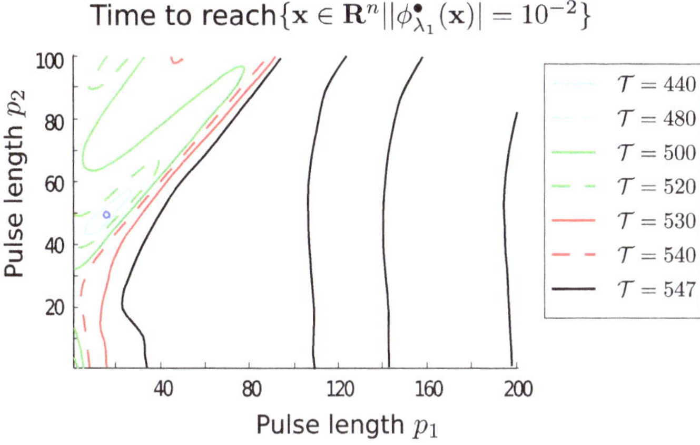


**Fig. 11.6** Level sets of the convergence time to the equilibrium $\mathbf{x}^{\bullet}$ (non-healthy state) for different times $p_1$, $p_2$. The convergence time is a continuous function (as it is a function of the $C^1$ dominant eigenfunction), and hence for the considered combinations of $p_1$ and $p_2$ the switch to the healthy state cannot be performed. Furthermore, numerical simulations suggest that the convergence time is bounded from above and hence the flow lies relatively far from the manifold separating the basins of attraction. This indicates that continuous therapies are not effective as an HIV treatment protocol

$$\mathbf{x}^u = \left(6.64 \cdot 10^5,\ 49.97,\ 1.21 \cdot 10^3,\ 11.33,\ 6.29 \cdot 10^3,\ 2.07 \cdot 10^5\right),$$
$$\mathbf{x}^{\bullet} = \left(1.63 \cdot 10^5,\ 4.99,\ 1.19 \cdot 10^4,\ 45.59,\ 6.39 \cdot 10^4,\ 23.54\right),$$
$$\mathbf{x}^{*} = \left(9.67 \cdot 10^5,\ 620.51,\ 76.01,\ 6.09,\ 415.37,\ 3.5311 \cdot 10^5\right).$$

The equilibrium $\mathbf{x}^u$ is a saddle with an unstable manifold of dimension 1; $\mathbf{x}^*$ corresponds to the "healthy" state, and $\mathbf{x}^{\bullet}$ corresponds to the "non-healthy" state. The equilibria $\mathbf{x}^{\bullet}, \mathbf{x}^*$ are exponentially stable with complex dominant eigenvalues, which rules out monotonicity. In [1], the authors considered an optimal control problem, which they have addressed using an approximated solution to the Pontryagin's maximum principle. The optimal therapies $u_1$ and $u_2$ take only values between zero and one, however, it is noticeable that these strategies are not pulses (Fig. 4 in [1]). We will show that switching from the unhealthy state to the healthy one is highly unlikely using temporal pulses.

In [1, 5], the optimal therapies were designed in such a way that the maximal values of $u_1$ and $u_2$ are equal to 0.7 and 0.3, respectively. Therefore, we assume that

$$u_1(t) = 0.7h(t, p_1),$$
$$u_2(t) = 0.3h(t, p_2).$$

This means that we have a vector $\mathbf{p}$ with two entries and $\tau = \max(p_1, p_2)$. Therefore, we can use the discussion above in order to compute the function $r(\mathbf{x}^{\bullet}, \mathbf{p}, \tau)$ defined using the eigenfunction $\phi_{\lambda_1}$ around the equilibrium $\mathbf{x}^{\bullet}$. Since $\tau$ is equal to one of the two $p_i$, we can plot the function $r$ in the parameter space $p_1$, $p_2$. The computation results are depicted in Fig. 11.6.

It is noticeable that the function $r$ is finite for all $\mathbf{p}$, which means that the switch to $\mathbf{x}^*$ is not occurring. Moreover, when $p_2 \to \infty$, the flow converges to the origin. As can be verified through simulation, small nonnegative state values of $\mathbf{x}$ around 0 lie in the basin of attraction of $\mathbf{x}^\bullet$. This means that the switching to $\mathbf{x}^*$ using these types of pulses is not possible. This validates the STI approach studied in [1], according to which both therapies have to be interrupted at specific times in order to reach the "healthy" state $\mathbf{x}^*$. The more interesting and important question of how to design (not necessarily optimally) the control signals $u_1$ and $u_2$ is currently hard to answer using the presented approach.

We note that some bistable systems with complex dominant eigenvalues of the Jacobian at the equilibria are amenable to switching using pulses. In fact, the behavior of these systems can be extremely complex. In [29], it was shown that such a control strategy is successful for switching between equilibria in a Lorentz system, which has chaotic attractors.

## 11.7 Conclusion

In this chapter, we discussed how to use the Koopman operator (and its eigenfunctions) to solve the problem of convergence to an equilibrium. The problem is cast using the concept of isostables—level sets of the dominant Koopman eigenfunction. For monotone systems, this reformulation allows rewriting of the optimal control problem as a static optimization program under additional regularity assumptions. The core idea of the method is to use isostables as a target set for the optimal control problem, which simplifies the computation of time-parametrized optimal control signals. This method can be applied to control bistable systems, if the control variable, considered as a constant parameter, forces a bifurcation in the system, and the resulting equilibrium is attractive and lies in the basin of attraction of the target equilibrium. Therefore, the approach can be applied to non-monotone systems, at the cost of a potential loss of optimality. For example, this approach was applied to the synchronization of cardiac cells [36], event-based regulation around a saddle point [33], switching under parametric uncertainty [33], and the problem of inducing oscillations as discussed above. At the moment optimality of solutions is guaranteed for the class of *monotone* systems, but as we discussed in Sect. 11.5.3 optimality results can potentially be extended to a larger class of systems. Another interesting question is whether Koopman eigenfunctions (or their properties) can be used to predict a parametric form of control signals that provide solutions to the convergence problem. Advances in the data-based Koopman operator algorithms such as dynamic mode decomposition (DMD), can potentially open the door for data-efficient data-based control methods.

**Acknowledgements** Most of the work was conducted when Dr. Sootla held F.R.S-FNRS fellowship at University of Liège. Currently, Dr. Sootla is supported by UK's Engineering and Physical Sciences (EPSRC) Grant EP/M002454/1. Dr. Stan gratefully acknowledges support from the U.K. EPSRC through the EPSRC Fellowship EP/M002187/1.

## References

1. Adams, B.M., Banks, H.T., Kwon, H.-D., Tran, H.T.: Dynamic multidrug therapies for HIV: optimal and STI control approaches. Math. Biosci. Eng. **1**(2), 223–241 (2004)
2. Angeli, D., Sontag, E.D.: Monotone control systems. IEEE Trans. Autom. Control **48**(10), 1684–1698 (2003)
3. Berman, A., Plemmons, R.J.: Nonnegative Matrices in the Mathematical Sciences, vol. 9. SIAM, Philadelphia (1994)
4. Brophy, J.A.N., Voigt, C.A.: Principles of genetic circuit design. Nat. Methods **11**(5), 508–520 (2014)
5. Ernst, D., Stan, G.-B., Goncalves, J., Wehenkel, L.: Clinical data based optimal STI strategies for HIV: a reinforcement learning approach. In: IEEE Conference Decision Control, pp. 667–672. IEEE, San Diego (2006)
6. Ernst, D., Geurts, P., Wehenkel, L.: Tree-based batch mode reinforcement learning. J. Mach. Learn. Res. **6**, 503–556 (2005)
7. Forni, F., Sepulchre, R.: Differentially positive systems. IEEE Trans. Autom. Control **61**(2), 346–359 (2016)
8. Freemont, P.S., Kitney, R.I.: Synthetic Biology-A Primer (revised Edition). World Scientific, Singapore (2015)
9. Hirsch, M.W., Smith, H., et al.: Monotone dynamical systems. In: Handbook of Differential Equations: Ordinary Differential Equations, vol. 2, pp. 239–357. Elsevier, Amsterdam (2005)
10. Hsiao, V., Swaminathan, A., Murray, R.M.: Control theory for synthetic biology: recent advances in system characterization, control design, and controller implementation for synthetic biology. IEEE Control Syst. **38**(3), 32–62 (2018)
11. Kahl, L.J., Endy, D.: A survey of enabling technologies in synthetic biology. J. Biol. Eng. **7**(1):13 (2013)
12. Kaiser, E., Kutz, J.N., Brunton, S.L.: Data-driven discovery of Koopman eigenfunctions for control (2017). arXiv:1707.01146
13. Kim, E.S., Arcak, M., Seshia, S.A.: Directed specifications and assumption mining for monotone dynamical systems. In: Proceedings of the Conference on Hybrid Systems: Computation Control, pp. 21–30. ACM, New York (2016)
14. Korda, M., Mezić, I.: Linear predictors for nonlinear dynamical systems: Koopman operator meets model predictive control. Automatica **93**, 149–160 (2018)
15. Lan, Y., Mezić, I.: Linearization in the large of nonlinear systems and Koopman operator spectrum. Phys. D **242**, 42–53 (2013)
16. Lugagne, J.-B., Carrillo, S.S., Kirch, M., Köhler, A., Batt, G., Hersen, P.: Balancing a genetic toggle switch by real-time feedback control and periodic forcing. Nat. Commun. **8**(1), 1671 (2017)
17. Mauroy, A., Forni, F., Sepulchre, R.: An operator-theoretic approach to differential positivity. In: IEEE Conference on Decision Control, pp. 7028–7033 (2015)
18. Mauroy, A.: Converging to and escaping from the global equilibrium: Isostables and optimal control. In: IEEE Conference on Decision Control, pp. 5888–5893 (2014)
19. Mauroy, A., Mezić, I.: Global stability analysis using the eigenfunctions of the Koopman operator. IEEE Trans. Autom. Control **61**(11), 3356–3369 (2016)
20. Mauroy, A., Mezić, I., Moehlis, J.: Isostables, isochrons, and Koopman spectrum for the action-angle representation of stable fixed point dynamics. Phys. D **261**, 19–30 (2013)

21. Menolascina, F., Di Bernardo, M., Di Bernardo, D.: Analysis, design and implementation of a novel scheme for in-vivo control of synthetic gene regulatory networks. Autom. Spec. Issue Syst. Biol. **47**(6), 1265–1270 (2011)
22. Milias-Argeitis, A., Summers, S., Stewart-Ornstein, J., Zuleta, I., Pincus, D., El-Samad, H., Khammash, M., Lygeros, J.: In silico feedback for in vivo regulation of a gene expression circuit. Nat. Biotechnol. **29**(12), 1114–1116 (2011)
23. Peitz, S., Klus, S.: Koopman operator-based model reduction for switched-system control of PDEs (2017). arXiv:1710.06759
24. Peitz, S.: Controlling nonlinear PDEs using low-dimensional bilinear approximations obtained from data (2018). arXiv:1801.06419
25. Purnick, P.E.M., Weiss, R.: The second wave of synthetic biology: from modules to systems. Nat. Rev. Mol. Cell Biol. **10**(6), 410–422 (2009)
26. Schmid, P.J.: Dynamic mode decomposition of numerical and experimental data. J. Fluid Mech. **656**, 5–28 (2010)
27. Sontag, E.D.: Mathematical Control Theory: Deterministic Finite Dimensional Systems, vol. 6. Springer, Berlin (2013)
28. Sontag, E.D.: Monotone and near-monotone biochemical networks. Syst. Synth. Biol. **1**(2), 59–87 (2007)
29. Sootla, A., Mauroy, A., Gonçalves, J.: Shaping pulses to control monotone bistable systems using Koopman operator. In: Proceedings of the Symposium on Nonlinear Control Systems, pp. 710–715 (2016)
30. Sootla, A., Mauroy, A.: Properties of isostables and basins of attraction of monotone systems. In: Proceedings of the American Control Conference, pp. 7365–7370 (2016)
31. Sootla, A., Strelkowa, N., Ernst, D., Barahona, M., Stan, G.-B.: On reference tracking using reinforcement learning with application to gene regulatory networks. In: Proceedings of the Conference on Decision Control, pp. 4086–4091, Florence (2013)
32. Sootla, A., Strelkowa, N., Ernst, D., Barahona, M., Stan, G.-B.: Toggling the genetic switch using reinforcement learning. In: Proceedings of the French Meeting on Planning, Decision Making and Learning, Liège (2014)
33. Sootla, A., Ernst, D.: Pulse-based control using Koopman operator under parametric uncertainty. IEEE Trans. Autom. Control **63**(3), 791–796 (2018)
34. Sootla, A., Mauroy, A.: Geometric properties and computation of isostables and basins of attraction of monotone systems. IEEE Trans. Autom. Control **62**(12), 6183–6194 (2017)
35. Sootla, A., Oyarzún, D., Angeli, D., Stan, G.-B.: Shaping pulses to control bistable systems analysis, computation and counterexamples. Automatica **63**, 254–264 (2016)
36. Sootla, A., Mauroy, A., Ernst, D.: An optimal control formulation of pulse-based control using Koopman operator. Automatica **91**, 217–224 (2018)
37. Strelkowa, N., Barahona, M.: Switchable genetic oscillator operating in quasi-stable mode. J. R. Soc. Interface **7**(48), 1071–1082 (2010)
38. Tu, J.H., Rowley, C.W., Luchtenburg, D.M., Brunton, S.L., Kutz, J.N.: On dynamic mode decomposition: theory and applications. J. Comput. Dyn. **1**(2), 391–421 (2014)
39. Uhlendorf, J., Miermont, A., Delaveau, T., Charvin, G., Fages, F., Bottani, S., Batt, G., Hersen, P.: Long-term model predictive control of gene expression at the population and single-cell levels. Proc. Nat. Acad. Sci. **109**(35), 14271–14276 (2012)
40. Wilson, D., Moehlis, J.: An energy-optimal methodology for synchronization of excitable media. SIAM J. Appl. Dyn. Syst. **13**(2), 944–957 (2014)

# Chapter 12
# Data-Driven Nonlinear Stabilization Using Koopman Operator

**Bowen Huang, Xu Ma and Umesh Vaidya**

**Abstract** We propose the application of Koopman operator theory for the design of stabilizing feedback controller for a nonlinear control system. The proposed approach is data-driven and relies on the use of time-series data generated from the control dynamical system for the lifting of a nonlinear system in the Koopman eigenfunction coordinates. In particular, a finite-dimensional bilinear representation of a control-affine nonlinear dynamical system is constructed in the Koopman eigenfunction coordinates using time-series data. Sample complexity results are used to determine the data required to achieve the desired level of accuracy for the approximate bilinear representation of the nonlinear system in Koopman eigenfunction coordinates. A control Lyapunov function-based approach is proposed for the design of stabilizing feedback controller. A systematic convex optimization-based formulation is proposed for the search of control Lyapunov function. Several numerical examples are presented to demonstrate the application of the proposed data-driven stabilization approach.

---

B. Huang
Department of Electrical and Computer Engineering, Iowa State University,
Ames, IA 50011, USA
e-mail: bowen@iastate.edu

X. Ma
Pacific Northwest National Laboratory, Richland, WA 99352, USA
e-mail: xu.ma@pnnl.gov

U. Vaidya (✉)
Department of Mechanical Engineering, Clemson University,
Clemson, SC 29634, USA
e-mail: uvaidya@clemson.edu



A. Mauroy et al. (eds.), *The Koopman Operator in Systems and Control*,
Lecture Notes in Control and Information Sciences 484,
https://doi.org/10.1007/978-3-030-35713-9_12

## 12.1 Introduction

Providing a systematic procedure for the design of stabilizing feedback control for a general nonlinear system will have a significant impact on a variety of application domains. The lack of proper structure for a general nonlinear system makes this design problem challenging. There have been several attempts to provide such a systematic approach, including convex optimization-based sum-of-squares (SoS) programming [12, 22] and differential geometric-based feedback linearization control [3, 27]. The introduction of operator theoretic methods from the ergodic theory of dynamical systems provides another opportunity for the development of systematic methods for the design of feedback controllers [19]. The operator theoretic methods provide a linear representation for a nonlinear dynamical system. This linear representation of the nonlinear system is made possible by shifting the focus from state space to space of functions using two linear and dual operators, namely, the Perron–Frobenius (P-F) and Koopman operators. The work involving the third author [8, 9, 14, 25, 35–37] provided a systematic linear programming-based approach involving transfer P-F operator for the optimal control of nonlinear systems. This contribution was made possible by exploiting the linearity and the positivity properties of the P-F operator.

More recently, there has been increased research activity on the use of Koopman operator for the analysis and control of nonlinear systems [5, 15, 20, 21, 23, 32–34]. This recent work is mainly driven by the ability to approximate the spectrum (i.e., eigenvalues and eigenfunctions) of the Koopman operator from time-series data [14, 26, 29, 38]. The data-driven approach for computing the spectrum of the Koopman operator is attractive as it opens up the possibility of employing operator theoretic methods for data-driven control. Research works in [1, 11, 15, 17, 18, 23, 31] are proposing to develop Koopman operator-based data-driven methods for the design of optimal control and model predictive control for nonlinear and partial differential equations as well. The existing approaches rely on identification of linear predictors and the use of linear control design techniques for Koopman-based control. However, the tightness of these linear predictors cannot be theoretically guaranteed. In comparison, this book chapter proposes data-driven identification and bilinear representation of nonlinear control systems in Koopman eigenfunction coordinates. The bilinear representation is tight and theoretically justified in the sense that in the limit as the number of basis function approaches infinity, the finite-dimensional bilinear representation will approach the true lifting of a control system in the function space. To address the control design problem of a more complex bilinear system, we propose a control Lyapunov function-based approach for feedback stabilization. Furthermore, sample complexity results from [6] are used to characterize the relationship between the amount of training data and the approximation error of our bilinear predictor. The work in this book chapter is the extended version of the work presented in [13], where the data-driven identification for control component is new.

The main contributions of the book chapter are as follows. We present a data-driven approach for feedback stabilization of a nonlinear system. Refer to Fig. 12.1

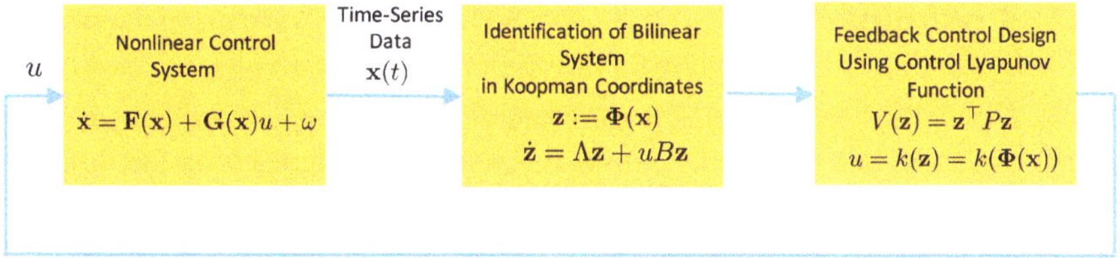


**Fig. 12.1** Data-driven identification and control of nonlinear system

for the schematic of data-driven nonlinear stabilization. We first show that the nonlinear control system can be identified from the time-series data generated by the system for two different input signals, namely, zero input and constant input. For this identification, we make use of linear operator theoretic framework involving the Fokker–Planck equation. Furthermore, sample complexity results developed in [6] are used to determine the data required to achieve the desired level for the approximation. This process of identification leads to a finite-dimensional bilinear representation of the nonlinear control system in Koopman eigenfunction coordinates. This finite-dimensional approximation of the bilinear system is used for the design of a stabilizing feedback controller. While the control design for a bilinear system is, in general, a challenging problem, we propose a systematic approach-based on the theory of control Lyapunov function (CLF) and inverse optimality for feedback control design [16]. The search for CLFs for a general nonlinear system is a difficult problem, we exploit the bilinear representation of the nonlinear control system in the Koopman eigenfunction space to search for a CLF for the bilinear system. By restricting the search of CLFs to a class of quadratic Lyapunov functions, we can provide a convex programming-based systematic approach for determining the CLF [4]. It is important to emphasize that while the CLF is quadratic in the lifted eigenfunction space it is, in fact, non-quadratic and contains higher order nonlinear terms in the original state-space coordinates. The principle of inverse optimality allows us to connect the CLF to an optimal cost function. The controller designed using CLF also optimizes an appropriate cost. Using this principle, we comment on the optimality of the controller designed using CLF.

This book chapter is organized as follows. In Sect. 12.2, we present some preliminaries on the Koopman operator, Fokker–Planck equation, and control Lyapunov functions. In Sect. 12.3, we present the identification scheme for the data-driven identification of a nonlinear control system as a bilinear system in Koopman eigenfunction coordinates. In Sect. 12.4, a convex optimization-based formulation is proposed to search for quadratic CLFs and for the design of stabilizing feedback controller. Simulation results are presented in Sect. 12.5, followed by conclusion in Sect. 12.6.

## 12.2 Preliminaries

In this section, we present some preliminaries on the Koopman operator, Fokker–Planck equation, and control Lyapunov function-based approach on the design of stabilizing feedback controllers for nonlinear systems.

### 12.2.1 Koopman Operator

Consider a continuous-time dynamical system of the form

$$\dot{\mathbf{x}} = \mathbf{F}(\mathbf{x}), \tag{12.1}$$

where $\mathbf{x} \in X \subset \mathbb{R}^n$ and the vector field $\mathbf{F}$ is assumed to be continuously differentiable. Let $\mathbf{S}(t, \mathbf{x}_0)$ be the solution of the system (12.1) starting from initial condition $\mathbf{x}_0$ and at time $t$. Let $\mathcal{O}$ be the space of all observables $f : X \to \mathbb{C}$.

**Definition 12.1** (*Koopman operator*) The Koopman semigroup of operators $U_t : \mathcal{O} \to \mathcal{O}$ associated with system (12.1) is defined by

$$[U_t f](\mathbf{x}) = f(\mathbf{S}(t, \mathbf{x})). \tag{12.2}$$

It is easy to observe that the Koopman operator is linear on the space of observables although the underlying dynamical system is nonlinear. In particular, we have

$$[U_t(\alpha f_1 + f_2)](\mathbf{x}) = \alpha[U_t f_1](\mathbf{x}) + [U_t f_2](\mathbf{x}).$$

Under the assumption that the function $f$ is continuously differentiable, the semigroup $[U_t f](\mathbf{x}) = p(\mathbf{x}, t)$ can be obtained as the solution of the following partial differential equation:

$$\frac{\partial p}{\partial t} = \mathbf{F} \cdot \nabla p =: Lp$$

with initial condition $p(\mathbf{x}, 0) = f(\mathbf{x})$. From the semigroup theory it is known [19] that the operator $L$ is the infinitesimal generator for the Koopman operator, i.e.,

$$Lp = \lim_{t \to 0} \frac{U_t p - p}{t}.$$

The linear nature of the Koopman operator allows us to define the eigenfunctions and eigenvalues of this operator as follows.

**Definition 12.2** (*Koopman eigenfunctions*) The eigenfunction of the Koopman operator is a function $\phi_\lambda \in \mathcal{O}$ that satisfies

$$[U_t\phi_\lambda](x) = e^{\lambda t}\phi_\lambda(x) \tag{12.3}$$

for some $\lambda \in \mathbb{C}$. The value $\lambda$ is the associated eigenvalue of the Koopman eigenfunction.

The spectrum of the Koopman operator is far more complex than simple point spectrum and could include continuous spectrum [21]. The eigenfunctions can also be expressed in terms of the infinitesimal generator of the Koopman operator $L$ as follows:

$$L\phi_\lambda = \lambda\phi_\lambda.$$

The eigenfunctions of the Koopman operator corresponding to the point spectrum are smooth functions and can be used as coordinates for linear representation of nonlinear systems.

### *12.2.2 Fokker–Planck Equation*

We need the preliminaries on Fokker–Planck equation for the purpose of data-driven identification of nonlinear control system. Consider a nonlinear dynamical system perturbed with white noise process

$$\dot{\mathbf{x}} = \mathbf{F}(\mathbf{x}) + \boldsymbol{\omega}, \tag{12.4}$$

where $\boldsymbol{\omega}$ is the white noise process with mean $\mu = 0$ and standard deviation $\sigma = 1$. The addition of noise term allows us to use the sample complexity results discovered in [6] to determine minimum data requirement for the data-driven approximation of nonlinear dynamics. The following assumption is made on the vector function $\mathbf{F}$.

**Assumption 12.1** *Let* $\mathbf{F} = (\mathbf{F}_1, \ldots, \mathbf{F}_n)^\top$. *We assume that the functions* $\mathbf{F}_i$ $i = 1, \ldots n$ *are* $\mathbf{C}^4$ *functions.*

We assume that the distribution of $\mathbf{x}(0)$ is absolutely continuous and has density $\rho_0(\mathbf{x})$. Then we know that $\mathbf{x}(t)$ has a density $\rho(\mathbf{x}, t)$ which satisfies the following Fokker–Planck (F-P) equation also known as Kolomogorov forward equation:

$$\frac{\partial\rho(\mathbf{x}, t)}{\partial t} = -\nabla \cdot (\mathbf{F}(\mathbf{x})\rho(\mathbf{x}, t)) + \frac{1}{2}\nabla^2\rho(\mathbf{x}, t). \tag{12.5}$$

Following Assumption 12.1, we know the solution $\rho(\mathbf{x}, t)$ to F-P equation exists and is differentiable (Theorem 11.6.1 [19]). Under some regularity assumptions on the coefficients of the F-P equation (Definition 11.7.6 [19]) it can be shown that the F-P admits a generalized solution. The generalized solution is used in defining stochastic semigroup of operators $\{\mathbb{P}_t\}_{t\geq 0}$ such that

$$[\mathbb{P}_t\rho_0](\mathbf{x}) = \rho(\mathbf{x}, t). \tag{12.6}$$

Furthermore, the right-hand side of the F-P equation is the infinitesimal generator for stochastic semigroup of operators $\mathbb{P}_t$, i.e.,

$$\mathbb{A}\varphi = \lim_{t\to 0} \frac{(\mathbb{P}_t^s - I)\varphi}{t}, \tag{12.7}$$

where

$$\mathbb{A}\varphi := -\boldsymbol{\nabla} \cdot ((\mathbf{F}(\mathbf{x})\varphi)) + \frac{1}{2}\nabla^2\varphi.$$

Let $\psi(x) \in \mathbf{C}^2(\mathbb{R}^n)$ be an observable. We have

$$\begin{aligned} \frac{d}{dt}\int \rho(\mathbf{x},t)\psi(\mathbf{x})dx &= \int \mathbb{A}\rho(\mathbf{x},t)\psi(\mathbf{x})d\mathbf{x} \\ &= \int \rho(\mathbf{x},t)\mathbb{A}^*\psi(\mathbf{x})d\mathbf{x}, \end{aligned} \tag{12.8}$$

where $\mathbb{A}^*$ is adjoint to $\mathbb{A}$ and is defined as

$$\mathbb{A}^*\psi = \mathbf{F}\cdot\nabla\psi + \frac{1}{2}\nabla^2\psi. \tag{12.9}$$

The semigroup corresponding to the operator $\mathbb{A}^*$ is given by

$$\mathbb{A}^*\psi = \lim_{t\to 0} \frac{(\mathbb{U}_t - I)\psi}{t}, \tag{12.10}$$

where

$$[\mathbb{U}_t\psi](\mathbf{x}) = \mathbb{E}[\psi(\mathbf{x}(t)) \mid \mathbf{x}(0) = \mathbf{x}]. \tag{12.11}$$

For the deterministic dynamical system $\dot{\mathbf{x}} = \mathbf{F}(\mathbf{x})$, i.e., in the absence of noise term, the above definitions of generators and semigroups reduces to Perron–Frobenius and Koopman operators. In particular, the propagation of probability density function capturing uncertainty in initial condition is given by the Perron–Frobenius (P-F) operator and is defined as follows.

**Definition 12.3** The P-F operator for a deterministic dynamical system $\dot{\mathbf{x}} = \mathbf{F}(\mathbf{x})$ is defined as follows:

$$[P_t\rho_0](\mathbf{x}) = \rho_0(\mathbf{S}(-t,\mathbf{x}))\left|\frac{\partial \mathbf{S}(-t,\mathbf{x})}{\partial \mathbf{x}}\right|, \tag{12.12}$$

where $\mathbf{S}(t,\mathbf{x})$ is the solution of the system (12.1) starting from initial condition $\mathbf{x}$ and at time $t$, and $|\cdot|$ stands for the determinant.

The infinitesimal generator for the P-F operator is given by

$$A\varphi := -\nabla \cdot (\mathbf{F}(\mathbf{x})\varphi) = \lim_{t\to 0} \frac{(P_t - I)\varphi}{t}. \tag{12.13}$$

### *12.2.3 Feedback Stabilization and Control Lyapunov Functions*

For the simplicity of the presentation, we will consider only the case of single input in this chapter. All the results carry over to the multi-input case in a straightforward manner. Consider a single input control-affine system of the form

$$\dot{\mathbf{x}} = \mathbf{F}(\mathbf{x}) + \mathbf{G}(\mathbf{x})u, \tag{12.14}$$

where $\mathbf{x}(t) \in \mathbb{R}^n$ denotes the state of the system, $u(t) \in \mathbb{R}$ denotes the single input of the system, and $\mathbf{F}, \mathbf{G} : \mathbb{R}^n \to \mathbb{R}^n$ are assumed to be continuously differentiable mappings. We assume that $\mathbf{F}(\mathbf{0}) = \mathbf{0}$ and the origin is an unstable equilibrium point of the uncontrolled system $\dot{\mathbf{x}} = \mathbf{F}(\mathbf{x})$.

The *state feedback stabilization* problem associated with system (12.14) seeks a possible feedback control law of the form

$$u = k(\mathbf{x}).$$

with $k : \mathbb{R}^n \to \mathbb{R}$ such that $\mathbf{x} = \mathbf{0}$ is asymptotically stable within some domain $\mathscr{D} \subset \mathbb{R}^n$ for the closed-loop system

$$\dot{\mathbf{x}} = \mathbf{F}(\mathbf{x}) + \mathbf{G}(\mathbf{x})k(\mathbf{x}). \tag{12.15}$$

One of the possible approaches for the design of stabilizing feedback controllers for the nonlinear system (12.14) is via control Lyapunov functions that are defined as follows:

**Definition 12.4** Let $\mathscr{D} \subset \mathbb{R}^n$ be a neighborhood that contains the equilibrium $\mathbf{x} = \mathbf{0}$. A *control Lyapunov function* (CLF) is a continuously differentiable positive definite function $V : \mathscr{D} \to \mathbb{R}_+$ such that for all $\mathbf{x} \in \mathscr{D} \setminus \{\mathbf{0}\}$ we have

$$\inf_u \left[ \frac{\partial V}{\partial x} \cdot \mathbf{F}(\mathbf{x}) + \frac{\partial V}{\partial x} \cdot \mathbf{G}(\mathbf{x})u \right] := \inf_u \left[ V_x \mathbf{F}(\mathbf{x}) + V_x \mathbf{G}(\mathbf{x})u \right] < 0.$$

It has been shown in [2, 30] that the existence of a CLF for system (12.14) is equivalent to the existence of a stabilizing control law $u = k(\mathbf{x})$ which is almost smooth everywhere except possibly at the origin $\mathbf{x} = \mathbf{0}$.

**Theorem 12.2** (see [3], Theorem 2) *There exists an almost smooth feedback* $u = k(\mathbf{x})$, *i.e., k is continuously differentiable for all* $\mathbf{x} \in \mathbb{R}^n \setminus \{\mathbf{0}\}$ *and continuous at* $\mathbf{x} = \mathbf{0}$, *which globally asymptotically stabilizes the equilibrium* $\mathbf{x} = \mathbf{0}$ *for system (12.14) if and only if there exists a radially unbounded CLF* $V(\mathbf{x})$ *such that*

1. *For all* $\mathbf{x} \neq \mathbf{0}$, $V_x\mathbf{G}(\mathbf{x}) = 0$ *implies* $V_x\mathbf{F}(\mathbf{x}) < 0$.
2. *For each* $\varepsilon > 0$, *there is a* $\delta > 0$ *such that* $\|\mathbf{x}\| < \delta$ *implies the existence of a* $|u| < \varepsilon$ *satisfying* $V_x\mathbf{F}(\mathbf{x}) + V_x\mathbf{G}(\mathbf{x})u < 0$.

In the theorem above, condition (2) is known as the small control property, and it is necessary to guarantee continuity of the feedback at $x \neq 0$. If both conditions (1) and (2) hold, an almost smooth feedback can be given by the so-called Sontag's formula

$$k(\mathbf{x}) := \begin{cases} -\frac{V_x\mathbf{F}+\sqrt{(V_x\mathbf{F})^2+(V_x\mathbf{G})^4}}{V_x\mathbf{G}} & \text{if } V_x\mathbf{G}(\mathbf{x}) \neq 0 \\ 0 & \text{otherwise.} \end{cases} \tag{12.16}$$

Besides Sontag's formula, we also have several other possible choices to design a stabilizing feedback control law based on the CLF given in Theorem 12.2. For instance, if we are not constrained to any specifications on the continuity or amplitude of the feedback, we may simply choose

$$k(\mathbf{x}) := -K \operatorname{sign}\left[V_x\mathbf{G}(\mathbf{x})\right], \tag{12.17}$$

$$k(\mathbf{x}) := -K V_x\mathbf{G}(\mathbf{x}) \tag{12.18}$$

with some constant gain $K > 0$. Then, differentiating the CLF with respect to time along trajectories of the closed loop (12.15) yields

$$\dot{V} = V_x\mathbf{F}(\mathbf{x}) - K\left|V_x\mathbf{G}(\mathbf{x})\right|,$$

$$\dot{V} = V_x\mathbf{F}(\mathbf{x}) - K(V_x\mathbf{G}(\mathbf{x}))^2.$$

Hence, by the stabilizability property of condition (1), there must exist some $K$ large enough such that $\dot{V} < 0$ for all $\mathbf{x} \neq \mathbf{0}$, because whenever $V_x\mathbf{F}(\mathbf{x}) \geq 0$ we have $V_x\mathbf{G}(\mathbf{x}) \neq 0$.

On the other hand, the CLFs also enjoy some optimality property using the principle of inverse optimal control. In particular, consider the following optimal control problem:

$$\begin{aligned} \text{minimize}_u \quad & \int_0^\infty (q(x) + u^\top u)dt \\ \text{subject to} \quad & \dot{\mathbf{x}} = \mathbf{F}(\mathbf{x}) + \mathbf{g}(\mathbf{x})u \end{aligned} \tag{12.19}$$

for some continuous, positive semidefinite function $q : \mathbb{R}^n \to \mathbb{R}$. Then the modified Sontag's formula

$$k(\mathbf{x}) := \begin{cases} -\frac{V_x\mathbf{F}+\sqrt{(V_x\mathbf{F})^2+q(x)(V_x\mathbf{G})^2}}{V_x\mathbf{G}} & \text{if } V_x\mathbf{G}(\mathbf{x}) \neq 0 \\ 0 & \text{otherwise.} \end{cases} \tag{12.20}$$

builds a strong connection with the optimal control. In particular, if the CLF has level curves that agree in shape with those of the value function associated with cost (12.19), then the modified Sontag's formula (12.20) will reduce to the optimal controller [10, 24].

## 12.3 Data-Driven Identification of Nonlinear System

In this section, we discuss the application of the linear operator theoretic framework for the identification of nonlinear dynamical system in the Koopman eigenfunctions space.

### *12.3.1 Infinite Dimensional Bilinear Representation*

Consider the control dynamical system perturbed by stochastic noise process

$$\dot{\mathbf{x}} = \mathbf{F}(\mathbf{x}) + \mathbf{G}(\mathbf{x})u + \boldsymbol{\omega}, \tag{12.21}$$

where $\boldsymbol{\omega} \in \mathbb{R}^n$ is the white noise process. As already discussed the presence of noise term will allow us to use sample complexity bounds from [6] to determine data requirement for the approximation. Sample complexity bounds can also be discovered without the additive noise term and is the topic of our current investigation.

**Assumption 12.3** *Let* $\mathbf{F} = (\mathbf{F}_1, \ldots, \mathbf{F}_n)^\top$ *and* $\mathbf{G} = (\mathbf{G}_1, \ldots, \mathbf{G}_n)^\top$. *We assume that the functions* $\mathbf{F}_i$ *and* $\mathbf{G}_i$ *for* $i = 1, \ldots n$ *are* $\mathbf{C}^4$ *functions.*

The objective is to identify the nonlinear vector fields $\mathbf{F}$ and $\mathbf{G}$ using the time-series data generated by the control dynamical system and arrive at a continuous-time dynamical system of the form

$$\dot{\mathbf{z}} = \Lambda\mathbf{z} + uB\mathbf{z}, \tag{12.22}$$

where $\mathbf{z} \in \mathbb{R}^N$ with $N \geq n$. We now make the following assumption on the control dynamical system (12.21).

**Assumption 12.4** *We assume that all the trajectories of the control dynamical system (12.21) starting from different initial conditions for control input* $u = 0$ *and for constant input remain bounded.*

*Remark 12.1* This assumption is essential to ensure that the control dynamical system can be identified from the time-series data generated by the system for two different input signals.

The goal is to arrive at a continuous-time bilinear representation of the nonlinear control system (12.21). Toward this goal we assume that the time-series data from the continuous-time dynamical system (12.21) is available for two different control inputs, namely, zero input and constant input. The discrete time-series data is generated from the continuous-time dynamical system with sufficiently small discretization time step $\Delta t$ and this time-series data is represented as

$$(\mathbf{x}_{k+1}^s, \mathbf{x}_k^s). \tag{12.23}$$

The subscript $s$ signifies that the data is generated by a dynamical system of the form

$$\dot{\mathbf{x}} = \mathbf{F}(\mathbf{x}) + \mathbf{G}(\mathbf{x})s + \boldsymbol{\omega}. \tag{12.24}$$

so that $s = 0$ and $s = 1$ correspond to the case of zero input and constant input, respectively. Let

$$\Psi = [\psi_1, \ldots, \psi_N]$$

be the set of observables with $\psi_i : \mathbb{R}^n \to \mathbb{R}$. The time evolution of these observables under the continuous-time control dynamical system with no noise can be written as

$$\begin{aligned} \frac{d\Psi}{dt} &= \mathbf{F}(\mathbf{x}) \cdot \nabla\Psi + u\mathbf{G}(\mathbf{x}) \cdot \nabla\Psi \\ &= \mathscr{A}\Psi + u\mathscr{B}\Psi, \end{aligned} \tag{12.25}$$

where $\mathscr{A}$ and $\mathscr{B}$ are linear operators. The objective is to construct the finite-dimensional approximation of these linear operators, $\mathscr{A}$ and $\mathscr{B}$, respectively, from time-series data to arrive at a finite-dimensional approximation of control dynamical system as in Eq. (12.22).

With reference to Eq. (12.9), let $\mathbb{A}_1^*$ and $\mathbb{A}_0^*$ be the generator corresponding to the control dynamical system with constant input, i.e., $s = 1$ and $s = 0$, respectively, in Eq. (12.24). We have

$$(\mathbb{A}_1^* - \mathbb{A}_0^*)\psi = \mathbf{G}(\mathbf{x}) \cdot \nabla\psi. \tag{12.26}$$

Under the assumption that the sampling time $\Delta t$ between the two consecutive time-series data point is sufficiently small, the generators $\mathbb{A}_s^*$ can be approximated as

$$\mathbb{A}_s^* \approx \frac{\mathbb{U}_{\Delta t}^s - I}{\Delta t}. \tag{12.27}$$

Substituting for $s = 1$ and $s = 0$ in (12.27) and using (12.26), we obtain

$$\frac{\mathbb{U}^1_{\Delta t} - \mathbb{U}^0_{\Delta t}}{\Delta t} \approx \mathbf{G}(\mathbf{x}) \cdot \nabla = \mathscr{B}. \tag{12.28}$$

and

$$\frac{\mathbb{U}^0_{\Delta t} - I}{\Delta t} \approx \mathbf{F}(\mathbf{x}) \cdot \nabla = \mathscr{A}. \tag{12.29}$$

Using the time-series data generated from dynamical system (12.24) for $s = 0$ and $s = 1$, it is possible to construct the finite-dimensional approximation of the operators $\mathbb{U}^0_{\Delta t}$ and $\mathbb{U}^1_{\Delta t}$, respectively, thereby approximating the operators $\mathscr{A}$ and $\mathscr{B}$, respectively. In the following, we explain the extended dynamic mode decomposition-based procedure for the approximation of these operators from time-series data.

### 12.3.2 Finite-Dimensional Approximation

We use extended dynamic mode decomposition (EDMD) algorithm for the approximation of $\mathbb{U}^1_{\Delta t}$ and $\mathbb{U}^0_{\Delta t}$ thereby approximating $\mathscr{A}$ and $\mathscr{B}$ in Eqs. (12.27) and (12.28), respectively [38]. For this purpose let the time-series data generated by the dynamical system (12.24) be given by

$$\overline{\mathbf{X}} = [\mathbf{x}^s_1, \mathbf{x}^s_2, \dots, \mathbf{x}^s_M], \ \overline{\mathbf{Y}} = [\mathbf{y}^s_1, \mathbf{y}^s_2, \dots, \mathbf{y}^s_M]., \tag{12.30}$$

where $\mathbf{y}^s_k = \mathbf{x}^s_{k+1}$ with $s = 0$ or $s = 1$, i.e., zero input and constant input. Furthermore, let $\mathscr{H} = \{\psi_1, \psi_2, \dots, \psi_N\}$ be the set of dictionary functions or observables and $\mathscr{G}_{\mathscr{H}}$ be the span of $\mathscr{H}$. The choice of dictionary functions is very crucial and it should be rich enough to approximate the leading eigenfunctions of the Koopman operator. Define vector-valued function $\Psi : X \to \mathbb{C}^N$

$$\Psi(\mathbf{x}) := \begin{bmatrix} \psi_1(\mathbf{x}) & \psi_2(\mathbf{x}) & \cdots & \psi_N(\mathbf{x}) \end{bmatrix}^\top. \tag{12.31}$$

In this application, $\Psi$ is the mapping from state space to function space. Any two functions $f$ and $\hat{f} \in \mathscr{G}_{\mathscr{H}}$ can be written as

$$f = \sum_{k=1}^{N} a_k h_k = \Psi^\top \mathbf{a}, \quad \hat{f} = \sum_{k=1}^{N} \hat{a}_k h_k = \Psi^\top \hat{\mathbf{a}}. \tag{12.32}$$

for some coefficients $\mathbf{a}$ and $\hat{\mathbf{a}} \in \mathbb{C}^N$. Let

$$\hat{f}(\mathbf{x}) = [U^s_{\Delta t} f](\mathbf{x}) + r,$$

where $r$ is a residual function that appears because $\mathscr{G}_{\mathscr{H}}$ is not necessarily invariant to the action of the Koopman operator. To find the optimal mapping which can minimize this residual, let $\mathbf{U}$ be the finite-dimensional approximation of the Koopman operator $U^s_{\Delta t}$. Then the matrix $\mathbf{U}^s$ is obtained as a solution of least-squares problem as follows:

$$\underset{\mathbf{U}^s}{\text{minimize}} \quad \|\mathbf{G}^s\mathbf{U}^s - \mathbf{A}^s\|_F, \tag{12.33}$$

where

$$\mathbf{G}^s = \frac{1}{M}\sum_{m=1}^{M}\Psi(\mathbf{x}^s_m)^\top\Psi(\mathbf{x}^s_m), \quad \mathbf{A}^s = \frac{1}{M}\sum_{m=1}^{M}\Psi(\mathbf{x}^s_m)^\top\Psi(\mathbf{y}^s_m) \tag{12.34}$$

with $\mathbf{U}^s, \mathbf{G}^s, \mathbf{A}^s \in \mathbb{C}^{N\times N}$. The optimization problem (12.33) can be solved explicitly with a solution in the following form:

$$\mathbf{U}^s = (\mathbf{G}^s)^\dagger\mathbf{A}^s., \tag{12.35}$$

where $(\mathbf{G}^s)^\dagger$ denotes the pseudoinverse of matrix $\mathbf{G}^s$.

Under the assumption that the leading Koopman eigenfunctions are contained within $\mathcal{G}_{\mathcal{H}}$, the eigenvalues of $\mathbf{U}$ are approximations of the Koopman eigenvalues. The right eigenvectors of $\mathbf{U}^{s=0}$ can be used then to generate the approximation of Koopman eigenfunctions. In particular, the approximation of Koopman eigenfunction is given by

$$\phi_j = \Psi^\top v_j, \quad j = 1, \ldots, N, \tag{12.36}$$

where $v_j$ is the $j$th right eigenvector of $\mathbf{U}^0$, and $\phi_j$ is the approximation of the eigenfunction of Koopman operator corresponding to the $j$th eigenvalue, $\lambda_j \in \mathbb{C}$.

The bilinear representation of nonlinear control dynamical system can be constructed either in the space of basis function $\Psi$ or the eigenfunctions of the Koopman operator $\Phi$, where

$$\Phi(\mathbf{x}) := [\phi_1(\mathbf{x}), \ldots, \phi_N(\mathbf{x})]^\top.$$

In this work, we constructed the bilinear representation in the Koopman eigenfunctions coordinates [20, 31]. Toward this goal, we define

$$\hat{\Phi}(\mathbf{x}) := [\hat{\phi}_1(\mathbf{x}), \ldots, \hat{\phi}_N(\mathbf{x})]^\top,$$

where $\hat{\phi}_i := \phi_i$ if $\phi_i$ is a real-valued eigenfunction and $\hat{\phi}_i := 2\mathrm{Re}(\phi)$, $\hat{\phi}_{i+1} := -2\mathrm{Im}(\phi_i)$, if $i$ and $i+1$ are complex conjugate eigenfunction pairs. Consider now the transformation $\hat{\Phi} : \mathbb{R}^n \to \mathbb{R}^N$ as

$$\mathbf{z} = \hat{\Phi}(\mathbf{x}).$$

Then in this new coordinates system Eq. (12.14) takes the following form:

$$\dot{\mathbf{z}} = \Lambda\mathbf{z} + uB\mathbf{z}, \tag{12.37}$$

where the matrix $\Lambda$ has a block diagonal form where the block corresponding to the eigenvalue $\hat{\lambda}_i$, such that $\Lambda_{(i,i)} = \hat{\lambda}_i$ if $\phi_i$ is real, and

$$\begin{bmatrix} \Lambda_{(i,i)} & \Lambda_{(i,i+1)} \\ \Lambda_{(i+1,i)} & \Lambda_{(i+1,i+1)} \end{bmatrix} = |\lambda_i| \begin{bmatrix} \cos(\angle\hat{\lambda}_i) & \sin(\angle\hat{\lambda}_i) \\ -\sin(\angle\hat{\lambda}_i) & \cos(\angle\hat{\lambda}_i) \end{bmatrix}. \tag{12.38}$$

if $\phi_i$ and $\phi_{i+1}$ are complex conjugate pairs. The value $\hat{\lambda}_i$ associated with the continuous-time system dynamics. The relationship between discrete-time Koopman eigenvalues $\lambda_i$ and continuous time $\hat{\lambda}_i$ can be written as $\hat{\lambda}_i = \log(\lambda_i)/\Delta t$.

Similarly, data generated using constant input for the control dynamical system is used to generate time-series data $\{\mathbf{x}_k^1\}$ and for the approximation of $\mathbf{U}^1$. The approximation of the operator $\mathscr{B}$ in the coordinates of basis functions, $\Psi(\mathbf{x})$ denoted by $\overline{B}$, and the eigenfunction coordinates $\hat{\Phi}(\mathbf{x})$ denoted by $B$ can be obtained as follows:

$$\overline{B} = \frac{\mathbf{U}^1 - \mathbf{U}^0}{\Delta t}, \qquad B = V^\top \overline{B} (V^\top)^{-1}, \tag{12.39}$$

where each column of $V$, $v_j$ is the $j$th eigenvector of $\mathbf{U}^0$.

There exist two sources of error in the approximation of Koopman operator and its spectrum, and both of them will be reflected in the bilinear representation of nonlinear system, namely, $\Lambda$ and $B$ matrices. The first source of error is due to a finite number of basis functions used in the approximation of the Koopman operator. Under the assumption that the choice of basis functions is sufficiently rich and $N$ is large, this approximation error is expected to be small. However, the selection of basis functions is an active research topic with no agreement on the best choice of basis functions for general nonlinear systems. The second source of error, which is more relevant to this work, arise due to the finite length of data used in the approximation of the Koopman operator. Sample complexity results for nonlinear stochastic dynamics using linear operator theory is developed in [6]. These results provide error bounds for the approximation of the Koopman operator as the function of finite data length under the assumption that the action of the Koopman operator is closed on the space of finite basis functions. In particular, for any given $\varepsilon > 0$ and $T > 2M + 2$, with probability at least $1 - \varepsilon$, the least-square estimator $\mathbf{U}^s$ in (12.35) will reconstructs the true Koopman operator $\mathbf{U}_{\text{true}}$ with the following error bound: [6]

$$\|\mathbf{U}^s - \mathbf{U}^s_{\text{true}}\|_F \leq \frac{c}{\varepsilon\sqrt{T}} \sqrt{\mathbb{E}\{\text{Tr}(\mathbf{G}^s)\}\mathbb{E}\{\|(\mathbf{G}^s)^{-1}\|_F^2\}}, \tag{12.40}$$

where $c$ is constant and is function of the additive noise variance and $\|\cdot\|_F$ stands for Frobenius norm. These sample complexity results are used to determine the data required to achieve the desired level of accuracy of the approximation.

## 12.4 Feedback Controller Design

The control Lyapunov function provides a powerful tool for the design of a stabilizing feedback controller which also enjoys some optimality property using the principle of inverse optimality. However, one of the main challenges is to provide a systematic procedure to find CLFs. For a general nonlinear system finding a CLF remains a challenging problem. We exploit the bilinear structure of the nonlinear system in the Koopman eigenfunction space to provide a systematic procedure for computing control Lyapunov function. We restrict the search for the control Lyapunov function to the class of quadratic Lyapunov function of the form $V(\mathbf{z}) = \mathbf{z}^\top P\mathbf{z}$. It is important to emphasize that although the Lyapunov function is restricted to be quadratic in Koopman eigenfunctions space $\mathbf{z}$, the Lyapunov function contains higher order nonlinearities in the original state space $\mathbf{x}$. Theorem 12.5 can be stated for the quadratic stabilization of the following bilinear control system:

$$\dot{\mathbf{z}} = \Lambda\mathbf{z} + uB\mathbf{z}. \tag{12.41}$$

In the sequel, if there exists a quadratic CLF for the bilinear system (12.41), then we will say that the system (12.41) is *quadratic stabilizable.*

**Theorem 12.5** *System (12.41) is quadratic stabilizable if and only if there exists an $N \times N$ symmetric positive definite $P$ such that for all nonzero $\mathbf{z} \in \mathbb{R}^N$ with $\mathbf{z}^\top(P\Lambda + \Lambda^\top P)\mathbf{z} \geq 0$, we have $\mathbf{z}^\top(PB + B^\top P)\mathbf{z} \neq 0$.*

*Proof* Sufficiency ($\Leftarrow$): Suppose there is a symmetric, positive definite $P$ that satisfies the condition of Theorem 12.5. We can use it to construct $V(\mathbf{z}) = \mathbf{z}^\top P\mathbf{z}$ as our Lyapunov candidate function, and the derivative of $V$ with respect to time along trajectories of (12.41) is given by

$$\begin{aligned}\dot{V} &= \mathbf{z}^\top P\dot{\mathbf{z}} + \dot{\mathbf{z}}^\top P\mathbf{z}\\ &= \mathbf{z}^\top(P\Lambda + \Lambda^\top P)\mathbf{z} + u\mathbf{z}^\top(PB + B^\top P)\mathbf{z}.\end{aligned}$$

Since for all $\mathbf{z} \neq 0$ we have $\mathbf{z}^\top(PB + B^\top P)\mathbf{z} \neq 0$ when $\mathbf{z}^\top(P\Lambda + \Lambda^\top P)\mathbf{z} \geq 0$, we can always find a control input $u(\mathbf{z})$ such that

$$\dot{V} < 0, \quad \forall\mathbf{z} \in \mathbb{R}^N \setminus \{0\}.$$

Therefore, $V(\mathbf{z})$ is indeed a CLF for system (12.41).

Necessity ($\Rightarrow$): We will prove this by contradiction. Suppose that system (12.41) has a CLF in the form of $V(\mathbf{z}) = \mathbf{z}^\top P\mathbf{z}$, where $P$ does not satisfy the condition of Theorem 12.5. That is, there exists some $\bar{\mathbf{z}} \neq 0$ such that $\bar{\mathbf{z}}^\top(P\Lambda + \Lambda^\top P)\bar{\mathbf{z}} \geq 0$ but $\bar{\mathbf{z}}^\top(PB + B^\top P)\bar{\mathbf{z}} = 0$. In this case, we have

$$\dot{V}(\bar{\mathbf{z}}) = \bar{\mathbf{z}}^\top(P\Lambda + \Lambda^\top P)\bar{\mathbf{z}} \geq 0$$

for any input $u$, which contradicts the definition of a CLF. This completes the proof. □

The following convex optimization formulation can be formulated to search for quadratic Lyapunov function for bilinear system without uncertainty in Eq. (12.41):

$$\begin{aligned} \underset{t>0,\ P=P^\top}{\text{minimize}} \quad & t - \gamma \text{Trace}(PB) \\ \text{subject to} \quad & tI - (P\Lambda + \Lambda^\top P) \succeq 0 \\ & c^{\max} I \succeq P \succeq c^{\min} I, \end{aligned} \tag{12.42}$$

where $c^{\max} > c^{\min} > 0$, respectively, are two given positive scalars forming bounds for the largest and the smallest eigenvalues of $P$. The variable $t$ here represents an epigraph form for the largest eigenvalue of $P\Lambda + \Lambda^\top P$.

Optimization (12.42) has combined two objectives. On the one hand, we minimize the largest eigenvalue of $P\Lambda + \Lambda^\top P$. On the other hand, we try to maximize the smallest singular value of $PB + B^\top P$ at the same time. Noticing that it may be difficult to maximize the smallest singular value of $PB + B^\top P$ directly, we maximize the trace of $PB$ instead and employ a parameter $\gamma > 0$ to balance these two objectives.

*Remark 12.2* When an optimal $P^\star$ is solved from (12.42), we still need to check whether it satisfies the condition of Theorem 12.5 or not. So if a matrix $P^\star$ fails the condition check, then we may tune the parameter $\gamma$ and solve the above optimization again until we obtain a correct $P^\star$. Nevertheless, we observe from simulations (see the multiple examples in our simulation section) that when we choose a value $\gamma = 2$, optimization (12.42) will always yield an optimal $P^\star$ that satisfies the condition of Theorem 12.5.

*Remark 12.3* We also need to point out that, compared to searching for a nonlinear CLF for the original nonlinear system (12.14), the procedure for seeking a quadratic CLF for the bilinear system (12.41) becomes quite easier and more systematic. Furthermore, a quadratic CLF for the bilinear system is, in fact, non-quadratic (i.e., contains higher order nonlinear terms) for the system (12.14).

Once a quadratic control Lyapunov function $V(\mathbf{z}) = \mathbf{z}^\top P\mathbf{z}$ is found for bilinear system (12.41), we have several choices for designing a stabilizing feedback control law. For instance, applying the control law (12.17) or (12.18) we can construct

$$k(\mathbf{z}) = -\beta_k \,\text{sign}\left[\mathbf{z}^\top (PB + B^\top P)\mathbf{z}\right]. \tag{12.43}$$

$$k(\mathbf{z}) = -\beta_k \mathbf{z}^\top (PB + B^\top P)\mathbf{z}. \tag{12.44}$$

Moreover, given a positive semidefinite cost $q(\mathbf{z}) \geq 0$, we may also apply the inverse optimality property to design an optimal control via Sontag's formula (12.20) to obtain

$$k(\mathbf{z}) = \begin{cases} -\frac{\mathbf{z}^\top (P\Lambda+\Lambda^\top P)\mathbf{z}+\sqrt{(\mathbf{z}^\top (P\Lambda+\Lambda^\top P)\mathbf{z})^2+q(x)(\mathbf{z}^\top (PB+B^\top P)\mathbf{z})^2}}{\mathbf{z}^\top (PB+B^\top P)\mathbf{z}} & \text{if } \mathbf{z}^\top (PB + B^\top P)\mathbf{z} \neq 0 \\ 0 & \text{otherwise.} \end{cases} \tag{12.45}$$

**Algorithm 12.1** Data-driven stabilizing controller design framework

**Data**: Given open-loop time-series data $\{\mathbf{x}_k^0\} = \{\mathbf{x}_0^0, \mathbf{x}_1^0, \ldots, \mathbf{x}_M^0\}$, and $\{\mathbf{x}_k^1\}$ with $s = 1$ in (12.24) both with Gaussian process noise added

**Result**: Feedback control $u = k(\mathbf{z})$

1 **Phase I: Modeling and Identification**

2 Choose $N$ dictionary functions $\Psi(\mathbf{x}) := \left[\psi_1(\mathbf{x})\ \psi_2(\mathbf{x})\ \cdots\ \psi_N(\mathbf{x})\right]^\top$.

3 **for** $\mathbf{x}_i,\ i = 0, 1, 2, \ldots, M$ **do**

4 $\Psi(\mathbf{x}_i) := \left[\psi_1(\mathbf{x}_i)\ \psi_2(\mathbf{x}_i)\ \cdots\ \psi_N(\mathbf{x}_i)\right]^\top$

5 **end**

6 Obtain $\mathbf{G}^0$ and $\mathbf{A}^0$ matrices $\mathbf{G}^0 = \frac{1}{M}\sum_{m=1}^{M}\Psi(\mathbf{x}_m)\Psi(\mathbf{x}_m)^\top$; $\mathbf{A}^0 = \frac{1}{M}\sum_{m=0}^{M-1}\Psi(\mathbf{x}_m)\Psi(\mathbf{x}_{m+1})^\top$.

7 Compute $\mathbf{U}^0 = (\mathbf{G}^0)^\dagger\mathbf{A}^0$, and its eigenfunctions $\phi_j = \Psi^\top v_j$, where $v_j$ is the $j$th eigenvector of $\mathbf{U}^0$ with respect to eigenvalue $\lambda_j$, $j = 1, 2, \ldots, N$.

8 Convert to continuous-time eigenvalues $\hat{\lambda}_i = \log(\lambda_i)/\Delta t$

9 Get $\Lambda = diag(\hat{\lambda}_1, \hat{\lambda}_2, \ldots, \hat{\lambda}_N)$ by block diagonalization of eigenvalues $\lambda_i$, use (12.38) if $i$, $i+1$ complex conjugate.

10 Obtain the new eigenfuntion $\hat{\Phi}(\mathbf{x})$ similarly, where $\hat{\phi}_i := \phi_i$ if $\phi_i$ is a real-valued and $\hat{\phi}_i := 2\mathrm{Re}(\phi)$, $\hat{\phi}_{i+1} := -2\mathrm{Im}(\phi_i)$, if $i$ and $i+1$ are complex conjugate.

11 Replace the dictionary function $\Psi(\mathbf{x})$ with $\mathbf{z} = \hat{\Phi}(\mathbf{x})$ and repeat Step 2 to 7 with the datasets $\{\mathbf{x}_k^0\}$ and $\{\mathbf{x}_k^1\}$ to get $\overline{\mathbf{U}}^0$ and $\overline{\mathbf{U}}^1$.

12 Get $B = (\overline{\mathbf{U}}^1 - \overline{\mathbf{U}}^0)/\Delta t$

13 **end**

14 **Phase II: Optimization**

15 Solve the following convex problem for optimal $P*$ with $\Lambda$ and $B$,

$$\begin{aligned} \underset{t>0,\ P=P^\top}{\text{minimize}} \quad & t - \gamma\,\mathrm{Trace}(PB) \\ \text{subject to} \quad & tI - (P\Lambda + \Lambda^\top P) \succeq 0 \\ & c^{\max} I \succeq P \succeq c^{\min} I, \end{aligned}$$

where $c^{\max} > c^{\min} > 0$, $\gamma > 0$ are chosen properly.

16 **end**

17 Feedback control $u = k(\mathbf{z}) = -\beta_k \mathbf{z}^\top(PB + B^\top P)\mathbf{z}$ or modified Sontag's formula,

$$k(\mathbf{z}) = \begin{cases} -\frac{\mathbf{z}^\top(P\Lambda+\Lambda^\top P)\mathbf{z}+\sqrt{(\mathbf{z}^\top(P\Lambda+\Lambda^\top P)\mathbf{z})^2+q(x)(\mathbf{z}^\top(PB+B^\top P)\mathbf{z})^2}}{\mathbf{z}^\top(PB+B^\top P)\mathbf{z}} & \text{if } \mathbf{z}^\top(PB+B^\top P)\mathbf{z} \neq 0 \\ 0 & \text{otherwise.} \end{cases}$$

The controller design framework is outlined in Algorithm 12.1 for the design of stabilizing feedback controller from time-series data.

## 12.5 Simulation Results

*Example 12.1* (*Duffing Oscillator*) The first example we present is the stabilization of Duffing oscillator. The controlled Duffing oscillator equation is written as follows:

$$\begin{aligned} \dot{x}_1 &= x_2, \\ \dot{x}_2 &= (x_1 - x_1^3) - 0.5x_2 + u. \end{aligned} \tag{12.46}$$

The uncontrolled equation for Duffing oscillator consists of three equilibrium points, two of the equilibrium points at $(\pm 1, 0)$ are stable, and one equilibrium point at the origin is unstable. For identification of the control system dynamics, we excite the system with white noise with zero mean and 0.01 variance. The continuous-time control equation is discretized with a sampling time of $\Delta t = 0.25s$. In Fig. 12.2a, we show the sampling complexity plot for the approximation error as the function of data length. As proved in [6], the error for the approximation of the $\Lambda$ and $B$ matrix decreases as $\frac{1}{\sqrt{T}}$, where $T$ is a data length. The error plot in Fig. 12.2a satisfies this rate of decay. The sample complexity results in Fig. 12.2a are obtained using ten randomly chosen initial conditions and generating time-series data over the different lengths of time ranging from 6-time steps to 30-time steps. For each fixed time step we compute the $\Lambda$ and $B$ matrices. The error $\| \Lambda - \overline{\Lambda} \|_2$ and $\| B - \overline{B} \|_2$ is computed at each fixed time step where $\overline{\Lambda}$ and $\overline{B}$ are computed using data collected over 50-time steps. The dictionary function used in the approximation of the Koopman operator has a maximum degree of five, i.e., 21 basis functions, $N = 21$. In particular, the following choice of dictionary function is made in the approximation:

$$\Psi(\mathbf{x}) = [1,\ x_1,\ x_2,\ x_1x_2,\ \ldots,\ x_1^5,\ x_1^4x_2,\ x_1^3x_2^2,\ x_1^2x_2^3,\ x_1x_2^4,\ x_2^5].$$

For control design, we use an approximation of $\Lambda$ and $B$ matrices computed over 30-time steps. The controller is designed using the Algorithm 12.1. For this Duffing oscillator example, we use a control design formula in Eq. (12.44). To verify the effectiveness of the designed controller we simulate the closed-loop system with the `ode15s` solver in **MATLAB** starting from 10 randomly chosen initial conditions within the region $[-1.5, 1.5] \times [-1, 1]$. In Fig. 12.2d, we show the closed-loop trajectories in red starting from different initial conditions overlaid on the open-loop trajectories in blue. We notice that the controller forces the trajectories of the closed-loop system along the stable manifold of the open-loop system before the trajectories slide to the origin. The time trajectories and control plots from different initial conditions are shown in Fig. 12.2b and c, respectively.

*Example 12.2* (*Lorenz System*) The second example we pick is that of Lorenz system. The control Lorenz system can be written as follows:

$$\begin{aligned} \dot{x}_1 &= \sigma(x_2 - x_1), \\ \dot{x}_2 &= x_1(\rho - x_3) - x_2 + u, \\ \dot{x}_3 &= x_1x_2 - \beta x_3, \end{aligned} \tag{12.47}$$

where $\mathbf{x} \in \mathbb{R}^3$ and $u \in \mathbb{R}$ is the single input. With the parameter values $\rho = 28$, $\sigma = 10, \beta = \frac{8}{3}$, and control input $u = 0$ the Lorenz system exhibits chaotic behavior. In this 3D example, we generated the time-series data from 1000 random chosen

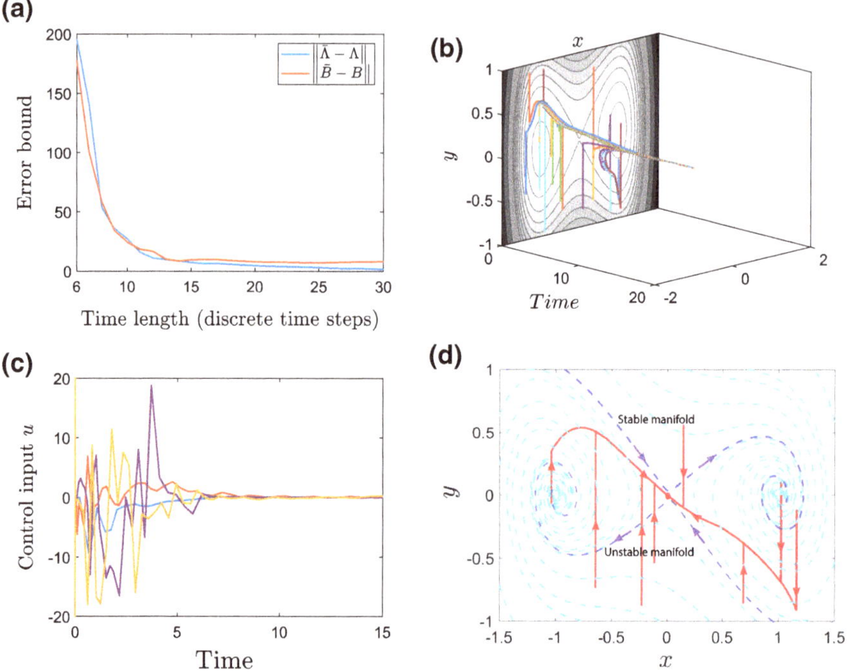


**Fig. 12.2** Data-driven stabilization of duffing oscillator. **a** Sample complexity error bounds for the approximation of $\Lambda$ and $B$ matrices as the function of data length; **b** Closed-loop trajectories versus time from multiple initial conditions; **c** Control value versus time from different initial conditions; **d** Comparison of closed-loop and open-loop trajectories in state space

initial conditions and propagate each of them for $T_{final} = 10$ s with sampling time $\Delta t = 0.001s$. For the purpose of identification the system is excited with white noise input with zero mean and 0.01 variance. The dictionary functions $\Psi(\mathbf{x})$ consist of 20 monomials of most degree $D = 3$

$$\Psi(\mathbf{x}) = [1,\ x_1,\ x_2,\ x_3,\ \ldots,\ x_1^3,\ x_1^2 x_2,\ x_1^2 x_3,\ x_1 x_2 x_3,\ \ldots\ x_3^3].$$

The objective is to stabilize one of the critical points $(\sqrt{\beta(\rho-1)}, \sqrt{\beta(\rho-1)}, \rho-1)$ of the Lorenz system. The system is stabilized using the control formula in Eq. (12.44). To validate the closed-loop control designed using the Algorithm 12.1, we perform the closed-loop simulation with five randomly chosen initial conditions in the domain $[-5, 5] \times [-5, 5] \times [0, 10]$ and solve the closed-loop system with `ode15s` solver in **MATLAB**. In Fig. 12.3a, we show the open-loop and closed-loop trajectories starting from five different initial conditions and the closed-loop trajectories are converging to the critical point. The time trajectories in Fig. 12.3b–d shows that all the initial conditions can be stabilized to the desired point within 4s.

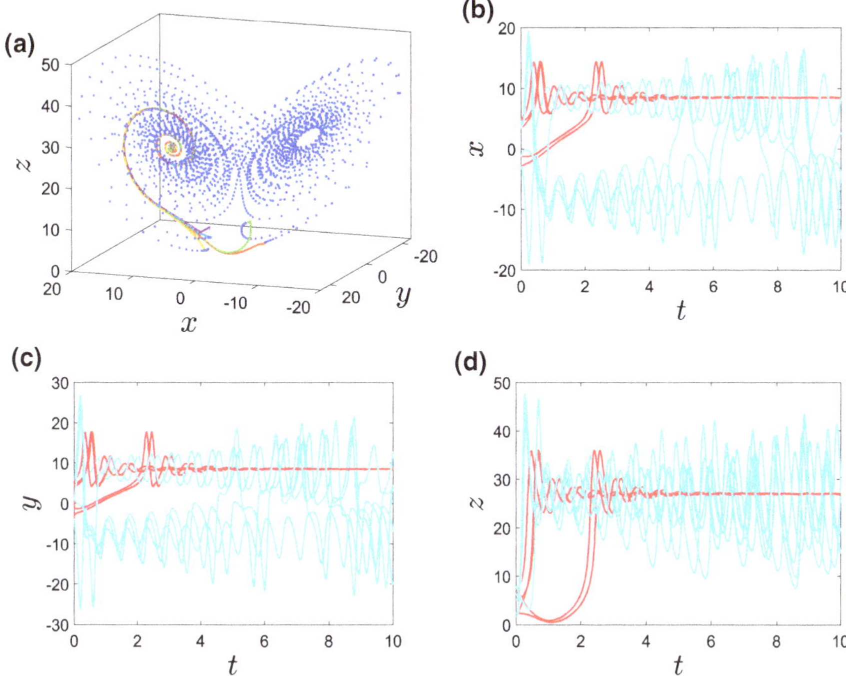


**Fig. 12.3** Feedback Stabilization of Lorenz system. **a** Comparison of open-loop and closed-loop trajectories in state space; **b** $x(t)$ *versus* time, open loop (blue) and closed loop (red); **c** $y(t)$ *versus* time, open loop (blue) and closed loop (red); **d** $z(t)$ *versus* time, open loop (blue) and closed loop (red)

*Example 12.3* (*IEEE 9 bus Power System*) In the last example, we consider the IEEE 9 bus system, the line diagram of which is shown in Fig. 12.4a. The model we are using is based on the modified 9-bus test system in [28]. The system consists of three synchronous machines(generators) with IEEE type-I exciters, loads, and transmission lines. The synthetic data is generated using Power System Toolbox (PST) in MATLAB [7]. The 9-bus power system network can be described by a set of differential algebraic equations (DAE). Consider a power system model with $n_g$ generator buses and $n_l$ load buses, the closed-loop generator dynamics for the $i$th generator bus can be represented as a second-order dynamical model with the control $u$:

$$\begin{aligned} \frac{d\delta_i}{dt} &= \omega_i - \omega_s, \\ \frac{d\omega_i}{dt} &= \frac{1}{M_i}\left(P_{m_i} - \sum_{j\in\mathcal{N}_i} \frac{E_i E_j}{X_{ij}} \sin(\delta_i - \delta_j) - D_i(\omega_i - \omega_s)\right) + u_i, \end{aligned} \tag{12.48}$$

where $\delta_i$, $\omega_i$ are the dynamic states of the generator and correspond to the generator rotor angle and the angular velocity of the rotor. The values for the other parameters are chosen as follows: $\omega_s = 1$, the generator mass $M_i = 23.64,\ 6.4,\ 3.1$, the internal damping $D_i = 0.05,\ 0.95,\ 0.05$, the generator power $P_{m_i} = 0.719,\ 1.63,\ 0.85$ for $i = 1, 2, 3$. The values of $X_{ij}$ are taken from the PST in MATLAB.

For the approximation of Koopman operator and eigenfunctions, the time-series data is generated from 100 initial conditions. Each initial condition is propagated for $T_{final} = 10s$ and $\Delta t = 0.01s$. The dictionary functions $H(x)$ in this example are chosen as 84 monomials of most degree $D = 3$. The data-driven stabilizing control is designed using modified Sontag's formula control in Eq. (12.20), where $q(x) = 10x^\top x$. The simulation results for this example are shown in Fig. 12.4. We notice that the open-loop system is marginally stable with sustained oscillations. The objective of the stabilizing controller is to stabilize the frequencies to $\omega_s = 1$ and the point for the stabilization of $\delta$ dynamics is determined by $P_{m_i}$. Simulation results in Fig. 12.4c and d show that the data-driven stabilizing controller is successful in stabilizing the power system dynamics.

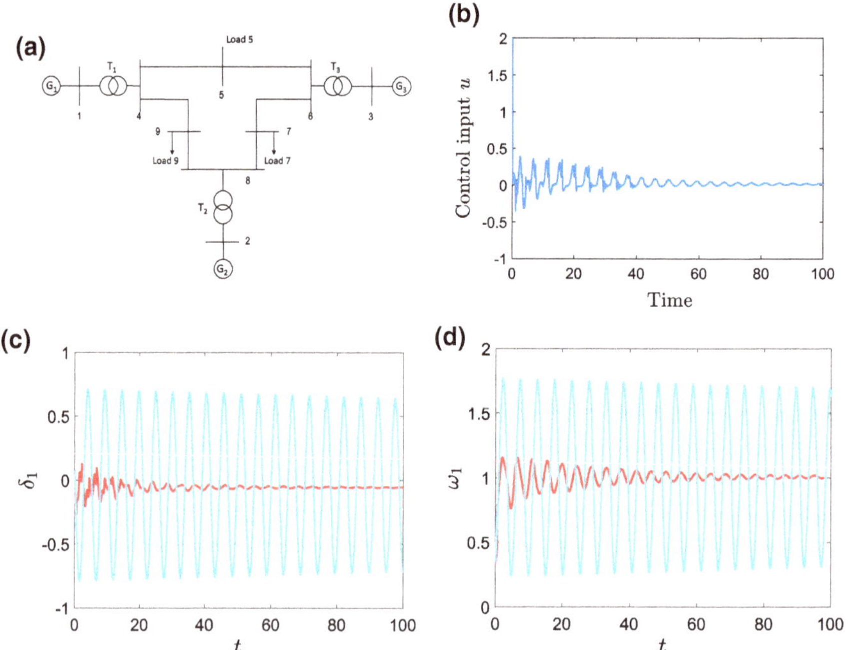


**Fig. 12.4** Stabilization of IEEE 9 bus system. **a** Line diagram for IEEE 9 bus system; **b** Control value versus time; **c** Comparison of open-loop and closed-loop trajectory for phase angle $\delta_1(t)$ of generator 1; **d** Comparison of open-loop and closed-loop trajectory for frequency $\omega_1(t)$ of generator 1

## 12.6 Conclusion

In this chapter, we provided a systematic approach for the data-driven feedback stabilization of nonlinear control systems. A data-driven approach is proposed for the identification of nonlinear control system and control Lyapunov function-based stabilizing feedback controller. The bilinear structure of the control system in Koopman eigenfunction coordinates is exploited to provide a convex optimization-based approach for the search of control Lyapunov function. Simulation results are presented to verify the applicability of the developed framework. In our future research efforts, we would account for the sample complexity-based error bound from the finite-dimensional approximation of the Koopman operator explicitly in the control design process for the nonlinear stabilization using tools from robust optimization.

## References


1. Arbabi, H., Korda, M., Mezić, I.: A data-driven Koopman model predictive control framework for nonlinear flows (2018). arXiv preprint arXiv:1804.05291
2. Artstein, Z.: Stabilization with relaxed controls. Nonlinear Anal.: Theory, Methods Appl. **7**(11), 1163–1173 (1983)
3. Astolfi, A.: Feedback stabilization of nonlinear systems. Encycl. Syst. Control, 437–447 (2015)
4. Boyd, S., Ghaoui, L.E., Feron, E., Balakrishnan, V.: Linear Matrix Inequalities in System and Control Theory. SIAM, Philadelphia (1994)
5. Budisic, M., Mohr, R., Mezić, I.: Applied Koopmanism. Chaos **22**(047), 510–532 (2012)
6. Chen, Y., Vaidya, U.: Sample complexity of nonlinear stochastic dynamics. In: American Control Conference (2019)
7. Chow, J.H., Cheung, K.W.: A toolbox for power system dynamics and control engineering education and research. IEEE Trans. Power Syst. **7**(4), 1559–1564 (1992). https://doi.org/10.1109/59.207380
8. Das, A.K., Huang, B., Vaidya, U.: Data-driven optimal control using transfer operators. In: 2018 IEEE Conference on Decision and Control (CDC), pp. 3223–3228. IEEE (2018)
9. Das, A.K., Raghunathan, A.U., Vaidya, U.: Transfer operator-based approach for optimal stabilization of stochastic systems. In: 2017 American Control Conference (ACC), pp. 1759–1764. IEEE (2017)
10. Freeman, R.A., Primbs, J.A.: Control Lyapunov functions: New ideas from an old source. In: Proceedings of the 35th IEEE Conference on Decision and Control, 1996, vol. 4, pp. 3926–3931. IEEE (1996)
11. Hanke, S., Peitz, S., Wallscheid, O., Klus, S., Böcker, J., Dellnitz, M.: Koopman operator based finite-set model predictive control for electrical drives (2018). arXiv preprint arXiv:1804.00854
12. Henrion, D., Garulli, A. (eds.): Positive Polynomials in Control. Lecture Notes in Control and Information Sciences, vol. 312. Springer, Berlin (2005)
13. Huang, B., Ma, X., Vaidya, U.: Feedback stabilization using Koopman operator. In: Proceedings of IEEE Control and Decision Conference, Miami (2018)
14. Huang, B., Vaidya, U.: Data-driven approximation of transfer operators: naturally structured dynamic mode decomposition. In: 2018 Annual American Control Conference (ACC), pp. 5659–5664. IEEE (2018)
15. Kaiser, E., Kutz, J.N., Brunton, S.L.: Data-driven discovery of Koopman eigenfunctions for control (2017). arXiv preprint arXiv:1707.01146

16. Khalil, H.K.: Nonlinear Systems. Prentice Hall, New Jersey (1996)
17. Korda, M., Mezić, I.: Linear predictors for nonlinear dynamical systems: Koopman operator meets model predictive control. Automatica **93**, 149–160 (2018)
18. Korda, M., Susuki, Y., Mezić, I.: Power grid transient stabilization using Koopman model predictive control (2018). arXiv preprint arXiv:1803.10744
19. Lasota, A., Mackey, M.C.: Chaos, Fractals, and Noise: Stochastic Aspects of Dynamics. Springer, New York (1994)
20. Mauroy, A., Mezić, I.: Global stability analysis using the eigenfunctions of the Koopman operator. IEEE Trans. Autom. Control **61**(11), 3356–3369 (2016)
21. Mezić, I.: Spectral properties of dynamical systems, model reductions and decompositions. Nonlinear Dyn. (2005)
22. Parrilo, P.A.: Structured semidefinite programs and semialgebraic geometry methods in robustness and optimization. Ph.D. thesis, California Institute of Technology, Pasadena (2000)
23. Peitz, S., Klus, S.: Koopman operator-based model reduction for switched-system control of pdes (2017). arXiv preprint arXiv:1710.06759
24. Primbs, J.A., Nevistić, V., Doyle, J.C.: Nonlinear optimal control: a control Lyapunov function and receding horizon perspective. Asian J. Control **1**(1), 14–24 (1999)
25. Raghunathan, A., Vaidya, U.: Optimal stabilization using Lyapunov measures. IEEE Trans. Autom. Control **59**(5), 1316–1321 (2014)
26. Rowley, C.W., Mezić, I., Bagheri, S., Schlatter, P., Henningson, D.S.: Spectral analysis of nonlinear flows. J. Fluid Mech. **641**, 115–127 (2009)
27. Sastry, S.: Nonlinear Systems: Analysis, Stability, and Control, vol. 10. Springer Science & Business Media, Berlin (2013)
28. Sauer, P.W., Pai, M.: Power system dynamics and stability. Urbana **51**, 61,801 (1997)
29. Schmid, P.J.: Dynamic mode decomposition of numerical and experimental data. J. Fluid Mech. **656**, 5–28 (2010)
30. Sontag, E.D.: A 'universal' construction of Artstein's theorem on nonlinear stabilization. Syst. Control. Lett. **13**(2), 117–123 (1989)
31. Sootla, A., Mauroy, A., Ernst, D.: Optimal control formulation of pulse-based control using Koopman operator. Automatica **91**, 217–224 (2018)
32. Surana, A.: Koopman operator framework for time series modeling and analysis. J. Nonlinear Sci. 1–34 (2018)
33. Surana, A., Banaszuk, A.: Linear observer synthesis for nonlinear systems using Koopman operator framework. In: Proceedings of IFAC Symposium on Nonlinear Control Systems. Monterey, California (2016)
34. Susuki, Y., Mezić, I.: Nonlinear Koopman modes and coherency identification of coupled swing dynamics. IEEE Trans. Power Syst. **26**(4), 1894–1904 (2011)
35. Vaidya, U., Mehta, P., Shanbhag, U.: Nonlinear stabilization via control Lyapunov measure. IEEE Trans. Autom. Control **55**, 1314–1328 (2010)
36. Vaidya, U., Mehta, P.G.: Computation of Lyapunov measure for almost everywhere stability. In: Proceedings of IEEE Conference on Decision and Control, pp. 5228–5233. San Diego (2006)
37. Vaidya, U., Mehta, P.G.: Lyapunov measure for almost everywhere stability. IEEE Trans. Autom. Control **53**, 307–323 (2008)
38. Williams, M.O., Kevrekidis, I.G., Rowley, C.W.: A data-driven approximation of the Koopman operator: Extending dynamic mode decomposition. J. Nonlinear Sci. **25**(6), 1307–1346 (2015)

# Chapter 13
# Parameter Estimation and Identification of Nonlinear Systems with the Koopman Operator

**Alexandre Mauroy and Jorge Goncalves**

**Abstract** We report on a novel framework for parameter estimation and nonlinear system identification. This framework is based on the key idea that nonlinear system identification in the state space is equivalent to linear identification of the Koopman operator in the space of observables. The proposed technique is divided into three steps. First, data is lifted in a higher dimensional space spanned by properly chosen basis functions. Second, a matrix representation of the Koopman operator is identified in this lifted space. This is equivalent to a linear identification problem. Finally, the infinitesimal generator of the Koopman semigroup is computed and used to identify the unknown vector field. Two methods (main and dual) are proposed in this framework and complemented with theoretical convergence results. The main method is applied to parameter estimation in the case of polynomial vector fields and both autonomous and input–output systems are considered. It is shown to be efficient with a general class of systems, including chaotic systems, and well suited to low sampling rate datasets. The other (dual) method is well suited to high-dimensional systems and is used in the context of network reconstruction.

## 13.1 Introduction

Identifying nonlinear dynamical systems from data is an important problem arising in various fields such as engineering, biology, finance, to name a few. Toward this end, black-box methods have been developed to describe nonlinear dynamics through input–output relationships (e.g., Wiener and Volterra series models [32], nonlinear auto-regressive models [14], neural network models [20], see also [26]).

A. Mauroy (✉)
Department of Mathematics, University of Namur, Namur, Belgium
e-mail: alexandre.mauroy@unamur.be

J. Goncalves
Luxembourg Centre for Systems Biomedecine, University of Luxembourg, Belvaux, Luxembourg
e-mail: jorge.goncalves@uni.lu



A. Mauroy et al. (eds.), *The Koopman Operator in Systems and Control*,
Lecture Notes in Control and Information Sciences 484,
https://doi.org/10.1007/978-3-030-35713-9_13

These methods are framed in the classical context of system identification [16] and are typically well suited to long and highly sampled time-series.

In a related context, parameter estimation methods aim at identifying the system dynamics with a (supposedly) known structure (see, e.g., [2, 3] and references therein). They typically seek for the best fit of the time derivatives of the state variables over a given set of library functions [31]. Recently, similar approaches were proposed, exploiting sparsity through, e.g., Bayesian learning [22] or pruning [7]. These abovementioned methods are *direct methods*, in the sense that they rely on the *a priori* knowledge of time derivatives. In contrast, *indirect methods* have also been developed, which rely on an initial value problem formulation and do not require the estimation of time derivatives [4]. If we draw a parallel with linear systems of the form $\dot{\mathbf{x}} = \mathbf{A}\mathbf{x}$, direct methods identify the matrix $\mathbf{A}$ from approximate time derivatives while indirect methods identify the exponential matrix $e^{\mathbf{A}T_s}$ from snapshot data (with sampling time $T_s$) and then recover the matrix $\mathbf{A}$ (see [35]). Direct and indirect methods have both advantages and drawbacks. On the one hand, direct methods are not well suited to noisy, low sampling rate datasets where the computation of time derivatives is prohibitive, but they are based on static linear regression techniques. On the other hand, indirect methods are adapted to low sampling rate datasets, but at the expense of solving a (nonconvex) nonlinear least-squares problem.

In this chapter, we present an identification method based on the Koopman operator framework. Initially motivated by network reconstruction problems [30], the proposed method is not developed in the vein of classical system identification and can deal with snapshot pairs of data instead of time-series data. It is an indirect method, therefore well-suited to low sampling, but at the same time relying on linear least-squares optimization. This method has been initially presented in [17, 18] and a similar approach, though based on nonconvex optimization, has also been developed in a stochastic context [25]. Although the proposed technique has been proposed for vector field identification and not for prediction (see instead [11]), it has recently been successfully used in the context of robot motion prediction [5].

The proposed method is based on the key idea that nonlinear system identification in the state space is equivalent to linear identification of the Koopman operator in the space of observables. The related numerical scheme is divided into three steps. First, data is lifted in a higher dimensional lifted space. Second, a linear identification problem is solved to compute the matrix representation of the Koopman operator in the lifted space. Two methods (hereafter called main and dual methods) are proposed, which consider the Koopman operator representation in the space of observables and in a "sample space", respectively. Third, the infinitesimal generator of the Koopman semigroup is computed through the matrix logarithm and connected to the unknown vector field. Note that the first step of the numerical scheme is directly connected to the so-called dynamic mode decomposition (EDMD) technique [33] (see also Chap. 1).

In the rest of this chapter, the identification problem is presented in Sect. 13.2 and the two methods are described in Sects. 13.3 (main method) and 13.4 (dual method).

Theoretical convergence results are provided in Sect. 13.5 and both methods are illustrated with several examples in Sect. 13.6. Concluding remarks are given in Sect. 13.7.

## 13.2 Problem Setting and Koopman Operator

### *13.2.1 Identification Problem*

Consider the dynamical system

$$\dot{\mathbf{x}} = \mathbf{F}(\mathbf{x}) = \sum_{k=1}^{N_F} \mathbf{w}_k \, h_k(\mathbf{x}) \,, \quad \mathbf{x} \in \mathbb{R}^n, \tag{13.1}$$

where $h_k : \mathbb{R}^n \to \mathbb{R}^n$ are known library functions (e.g., monomials). We aim at identifying the unknown coefficients $\mathbf{w}_k = (w_k^1 \; \cdots \; w_k^n) \in \mathbb{R}^n$ from $M$ snapshot pairs

$$(\mathbf{x}_k, \mathbf{y}_k) = (\mathbf{x}_k, \mathbf{S}(T_s, \mathbf{x}_k)) \in \mathbb{R}^{2n}, \tag{13.2}$$

where $\mathbf{S} : \mathbb{R}^+ \times \mathbb{R}^n \to \mathbb{R}^n$ is the flow induced by (13.1) and $T_s$ is the sampling time. These data pairs can possibly be affected by measurement errors (e.g., measurement noise).

### *13.2.2 Koopman Operator and Lifting Method*

In the context of the identification problem, we aim at "identifying" the Koopman semigroup of operators

$$U^t f = f \circ \mathbf{S}^t \quad f \in \mathscr{F}, \tag{13.3}$$

where $\circ$ denotes the composition of functions and $\mathbf{S}^t(\cdot) = \mathbf{S}(t, \cdot)$. Note that the space of functions $\mathscr{F}$ is assumed to be such that the Koopman semigroup is strongly continuous (e.g., $\mathscr{F} = L^2$). When the Koopman semigroup is identified, one can compute its infinitesimal generator

$$Lf = \lim_{t \downarrow 0} \frac{U^t f - f}{t} = \mathbf{F} \cdot \nabla f, \tag{13.4}$$

which is directly related to the unknown vector field. This yields a *lifting* technique that consists in three steps:

1. **Lifting:** Data points are "lifted" to a high-dimensional space that approximates the (infinite-dimensional) space $\mathscr{F}$.
2. **Linear identification:** The Koopman operator (13.3) and its infinitesimal generator (13.4) are identified in the lifted space through linear least-squares regression.
3. **"Lifting back":** The vector field of (13.1) is computed from the identified infinitesimal generator.

Depending on the choice of lifted space, these three steps lead to two identification methods described in the following sections: a main method and a dual method.

## 13.3 Main Method

In this section, we describe the main identification method based on the identification of the Koopman operator and discuss a few extensions.

### *13.3.1 Description of the Main Method*

The first step and a part of the second step are closely related to the so-called extended dynamic mode decomposition (EDMD) algorithm ([33], see also Chap. 1).

#### 13.3.1.1 First Step: Lifting

We consider a subspace $\mathscr{F}_N \subset \mathscr{F}$ spanned by the basis functions $\psi_k, k = 1, \cdots, N$, such that

$$\{\psi_k\}_{k=1}^N \supseteq \left(\{h_k\}_{k=1}^{N_F} \cup \{\mathrm{id}_j\}_{j=1}^n\right) \tag{13.5}$$

with the identity functions $\mathrm{id}_j$ mapping $\mathbf{x}$ to its $j$th component $x_j$. Note that this basis is said to be state-inclusive [10]. There is no general rule for the choice of basis functions, which might affect the quality of the results. However, we will generally choose a polynomial basis. The number of basis functions must be larger than (or equal to) the number of data points (i.e., $N \geq M$). Data points $\mathbf{x}_k \in \mathbb{R}^n$ are somehow lifted to the space $\mathscr{F}_N$ and transformed to new data points $(\psi_1(\mathbf{x}_k), \ldots, \psi_N(\mathbf{x}_k)) \in \mathbb{R}^N$. In particular, we construct the $M \times N$ data matrices

$$\mathbf{P_x} = \begin{pmatrix} \psi_1(\mathbf{x}_1) & \psi_2(\mathbf{x}_1) & \cdots & \psi_N(\mathbf{x}_1) \\ \psi_1(\mathbf{x}_2) & \psi_2(\mathbf{x}_2) & \cdots & \psi_N(\mathbf{x}_2) \\ \vdots & \vdots & \ddots & \vdots \\ \psi_1(\mathbf{x}_M) & \psi_2(\mathbf{x}_M) & \cdots & \psi_N(\mathbf{x}_M) \end{pmatrix} \quad \mathbf{P_y} = \begin{pmatrix} \psi_1(\mathbf{y}_1) & \psi_2(\mathbf{y}_1) & \cdots & \psi_N(\mathbf{y}_1) \\ \psi_1(\mathbf{y}_2) & \psi_2(\mathbf{y}_2) & \cdots & \psi_N(\mathbf{y}_2) \\ \vdots & \vdots & \ddots & \vdots \\ \psi_1(\mathbf{y}_M) & \psi_2(\mathbf{y}_M) & \cdots & \psi_N(\mathbf{y}_M) \end{pmatrix}. \tag{13.6}$$

#### 13.3.1.2 Second Step: Identification of the Operators

The Koopman matrix $\mathbf{U} \in \mathbb{R}^{N \times N}$ is computed with the data matrices

$$\mathbf{U} = \mathbf{P}_{\mathbf{x}}^{\dagger} \mathbf{P}_{\mathbf{y}}, \tag{13.7}$$

where $^{\dagger}$ denotes the pseudoinverse. This matrix is the representation of the operator $\Pi_N U^{T_s}$ (restricted to $\mathscr{F}_N$), where $\Pi_N : \mathscr{F} \to \mathscr{F}_N$ is the discrete orthogonal projection

$$\Pi_N f = \underset{\tilde{f} \in \text{span}\{\psi_1, \ldots, \psi_N\}}{\text{argmin}} \sum_{k=1}^{M} |\tilde{f}(\mathbf{x}_k) - f(\mathbf{x}_k)|^2 \,. \tag{13.8}$$

Considering the vector of basis functions $\Psi(\mathbf{x}) = (\psi_1(\mathbf{x}), \ldots, \psi_N(\mathbf{x}))^{\mathbf{T}}$, we have $U^{T_s} f \approx \Pi_N U^{T_s} f = \Psi^{\mathbf{T}}(\mathbf{Ua})$ for all $f \in \mathscr{F}_N$ such that $f = \Psi^{\mathbf{T}}\mathbf{a}$.

Next, we can compute the matrix representation $\mathbf{L}$ of the projected infinitesimal generator $\Pi_N L$. Assuming that $\mathbf{U} \approx e^{\mathbf{L} T_s}$ (see Sect. 13.5), we obtain

$$\mathbf{L} \approx \frac{1}{T_s} \log \mathbf{U} \triangleq \hat{\mathbf{L}}, \tag{13.9}$$

where "log" denotes the matrix logarithm. Note that we have

$$Lf \approx \Pi_N L f = \Psi^{\mathbf{T}}(\mathbf{La}) \tag{13.10}$$

for all $f = \Psi^{\mathbf{T}}\mathbf{a}$.

#### 13.3.1.3 Third Step: Identification of the Vector Field

Using (13.4), we observe that $L\,\text{id}_j = F_j$, which is the $j$th component of the vector field. Moreover, $\text{id}_j = \psi_l$ for some $l$, or equivalently $\text{id}_j = \Psi^{\mathbf{T}}\mathbf{e}_l$, where $\mathbf{e}_l$ is the $l$th unit vector. It follows from (13.10) that $F_j \approx \Psi^{\mathbf{T}}(\mathbf{L}\mathbf{e}_l)$ so that the $l$th column of $\mathbf{L}$ is the expansion of $F_j$ in the basis of functions $\psi_k$. Since the library functions $\{h_k\}_{k=1}^{N_F}$ of the vector field are in the set of basis functions, we have in particular $w_k^j = \mathbf{L}_{ml}$ with $\psi_m = h_k$ and $\psi_l = \text{id}_j$. Finally, since $\hat{\mathbf{L}} \approx \mathbf{L}$, we obtain the estimates $\hat{w}_k^j = \hat{\mathbf{L}}_{ml}$.

### 13.3.2 Algorithm

The main method is summarized in Algorithm 13.1. A code can be found at www.github.com/AlexMauroy/Koopman-identification/tree/master/main.

**Algorithm 13.1** Main Lifting Method

**Input:** Snapshot pairs $\{(\mathbf{x}_k, \mathbf{y}_k)\}_{k=1}^M$, with $\mathbf{x}_k \in \mathbb{R}^n$; sampling period $T_s$; vector field library functions $\{h_k\}_{k=1}^{N_F}$.

**Output:** Estimates $\hat{w}_k^j$.

1: Choose $N \geq M$ basis functions $\{\psi_k\}_{k=1}^N$ such that

$$\left(\{h_k\}_{k=1}^{N_F} \cup \{\mathrm{id}_j\}_{j=1}^n\right) \subseteq \{\psi_k\}_{k=1}^N$$

2: Construct the $M \times N$ matrices $\mathbf{P_x}$ and $\mathbf{P_y}$ defined in (13.6)

3: Compute the $N \times N$ Koopman matrix $\mathbf{U} = \mathbf{P_x^\dagger} \mathbf{P_y}$

4: Compute the $N \times N$ matrix $\hat{\mathbf{L}} = 1/T_s \log \mathbf{U}$

5: $\hat{w}_k^j := \hat{\mathbf{L}}_{ml}$, with $\psi_m = h_k$ and $\psi_l = \mathrm{id}_j$

## 13.3.3 Variants and Extensions

The numerical method presented in this section has several variants and generalizations.

### 13.3.3.1 Optimization Methods

In the proposed method, we obtain the estimates $\hat{w}_j^k$ with only $n$ columns of $\hat{\mathbf{L}}$, which are related to specific basis functions $\mathrm{id}_j$. Instead, one could consider other columns of the matrix, yielding additional linear equations on the estimates $\hat{w}_j^k$. Since

$$L = \sum_{j=1}^{n} \sum_{k=1}^{N_F} w_k^j L_k^j$$

with the operators $L_k^j = h_k \frac{\partial}{\partial x_j}$, we obtain the matrix equality

$$\hat{\mathbf{L}} \approx \mathbf{L} = \sum_{j=1}^{n} \sum_{k=1}^{N_F} \hat{w}_k^j \mathbf{L}_k^j,$$

where $\mathbf{L}_k^j$ is the matrix representation of $\Pi_N L_k^j$. This leads to an overconstrained problem, which could be solved by promoting sparsity (e.g., Lasso method). However numerical results suggest that this variant of the method does not improve the results.

Alternatively, one can avoid the computation of the matrix logarithm (in step 2) and consider the equality $\mathbf{U} = e^{\mathbf{L}T_s}$, where $\mathbf{U}$ is obtained from data and $\mathbf{L}$ depends on the coefficients $w_k^j$. This amounts to solving a (nonconvex) nonlinear least-squares problem. This alternative method is considered in [25].

#### 13.3.3.2 Systems with Inputs

The main identification method can be extended to systems with inputs of the form

$$\dot{\mathbf{x}} = \mathbf{F}(\mathbf{x}, \mathbf{u}(t)), \tag{13.11}$$

where $\mathbf{x} \in \mathbb{R}^n$ and $\mathbf{u} \in \mathscr{U} : \mathbb{R}^+ \to \mathbb{R}^p$. The flow $\mathbf{S} : \mathbb{R}^+ \times \mathbb{R}^n \times \mathscr{U}$ is such that $\mathbf{S}(\cdot, \mathbf{x}, \mathbf{u}(\cdot))$ is the solution to (13.11) associated with the initial condition $\mathbf{x}$ and the input $\mathbf{u}(\cdot)$. Following the generalization proposed in [6, 23], we consider the input $\mathbf{u}$ as additional state variables and the observables $f : \mathbb{R}^n \times \mathbb{R}^p \to \mathbb{R}$. This allows to define the Koopman semigroup of operators

$$U^t f(\mathbf{x}, \mathbf{u}) = f(\mathbf{S}^t(\mathbf{x}, \mathbf{u}(\cdot) = \mathbf{u}), \mathbf{u}), \tag{13.12}$$

where $\mathbf{u}(\cdot) = \mathbf{u}$ is a constant input. The above operator is the Koopman semigroup of operators associated with the augmented system $\dot{\mathbf{x}} = \mathbf{F}(\mathbf{x}, \mathbf{u})$, $\dot{\mathbf{u}} = \mathbf{0}$. In particular, the infinitesimal generator is still given by (13.4).

The operator (13.12) can be identified from data provided that the zero-order hold assumption is satisfied, that is

$$\mathbf{S}^{T_s}(\mathbf{x}, \mathbf{u}(\cdot)) \approx \mathbf{S}^{T_s}(\mathbf{x}, \mathbf{u}(\cdot) = \mathbf{u}(0)) \ .$$

This holds in particular if the sampling rate is high enough. We note, however, that this assumption might not always be satisfactory in the context of sampled-data control theory [1]. The Koopman matrix $\mathbf{U}$ is obtained with snapshot pairs $([\mathbf{x}_k, \mathbf{u}_k], [\mathbf{y}_k, \mathbf{u}_k]) \in \mathbb{R}^{(n+p)\times 2}$ and the procedure follows similar lines, as described in Sect. 13.3.1, but with the augmented state space $\mathbb{R}^{n+p}$. It is noticeable that the identification method not only provides the vector field coefficients associated with the state $\mathbf{x}$, but also those associated with the input $\mathbf{u}$.

#### 13.3.3.3 Stochastic Systems

We consider a system described by the stochastic differential equation

$$d\boldsymbol{x} = \boldsymbol{F}(\mathbf{x})dt + \sigma\, d\boldsymbol{w}(t), \tag{13.13}$$

where $\boldsymbol{w}(t)$ is a Wiener process. The flow $\mathbf{S} : \mathbb{R}^+ \times \mathbb{R}^n \times \Omega \to \mathbb{R}^n$, where $\Omega$ is the probability space, is such that $\mathbf{S}(\cdot, \mathbf{x}, \omega)$ is a realization of the stochastic process which is solution to (13.13) and associated with the initial condition $\mathbf{x}$. The Koopman semigroup of operators is defined by (see, e.g., [19] and Chap. 6)

$$U^t_{stoc} f(\mathbf{x}) = \mathbb{E}[f(\mathbf{S}(t, \mathbf{x}, \omega))] \tag{13.14}$$

and its infinitesimal generator is given by

$$L_{stoc} f = \mathbf{F} \cdot \nabla f + \frac{\sigma^2}{2} \Delta f, \tag{13.15}$$

where $\Delta = \sum_k \partial^2 / \partial x_k^2$ denotes the Laplacian operator.

The identification method is similar to the deterministic case. As shown in [33], the Koopman matrix $\mathbf{U}_{stoc}$ associated with (13.14) is still computed with (13.7) and the matrix $\hat{\mathbf{L}}_{stoc}$ is obtained with the matrix logarithm as in (13.9). To compute the estimates $\hat{w}_k^j$, we need to use the approximation $\hat{\mathbf{L}}_{stoc} \approx \mathbf{L} + \mathbf{D}\sigma^2/2$, where $\mathbf{L}$ represents the infinitesimal generator $L = \mathbf{F} \cdot \partial / \partial x$ of the deterministic case and $\mathbf{D}$ represents the Laplacian operator. For all $\psi_l = \mathrm{id}_j$, we have $\Delta \psi_l = 0$, so that the $l$th columns of $\mathbf{D}$ contain only zeros. It follows that $w_k^j = \mathbf{L}_{ml} = [\mathbf{L}_{stoc}]_{ml}$ and therefore we have again $\hat{w}_k^j = [\hat{\mathbf{L}}_{stoc}]_{ml}$.

We also refer to the recent work [25] for another lifting identification method in the case of stochastic systems.

## 13.4 Dual Method

The main method described above is not suited to high-dimensional systems. In the case of high-dimensional systems, the number $N_F$ of candidate library functions is typically large and implies that $N \geq N_F$ is large. In this context, the main method becomes computationally intractable and requires a prohibitive number $M \geq N$ of data points. To tackle these issues, one can use the dual method presented in this section. This dual method requires $M \leq N$ by using a $M$-dimensional "*sample space*" instead of the $N$-dimensional functional space.

### *13.4.1 Description of the Method*

We now describe the three steps of the dual method.

#### 13.4.1.1 First Step: Lifting

The first step is the same as the first step of the main method (see Sect. 13.3.1.1). However, in contrast to the main method, there is no constraint on the choice of basis functions (i.e., the set of basis functions do not need to contain the vector field functions $h_k$ and the identity functions). We suggest to consider radial basis functions centered at the data points (e.g., Gaussian basis functions $\psi_k(\mathbf{x}) = \exp(-\gamma \|\mathbf{x} - \mathbf{x}_k\|^2)$, with $\gamma > 0$), although there is no general rule to choose the basis.

#### 13.4.1.2 Second Step: Identification of the Operators

Instead of the Koopman matrix $\mathbf{U} = \mathbf{P}_\mathbf{x}^\dagger \mathbf{P}_\mathbf{y}$ used in the main method, we construct the matrix $\widetilde{\mathbf{U}} \in \mathbb{R}^{M \times M}$ with

$$\widetilde{\mathbf{U}} = \mathbf{P}_\mathbf{y}\,\mathbf{P}_\mathbf{x}^\dagger.$$

Note that we must have $N \geq M$. Since $\widetilde{\mathbf{U}} = \mathbf{P}_\mathbf{x}\mathbf{U}\mathbf{P}_\mathbf{x}^\dagger$, the data matrix $\mathbf{P}_\mathbf{x}$ can be interpreted as a change of coordinates, from the $N$-dimensional subspace spanned by the basis functions $\psi_k$ to the $M$-dimensional "*sample space*" where a function $f$ is represented by the vector $(f(\mathbf{x}_1) \cdots f(\mathbf{x}_M))^\mathrm{T}$. The matrix $\widetilde{\mathbf{U}}$, therefore, represents the Koopman operator $U^{T_s}$ in the sample space: for all $f \in \mathscr{F}_N$, we have

$$\begin{pmatrix} U^{T_s} f(\mathbf{x}_1) \\ \vdots \\ U^{T_s} f(\mathbf{x}_M) \end{pmatrix} \approx \widetilde{\mathbf{U}} \begin{pmatrix} f(\mathbf{x}_1) \\ \vdots \\ f(\mathbf{x}_M) \end{pmatrix}. \tag{13.16}$$

In fact, the basis functions $\psi_k$ play the role of test functions to compute the matrix $\widetilde{\mathbf{U}}$ satisfying (13.16) in the least-squares sense. The matrix $\widetilde{\mathbf{U}}$ can also be interpreted as the (discrete) projection of the dual Perron–Frobenius operator in a basis of $M$ evaluation functionals $\xi_k : f \mapsto f(\mathbf{x}_k)$, evaluated at the functions $\psi_j$. In that sense, this proposed method is truly a dual method.

Now we can compute the matrix representation $\widetilde{\mathbf{L}}$ of $L$ in the sample space, which satisfies

$$\begin{pmatrix} Lf(\mathbf{x}_1) \\ \vdots \\ Lf(\mathbf{x}_M) \end{pmatrix} \approx \widetilde{\mathbf{L}} \begin{pmatrix} f(\mathbf{x}_1) \\ \vdots \\ f(\mathbf{x}_M) \end{pmatrix}. \tag{13.17}$$

Similarly to the main method, an approximation of $\widetilde{\mathbf{L}}$ is given by

$$\widetilde{\mathbf{L}} \approx \frac{1}{T_s} \log \widetilde{\mathbf{U}} \triangleq \hat{\mathbf{L}}.$$

#### 13.4.1.3 Third Step: Identification of the Vector Field

Since $L\,\mathrm{id}_j = F_j$, it follows from (13.17) that we can obtain an estimate $\hat{\mathbf{F}}$ of the vector field at the sample points

$$\begin{pmatrix} \hat{\mathbf{F}}(\mathbf{x}_1)^\mathrm{T} \\ \vdots \\ \hat{\mathbf{F}}(\mathbf{x}_M)^\mathrm{T} \end{pmatrix} \approx \widetilde{\mathbf{L}} \begin{pmatrix} \mathbf{x}_1^\mathrm{T} \\ \vdots \\ \mathbf{x}_M^\mathrm{T} \end{pmatrix} \approx \hat{\mathbf{L}} \begin{pmatrix} \mathbf{x}_1^\mathrm{T} \\ \vdots \\ \mathbf{x}_M^\mathrm{T} \end{pmatrix}. \tag{13.18}$$

Once the value of the vector field is estimated at every data points, we can compute the estimates $\hat{w}_j^l$ of the coefficients $w_j^l$ by solving a static regression problem. For each $j = 1, \dots, n$, one has to solve

$$\hat{F}_j(\mathbf{x}_k) = \sum_{j=1}^{N_F} \hat{w}_l^j \, h_l(\mathbf{x}_k) \qquad k = 1, \dots, M$$

or equivalently

$$\begin{pmatrix} \hat{F}_j(\mathbf{x}_1) \\ \vdots \\ \hat{F}_j(\mathbf{x}_M) \end{pmatrix} = \mathbf{H}_\mathbf{x} \begin{pmatrix} \hat{w}_1^j \\ \vdots \\ \hat{w}_{N_F}^j \end{pmatrix} \tag{13.19}$$

with the matrix

$$\mathbf{H}_\mathbf{x} = \begin{pmatrix} \mathbf{h}(\mathbf{x}_1)^\mathrm{T} \\ \vdots \\ \mathbf{h}(\mathbf{x}_M)^\mathrm{T} \end{pmatrix}. \tag{13.20}$$

In most cases, it is reasonable to assume that most coefficients are zero, so that one can use a sparsity promoting method such as Lasso regularization [29]:

$$\min_{\mathbf{w} \in \mathbb{R}^{N_F}} \frac{1}{M} \left\| \mathbf{H}_\mathbf{x} \mathbf{w} - \begin{pmatrix} \hat{F}_j(\mathbf{x}_1) \\ \vdots \\ \hat{F}_j(\mathbf{x}_M) \end{pmatrix} \right\|_2^2 + \rho \|\mathbf{w}\|_1, \tag{13.21}$$

where $\rho$ is a positive regularization parameter.

### *13.4.2 Algorithm*

The dual method is summarized in Algorithm 13.2. A code can be found at www.github.com/AlexMauroy/Koopman-identification/tree/master/dual.

## 13.5 Convergence Results

In this section, we present convergence results showing that exact estimates of the vector field are obtained with both methods in optimal conditions.

**Algorithm 13.2** Dual Lifting Method

**Input:** Snapshot pairs $\{(\mathbf{x}_k, \mathbf{y}_k)\}_{k=1}^{M}$, $\mathbf{x}_k \in \mathbb{R}^n$; basis functions $\{\psi_k\}_{k=1}^{N}$ (with $N \geq M$); library functions $\{h_k\}_{k=1}^{N_F}$.

**Output:** Estimates $\hat{\mathbf{F}}(\mathbf{x}_k)$ and $\hat{w}_k^j$.

1: Choose $N \leq M$ basis functions $\{\psi_k\}_{k=1}^{N}$ (e.g., Gaussian radial basis functions)
2: Construct the $M \times N$ matrices $\mathbf{P_x}$ and $\mathbf{P_y}$ defined in (13.6)
3: Compute the $M \times M$ matrix $\widetilde{\mathbf{U}} = \mathbf{P_y}\,\mathbf{P_x}^{\dagger}$
4: Compute the $M \times M$ matrix $\hat{\mathbf{L}} = 1/T_s \log \widetilde{\mathbf{U}}$
5: Obtain $\hat{\mathbf{F}}(\mathbf{x}_k)$ with (13.18)
6: For each $j$, solve the regression problem (13.19) (e.g., solve the Lasso problem (13.21)) to obtain $\hat{w}_k^j$

## 13.5.1 Convergence of the Main Method

The main method yields exact estimates $\hat{w}_k^j$ of the vector field coefficients as the sampling time goes to zero.

**Theorem 13.1** *Assume that the sample points $\mathbf{x}_k$ are uniformly randomly distributed in a compact forward or backward invariant set $X \subset [-1, 1]^n$, and consider $\mathbf{y}_k = \mathbf{S}^{T_s}(\mathbf{x}_k)$ where $\mathbf{S}^t$ is an invertible and nonsingular flow generated by the dynamics* (13.1)*. If Algorithm 13.1 is used with the data pairs $\{\mathbf{x}_k, \mathbf{y}_k\}_{k=1}^{M}$ and with N basis functions* (13.5) *(with $N_F \leq N \leq K$), then*

$$\lim_{T_s \to 0} \hat{w}_k^j = w_k^j$$

*for all $j = 1, \ldots, n$ and $k = 1, \ldots, N_F$.*

*Proof* We refer to [18] for a detailed proof of the result and provide below a brief sketch of the proof. Consider the finite-dimensional operators $U_N = \Pi_N U^{T_s}|_{\mathscr{F}_N}$ and $A_N = e^{P_N L P_N T_s}$ (whose matrix representations are $\mathbf{U}$ and $e^{\mathbf{L}T_s}$, respectively). It can be shown that

$$\lim_{T_s \to 0} \frac{1}{T_s} \|(\log U_N - \log A_N) f\|_{L^2(X)} = 0$$

for all $f \in \mathscr{F}_N$ and we have

$$\begin{aligned} \lim_{T_s \to 0} \left\| \hat{F}_j - \Pi_N F_j \right\|_{L^2(X)} &= \lim_{T_s \to 0} \left\| \left( \frac{\log U_N}{T_s} - \Pi_N L \Pi_N \right) \mathrm{id}_j \right\|_{L^2(X)}, \\ &= \lim_{T_s \to 0} \frac{1}{T_s} \left\| (\log U_N - \log A_N)\, \mathrm{id}_j \right\|_{L^2(X)} = 0, \end{aligned} \tag{13.22}$$

where we used the fact that the identity function $\mathrm{id}_j \in \mathscr{F}_N$ satisfies $\Pi_N \mathrm{id}_j = \mathrm{id}_j$ and $L\mathrm{id}_j = F_j$. Since $F_j \in \mathscr{F}_N$, we finally obtain that

$$\lim_{T_s \to 0} \left\| \hat{F}_j - F_j \right\|_{L^2(X)} = 0$$

so that $\lim_{T_s \to 0} \hat{w}_k^j = w_k^j$. □

The convergence property is illustrated in Fig. 13.1a. Intuitively, the limit $T_s \to 0$ is required to capture the infinite set of frequencies of a nonlinear system and is related to the so-called system aliasing [35]. In (13.22), we used the equality $\log e^{\Pi_N L \Pi_N T_s} = \Pi_N L \Pi_N T_s$, which is true provided that the eigenvalues of $\Pi_N L \Pi_N$ lie in the strip $\{z \in \mathbb{C} : |\mathrm{Im}\{z\}| < \pi / T_s\}$. The limit $T_s \to 0$ ensures that this condition is satisfied (for arbitrarily large values $N$). It is also noticeable that the algorithm is equivalent to a direct computation of time derivatives in the limit $T_s \to 0$, since $\log U_N \mathrm{id}_j / T_s \approx (I - U_N)\mathrm{id}_j / T_s \approx F_j$. This explains that the number $N$ of basis functions does not need to tend to infinity. However, choosing large values $N$ can be useful. It can be shown by the Trotter–Kato theorem [8] that

$$\lim_{N \to \infty} \|(U_N - A_N) f\|_{L^2(X)} \to 0$$

for all $f \in \mathscr{F}_N$. It follows that one expects that, for nonzero values $T_s$, the error $1/T_s \| (\log U_N - \log A_N) f \|$ (and the estimation error) decreases as $N$ grows. This property is shown in Fig. 13.1b. However it only holds below some threshold value $N$ which depends on the sampling time $T_s$ (see the case $T_s = 0.5$).

When the vector field is not contained in the space $\mathscr{F}_N$ spanned by the basis functions, the $L^2$ estimation error on the vector field goes to zero as $N$ tends to infinity. This is summarized in the following corollary.

**Corollary 13.1** *Consider the assumptions of Theorem 13.1, but with vector field components that are not necessarily in $\mathscr{F}_N$. If Algorithm 13.1 is used with the data*

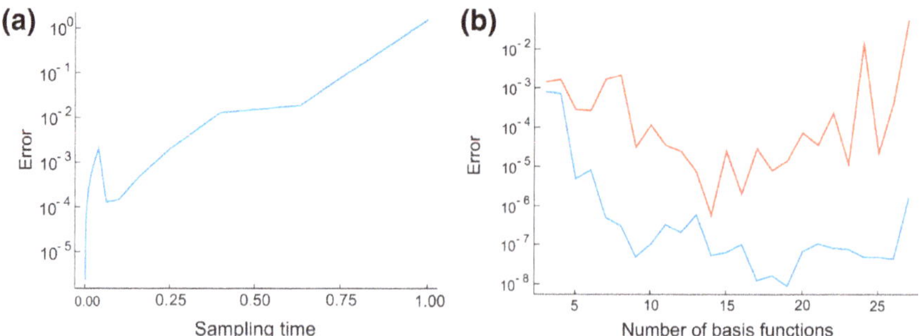


**Fig. 13.1** Convergence of the error for the main method used with the Van der Pol dynamics $\dot{x}_1 = x_2$, $\dot{x}_2 = (1 - x_1^2)x_2 - x_1$. **a** Error on the estimation of the coefficient associated with the monomial $x_1^2 x_2$ as a function of the sampling time (number of samples $M = 100$, number of basis functions $N = N_F = 3$). **b** Error on the estimation of the coefficient associated with the monomial $x_1^2 x_2$ as a function of the number of basis functions (number of samples $M = 600$). Blue: $T_s = 0.3$, red: $T_s = 0.5$

*pairs* $\{\mathbf{x}_k, \mathbf{y}_k\}_{k=1}^{M}$ *and with a set of basis functions* $\{\psi_k\}_{k=1}^{\infty}$ *which contains* $id_j$ *and whose span is dense in* $L^2(X)$, *then the estimated vector field* $\hat{F}_j = \sum_{k=1}^{N} \hat{w}_k^j \psi_k$ *satisfies*

$$\lim_{N\to\infty} \lim_{M\to\infty} \lim_{T_s\to 0} \|\hat{F}_j - F_j\|_{L^2(X)} = 0$$

*with probability one.*

*Proof* We note that the discrete orthogonal projection converges in the strong operator topology (in $L^2(X)$) to the orthogonal $L^2$ projection as $M \to \infty$ (see, e.g., [12]). Since the basis is complete in $L^2(X)$, the orthogonal projection converges in the strong operator topology to the identity operator as $N \to \infty$, with probability one. Then we have

$$\begin{aligned} &\lim_{N\to\infty} \lim_{K\to\infty} \lim_{T_s\to 0} \left\| \hat{F}_j - F_j \right\|_{L^2(X)} \\ &\leq \lim_{N\to\infty} \lim_{K\to\infty} \left( \lim_{T_s\to 0} \left\| \hat{F}_j - \Pi_N F_j \right\|_{L^2(X)} + \left\| \Pi_N F_j - F_j \right\|_{L^2(X)} \right) = 0, \end{aligned}$$

where we used (13.22). □

## 13.5.2 Convergence of the Dual Method

The dual method provides exact estimates of the vector field values $\hat{\mathbf{F}}(\mathbf{x}_k)$ at the sample points as the sampling time goes to zero and additionally, depending on the assumptions, as the number of sample points grows to infinity.

Before stating the main result, we need some preliminaries. We consider the space $\mathscr{F} = C(X)$ and the conjugate space $\mathscr{F}^*$ of linear functionals. Denoting by $\widetilde{\mathscr{F}}_M^*$ the subspace spanned by the evaluation functionals $\xi_k : f \mapsto f(\mathbf{x}_k)$, we define the discrete projection $\widetilde{\Pi}_M^* : \mathscr{F}^* \to \widetilde{\mathscr{F}}_M^*$ by

$$\widetilde{\Pi}_M^* \xi = \underset{\tilde{\xi} \in \text{span}\{\xi_1,\ldots,\xi_M\}}{\text{argmin}} \sum_{j=1}^{N} |\xi(\psi_j) - \tilde{\xi}(\psi_j)|^2 . \tag{13.23}$$

Then we can also define its adjoint projection operator $\widetilde{\Pi}_M : \mathscr{F} \to \widetilde{\mathscr{F}}_M$ (where the subspace $\widetilde{\mathscr{F}}_M$ is contained in the span of the functions $\psi_j$), which is obtained through interpolation and satisfies $\widetilde{\Pi}_M f(\mathbf{x}_k) = f(\mathbf{x}_k)$ for all $k$ (see [18] for the details).

**Theorem 13.2** *Assume that the sample points* $\mathbf{x}_k$ *are uniformly randomly distributed in a compact forward invariant set* $X$, *consider* $\mathbf{y}_k = \mathbf{S}^{T_s}(\mathbf{x}_k)$ *where* $\mathbf{S}^t$ *is an invertible and nonsingular flow generated by the dynamics* $\dot{\mathbf{x}} = \mathbf{F}(\mathbf{x})$ *and suppose that Algorithm 13.2 is used with the data pairs* $\{\mathbf{x}_k, \mathbf{y}_k\}_{k=1}^{M}$ *and with* $N$ *basis functions* $\psi_k \in C^1(X)$ *such that*

$$\mu\left\{\mathbf{x} \in X \mid \sum_{k=1}^{\infty} c_l \psi_k(\mathbf{x}) = 0\right\} = 0 \quad \forall (c_1, c_2, \dots) \neq \mathbf{0} \tag{13.24}$$

*where $\mu$ is the Lebesgue measure.*

1. *If $N = M$ and the identity function $id_j(\mathbf{x}) = x_j$ is in the span of $\{\psi_k\}_{k=1}^N$, then*

$$\lim_{T_s \to 0} \left| F_j(\mathbf{x}_k) - \hat{F}_j(\mathbf{x}_k) \right| = 0 \quad \forall k$$

*with probability one.*

2. *If $\lim_{M\to\infty} \lim_{N\to\infty} \|\widetilde{\Pi}_M id_j - id_j\|_L = 0$ (with the graph norm $\|f\|_L = \|f\|_{L^\infty(X)} + \|Lf\|_{L^\infty(X)}$), then*

$$\lim_{M\to\infty} \lim_{N\to\infty} \lim_{T_s \to 0} \left| F_j(\mathbf{x}_k) - \hat{F}_j(\mathbf{x}_k) \right| = 0 \quad \forall k$$

*with probability one.*

*Proof* We refer to [18] for a detailed proof of this result and provide below the sketch of the proof. It can be shown that $\widetilde{\mathbf{U}}$ and $e^{\widetilde{\mathbf{L}}T_s}$ are the matrix representations of the finite-dimensional operators $\widetilde{U}_M = \widetilde{\Pi}_M U^{T_s}|_{\widetilde{\mathscr{F}}_M}$ and $\widetilde{A}_M = e^{\widetilde{\Pi}_M L \widetilde{\Pi}_M T_s}$.

Similarly to the proof of Theorem 13.1, it can also be shown that

$$\lim_{T_s \to 0} \frac{1}{T_s} \|(\log \widetilde{U}_M - \log \widetilde{A}_M) f\|_{L^\infty(X)} = 0$$

for all $f \in \widetilde{\mathscr{F}}_M$ and we have

$$\begin{aligned} \lim_{T_s \to 0} \left| \hat{F}_j(\mathbf{x}_k) - L\widetilde{\Pi}_M \mathrm{id}_j(\mathbf{x}_k) \right| &\leq \lim_{T_s \to 0} \left\| \left( \frac{\log \widetilde{U}_M}{T_s} - \widetilde{\Pi}_M L \right) \widetilde{\Pi}_M \mathrm{id}_j \right\|_{L^\infty(X)}, \\ &= \lim_{T_s \to 0} \frac{1}{T_s} \left\| \left( \log \widetilde{U}_M - \log \widetilde{A}_M \right) \widetilde{\Pi}_M \mathrm{id}_j \right\|_{L^\infty(X)} = 0, \end{aligned} \tag{13.25}$$

where we used $\widetilde{\Pi}_M f(\mathbf{x}_k) = f(\mathbf{x}_k)$.

1. If $M = N$ and (13.24) holds, it can be shown that $\widetilde{\Pi}_M \psi_k = \psi_k$, so that $\widetilde{\Pi}_M \mathrm{id}_j = \mathrm{id}_j$ if $\mathrm{id}_j \in \mathrm{span}\{\psi_k\}_{k=1}^N$, which implies with (13.25) that

$$\lim_{T_s \to 0} \left| \hat{F}_j(\mathbf{x}_k) - F_j(\mathbf{x}_k) \right| = \lim_{T_s \to 0} \left| \hat{F}_j(\mathbf{x}_k) - L\widetilde{\Pi}_M \mathrm{id}_j(\mathbf{x}_k) \right| = 0\,.$$

2. If $\lim_{M\to\infty} \lim_{N\to\infty} \|\widetilde{\Pi}_M f_j - f_j\|_L = 0$, it follows from (13.25) that

$$\begin{aligned}
&\lim_{M\to\infty}\lim_{N\to\infty}\lim_{T_s\to 0}\left|\hat{F}_j(\mathbf{x}_k)-F_j(\mathbf{x}_k)\right| \\
&\leq \lim_{M\to\infty}\lim_{N\to\infty}\lim_{T_s\to 0}\left|\hat{F}_j(\mathbf{x}_k)-L\widetilde{\Pi}_M\mathrm{id}_j(\mathbf{x}_k)\right| \\
&+ \lim_{M\to\infty}\lim_{N\to\infty}\lim_{T_s\to 0}\left\|L\widetilde{\Pi}_M\mathrm{id}_j - L\mathrm{id}_j\right\|_{L^\infty(X)} \\
&= 0\,.
\end{aligned}$$

□

Convergence of Algorithm 13.2 is obtained for finite values $M$ and $N$ as the sampling time goes to zero provided that the identity function is contained in the span of test functions $\psi_k$ (case 1). We will use in practice a set of Gaussian radial basis functions that does not contain the identity function. In this case, it can be shown that the convergence of $\widetilde{\Pi}_M \mathrm{id}_j$ to $\mathrm{id}_j$ is uniform with all the derivatives (case 2) (see, e.g., [9]). The set of Gaussian radial basis functions also satisfies the independence condition (13.24) (see, e.g., [12]).

## 13.6 Examples

In this section, we apply the two methods on several examples, for illustrative purposes only. Additional examples can be found in [18].

Data is generated by numerical integration of the set of ODEs and scaled to ensure that the data points lie in $[-1, 1]^n$. Measurement noise $\eta$ is added to the data. Noise amplitude is normally distributed, with zero mean and standard deviation proportional to the state (i.e., $\eta \sim \mathcal{N}(\mathbf{0}, \sigma\mathbf{x}/\|\mathbf{x}\|)$.

The performance of the methods is evaluated by considering the estimation of the vector field coefficients $w_k^i$. In particular, we compute the root-mean-square error

$$\mathrm{RMSE} = \sqrt{\frac{1}{nN_F}\sum_{j=1}^{n}\sum_{k=1}^{N_F}\left((w_k^j)-(\hat{w}_k^j)\right)^2}$$

and the normalized root-mean-square error $\mathrm{NRMSE} = \mathrm{RMSE}/\overline{w}$, where $\overline{w}$ is the average value of the nonzero coefficients $|w_k^j|$.

All simulations have been performed in Julia, with additional Julia packages: `LinearAlgebra.jl`, `Random.jl`, `Plots.jl`, `Lasso.jl`, `DifferentialEquations.jl` [24].

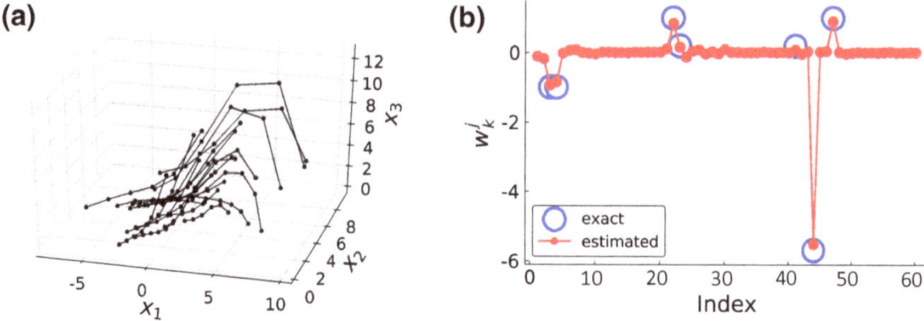


**Fig. 13.2** The main method is used to estimate the vector field coefficients of the Rössler system. **a** Data points generated by the dynamics (the trajectories are represented by black lines). **b** Exact and estimated coefficients of the vector field

## 13.6.1 Main Method

We first consider the main method to identify the chaotic Rössler system, the genetic toggle switch, and the forced inversed pendulum.

### 13.6.1.1 Rössler System

We use the Rössler dynamics

$$\begin{aligned}\dot{x}_1 &= -x_2 - x_3,\\ \dot{x}_2 &= x_1 + 0.2\,x_2,\\ \dot{x}_3 &= 0.2 + x_3\,(x_1 - 5.7)\end{aligned}$$

to generate $M = 100$ snapshot data pairs taken from 20 trajectories, with a sampling period $T_s = 0.2$ (see Fig. 13.2a). Initial conditions are randomly distributed in $[0, 10]^3$ and measurement noise is added with $\sigma = 0.01$. The basis functions are monomials with total degree less or equal to 3. The estimation of the vector field coefficients is shown in Fig. 13.2b. The NRMSE is equal to 0.06 (averaged over 10 simulations).

### 13.6.1.2 Genetic Toggle Switch

We consider the genetic toggle switch dynamics

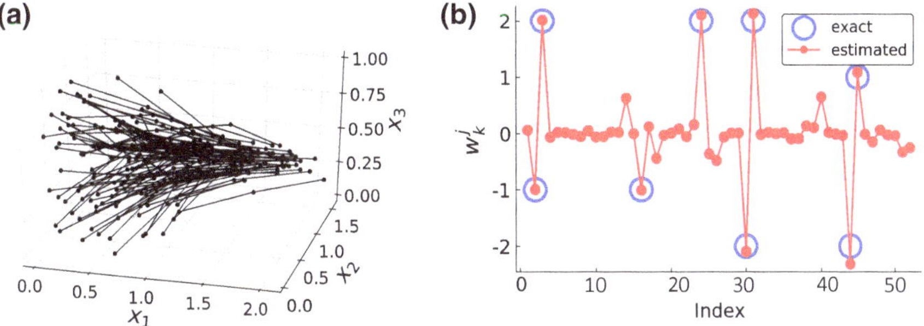


**Fig. 13.3** The main method is used to estimate the vector field coefficients of a toggle switch system. **a** Data points generated by the dynamics (the trajectories are represented by black lines). **b** Exact and estimated coefficients of the vector field

$$
\begin{aligned}
\dot{x}_1 &= -x_1 + 2\,x_2, \\
\dot{x}_2 &= -x_2 + \frac{2}{1 + x_3^2}, \\
\dot{x}_3 &= -2\,x_3 + 2\,x_4, \\
\dot{x}_4 &= -2\,x_4 + \frac{1}{1 + x_1}
\end{aligned}
$$

to generate $M = 200$ snapshot data pairs taken from 100 trajectories, with a sampling period $T_s = 0.5$ (see Fig. 13.3a). Initial conditions are randomly distributed in $[0, 1]^4$ and measurement noise is added with $\sigma = 0.001$. The basis functions are the constant function, the identity functions $\mathrm{id}_j(\mathbf{x}) = x_j$, $j = 1, \ldots, 4$ and 8 Hill functions

$$
\frac{1}{1 + x_k^l} \quad k = \{1, 2, 3, 4\}, \quad l = \{1, 2\}.
$$

The estimation of the vector field coefficients is shown in Fig. 13.3b. The NRMSE is equal to 0.09 (averaged over 10 simulations).

#### 13.6.1.3 Forced Pendulum

We consider the forced, damped (inverted) pendulum

$$
\begin{aligned}
\dot{\theta}_1 &= \theta_2, \\
\dot{\theta}_2 &= 2\,\sin\theta_1 - 0.5\,\theta_2 + 0.5\,\cos\theta_1\,u(t)
\end{aligned}
$$

with the input $u(t) = \sin(t)$. We generate $M = 600$ snapshot data pairs taken from 20 trajectories, with a sampling period $T_s = 0.3$ and initial conditions randomly distributed in $[-\pi/2, \pi/2]^2$ (see Fig. 13.3a). Measurement noise is characterized

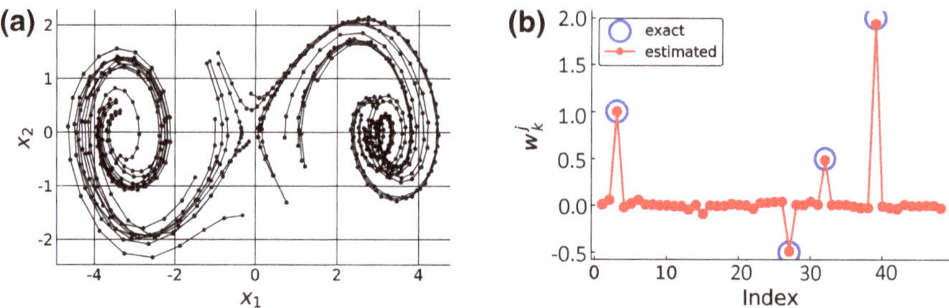


**Fig. 13.4** The main method is used to estimate the vector field coefficients of a forced inverted pendulum. **a** Data points generated by the dynamics (the trajectories are represented by black lines). **b** Exact and estimated coefficients of the vector field (including the forcing term)

by $\sigma = 0.01$. The basis functions are the constant function, the identity functions $\text{id}_1(\theta_1, \theta_2, u) = \theta_1$, $\text{id}_2(\theta_1, \theta_2, u) = \theta_2$, $\text{id}_3(\theta_1, \theta_2, u) = u$, and 20 additional basis functions of the form

$$\cos(\theta_1)\, \theta_1^{n_1} \theta_2^{n_2} u^{n_3} \quad \text{and} \quad \sin(\theta_1)\, \theta_1^{n_1} \theta_2^{n_2} u^{n_3}$$

with $n_1, n_2, n_3 \in \{0, 1, 2\}$ and $n_1 + n_2 + n_3 \leq 2$. The estimation of the vector field coefficients is shown in Fig. 13.4b. Note that the input term $0.5 \cos(\theta_1)\, u(t)$ is also recovered. The NRMSE is equal to 0.05 (averaged over 10 simulations).

## 13.6.2 *Dual Method*

We now apply the dual method on higher dimensional systems. We will consider a power grid system and a nonlinear network.

### 13.6.2.1 Power Grid System

We apply the dual identification method on a power grid model.[1] The swing dynamics of $n_g$ generators are described by the classical model (see, e.g., [13, 27])

$$\begin{aligned} \dot{\delta}_i &= \omega_i \\ \dot{\omega}_i &= \frac{1}{M_i} \left( -D_i \omega_i + P_{m,i} - E_i \sum_{j=1}^{n} E_j \left( G_{ij} \cos(\delta_i - \delta_j) + B_{ij} \sin(\delta_i - \delta_j) \right) \right), \end{aligned}$$

[1] Another illustration of Koopman operator techniques in the context of power grids can be found in Chap. 19.

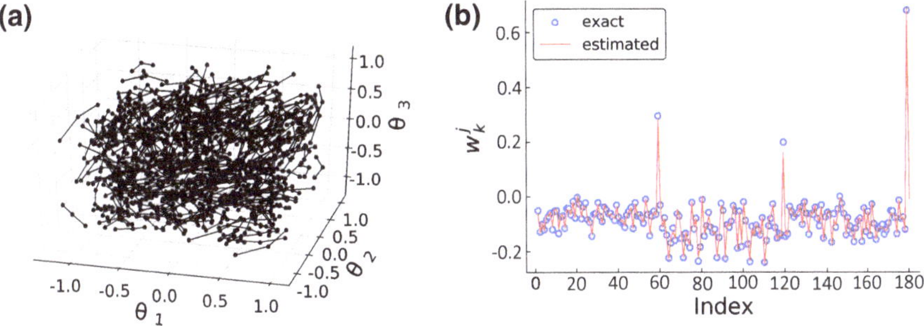


**Fig. 13.5** The dual method is used to estimate the parameters of a power grid dynamics with $n_g = 30$ generators. **a** Data points generated by the dynamics (the rotation angles of three generators $i = 1, 2, 3$ are shown; the trajectories are represented by black lines). **b** Exact and estimated coefficients of the vector field (only the parameters associated with the generators $i = 1, 2, 3$ are shown)

where $\delta_i$ and $\omega_i$ are the rotor angle and speed for generator $i$ (with $i = 1, \ldots, n_g$), respectively. The parameters $M_i$, $D_i$, and $P_{m,i}$ are the rotational inertia, the damping and the mechanical power input, respectively, of the generator. The value $G_{ij} + \mathrm{i}B_{ij}$ represents the transfer impedance between generators $i$ and $j$ (in the network-reduction representation).

We consider the case of $n_g = 30$ generators, so that $n = 60$. The parameters are uniformly randomly distributed (over $[0, 0.1]$ for $G_{ij}$, $B_{ij}$ and $D_i$; $[0, 1]$ for $P_i$; $[0.5, 1.5]$ for $M_i$; $[1, 2]$ for $E_i$). We generate $M = 1000$ samples taken from $M/2$ trajectories, with $T_s = 0.2$ and initial conditions in $[-1, 1]^n$. Measurement noise is added with $\sigma = 0.01$. Using $N = M$ Gaussian radial basis functions $\psi_k(\mathbf{x}) = e^{-0.01\|\mathbf{x}-\mathbf{x}_k\|^2}$ and solving the regression problem (13.21) (with $\lambda = 0$) with $n$ library functions

$$\left\{1, \omega_i, \cos(\delta_i - \delta_j) \text{ for } j \neq i, \sin(\delta_i - \delta_j) \text{ for } j \neq i\right\}$$

for the dynamics associated with $\omega_i$, we recover all the parameters for the 30 generators (Fig. 13.5). The NRMSE is equal to 0.0689 (averaged over 5 simulations).

**Table 13.1** NRMSE for the reconstruction of a network with quadratic and cubic interactions

| $n$ | $N_F$ | $n_{inter}$ | $M$ | NRMSE |
|---|---|---|---|---|
| 20 | 1771 | 5 | 200 | 0.016 |
| 50 | 23426 | 15 | 600 | 0.012 |
| 100 | 176851 | 10 | 1000 | 0.004 |

#### 13.6.2.2 Networks with Nonlinear Couplings

We consider a network of interaction between states, where each state is directly influenced by other states through $n_{inter}$ quadratic and cubic nonlinearities. The dynamics of the system are described by

$$\dot{x}_j = w_1^j x_j + \sum_{k=2}^{N_F} w_k^j h_k \qquad j = 1, \ldots, n,$$

where the functions $h_k$ are monomials with total degree equal to 2 or 3 and where only $n_{inter}$ coefficients are nonzero for each individual state dynamics. Moreover, we choose $w_1^j \in [-1, 0]$. For several network sizes, we generate $M$ samples taken from $M/2$ trajectories, with $T_s = 0.5$ and initial conditions in $[-0.5, 0.5]^n$. Measurement noise is added with $\sigma = 0.01$. Using $N = M$ Gaussian radial basis functions $\psi_k(\mathbf{x}) = e^{-0.01\|\mathbf{x}-\mathbf{x}_k\|^2}$ and solving the Lasso regression problem (13.21) (with $\lambda = 2 \times 10^{-4}$) with the monomials $h_k$, we obtain the results summarized in Table 13.1. As shown in Fig. 13.6, the method is less accurate as the dimension $n$ grows where in particular some nonzero coefficients are estimated to zero. However, most coefficients have a zero value and are correctly estimated, which explains that the NRMSE remains small.

In the context of network reconstruction, we consider that there is a link between state $j$ and state $i$ if there is at least one nonzero coefficient $w_k^j$ such that $h_k$ depends on $x_i$. As illustrated in Fig. 13.7, the dual method is efficient to infer the network structure from a small dataset and is characterized by very few false positives.

## 13.7 Conclusion

Attracting growing interest in nonlinear control theory, the Koopman operator framework is applied to an increasing number of problems. This chapter reports on another contribution in this context. It demonstrates that the Koopman operator can be used in system identification and parameter estimation, somehow bridging two sets of Koopman-based methods: theoretical methods focusing on vector fields and data-driven methods. The proposed approach yields novel numerical techniques that allow to identify a nonlinear system through linear algebraic methods. Moreover, a dual method has been presented, which is well suited to high-dimensional systems and small datasets. The efficiency of both methods have been demonstrated on several examples, including chaotic systems and networked systems.

These results also offer some perspectives and future research directions, such as combining the identification lifting methods with dictionary learning techniques [15, 34], developing the framework to estimate parameters that appear nonlinearly in the vector field [21], investigating the robustness of the method on real datasets (e.g.,

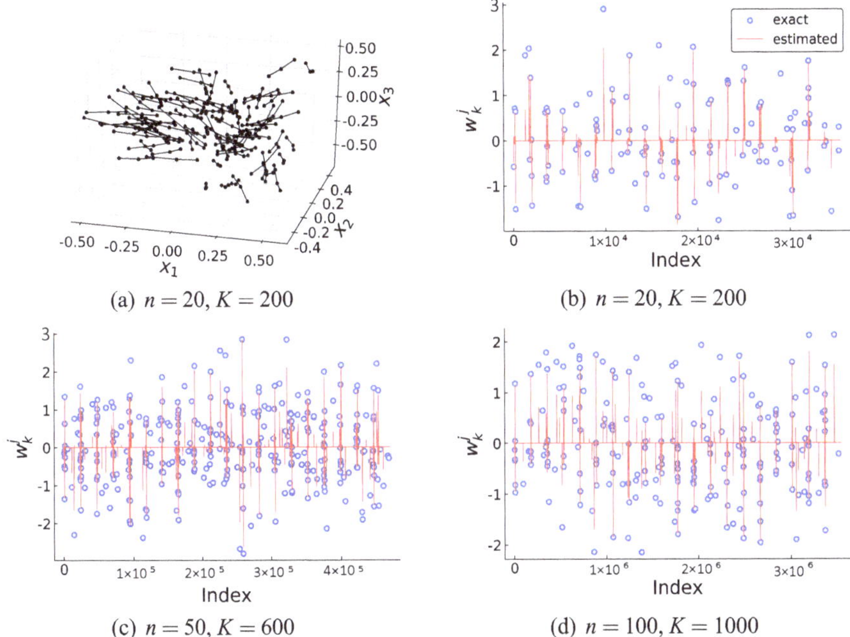


**Fig. 13.6** The dual method is used to estimate the vector field coefficients of a network with nonlinear coupling terms. **a** Data points generated by the dynamics (the first three components are shown; the trajectories are represented by black lines). **b–d** Exact and estimated coefficients of the vector field (only the coefficients of the first 20 components of the vector field are shown)

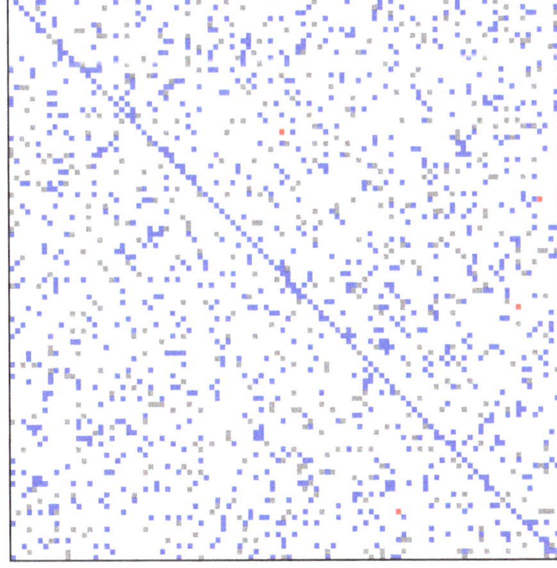

**Fig. 13.7** The dual method is efficient to estimate the structure of the network with nonlinear couplings and generate few false positives (red). Blue: correctly identified links (true positives); white: correct missing links (true negatives); gray: unidentified links (false negatives); red: incorrectly identified links (false positives)

power grids [28]), and addressing the classical identification problem in the case of nonlinear input–output systems.

**Acknowledgements** A.M. acknowledges J. Winkin, F. Lamoline, A. Hastir, and Y. Susuki for fruitful discussions and suggestions. The authors acknowledge support from the Luxembourg National Research Fund.

## References

1. Ackermann, J.: Sampled-Data Control Systems: Analysis and Synthesis, Robust System Design. Springer Science & Business Media, Berlin (2012)
2. Åström, K.J., Eykhoff, P.: System identification: a survey. Automatica **7**(2), 123–162 (1971)
3. Bard, Y.: Nonlinear Parameter Estimation. Academic, Cambridge (1974)
4. Bock, H.G.: Recent advances in parameter identification techniques for ode. In: Numerical Treatment of Inverse Problems in Differential and Integral Equations, pp. 95–121. Springer (1983)
5. Bruder, D., Remy, C.D., Vasudevan, R.: Nonlinear system identification of soft robot dynamics using Koopman operator theory (2018). arXiv:1810.06637
6. Brunton, S.L., Proctor, J.L., Kutz, J.N.: Sparse identification of nonlinear dynamics with control (SINDYc). Proc. IFAC Conf. **49**, 710–715 (2016)
7. Brunton, S.L., Proctor, L.P., Kutz, J.N.: Discovering governing equations from data by sparse identification of nonlinear dynamical systems. Proc. Natl. Acad. Sci. **113**(15), 3932–3937 (2016)
8. Engel, K.J., Nagel, R.: One-Parameter Semigroups for Linear Evolution Equations, vol. 194. Springer Science & Business Media, Berlin (1999)
9. Ferrari-Trecate, G., Rovatti, R.: Fuzzy systems with overlapping Gaussian concepts: approximation properties in Sobolev norms. Fuzzy Sets Syst. **130**(2), 137–145 (2002)
10. Johnson, C., Yeung, E.: A class of logistic functions for approximating state-inclusive Koopman operators (2017). arXiv:1712.03132
11. Korda, M., Mezić, I.: Linear predictors for nonlinear dynamical systems: Koopman operator meets model predictive control. Automatica **93**, 149–160 (2018)
12. Korda, M., Mezić, I.: On convergence of extended dynamic mode decomposition to the Koopman operator. J. Nonlinear Sci. **28**(2), 687–710 (2018)
13. Kundur, P.: Power System Stability and Control. McGraw-Hill, New York (1994)
14. Leontaritis, I., Billings, S.A.: Input-output parametric models for non-linear systems part I: deterministic non-linear systems. Int. J. Control **41**(2), 303–328 (1985)
15. Li, Q., Dietrich, F., Bollt, E.M., Kevrekidis, I.G.: Extended dynamic mode decomposition with dictionary learning: a data-driven adaptive spectral decomposition of the Koopman operator. Chaos: Interdiscip. J. Nonlinear Sci. **27**, 103,111 (2017)
16. Ljung, L.: System identification. In: Signal Analysis and Prediction, pp. 163–173. Springer (1998)
17. Mauroy, A., Goncalves, J.: Linear identification of nonlinear systems: a lifting technique based on the Koopman operator. In: Proceedings of the 55th IEEE Conference on Decision and Control, pp. 6500–6505 (2016)
18. Mauroy, A., Goncalves, J.: Koopman-based lifting techniques for nonlinear systems identification. In: IEEE Transactions on Automatic Control (2017). https://ieeexplore.ieee.org/abstract/document/8836606, arXiv:1709.02003
19. Mezić, I.: Spectral properties of dynamical systems, model reduction and decompositions. Nonlinear Dyn. **41**(1–3), 309–325 (2005)
20. Narendra, K.S., Parthasarathy, K.: Identification and control of dynamical systems using neural networks. IEEE Trans. Neural Netw. **1**(1), 4–27 (1990)

21. Pan, W., Menolascina, F., Stan, G.B.: Online model selection for synthetic gene networks. In: Proceedings of the 55th IEEE Conference on Decision and Control, pp. 776–782. IEEE (2016)
22. Pan, W., Yuan, Y., Goncalves, J., Stan, G.B.: A sparse Bayesian approach to the identification of nonlinear state-space systems. IEEE Trans. Autom. Control **61**(1), 182–187 (2016)
23. Proctor, L.P., Brunton, S.L., Kutz, J.N.: Generalizing Koopman operator theory to allow for inputs and control. SIAM J. Appl. Dyn. Syst. **17**(1), 909–930 (2018)
24. Rackauckas, C., Nie, Q.: DifferentialEquations.jl - A Performant and feature-rich ecosystem for solving differential equations in Julia. J. Open Res. Softw. **5**(1), 5 (2017)
25. Riseth, A.N., Taylor-King, J.P.: Operator fitting for parameter estimation of stochastic differential equations (2017). arXiv:1709.05153
26. Sjöberg, J., Zhang, Q., Ljung, L., Benveniste, A., Delyon, B., Glorennec, P.Y., Hjalmarsson, H., Juditsky, A.: Nonlinear black-box modeling in system identification: a unified overview. Automatica **31**(12), 1691–1724 (1995)
27. Susuki, Y., Mezic, I.: Nonlinear Koopman modes and coherency identification of coupled swing dynamics. IEEE Trans. Power Syst. **26**(4), 1894–1904 (2011)
28. Susuki, Y., Mezic, I., Raak, F., Hikihara, T.: Applied Koopman operator theory for power systems technology. Nonlinear Theory Appl. IEICE **7**(4), 430–459 (2016)
29. Tibshirani, R.: Regression shrinkage and selection via the lasso. J. R. Stat. Soc. Ser. B (Methodological) **58**, 267–288 (1996)
30. Timme, M., Casadiego, J.: Revealing networks from dynamics: an introduction. J. Phys. A: Math. Theor. **47**(34), 343,001 (2014)
31. Varah, J.M.: A spline least squares method for numerical parameter estimation in differential equations. SIAM J. Sci. Stat. Comput. **3**(1), 28–46 (1982)
32. Wiener, N.: Nonlinear problems in random theory. In: Wiener, N. (ed.) Nonlinear Problems in Random Theory, p. 142. The MIT Press, Cambridge (1966). ISBN 0-262-73012-X
33. Williams, M.O., Kevrekidis, I.G., Rowley, C.W.: A data-driven approximation of the Koopman operator: extending dynamic mode decomposition. J. Nonlinear Sci. **25**(6), 1307–1346 (2015)
34. Yeung, E., Kundu, S., Hodas, N.: Learning deep neural network representations for Koopman operators of nonlinear dynamical systems (2017). arXiv:1708.06850
35. Yue, Z., Thunberg, J., Ljung, L., Goncalves, J.: Identification of Sparse Continuous-Time Linear Systems with Low Sampling Rate: Exploring Matrix Logarithms. arXiv:1605.08590

# Chapter 14
# Manifold Learning for Data-Driven Dynamical System Analysis

**Tal Shnitzer, Ronen Talmon and Jean-Jacques Slotine**

**Abstract** High-dimensional signals generated by dynamical systems arise in many fields of science. For example, many biomedical signals can be modeled by a few latent physiologically related variables measured indirectly through a large set of noisy sensors. In such applications, a notable challenge in analyzing and processing the observed data is that the generating system is typically unknown. In this chapter, this problem is addressed through geometric analysis by applying manifold learning. Specifically, we show how using manifold learning in a purely data-driven manner, with minimal prior knowledge or model assumptions, we can both discover the hidden state, dynamics, and observation function, and also attain a compact linear description of the full system. The main assumption is that the accessible high-dimensional data (the observations of the system) lie on an underlying nonlinear manifold of lower dimensions. Furthermore, when applying manifold learning to time series, critical information is typically overlooked. Time series are processed as data sets of samples, ignoring their embodied dynamics and temporal order. We address the challenge of incorporating the time dependencies into manifold learning and present a purely data-driven scheme. First, an intrinsic representation is derived without prior knowledge of the system by applying diffusion maps. Second, we show that even for highly nonlinear systems, the dynamics of the constructed representation is approximately linear, and present accordingly two filtering frameworks, one based on a linear observer and another based on the Kalman filter. These filtering methods enable us to directly incorporate the inherent dynamics and time dependencies between consecutive system observations into the diffusion maps coordinates.

T. Shnitzer (✉) · R. Talmon
Viterbi Faculty of Electrical Engineering, Technion - Israel Institute of Technology, Haifa, Israel
e-mail: shnitzer@campus.technion.ac.il

R. Talmon
e-mail: ronen@ee.technion.ac.il

J.-J. Slotine
Nonlinear Systems Laboratory, Massachusetts Institute of Technology, Cambridge, MA, USA
e-mail: jjs@mit.edu



A. Mauroy et al. (eds.), *The Koopman Operator in Systems and Control*,
Lecture Notes in Control and Information Sciences 484,
https://doi.org/10.1007/978-3-030-35713-9_14

We show that this approach is analogous to Koopman spectral analysis, since applying diffusion maps to the given measurements generates a parametrization of the state space in a known analytic form of the dynamics with a linear drift. We demonstrate the benefits of the approach on simulated data and on real recordings from various applications.

## 14.1 Introduction

Analyzing observed data from dynamical systems is typically challenging, especially when the dynamical system is unknown or does not have a definitive model. Analyzing such data often requires many modeling assumptions and the estimation of a large number of parameters. In this chapter, we present two nonparametric filtering frameworks, first presented in [15, 16], which require only minimal model assumptions and recover the system model from the observations in a data-driven manner.

Using diffusion maps , a particular manifold learning technique [6], we form new system coordinates, which serve as an intrinsic representation of a class of dynamical systems, particularly, of gradient flows with isotropic diffusion. More concretely, a new *linear* data-driven state-space model of the dynamical system is constructed, by exploiting the theoretical dynamics of the diffusion maps coordinates, which were shown to evolve according to a linear drift [5]. Due to the linearity of the model, we obtain a setting, which is highly related to the Koopman operator, i.e., given the measurements of some nonlinear system, we obtain coordinates in which the evolution is linear. Two filtering frameworks are then constructed by incorporating this linear model into two standard algorithms, a contracting observer and the Kalman filter. These frameworks give rise to completely data-driven filters, which are adapted for the dynamical system at hand, and provide a new representation of the system. We demonstrate these frameworks on an object tracking problem and on music signals and we show that the new representations obtained by the two filtering frameworks reveal meaningful properties of these systems.

This chapter is organized as follows. Section 14.2 reviews related work. In Sect. 14.3, we present the general setting of the problem and elaborate on the relation of the method to the Koopman operator. In Sect. 14.4, a method for recovering a linear system model based on the diffusion maps algorithm is introduced. Section 14.5 presents the two filtering frameworks, which are based on the devised linear system model, and in Sect. 14.6, the properties of these frameworks are illustrated using two different applications.

## 14.2 Background

In this work, we focus on nonparametric methods, which allow for dynamical system analysis and filtering with minimal model assumptions, and provide a flexible framework. For example, [8, 23] propose to use Gaussian Processes for describing the dynamics and measurement functions of nonlinear dynamical systems, which are then utilized for filtering. Such methods commonly require a set of corresponding state and measurement pairs, which are often unavailable.

Recently, several nonparametric frameworks which encompass concepts from Koopman theory have been proposed [4, 19, 24]. In these methods, the Koopman operator is empirically approximated and employed for establishing a new space in which the propagation of dynamical systems is linear. One example is Extended Dynamic Mode Decomposition (EDMD) [4, 24], which is a method for learning the eigenfunctions, eigenvalues, and modes of the Koopman operator from measurements and a set of empirically chosen dictionary elements. In [19], a Kalman filtering framework was proposed based on the eigenfunctions of the Koopman operator, approximated using EDMD. In this framework, a linear model was constructed using the linearity of the Koopman eigenfunctions, and then implemented as part of a linear Kalman filter. An extension for the Koopman operator to stochastic dynamical systems was presented in [7, 12], which defined the stochastic Koopman operator. In [7], different forms of stochastic dynamical systems were analyzed in the context of this operator, and a method for approximating the eigenfunctions and eigenvalues of the operator was presented. However, when observation noise is present, in addition to the process noise of stochastic dynamical systems, common existing methods fail to recover a good approximation of the stochastic Koopman operator [21]. Two recent papers, [18, 21], address this challenge and propose two methods for approximating the stochastic Koopman operator based on noisy observations. In [21], subspace dynamical mode decomposition (subspace DMD) is presented, which assumes access to a set of observables that spans an invariant subspace of the stochastic Koopman operator. An extension of DMD is proposed and it is shown that this extension converges to the stochastic Koopman operator, under certain assumptions, even in the presence of observation noise. In [18], an extension of common algorithms, such as DMD and EDMD, is presented, which provides a robust approximation of the stochastic Koopman operator in the presence of noise. However, both methods suffer from the same limitations as DMD and EDMD, where a careful choice of observables or dictionary elements is required. One example of a scheme for constructing suitable observables is presented in [20], which proposes a method for recovering observables spanning a Koopman invariant subspace, as required in [21], using neural networks. However, the process and observation noise are not addressed in this framework.

A different class of related work, addressing the nonparametric analysis of dynamical systems, relies on diffusion maps [6], which is a manifold learning technique. Using diffusion maps, a new set of coordinates can be constructed in a data-driven manner, which captures the geometrical properties of the data and provides a new representation of the system. It was shown in [5] that these coordinates approximate functions that evolve according to known dynamics, for a specific class of dynamical systems. In [1, 2], these known dynamics are exploited for propagating dynamical systems. There, the dynamical system is projected onto the new set of coordinates, where the probability density of the system state is propagated in time in a data-driven manner. However, these frameworks assume that the state of the system is accessible. Another work [9] combines concepts from Koopman theory and diffusion maps and proposes a method for approximating the eigenfunctions of the Koopman generator using the coordinates obtained from diffusion maps. In addition, they rigorously analyze this relationship between diffusion maps and the Koopman generator.

## 14.3 The Considered Class of Dynamical Systems

Consider the following dynamical system:

$$\dot{\mathbf{x}}(t) = F\left(\mathbf{x}\left(t\right), \dot{\boldsymbol{\omega}}(t)\right), \tag{14.1}$$

$$\mathbf{z}(t) = G\left(\mathbf{x}\left(t\right), \mathbf{v}\left(t\right)\right), \tag{14.2}$$

where $\mathbf{x}(t) \in \mathbb{R}^r$ is the state of the system, $\boldsymbol{\omega}(t)$ is Brownian motion and $\dot{\boldsymbol{\omega}}(t)$ denotes its derivative, $\mathbf{z}(t) \in \mathbb{R}^n$ is a measurement of the system, $\mathbf{v}(t)$ is a noise term, and $F$ and $G$ are some nonlinear functions.

We focus on dynamical systems with state dynamics of gradient flows with isotropic diffusion, in which the nonlinear functions, $F$ and $G$, are modeled as follows:

$$\dot{\mathbf{x}}(t) = -\nabla Q\left(\mathbf{x}(t)\right) + \sqrt{\frac{2}{\beta}}\dot{\boldsymbol{\omega}}(t), \tag{14.3}$$

$$\mathbf{z}(t) = h\left(\mathbf{x}(t)\right) + \mathbf{v}(t), \tag{14.4}$$

where $Q\left(\mathbf{x}(t)\right)$ is a smooth potential function, defining a deterministic drift and $\sqrt{2/\beta}$ is a constant diffusion coefficient. The restriction of the work presented in this chapter to such system dynamics will be explained in Sect. 14.4.2.

Our goal here is to recover a new data-driven representation for such a dynamical system, which captures its inherent properties, given only the measurements $\mathbf{z}(t)$, and without additional model assumptions. In this setting, the state $\mathbf{x}(t)$, the potential function $Q\left(\mathbf{x}(t)\right)$, the diffusion coefficient $\beta$, the measurement function $h\left(\mathbf{x}(t)\right)$, and the dimensionality of the underlying state $r$, are all unknown.

The methods presented in this chapter are highly related to the stochastic Koopman operator [7, 12], which is defined by

$$\left(U_{st}^{t} f\right)\left(\mathbf{x}(t)\right) = \mathbb{E}\left[f \circ S^{t}\left(\mathbf{x}(t), \dot{\boldsymbol{\omega}}(t)\right)\right], \tag{14.5}$$

where $S^t$ is the flow induced by $F$, $\circ$ is the composition operator, $f$ are some observables, and $\mathbb{E}$ denotes expectation.

The stochastic Koopman operator describes the average evolution in time of stochastic dynamical systems. Moreover, it is a linear operator, even for highly nonlinear systems, which propagates observables of the underlying state, i.e., $f(\mathbf{x}(t))$, where the space of observables is commonly infinite dimensional. Specifically, it was shown in [7] that for autonomous nonlinear stochastic differential equations of the form (14.3), under the assumption that the state is 1-dimensional, the eigenfunctions of the stochastic Koopman operator evolve according to the following stochastic differential equation:

$$\dot{\tilde{\phi}}(x(t)) = \tilde{\lambda}\tilde{\phi}(x(t)) + \sqrt{\frac{2}{\beta}}\frac{d\tilde{\phi}(x(t))}{dx}\dot{\omega}(t), \tag{14.6}$$

where $\tilde{\phi}(x(t))$ and $\tilde{\lambda}$ denote the eigenfunctions and eigenvalues of the stochastic Koopman operator, respectively, $x(t)$ is a 1-dimensional underlying system state and $\omega(t)$ denotes Brownian motion. Based on these properties of the stochastic Koopman operator, the average propagation of dynamical systems can be described using the eigenfunctions and eigenvalues of the operator.

In [15, 16], a related data-driven framework using diffusion maps was devised. It was shown that a linear model for the system in (14.3) and (14.4) can be recovered based on the eigenvectors and eigenvalues obtained from diffusion maps. Furthermore, these eigenvectors approximate eigenfunctions that evolve according to an equation similar to (14.6).

Figure 14.1 illustrates the data-driven system model presented in [15, 16], and emphasizes its relation to the stochastic Koopman operator. Based on the measurements, $\mathbf{z}(t)$, a new representation of the dynamical system is obtained, denoted by $\boldsymbol{\Psi}$, using the diffusion maps algorithm. By exploiting certain properties of this new representation, specifically, a linear drift as in (14.6), a linear propagation model and lift function can be constructed, denoted by $\tilde{\mathbf{F}}$ and $\boldsymbol{\alpha}$, respectively. Therefore, similarly to the Koopman operator scheme, instead of the nonlinear propagation of the measurements, the new representation $\boldsymbol{\Psi}$ can be linearly propagated, and then, lifted back to the measurement space.

The relation of diffusion maps to the Koopman operator is further analyzed in [9]. Specifically, a representation of the generator of the Koopman operator is constructed based on diffusion maps. In addition, it is shown that the Laplace–Beltrami operator, which is approximated by diffusion maps, and the generator of the Koopman operator have common eigenfunctions, when using a particular data representation (delay-embedding [22]) in the construction of diffusion maps.

In [15, 16], the derived data-driven system model was incorporated into two filtering frameworks. This yielded two main benefits. First, in these filtering frameworks, the measurements are taken into account, allowing for estimation of specific trajectories, instead of the expected propagation in the stochastic Koopman operator model. Second, due to the use of diffusion maps, a completely data-driven representation is

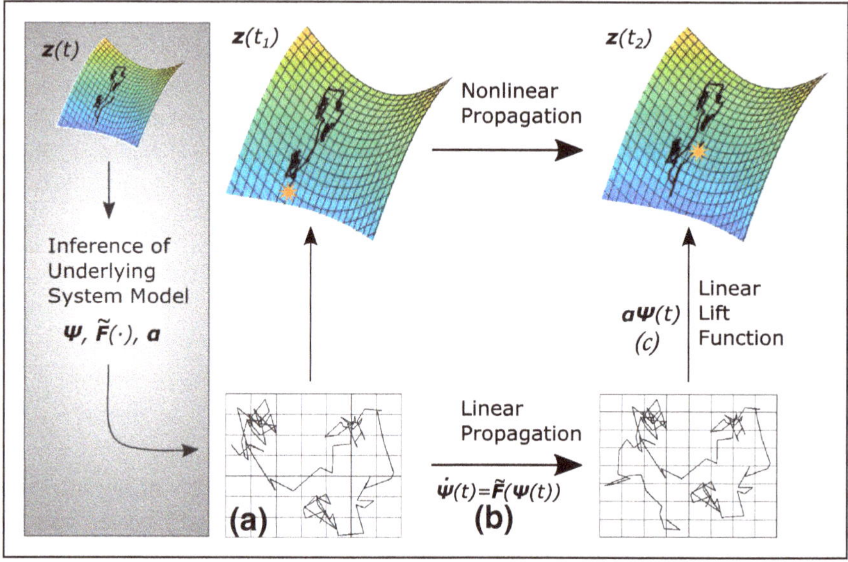


**Fig. 14.1** Illustration of the data-driven model and state derivation framework

obtained, which in many cases, can be well approximated by a low-dimensional set of coordinates [6].

## 14.4 Data-Driven System Model Inference

In this section, we describe a method for recovering a data-driven system model and system representation. We show that using diffusion maps [6], a new intrinsic representation of the system is obtained, as well as a new system model, based solely on the measurements. Moreover, we show that the recovered system dynamics contain a linear drift component and that the mapping between the new state representation and the measurements is linear as well. In Sect. 14.5, the linearity of the recovered system model is exploited to construct filtering frameworks, which are based on well-established linear algorithms: a linear contracting observer [11] and the (linear) Kalman filter [10].

The outline of this section is illustrated in Fig. 14.1. In Sect. 14.4.1, we elaborate on the construction of the new representation of the system, marked by (a) in the figure. In Sect. 14.4.2, we present the recovered dynamics, marked by (b) in the figure. In Sect. 14.4.3, a method for constructing the linear lift function, marked by (c), is presented. This lift function maps between the new system representation and the measurements, and completes the data-driven system model.

### 14.4.1 New State Representation

We briefly present the diffusion maps algorithm, which allows for the construction of a new representation of the system state based on the measurements $\mathbf{z}(t)$. We assume here that the measurements are given by some nonlinear function applied to the underlying system state of the form (14.3), and disregard the noise term in (14.4). This additional noise term will be taken into account in the definition of the filtering frameworks in Sect. 14.5.

Given the measurements, an affinity kernel is constructed by

$$k_{\varepsilon}(t, s) = \exp\left\{-\frac{d(\mathbf{z}(t), \mathbf{z}(s))^2}{2\varepsilon^2}\right\}, \tag{14.7}$$

where $d(\cdot, \cdot)$ denotes a distance function and $\varepsilon$ is a kernel scale that induces a notion of locality, i.e., when $d(\mathbf{z}(t), \mathbf{z}(s))^2 \gg \varepsilon^2$, the affinity kernel is approximately zero.

The kernel is then normalized as follows:

$$p_{\varepsilon}(t, s) = \frac{k_{\varepsilon}(t, s)}{w_{\varepsilon}(t)}, \tag{14.8}$$

where $w_{\varepsilon}(t) = \int k_{\varepsilon}(t, s) q(s) ds$, and $q(s)$ is the probability density of the data.

Based on $p_{\varepsilon}$, the following operator is defined for any real function of the data $g(\cdot)$:

$$(P_{\varepsilon} g)(t) = \int p_{\varepsilon}(t, s) g(s) q(s) ds \tag{14.9}$$

In [14] it was shown that the operator $(P_{\varepsilon} - \mathrm{I})/\varepsilon$ converges in the limit $\varepsilon \to 0$ to the backward Fokker–Plank operator, which has a discrete set of decreasing eigenvalues and corresponding eigenfunctions, denoted by $\{\lambda_{\ell}\}_{\ell=0}^{\infty}$ and $\{\phi_{\ell}(t)\}_{\ell=0}^{\infty}$, respectively. This convergence implies that the eigenfunctions of $P_{\varepsilon}$, which are denoted by $\{\psi_{\ell}(t)\}_{\ell=0}^{\infty}$, approximate the eigenfunctions of the backward Fokker–Planck operator.

We can now construct a new representation of the data using the eigenfunctions of $P_{\varepsilon}$. However, there is an infinite number of such eigenfunctions. To obtain a finite-dimensional coordinate system, we follow [6], which shows that when there exists a spectral gap in $\{\lambda_{\ell}\}_{\ell=0}^{\infty}$, the data can be well represented by a finite set of eigenfunctions, corresponding to the largest eigenvalues. The new finite coordinate system is then given by

$$\mathbf{z}(t) \mapsto \boldsymbol{\Psi}(t) = [\psi_1(t), \psi_2(t), ..., \psi_m(t)], \tag{14.10}$$

where $\psi_1(t), \ldots, \psi_m(t)$ are the eigenfunctions corresponding to the $m$ largest eigenvalues. Note that $m$ is the dimension of the new coordinate system.

At this point, the resulting diffusion maps coordinates represent the measurements $\mathbf{z}(t)$, which are given by some nonlinear function distorting the underlying system state. In order to obtain a more accurate representation of the system state using the diffusion maps coordinates, we construct the affinity kernel in (14.7) using a variant of the Mahalanobis distance [17]:

$$d\left(\mathbf{z}(t), \mathbf{z}(s)\right) = \frac{1}{2}\left(\mathbf{z}(t) - \mathbf{z}(s)\right)\left(C^{-1}(t) + C^{-1}(s)\right)\left(\mathbf{z}(t) - \mathbf{z}(s)\right)^T, \quad (14.11)$$

where $C^{-1}(t)$ denotes the inverse of the covariance matrix of the measurements at time $t$. It was shown in [17], that by constructing the diffusion maps kernel using this modified Mahalanobis distance, the diffusion maps coordinates in (14.10) approximate the eigenfunctions of the backward Fokker–Planck operator defined on the *underlying data* $\mathbf{x}(t)$, and therefore can be used to represent the underlying system state. In other words, the use of this Mahalanobis distance partially solves the inverse problem of recovering the (Euclidean) distances between samples of the system state $\mathbf{x}(t)$ (rather than the samples themselves) from the measurements $\mathbf{z}(t)$.

### 14.4.2 State Dynamics

We now have a new set of coordinates that represent the underlying system state, which are based on an approximation of the eigenfunctions of the backward Fokker–Planck operator defined on $\mathbf{x}(t)$. In this subsection, we show that these new coordinates also induce a data-driven model for the system dynamics.

The eigenfunctions of the operator $P_\varepsilon$, describing the system (14.3), are a function of the underlying stochastic system state, $\mathbf{x}(t)$. Therefore, based on Itô's lemma, the evolution in time of these eigenfunctions is given by

$$\dot{\psi}_\ell\left(\mathbf{x}(t)\right) = -\lambda_\ell \psi_\ell\left(\mathbf{x}(t)\right) + \sqrt{\frac{2}{\beta}}\left\|\nabla_{\mathbf{x}(t)}\psi_\ell\left(\mathbf{x}(t)\right)\right\|_2 \dot{\omega}_\ell(t), \;\; \ell = 0, 1, 2 \ldots, \quad (14.12)$$

where $\{\psi_\ell\left(\mathbf{x}(t)\right)\}_{\ell=1}^{\infty}$ and $\{\lambda_\ell\}_{\ell=1}^{\infty}$ are the eigenfunctions and eigenvalues of the operator $P_\varepsilon$, respectively, and $\boldsymbol{\omega}_\ell(t)$ denotes Brownian motion [5]. This equation implies that the eigenfunctions of the operator $P_\varepsilon$ evolve according to a *linear drift*, given by the eigenvalues, $\{\lambda_\ell\}_{\ell=1}^{\infty}$, and an additional diffusion term.

Note that we restrict the work presented in this chapter to systems with state dynamics of gradient flows with constant diffusion, as in (14.3), since such underlying state dynamics lead to the linear drift term of the eigenfunction dynamics in (14.12).

We exploit this property and set the dynamics of the data-driven system model to be a linear drift, given by the eigenvalues of $(P_\varepsilon - \mathrm{I})/\varepsilon$, which approximate the eigenvalues of the backward Fokker–Planck operator. Note that by doing so we disregard the stochastic diffusion component of (14.12). This stochastic term will be compensated by using a filtering framework which takes into account the measurements as well, as will be described in Sect. 14.5.

### 14.4.3 Lift Function

To complete the data-driven framework presented in Fig. 14.1, a mapping between the new representation of the system state $\boldsymbol{\Psi}(t)$, and the measurements $\mathbf{z}(t)$, is required (marked by (c) in Fig. 14.1). We show that using the properties of the recovered representation, this mapping can be obtained in a data-driven manner.

The eigenfunctions of the backward Fokker–Planck operator form a basis for all real functions of the underlying state $\mathbf{x}(t)$, with respect to the inner product $\langle f, g\rangle_q = \int_{-\infty}^{\infty} f(t)g(t)q(t)dt$, where $q(t)$ is the probability density of the data [13]. Therefore, any function of $\mathbf{x}(t)$ can be represented using these eigenfunctions. Specifically, the measurements are a function of $\mathbf{x}(t)$ and can be represented by

$$z_j(t) = \sum_{\ell=1}^{\infty} \alpha_{j,\ell}\phi_\ell\left(\mathbf{x}(t)\right), \quad j = 1, \ldots, n, \tag{14.13}$$

where $\alpha_{j,\ell} = \langle z_j, \phi_\ell\rangle_q = \int_{-\infty}^{\infty} z_j(t)\phi_\ell\left(\mathbf{x}(t)\right) q(t)dt$.

Since the eigenfunctions of $P_\varepsilon$ approximate the eigenfunctions of the backward Fokker–Planck operator, we can write an approximate equation for $z_j(t)$ as follows:

$$z_j(t) \approx \sum_{\ell=1}^{\infty} \alpha_{j,\ell}\psi_\ell(t), \quad j = 1, \ldots, n, \tag{14.14}$$

where $\alpha_{j,\ell} = \langle z_j, \psi_\ell\rangle_q$.

In addition, as described in Sect. 14.4.1, when there exists a spectral gap in $\{\lambda_\ell\}_{\ell=1}^{\infty}$, the infinite sum in (14.14) can be approximated by the following finite sum:

$$z_j(t) \approx \sum_{\ell=1}^{m} \alpha_{j,\ell}\psi_\ell(t). \tag{14.15}$$

In such cases, the lift function between the eigenfunctions and the measurements can be approximated by the following $n \times m$ matrix:

$$\mathbf{z}(t) = g(\boldsymbol{\Psi}(t)) \simeq \boldsymbol{\alpha}\boldsymbol{\Psi}(t), \tag{14.16}$$

where $(\boldsymbol{\alpha})_{j,\ell} = \alpha_{j,\ell}$.

We conclude this section by emphasizing that the presented method provides a new representation for the system, along with a linear data-driven model, even for highly nonlinear systems. This method does not assume access to the true underlying system state and requires only a basic assumption regrading the general form of the system dynamics.

## 14.5 Diffusion Filtering

In Sect. 14.4 we presented a data-driven derivation of a new coordinate system, which is based on diffusion maps, along with a linear system model. Namely, we showed how to obtain a new meaningful representation for the system in a data-driven manner. Yet, in the construction of the diffusion maps coordinates the inherent time dependencies between consecutive samples of the dynamical system are disregarded and the measurements are treated as a collection of independent samples. This is especially apparent in the discrete setting of the diffusion maps algorithm that will be presented next in Sect. 14.5.1. In addition, as mentioned above, in the construction of the data-driven model and system representation, possible measurement noise is disregarded, which could lead to a degraded system representation.

In this section, we show how the system representation obtained from diffusion maps could be improved. More concretely, by exploiting the properties of the eigenfunctions of the Fokker–Planck operator, presented in Sects. 14.4.2 and 14.4.3, we show how to derive a linear system model. Based on this derived model, we construct two linear filtering frameworks using a contracting observer in Sect. 14.5.2 and using the Kalman filter in Sect. 14.5.3. Generally, the two filtering alternatives can be described by the following scheme:

$$\dot{\hat{\boldsymbol{\Psi}}}(t) = \tilde{F}\left(\hat{\boldsymbol{\Psi}}(t), \boldsymbol{\alpha}^{\dagger}\mathbf{z}(t)\right), \tag{14.17}$$

$$\hat{\mathbf{z}}(t) = \boldsymbol{\alpha}\hat{\boldsymbol{\Psi}}(t), \tag{14.18}$$

where $\tilde{F}(\cdot,\cdot)$ denotes a linear function that updates the state estimation at time $t$, denoted by $\hat{\boldsymbol{\Psi}}(t)$, according to the recovered linear system model from Sect. 14.4 and the measurement at time $t$, denoted by $\mathbf{z}(t)$. Additionally, $\boldsymbol{\alpha}$ denotes the lift function from the state estimation to the measurements, and $\alpha^{\dagger}$ denotes its pseudoinverse. The filtered measurements are denoted by $\hat{\mathbf{z}}(t)$.

We note that when the measurement noise in (14.4) is significant, the derived model is no longer accurate, since the theoretical derivations in Sect. 14.4 no longer hold. This hampers the performance of the presented methods. However, the experimental results in Sect. 14.6 empirically show that we can still obtain a relatively good representation for the system using the constructed filtering frameworks.

Figure 14.2 illustrates this filtering scheme. Based on the measurements, the diffusion maps coordinates, denoted by $\boldsymbol{\Psi}$ in the figure, are constructed, and the linear system model is derived. These coordinates are used for initialization of the filtering scheme. Then, the new system representation, denoted by $\hat{\boldsymbol{\Psi}}$, is obtained by applying a linear filter to the measurements with the state initialization. The filter is based on the derived state propagation model in (14.12) and lift function in (14.16), denoted by $\tilde{F}(\cdot,\cdot)$ and $\boldsymbol{\alpha}$, respectively.

The presented filtering scheme incorporates the inherent time dependencies into the diffusion maps coordinates and provides an improved representation of the system. In addition, the system model that is used in this scheme relies on the intrinsic

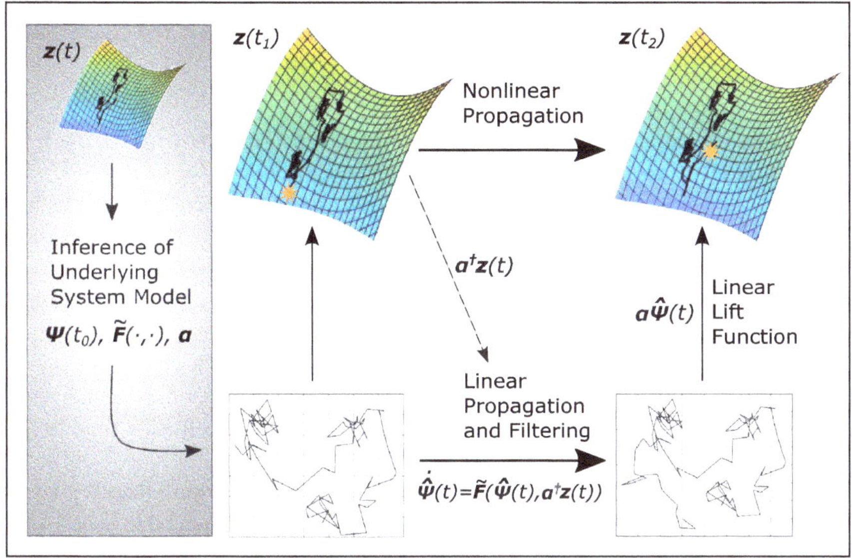


**Fig. 14.2** Outline of the presented filtering frameworks

properties of the system, recovered by geometry analysis and learning, specifically, conveyed by diffusion maps. Therefore, this scheme can be interpreted as performing "geometrically informed" filtering of the measurements.

This section is organized as follows. In Sect. 14.5.1, we present the discrete counterpart of the derivation of the diffusion maps coordinates and system model. In Sect. 14.5.2, we present a filtering framework which is based on a contracting observer, first shown in [16]. In Sect. 14.5.3, we present an improved framework based on the Kalman filter, presented in [15].

## *14.5.1 Discrete Setting*

**Diffusion Maps Coordinates.** Consider a discrete set of $N$ measurements, denoted by $\{\mathbf{z}(t_i)\}_{i=0}^{N-1}$, where $t_i = i\Delta t$, $\Delta t > 0$. Throughout this section, we take $\Delta t = 1$ for simplicity. Based on the measurements, the diffusion maps coordinates are constructed as follows. The $N \times N$ affinity matrix is constructed by

$$K_{i,j} = \exp\left(-\frac{d\left(\mathbf{z}\left(t_i\right), \mathbf{z}(t_j)\right)^2}{2\varepsilon^2}\right), \quad i, j = 1, \ldots, N \tag{14.19}$$

where $d\left(\mathbf{z}\left(t_i\right), \mathbf{z}(t_j)\right)$ is the modified Mahalanobis distance.

The matrix $\mathbf{K}$ is then normalized to create a row stochastic matrix:

$$P_{i,j} = \frac{K_{i,j}}{W_i}, \tag{14.20}$$

where $W_i = \sum_{j=0}^{N-1} K_{i,j}$.

The eigenvalues and eigenvectors of $\mathbf{P}$ are computed, denoted by $\mu_0, \ldots, \mu_{N-1}$ and $\psi_0, \ldots, \psi_{N-1}$, respectively. We order the eigenvalues such that $1 = \mu_0 \geq \mu_1 \geq \cdots \geq \mu_{N-1} \geq 0$ and take the first $m$ eigenvectors corresponding to the largest eigenvalues. The diffusion maps coordinates are given by

$$\mathbf{z}(t_i) \mapsto \boldsymbol{\Psi}(t_i) = [\psi_1(t_i), \ldots, \psi_m(t_i)]. \tag{14.21}$$

Note that we discard the first eigenvector, $\psi_0$, corresponding to the eigenvalue $\mu_0 = 1$ since it is a constant vector (the trivial eigenvector). We also omit the dependence of the coordinates on the eigenvalues, essentially by setting the diffusion time to zero (see [6] for details).

**Coordinate Dynamics.** The linear drift of the dynamics of the coordinates is given by the eigenvalues of the backward Fokker–Planck operator, as presented in Sect. 14.4.2. We approximate these eigenvalues, denoted by $\{\lambda_\ell\}_{\ell=1}^m$, using the eigenvalues of $\mathbf{P}$, as follows:

$$\lambda_\ell = -\frac{2}{\varepsilon^2} \log \mu_\ell, \tag{14.22}$$

where $\mu_\ell$ are the eigenvalues of $\mathbf{P}$ and $\varepsilon^2$ is the kernel scale in (14.19), for details, we refer the reader to [17].

**Lift Function.** In Sect. 14.4.3, it was shown that the lift function can be approximated by the inner product between the measurements and the recovered eigenfunctions, i.e., $\alpha_{j,\ell} = \langle z_j, \psi_\ell \rangle_q$. In the discrete setting, this inner product is approximated by

$$\alpha_{j,\ell} = \langle z_j, \psi_\ell \rangle = \sum_{i=0}^{N-1} z_j(t_i) \psi_\ell(t_i), \; \ell = 1, \ldots, m \;, j = 1, \ldots, n. \tag{14.23}$$

Therefore, the lift function between the new state representation, $\boldsymbol{\Psi}$, and the measurements is approximated simply by a matrix multiplication:

$$\mathbf{z}(t_i) = \boldsymbol{\alpha} \boldsymbol{\Psi}(t_i). \tag{14.24}$$

### *14.5.2 Contracting Observer*

Here we present a filtering framework which is based on a linear contracting observer [11]. A new representation of the system is constructed, denoted by $\hat{\boldsymbol{\Psi}}(t_i) \in \mathbb{R}^m$, $i = 0, \ldots, N-1$, using the following recursive observer equation:

$$\hat{\boldsymbol{\Psi}}(t_{i+1}) - \hat{\boldsymbol{\Psi}}(t_i) = \Lambda \hat{\boldsymbol{\Psi}}(t_i) + \boldsymbol{\kappa}\left(\mathbf{z}(t_i) - \hat{\mathbf{z}}(t_i)\right), \tag{14.25}$$

where $\mathbf{z}(t_i)$ is the measurement at time $t_i$, $\hat{\mathbf{z}}(t_i) = \boldsymbol{\alpha}\hat{\boldsymbol{\Psi}}(t_i)$ is the reconstruction of the measurement from the current state estimation where $\boldsymbol{\alpha}$ is the lift function, $\Lambda \in \mathbb{R}^{m\times m}$ is a diagonal matrix with $\Lambda_{\ell\ell} = -\lambda_\ell$, and $\lambda_\ell$ is the approximation of the $\ell$th eigenvalue of the backward Fokker–Planck operator.

By substituting the definition of $\hat{\mathbf{z}}(t_i)$ into (14.25), the following update equation is obtained:

$$\hat{\boldsymbol{\Psi}}(t_{i+1}) - \hat{\boldsymbol{\Psi}}(t_i) = (\Lambda - \boldsymbol{\kappa}\boldsymbol{\alpha})\,\hat{\boldsymbol{\Psi}}(t_i) + \boldsymbol{\kappa}\mathbf{z}(t_i). \tag{14.26}$$

The Jacobian of the system $\mathbf{J} = (\Lambda - \boldsymbol{\kappa}\boldsymbol{\alpha})$ is set to be negative by taking

$$\boldsymbol{\kappa} = \gamma \Lambda \boldsymbol{\alpha}^{\dagger}, \tag{14.27}$$

where $\gamma \in [0, 1]$ is an empirically chosen tunable parameter and $\boldsymbol{\alpha}^{\dagger}$ denotes the pseudoinverse. By substituting this choice of the Jacobian into (14.26), the update equation can be recast as

$$\hat{\boldsymbol{\Psi}}(t_{i+1}) = \left[\mathrm{I} + (1-\gamma)\Lambda\right]\hat{\boldsymbol{\Psi}}(t_i) + \gamma \Lambda \boldsymbol{\alpha}^{\dagger}\mathbf{z}(t_i), \tag{14.28}$$

where I is the identity matrix. In the resulting update equation, $\gamma$ serves as a weighting parameter, which controls the relative influence of the derived propagation model $\Lambda\hat{\boldsymbol{\Psi}}(t_i)$ on the resulting state estimation, compared with the measurement agreement term $\Lambda\boldsymbol{\alpha}^{\dagger}\mathbf{z}(t_i)$.

Note that $\boldsymbol{\kappa}$ in (14.27) was set empirically, and this particular choice was driven by two main considerations: it inverts the lift function, and it facilitates a contracting observer. In Sect. 14.5.3, we will revisit the choice of $\boldsymbol{\kappa}$ and present a filtering framework based on the Kalman filter, in which $\boldsymbol{\kappa}$ is optimal.

#### 14.5.2.1 The Observer as a Data-Driven Filter

Further insight about the role of $\gamma$ and the behavior of the observer framework is gained by substituting $\boldsymbol{\Psi}(t_i) = \boldsymbol{\alpha}^{\dagger}\mathbf{z}(t_i)$ into (14.28) and recursively unfolding the equation, starting from $\hat{\boldsymbol{\Psi}}(t_0) = \boldsymbol{\Psi}(t_0)$ [16]. This yields the following state estimation equation:

$$\hat{\boldsymbol{\Psi}}(t_{k+1}) = \left[\mathrm{I} + (1-\gamma)\Lambda\right]^{k+1}\boldsymbol{\Psi}(t_0) + \gamma\Lambda\sum_{i=0}^{k}\left[\mathrm{I} + (1-\gamma)\,\Lambda\right]^{k-i}\boldsymbol{\Psi}(t_i), \tag{14.29}$$

where $\gamma$ is presented as a weighting parameter. On the one hand, when $\gamma$ is close to 0, the state estimation in (14.29) is mostly influenced by $[\mathrm{I} + \Lambda]^{k+1}\,\boldsymbol{\Psi}(t_0)$, in which the propagation in time is based only on the data-driven dynamics $\Lambda$. On the other hand, when $\gamma$ is close to 1, the state estimation is mostly influenced by $\boldsymbol{\Psi}(t_0) +$

$\Lambda \sum_{i=0}^{k} \boldsymbol{\Psi}(t_i)$, which estimates the state by summing over increments determined by propagating each measurement one step forward, since $\boldsymbol{\Psi}(t_i) = \boldsymbol{\alpha}^{\dagger}\mathbf{z}(t_i)$. The latter term takes into account the current realization (measurements) and can be viewed as a correction for the derived model propagation term.

Another interpretation for the observer arises when initializing the observer equation with $\hat{\boldsymbol{\Psi}}(t_0) = 0$ and placing $\boldsymbol{\Psi}(t_i) = \boldsymbol{\alpha}^{\dagger}\mathbf{z}(t_i)$:

$$\hat{\boldsymbol{\Psi}}(t_{k+1}) = \gamma \Lambda \sum_{i=0}^{k} \left[\mathrm{I} + (1-\gamma)\,\Lambda\right]^{k-i} \boldsymbol{\alpha}^{\dagger}\mathbf{z}(t_i). \tag{14.30}$$

This equation describes a discrete convolution between the transformed measurements $\boldsymbol{\alpha}^{\dagger}\mathbf{z}(t_i)$, and a weighted filter $\left[\mathrm{I} + (1-\gamma)\,\Lambda\right]^{k}$, with weights that are determined by the recovered dynamics $\Lambda$ and the tuning parameter $\gamma$. Therefore, the observer equation can be interpreted as a weighted moving average filter of the measurements, which is adapted to the system at hand in a data-driven manner. A demonstration of this property was presented in [16], where the observer framework was compared to moving average filters with different fixed window sizes, in an object tracking problem. It was shown there that the estimation of the position using the observer is better than the best moving average filter in a broad range of system regimes.

### *14.5.3 Diffusion Maps Kalman Filter*

The observer presented in Sect. 14.5.2 suffers from two main shortcomings. First, the gain parameter $\boldsymbol{\kappa}$ in (14.25) is chosen empirically without optimality considerations. Second, in the construction of the observer equation (14.28), the focus is on the deterministic linear component of the dynamics in (14.12) and the stochastic component is disregarded.

Here, we present a different filtering framework based on the Kalman filter, which addresses these two issues [15]. The Kalman filter framework provides an optimal time-varying choice of the gain parameter $\boldsymbol{\kappa}$. In addition, the state formulation of the Kalman filter naturally includes a stochastic component, which is more suitable to the theoretic properties of the eigenfunctions of the Fokker–Planck operator (14.12). Note that the stochastic term in the dynamics of the eigenfunctions, i.e., $\sqrt{2/\beta}\,\left\|\nabla_{\mathbf{x}(t)}\phi_{\ell}(t)\right\|_2 \dot{\omega}_{\ell}(t)$, is not fully compatible with the Kalman filter model. This stochastic term is dependent on the system state, and therefore induces dependencies between different samples of the process noise in the derived system model. However, when the gradient of the new state representation, which approximates $\nabla_{\mathbf{x}(t)}\phi_{\ell}(t)$, is sufficiently small, the deviation from the model can be negligible. Therefore, by taking the *leading* diffusion maps coordinates, which are typically slowly varying functions of $\mathbf{x}(t)$, the gradient is approximately constant and the Kalman filter formulation is appropriate.

Based on the derived system model and the observer in Sect. 14.5.2, we define the following state-space equations of the system:

$$\boldsymbol{\Psi}(t_i) = (\mathrm{I} + \Lambda)\,\boldsymbol{\Psi}(t_{i-1}) + \mathbf{w}(t_{i-1}), \tag{14.31}$$

$$\mathbf{z}(t_i) = \boldsymbol{\alpha}\boldsymbol{\Psi}(t_i) + \mathbf{v}(t_i), \tag{14.32}$$

where $\Lambda$ is the linear dynamics matrix with the eigenvalues $\{-\lambda_\ell\}_{\ell=1}^{m}$ on its diagonal, I is the identity matrix, $\boldsymbol{\alpha}$ is the recovered lift function, and $\mathbf{w}(t_i)$ and $\mathbf{v}(t_i)$ are independent Gaussian noises with covariance matrices denoted by $\mathbf{Q}(t_i)$ and $\mathbf{R}(t_i)$, respectively.

Based on this system formulation, we construct the Kalman filter update equations as follows:

$$\begin{aligned}
\hat{\boldsymbol{\Psi}}(t_i) &= \mathbf{F}\hat{\boldsymbol{\Psi}}(t_{i-1}) + \kappa(t_i)\left(\mathbf{z}(t_i) - \mathbf{H}\mathbf{F}\hat{\boldsymbol{\Psi}}(t_{i-1})\right), \\
\mathbf{P}(t_i) &= (\mathrm{I} - \kappa(t_i)\mathbf{H})\left(\mathbf{F}\mathbf{P}(t_{i-1})\mathbf{F}^T + \mathbf{Q}(t_i)\right), \\
\kappa(t_i) &= \left(\mathbf{F}\mathbf{P}(t_{i-1})\mathbf{F}^T + \mathbf{Q}(t_i)\right)\mathbf{H}^T \\
&\quad \left(\mathbf{H}\mathbf{F}\mathbf{P}(t_{i-1})\mathbf{F}^T\mathbf{H}^T + \mathbf{H}\mathbf{Q}(t_i)\mathbf{H}^T + \mathbf{R}(t_i)\right)^{-1},
\end{aligned} \tag{14.33}$$

where $\hat{\boldsymbol{\Psi}}(t_i)$ and $\mathbf{z}(t_i)$ are the state estimation and the measurements at time $t_i$, respectively, $\mathbf{F} = \mathrm{I} + \Lambda$ denotes the dynamics of the new system state representation, $\mathbf{H} = \boldsymbol{\alpha}$ is the lift function, and $\mathbf{Q}(t_i)$ and $\mathbf{R}(t_i)$ are the covariance matrices of the process noise and the measurement noise, respectively. In the sequel, we will evaluate the performance of this Kalman filter, demonstrating its capability of filtering complex observations in a data-driven manner without explicit ground-truth or prior model knowledge.

## 14.6 Experimental Results

In this section, we apply the presented filtering frameworks to a simulated tracking problem, in Sect. 14.6.1, and to music analysis in Sect. 14.6.2.

### *14.6.1 Tracking Example*

We present a 2-dimensional simulated tracking problem and show that the data-driven observer and Kalman filter are well suited for noise reduction in nonlinear systems of the form (14.3), in which the system model and the underlying state are completely unknown.

Consider a moving object whose 2-dimensional position is measured through its radius and azimuth with respect to a reference (measurement) position. The 2-

dimensional position of the objected in Cartesian coordinates at time $t$, denoted by $(x_1(t), x_2(t))$, is simulated using the following nonlinear discrete stochastic equations:

$$\Delta x_1(t_{i+1}) = -\nabla Q^{(1)}(x_1(t_i)) + \sqrt{2}\omega_1(t_i), \quad (14.34)$$

$$\Delta x_2(t_{i+1}) = -\nabla Q^{(2)}(x_2(t_i)) + \sqrt{2}, \omega_2(t_i), \quad (14.35)$$

where $Q^{(1)}(x) = \frac{1}{8}(x-1)^4 - \frac{1}{2}(x-1)^2$ and $Q^{(2)}(x) = \frac{1}{8}(x-6)^4 - \frac{1}{2}(x-6)^2$ are double-well potential equations and $\omega_1(t_i)$, $\omega_2(t_i)$ denote standard Gaussian noises.

The accessible noisy measurements are given by

$$r(t_i) = \arctan\left(\frac{x_1(t_i)}{x_2(t_i)}\right) + v_1(t_i), \quad (14.36)$$

$$\phi(t_i) = \sqrt{x_1^2(t_i) + x_2^2(t_i)} + v_2(t_i), \quad (14.37)$$

where $v_1(t_1)$ and $v_2(t_i)$ are Gaussian noises. We denote the measurements by $\mathbf{z}(t_i) = [r(t_i), \phi(t_i)]$.

We simulated position trajectories of $N = 1000$ samples with a time step of $\Delta t = 0.01$ and with different measurement noise levels. An example of the resulting underlying state is presented in Fig. 14.3.

Given the measurements $\{\mathbf{z}(t_i)\}_{i=0}^{N-1}$, we applied the filtering frameworks described in Sect. 14.5, where $\varepsilon$ in (14.19) was set to be the median of the distances $d\left(\mathbf{z}(t_i), \mathbf{z}(t_j)\right)$, the dimensionality of the state coordinates in (14.21) was set to $m = 2$, and the covariance matrices for the modified Mahalanobis distance in (14.11) were computed by calculating the covariance of the measurements in a running window of 15 samples. In the observer framework, the gain parameter was empirically set to $\gamma = 0.01$, for which the best results were obtained. In the Kalman filter

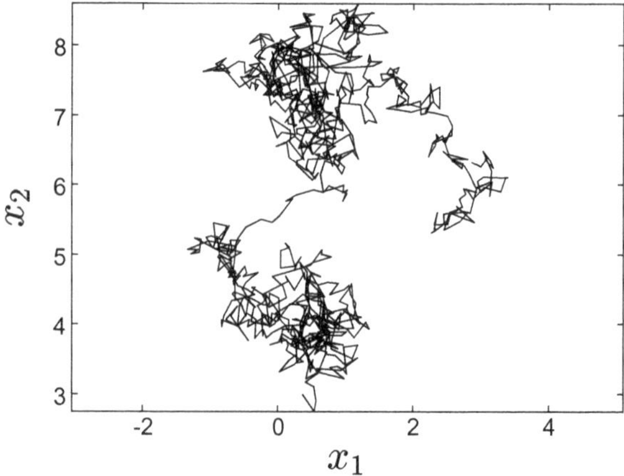


**Fig. 14.3** Example of a 2-dimensional trajectory created using Eqs. (14.36) and (14.37)

framework, the process covariance matrix, $\mathbf{Q}$ in (14.33), was set to be a diagonal matrix, with the variance of the recovered coordinate time steps on its diagonal, i.e., $Q_{k,k} = var\ (\lambda_k \psi_k)$. This empirical choice led to good results in all examined applications; however, it is possibly an underestimation of the covariance and could be further improved. The measurement noise covariance matrix, $\mathbf{R}$ in (14.33), was set to be a diagonal matrix with the variance of the measurements on its diagonal. We note that these choices are not optimal, and a better result may be obtained by estimating the covariance matrices as described in [3], which presents an adaptive estimation of these matrices for systems with an unknown model.

In this application, the goal is to recover the position of an object. Since the filtering frameworks provide new representations of the system state and not necessarily the object position, the recovered state coordinates cannot directly be used for tracking. Therefore, we use the new representations for denoising. This is obtained by first constructing the new representations and then lifting them to the measurement space using (14.24).

Figure 14.4 presents the resulting normalized root mean square error (RMSE) between the denoised measurements and the clean reference measurements. The RMSE was calculated for different signal-to-noise ratios (SNR) and averaged over 50 realizations of the underlying state (object position). In this figure, we denote the observer and Kalman filter frameworks by "Observer" and "DMK", respectively. We compare the presented filtering frameworks to two other algorithms: a particle filter, which is constructed using the *real hidden system model*, denoted by "PF" in the figure and will be referred to as the "optimal particle filter", and the algorithm presented in [19], denoted by "KKF" in the figure. In addition, we present the RMSE of the noisy measurements as an upper bound, denoted by "Meas". in the figure.

The KKF algorithm was implemented according to [19], which constructs a Kalman filtering framework based on the Koopman operator. There, the Koopman operator is approximated by EDMD and used for recovering a new linear system model. In contrast to the filtering frameworks presented in this chapter, this algorithm requires a set of state and measurement pairs, as well as samples of the underlying state in consecutive time steps. We implemented the KKF algorithm with 21 such reference pairs. This number of pairs was chosen empirically and led to the best results. In addition, the EDMD method requires the choice of a kernel. Here, we implemented the KKF with the kernel used by the examples in [19], i.e., $K\ (\mathbf{x}, \mathbf{y}) = \left(1 + \sqrt{5}d/l + 5d^2/3l^2\right) \exp\left(-\sqrt{5}d/l\right)$, where $d = \|\mathbf{x} - \mathbf{y}\|$ and $l$ was empirically set to 2.

Figure 14.4 depicts that both the observer framework and the Kalman filter framework indeed provide a denoised version of the measurements. However, the Kalman filter obtained significantly better results. This is mainly due to the adaptive optimal gain of the Kalman filter, in contrast to the fixed empirically chosen gain of the observer framework. In addition, the Kalman filter framework leads to significantly smaller errors than the KKF algorithm. Moreover, it obtains results that are close to the optimal particle filter, which was constructed using the true (hidden) model, especially for high SNR (SNR of $0.66 - 1$). We note that in low SNR the perfor-

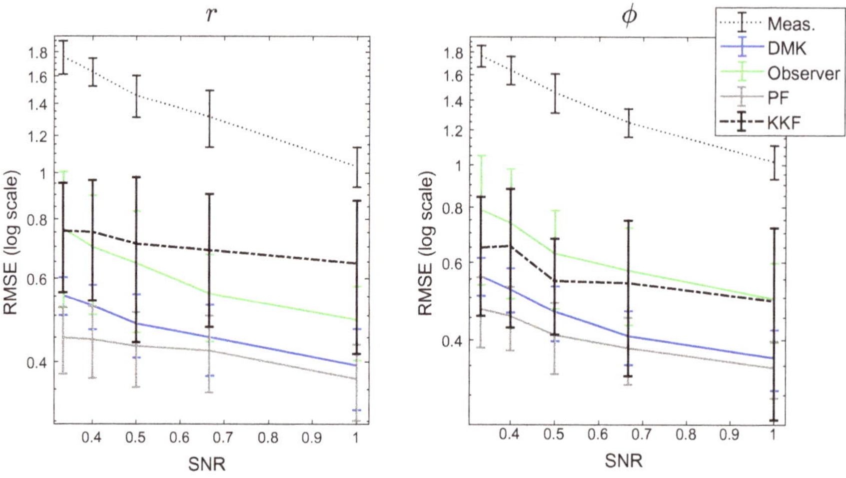


**Fig. 14.4** Average and standard deviation of the RMSE of the measurement estimation, calculated over 50 realizations, and presented as a function of the SNR. A comparison between 4 different algorithms is presented: (1) the Kalman filter framework from Sect. 14.5.3, denoted by "DMK", (2) the observer framework from Sect. 14.5.2, denoted by 'Observer', (3) the optimal particle filter, denoted by "PF", and (4) the method proposed in [19], denoted by "KKF". The RMSE of the noisy measurements is presented as an upper bound, denoted by "Meas"

mance of both filtering frameworks degrade, as can be seen in the figure. Since the theoretical derivation of the system model in Sect. 14.4 did not include measurement noise, significant noise leads to model errors, which hamper the performance.

The similarity in the performance of the Kalman filter and the optimal particle filter for high SNR is highlighted in Fig. 14.5. This figure presents an example of a measurement trajectory, created with SNR $= 1$, which was filtered using the presented Kalman filter, denoted by "DMK", and compared to the measurement trajectory obtained by the optimal particle filter, denoted by "PF". We present the clean measurement trajectory, marked by a dotted black curve, and the noisy measurements, marked by gray "x", for reference. We observe that almost everywhere the measurement estimation obtained by the Kalman filter closely follows the estimation obtained by the optimal particle filter. Remarkably, this comparable performance was obtained by the Kalman filter in a purely data-driven manner without any knowledge on the underlying system model, whereas the optimal particle filter was constructed using the true system model.

A MATLAB implementation of the observer and the Kalman filtering frameworks and the tracking problem is available on https://github.com/shnitzer/Manifold-Learning-for-Data-Driven-Dynamical-System-Analysis.git.

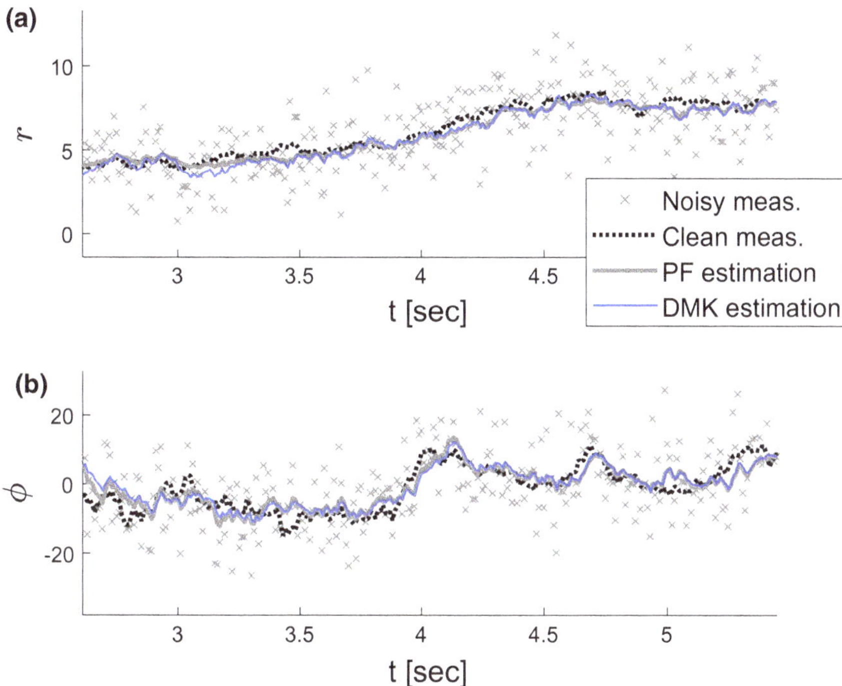


**Fig. 14.5** Example of measurement filtering for SNR = 1 using the Kalman filter framework, denoted by "DMK", compared with the optimal particle filter, denoted by "PF". The noisy measurements are marked by gray "x" and the clean measurements are denoted by a dotted black line

## *14.6.2 Music Analysis*

We demonstrate the presented approach in a music analysis application. We show that the filtering frameworks provide a new meaningful representation for music signals, which captures their inherent properties. Specifically, the constructed coordinates distinguish between different musical notes, as well as different musical instruments.

We apply both frameworks presented in Sect. 14.5 to 20–40 s segments of the theme song of "The good, the bad and the ugly" by Ennio Morricone, which is sampled at 44.1 kHz. First, the short-time Fourier transform (STFT) is applied to the music signal with a window of 23[msec]. Note that due to the use of the STFT, the input to the filtering frameworks, i.e., the system measurements, are high dimensional. Second, the diffusion maps coordinates and linear model are constructed based on the absolute value of the STFT of the signal, as described in Sect. 14.5.1, by treating each frequency vector in the STFT (one time frame) as one sample. For this purpose, the measurement covariance matrices, used in the calculation of the modified Mahalanobis distance (14.11), were estimated based on short time windows of 15 consecutive samples. In addition, we set $\varepsilon$ in (14.19) to be the median of the distances, which is common practice. Third, we construct the two filtering frameworks, the contracting observer, presented in Sect. 14.5.2, and the Kalman filter, presented in Sect. 14.5.3. The tunable parameter in the observer framework was set to be $\gamma = 0.1$, which empirically led to the best results. The dimension of the diffusion maps coordinates, denoted by $m$ in (14.21), was set to be 3 in both frameworks. Other choices of $m$ led to comparable results. The process and measurement noise covariance matrices in the Kalman filter framework were calculated as in Sect. 14.6.1.

Figure 14.6 presents the resulting system representation obtained using the observer framework, compared with the diffusion maps coordinates without additional processing. Plot (a) in this figure displays a 20 s segment of the spectrogram of the song, along with the audio waveform. In this song part, there is an alternation between two different musical instrument groups, which play in nonoverlapping segments. We mark these different segments by 1 and 2 on the waveform. The dashed black lines denote the transitions between the two instrument sets. In addition, we mark the musical notes played by instrument set 2 on the spectrogram using colored rectangles, where $Do^h$ denotes that the musical note "$Do$" is played in a higher octave. Plots (b)–(e) in Fig. 14.6 present the coordinates obtained by diffusion maps with no additional processing (plots (b) and (d)) and the coordinates obtained by the observer framework (plots (c) and (e)). The colors in plots (b) and (c) represent the different musical instrument sets and correspond to the colors of the instrument sets marked on the waveform in plot (a), i.e., blue for set 1 and green for set 2. Plots (d) and (e) are colored according to the musical notes, which are marked on the spectrogram in plot (a). The gray points in plots (d) and (e) correspond to time frames in which the musical notes were not marked, i.e., when instrument set 1 is playing.

The plots show that the observer framework provides a better representation for the music signal segment, compared with diffusion maps. Due to the incorporation of the dynamics, the separation between the two musical instrument sets in plot (c) and between different musical notes in plot (e) is more distinct than in plots (b) and (d), respectively. Importantly, note that in the coordinates constructed using the observer framework in plot (e), similar musical notes from different points in time are located in the same region in the new coordinate space. For example, the two instances of "Sol" that are marked on the spectrogram and colored in dark green are clustered in plot (c). This indicates that the observer framework indeed represents meaningful features in this application.

The Kalman filter framework led to comparable plots, which were omitted for brevity. We note that the main advantage of the Kalman filter framework over the observer in this application is that the Kalman gain is optimally calculated and does not require parameter tuning. In the observer framework, the gain parameter $\gamma$ was empirically chosen to provide the best results, whereas the Kalman filter obtained comparable results without such tuning.

Table 14.1 further demonstrates that the obtained coordinates provide a good representation for the music signal. This table presents the success rates of musical note classification performed on different representations of the song at times 6.2–45 s. Only musical notes played by instrument set 2 were manually labeled and used in the classification, resulting in 1466 samples and 7 classes (i.e., 7 different musical

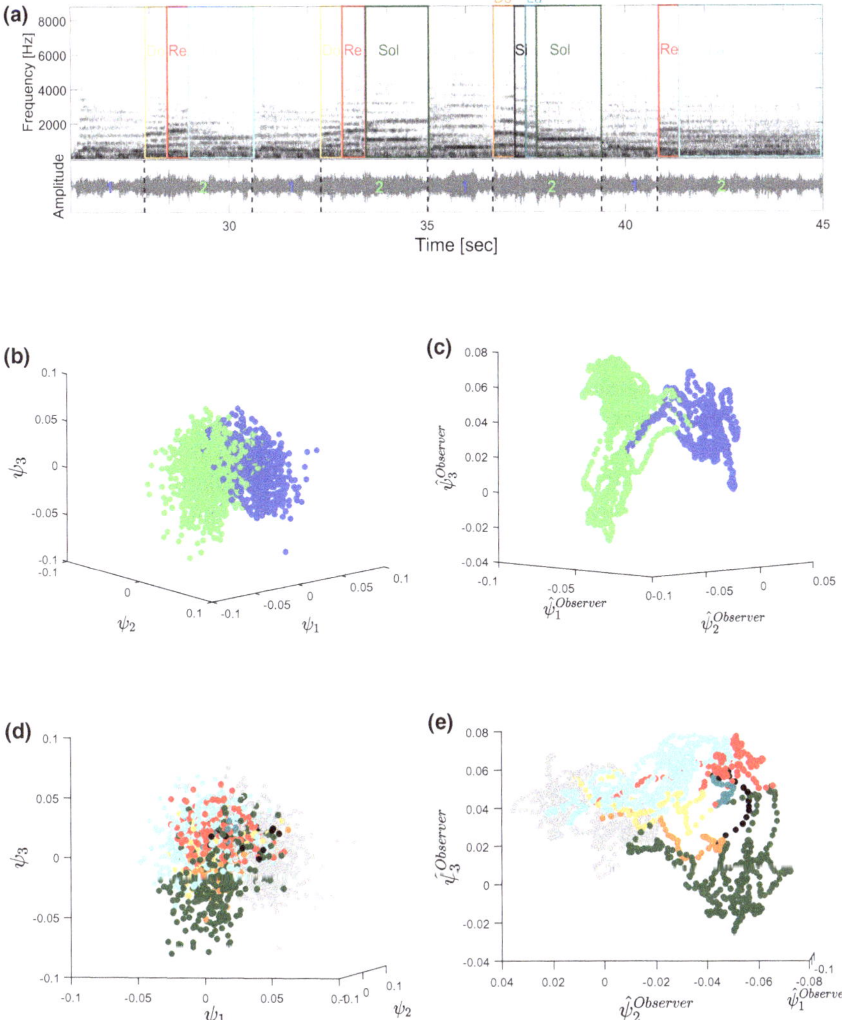


**Fig. 14.6** A comparison between the coordinates obtained by the observer framework and by diffusion maps for a 20 s segment of "The good, the bad and the ugly" theme song. **a** A spectrogram and waveform of the song segment. The two musical instrument sets are marked by 1 and 2 on the waveform and transitions between them are marked by black dashed lines. Musical notes are marked by colored rectangles on the spectrogram. **b** and **d** Diffusion maps coordinates, colored according to the musical instrument sets and according to the played musical notes, respectively. **c** and **e** Observer framework coordinates colored according to the musical instrument sets and according to the musical notes, respectively

**Table 14.1** Musical notes classification success rates and standard deviation

| $PCA$ | $\psi^{DM}$ | $\hat{\psi}^{Observer}$ | $\hat{\psi}^{Kalman}$ |
|---|---|---|---|
| 74.2% ± 2.2% | 46.6% ± 2.5% | 83.6% ± 2.2% | 89.6% ± 1.9% |

notes). We elaborate on the reason for this exclusion in the subsequent paragraph. We compared 4 algorithms: diffusion maps with no further processing, denoted by $\psi^{DM}$ in the table, the linear observer, denoted by $\hat{\psi}^{Observer}$, the Kalman filter, denoted by $\hat{\psi}^{Kalman}$, and principal component analysis (PCA), which was applied to the spectrogram, denoted by $PCA$ in the table. For each algorithm, 3 coordinates were used as input to the classifier and each 3-dimensional point, representing one (high dimensional) time sample in the STFT, was classified separately. The classification was performed using a support vector machine (SVM) with a radial basis function kernel (RBF). The data points were divided into a training set, which included 75% of the samples, and a test set, which included 25% of the samples. Cross-validation was performed and averaged over 50 random partitioning of the data.

We note that the segments played by musical instrument set 1 included many rapid note changes. These fast changes were not captured properly by any of the coordinates and were, therefore, ignored in the musical notes classification. Such segments can be separately analyzed by taking different values of $\varepsilon$ or by changing the considered time steps $\Delta t_i = t_{i+1} - t_i$, as described in [16].

To conclude, we showed that the obtained data-driven representation captures relevant features and characterizes the different musical notes and different instrument sets in music. These new representations improved the diffusion maps coordinates by incorporating time dependencies and by further adaptive filtering of the measurements, even though the validity of the state model in (14.3) is not guaranteed and can only be assumed in this application.

## 14.7 Conclusion

In this chapter, we presented two nonparametric and completely data-driven linear filtering frameworks based on a linear observer and the Kalman filter, which learn the system model by exploiting the properties of the diffusion maps embedding. We showed that these frameworks recover a new meaningful representation of the system, given only a set of noisy measurements, even for nonlinear systems in which the model is completely unknown.

Using a simulated tracking problem consisting of stochastic state equations and measurement noise, we demonstrated that the Kalman filter framework, which is more suitable for such problems, outperforms other nonparametric methods. Moreover, the Kalman framework obtained results which are comparable to the parametric particle filter (equipped with the complete system model) for high signal-to-noise rations. The properties and strengths of the filtering frameworks were further demonstrated on real musical data, where the recovered coordinates captured intrinsic information related to musical notes and instruments.

We conclude by noting that in these frameworks, the new recovered representation is not necessarily easily related to the true system state, however, we showed that it captures meaningful properties and can be leveraged to obtain an estimation of the "clean" system measurements, as illustrated in the simulated problem.

## References


1. Berry, T., Giannakis, D., Harlim, J.: Nonparametric forecasting of low-dimensional dynamical systems. Phys. Rev. E **91**(3), 032,915 (2015)
2. Berry, T., Harlim, J.: Nonparametric uncertainty quantification for stochastic gradient flows. SIAM/ASA J. Uncertain. Quantif. **3**(1), 484–508 (2015)
3. Berry, T., Sauer, T.: Adaptive ensemble Kalman filtering of non-linear systems. Tellus A: Dyn. Meteorol. Oceanogr. **65**(1), 20,331 (2013)
4. Budišić, M., Mohr, R., Mezić, I.: Applied Koopmanism a. Chaos: Interdiscip. J. Nonlinear Sci. **22**(4), 047,510 (2012)
5. Coifman, R., Kevrekidis, I., Lafon, S., Maggioni, M., Nadler, B.: Diffusion maps, reduction coordinates, and low dimensional representation of stochastic systems. Multiscale Model. Simul. **7**(2), 842–864 (2008)
6. Coifman, R., Lafon, S.: Diffusion maps. Appl. Comput. Harmon. Anal. **21**, 5–30 (2006)
7. Črnjarić-Žic, N., Maćešić, S., Mezić, I.: Koopman operator spectrum for random dynamical system (2017). arXiv:1711.03146
8. Deisenroth, M.P., Turner, R.D., Huber, M.F., Hanebeck, U.D., Rasmussen, C.E.: Robust filtering and smoothing with gaussian processes. IEEE Trans. Autom. Control **57**(7), 1865–1871 (2012)
9. Giannakis, D.: Data-driven spectral decomposition and forecasting of ergodic dynamical systems. Appl. Comput. Harmonic Anal. (2017)
10. Kalman, R.: A new approach to linear filtering and prediction problems. Trans. ASME J. Basic Eng. **82**, 34–45 (1960)
11. Lohmiller, W., Slotine, J.: On contraction analysis for non-linear systems. Automatica **34**(6), 683–696 (1998)
12. Mezić, I.: Spectral properties of dynamical systems, model reduction and decompositions. Nonlinear Dyn. **41**(1), 309–325 (2005)
13. Nadler, B., Lafon, S., Coifman, R., Kevrekidis, I.G.: Diffusion maps, spectral clustering and eigenfunctions of Fokker-Planck operators. In: Neural Information Processing Systems (NIPS), vol. 18 (2005)
14. Nadler, B., Lafon, S., Coifman, R., Kevrekidis, I.G.: Diffusion maps, spectral clustering and reaction coordinates of dynamical systems. Appl. Comput. Harmon. Anal. **21**, 113–127 (2006)
15. Shnitzer, T., Talmon, R., Slotine, J.J.: Diffusion maps Kalman filter (2017). arXiv:1711.09598
16. Shnitzer, T., Talmon, R., Slotine, J.J.E.: Manifold learning with contracting observers for data-driven time-series analysis. IEEE Trans. Signal Process. **65**(4), 904–918 (2017)
17. Singer, A., Coifman, R.: Non-linear independent component analysis with diffusion maps. Appl. Comput. Harmon. Anal. **25**, 226–239 (2008)
18. Sinha, S., Bowen, H., Vaidya, U.: On robust computation of Koopman operator and prediction in random dynamical systems (2018). arXiv:1803.08562
19. Surana, A., Banaszuk, A.: Linear observer synthesis for nonlinear systems using Koopman operator framework. IFAC-PapersOnLine **49**(18), 716–723 (2016)

20. Takeishi, N., Kawahara, Y., Yairi, T.: Learning Koopman invariant subspaces for dynamic mode decomposition. In: Advances in Neural Information Processing Systems, pp. 1130–1140 (2017)
21. Takeishi, N., Kawahara, Y., Yairi, T.: Subspace dynamic mode decomposition for stochastic Koopman analysis. Phys. Rev. E **96**(3), 033,310 (2017)
22. Takens, F.: Dynamical systems and turbulence (detecting strange attractors in fluid turbulence) (1981)
23. Wang, Y., Brubaker, M.A., Chaib-draa, B., Urtasun, R.: Bayesian filtering with online Gaussian process latent variable models. UAI pp. 849–857 (2014)
24. Williams, M.O., Kevrekidis, I.G., Rowley, C.W.: A data-driven approximation of the Koopman operator: extending dynamic mode decomposition. J. Nonlinear Sci. **25**(6), 1307–1346 (2015)

# Chapter 15
# Phase-Amplitude Reduction of Limit Cycling Systems

**Sho Shirasaka, Wataru Kurebayashi and Hiroya Nakao**

**Abstract** Optimization and control of collective dynamics in rhythmic systems have attracted increasing interest, and various methods have been developed on the basis of the phase reduction framework for limit cycle oscillators. The phase reduction relies on the notion of the isochron, which represents the set of system states that share the same asymptotic phase. However, the phase reduction does not take into account the amplitude degrees of freedom representing deviations of the system state from the limit cycle attractor, to which some rich and nontrivial transient behaviors of the oscillators are attributed. In this chapter, after a brief introduction of the phase reduction framework, a phase-amplitude reduction framework that is applicable to transient dynamics far from the limit cycle attractor is formulated. A rigorous theoretical background for defining the amplitudes is provided by the notion of the isostable, which naturally complements the isochron in the sense that both of them can be understood from a unified viewpoint of the spectral properties of the Koopman operator. The utility of the proposed phase-amplitude reduction framework is illustrated by evaluating the optimal injection timing of a weak control input that efficiently suppresses deviations of the system state from the limit cycle attractor.

---

The text of Sect. 15.3 is adapted from [78], with permission of AIP Publishing.

---

S. Shirasaka (✉)
Department of Information and Physical Sciences, Osaka University, 1-5 Yamadaoka, Suita 565-0871, Japan
e-mail: shirasaka@ist.osaka-u.ac.jp

W. Kurebayashi
Center for Data Science Education and Research, Shiga University, 1-1-1 Banba, Hikone 522-8522, Japan
e-mail: wataru-kurebayashi@biwako.shiga-u.ac.jp

H. Nakao
Department of Systems and Control Engineering, Tokyo Institute of Technology, 2-12-1 O-okayama, Meguro 152-8552, Japan
e-mail: nakao@sc.e.titech.ac.jp



A. Mauroy et al. (eds.), *The Koopman Operator in Systems and Control*,
Lecture Notes in Control and Information Sciences 484,
https://doi.org/10.1007/978-3-030-35713-9_15

## 15.1 Introduction

The *phase reduction* provides a general framework of dimensionality reduction for stable rhythmic systems, which approximately simplifies a complex, multi-dimensional limit cycling system describing a persistent spontaneous oscillatory activity to a one-dimensional phase equation evolving on a circle [4, 17, 26, 41, 60, 105]. It has been successfully used to clarify the synchronization phenomena of weakly interacting rhythmic elements in physical, chemical, biological, and engineered systems [4, 12, 17, 26, 41, 60, 66, 74, 84, 88, 89, 105]. Various methods to optimize and control synchrony of oscillators have also been developed on the basis of the phase reduction framework [24, 34, 57, 108, 109].

In the phase reduction framework, the system states that converge to the same state on the limit cycle attractor are identified and the same phase value is assigned to them; such sets of system states with identical phase values are called *isochrons* [17, 26, 41, 60, 66, 105]. By focusing only on the phase value of the system state, the dynamics of the system is approximately represented by a simple one-dimensional phase equation. However, to describe the system dynamics away from the limit cycle, such as the transient relaxation of the system state to the limit cycle, the amplitude degrees of freedom should also be taken into consideration. In this chapter, we briefly review the phase reduction framework and then formulate a *phase-amplitude reduction* framework that can be applied to excursions of the system far off the limit cycle attractor by extending the preceding studies.

The roles of the amplitude degrees of freedom in limit cycling systems, which represent deviations of the system states from the limit cycle and are eliminated in the phase reduction framework, have been studied extensively in the literature, because they are rich sources of intriguing oscillator dynamics at individual [4, 20, 25, 52, 66, 93, 107] and ensemble [3, 4, 10, 36, 41, 48, 61, 66] levels. In most studies, however, the analysis is restricted to the vicinity of a supercritical Hopf bifurcation, where a simple normal form (often called the Stuart–Landau equation) of the oscillator dynamics can be obtained by the center manifold reduction [22, 41, 85]. In this case, the system state is approximately described by a single complex variable, whose argument gives the phase of the oscillation and whose modulus represents the oscillation amplitude measured from the unstable fixed point at the center of the limit cycle. Synchronization dynamics of limit cycle oscillators near a Hopf bifurcation caused by a periodic input or mutual coupling with other oscillators have been analyzed on the basis of the Stuart–Landau equation.

For general limit cycle oscillators not necessarily near the Hopf bifurcation, some studies have introduced a moving orthonormal coordinate frame along the limit cycle in order to define the amplitudes of the oscillation [4, 25, 93], where the amplitude variables represent deviations of the system state from the limit cycle. It enables us to quantitatively analyze the amplitude dynamics of the oscillators far from the bifurcation point. However, in general, those amplitude variables interact nonlinearly with the phase and other amplitude variables, which prevents simplification of the system description. Thus, it is desirable to establish a framework for a quantitative reduced description of limit cycling systems that is applicable to the transient dynamics far from the limit cycle. Such a framework would facilitate in-depth studies of the roles played by the amplitude degrees of freedom of limit cycling systems in realistic settings.

The key idea in the phase reduction is to assign the same phase value to the set of initial conditions that share the same asymptotic behavior by using the notion of *isochrons* [17, 26, 41, 60, 66, 105]. Analogously, in a recent work [51], Mauroy, Mezić, and Moehlis have introduced the notion of *isostables* by identifying the initial conditions that share the same relaxation property. It has also been shown [51] that the isochrons and isostables can be understood from a unified viewpoint of the spectral properties of the Koopman operator [7]; for each characteristic decay rate of the system state toward the attractor, a set of isostables representing an amplitude degree of freedom can be introduced, which evolves independently from the phase and the other amplitude degrees of freedom when the system is autonomous. By retaining a small number of amplitude variables representing the dominant (slowly decaying) components of the transient dynamics, approximate reduced description of the system dynamics can be derived. The Koopman operator viewpoint has attracted broad interest recently, because it is closely related to the rapidly developing data-driven approach to complex nonlinear systems, called the dynamic mode decomposition [7, 69, 71, 73, 91, 95, 96] (see also Chap. 1).

By making use of both the isochrons and isostables, phase-amplitude reduction frameworks for limit cycling systems, which generalize the classical phase reduction framework, have been developed recently. Wilson et al. have developed phase-amplitude reduction frameworks using the first-order [103] and second-order [98] response properties of the amplitudes evaluated on the limit cycle [16, 99]. Castejón et al. [8], Mauroy and Mezić [50], and Shirasaka et al. [78] have considered the response properties of the phase and amplitudes evaluated on the transient orbits relaxing toward the limit cycle. By using the phase-amplitude reduction frameworks, optimal control strategies for modifying the period of the oscillator [98] and for suppressing the transient excursions of the system state from the limit cycle [78] have been developed. Also, amplitude reduction frameworks for non-oscillatory systems near a stable equilibrium based on the notion of isostables have been established for multi-dimensional [51, 53, 101, 104] and infinite-dimensional systems [102], and have been used to formulate optimal control problems [53, 101, 102, 104].

In this chapter, we formulate a phase-amplitude reduction framework for a stable limit cycle oscillator subjected to weak perturbations, which is also applicable to the transient dynamics far from the limit cycle. We introduce a systematic numerical method developed in [78] to accurately evaluate the fundamental quantities for the phase-amplitude reduction, that is, the first order sensitivity functions of the phase and amplitudes to perturbations along a given orbit, which is not necessarily the limit cycle itself. These sensitivity functions can be interpreted as an extension of the adjoint Floquet vectors (also called the adjoint or dual covariant Lyapunov vectors) [39, 68, 90] to the transient dynamics of limit cycling systems. We also discuss a simplified version of the numerical method to evaluate the phase-amplitude sensitivity functions. The utility of the proposed framework is illustrated by estimating an optimal injection timing of an external control signal that realizes the optimal suppression of the most persistent (least-decaying) amplitude degree of freedom.

This chapter is organized as follows: In Sect. 15.2, a brief review of the classical phase reduction framework for limit cycling systems is provided. In Sect. 15.3, the phase-amplitude reduction framework for limit cycling systems is introduced from the viewpoint of the Koopman operator theory. Numerical methods to calculate the sensitivity functions of the phase and amplitudes using the bi-orthogonality of the sensitivity functions and their dual vectors are also introduced. In Sect. 15.4, an analytically tractable example of a limit cycle oscillator is introduced and the numerical bi-orthogonalization method is verified by comparing the phase and amplitude sensitivity functions with the exact results. Also, an optimal control problem for the oscillator is introduced and is analyzed by using the phase-amplitude reduction framework. Section 15.5 summarizes the results and presents some recent related developments in the phase-amplitude reduction and future directions.

## 15.2 Brief Review of the Classical Phase Reduction Framework

In this section, we briefly review the classical phase reduction framework for weakly perturbed limit cycle oscillators, including a simple application to optimal control of synchronization.

### *15.2.1 Background*

Synchronization of rhythmic systems often plays important functional roles in real-world systems, such as the synchronized secretion of insulin from pancreatic $\beta$-cells [33] and synchronized operation of power grids [12]. Many of such rhythmic systems are modeled as limit cycle oscillators, namely, by using continuous dynamical systems that possess stable limit cycle solutions. When a limit cycle oscillator is driven by a periodic external input whose frequency is close to the natural frequency of the oscillator, the oscillator can be synchronized, or entrained, to the periodic input. When a pair of limit cycle oscillators are mutually coupled, they can exhibit mutual synchronization if their natural frequencies are close to each other. A population of coupled limit cycle oscillators can undergo collective synchronization transition even if their properties are slightly heterogeneous, which leads to macroscopic collective oscillations of the whole population where individual oscillators behave in unison with other oscillators.

Because limit cycle oscillations can only arise in multi-dimensional nonlinear dynamical systems, it is difficult to study the dynamics of coupled limit cycle oscillators by straightforward analytical methods. In particular, it is generally impossible to write down the analytical expressions of limit cycle solutions or to decompose their dynamics into linear independent modes. To overcome this difficulty, reduction frameworks for limit cycle oscillators have been developed, which approximately reduce the dimensionality of the original system and transform into simpler dynamical equations. As the result of reduction, lower dimensional dynamical equations with higher symmetry are generally obtained, which facilitates detailed theoretical analysis of coupled limit cycle oscillators. Moreover, essentially the same equations are often obtained from physically distinct systems through the reduction, which leads us to universal understanding of complex oscillatory dynamics independent of the detailed material properties of individual systems. Two major reduction frameworks for the limit cycle oscillators are center manifold reduction and phase reduction.

The center manifold reduction is valid only near the bifurcation point of limit cycle oscillations. For example, in the case of a supercritical Hopf bifurcation, it yields a two-dimensional normal form of the system called the Stuart–Landau oscillator [41, 66], and in the case of a saddle-node on an invariant circle bifurcation, a one-dimensional normal form called the theta model is obtained [17, 26]. In contrast, the phase reduction framework can generally be applied to stable limit cycle oscillators, provided that the perturbations given to the oscillators are sufficiently weak, and the system dynamics is approximated by a one-dimensional semi-linear phase equation [4, 17, 26, 41, 60, 105]. In the classical phase reduction framework, the deviations of the system state from the unperturbed limit cycle are eliminated and only the oscillator phase is retained.

The phase reduction framework has a long history, beginning with Malkin [46] and Winfree [105]. It has been widely applied to the analysis of spatiotemporal patterns in oscillatory media and collective dynamics in networks of coupled oscillators. Among them, Kuramoto's analysis of coupled phase equations that undergo collective synchronization transition is particularly well known [40, 83]. Recent studies include generalizations of the phase reduction framework to nonconventional oscillatory systems such as hybrid systems [65, 78, 99], time-delayed systems [37, 63], reaction–diffusion systems [62], and fluid systems [30, 86], applications of the reduced phase equations in engineering problems such as control, optimization, and design of oscillators [11, 13, 24, 31, 57, 59, 67, 76, 79, 80, 100, 108, 109], and developments of inverse methods to infer coupled phase equations from observed multivariate experimental data [82]. Inclusion of amplitude degrees of freedom representing deviations of the system state from the limit cycle has also been discussed in the literature [4, 25, 93], and recent introduction of the Koopman operator viewpoint [51] has provided a unified formulation to generalize the phase reduction to phase-amplitude reduction, as we discuss in the next section.

### 15.2.2 Phase Reduction

In the classical phase reduction framework, the starting point is to define a phase variable, known as the *asymptotic phase* [23, 105], for the system state of a limit cycle oscillator. Consider a limit cycle oscillator whose $N$-dimensional state $\mathbf{x}$ obeys a sufficiently smooth autonomous dynamical system

$$\dot{\mathbf{x}} = \mathbf{F}(\mathbf{x}), \qquad \mathbf{x} \in \mathbb{R}^N, \tag{15.1}$$

which has an exponentially stable limit cycle solution $\mathbf{x}_0(t)$ with natural period $T$ and frequency $\omega = 2\pi/T$, satisfying $\mathbf{x}_0(t) = \mathbf{x}_0(t+T)$. We denote this limit cycle by $\chi$ and its basin of attraction by $B \subset \mathbb{R}^N$. We first introduce a phase $\theta \in [0, 2\pi)$ on $\chi$, where 0 and $2\pi$ are considered identical (see Fig. 15.1a for a schematic illustration). We choose a state point $\mathbf{x}_0(0)$ on $\chi$ as the origin of phase, $\theta(\mathbf{x}_0(0)) = 0$, and define the phase of $\mathbf{x}_0(t)$ as $\theta(\mathbf{x}_0(t)) = \omega t$. With this definition, it is clear that $\dot{\theta} = \omega$ holds as long as $\mathbf{x}$ evolves on $\chi$. In the following, we parametrize a point $\mathbf{x}_0(\theta)$ on $\chi$ as a function of the phase $\theta \in [0, 2\pi)$ rather than the time $t$.

Next, we extend the definition of the phase to the basin of attraction $B$, because the system state may be perturbed and kicked out from $\chi$. In the phase reduction framework, this is accomplished by introducing the notion of asymptotic phase. We consider a set of states $\{\mathbf{x}\} \subset B$ that converge to the same state as that started from $\mathbf{x}_0(\theta)$ on $\chi$,

$$I(\theta) = \{\mathbf{x} \in B \mid \lim_{t\to\infty} \|\mathbf{S}(t, \mathbf{x}) - \mathbf{S}(t, \mathbf{x}_0(\theta))\| = 0\}, \tag{15.2}$$

for $0 \le \theta < 2\pi$, where $\|\cdot\|$ is the Euclidean norm and $\mathbf{S}(t, \mathbf{x})$ is the flow of Eq. (15.1). Here and in the following discussions, we also denote the flow simply by $\mathbf{x}(t) = \mathbf{S}(t, \mathbf{x}(0))$ when the initial condition $\mathbf{x}(0)$ is implicitly assumed. Such a set $I(\theta)$ is called the isochron of $\chi$ (see Fig. 15.1b for a schematic illustration). We assign the same phase value $\theta_*$ to $\mathbf{x} \in I(\theta_*)$, i.e., we introduce a phase function $\theta(\mathbf{x}) : B \to [0, 2\pi)$ that maps the system state to a phase value as

$$\theta(\mathbf{x}) = \theta(\mathbf{x}_0(\theta_*)) = \theta_*, \quad \mathbf{x} \in I(\theta_*), \tag{15.3}$$

where we use the asterisk to make a distinction between the actual phase value and the phase function. It is assumed that $\theta(\mathbf{x})$ is a sufficiently smooth function of $\mathbf{x}$.

The asymptotic phase obeys

$$\dot{\theta}(\mathbf{x}(t)) = \lim_{\tau\to 0} \frac{\theta(\mathbf{x}(t+\tau)) - \theta(\mathbf{x}(t))}{\tau} = \mathbf{F}(\mathbf{x}(t)) \cdot \nabla\theta(\mathbf{x}(t)) = \omega, \tag{15.4}$$

where $\nabla$ denotes the gradient and $\cdot$ is the dot product of vectors, and we have used

$$\mathbf{x}(x+\tau) = \mathbf{x}(t) + \mathbf{F}(\mathbf{x}(t))\tau + O(\tau^2) \tag{15.5}$$

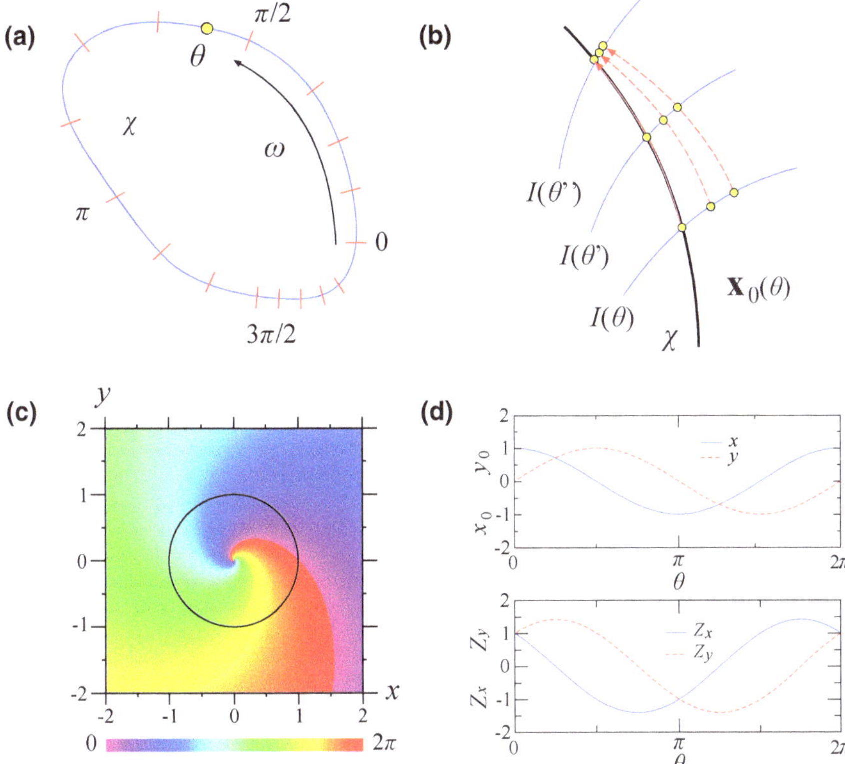


**Fig. 15.1** Asymptotic phase of a limit cycle oscillator. **a** Definition of the phase along the limit cycle. **b** Isochrons of the limit cycle. **c** Asymptotic phase of the Stuart–Landau oscillator. **d** Limit cycle solution $\mathbf{x}_0(\theta)$ and phase sensitivity function $\mathbf{Z}(\theta)$ of the Stuart–Landau oscillator. Adapted from [60] by permission of Taylor & Francis Ltd

and expanded $\theta(\mathbf{x})$ in Taylor series assuming its smoothness. Thus, the phase $\theta$ defined in this way always increases with a constant frequency $\omega$ as long as the state $\mathbf{x}$ evolves in $B$. The above definition is equivalent to introducing the phase function $\theta(\mathbf{x})$ such that $\theta(\mathbf{x}_0(\theta_*)) = \theta_*$ and

$$\mathbf{F}(\mathbf{x}) \cdot \nabla\theta(\mathbf{x}) = \omega, \quad \forall \mathbf{x} \in B, \tag{15.6}$$

hold. In the next section, we will see that $e^{i\theta(\mathbf{x})}$ is actually a Koopman eigenfunction associated with eigenvalue $i\omega$ of the infinitesimal generator of the Koopman operator semigroup, $\mathbf{F}(\mathbf{x}) \cdot \nabla$, where i is the imaginary unit.

As an example, Fig. 15.1c shows the phase function $\theta(x, y)$ of a two-dimensional Stuart–Landau oscillator, given by

$$\mathbf{x} = \begin{pmatrix} x \\ y \end{pmatrix}, \quad \mathbf{F}(\mathbf{x}) = \begin{pmatrix} x - \alpha y - (x - \beta y)(x^2 + y^2) \\ \alpha x + y - (\beta x + y)(x^2 + y^2)) \end{pmatrix}, \tag{15.7}$$

with $\alpha = 1$ and $\beta = -1$. This model has a stable limit cycle solution of frequency $\omega = \alpha - \beta$, whose basin of attraction is the whole $xy$ plane except the origin $(0, 0)$. The limit cycle is a unit circle and is analytically given by $\mathbf{x}_0(\theta) = (x_0(\theta), y_0(\theta))^\dagger = (\cos\theta, \sin\theta)^\dagger$, where $\dagger$ denotes the matrix transpose. For this model, the phase function can be obtained explicitly as

$$\theta(x, y) = \arctan(y/x) - (\beta/2)\ln(x^2 + y^2). \tag{15.8}$$

Using the asymptotic phase $\theta(\mathbf{x})$, we can introduce two important functions that characterize response properties of the oscillator phase to applied perturbations. First, the *phase response function* (a.k.a. phase response curve or phase resetting curve) [17, 105] is defined as

$$g(\theta_*; \mathbf{k}) = \theta(\mathbf{x}_0(\theta_*) + \mathbf{k}) - \theta_*. \tag{15.9}$$

This function quantifies the asymptotic phase shift of the oscillator caused by an impulsive perturbation $\mathbf{k}$ given to the system state $\mathbf{x}_0(\theta_*)$ at phase $\theta_*$ on $\chi$, which instantaneously kicks the system state from $\mathbf{x}_0(\theta_*)$ to $\mathbf{x}_0(\theta_*) + \mathbf{k}$.

Second, when the perturbation $\mathbf{k}$ is sufficiently weak, $g(\theta_*; \mathbf{k})$ can be linearly approximated by neglecting higher order terms in $\mathbf{k}$ as

$$g(\theta_*; \mathbf{k}) = \nabla\theta(\mathbf{x}_0(\theta_*)) \cdot \mathbf{k}, \tag{15.10}$$

where we have used $\theta(\mathbf{x}_0(\theta_*)) = \theta_*$. The gradient $\nabla\theta(\mathbf{x}_0(\theta_*))$ of $\theta$, evaluated at $\mathbf{x} = \mathbf{x}_0(\theta_*)$, is called the *phase sensitivity function* (a.k.a. infinitesimal phase resetting curve) [17, 41, 105], which characterizes linear response property of the oscillator phase to weak perturbations. The phase response and sensitivity functions have been central in the analysis of limit cycle oscillations and measured experimentally in various chemical [34, 56], biological [105], or physiological systems [74, 88]. The phase sensitivity function plays an essential role in the following formulation for weakly perturbed limit cycle oscillators. In this section, we denote the phase sensitivity function evaluated at $\mathbf{x} = \mathbf{x}_0(\theta_*)$ as $\mathbf{Z}(\theta_*) = \nabla\theta(\mathbf{x}_0(\theta_*))$.

Now, let us consider a limit cycle oscillator subjected to a weak perturbation, described by

$$\dot{\mathbf{x}} = \mathbf{F}(\mathbf{x}) + \varepsilon\mathbf{p}, \qquad \mathbf{p} \in \mathbb{R}^N, \tag{15.11}$$

where $\varepsilon\mathbf{p}$ describes the perturbation given to the oscillator of tiny magnitude $\varepsilon$, i.e., $0 < \varepsilon \ll 1$. The dynamics of the phase $\theta(\mathbf{x}(t))$ is given by

$$\dot{\theta} = \nabla\theta(\mathbf{x}) \cdot \{\mathbf{F}(\mathbf{x}) + \varepsilon\mathbf{p}\} = \omega + \varepsilon\nabla\theta(\mathbf{x}) \cdot \mathbf{p}. \tag{15.12}$$

This equation is not yet closed in phase $\theta$ because $\nabla\theta(\mathbf{x})$ depends on $\mathbf{x}$. In order to obtain an equation for $\theta$, we use the fact that the perturbation is small and $O(\varepsilon)$, and hence the deviation of the state $\mathbf{x}$ from $\chi$ is also small and $O(\varepsilon)$, i.e., $\mathbf{x} = \mathbf{x}_0(\theta_*) + O(\varepsilon)$, where $\theta_* = \theta(\mathbf{x})$. The gradient $\nabla\theta$ at $\mathbf{x}$ can then be expressed as

$$\nabla\theta(\mathbf{x}) = \nabla\theta(\mathbf{x}_0(\theta_*)) + O(\varepsilon) \tag{15.13}$$

and, by plugging into Eq. (15.12), we obtain an approximate phase equation for $\theta$,

$$\dot{\theta} = \omega + \varepsilon\mathbf{Z}(\theta) \cdot \mathbf{p}, \tag{15.14}$$

by neglecting terms of $O(\varepsilon^2)$. This phase equation is now closed in $\theta$ and can be solved for $\theta$ when the phase sensitivity function $\mathbf{Z}$ and perturbation $\varepsilon\mathbf{p}$ are given.

In practice, it is generally impossible to obtain the analytical expression for the asymptotic phase function $\theta(\mathbf{x})$. However, it is known that the gradient $\nabla\theta$ on $\chi$ can be obtained by calculating a $2\pi$-periodic solution to the following equation, which is an adjoint system to Eq. (15.1) linearized around $\chi$:

$$\omega\frac{\mathrm{d}}{\mathrm{d}\theta}\mathbf{Z}(\theta) = -\mathrm{D}\mathbf{F}(\mathbf{x}_0(\theta))^{\dagger}\mathbf{Z}(\theta), \tag{15.15}$$

where $\mathrm{D}\mathbf{F}(\mathbf{x}_0(\theta))$ is the Jacobian matrix of $\mathbf{F}(\mathbf{x})$ evaluated at $\mathbf{x} = \mathbf{x}_0(\theta)$ and the resulting $\mathbf{Z}(\theta)$ should be normalized to satisfy Eq. (15.6), i.e., $\mathbf{Z}(\theta) \cdot \mathbf{F}(\mathbf{x}_0(\theta)) = \omega$. See the next section and also Ermentrout [15] and Brown et al. [5] for the derivation of the adjoint equation and its relation to the classical Floquet theory.

### 15.2.3 Entrainment by a Periodic External Input

In this subsection, as the simplest example of the phase reduction framework, we analyze a single limit cycle oscillator subjected to a weak external periodic input. For details see, e.g., Kuramoto [41], Hoppensteadt and Izhikevich [26], and Ermentrout and Terman [17].

In this case, the perturbation $\varepsilon\mathbf{p}(t)$ is a periodic signal of period $T'$ and frequency $\Omega = 2\pi/T'$, satisfying $\varepsilon\mathbf{p}(t) = \varepsilon\mathbf{p}(t + T')$. We assume that the frequency $\Omega$ of the periodic input is close to the natural frequency $\omega$ of the oscillator in the sense that their frequency mismatch is of $O(\varepsilon)$, i.e.,

$$\omega - \Omega = \varepsilon\Delta, \tag{15.16}$$

where $\Delta$ is a constant of $O(1)$. The approximate phase dynamics is given by Eq. (15.14), which is a nonautonomous equation. To proceed, we consider the phase difference between the oscillator phase $\theta$ and the phase $\theta_{\mathrm{ext}} := \Omega t$ of the periodic input,

$$\varphi(t) = \theta(t) - \Omega t. \tag{15.17}$$

From Eq. (15.14), this phase difference $\varphi(t)$ obeys

$$\dot{\varphi}(t) = \varepsilon\{\Delta + \mathbf{Z}(\varphi(t) + \Omega t) \cdot \mathbf{p}(t)\}. \tag{15.18}$$

Here, because the right-hand side of this equation is $O(\varepsilon)$ and small, $\varphi(t)$ evolves slowly while the quantity $\varphi(t) + \Omega t$ varies quickly and approximately uniformly over $[0, 2\pi)$. In this case, we can invoke the averaging approximation [27, 32, 72], that is, we can average the right-hand side over one period of oscillation while fixing $\varphi(t)$ to obtain an approximate, autonomous phase equation. We thus obtain

$$\dot{\varphi}(t) = \varepsilon\{\Delta + \Gamma(\varphi)\}, \tag{15.19}$$

where we have introduced the *phase coupling function* as

$$\begin{aligned} \Gamma(\varphi) &= \frac{1}{T'} \int_0^{T'} \mathbf{Z}(\varphi + \Omega\tau) \cdot \mathbf{p}(\tau) d\tau \\ &= \frac{1}{2\pi} \int_0^{2\pi} \mathbf{Z}(\varphi + \theta_{\text{ext}}) \cdot \mathbf{p}\left(\frac{\theta_{\text{ext}}}{\Omega}\right) d\theta_{\text{ext}}. \end{aligned} \tag{15.20}$$

Precisely speaking, the variable $\varphi(t)$ in Eq. (15.19) after averaging and the original $\varphi(t)$ in Eq. (15.17) before averaging are slightly different and are related by a near-identity transform, but we use the same notation. The error caused by this averaging approximation is also $O(\varepsilon^2)$ and thus ignored.

We can use Eq. (15.18) to analyze the entrainment of the oscillator to the periodic input. If Eq. (15.18) has a stable fixed point $\varphi^*$ satisfying $\Delta + \Gamma(\varphi^*) = 0$ and $\Gamma'(\varphi^*) < 0$, the phase difference $\varphi(t)$ can converge to this point. The averaged frequency of the oscillator becomes $\Omega$ and the oscillator phase $\theta$ keeps a fixed relation with the phase $\theta_{\text{ext}}$ of the periodic input on average, that is, the oscillator is entrained to the periodic input. Because $\Gamma(\varphi)$ is $2\pi$-periodic in $\varphi$ and generally nonconstant, this occurs when $\min_\varphi \Gamma(\varphi) < -\Delta < \max_\varphi \Gamma(\varphi)$.

For example, for the Stuart–Landau oscillator, the phase sensitivity function can be explicitly obtained as $\mathbf{Z}(\theta) = (-\sin\theta - \beta\cos\theta, \cos\theta - \beta\sin\theta)^\dagger$ by calculating the gradient of the phase function $\theta(x, y)$ at $\mathbf{x}_0(\theta)$, where $\beta$ is a real parameter (see Fig. 15.1d). If the external periodic input is given to the first component as $\varepsilon\mathbf{p}(t) = \varepsilon(K\cos\Omega t, 0)^\dagger$, where $K > 0$ characterizes the intensity of the input, the phase coupling function can be calculated as

$$\Gamma(\varphi) = -\frac{1}{2}K(\beta\cos\varphi + \sin\varphi) = -\frac{1}{2}K\sqrt{1+\beta^2}\sin(\varphi + \delta), \tag{15.21}$$

where $\delta$ is a constant satisfying $\cos\delta = 1/\sqrt{1+\beta^2}$ and $\sin\delta = \beta/\sqrt{1+\beta^2}$. Thus, if the frequency mismatch $\Delta = (\omega - \Omega)/\varepsilon$ satisfies

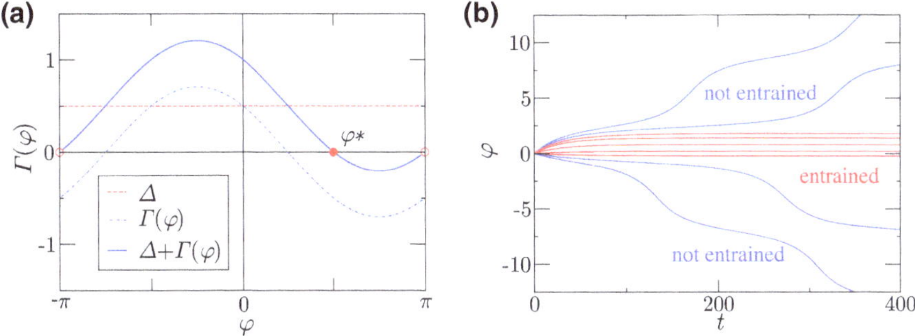


**Fig. 15.2** Entrainment of a limit cycle oscillator to a periodic input. **a** Phase coupling function and fixed point of the phase difference. $\Delta = 0.5$ and $\beta = -1$. **b** Evolution of the phase difference. Cumulative phase differences without taking mod $2\pi$ are plotted for varying values of $\Delta$. Adapted from [60] by permission of Taylor & Francis Ltd

$$|\Delta| < \frac{1}{2}K\sqrt{1+\beta^2}, \tag{15.22}$$

there exists a single stable fixed point $\varphi^*$ of Eq. (15.19) and the oscillator is entrained to the periodic input with this phase shift.

Figure 15.2 shows $\Delta + \Gamma(\varphi)$ in the right-hand side of Eq. (15.19) and evolution of the phase difference for varying values of $\Omega$. The parameters of the oscillator are $\alpha = 1$, $\beta = -1$, which gives $\omega = 2$. The small parameter $\varepsilon$ is fixed at $\varepsilon = 0.05$, and the intensity of the periodic input is fixed at $K = 1$. The driving frequency $\Omega$ is varied around $\omega = 2$. When the frequency mismatch $\Delta$ satisfies the condition derived above, the oscillator is entrained to the periodic input and the phase difference converges to a constant. When $|\Delta|$ is too large, the entrainment cannot occur and the phase difference increases or decreases indefinitely. If we draw the region where the entrainment takes place on the $(\Delta, \varepsilon)$ plane using $\Gamma$, we obtain a triangular parameter region. When $\varepsilon$ is sufficiently small, this region well approximates the vertical-tongue-shaped frequency locking region of the original system (15.11) known as the Arnold tongue [26, 66].

### 15.2.4 Optimal Periodic Input for Stable Entrainment

In this subsection, we briefly discuss an application of the phase reduction to the control of limit cycle oscillators. Due to the nonlinearity of the governing equations, it is generally not easy to develop systematic methods to control the dynamics of limit cycle oscillators. However, when the input to the oscillator is sufficiently weak, we can reduce the oscillator dynamics to a phase equation and use it to derive simple control methods [24, 34, 57, 108, 109]. A typical example of such studies is the optimal entrainment of a limit cycle oscillator considered by Zlotnik et al. [108], where an oscillator driven by a weak periodic input is considered and the periodic input is optimized to achieve the largest linear stability of the entrained state.

For simplicity, let us assume that the frequency $\Omega$ of the periodic input is equal to the natural frequency $\omega$ of the oscillator. From Sect. 15.2.3, the dynamics of the phase difference $\varphi$ between the oscillator and the periodic forcing is given by $\dot{\varphi}(t) = \varepsilon \Gamma(\varphi)$, where the frequency mismatch is $\Delta = 0$ by assumption and the phase coupling function $\Gamma(\varphi)$ is given by Eq. (15.20) with $T' = T$. By appropriately choosing the origin of the periodic input, we can always assume that the entrainment occurs at $\varphi = 0$, and the linear stability of this point is given by $-\Gamma'(0) = -\mathrm{d}\Gamma(\varphi)/\mathrm{d}\varphi|_{\varphi=0}$. To derive the optimal periodic input $\mathbf{p}(t)$ for realizing stable entraiment, we consider the following maximization problem:

$$\text{maximize} \quad -\Gamma'(0) \quad \text{subject to} \quad \frac{1}{T}\int_0^T \|\mathbf{p}(\tau)\|^2 d\tau = P, \tag{15.23}$$

where the constraint indicates that the mean power of the input $\mathbf{p}(t)$ is fixed at $P > 0$. Using a Lagrange multiplier $\nu$, we introduce a cost functional

$$S\{\mathbf{p}(\cdot), \nu\} = -\Gamma'(0) + \nu \left( \frac{1}{T}\int_0^T \|\mathbf{p}(\tau)\|^2 d\tau - P \right), \tag{15.24}$$

and by solving the conditions for the extremum of $S$, i.e., $\delta S/\delta \mathbf{p}(t) = 0$ and $\partial S/\partial \nu = 0$, we can obtain the optimal functional form for $\mathbf{p}(\tau)$ as

$$\mathbf{p}(t) = -\frac{1}{2\nu} \left. \frac{\mathrm{d}\mathbf{Z}(\varphi)}{\mathrm{d}\varphi} \right|_{\varphi=\omega t}, \tag{15.25}$$

where the Lagrange multiplier $\nu > 0$ is determined from the constraint.

Thus, by using the phase reduction framework, the optimal periodic input for realizing stable entrainment can be systematically derived in a concise form. It is notable that the optimal input is directly related to the phase sensitivity function of the oscillator. In [108], the authors have experimentally validated the above result by using an electrochemical oscillator and demonstrated that faster entrainment is realized with the optimized input. They further generalized the method to realize complex entrainment patterns in an ensemble of oscillators in [109]. Similarly, the phase reduction framework can also be used to find the optimal coupling schemes for mutual synchronization between weakly coupled limit cycle oscillators [76, 92].

### 15.2.5 Networks of Coupled Phase Oscillators

Using the phase reduction, we can also analyze networks of coupled limit cycle oscillators [41]. Consider a system of $M$ nearly identical oscillators described by

$$\dot{\mathbf{x}}_i = \mathbf{F}(\mathbf{x}_i) + \varepsilon \mathbf{f}_i(\mathbf{x}_i) + \varepsilon \sum_{j=1}^{M} \mathbf{G}_{ij}(\mathbf{x}_i, \mathbf{x}_j), \quad (i = 1, \ldots, M), \tag{15.26}$$

where $\mathbf{x}_1, ..., \mathbf{x}_M$ are the oscillator states, $\mathbf{F}$ represents the common part of their dynamics, $\varepsilon \mathbf{f}_i$ represents tiny heterogeneity of individual dynamics from the common part, and $\varepsilon \mathbf{G}_{ij}$ represents a pairwise interaction between oscillators $i$ and $j$. As before, we assume that the common part $\mathbf{F}$ has a stable limit cycle $\mathbf{x}_0(t) = \mathbf{x}_0(t + T)$ of period $T$ and frequency $\omega = 2\pi/T$. We also assume that the heterogeneity and mutual coupling are sufficiently weak and $O(\varepsilon)$, and treat them as weak perturbations.

By introducing the phase of each oscillator as $\theta_i(t) = \theta(\mathbf{x}_i(t))$, this model can be approximately reduced to a coupled phase equation of the form

$$\dot{\theta}_i = \omega + \varepsilon \mathbf{Z}(\theta_i) \cdot \left\{ \mathbf{f}_i(\theta_i) + \sum_{j=1}^{M} \mathbf{G}_{ij}(\theta_i, \theta_j) \right\}, \quad (i = 1, ..., M), \tag{15.27}$$

where $\mathbf{f}_i(\theta_i) = \mathbf{f}_i(\mathbf{x}_0(\theta_i))$ and $\mathbf{G}_i(\theta_i, \theta_j) = \mathbf{G}_{ij}(\mathbf{x}_0(\theta_i), \mathbf{x}_0(\theta_j))$. We can further average these equations as before and obtain simpler coupled phase equations,

$$\dot{\theta}_i = \omega + \varepsilon \cdot \left\{ \Delta_i + \sum_{j=1}^{M} \Gamma_{ij}(\theta_i - \theta_j) \right\}, \quad (i = 1, ..., M), \tag{15.28}$$

where

$$\Delta_i = \frac{1}{2\pi} \int_0^{2\pi} \mathbf{Z}(\theta_i) \cdot \mathbf{f}_i(\theta_i) d\theta_i, \quad (i = 1, ..., M) \tag{15.29}$$

gives an average frequency shift of oscillator $i$ due to tiny heterogeneity and

$$\Gamma_{ij}(\varphi) = \frac{1}{2\pi} \int_0^{2\pi} \mathbf{Z}(\theta + \varphi) \cdot \mathbf{G}_{ij}(\theta + \varphi, \theta) d\theta, \quad (i, j = 1, ..., M) \tag{15.30}$$

is the phase coupling function between the oscillators $i$ and $j$.

A particularly well-known case is $\varepsilon \Gamma_{ij}(\varphi) = -K/M \sin \varphi$ considered by Kuramoto [40], which gives

$$\dot{\theta}_i = \omega_i + \frac{K}{M} \sum_{j=1}^{M} \sin(\theta_j - \theta_i), \quad (i = 1, ..., M), \tag{15.31}$$

where $\omega_i = \omega + \varepsilon \Delta_i$ is the natural frequency of oscillator $i$ and $K$ represents the coupling intensity between the oscillators. This model with a sinusoidal phase coupling function can actually be derived by the phase reduction (and rescaling) from a system of globally coupled Stuart–Landau oscillators [60]. It is often assumed that the frequency mismatch $\omega_i$ obeys a one-humped symmetric frequency distribution and the limit of $M \to \infty$ is considered. This model then exhibits a collective synchronization transition as the coupling intensity $K$ is increased past a critical value $K_c$. The degree of synchronization is quantified by using a complex order parameter,

$$Re^{\mathrm{i}\Phi} = \frac{1}{M}\sum_{i=1}^{M} e^{\mathrm{i}\theta_i}, \tag{15.32}$$

where the modulus $R$ gives the amplitude of the collective oscillation and the argument $\Phi$ gives the collective phase. When the oscillators are desynchronized, $R$ takes a small value of $O(1/\sqrt{M})$, and when they are synchronized, $R$ takes a value close to 1. Because of its simplicity, this model allows detailed analytical treatment of the synchronization transition and has been extensively studied in the literature [2, 13, 41, 70, 83].

Typical behavior of the Kuramoto model is shown in Fig. 15.3. Figure 15.3a shows typical dynamics of the phase variables $\theta_i$ of the oscillators for $K > K_c$, showing mutual synchronization. Note that complete synchronization does not occur because of the frequency mismatch between the oscillators. Figure 15.3b plots the modulus $R$ of the complex order parameter as a function of the coupling intensity $K$. It can be seen that, when $K$ exceeds a critical value $K_c$, the oscillators exhibit sudden transition to collectively synchronized state. The inset of Fig. 15.3b shows the snapshots of the oscillators on the unit circle below and above the synchronization transition.

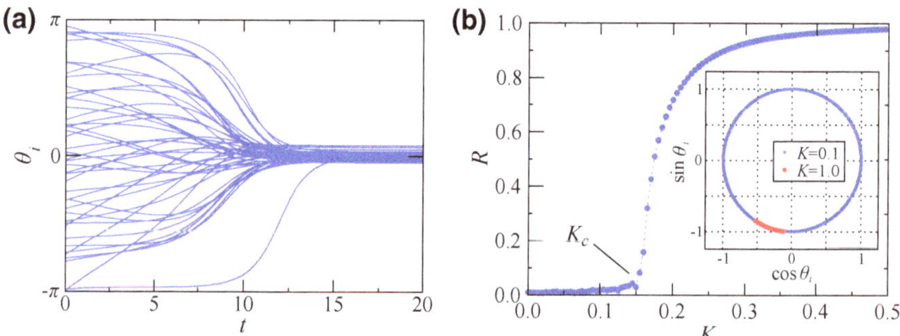


**Fig. 15.3** Collective synchronization transition of coupled phase oscillators. The number of oscillators are $M = 12000$ and the frequencies of the oscillators are randomly drawn from a Gaussian distribution of mean 1 and variance 0.1. **a** Dynamics of individual oscillators (only a few dozen of oscillators are shown). **b** Modulus $R$ of the complex order parameter versus coupling intensity $K$. The inset shows snapshots of the oscillator distributions at $K = 0.1$ (below transition) and at $K = 1.0$ (above transition). Adapted from [60] by permission of Taylor & Francis Ltd

## 15.3 Phase-Amplitude Reduction Framework

In this section, a phase-amplitude reduction framework for limit cycling systems is introduced from the viewpoint of the Koopman operator theory. Numerical methods to calculate the sensitivity functions of the phase and amplitudes are also developed.

### *15.3.1 Phase, Amplitudes and the Koopman Operator*

In this subsection, we reformulate the asymptotic phase given in the previous section in terms of the Koopman operator framework. We will see that we can further introduce amplitude variables and *isostables*, which are the level sets of the amplitudes, in such a way that they naturally complement the phase variables and isochrons in this framework [49, 51].

As before, we consider a $N$-dimensional autonomous dynamical system (15.1) that has an exponentially stable limit cycle solution $\chi : \mathbf{x}_0(t)$ with period $T$. The stability of the limit cycle $\chi$ is characterized by the *Floquet multipliers* $\mu_i$ $(i = 1, \dots, N)$ [22], which are the eigenvalues of the time-$T$ flow linearized around a point $\mathbf{x}_0(t_*)$ on the orbit $\chi$ (also called the *monodromy matrix*) $\mathbf{M}(\mathbf{x}_0(t_*)) = \partial \mathbf{S}(T, \mathbf{x})/\partial \mathbf{x}|_{\mathbf{x}=\mathbf{x}_0(t_*)}$. The exponential stability of $\chi$ means that the relation $1 = \mu_1 > |\mu_2| \geq \cdots \geq |\mu_N|$ holds. For simplicity, we hereafter assume that the Floquet multipliers $\mu_i$ are positive, real, and simple. Extension of the argument to the case with complex conjugate multipliers can be performed in a parallel way to the analysis of stable equilibria [51, 97, 102]. We consider the dynamics of the system in the basin of attraction $B \subset \mathbb{R}^N$ of $\chi$.

Let us consider the Koopman operator semigroup $U^t$ acting on an observable function $f : B \to \mathbb{C}$, defined as $(U^t f)(\mathbf{x}) = f \circ \mathbf{S}(t, \mathbf{x})$, where $\circ$ represents composition of functions. For the exponentially stable limit cycle $\chi$, the Koopman eigenvalues include its *Floquet exponents* [22], hence they reflect the linear stability of $\chi$. That is, the operator $U^t$ has $N$ eigenfunctions $\phi_{\lambda_i}(\mathbf{x})$ $(i = 1, \dots, N)$ associated with eigenvalues $\lambda_i$ $(i = 1, \dots, N)$ (we use the term eigenvalue to indicate $\lambda_i$ rather than $e^{\lambda_i t}$) [43, 54], satisfying

$$(U^t \phi_{\lambda_i})(\mathbf{x}) = e^{\lambda_i t} \phi_{\lambda_i}(\mathbf{x}), \tag{15.33}$$

where $\lambda_1 = \mathrm{i}\omega$, $\omega \equiv 2\pi/T$, and $\lambda_i = \log(\mu_i)/T$ $(i = 2, \dots, N)$ are the Floquet exponents of $\chi$ (we adopt $\lambda_1 = \mathrm{i}\omega$ rather than the ordinary $\lambda_1 = 0$). These Koopman eigenfunctions associated with the Floquet exponents are called the principal Koopman eigenfunctions [55, 58].

The eigenfunctions $\phi_{\lambda_i}$ are also eigenfunctions of the infinitesimal generator of the Koopman operator semigroup $L$, namely,

$$(L\phi_{\lambda_i})(\mathbf{x}) = \lambda_i \phi_{\lambda_i}(\mathbf{x}), \tag{15.34}$$

where

$$(Lf)(\mathbf{x}) := \left(\lim_{\Delta t \to 0} \frac{U^{\Delta t} f - f}{\Delta t}\right)(\mathbf{x}) = \mathbf{F}(\mathbf{x}) \cdot \nabla f(\mathbf{x}). \tag{15.35}$$

In addition to the principal eigenfunctions, the Koopman operator generally has infinitely many eigenfunctions, which are obtained as the pointwise products of the principal eigenfunctions [7]. The principal Koopman eigenfunctions associated with the Floquet exponents provide a globally linearizing transformation of the nonlinear system (15.1) in the whole basin of attraction $B$ as discussed below [43].

We hereafter assume that the Koopman eigenfunctions $\phi_{\lambda_i}(\mathbf{x})$ are continuously differentiable with respect to $\mathbf{x}$ and their first derivatives are Lipschitz continuous on $B$, which is required for the perturbative analysis. *Nonresonant* analyticity of $\mathbf{F}$, which holds in many practical situations, is sufficient for the Lipschitz continuity, because this ensures that $\phi_{\lambda_i}$ is analytic [55]. See [14, 42, 43] for the results on sufficient conditions for nonanalytic regularities of the Koopman eigenfunctions.

By using the first principal Koopman eigenfunction $\phi_{\lambda_1}(\mathbf{x})$, we can introduce a phase function of $\mathbf{x}$ by $\theta(\mathbf{x}) := \arg(\phi_{\lambda_1}(\mathbf{x}))$, where $\arg(z)$ is the argument of the complex number $z$, whose range is defined as the interval $[0, 2\pi)$. Indeed, because $\lambda_1 = \mathrm{i}\omega$, $\theta$ obeys

$$\dot{\theta}(\mathbf{x}) = \left(\lim_{\Delta t \to 0} \frac{\arg \circ U^{\Delta t} \phi_{\lambda_1} - \arg \circ \phi_{\lambda_1}}{\Delta t}\right)(\mathbf{x}) = \omega. \tag{15.36}$$

This definition of the phase by using the Koopman eigenfunction coincides with that of the asymptotic phase in the conventional phase reduction framework given in the previous section. Therefore, the level sets of $\arg(\phi_{\lambda_1})$ provide the isochrons. Analogously, let us introduce amplitudes of the system state $\mathbf{x}$ by $r_i(\mathbf{x}) := \phi_{\lambda_i}(\mathbf{x})$ $(i = 2, \ldots, N)$. Each $r_i$ then obeys a linear equation,

$$\dot{r}_i(\mathbf{x}) = (Lr_i)(\mathbf{x}) = \lambda_i r_i(\mathbf{x}). \tag{15.37}$$

The level sets of $r_i$ are called isostables. The isostables, associated with one of the exponents $\lambda_i$, foliate the basin of attraction of the limit cycle, and each leaf of this foliation provides a level set of the amplitude associated with this exponent. From Eq. (15.37), we can see that initial conditions on the same isostable share the same decay rate and show the same asymptotic convergence toward the limit cycle.

Note that the linear form (15.36), (15.37), which is valid in the entire basin of attraction [43], is not necessarily derived by the perturbative power series approach based on the *Poincaré-Dulac normal form theory* and its extensions [19, 22, 72, 75, 94]. Therefore, we do not need to assume the non-resonance condition usually required for a complete linearization in the Poincaré-Dulac type scheme. See Sect. 3.2 of Lan and Mezić's work [43] for an example with resonance that can be linearized by using the Koopman eigenfunctions including nonanalytic (trans)monomials.

The phase and amplitudes defined above evolve independently under linear time-invariant dynamics. Thus, they provide simpler description of the dynamics around the limit cycle than those introduced using the moving orthonormal frames [4, 25, 93], in which different dynamical components interact nonlinearly. Also, we do not need to assume that the oscillatory dynamics are close to a bifurcation point in order to obtain the simple form.

### 15.3.2 Phase-Amplitude Reduction

Suppose now that a tiny perturbation $\varepsilon \mathbf{p}(t)$, where $0 < \varepsilon \ll 1$ characterizes its magnitude, is introduced to the oscillator (15.1) as

$$\dot{\mathbf{x}} = \mathbf{F}(\mathbf{x}) + \varepsilon \mathbf{p}(t). \tag{15.38}$$

We denote a coordinate transformation $\mathbf{x} \mapsto \Theta$ from the system state to the phase and amplitudes by $\mathbf{x} = \mathbf{h}(\Theta)$, where $\Theta := (\theta, r_2, \ldots, r_N)^{\dagger}$. In this phase-amplitude coordinate, the perturbed system (15.38) takes the following form:

$$\dot{\theta} = \omega + \varepsilon \nabla \theta(\mathbf{h}(\Theta)) \cdot \mathbf{p}(t), \tag{15.39}$$

$$\dot{r_i} = \lambda_i r_i + \varepsilon \nabla r_i(\mathbf{h}(\Theta)) \cdot \mathbf{p}(t), \quad (i = 2, \ldots, N). \tag{15.40}$$

These equations are exact and provide a semi-linear representation of the system dynamics.

Though the above phase-amplitude equations are exact, the phase and amplitude variables still interact with each other. Also, in order to actually use these equations for the analysis of the system dynamics, we need to know all the principal Koopman eigenfunctions beforehand, which is generally not easy. However, if we are interested only in the effect of the perturbation on the system dynamics along a certain transient orbit that is converging to the limit cycle, we may linearize the above equations around this orbit and derive simpler expressions, which do not require the whole principal eigenfunctions but only the *sensitivity functions* of phase and amplitudes along this orbit.

Consider a solution $\chi^* : \mathbf{x}^*(t)$ of the unperturbed system (15.1) with an initial condition $\mathbf{x}^*(0)$ taken arbitrarily in the basin of attraction $B$. We can then expand the gradients using the Lipschitz continuity as $\nabla \theta(\mathbf{h}(\Theta)) = \nabla \theta(\mathbf{x}^*(t)) + O(\varepsilon)$ and $\nabla r_i(\mathbf{h}(\Theta)) = \nabla r_i(\mathbf{x}^*(t)) + O(\varepsilon)$ in Eqs. (15.39), (15.40). Thus, we can approximate Eqs. (15.39), (15.40) as

$$\dot{\theta} = \omega + \varepsilon \nabla \theta(\mathbf{x}^*(t)) \cdot \mathbf{p}(t), \tag{15.41}$$

$$\dot{r_i} = \lambda_i r_i + \varepsilon \nabla r_i(\mathbf{x}^*(t)) \cdot \mathbf{p}(t), \quad (i = 2, \ldots, N), \tag{15.42}$$

by neglecting the terms of $O(\varepsilon^2)$. Note that only the gradients $\theta(\mathbf{x}^*(t))$ and $r_i(\mathbf{x}^*(t))$ along $\mathbf{x}^*(t)$, not the whole functions $\theta$ and $r_i$, are needed. These equations are completely decoupled from each other and we can adopt combinations of these $N$ equations (15.41), (15.42) as a reduced form of the system dynamics in the close-enough neighborhood of the transient orbit $\chi^*$. In many cases, only the first $J$ equations of (15.41), (15.42) for some $J(\ll N)$ are of interest to us, because they describe relatively persistent, slowly decaying modes.

We hereafter call the gradients $\nabla\theta(\mathbf{x}^*(t))$ and $\nabla r_i(\mathbf{x}^*(t))$ the sensitivity functions of the phase and amplitudes, because they characterize linear response properties of the phase and amplitude variables to the tiny perturbation $\varepsilon\mathbf{p}(t)$ given at $\mathbf{x}^*(t)$. These are the fundamental quantities in the present phase-amplitude reduction framework. Note that $\nabla\theta(\mathbf{x}^*(t))$ is equivalent to the phase sensitivity function $\mathbf{Z}(\theta)$ in the previous section if the initial condition $\mathbf{x}^*(0)$ is taken on the limit cycle $\chi$. Thus, the above result is a generalization of the classical phase reduction framework to transient orbits. In the following subsections, we discuss the properties of these sensitivity functions and numerical methods to evaluate them.

Before proceeding, let us briefly discuss the validity of the approximate phase-amplitude equations derived above. Let $\chi_{\mathrm{p}}^* : \mathbf{x}_{\mathrm{p}}^*(t)$ be a solution of the perturbed system (15.38) with the same initial condition $\mathbf{x}_{\mathrm{p}}^*(0) = \mathbf{x}^*(0)$ as the unperturbed system. As is known in the regular perturbation theory [27, 72], we can show by the *Grönwall–Bellman inequality* that the magnitude of the error $\|\mathbf{x}_{\mathrm{p}}^*(t) - \mathbf{x}^*(t)\|$ is bounded by $b\varepsilon(e^{at} - 1)/a$, where $a$ and $b$ are positive constants. This means that $\mathbf{x}_{\mathrm{p}}^*(t)$ is in a ball of radius $\varepsilon$ centered at $\mathbf{x}^*(t)$ within a finite time interval of length $O(1)$. This does not imply the breakdown of the continuous dependence of the solutions on $\varepsilon$ within a specific, fixed finite time interval (as long as the unperturbed solution exists on the entire half line, which is the case here). In fact, once we fix an arbitrary large finite length interval $[0, T_{\mathrm{f}}]$, we can assume that $\mathbf{x}_{\mathrm{p}}^*(t)$ is in a ball of radius $\varepsilon$ centered at $\mathbf{x}^*(t)$ on this interval by taking appropriately small $\varepsilon$, because $T_{\mathrm{f}}$ is independent of $\varepsilon$. The fact that the length of this interval is $O(1)$ means that the convergence of $\mathbf{x}_{\mathrm{p}}^*(t)$ to $\mathbf{x}^*(t)$ is *nonuniform* on an $\varepsilon$-dependent interval $[0, \varepsilon^c)$ for any $c < 0$, i.e., the limiting passages $t \to \varepsilon^c$ and $\varepsilon \to +0$ cannot be interchanged. This does not affect our analysis here, because no asymptotic properties of the perturbed dynamics are discussed. Thus, the approximate phase-amplitude Eqs. (15.41), (15.42) are valid in this interval.

Note that we can also analyze the asymptotic properties of the dynamics using the perturbed system of the phase-amplitude Eqs. (15.39), (15.40) under a suitable setting. For example, in the work of Wilson and Ermentrout [98], entrainment of an oscillator to a periodic input of large magnitudes is discussed using the approximated phase-amplitude equations, which are expanded around the limit cycle orbit, not around a transient orbit.

### *15.3.3 Properties of the Sensitivity Functions*

In this subsection, we discuss fundamental properties of the sensitivity functions of the phase and amplitudes, which can be seen as an extension of the covariant properties of the Lyapunov vectors defined on attractors, and derive the adjoint equations describing the sensitivity functions. These properties will be used to develop numerical methods for computing the sensitivity functions in the next subsection.

First, we evaluate the gradients of the phase and amplitudes on the periodic orbit $\chi$. Consider an initial condition slightly deviated from $\chi$, $\mathbf{h}_\mathrm{p} \equiv \mathbf{h}(\Theta_1) + \delta\mathbf{x}$, where we defined $\Theta_1 = (\theta, 0, \ldots, 0)^\dagger$. Then, from Eq. (15.37), each amplitude $r_i$ $(i = 2, \ldots, N)$ of the system state obeys

$$(U^T r_i)(\mathbf{h}_\mathrm{p}) = e^{\lambda_i T} r_i(\mathbf{h}(\Theta_1) + \delta\mathbf{x}) \tag{15.43}$$

when the perturbation is absent. Using the time-$T$ flow, we can also express $(U^T r_i)(\mathbf{h}_\mathrm{p})$ by using the monodromy matrix as

$$(U^T r_i)(\mathbf{h}_\mathrm{p}) = r_i(\mathbf{h}(\Theta_1) + \mathbf{M}(\mathbf{h}(\Theta_1))\delta\mathbf{x} + O(\|\delta\mathbf{x}\|^2)). \tag{15.44}$$

Equating the RHSs of Eqs. (15.43), (15.44), Taylor expanding $r_i$ around $\mathbf{h}(\Theta_1)$, considering that $r_i(\mathbf{h}(\Theta_1)) = 0$ and that the direction of $\delta\mathbf{x}$ is arbitrary, and finally taking the limit $\|\delta\mathbf{x}\| \to 0$, we find that

$$\nabla r_i^\dagger(\mathbf{h}(\Theta_1))\mathbf{M}(\mathbf{h}(\Theta_1)) = e^{\lambda_i T}\nabla r_i^\dagger(\mathbf{h}(\Theta_1)). \tag{15.45}$$

Similarly, for the phase $\theta$, we obtain

$$\nabla\theta^\dagger(\mathbf{h}(\Theta_1))\mathbf{M}(\mathbf{h}(\Theta_1)) = \nabla\theta^\dagger(\mathbf{h}(\Theta_1)). \tag{15.46}$$

Thus, the sensitivity functions, which are the gradient vectors of the phase and amplitudes evaluated on the limit cycle orbit $\chi$, are left eigenvectors of the monodromy matrix, which are called the adjoint covariant Lyapunov vectors [39, 68, 90]. These vectors can be numerically obtained by the *QR-decomposition*-based methods [39, 68] or by the *spectral dichotomy* approaches [18, 28].

Next, we seek the equations for the gradients of the phase and amplitudes on the transient orbit $\chi^* : \mathbf{x}^*(t)$. Here, we introduce logarithmic amplitudes $\eta_i(\mathbf{x}) := \ln(|r_i(\mathbf{x})|)$ $(i = 2, \ldots, N)$ in order to make the following treatment of the gradients simple and parallel with the standard arguments in the conventional phase reduction framework. For convenience of notation, let $\eta_1(\mathbf{x}) = \theta(\mathbf{x})$. We evaluate the gradient vectors of $\eta_i$, whose directions coincide with those of $\theta$ and $r_i$. The gradients $\nabla\theta$ and $\nabla r_i$ can be calculated from $\nabla\eta_i$ by rescaling, where the following normalization conditions should be satisfied:

$$\nabla r_i(\mathbf{x}^*(t)) \cdot \mathbf{F}(\mathbf{x}^*(t)) = \lambda_i r_i, \tag{15.47}$$

$$\nabla \theta(\mathbf{x}^*(t)) \cdot \mathbf{F}(\mathbf{x}^*(t)) = \omega. \tag{15.48}$$

These normalization conditions are equivalent to Eqs. (15.36), (15.37).

We can derive the adjoint equations for the gradients by using the same argument as the conventional derivation of the adjoint equation Eq. (15.15) for the phase sensitivity function. Consider an infinitesimal error $\delta\mathbf{x}(0)$ introduced at $t = 0$ between two unperturbed solutions $\mathbf{x}^*(t) + \delta\mathbf{x}(t)$ and $\mathbf{x}^*(t)$, which satisfies the *variational equation* [22, 72, 75] $\mathrm{d}(\delta\mathbf{x}(t))/\mathrm{d}t = \mathbf{DF}(\mathbf{x}^*(t))\delta\mathbf{x}(t)$. Because each logarithmic amplitude $\eta_i$ increases constantly as $\dot{\eta}_i(\mathbf{x}(t)) = \nabla\eta_i(\mathbf{x}(t)) \cdot \dot{\mathbf{x}}(t) = \lambda_i$ in the absence of perturbation, the error in this logarithmic amplitude coordinate $\eta_i(\mathbf{x}^*(t) + \delta\mathbf{x}(t)) - \eta_i(\mathbf{x}^*(t)) = \nabla\eta_i(\mathbf{x}^*(t)) \cdot \delta\mathbf{x}(t)$ should be independent of time, i.e., $\mathrm{d}(\nabla\eta_i(\mathbf{x}^*(t)) \cdot \delta\mathbf{x}(t))/\mathrm{d}t = 0$. This yields

$$\begin{aligned}
\frac{\mathrm{d}\nabla\eta_i(\mathbf{x}^*(t))}{\mathrm{d}t} \cdot \delta\mathbf{x}(t) &= -\nabla\eta_i(\mathbf{x}^*(t)) \cdot \frac{\mathrm{d}(\delta\mathbf{x}(t))}{\mathrm{d}t} \\
&= -\nabla\eta_i(\mathbf{x}^*(t)) \cdot \mathbf{DF}(\mathbf{x}^*(t))\delta\mathbf{x}(t) \\
&= -\mathbf{DF}^{\dagger}(\mathbf{x}^*(t))\nabla\eta_i(\mathbf{x}^*(t)) \cdot \delta\mathbf{x}(t).
\end{aligned} \tag{15.49}$$

Here, we used the variational equation and the definition of the adjoint matrix. We can take $N$ linearly independent initial errors $\delta\mathbf{x}_i(0) = \varepsilon'\mathbf{e}_i$, where $0 < \varepsilon' \ll 1$ and $\mathbf{e}_i$ is the $i$th unit vector, and define the *fundamental solution matrix* $\Xi(t)$ of the variational equation as $\Xi(t) = (\delta\mathbf{x}_1(t), \delta\mathbf{x}_2(t), \ldots, \delta\mathbf{x}_N(t))$. The sign of the determinant of the fundamental solution matrix, i.e., the *Wronskian*, is time-invariant due to *Liouville's trace formula* [9, 94]. Because $\det(\Xi(0)) = (\varepsilon')^N > 0$, we obtain $\det(\Xi(t)) > 0$ for all $t$, and thus the fundamental solution matrix is always invertible. From a matrix form of Eq. (15.49), $\mathrm{d}(\nabla\eta_i(\mathbf{x}^*(t))/\mathrm{d}t)\Xi(t) = -\mathbf{DF}^{\dagger}(\mathbf{x}^*(t))\nabla\eta_i(\mathbf{x}^*(t))\Xi(t)$, we can eliminate $\Xi(t)$ by multiplying its inverse from the right on both sides of this equation and derive the adjoint equation,

$$\frac{\mathrm{d}\nabla\eta_i(\mathbf{x}^*(t))}{\mathrm{d}t} = -\mathbf{DF}^{\dagger}(\mathbf{x}^*(t))\nabla\eta_i(\mathbf{x}^*(t)), \tag{15.50}$$

which should be satisfied by $\nabla\eta_i(\mathbf{x}^*(t))$.

Note that this equation should be solved with an appropriate end condition. We can approximately take this end condition as $\nabla\eta_i(\mathbf{x}^*(\tau)) \parallel \nabla r_i(\mathbf{h}(\Theta_1))|_{\theta=\theta_*}$ for some $t = \tau$, where $\theta_* \simeq \theta(\mathbf{x}^*(\tau))$ and the symbol $\parallel$ between the two vectors indicates that they are parallel to each other, because the gradient field $\nabla r_i(\mathbf{x})$ is continuous and the transient orbit eventually converges to the limit cycle.

Let us consider the adjoint tangent *propagator* $\mathscr{G}(t_1, t_2) := \mathbf{N}(t_2)\mathbf{N}^{-1}(t_1)$, where $\mathbf{N}(t)$ is a fundamental solution matrix of the linear system given by Eq. (15.50), which maps $\nabla\eta_i(\mathbf{x}^*(t_1))$ to $\nabla\eta_i(\mathbf{x}^*(t_2))$. Since $\nabla\theta(\mathbf{x}^*(t_2)) \parallel \mathscr{G}(t_1, t_2)\nabla\theta(\mathbf{x}^*(t_1))$ and $\nabla r_i(\mathbf{x}^*(t_2)) \parallel \mathscr{G}(t_1, t_2)\nabla r_i(\mathbf{x}^*(t_1))$ hold, the gradient vectors of the phase and amplitudes are covariant with respect to the action of the propagator $\mathscr{G}$. Thus, they can be interpreted as an extension of the adjoint covariant Lyapunov vectors to transient regimes. Note that the adjoint covariant Lyapunov vectors evaluated on the limit cycle, given by Eqs. (15.45, 15.46), are covariant with respect to the action of the adjoint of the monodromy matrix, which is the one period (time-$T$) propagator.

### 15.3.4 *Numerical Methods for the Sensitivity Functions*

In this subsection, we discuss numerical methods to calculate the sensitivity functions using the results of the previous subsection.

#### 15.3.4.1 Sensitivity Functions and Bi-orthogonal Vectors

In the numerical evaluation of the phase sensitivity function, $\nabla\theta$ (or $\nabla\eta_1$), the standard method is to integrate the adjoint equation backward-in-time, while renormalizing $\nabla\theta$ occasionally so that the normalization condition (15.48) is satisfied [17]. This is because $\nabla\theta$ corresponds to the neutrally stable component ($\mathrm{Re}(\lambda_1) = 0$) of the linearized system while the other components have negative growth rates ($\lambda_{2,\ldots,N} < 0$).

In the present case, however, such a naive backward integration does not provide correct results for the sensitivity functions for the amplitudes, $\nabla\eta_{2,\ldots,N}$, because the vector components arising from tiny numerical errors in the relatively (backward-in-time) unstable covariant subspaces grow and accumulate. We thus need to develop a method to eliminate such unnecessary components for the correct evaluation of the sensitivity functions. Note that the standard QR-decomposition-based methods [39, 68] for obtaining the covariant subspace require the ergodicity of the underlying dynamical process, and hence they cannot be directly applied to the transient orbits far from attractors. Note also that the spectral dichotomy techniques [18, 28] for evaluating them may not work appropriately near the left boundary of the time evolution (See Sects. 2.6 and 2.7 of Hüls's work [28]).

In order to develop a numerical method, we introduce dual vectors $\boldsymbol{\gamma}_i$ of $\nabla\eta_i$, which satisfy the bi-orthogonal relation,

$$\boldsymbol{\gamma}_i(\mathbf{x}^*(t)) \cdot \nabla\eta_j(\mathbf{x}^*(t)) = \delta_{ij}, \tag{15.51}$$

where $\delta_{ij}$ is the Kronecker's delta. By using $\boldsymbol{\gamma}_i(\mathbf{x}^*(t))$, we can subtract the vector component in the covariant subspace $\nabla\eta_i(\mathbf{x}^*(t))$ from the solution $\mathbf{z}(t)$ of the linear system of the form Eq. (15.50) by projecting $z(t)$ onto this subspace as

$$(\boldsymbol{\gamma}_i(\mathbf{x}^*(t)) \cdot \mathbf{z}(t))\nabla\eta_i(\mathbf{x}^*(t)). \tag{15.52}$$

Differentiating Eq. (15.51) by $t$, we obtain

$$(\dot{\boldsymbol{\gamma}}_i(\mathbf{x}^*(t)) - \mathrm{D}\mathbf{F}(\mathbf{x}^*(t))\boldsymbol{\gamma}_i(\mathbf{x}^*(t))) \cdot \nabla\eta_j(\mathbf{x}^*(t)) = 0. \tag{15.53}$$

The sign of the Wronskian of Eq. (15.50) is also time-invariant due to Liouville's trace formula. By using this fact and the linear independence of the left eigenvectors of the monodromy matrix, we can show the linear independence of $\{\nabla\eta_i(\mathbf{x})\}_{i=1}^N$ at every point $\mathbf{x}$ in the whole basin of attraction $B$. Thus, we obtain

$$\dot{\boldsymbol{\gamma}}_i(\mathbf{x}^*(t)) = \mathbf{DF}(\mathbf{x}^*(t))\boldsymbol{\gamma}_i(\mathbf{x}^*(t)). \tag{15.54}$$

The vectors $\{\boldsymbol{\gamma}_i\}_{i=1}^N$ are covariant with respect to the action of the propagator $\mathscr{F}(= (\mathscr{G}^\dagger)^{-1})$ of the linear system (15.54), and hence they can be seen as covariant Lyapunov vectors extended to transient regimes. The relative stability relation of the covariant subspace of Eq. (15.54) forward-in-time coincides with that of Eq. (15.50) backward-in-time. In order to subtract the unstable components using the projection (15.52), the system (15.54) should be solved forward-in-time with an approximate initial condition $\boldsymbol{\gamma}_i(\mathbf{x}^*(0))$. The vectors $\{\nabla\eta_i(\mathbf{x}^*(0))\}_{i=1}^N$ can be approximated by direct numerical simulation of the dynamics using the *Fourier average* and the *generalized Laplace average* [49, 54] (See appendix for details). Then, $\boldsymbol{\gamma}_i(\mathbf{x}^*(0))$ can be obtained by using the bi-orthogonality relation (15.51).

#### 15.3.4.2 Bi-orthogonalization Method

We here introduce a bi-orthogonalization method to obtain the sensitivity functions of the phase and amplitudes up to the $J$th unstable mode. The procedure is as follows:

(a) Evaluate the adjoint Lyapunov vectors on the limit cycle $\chi$ and the Floquet exponents.
(b) Calculate $\{\boldsymbol{\gamma}_i(\mathbf{x}^*(0))\}_{i=1}^J$ from $\{\nabla\eta_i(\mathbf{x}^*(0))\}_{i=1}^N$ by using the bi-orthogonality relation (15.51).
(c) Obtain $\nabla\eta_1(\mathbf{x}^*(t))$ by backward integration of Eq. (15.50).
(d) Obtain $\boldsymbol{\gamma}_1(\mathbf{x}^*(t))$ by forward integration of Eq. (15.54).
(e) Obtain $\nabla\eta_2(\mathbf{x}^*(t))$ by backward integration of Eq. (15.50) while subtracting relatively unstable component $\nabla\eta_1(\mathbf{x}^*(t))$ by using the projection (15.52).
(f) Obtain $\boldsymbol{\gamma}_2(\mathbf{x}^*(t))$ by forward integration of Eq. (15.54) while subtracting relatively unstable component $\boldsymbol{\gamma}_1(\mathbf{x}^*(t))$ by using the projection

$$(\nabla\eta_i(\mathbf{x}^*(t)) \cdot \boldsymbol{\xi}(t))\boldsymbol{\gamma}_i(\mathbf{x}^*(t)), \tag{15.55}$$

where $\boldsymbol{\xi}(t)$ is a solution of the linear system of the form Eq. (15.54).
(g) Perform (e) and (f) consecutively to obtain $\{\nabla\eta_i(\mathbf{x}^*(t))\}_{i=3}^J$ and $\{\boldsymbol{\gamma}_i(\mathbf{x}^*(t))\}_{i=3}^J$ (note that all relatively unstable modes should be subtracted during integration).
(h) Obtain $\nabla\theta$ and $\nabla r_i$ $(i = 2, \ldots, J)$ using the normalization conditions (15.47, 15.48), where $r_i(\mathbf{x}^*(t))$ on the transient orbit $\chi^*$ is evaluated using Eq. (15.37) with the initial condition $r_i(\mathbf{x}^*(0))$, which is calculated in (b) by the direct numerical simulation.

This method has a significant computational advantage in evaluating the sensitivity functions. To calculate the sensitivity functions $\{\nabla \eta_i\}_{i=1}^{J}$ at $m$ points on the transient orbit $\chi^*$, it is necessary to repeat a long-time evolution of the system $mJ(N+1)$ times if we evaluate them by direct numerical integration. In contrast, we need only $J(N+1)+2J$ long-time evolutions in the bi-orthogonalization method. The above numerical method was introduced and discussed in [78].

#### 15.3.4.3 Simplified Bi-orthogonalization Method

Since the generalized Laplace average involves unstable computations, evaluation of the complete set of the basis vectors $\{\nabla \eta_i(\mathbf{x}^*(0))\}_{i=1}^{N}$ in the numerical procedure (b) above can be a very delicate task. Here, we discuss a slightly simplified version of the bi-orthogonalization method, which uses only an incomplete set of the basis vectors $\{\nabla \eta_i(\mathbf{x}^*(0))\}_{i=1}^{J}$. This is particularly useful when $\mathbf{F}$ is a nonresonant analytic vector field and the $J$th Floquet exponent satisfies $\lambda_J > 2\lambda_2$, because we can use a simpler version of the Laplace average in this case (see appendix). In this method, we make use of a set of vectors $\{\tilde{\boldsymbol{\gamma}}_i(\mathbf{x}^*(0))\}_{i=1}^{J}$ given by

$$
\begin{aligned}
&(\tilde{\boldsymbol{\gamma}}_1(\mathbf{x}^*(0)), \tilde{\boldsymbol{\gamma}}_2(\mathbf{x}^*(0)), \ldots, \tilde{\boldsymbol{\gamma}}_J(\mathbf{x}^*(0)))^\dagger \\
=&(\nabla \eta_1(\mathbf{x}^*(0)), \nabla \eta_2(\mathbf{x}^*(0)), \ldots, \nabla \eta_J(\mathbf{x}^*(0)))^+,
\end{aligned} \tag{15.56}
$$

where + indicates the Moore-Penrose generalized inverse. We can use $\{\tilde{\boldsymbol{\gamma}}_i(\mathbf{x}^*(0))\}_{i=1}^{J}$ and the solution of the linear system of the form Eq. (15.54) from these initial conditions for the bi-orthogonalization procedure in Sect. 15.3.4.2 instead of $\{\boldsymbol{\gamma}_i\}_{i=1}^{J}$.

Let us briefly explain how this works. Consider the case where we are estimating $\nabla \eta_{k+1}(\mathbf{x}^*(t))$. It is sufficient to show that we can successfully subtract the relatively unstable modes by using the projection (15.52), i.e., $\tilde{\boldsymbol{\gamma}}_i(\mathbf{x}^*(t)) \cdot \mathbf{z}(t)$ well approximates $\boldsymbol{\gamma}_i(\mathbf{x}^*(t)) \cdot \mathbf{z}(t)$ for $i = 1, \ldots, k$. We use the fact that $\tilde{\boldsymbol{\gamma}}_i(\mathbf{x}^*(t))$ can be written as $\boldsymbol{\gamma}_i(\mathbf{x}^*(t)) + \mathbf{R}_i(t)$, where $\mathbf{R}_l(t)$ is a linear combination of $\{\boldsymbol{\gamma}_j(\mathbf{x}^*(t))\}_{j=J+1}^{N}$ (this can be shown by recursion). In the backward integration of the adjoint system (15.50), the vector components of the solution $\mathbf{z}(t)$ in the directions of $\{\nabla \eta_j(\mathbf{x}^*(t))\}_{j=J+1}^{N}$ decay faster than the first $k$ components in the directions of $\{\nabla \eta_j(\mathbf{x}^*(t))\}_{j=1}^{k}$. Therefore, $\mathbf{R}_i \cdot \mathbf{z}(t)$ gives only a tiny contribution to $\tilde{\boldsymbol{\gamma}}_i(\mathbf{x}^*(t)) \cdot \mathbf{z}(t)$. Thus, $\tilde{\boldsymbol{\gamma}}_i(\mathbf{x}^*(t)) \cdot \mathbf{z}(t)$ well approximates $\boldsymbol{\gamma}_i(\mathbf{x}^*(t)) \cdot \mathbf{z}(t)$.

## 15.4 Example

In this section, we analyze an analytically tractable model of a limit cycle oscillator to illustrate the phase-amplitude reduction framework. As an application, we also consider an optimal control problem and solve it by using the proposed reduction framework.

### 15.4.1 Model

We consider a limit cycle oscillator described by a three-dimensional state $\mathbf{x} = (x, y, z)^\dagger \in \mathbb{R}^3 \setminus \{0\} \times \{0\} \times \mathbb{R}$, which obeys the following dynamics:

$$\begin{cases} \dot{x} &= x - ay - (x^2+y^2+z^2)(x-by) + \dfrac{cxz^2}{\sqrt{x^2+y^2}\sqrt{x^2+y^2+z^2}}, \\ \dot{y} &= ax + y - (x^2+y^2+z^2)(bx+y) + \dfrac{cyz^2}{\sqrt{x^2+y^2}\sqrt{x^2+y^2+z^2}}, \\ \dot{z} &= \left(1 - (x^2+y^2+z^2) - c\dfrac{\sqrt{x^2+y^2}}{\sqrt{x^2+y^2+z^2}}\right) z, \end{cases} \tag{15.57}$$

where $a, b$, and $c$ are real parameters. We hereafter assume $c > 2$. This model is a generalization of the Stuart–Landau oscillator and has a stable limit cycle described as $\mathbf{x}_0(\theta) = (\cos\theta, \sin\theta, 0)^\dagger$. In a spherical coordinate system $(r_s, \theta_s, \phi_s)$, this model is written as

$$\dot{r}_s = r_s - r_s^3, \quad \dot{\theta}_s = c\cos\theta_s, \quad \dot{\phi}_s = a - br_s^2. \tag{15.58}$$

The initial value problem for Eq. (15.58) can be solved as

$$\begin{cases} r_s(t) = \left(1 + \frac{1-r_s(0)^2}{r_s(0)^2} e^{-2t}\right)^{-1/2}, \\ \theta_s(t) = 2\arctan\left(\tanh\left(\frac{1}{2}(ct+k_0)\right)\right), \\ \phi_s(t) = \phi_s(0) + (a-b)t - \frac{1}{2}\ln\left(r_s(0)^2 + (1-r_s(0)^2)e^{-2t}\right), \end{cases} \tag{15.59}$$

where $k_0 = 2\,\mathrm{arctanh}\,(\tan(\theta_s(0)/2))$.

Considering that the dynamics of $(r_s, \phi_s)$ are decoupled from $\theta_s$ and obey the Stuart–Landau equation, the principal eigenvalues of the infinitesimal generator of the Koopman operator semigroup are obtained as $\lambda_1 = \mathrm{i}(a-b)$, $\lambda_2 = -2$ and $\lambda_3 = -c$. The associated principal Koopman eigenfunctions are given by

$$\phi_{\lambda_1} = e^{\mathrm{i}(\phi_s - b\ln r_s)}, \quad \phi_{\lambda_2} = \frac{r_s^2 - 1}{r_s^2}, \quad \phi_{\lambda_3} = \frac{1 - \tan(\theta_s/2)}{1 + \tan(\theta_s/2)}. \tag{15.60}$$

Therefore, the isochrons are given by $\phi_s - b\ln r_s = \text{const.}$, and the isostables are given by $r_s = \text{const.}$ and $\theta_s = \text{const.}$ Fig. 15.4 shows the isochrons and isostables of this model. The isochrons are the logarithmic spiral surfaces, the first isostables are the concentric spheres, and the second isostables are the coaxial cones.

In the following, we consider the dynamics around a transient orbit $\mathbf{x}^*(t)$ starting from an initial condition $\mathbf{x}^*(0) = (x^*(0), y^*(0), z^*(0))^\dagger = (2.0, 1.5, 1.0)^\dagger$ and reduce it to a $J = 2$ dimensional system. See Fig. 15.5a for the time series of $\mathbf{x}^*(t)$.

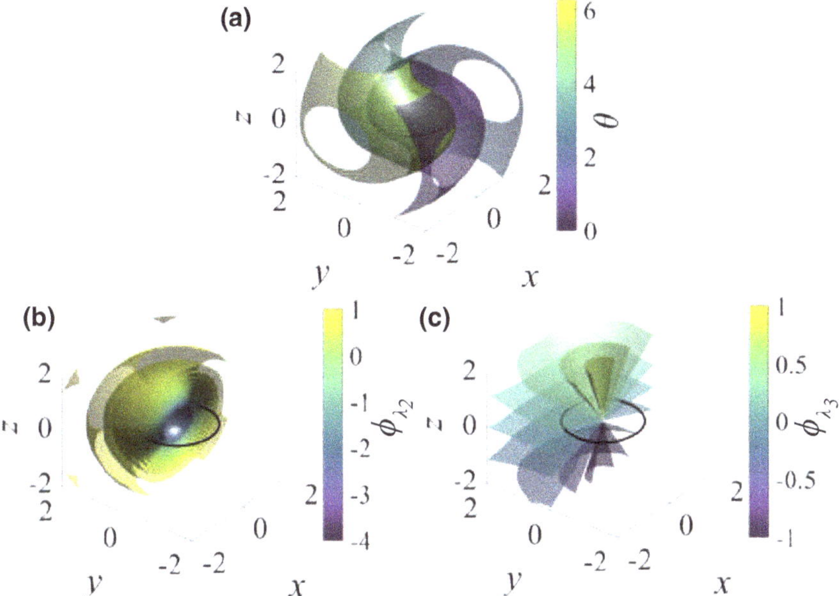


**Fig. 15.4** Isochrons and isostables of the analytically tractable model (15.57). The black solid lines show the periodic solution. **a** Isochrons corresponding to $\theta \in \{\pi/4, 3\pi/4, 5\pi/4, 7\pi/4\}$. **b**, **c** Isostables of the model in the $y \geq 0$ domain. Both of them are symmetric with respect to the $xz$-plane

## 15.4.2 Sensitivity Functions of the Phase and Amplitude

From the first two principal Koopman eigenfunctions in Eq. (15.60), the gradients of the phase $\theta = \arg(\phi_{\lambda_1})$ and the first amplitude $r_2 = \phi_{\lambda_2}$ are analytically given by

$$\nabla\theta(\mathbf{x}) = -\frac{b}{x^2 + y^2 + z^2}(x, y, z)^{\dagger} - \frac{1}{x^2 + y^2}(y, -x, 0)^{\dagger}, \tag{15.61}$$

$$\nabla r_2(\mathbf{x}) = \frac{2}{(x^2 + y^2 + z^2)^2}(x, y, z)^{\dagger}. \tag{15.62}$$

To test the validity of the numerical algorithm, we also evaluated the phase-amplitude sensitivity functions by using the simplified bi-orthogonalization method in Sect. 15.3.4.3. Here, in order to consider a practical situation, we numerically analyzed Eq. (15.57), not Eq. (15.58), assuming that the symmetries of the system are unknown. We set the end time as $\tau = 3.4$ for the backward integration of the adjoint equation (15.50).

In Fig. 15.5b, c, the phase-amplitude sensitivity functions $\nabla\theta(\mathbf{x}^*(t))$, $\nabla r_2(\mathbf{x}^*(t))$ along the orbit $\mathbf{x}^*(t)$ obtained analytically from Eqs. (15.61), (15.62) are compared with the result of the simplified bi-orthogonalization method. We can confirm that

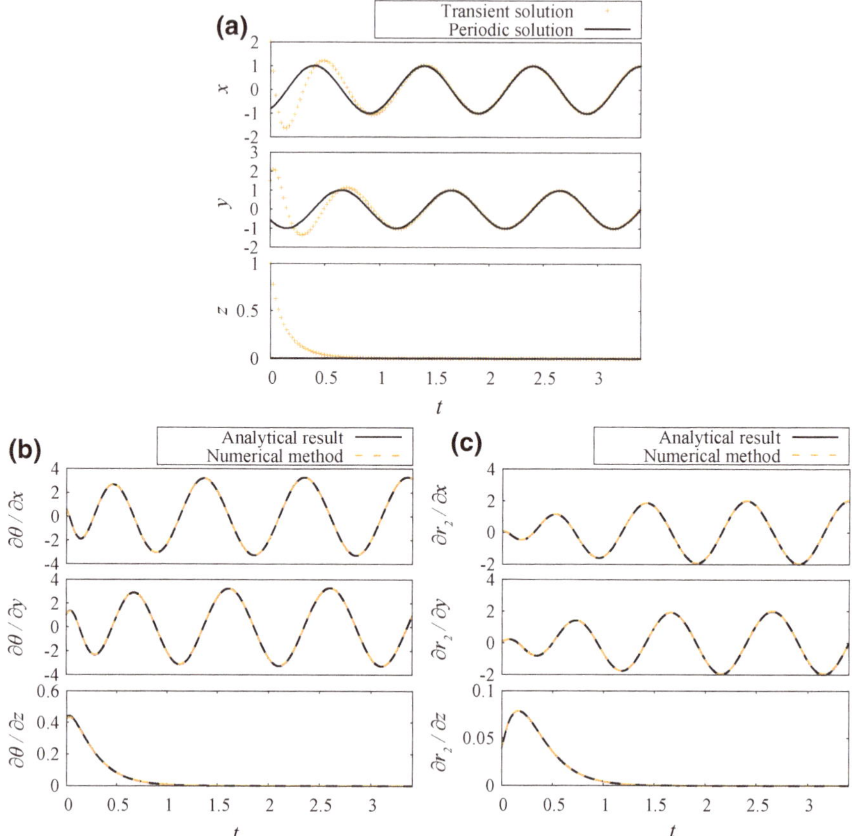


**Fig. 15.5** Phase-amplitude sensitivity functions of the analytically tractable model. **a** The stable periodic solution of the model (lines) and the transient solution $\mathbf{x}^*(t)$ (plus signs). **b** Three components of the phase sensitivity function $\nabla\theta(\mathbf{x}^*(t))$ obtained analytically (solid lines) and numerically by the simplified bi-orthogonalization method (dashed lines). **c** Three components of the sensitivity function of the second amplitude $\nabla r_2(\mathbf{x}^*(t))$ obtained analytically (solid lines) and numerically by the simplified bi-orthogonalization method (dashed lines)

the results agree well with each other and thus the proposed numerical algorithm works appropriately.

### *15.4.3 Optimal Control Based on the Phase-Amplitude Reduction*

In this subsection, we illustrate the utility of the reduced amplitude equation (15.42) by solving an optimal control problem regarding the transient dynamics converging to the limit cycle attractor.

When designing a control strategy that efficiently drives the system state toward an attractor, how to measure the deviation of the state from the attractor can be a problem. Distances introduced in the state space in ad hoc ways might not be satisfactory since they are not likely to reflect the nonlinear dynamical properties of the system. For example, consider a spike-like excursion of the system state from a stable equilibrium in an excitable dynamical system such as neurons [17]. If we use the Euclidean distance defined in the state space, we may conclude that the state corresponding to the tip of the spike is farther from the equilibrium than the state at the onset of the spike. However, this may not be suitable for our purpose, because the state at the tip of the spike relaxes to the equilibrium earlier than the state at the onset of the spike, and therefore the former state may be considered closer to the equilibrium than the latter. Thus, the Euclidean distance can fail to capture the nonlinear dynamics of the system.

The isostables associated with each decay rate provide a natural measure of the distance of the system state from the attractor, because they quantify the time necessary for the state to converge to the attractor, and also because they are uniquely defined in the entire basin of attraction. Moreover, by using the isostable coordinates, dynamics of the nonlinear system can be transformed to a simple linear form with reduced dimensions as described in the previous section.

An approach to the optimal control problem using the isostable coordinates was first formulated by Mauroy [53]. Control theories of nonlinear systems that are based on the Koopman eigenfunctions have been rapidly developing since then [1, 6, 29, 35, 81, 101, 104] (see also Chaps. 8, 9, 11, and 16). These results are useful in practice because they can be implemented in a data-driven way with the aid of the dynamic mode decomposition and thus can be applied to complex nonlinear systems in realistic settings without knowing their detailed mathematical models. It is also convenient that the isostable coordinates can be constructed without using the polynomial basis [21] (see also Chap. 4), in contrast to the conventional technique called the *Carleman linearization* [38]. Thus, the isostable coordinates enable us to obtain simple expressions of nonlinear dynamical systems by using only a small set of non-polynomial bases, which can also be evaluated in a data-driven way, for example, by using machine learning methods [44, 45, 47, 64, 87, 106].

Here, in the spirit of the pioneering study by Mauroy [53], we consider an optimal control problem on a fixed time horizon of minimizing the deviation of the system state from the limit cycle attractor at the end of the time horizon, which is measured in terms of the amplitude coordinates introduced by using the isostables.

We consider a stable limit cycling system subjected to a transient control input $\varepsilon \mathbf{p}(t)$, described by Eq. (15.38). We focus only on the amplitude $r_2$ of the system, because it is the most persistent component among all the amplitudes. As for the transient control input $\varepsilon \mathbf{p}(t)$, we assume a fixed waveform $\mathbf{w}(t)$ of a fixed duration $\tau_*$, i.e., $\mathbf{p}(t) = \mathbf{w}(t - s)$, where $\mathbf{w}(t)$ is nonzero only on $t \in [0, \tau_*]$ and the time $s$ determines the injection timing of the input. We denote the solution of the system subjected to the external control input by $\mathbf{x}_{\mathrm{p}}^*(t)$. The initial condition is taken arbitrarily in the basin of attraction of the limit cycle.

This control problem on a finite time horizon $[0, T_e]$ can be formulated as follows: find the injection timing $s_*$ such that

$$s_* = \arg\min_{s \in \mathscr{I}} |r_2(\mathbf{x}_p^*(T_e))|, \tag{15.63}$$

where $\mathscr{I} := [0, T_e - \tau_*]$. That is, we evaluate the optimal injection timing of a weak control signal to suppress the most persistent component $r_2$ of the amplitudes. When the magnitude of the input $\varepsilon$ is sufficiently small, the evolution of the amplitude $r_2$ is approximated by the reduced amplitude equation (15.42). Then, by using an analytical solution of the linear one-dimensional non-homogeneous differential equation (15.42) for $r_2$, the optimal control problem (15.63) can be approximated by a problem of finding $s_*$ such that

$$\mathrm{sgn}(r_2(\mathbf{x}^*(0))) \int_0^{T_e} \mathbf{p}(t) \cdot \nabla r_2(\mathbf{x}^*(t)) e^{\lambda_2(T_e - t)} \mathrm{d}t \tag{15.64}$$

is minimized. We analyze this approximate problem for the limit cycle oscillator described by Eq. (15.57) along the transient orbit $\mathbf{x}^*(t)$ given in Sect. 15.4.1. We set $T_e = 0.7$, $\tau_* = 0.035$, and $\mathbf{w}(t) = (1, 0, 0)^\dagger$.

Figure 15.6 shows the effect of the control input on the amplitude $r_2(\mathbf{x}_p^*(T_e))$ at time $T_e$. The result obtained by using the reduced amplitude equation (15.42) is compared with the result of direct numerical simulations, showing good agreement for the sufficiently weak input with $\varepsilon = 0.1$ or $1.0$. This verifies the validity of the approximate amplitude equation for the present model. The optimal injection timing of the control input can be theoretically predicted using Eq. (15.64), because it is essentially equivalent to solving the approximate amplitude equation (15.42) directly. In this case, the initial value of the amplitude is positive, i.e., $r_2(\mathbf{x}^*(0)) > 0$. Hence, the optimal injection timing $s_*$ of the control input can be evaluated by finding the minimum of the waveform in Fig. 15.6, which gives $s_* = 0.183$ in the present case. Note that when the magnitude of the control input is too large ($\varepsilon = 10.0$), the approximation by the amplitude equation (15.42) fails and the results deviate from each other considerably. Thus, when the control input is not too large, the phase-amplitude reduction framework is useful in analyzing the optimal control problem for limit cycle oscillators.

## 15.5 Concluding Remarks

In this chapter, we formulated a phase-amplitude reduction framework for stable limit cycling systems, which can be applied to transient dynamical regimes far from attractors in high-dimensional systems. We also developed a bi-orthogonalization method and its simplified variant for numerical evaluation of the sensitivity functions of the phase and amplitudes, which provide accurate results. As an application, we

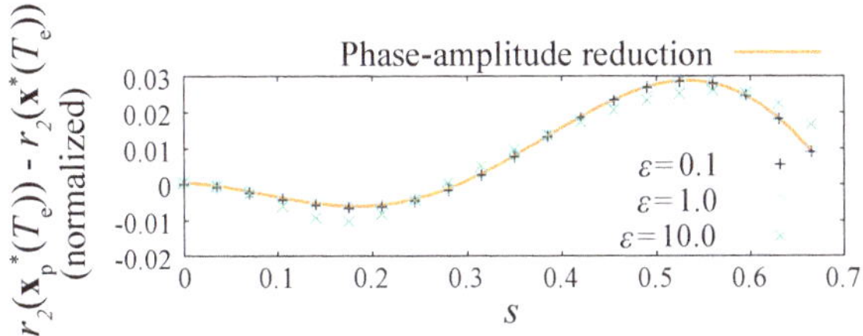


**Fig. 15.6** Optimal control problem for the limit cycle oscillator described by Eq. (15.57). Effect of the control input on the amplitude at a given time $r_2(\mathbf{x}_{\mathrm{p}}^*(T_{\mathrm{e}}))$, obtained by the reduced amplitude equation (line) and by the direct numerical simulations for 20 different injection timings (symbols) for three different magnitudes of the control input: $\varepsilon = 0.1$ (black plus symbols), $\varepsilon = 1.0$ (blue circles) and $\varepsilon = 10.0$ (green crosses). The results are normalized so that the $l_2$ norms of the waveforms are the same

considered a problem of finding the optimal injection timing for suppressing the amplitude (deviation) of the system from the limit cycle attractor, and showed that the reduced amplitude equation accurately predicts the optimal injection timing of the control input. The proposed theory would be useful in analyzing and controlling response properties of high-dimensional rhythmic systems.

In the optimization problem, we assumed that the external control input is sufficiently weak, because we considered the situation where only the sensitivity functions along a given transient orbit are given. This limitation can be removed when the information of the full isostables is available. For the dynamics around a stable equilibrium, an optimal control problem with a strong external control input is formulated and solved on the basis of the isostables in [53]. Such kind of studies would be of practical importance, e.g., in biological systems [59].

Recently, further extensions of the phase-amplitude reduction framework have been developed for piecewise-smooth systems [97] and for quasiperiodic systems [55]. Extending the present framework to other complex systems, such as stochastic, networked, time delayed, reaction–diffusion, and fluid systems, would be a promising future direction of the study.

**Acknowledgements** S. S. acknowledges financial support from Japan Society for the Promotion of Science (JSPS) KAKENHI Grant No. 18H06478. W. K. acknowledges financial support from JSPS KAKENHI Grants No. 16K16125 No. 17H03279. H. N. acknowledges financial support from JSPS KAKENHI Grants No. 16K13847, No. 17H03279, No. 18K03471, No. 18H03287, and JST CREST Grant No. JPMJCR1913.

## Appendix: Laplace Average

In this appendix, we propose a simpler version of the Laplace average [49, 54]. The following results also hold for complex Koopman eigenvalues $\lambda$. When $\mathbf{F}$ is a nonresonant analytic vector field, a generic scalar-valued observable function $f$ can be expressed as [55]

$$f(\mathbf{x}) = \sum_{k \in \mathbb{Z},\ \mathbf{m} \in \mathbb{Z}_{\geq 0}^{N-1}} a_{k,\mathbf{m}}(f)\phi_{\lambda_{k,\mathbf{m}}}(\mathbf{x}), \tag{15.65}$$

where $\mathbf{m} = (m_2, m_3, \ldots, m_N)$. The complex coefficients $a_{k,\mathbf{m}}(f) \in \mathbb{C}$ are called the Koopman modes, $\lambda_{k,\mathbf{m}} = k\lambda_1 + \sum_{i=2}^{N} m_i\lambda_i$ are the eigenvalues of the infinitesimal generator of the Koopman operator semigroup, and

$$\phi_{\lambda_{k,\mathbf{m}}}(\mathbf{x}) := e^{\mathrm{i}k\theta(\mathbf{x})}\phi_{\lambda_2}^{m_2}(\mathbf{x})\phi_{\lambda_3}^{m_3}(\mathbf{x}) \cdots \phi_{\lambda_N}^{m_N}(\mathbf{x}) \tag{15.66}$$

are their associated eigenfunctions. When the observable can be expanded as (15.65), we can evaluate the value of a Koopman eigenfunction at point $\mathbf{x}$ by using the generalized Laplace average [54] as

$$\begin{aligned} &a_{k,\mathbf{m}}(f)\phi_{\lambda_{k,\mathbf{m}}}(\mathbf{x}) \\ &= \lim_{s\to\infty} \frac{1}{s} \int_0^s \left[ f \circ \mathbf{S}(t,\mathbf{x}) - \sum_{\mathrm{Re}(\lambda_{k,\mathbf{m}'}) \geq \mathrm{Re}(\lambda_{k,\mathbf{m}})} a_{k,\mathbf{m}'}(f)\phi_{\lambda_{k,\mathbf{m}'}}(\mathbf{x})e^{\lambda_{k,\mathbf{m}'}t} \right] e^{-\lambda_{k,\mathbf{m}}t}\mathrm{d}t, \end{aligned} \tag{15.67}$$

where $\mathrm{Re}(\lambda)$ is the real part of $\lambda$.

If we evaluate the value of the $J$th principal Koopman eigenfunction associated with the Floquet exponent $\mathrm{Re}(\lambda_J)$ ($> 2\mathrm{Re}(\lambda_2)$), we can simplify the generalized Laplace average using an observable $g_J$ defined as

$$g_J(\mathbf{x}) = \nabla r_J(\mathbf{x}_0(\theta_*)) \cdot (\mathbf{x} - \mathbf{x}_0(\theta_*)), \tag{15.68}$$

where $\theta_* = \theta(\mathbf{x})$. Since this observable converges to zero along a transient orbit, we obtain $a_{k,\mathbf{0}}(g_J) = 0$ for all $k \in \mathbb{Z}$, where $\mathbf{0}$ is a zero vector. The gradient of $g_J$ evaluated on the periodic orbit $\chi$ is given by

$$\begin{aligned} \nabla g_J(\mathbf{x}_0(\theta_*)) &= (\mathbf{I} - \nabla\theta(\mathbf{x}_0(\theta_*))\mathbf{F}(\mathbf{x}_0(\theta_*))^\dagger)\nabla r_J(\mathbf{x}_0(\theta_*)) \\ &= \nabla r_J(\mathbf{x}_0(\theta_*)), \end{aligned} \tag{15.69}$$

where $\mathbf{I}$ is the identity matrix and we used the fact [41, Sect. 3.4] that $\mathbf{F}(\mathbf{x}_0(\theta_*))$ is parallel to $\boldsymbol{\gamma}_1(\mathbf{x}_0(\theta_*))$ and the bi-orthogonal relation (15.51).

Let us denote a vector which has a nonzero element of value $q$ only at the $(p-1)$th component as $\mathbf{m}_{p,q}$. Because $r_p = 0$ on $\chi$, the gradient of $f$ evaluated on $\chi$ contains the following terms corresponding to $\mathbf{m}_{p,1}$:

$$\sum_{k\in\mathbb{Z}} a_{k,\mathbf{m}_{p,1}}(f)e^{\mathrm{i}k\theta_*}\nabla r_p(\mathbf{x}_0(\theta_*)). \tag{15.70}$$

It can be easily shown that all other components corresponding to $|\mathbf{m}|_1 \geq 2$, where $|\cdot|_1$ is the $l^1$-norm, do not contribute to the gradient $\nabla f$ on $\chi$ since $r_p = 0$ on the periodic orbit. We can then obtain the Koopman modes $a_{k,\mathbf{m}_{p,1}}(f)$ for the generic observable $f$ from its gradient evaluated on the periodic orbit $\chi$ as follows. First, we take a phase-wise inner product of $\nabla f(\mathbf{x}_0(\theta_*))$ and $e^{-ik\theta_*}\boldsymbol{\gamma}_p(\mathbf{x}_0(\theta_*))$, which we denote by $\tilde{z}(\theta_*)$. Then we evaluate the average of $\tilde{z}(\theta_*)$ over the circle. From Eq. (15.69), there is only one nonzero Koopman mode for $|\mathbf{m}|_1 \leq 1$, i.e., $a_{0,\mathbf{m}_{J,1}}(g_J) = 1$ for the observable $g_J$. In general, the Koopman modes associated with the non-principal eigenfunctions with $|\mathbf{m}|_1 \geq 2$ are nonzero even for the observable $g_J$. However, when $\mathrm{Re}(\lambda_J) > 2\mathrm{Re}(\lambda_2)$ holds, all the Koopman modes $a_{k,\mathbf{m}'}(g_J)$ involved in the generalized Laplace average (15.67) are zero.

Thus, we can replace the generalized Laplace average with the following Laplace average when we evaluate the $J$th principal Koopman eigenfunction:

$$a_{0,\mathbf{m}_{J,1}}(g_J)\phi_{\lambda_J}(\mathbf{x}) = \lim_{s\to\infty}\frac{1}{s}\int_0^s g_J \circ \mathbf{S}(t,\mathbf{x})e^{-\lambda_J t}\mathrm{d}t. \tag{15.71}$$

## References


1. Abraham, I., De la Torre, G., Murphey, T.D.: Model-based control using Koopman operators. In: Proceedings of Robotics: Science and Systems (2017)
2. Acebrón, J.A., Bonilla, L.L., Pérez Vincente, C.J., Ritort, F., Spigler, R.: The Kuramoto model: a simple paradigm for synchronization phenomena. Rev. Mod. Phys. **77** (2005)
3. Aranson, I.S., Kramer, L.: The world of the complex Ginzburg-Landau equation. Rev. Mod. Phys. **74** (2002)
4. Ashwin, P., Coombes, S., Nicks, R.: Mathematical frameworks for oscillatory network dynamics in neuroscience. J. Math. Neurosci. **6** (2016)
5. Brown, E., Moehlis, J., Holmes, P.: On the phase reduction and response dynamics of neural oscillator populations. Neural Comput. **16** (2004)
6. Brunton, S.L., Brunton, B.W., Proctor, J.L., Kutz, J.N.: Koopman invariant subspaces and finite linear representations of nonlinear dynamical systems for control. PLoS ONE **11** (2016)
7. Budišić, M., Mohr, R., Mezić, I.: Applied Koopmanism. Chaos **22** (2012)
8. Castejón, O., Guillamon, A., Huguet, G.: Phase-Amplitude response functions for transient-state stimuli. J. Math. Neurosci. **3** (2013)
9. Colonius, F., Kliemann, W.: Dynamical Systems and Linear Algebra. American Mathematical Society, Providence (2014)
10. Cross, M.C., Hohenberg, P.C.: Pattern formation outside of equilibrium. Rev. Mod. Phys. **65** (1993)
11. Dasanayake, I., Li, J.-S.: Optimal design of minimum-power stimuli for phase models of neuron oscillators. Phys. Rev. E **83** (2011)
12. Dörfler, F., Chertkov, M., Bullo, F.: Synchronization in complex oscillator networks and smart grids. Proc. Natl. Acad. Sci. USA **110** (2013)
13. Dörfler, F., Bullo, F.: Synchronization in complex networks of phase oscillators: a survey. Proc. Autom. **50** (2014)
14. Eldering, J., Kvalheim, M., Revzen, S.: Global linearization and fiber bundle structure of invariant manifolds. Nonlinearity **31** (2018)
15. Ermentrout, G.B.: Type I membranes, phase resetting curves, and synchrony. Neural Comput. **8** (1996)

16. Ermentrout, B., Park, T., Wilson, D.: Recent advances in coupled oscillator theory. Philos. Trans. R. Soc. A **377** (2019)
17. Ermentrout, G.B., Terman, D.H.: Mathematical Foundations of Neuroscience. Springer, New York (2010)
18. Froyland, G., Hüls, T., Morriss, G.P., Watson, T.M.: Computing covariant Lyapunov vectors, Oseledets vectors, and dichotomy projectors: a comparative numerical study. Phys. D **247** (2013)
19. Gaeta, G.: Poincaré Normal and Renormalized Forms. Acta Appl. Math. **70** (2002)
20. Goldobin, D.S., Teramae, J.-N., Nakao, H., Ermentrout, G.B.: Dynamics of limit-cycle oscillators subject to general noise. Phys. Rev. Lett. **105** (2010)
21. Goswami, D., Paley, D.A.: Global bilinearization and controllability of control-affine nonlinear systems: a Koopman spectral approach. In: 56th Annual Conference on Decision and Control pp. 6107–6112 (2017)
22. Guckenheimer, J., Holmes, P.J.: Nonlinear Oscillations, Dynamical Systems, and Bifurcations of Vector Fields. Springer, New York (1983)
23. Hale, J.K.: Ordinary Differential Equations. Dover Publications, New York (2009)
24. Harada, T., Tanaka, H.-A., Hankins, M.J., Kiss, I.Z.: Optimal waveform for the entrainment of a weakly forced oscillator. Phys. Rev. Lett. **105** (2010)
25. Hitczenko, P., Medvedev, G.S.: The Poincaré map of randomly perturbed periodic motion. J. Nonlinear Sci. **23** (2013)
26. Hoppensteadt, F.C., Izhikevich, E.M.: Weakly Connected Neural Networks. Springer, New York (1997)
27. Hoppensteadt, F.C.: Analysis and Simulation of Chaotic Systems. Springer, New York (2000)
28. Hüls, T.: Computing stable hierarchies of fiber bundles. Discret. Contin. Dyn. Syst. Ser. B **22** (2017)
29. Kaiser, E., Kutz, J.N., Brunton, S.L.: Data-driven discovery of Koopman eigenfunctions for control (2017). arXiv:1707.01146
30. Kawamura, Y., Nakao, H.: Collective phase description of oscillatory convection. Chaos **23**D (2013)
31. Kawamura, Y., Shirasaka, S., Yanagita, T., Nakao, H.: Optimizing mutual synchronization of rhythmic spatiotemporal patterns in reaction-diffusion systems. Phys. Rev. E **96** (2017)
32. Keener, J.P.: Principles of Applied Mathematics. Westview Press, Boulder (2001)
33. Keener, J., Sneyd, J.: Mathematical Physiology I: Cellular Physiology. Springer, New York (2009)
34. Kiss, I.Z., Rusin, C.G., Kori, H., Hudson, J.L.: Engineering complex dynamical structures: sequential patterns and desynchronization. Science **316** (2007)
35. Korda, M., Mezic, I.: Linear predictors for nonlinear dynamical systems: Koopman operator meets model predictive control. Automatica **93** (2018)
36. Koseska, A., Volkov, E., Kurths, J.: Oscillation quenching mechanisms: amplitude vs. oscillation death. Phys. Rep. **531** (2013)
37. Kotani, K., Yamaguchi, I., Ogawa, Y., Jimbo, Y., Nakao, H., Ermentrout, G.B.: Adjoint method provides phase response functions for delay-induced oscillations. Phys. Rev. Lett. **109** (2012)
38. Kowalski, K., Steeb, W.H.: Nonlinear Dynamical Systems and Carleman Linearization. World Scientific, Singapore (2011)
39. Kuptsov, P.V., Parlitz, U.: Theory and computation of covariant Lyapunov vectors. J. Nonlinear Sci. **22** (2012)
40. Kuramoto, Y.: Self-entrainment of a population of coupled non-linear oscillators. In: Arakaki, H. (ed.) International Symposium on Mathematical Problems in Theoretical Physics. Lecture Notes in Physics, vol. 39. Springer, New York (1975)
41. Kuramoto, Y.: Chemical Oscillations, Waves, and Turbulence. Springer, Berlin (1984)
42. Kvalheim, M.D., Revzen, S.: Existence and uniqueness of global Koopman eigenfunctions for stable fixed points and periodic orbits (2019). arXiv:1911.11996
43. Lan, Y., Mezić, I.: Linearization in the large of nonlinear systems and Koopman operator spectrum. Phys. D **242** (2013)

44. Li, Q., Dietrich, F., Bollt, E.M., Kevrekidis, I.G.: Extended dynamic mode decomposition with dictionary learning: a data-driven adaptive spectral decomposition of the Koopman operator. Chaos **27** (2017)
45. Lusch, B., Kutz, J.N., Brunton, S.L.: Deep learning for universal linear embeddings of nonlinear dynamics. Nat. Commun. **9** (2018)
46. Malkin, I.G.: Some Problems in Nonlinear Oscillation Theory. Gostexizdat, Moskow (1956). (in Russian)
47. Mardt, A., Pasquali, L., Wu, H., Noé, F.: VAMPnets for deep learning of molecular kinetics. Nat. Commun. **9** (2018)
48. Matthews, P.C., Mirollo, R.E., Strogatz, S.H.: Dynamics of a large system of coupled nonlinear oscillators. Phys. D **52** (1991)
49. Mauroy, A., Mezić, I.: On the use of Fourier averages to compute the global isochrons of (quasi)periodic dynamics. Chaos **22** (2012)
50. Mauroy, A., Mezić, I.: Global computation of phase-amplitude reduction for limit-cycle dynamics. Chaos **28** (2018)
51. Mauroy, A., Mezić, I., Moehlis, J.: Isostables, isochrons, and Koopman spectrum for the action-angle representation of stable fixed point dynamics. Phys. D **261** (2013)
52. Mauroy, A., Rhoads, B., Moehlis, J., Mezic, I.: Global isochrons and phase sensitivity of bursting neurons. SIAM J. Appl. Dyn. Syst. **13** (2014)
53. Mauroy, A.: Converging to and escaping from the global equilibrium: isostables and optimal control. In: Proceedings of the 53rd IEEE Conference on Decision and Control, pp. 5888–5893 (2014)
54. Mezić, I.: Analysis of fluid flows via spectral properties of the Koopman operator. Ann. Rev. Fluid Mech. **45** (2013)
55. Mezić, I.: Spectum of the Koopman Operator, Spectral Expansions in Functional Spaces, and State-Space Geometry. J. Nonlinear Sci. (2019)
56. Mikhailov, A.S., Ertl, G.: Chemical Complexity. Springer International Publishing, Cham (2017)
57. Moehlis, J., Shea-Brown, E., Rabitz, H.: Optimal inputs for phase models of spiking neurons. J. Comput. Nonlinear Dyn. **1** (2006)
58. Mohr, R., Mezić, I.: Construction of eigenfunctions for scalar-type operators via Laplace averages with connections to the Koopman operator (2014). arXiv:1403.6559
59. Monga, B., Wilson, D., Matchen, T., Moehlis, J.: Phase reduction and phase-based optimal control for biological systems: a tutorial. Biol. Cybern. (2018)
60. Nakao, H.: Phase reduction approach to synchronisation of nonlinear oscillators. Contemp. Phys. **57** (2016)
61. Nakao, H., Mikhailov, A.S.: Diffusion-induced instability and chaos in random oscillator networks. Phys. Rev. E **79** (2009)
62. Nakao, H., Yanagita, T., Kawamura, Y.: Phase reduction approach to synchronization of spatiotemporal rhythms in reactiondiffusion systems. Phys. Rev. X **4** (2014)
63. Novičenko, V., Pyragas, K.: Phase reduction of weakly perturbed limit cycle oscillations in time-delay systems. Phys. D **241** (2012)
64. Otto, S.E., Rowley, C.W.: Linearly recurrent autoencoder networks for learning dynamics. SIAM J. Appl. Dyn. Syst. **18** (2019)
65. Park, Y., Shaw, K.M., Chiel, H.J., Thomas, P.J.: The infinitesimal phase response curves of oscillators in piecewise smooth dynamical systems. Eur. J. Appl. Math. **29** (2018)
66. Pikovsky, A., Rosenblum, M., Kurths, J.: Synchronization: A Universal Concept in Nonlinear Sciences. Cambridge University Press, Cambridge (2003)
67. Pikovsky, A.: Maximizing coherence of oscillations by external locking. Phys. Rev. Lett. **115** (2015)
68. Pikovsky, A., Politi, A.: Lyapunov Exponents: A Tool to Explore Complex Dynamics. Cambridge University Press, Cambridge (2015)
69. Proctor, J.L., Brunton, S.L., Kutz, J.N.: Dynamic mode decomposition with control. SIAM J. Appl. Dyn. Syst. **15** (2016)

70. Rodrigues, F.A., Peron, T.K.DM., Ji, P., Kurths, J.: The Kuramoto model in complex networks. Phys. Rep. **610** (2016)
71. Rowley, C.W., Mezić, I., Bagheri, S., Schlatter. P., Henningson, D.S.: Spectral analysis of nonlinear flows. J. Fluid Mech. **641** (2009)
72. Sanders, J.A., Verhulst, F., Murdock, J.: Averaging Methods in Nonlinear Dynamical Systems. Springer, New York (2007)
73. Schmid, P.J.: Dynamic mode decomposition of numerical and experimental data. J. Fluid Mech. **656** (2010)
74. Schultheiss, N.W., Prinz, A.A., Butera, R.J. (eds.): Phase Response Curves in Neuroscience: Theory, Experiment, and Analysis. Springer, New York (2011)
75. Shilnilov, L.P., Shilnilov, A.L., Turaev, D.V., Chua, L.O.: Methods of Qualitative Theory in Nonlinear Dynamics. Part I. World Scientific, Singapore (1998)
76. Shirasaka, S., Watanabe, N., Kawamura, Y., Nakao, H.: Optimizing stability of mutual synchronization between a pair of limit-cycle oscillators with weak cross coupling. Phys. Rev. E **96** (2017)
77. Shirasaka, S., Kurebayashi, W., Nakao, H.: Phase reduction theory for hybrid nonlinear oscillators. Phys. Rev. E **95** (2017)
78. Shirasaka, S., Kurebayashi, W., Nakao, H.: Phase-amplitude reduction of transient dynamics far from attractors for limit-cycling systems. Chaos **27** (2017)
79. Skardal, P.S., Taylor, D., Sun, J.: Optimal synchronization of complex networks. Phys. Rev. Lett. **113** (2014)
80. Skardal, P.S., Taylor, D., Sun, J.: Optimal synchronization of directed complex networks. Chaos **26** (2016)
81. Sootla, A., Mauroy, A., Ernst, D.: Optimal control formulation of pulse-based control using Koopman operator. Automatica **91** (2018)
82. Stankovski, T., Pereira, T., McClintock, P.V.E., Stefanovska, A.: Coupling functions: universal insights into dynamical interaction mechanisms. Rev. Mod. Phys. **89** (2017)
83. Strogatz, S.H.: From Kuramoto to Crawford: exploring the onset of synchronization in populations of coupled oscillators. Phys. D **143** (2000)
84. Strogatz, S.H., Abrams, D.M., McRobie, A., Eckhardt, B., Ott, E.: Crowd synchrony on the Millennium bridge. Nature **438** (2005)
85. Strogatz, S.H.: Nonlinear Dynamics and Chaos: With Applications to Physics, Biology, Chemistry, and Engineering. Westview press, Boulder (2014)
86. Taira, K., Nakao, H.: Phase-response analysis of synchronization for periodic flows. J. Fluid. Mech. **846** (2018)
87. Takeishi, N., Kawahara, Y., Yairi, Y.: Learning koopman invariant subspaces for dynamic mode decomposition. In: Advances in Neural Information Processing Systems, pp. 1130–1140 (2017)
88. Tass, P.A.: Phase Resetting in Medicine and Biology: Stochastic Modelling and Data Analysis. Springer, Berlin (2007)
89. Tinsley, M.R., Nkomo, S., Showalter, K.: Chimera and phase-cluster states in populations of coupled chemical oscillators. Nat. Phys. **8** (2012)
90. Traversa, F.L., Bonnin, M., Corinto, F., Bonani, F.: Noise in oscillators: a review of state space decomposition approaches. J. Comput. Electron. **14** (2015)
91. Tu, J.H., Rowley, C.W., Luchtenburg, D.M., Brunton, S.L., Kutz, J.N.: On dynamic mode decomposition: theory and applications. J. Comput. Dyn. **1** (2014)
92. Watanabe, N., Kato, Y., Shirasaka, S., Nakao, H.: Optimization of linear and nonlinear interaction schemes for stable synchronization of weakly coupled limit-cycle oscillators. Phys. Rev. E **100** (2019)
93. Wedgwood, K.C.A., Lin, K.K., Thul, R., Coombes, S.: Phase-amplitude descriptions of neural oscillator models. J. Math. Neurosci. **3** (2013)
94. Wiggins, S.: Introduction to Applied Nonlinear Dynamical Systems and Chaos. Springer, New York (2003)
95. Williams, M.O., Kevrekidis, I.G., Rowley, C.W.: A DataDriven approximation of the Koopman operator: extending dynamic mode decomposition. J. Nonlinear Sci. **25** (2015)

96. Williams, M.O., Rowley, C.W., Kevrekidis, I.G.: A kernel-based method for data-driven koopman spectral analysis. J. Comput. Dyn. **2015** (2015)
97. Wilson, D.: Isostable reduction of oscillators with piecewise smooth dynamics and complex Floquet multipliers. Phys. Rev. E **99** (2019)
98. Wilson, D., Ermentrout, G.B.: Greater accuracy and broadened applicability of phase reduction using isostable coordinates. J. Math. Biol. **76** (2018)
99. Wilson, D., Ermentrout, B.: Phase Models Beyond Weak Coupling. Phys. Rev. Lett. **123** (2019)
100. Wilson, D., Holt, A.B., Netoff, T.I., Moehlis, J.: Optimal entrainment of heterogeneous noisy neurons. Front. Neurosci. **29** (2015)
101. Wilson, D., Moehlis, J.: Extending phase reduction to excitable media: theory and applications. SIAM Rev. **57** (2015)
102. Wilson, D., Moehlis, J.: Isostable reduction with applications to time-dependent partial differential equations. Phys. Rev. E **94** (2016)
103. Wilson, D., Moehlis, J.: Isostable reduction of periodic orbits. Phys. Rev. E **94** (2016)
104. Wilson, D., Moehlis, J.: Spatiotemporal control to eliminate cardiac alternans using isostable reduction. Phys. D **342** (2017)
105. Winfree, A.T.: The Geometry of Biological Time. Springer, New York (2001)
106. Yeung, E., Kundu, S., Hodas, N.: Learning Deep Neural Network Representations for Koopman Operators of Nonlinear Dynamical Systems. In: 2019 American Control Conference, pp. 4832–4839 (2019)
107. Yoshimura, K., Arai, K.: Phase reduction of stochastic limit cycle oscillators. Phys. Rev. Lett. **101** (2008)
108. Zlotnik, A., Chen, Y., Kiss, I.Z., Tanaka, H.-A., Li, J.-S.: Optimal waveform for fast entrainment of weakly forced nonlinear oscillators. Phys. Rev. Lett. **111** (2013)
109. Zlotnik, A., Nagao, R., Kiss, I.Z., Li, J.-S.: Phase-selective entrainment of nonlinear oscillator ensembles. Nat. Commun. **7** (2016)

# Part III
# Applications

# Chapter 16
# Experimental Applications of the Koopman Operator in Active Learning for Control

**Thomas A. Berrueta, Ian Abraham and Todd Murphey**

**Abstract** Experimentation as a setting for learning demands adaptability on the part of the decision-making system. It is typically infeasible for agents to have complete a priori knowledge of an environment, their own dynamics, or the behavior of other agents. In order to achieve autonomy in robotic applications, learning must occur incrementally, and ideally as a function of decision-making by exploiting the underlying control system. Most artificial intelligence techniques are ill suited for experimental settings because they either lack the ability to learn incrementally or do not have information measures with which to guide their learning. This chapter examines the Koopman operator, its application in active learning, and its relationship to alternative learning techniques, such as Gaussian processes and kernel ridge regression. Additionally, examples are provided from a variety of experimental applications of the Koopman operator in active learning settings.

## 16.1 Introduction

Machine learning techniques describe complex patterns in data. Many learning methods, such as neural networks and convolutional networks, process large datasets in order to develop models that describe causal relationships between data. However, what happens when one must gather data as a part of a real-time process? How should one incorporate new data into models, and then leverage agency to further improve them? The answers to these questions form the basis of the robotic paradigm—*sense, plan, act*—and are at the heart of what defines autonomy.

T. A. Berrueta (✉) · I. Abraham · T. Murphey
Department of Mechanical Engineering, Northwestern University,
Evanston, IL, USA
e-mail: tberrueta@u.northwestern.edu

I. Abraham
e-mail: i-abr@u.northwestern.edu

T. Murphey
e-mail: t-murphey@northwestern.edu



A. Mauroy et al. (eds.), *The Koopman Operator in Systems and Control*,
Lecture Notes in Control and Information Sciences 484,
https://doi.org/10.1007/978-3-030-35713-9_16

In robotics, one is interested in endowing physical agents with the ability to reason about uncertainty and to *act* in order to achieve a goal. The experimental nature of this process necessitates learning frameworks capable of incremental adaptation, such that an agent is able to gather new information and use it to its advantage. We refer to the process in which an agent selectively seeks out new information relevant to better achieving a goal as *active learning*.

Not all machine learning techniques are well suited for active learning. Batch learning processes requiring large amounts of training data are not useful in real-time settings where learning is occurring at all times. For active learning purposes, we need the ability to incrementally update models to avoid recomputing the learned model from scratch, which can be prohibitive. In order to develop principled approaches for incremental information acquisition, we need to quantify the model's information gain. We use the approximate Koopman operator as formulated in [41] and used in our earlier work [2] as a model of choice, and apply it in active learning settings.

While Koopman operators are the primary focus of this chapter, techniques such as generalized linear regression, kernel ridge regression, and Gaussian processes are often also applied as online learning tools in experiments. Specifying the theoretical and practical relationships between these methods is important to understanding the qualities of good active learning techniques. By developing a set of assumptions under which these methods are equivalent, one can transfer techniques developed for one method to another.

In Sect. 16.2, we describe the Koopman operator as a technique for active learning. In Sect. 16.3, we consider generalized linear regression, kernel ridge regression, and Gaussian processes as alternative techniques for experimental learning, and compare them to approximate Koopman operators, both in theoretical terms and computational implications. In Sect. 16.4, we illustrate the Koopman operator's use in experimental settings. In Sect. 16.5, we conclude with a discussion of what makes the Koopman operator a good system representation in active learning settings.

## 16.2 Koopman Operator

The Koopman operator is an infinite-dimensional linear operator capable of describing the time evolution of any observable, measure-preserving dynamical system [22]. We consider discrete-time dynamical systems described by

$$\mathbf{x}_{k+1} = S(\mathbf{x}_k) = \mathbf{x}_k + \int_{t_k}^{t_k+\Delta t} F(\mathbf{x}(\tau))d\tau, \tag{16.1}$$

where the $F$ is the vector field of the continuous dynamical system, which evolves on a state-space manifold $\mathcal{M}$ such that $S : \mathcal{M} \to \mathcal{M}$. The right-hand side is a continuous-time system discretized with sampling interval $\Delta t$.

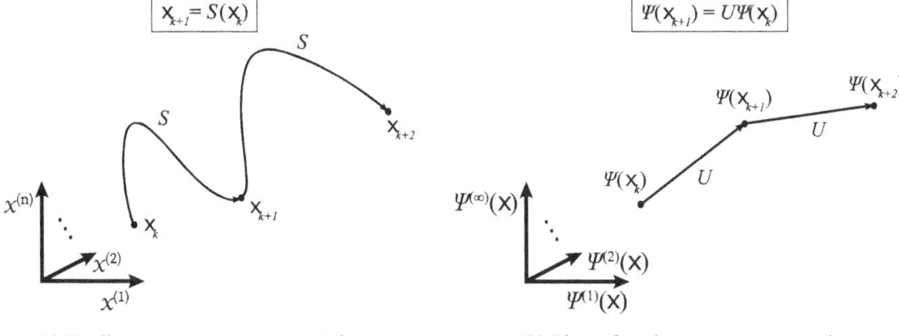


(a) Nonlinear state-space representation (b) Linear function-space representation

**Fig. 16.1** Schematic of the relationship between a nonlinear dynamical system and its corresponding Koopman operator representation. Given a nonlinear system in $n$-dimensional state space as shown in (**a**), the Koopman operator is a linear representation of the system in an infinite-dimensional function space, shown in (**b**). The map $\Psi$ lifts the original system onto an infinite-dimensional function space. In this function space, the Koopman operator, $U$, is a linear transformation representing the system dynamics

The Koopman operator is a linear operator that describes the time evolution of basis functions instead of states. This shifts the system representation from state-space to an infinite-dimensional function space. Figure 16.1 shows the relationship between a dynamical system $S(\mathbf{x})$ and its corresponding Koopman operator $U$. For a detailed analysis of the Koopman operator and its properties, we refer the reader to Chap. 1, as well as [7].

### *16.2.1 EDMD Approximation*

The Koopman operator itself is not a learning tool; it is an alternative representation of dynamical systems. However, numerically synthesizing an approximation of this representation in finite dimensions from data is a learning problem. In order to implement Koopman operators in computational applications, one generates finite-dimensional approximations. While certain systems can be represented by an exact, closed-form, finite-dimensional Koopman operator, this is generally not the case [20]. Dynamic mode decomposition (DMD) [33] and extended dynamic mode decomposition (EDMD) [41] are methods developed for approximating Koopman operators in finite dimensions and are commonly applied in fluid dynamics as analytical tools [26].

EDMD synthesizes a data-driven Koopman operator by developing a mapping that tries to span a finite-dimensional subspace of a given function space using a finite set of basis functions. Given basis functions $\Psi(\mathbf{x}) = [\psi_1(\mathbf{x}), \ldots, \psi_N(\mathbf{x})]^T$, $s.t.$ $\{\psi_i(\mathbf{x}) : \mathcal{M} \to \mathbb{R},\ \forall i\}$ and a dataset of sequential observations $\mathbf{X} = [\mathbf{x}_1, \ldots, \mathbf{x}_M]$, we can generate an approximate Koopman operator, $\mathbf{U}$, to describe the evolution of the lifted system

$$\Psi(\mathbf{x}_{k+1}) = \mathbf{U}\Psi(\mathbf{x}_k) + \varepsilon, \tag{16.2}$$

such that $\varepsilon \sim \mathcal{N}(\mathbf{0}, \sigma^2)$, where $\mathcal{N}$ is a multivariate Gaussian distribution with mean and variance $\mathbf{0}, \sigma^2 \in \mathbb{R}^N$. The corresponding approximate Koopman operator is given by the solution to the optimization

$$\min_{\mathbf{U}} \frac{1}{2} \sum_{k=1}^{M-1} ||\Psi(\mathbf{x}_{k+1}) - \mathbf{U}\Psi(\mathbf{x}_k)||^2 \tag{16.3}$$

which has closed-form solution

$$\mathbf{U} = \mathbf{A}\mathbf{G}^{\dagger}, \tag{16.4}$$

where $\dagger$ denotes the Moore–Penrose pseudoinverse and the individual matrix components are

$$\begin{aligned} \mathbf{G} = \mathbf{G}[M] &= \frac{1}{M} \sum_{k=1}^{M-1} \Psi(\mathbf{x}_k)\Psi(\mathbf{x}_k)^T, \\ \mathbf{A} = \mathbf{A}[M] &= \frac{1}{M} \sum_{k=1}^{M-1} \Psi(\mathbf{x}_{k+1})\Psi(\mathbf{x}_k)^T. \end{aligned} \tag{16.5}$$

### 16.2.2 *Incremental Updates*

The closed-form solution in (16.5) is of particular interest because it can be modified for incremental updates [1, 18]. We can define a set of difference equations describing an incremental update to the Koopman operator after a new observation

$$\begin{aligned} \mathbf{G}[M+1] &= \mathbf{G}[M] + \frac{1}{M+1}(\Psi(\mathbf{x}_M)\Psi(\mathbf{x}_M)^T - \mathbf{G}[M]), \\ \mathbf{A}[M+1] &= \mathbf{A}[M] + \frac{1}{M+1}(\Psi(\mathbf{x}_{M+1})\Psi(\mathbf{x}_M)^T - \mathbf{A}[M]), \end{aligned} \tag{16.6}$$

where we can calculate the updated Koopman operator with

$$\mathbf{U}[M+1] = \mathbf{A}[M+1](\mathbf{G}[M+1])^{\dagger}. \tag{16.7}$$

This formulation does not scale in complexity with the number of measurements since the pseudoinverse computation complexity scales with respect to the number of basis functions considered. The difference equation (16.6) describes an implicit cumulative average of all measurements. However, it is possible to perform incremental updates to the Koopman operator using other difference equations, such as moving average filters, exponential moving average filters, or Kalman filters [14, 16].

### *16.2.3 Formulation for Control*

In order to apply the Koopman operator in experimental settings, we can formulate the operator for real-time control [6]. By considering control inputs as a part of the measurement vector, we can split the basis functions $\Psi(\mathbf{x}, \mathbf{u})$ into $\Psi(\mathbf{x}, \mathbf{u}) = [\Psi_x(\mathbf{x})^T, \Psi_u(\mathbf{x}, \mathbf{u})^T]^T$, where $\{\Psi_x(\mathbf{x}) : \mathcal{M} \rightarrow \mathbb{R}^{N_x}\}$ and $\{\Psi_u(\mathbf{x}, \mathbf{u}) : \mathcal{M} \rightarrow \mathbb{R}^{N_u}\}$ such that $N = N_x + N_u$. As a result, the corresponding data-driven Koopman operator can be split into submatrices

$$\mathbf{U} = \begin{bmatrix} \mathbf{K}_x & \mathbf{K}_u \\ \mathbf{K}_{ux} & \mathbf{K}_{uu} \end{bmatrix}. \tag{16.8}$$

Using these submatrices we can formulate the time evolution of state observables with control [2]

$$\Psi_x(\mathbf{x}_{k+1}) = \mathbf{K}_x \Psi_x(\mathbf{x}_k) + \mathbf{K}_u \Psi_u(\mathbf{x}_k, \mathbf{u}_k). \tag{16.9}$$

Additionally, when considering the evolution of observables that depend on input, the following equation holds

$$\Psi_u(\mathbf{x}_{k+1}, \mathbf{u}_{k+1}) = \mathbf{K}_{ux} \Psi_x(\mathbf{x}_k) + \mathbf{K}_{uu} \Psi_u(\mathbf{x}_k, \mathbf{u}_k). \tag{16.10}$$

It is worth noting a couple of special cases of (16.9). For the case in which $\Psi_u(\mathbf{x}, \mathbf{u})$ is linear in $\mathbf{u}$, the equation becomes

$$\Psi_x(\mathbf{x}_{k+1}) = \mathbf{K}_x \Psi_x(\mathbf{x}_k) + \hat{\mathbf{K}}_u \mathbf{u}_k, \tag{16.11}$$

where $\hat{\mathbf{K}}_u = \mathbf{K}_u \frac{\partial \Psi_u(\mathbf{x}_k, \mathbf{u}_k)}{\partial \mathbf{u}}$. Then for the case in which $\Psi_u(\mathbf{x}, \mathbf{u})$ is linear in $\mathbf{u}$ and does not depend on the state, the equation is

$$\Psi_x(\mathbf{x}_{k+1}) = \mathbf{K}_x \Psi_x(\mathbf{x}_k) + \mathbf{K}_u \mathbf{u}_k. \tag{16.12}$$

This formulation in (16.12) of the Koopman operator for control systems enables us to easily incorporate it into classical control schemes such as optimal control, or other model-predictive frameworks [23]. Alternative data-driven formulations of the Koopman operator for control have been described in the work of [20].

In optimal control, we seek to specify control laws that best drive systems toward a desired goal state with respect to an objective. For a discrete-time linear quadratic (LQ) control problem with goal state $\mathbf{x}_d$, we can define an objective function of the following form over the iterate $k$

$$J = \sum_{k=0}^{\infty} (\Psi_x(\mathbf{x}_k) - \Psi_x(\mathbf{x}_d))^T \mathbf{Q} (\Psi_x(\mathbf{x}_k) - \Psi_x(\mathbf{x}_d)) + \mathbf{u}_k^T \mathbf{R} \mathbf{u}_k, \tag{16.13}$$

where $\mathbf{Q}$ and $\mathbf{R}$ are positive semidefinite matrices that specify weights on system states and control effort, respectively. Deriving an optimal solution to this LQ problem leads to a feedback law

$$\mathbf{u}_{LQ_K} = -\mathbf{F}_{LQ_K}(\Psi_x(\mathbf{x}) - \Psi_x(\mathbf{x}_d)), \tag{16.14}$$

such that the optimal feedback gain $\mathbf{F}_{LQ_K}$ is

$$\mathbf{F}_{LQ_K} = (\mathbf{R} + \mathbf{K}_u^T\mathbf{P}\mathbf{K}_u)^{-1}\mathbf{K}_u^T\mathbf{P}\mathbf{K}_x, \tag{16.15}$$

where $\mathbf{P}$ is the solution to the discrete-time algebraic Ricatti equation. Through this process, we can solve optimal control problems with data-driven Koopman models [6, 20].

Additionally, given Koopman operator dynamics of the same form as (16.12), we note that system derivatives with respect to basis functions of state and control are linear:

$$\frac{d}{d\Psi_x}\big(\Psi_x(\mathbf{x}_{k+1})\big) = \mathbf{K}_x, \quad \frac{d}{d\mathbf{u}}\big(\Psi_x(\mathbf{x}_{k+1})\big) = \mathbf{K}_u. \tag{16.16}$$

The quality of a data-driven model is dependent on the amount of information captured about the underlying system. In order to determine how much information is captured by system models, we must develop a means of quantifying information with respect to the model.

### *16.2.4 Information Measures*

Active learning is a process by which an agent can acquire more information about a task and improve its performance. Through experimental learning, we are interested in improving our system models such that we maximize task performance. In order to actively acquire information, we must develop information measures with respect to our system model. Fisher information provides a means of measuring the amount of information that a random variable carries about parameters [10]. Given a Koopman operator submatrix $\mathbf{K}_x$ from a system specified as (16.9), we model each of its elements as a normally distributed parameter $\mathbf{K}_x^{(i,j)} \sim \mathcal{N}(\boldsymbol{\mu}_{ij}, \Sigma_{ij}^2)$ and compute the Fisher information

$$I(\Psi_x(\mathbf{x})) = \frac{\partial \Gamma^T}{\partial \bar{k}}\Sigma^{-1}\frac{\partial \Gamma}{\partial \bar{k}}, \tag{16.17}$$

where $\bar{k} = \{\mathbf{K}_x^{(i,j)} : (i, j) \in \{1, \ldots, N_x\} \times \{1, \ldots, N_x\}\}$, $\Gamma \in \mathbb{R}^{N_x}$ is the measurement model defined by $\Gamma = \mathbf{K}_x\Psi_x(\mathbf{x})$, and $\Sigma \in \mathbb{R}^{N_x \times N_x}$ is assumed to be constant. The matrix $\Sigma$ is the measurement covariance that can be empirically estimated from sample data a priori. The matrix $\frac{\partial \Gamma}{\partial \bar{k}} \in \mathbb{R}^{N_x \times (N_x \cdot N_x)}$ encodes measurement sensitivities to all elements of the Koopman operator. The sensitivity matrix takes the form

$$\frac{\partial \Gamma}{\partial \bar{k}} = \begin{bmatrix} \Psi_x(\mathbf{x})^T & \mathbf{0} & \dots & \mathbf{0} \\ \mathbf{0} & \Psi_x(\mathbf{x})^T & \dots & \mathbf{0} \\ \vdots & & \ddots & \\ \mathbf{0} & \dots & & \Psi_x(\mathbf{x})^T \end{bmatrix}, \tag{16.18}$$

where $\mathbf{0} \in \mathbb{R}^{1 \times N_x}$. We now optimize the trace of the Fisher information matrix, also known as T-optimality [36], to approximate information gain

$$\operatorname{tr}(I(\Psi_x(\mathbf{x}))) = \sum_{i=1}^{N_x} ||\Psi_x(\mathbf{x})||^2_{\Sigma_{ii}^{-1}} . \tag{16.19}$$

We introduce (16.19) into a control objective by assigning a cost $q$ to the inverse of the Fisher information

$$J = \sum_{k=1}^{\infty} q \operatorname{tr}(I(\Psi_x(\mathbf{x}_k)))^{-1} + ||\Psi_x(\mathbf{x}_k) - \Psi_x(\mathbf{x}_d)||^2_{\mathbf{Q}} + ||\mathbf{u}_k||^2_{\mathbf{R}} . \tag{16.20}$$

Using the cost functional in (16.20) we synthesize information maximizing trajectories and continuously improve our Koopman operator representation of the dynamics, as shown in [1]. The simplicity of the Koopman operator's measurement model $\Gamma$ makes it easy to develop information measures with respect to the generated model.

### 16.2.5 Formulation for Hybrid Systems

Robotic systems often have discontinuous dynamics because of impacts and intermittent contact, leading to hybrid dynamics. Hybrid system dynamics are defined piecewise, where an indicator function determines which hybrid mode will evolve the current state. These dynamics are formulated as

$$S(\mathbf{x}) = \begin{cases} S_1(\mathbf{x}), & \text{if } \Phi(\mathbf{x}) = 1 \\ \vdots & \vdots \\ S_i(\mathbf{x}), & \text{if } \Phi(\mathbf{x}) = i \\ \vdots & \vdots \\ S_B(\mathbf{x}), & \text{otherwise} \end{cases} \tag{16.21}$$

where each $S_i(\mathbf{x})$ represents the dynamics of states $\mathbf{x} \in \mathbb{R}^n$ at mode $\Phi(\mathbf{x}) = i$. Additionally, at each hybrid mode boundary there is a discontinuous map between modes. Assuming that $\Phi(\mathbf{x})$ is known a priori, it is possible to compute a Koopman operator $\mathbf{U}_i$ to represent each hybrid mode. We can specify a hybrid Koopman operator using

$$\overline{\mathbf{U}}(\mathbf{x}), \overline{\Psi}(\mathbf{x}) = \begin{cases} \mathbf{U}_1(\mathbf{x}), \Psi_1, & \text{if } \Phi(\mathbf{x}) = 1 \\ \vdots & \vdots \\ \mathbf{U}_i(\mathbf{x}), \Psi_i, & \text{if } \Phi(\mathbf{x}) = i \\ \vdots & \vdots \\ \mathbf{U}_B(\mathbf{x}), \Psi_B, & \text{otherwise} \end{cases} \tag{16.22}$$

where we characterize the discontinuous jumps between hybrid modes by collecting data at the mode boundaries and using $\Psi_i^*(\mathbf{x}_k^+) = \Psi_i^*(\mathbf{x}_k^-)\mathbf{U}_i^*$. Given that these jumps in state often take place over very small timescales, it may be difficult to gather enough data to provide a robust Koopman operator for that mapping. Active excitation methods might be necessary in order to properly characterize transitions. Alternative treatments of the Koopman operator have looked into modeling and analyzing switched [31], hybrid [15], and distributed systems [17].

The Koopman operator's linearity, ability to update incrementally, and ease of generating information measures make it a good choice of data-driven model representation for experimental settings. However, the operator's performance is dependent on the suitability of the chosen basis functions, which may not be known ahead of time.

Though the model representation is different, data-driven Koopman operator synthesis, as formulated in EDMD, solves a similar least-squares optimization as alternative regression models, such as generalized linear regression, kernel ridge regression, and Gaussian processes. Characterizing the conditions under which these models are equivalent to one another can help expand understanding of both Koopman operator synthesis and alternative learning methods. Under certain sets of assumptions, methods developed for Koopman operators synthesized via EDMD apply to the alternatives and vice versa.

## 16.3 Alternative Methods

While the Koopman operator is neither a function approximation technique nor a regression tool, EDMD, as formulated in Sect. 16.2.1, is both [41]. In Sect. 16.2 we proposed that the EDMD approximation of the Koopman operator is well suited for active learning. In principle, other function approximation techniques can be applied to generate finite Koopman operators. However, EDMD is the most common Koopman operator synthesis technique, and as such it is of interest to explicitly relate it to alternative machine learning techniques.

We show that under certain assumptions generalized linear regression, kernel ridge regression, and Gaussian processes solve the same underlying optimization problem as EDMD and can be formulated similarly. By understanding these relationships, we can apply techniques formally developed for one method to others, and vice versa. Through these equivalences, we can extend the active learning techniques developed for the Koopman operator in Sect. 16.2 to other methods. We first provide a review of generalized linear regression, kernel ridge regression, and Gaussian processes, then establish the relationships between them.

### 16.3.1 Generalized Linear Regression

Linear regression and online linear regression have long been applied for statistical analysis of sequential models [34]. Linear regression models are typically limited to describing a function $\mathbf{y} = f(\mathbf{x}),\ s.t.\ \mathbf{x}, \mathbf{y} \in \mathbb{R}^S$ as linear combinations of inputs $\mathbf{y} = \mathbf{W}^T\mathbf{x}$, where we find $\mathbf{W} \in \mathbb{R}^{S\times S}$ to minimize an error metric. Generalized linear regression extends this approach to consider nonlinear functions of state, such that we are performing a linear regression in a given feature space defined by the basis functions, where the features are the basis functions. Given a dataset with inputs $\mathbf{X} = [\mathbf{x}_1, \ldots, \mathbf{x}_M]$, outputs $\mathbf{Y} = [\mathbf{y}_1, \ldots, \mathbf{y}_M]$, and a set of basis functions $\Psi(\mathbf{x}) = [\psi_1(\mathbf{x}), \ldots, \psi_N(\mathbf{x})]^T\ s.t.\ \{\psi_i(\mathbf{x}) : \mathbb{R}^S \to \mathbb{R},\ \forall i\}$, we consider the set of candidate functions defined by linear combinations of the basis functions subject to zero-mean Gaussian noise $\varepsilon$ with variance $\sigma^2$

$$\mathbf{y} = \mathbf{W}^T\Psi(\mathbf{x}) + \varepsilon. \tag{16.23}$$

We can minimize over the least-squares error functional

$$\min_{\mathbf{W}} \frac{1}{2}\sum_{k=1}^{M-1} ||\mathbf{y}_k - \mathbf{W}^T\Psi(\mathbf{x}_k)||^2 \tag{16.24}$$

to obtain a closed-form solution using maximum likelihood estimation [5]. We define the design matrix as

$$\Phi = [\Psi(\mathbf{x}_1), \ldots, \Psi(\mathbf{x}_M)]^T \in \mathbb{R}^{M\times N}, \tag{16.25}$$

which allows us to express the closed-form solution of (16.24) as

$$\mathbf{W} = (\Phi^T\Phi)^{-1}\Phi^T\mathbf{Y}, \tag{16.26}$$

where $\mathbf{Y} = [\mathbf{y}_1, \ldots, \mathbf{y}_M]^T$. We note that (16.26) is equivalent to

$$\mathbf{W} = \sum_{k=1}^{M-1}(\Psi(\mathbf{x}_k)\Psi(\mathbf{x}_k)^T)^{-1}(\Psi(\mathbf{x}_k)^T\mathbf{y}_k). \tag{16.27}$$

We also note that for autoregressive models where $\mathbf{y}_k = \Psi(\mathbf{x}_{k+1})$,

$$\Psi(\mathbf{x}_{k+1}) = \mathbf{W}^T\Psi(\mathbf{x}_k) + \varepsilon. \tag{16.28}$$

### 16.3.2 Kernel Ridge Regression

Kernel ridge regression reframes the generalized linear regression problem by considering infinite-dimensional feature spaces specified by kernels. Features are observed properties or characteristics of a phenomenon [5]. A feature space is then the space formed by all features considered. A kernel is a generalized inner product between vectors in a feature space. Finite-dimensional kernels can be specified directly in terms of a choice of basis functions

$$k(\mathbf{x}, \mathbf{x}') = \langle \Psi(\mathbf{x}), \Psi(\mathbf{x}') \rangle, \tag{16.29}$$

where the basis functions are a nonlinear transformation into a feature space. Kernels can implicitly define inner products in infinite-dimensional spaces without having to directly project coordinates into the feature space. The most common infinite-dimensional kernel of choice is the radial basis function kernel, also known as the Gaussian kernel [5]

$$k(\mathbf{x}, \mathbf{x}') = e^{-\frac{1}{2\sigma^2}||\mathbf{x}-\mathbf{x}'||^2}. \tag{16.30}$$

The sense in which the Gaussian kernel represents an inner product in an infinite-dimensional space can be understood by examining its Taylor series expansion. The Gaussian kernel is shown below as an infinite-dimensional polynomial kernel, with $\sigma^2 = \frac{1}{2}$ for simplicity:

$$\begin{aligned} k(\mathbf{x}, \mathbf{x}') &= e^{-||\mathbf{x}-\mathbf{x}'||^2} \\ &= e^{-(\mathbf{x}^T\mathbf{x}+2(\mathbf{x}^T\mathbf{x}')+(\mathbf{x}')^T\mathbf{x}')} \\ &= (e^{-\mathbf{x}^T\mathbf{x}})(e^{(\mathbf{x}')^T\mathbf{x}'})\sum_{n=0}^{\infty}\frac{(2\mathbf{x}^T\mathbf{x}')^n}{n!} \\ &= C(\mathbf{x}, \mathbf{x}')\sum_{n=0}^{\infty}\frac{2^n}{n!}\langle \mathbf{x}, \mathbf{x}' \rangle^n, \end{aligned} \tag{16.31}$$

where $C(\mathbf{x}, \mathbf{x}')$ is some constant value dependent on $\mathbf{x}$ and $\mathbf{x}'$ that does not affect the Taylor expansion. Equation (16.31) shows that the Gaussian kernel is an inner product over the infinite-dimensional space of polynomial functions, where polynomial functions of the input vector are considered as features.

Ridge regression, also known as $L_2$-norm regularized generalized linear regression, extends the least-squares residual minimization problem described in (16.24) by including a regularization term over the model weights. Solutions with low weight magnitudes are given preference to prevent overfitting of the model. For a dataset with inputs $\mathbf{X} = [\mathbf{x}_1, \ldots, \mathbf{x}_M]$ and outputs $\mathbf{Y} = [\mathbf{y}_1, \ldots, \mathbf{y}_M]$, the ridge regression optimization is given by

$$\min_{\mathbf{W}} \frac{1}{2}\sum_{k=1}^{M-1} ||\mathbf{y}_k - \mathbf{W}^T \Psi(\mathbf{x}_k)||^2 + \frac{\lambda}{2}||\mathbf{W}||_F^2, \tag{16.32}$$

where $\lambda \geqslant 0$ is a regularization parameter specifying a cost on the magnitude of the model weights, and $||\cdot||_F$ represents the Frobenius matrix norm.

Using the design matrix $\Phi$ in (16.25), we can write down a closed-form solution to the minimization in (16.32)

$$\mathbf{W} = (\Phi^T \Phi + \lambda \mathbf{I})^{-1} \Phi^T \mathbf{Y}. \tag{16.33}$$

We refer to this solution as the primal solution to the ridge regression estimation problem.

Kernel ridge regression is based on an alternative representation of the solution to the ridge regression problem that enables application with infinite-dimensional kernels. The primal solution to the ridge regression problem scales in complexity with the number of basis functions used, which makes it unable to consider high-dimensional feature spaces. However, we can manipulate (16.33) to get an equivalent representation, which we will refer to as the dual solution

$$\mathbf{W} = \Phi^T (\Phi \Phi^T + \lambda \mathbf{I})^{-1} \mathbf{Y}. \tag{16.34}$$

We define $\mathbf{B} = (\mathbf{G_r} + \lambda \mathbf{I})^{-1}\mathbf{Y}$ where $\mathbf{G_r}$ is the Gram matrix, such that $\mathbf{W} = \Phi^T \mathbf{B}$. The Gram matrix is a matrix containing all inner products between points in a dataset as specified by a given kernel $k(\mathbf{x}, \mathbf{x}')$, such that $\mathbf{G_r}^{(i,j)} = k(\mathbf{x}_i, \mathbf{x}_j),\ \forall i, j \in \{1, \ldots, M\}$. In this case, $\mathbf{G_r} = \Phi\Phi^T$ due to the choice of kernel in (16.29). If we consider $\mathbf{W}$ as the primal weight matrix, we can refer to $\mathbf{B}$ as the dual weight matrix and use it to formulate our estimate as a function of the kernel choice [27]

$$\begin{aligned} \hat{\mathbf{y}} &= \Psi(\mathbf{x}')^T \mathbf{W} && (16.35) \\ &= \Psi(\mathbf{x}')^T \Phi^T \mathbf{B} \\ &= \mathbf{k}(\mathbf{x}')^T \mathbf{B} \\ &= \mathbf{k}(\mathbf{x}')^T (\mathbf{G_r} + \lambda \mathbf{I})^{-1} \mathbf{Y}, && (16.36) \end{aligned}$$

where $\mathbf{k}(\mathbf{x}') = \Phi\Psi(\mathbf{x}') = [k(\mathbf{x}_1, \mathbf{x}'), \ldots, k(\mathbf{x}_M, \mathbf{x}')]^T$. The kernel ridge regression function prediction scales in complexity with the number of data points, and as such is not suitable for incremental learning. Equation (16.36) enables us to analyze the models in infinite-dimensional feature spaces using kernels, which is beneficial in different settings. This kernel-based representation is how kernel ridge regression is most often applied. However, kernel ridge regression can also be formulated recursively and applied in online learning [9, 12].

### 16.3.3 Gaussian Process Regression

Until this point, we have only considered non-probabilistic models for regression in active learning. In some settings, randomness and uncertainty are inherently present in state measurements and as such require models that incorporate uncertainty into their predictions. Stochastic processes, such as Gaussian processes, specialize in describing systems that probabilistically evolve over time [14].

A Gaussian process is a stochastic process where any collection of random variables drawn are jointly Gaussian. This is to say that any finite set of random variable observations will be distributed according to $f(\mathbf{x}_1, \ldots, \mathbf{x}_M) \sim \mathcal{N}(\mu, \Sigma),\ s.t\ \mu \in \mathbb{R}^M,\ \Sigma \in \mathbb{R}^{M \times M}$. We specify a Gaussian process by defining its mean and covariance functions, $m(\mathbf{x})$ and $k(\mathbf{x}, \mathbf{x}')$. Covariance functions describe a similarity measure between features in an infinite-dimensional function space, and are equivalent to positive-definite kernel functions [32]. Given a Gaussian process $f(\mathbf{x})$

$$\begin{aligned} f(\mathbf{x}) &\sim \mathcal{GP}(m(\mathbf{x}), k(\mathbf{x}, \mathbf{x}')) \\ m(\mathbf{x}) &= \mathbb{E}[f(\mathbf{x})] \\ k(\mathbf{x}, \mathbf{x}') &= \mathbb{E}[(f(\mathbf{x}) - m(\mathbf{x}))^T (f(\mathbf{x}') - m(\mathbf{x}'))]. \end{aligned} \tag{16.37}$$

Gaussian process regression specifies an approximated function by expressing a predictive distribution over a function space based on observations of realized random variables. This predictive distribution $p(f(\mathbf{x}')|\mathbf{X}, \mathbf{Y})$ is also known as a posterior, and can be computed given inputs $\mathbf{X} = [\mathbf{x}_1, \ldots, \mathbf{x}_M]$ and outputs $\mathbf{Y} = [\mathbf{y}_1, \ldots, \mathbf{y}_M]$ through a Bayesian inference framework

$$\begin{aligned} \text{posterior} &= \frac{\text{prior} \times \text{likelihood}}{\text{marginal likelihood}} \\ p(f(\mathbf{x})|\mathbf{X}, \mathbf{Y}) &= \frac{p(f(\mathbf{x}))p(\mathbf{Y}|\mathbf{X}, f(\mathbf{x}))}{\int p(\mathbf{Y}|f(\mathbf{x}), \mathbf{X})p(f(\mathbf{x})|\mathbf{X})df(\mathbf{x})}. \end{aligned} \tag{16.38}$$

By exploiting properties of the Gaussian distribution, one can avoid explicitly computing the posterior in (16.38). We use the fact that conditioning and marginalizations of Gaussian distributions are other Gaussians with closed-form solutions [32]. Without imposing any prior belief on the function space, this problem is ill posed. We will consider processes of the form $\mathbf{y}_k = f(\mathbf{x}_k) + \varepsilon$ with $\varepsilon \sim \mathcal{N}(\mathbf{0}, \sigma^2)$. As a prior, we restrict the set of functions considered to $\mathbf{y} = \mathbf{W}^T \Psi(\mathbf{x}) + \varepsilon$ such that $\mathbf{W} \sim \mathcal{N}(0, \Sigma_{\mathbf{W}})$ with zero mean for notational simplicity. We can see that the mean and covariance functions are then

$$\begin{aligned} m(\mathbf{x}) &= \mathbb{E}[f(\mathbf{x})] = \mathbb{E}[\mathbf{W}^T]\Psi(\mathbf{x}) + \mathbb{E}[\varepsilon] = \mathbf{0} \\ k(\mathbf{x}, \mathbf{x}') &= \mathbb{E}[f(\mathbf{x})^T f(\mathbf{x}')] = \Psi(\mathbf{x})^T \mathbb{E}[\mathbf{W}\mathbf{W}^T]\Psi(\mathbf{x}') + \mathbb{E}[\varepsilon^T \varepsilon] = \Psi(\mathbf{x})^T \Sigma_{\mathbf{W}} \Psi(\mathbf{x}') + \sigma^2. \end{aligned} \tag{16.39}$$

Given a design matrix as defined in (16.25), one can use the mean and covariance functions to define a Gaussian processes $f(\mathbf{x}) \sim \mathcal{GP}(m(\mathbf{x}), k(\mathbf{x}, \mathbf{x}'))$. To generate a predictive distribution, we then define a joint Gaussian distribution between the input dataset's design matrix $\Phi$ and a new observation $\Psi(\mathbf{x}')$

$$\begin{bmatrix} f(\mathbf{X}) \\ f(\mathbf{x}') \end{bmatrix} \sim \mathcal{N}\left(\mathbf{0}, \begin{bmatrix} \mathbf{k}(\Phi, \Phi) & \mathbf{k}(\Phi, \Psi(\mathbf{x}')) \\ \mathbf{k}(\Phi, \Psi(\mathbf{x}'))^T & k(\Psi(\mathbf{x}'), \Psi(\mathbf{x}')) \end{bmatrix}\right), \tag{16.40}$$

where the design matrix autocovariance is $\mathbf{k}(\Phi, \Phi) \in \mathbb{R}^{M \times M}$, and the covariance between the design matrix and the new observation $\mathbf{k}(\Phi, \Psi(\mathbf{x}')) \in \mathbb{R}^{M}$. We then express these covariances in closed form

$$\begin{bmatrix} f(\mathbf{X}) \\ f(\mathbf{x}') \end{bmatrix} \sim \mathcal{N}\left(\mathbf{0}, \begin{bmatrix} \Phi \Sigma_{\mathbf{W}} \Phi^T + \sigma^2 \mathbf{I} & \Phi \Sigma_{\mathbf{W}} \Psi(\mathbf{x}') + \sigma^2 \mathbf{1} \\ (\Phi \Sigma_{\mathbf{W}} \Psi(\mathbf{x}'))^T + \sigma^2 \mathbf{1}^T & \Psi(\mathbf{x}')^T \Sigma_{\mathbf{W}} \Psi(\mathbf{x}') + \sigma^2 \end{bmatrix}\right), \tag{16.41}$$

where $\mathbf{1} \in \mathbb{R}^M$, and $\mathbf{I}$ is the identity matrix. Equation (16.41) allows us to calculate a predictive posterior distribution for new observations of $f(\cdot)$ by marginalizing over the design matrix $\Phi$ and outputs $\mathbf{Y}$ [39]. The predictive posterior distribution is then of the form

$$\begin{aligned} f(\mathbf{x}')|\Phi, \mathbf{Y} &\sim \mathcal{N}(\mathbb{E}[f(\mathbf{x}')], \mathbb{V}[f(\mathbf{x}')]) \\ \mathbb{E}[f(\mathbf{x}')] &= \Psi(\mathbf{x}')^T \Sigma_{\mathbf{W}} \Phi^T (\Phi \Sigma_{\mathbf{W}} \Phi^T + \sigma^2 \mathbf{I})^{-1} \mathbf{Y} \\ \mathbb{V}[f(\mathbf{x}')] &= \Psi(\mathbf{x}')^T \Sigma_{\mathbf{W}} \Psi(\mathbf{x}') - \Psi(\mathbf{x}')^T \Sigma_{\mathbf{W}} \Phi^T (\Phi \Sigma_{\mathbf{W}} \Phi^T + \sigma^2 \mathbf{I})^{-1} \Phi \Sigma_{\mathbf{W}} \Psi(\mathbf{x}'). \end{aligned} \tag{16.42}$$

This posterior distribution enables us to generate predictions for new observations, as well as specify a covariance for each point [32].

Although we have constructed our derivation using a finite-dimensional basis representation, Gaussian processes are typically implemented using infinite-dimensional kernels. By following the same procedure shown in Sect. 16.3.2, kernelization is trivial and results in the following predictive distribution in terms of a specified kernel:

$$\begin{aligned} f(\mathbf{x}')|\Phi, \mathbf{Y} &\sim \mathcal{N}(\mathbb{E}[f(\mathbf{x}')], \mathbb{V}[f(\mathbf{x}')]) \\ \mathbb{E}[f(\mathbf{x}')] &= m(\mathbf{x}') + \mathbf{k}(\mathbf{x}')^T (\mathbf{G_r} + \sigma^2 \mathbf{I})^{-1} (\mathbf{Y} - m(\mathbf{X})) \\ \mathbb{V}[f(\mathbf{x}')] &= k(\mathbf{x}', \mathbf{x}') - \mathbf{k}(\mathbf{x}')^T (\mathbf{G_r} + \sigma^2 \mathbf{I})^{-1} \mathbf{k}(\mathbf{x}'), \end{aligned} \tag{16.43}$$

where we have added the mean function $m(\mathbf{x})$ for generality. The case in which $m(\mathbf{x}) = \mathbf{0}$ is referred to as a *centered* Gaussian process. This kernelized representation of the Gaussian process is not well suited for active learning settings. In order to overcome this limitation, there has been work in deriving local models for real-time applications [29].

### 16.3.4 Relationships Between Learning Techniques

Now that we have reviewed the foundations of generalized linear regression, kernel ridge regression, and Gaussian processes, we can examine their relationships to EDMD approximations of Koopman operators, as well as to one another. We develop assumptions for which these techniques are equivalent to one another.

#### 16.3.4.1 Generalized Linear Regression

For finite-dimensional bases, generalized linear regression is most closely related to EDMD. By limiting the functions we are willing to model to autoregressive functions, we can make the substitution $\mathbf{y}_k = \Psi(\mathbf{x}_{k+1})$ in (16.27). Then, if we note that $\mathbf{A}^\dagger = \mathbf{A}^{-1}$ when $\mathbf{A}$ is square, we can restate (16.27) in relation to the Koopman operator matrix components in (16.5)

$$
\begin{aligned}
\mathbf{W} &= \sum_{k=1}^{M-1} (\Psi(\mathbf{x}_k)\Psi(\mathbf{x}_k)^T)^\dagger (\Psi(\mathbf{x}_k)^T \Psi(\mathbf{x}_{k+1})) \\
&= ((\mathbf{G}[M])^\dagger)^T (\mathbf{A}[M])^T = (\mathbf{A}[M](\mathbf{G}[M])^\dagger)^T \\
&= \mathbf{U}^T. \qquad (16.44)
\end{aligned}
$$

Thus, generalized linear regression models are equivalent to finite-dimensional Koopman operators synthesized through EDMD for the same set of basis functions. This should not be surprising since Koopman operator synthesis via DMD [33, 38] and EDMD [41] is formulated as a regression problem.

The correspondence between these techniques indicates that all methods developed in Sect. 16.2 apply to generalized linear regression and vice versa. This means that generalized linear regression models can be applied in control, represent hybrid systems, update incrementally, and have information measures easily generated, which makes them suitable for experimental settings.

#### 16.3.4.2 Kernel Ridge Regression

Kernel ridge regression models are closely related to generalized linear regression models, and consequently Koopman operators. Ridge regression solves a slightly different optimization problem than EDMD, as shown in (16.32). However, for the case of $\lambda = 0$ the ridge regression optimization is equivalent to that of generalized linear regression and EDMD, as shown in (16.24).

For finite dimensional kernels of the form $k(\mathbf{x}, \mathbf{x}') = \langle \Psi(\mathbf{x}), \Psi(\mathbf{x}') \rangle,\ s.t.\ \Psi(\mathbf{x}) \in \mathbb{R}^n$, we have shown that the solutions in (16.34) and (16.33) are equivalent in Sect. 16.3.2. For $\lambda = 0$, (16.33) is equivalent to the generalized linear regression solution in (16.26), which we have shown in the previous section to be the same as that of EDMD. This is to say that the kernel ridge regression estimate can be equivalent to the Koopman operator's estimate derived in Sect. 16.2.1 for finite-dimensional kernels with a regularization coefficient of zero. In this limited representation, kernel ridge regression models can be computed in real time. However, it is important to note that kernel ridge regression without infinite-dimensional kernels or regularization is not how the technique is typically implemented, and does not take advantage of the compactness of kernel representations for high-dimensional data.

As shown by the work of [42], Koopman operators can be synthesized using infinite-dimensional kernel methods at the expense of growing computational complexity with additional samples. In this kernel-based representation, the Koopman operator estimate is equivalent to (16.36) for $\lambda = 0$. While the choice of finite-dimensional or infinite-dimensional kernel impacts online learning, the regularization coefficient does not. Hence, regularization coefficients may be used to promote sparsity in Koopman operators, which may be desirable in some settings.

#### 16.3.4.3 Gaussian Processes

Despite Gaussian processes being stochastic, they are closely related to kernel ridge regression. By examining the Gaussian processes' mean function in (16.43), we see that it is equivalent to the kernel ridge regression prediction in (16.35) given that we set $\sigma^2 = \lambda$, $\Sigma_W = \mathbf{I}$, and assume the Gaussian process is centered, with $\mathbf{W}$ as defined in (16.34). In order for the equivalence to hold between typical kernel ridge regression models and Gaussian processes, we ignore the process covariance estimate and sequential Bayesian model updates. However, recent work in Bayesian kernel ridge regression models has shown that they are equivalent to Gaussian processes when the Gaussian process covariance estimate is fixed [40].

The correspondence between Gaussian process regression models and Koopman operator synthesis follows from EDMD's relationship to kernel ridge regression as described in the previous section. By setting $\sigma^2 = 0$ and $\Sigma_W = \mathbf{I}$, the mean estimate from the Gaussian process regression matches the EDMD solution. However, by doing this we remove all non-determinism from the function approximation which makes it fundamentally *not* a Gaussian process. We note that the role $\sigma^2$ plays is very similar to that of $\lambda$ in kernel ridge regression, which indicates that it may be incorporated in the EDMD Koopman operator synthesis process. By increasing $\sigma^2$ one would prevent overfitting in the solution by promoting sparsity, and consequently admit more variance. Typical implementations of Gaussian processes scale in complexity with the number of data observed, and as such are not suitable for settings that demand incremental learning. Nonetheless, one can generate information measures for Gaussian processes with stationary mean and covariance functions [11].

Alternatively, generalized linear regression, kernel ridge regression, and DMD can be derived through a Bayesian inference framework [5, 37], which suggests that EDMD can be subject to a Bayesian treatment as well. It would be of interest to show a direct correspondence between Bayesian representations of the Koopman operator and Gaussian processes because of their relationship to neural networks.

Given a Gaussian prior over model weights, [28] showed that a single-layer neural network with infinite bandwidth approximates a Gaussian process as a result of the central limit theorem. This work was recently extended to describe the connection between Gaussian processes and deep neural networks in [24]. Moreover, neural networks have been applied both to Koopman operator synthesis [43], as well as basis function dictionary learning [25]. The relationships outlined in this section between Gaussian processes, neural networks and Koopman operators suggest a deeper connection between these methods which remains an open problem.

#### 16.3.4.4 Summary

The assumptions under which generalized linear regression, kernel ridge regression, and Gaussian processes produce the same solutions to an estimation problem as EDMD are summarized in Table 16.1. Approximate Koopman operators as formulated in Sect. 16.2.1 are equivalent to generalized linear regression models. Kernel ridge regression models can be made equivalent to a Koopman operator model under a set of simplifying assumptions for finite kernel representations, as well as for infinite-dimensional kernels by sacrificing online computation. Finally, the relationship between Gaussian processes and EDMD approximations of Koopman operators follows from its correspondence with kernel ridge regression in deterministic settings.

Although Gaussian processes and kernel ridge regression are often applied in real-time settings, in their typical formulations they are unable to be updated incrementally and are ill suited for active learning problems. However, generalized linear regression models are compatible with experimental settings because of their direct correspondence with EDMD. The correspondences between methods shown throughout this section imply that there should exist settings in which learning techniques such as Gaussian processes and deep neural networks can be made to learn incrementally and with respect to an information measure. Additionally, the relationship between

**Table 16.1** Assumptions under which each alternative method presented in Sect. 16.3 is equivalent to the EDMD estimate of a finite Koopman operator. As formulated, generalized linear regression models are equivalent to the synthesized Koopman operator matrix. The kernel ridge regression prediction is equivalent to that of EDMD when the regression regularization coefficient is 0, and only for finite-dimensional kernels. The mean function estimate of a Gaussian process is equivalent to an EDMD Koopman operator prediction when $\sigma^2 = 0$, $\Sigma_{\mathbf{W}} = \mathbf{I}$, for finite-dimensional kernels

| | Generalized linear regression | Kernel ridge regression | Gaussian processes |
|---|---|---|---|
| Koopman operators (EDMD) | $\mathbf{U} = \mathbf{W}^T$ | $\lambda = 0$<br>$k(\mathbf{x}, \mathbf{x}') = \langle \Psi(\mathbf{x}), \Psi(\mathbf{x}') \rangle$ | $\sigma^2 = 0$, $\Sigma_{\mathbf{W}} = \mathbf{I}$<br>$k(\mathbf{x}, \mathbf{x}') = \langle \Psi(\mathbf{x}), \Psi(\mathbf{x}') \rangle$<br>Ignore covariance estimate |

Gaussian processes and EDMD suggest that there should be a way to formulate Koopman operator synthesis to incorporate uncertainty estimates for predictions.

### *16.3.5 Example*

Though the methods presented in this section are formulated through alternative means, they all solve similar optimization problems. The primary difference between these methods is representation, which can lead to *the formal optimization being the same, but the implementation being different*. Under a set of assumptions we are able to relate the model representations of kernel ridge regression, Gaussian processes, and EDMD. Although this is technically the case, these assumptions do not represent typical use cases for these methods.

As a comparison between standard implementations of these methods, we generate three independent data-driven models of the dynamics of a double pendulum system—one with Koopman operators, one with Gaussian processes, and one with kernel ridge regression. We do not implement generalized linear regression because we showed it to produce an equivalent model representation to the EDMD solution in Sect. 16.3.4.1. We collected a training dataset consisting of a single 4 s trajectory taken from the double pendulum free dynamics sampled at 100 Hz with an initial condition of $(\theta_1, \theta_2, \dot{\theta}_1, \dot{\theta}_2) = (0.8, 0, 0, 0)$. We use the training dataset to instantiate a Koopman operator, kernel ridge regression model, and a Gaussian process model. The Koopman operator model is calculated using a second-order polynomial basis of the double pendulum states, while both the kernel ridge regression model and the Gaussian process use the Gaussian kernel shown in (16.30). The regularization and variance parameters were selected by the Scikit-Learn software package [30] via Bayesian hyperparameter optimization [35]. Each model is then used to predict double pendulum trajectories for a horizon of 3 s from each $(\theta_1, \theta_2)$ initial condition over the $\{(\theta_1, \theta_2) : [-1, 1] \times [-1, 1]\}$ domain, with zero initial velocity. Then, we calculate the integrated mean square error between each model's prediction and the actual dynamics over the entire horizon for each initial condition.

Figure 16.2 depicts the results of the comparison. The Koopman operator model prediction has the lowest average error of the three models. We observe that both the Gaussian process model and kernel ridge regression model prediction errors are lowest at the initial conditions of the training trajectory $(\theta_1, \theta_2) = (0.8, 0)$ while the Koopman operator model generalizes more easily over the domain. The purpose of this comparison is to show that while these methods can be shown to be closely related, standard implementations will produce very different results.

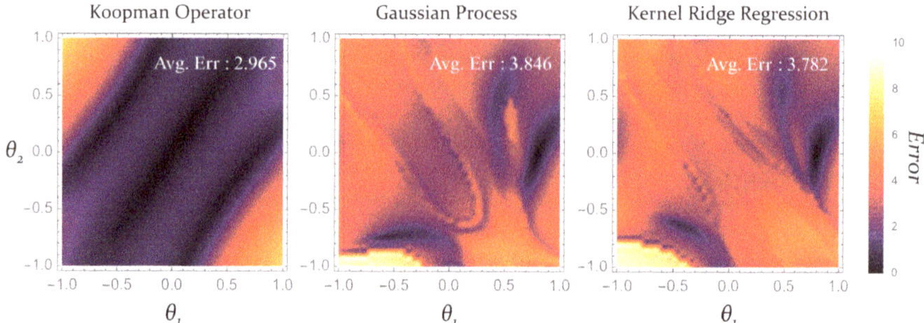


**Fig. 16.2** Performance comparison between EDMD, Gaussian processes, and kernel ridge regression. Each model was trained on the same dataset collected from a 4 s trajectory of the free dynamics of a double pendulum system with initial condition $(\theta_1, \theta_2, \dot{\theta}_1, \dot{\theta}_2) = (0.8, 0, 0, 0)$. Then, we calculated the integrated mean squared error over a 3 s prediction from each model of the double pendulum dynamics over the entire $\{(\theta_1, \theta_2) : [-1, 1] \times [-1, 1]\}$ domain. Standard implementations of each model were used so as to compare performance under typical usage

## 16.4 Experimental Applications

In this section, we demonstrate experimental applications of the Koopman operator in a variety of examples. We highlight applications in which active learning enables rapid system identification through information-driven incremental model updates. Additionally, we show how the Koopman operator's ability to represent hybrid systems enables the data-driven modeling of finite automata in experimental settings.

### *16.4.1 Learning Sphero SPRK Dynamics in Sand*

We are interested in demonstrating the use of a Koopman operator applied to learning complex nonlinear dynamics in a real-time experimental setting. We use the Sphero SPRK shown in Fig. 16.3, which is a programmable robotic toy shaped like a small ball that can be teleoperated using a controller or application. It is actuated through an internal mechanism designed for rolling on flat ground. If we change the system dynamics by switching the rolling surface to sand, the SPRK is incapable of rolling with its default controller. To enable the SPRK to roll on sand, we use the Koopman operator for active identification of the robot's dynamics on sand via real-time information maximization as described in Sect. 16.2.4, and shown in [1].

We use a nonlinear model-predictive controller known as sequential action control (SAC) [3] to generate information maximizing trajectories with respect to the information objective (16.20), which we use to update our Koopman operator model in real time using (16.7). Over a 20 s period, we synthesize information maximizing trajectories in order to characterize system dynamics on sand. After this period, we switch objectives to a tracking objective such as (16.13) where we attempt to trace

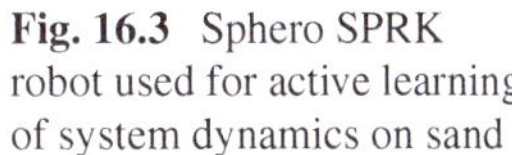

**Fig. 16.3** Sphero SPRK robot used for active learning of system dynamics on sand

a "figure 8" trajectory in a sandbox while continuing to update the Koopman operator. We compare the actively learned Koopman operator performance to a passive Koopman operator. The passive Koopman operator was computed from 20 s of data from tracing the "figure 8" trajectory with open-loop control.

The experiment consists of a rectangular sandbox, with an Xbox Kinect mounted overhead equipped with OpenCV providing odometry information. The system states collected are $\mathbf{x} = [x, y, \dot{x}, \dot{y}]^T$ where $(x, y)$ are coordinates in the Kinect field of vision and the corresponding velocities are estimated from finite differencing, and we define our control signals within this representation as $\mathbf{u} = [u_x, u_y]^T$. The set of basis functions used to model the SPRK robot were

$$
\begin{aligned}
\Psi_x(\mathbf{x}) &= [x, y, \dot{x}, \dot{y}, 1, \dot{x}^2, \dot{y}^2, \dot{x}^2\dot{y}, \ldots, \dot{x}^3\dot{y}^3]^T \in \mathbb{R}^{18}, \\
\Psi_u(\mathbf{u}) &= [u_x, u_y]^T \in \mathbb{R}^2,
\end{aligned}
\tag{16.45}
$$

which are linear combinations of third-order polynomial basis functions of the velocity states. The basis functions of control were chosen to be linear in $\mathbf{u}$ and not dependent on state.

In Fig. 16.4 we show the experimental results resulting from the active and passive Koopman operator learning approaches. The online learning approach is much more effective at tracking the desired trajectory than the fixed learning approach. Moreover, we notice that as the experiment progresses, the active Koopman operator trajectories continue to improve. The Koopman operator synthesized through active learning required no a priori knowledge of the system dynamics, bootstrapped initial model guess, or specification of granular media physics. Despite the fact that this problem could have been solved by other means, this experimental application is facilitated by the Koopman operator model representation.

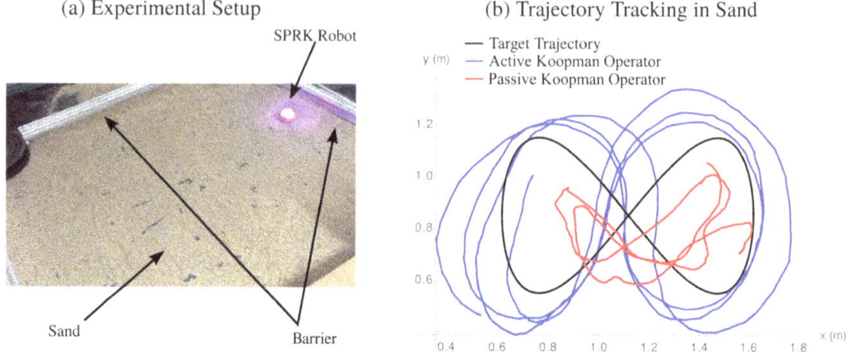


**Fig. 16.4** Experimental application of active learning with Koopman operators for online learning of the Sphero SPRK dynamics on sand. **a** Experimental setup consisting of a boxed area covered in sand, with an overhead Xbox Kinect using OpenCV reporting state information. **b** Results from trajectory tracking experiments in sand. The Sphero SPRK was used to track a "figure 8" trajectory (black), using both an active learning approach with a continuously updating Koopman operator (blue), and a precomputed passive Koopman operator (red). While the fixed model establishes a baseline performance, the active learning Koopman operator model continuously improves tracking performance as the experiment progresses

## *16.4.2 Rapid Quadcopter Stabilization with Active Learning*

The Koopman operator's ability to actively acquire information in real-time allows us to approach problems that demand rapid reactive efforts. For settings in which large disturbances permanently affect system dynamics, we must be able to quickly develop new models to adapt to the disturbance while still achieving an underlying goal. We demonstrate the Koopman operator's quick adaptability by stabilizing a simulated quadcopter with a motor malfunction. We define the quadcopter malfunction as a single motor operating at 80% capacity. However, the malfunction is unmodeled and unknown to the Koopman operator ahead of time, which means it must be learned in real time in order to stabilize the quadcopter and prevent it from falling, as shown in [1]. The quadcopter configuration is shown in Fig. 16.5, and the system dynamics used to carry out this simulated experiment are

$$\begin{aligned} \dot{\mathbf{h}} &= \mathbf{h} \begin{bmatrix} \hat{\boldsymbol{\omega}} & \mathbf{v} \\ \mathbf{0} & 0 \end{bmatrix} \\ \mathbf{J}\dot{\boldsymbol{\omega}} &= \mathbf{M} + \mathbf{J}\boldsymbol{\omega} \times \boldsymbol{\omega} \\ \dot{\mathbf{v}} &= \frac{1}{m} F \mathbf{e}_3 - \boldsymbol{\omega} \times \mathbf{v} - g \mathbf{R}^T \mathbf{e}_3, \end{aligned} \tag{16.46}$$

where the dynamics are defined on the Lie group $\mathbf{h} = (\mathbf{R}, \mathbf{p}) \in SE(3)$ in body-fixed coordinates, where $\hat{\boldsymbol{\omega}}$ is the angular velocity tensor of the vector $\boldsymbol{\omega}$, $\mathbf{v}$ is the vector of linear velocities. The matrix $\mathbf{J}$ is the inertia tensor, $m$ is the quadcopter mass, $g$ is acceleration due to gravity, and $\mathbf{e}_3$ is a body-fixed vector shown in Fig. 16.5.

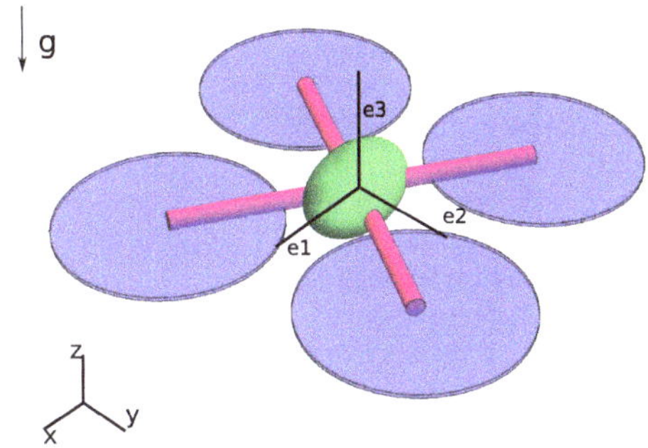


**Fig. 16.5** Quadcopter vehicle model configuration in body-fixed frame for active learning of vehicle dynamics under emergency conditions. The quadcopter dynamics are defined such that one motor has an unspecified malfunction and can only work at 80% capacity

The inputs to the system are a control force $F$ and control torque $\mathbf{M}$, each of which depends on the control vector $\mathbf{u} = [u_1, u_2, u_3, u_4]$ according to

$$F = k_t(u_1^2 + u_2^2 + u_3^2 + u_4^2),$$
$$\mathbf{M} = \begin{bmatrix} k_t l(u_2^2 - u_4^2) \\ k_t l(u_3^2 - u_1^2) \\ k_m(u_1^2 - u_2^2 + u_3^2 - u_4^2) \end{bmatrix}$$

with $k_t$, $k_m$, and $l$ as model parameters. Details regarding implementation of the quadcopter dynamics are covered in [13].

The set of basis functions chosen directly embed the quadcopter dynamics, such that the Koopman operator has a bootstrapped model of the nominal system dynamics

$$\Psi_x(\mathbf{x}) = [\mathbf{a_g}, \omega, \mathbf{v}, g(\mathbf{v}, \omega)]^T \in \mathbb{R}^{18},$$
$$\Psi_u(\mathbf{u}) = \mathbf{u} \in \mathbb{R}^4,$$

where $\mathbf{a_g} \in \mathbb{R}^3$ is the body-centered gravity vector, and $\mathbf{v}$, $\omega$ are as defined above. The additional basis functions $g(\mathbf{v}, \omega)$ consist of all elements of the outer product between the linear and angular velocity vectors, and were chosen without any a priori knowledge of the motor malfunction specification

$$\begin{aligned} g(\mathbf{v}, \omega) = [&v_1\omega_1, v_1\omega_2, v_1\omega_3, \\ &v_2\omega_1, v_2\omega_2, v_2\omega_3, \\ &v_3\omega_1, v_3\omega_2, v_3\omega_3]. \end{aligned}$$

The experimental procedure consists of setting the quadcopter in free fall from some initial condition and for a period of 1 s learning a Koopman operator online with respect to a specified objective. Active learning trials applied an information maximizing objective during the learning period, then switched to a stabilizing objective afterward. Passive learning trials followed a stabilizing objective for the entirety of each trial. After the 1 s learning period, we did not continue to update the operator online for either set of trials. A set of 100 Monte Carlo trials over uniformly dis-

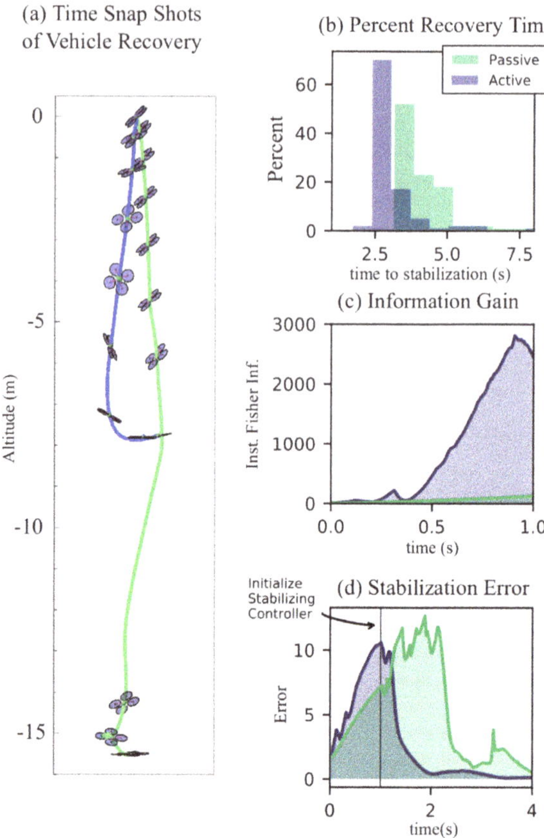


**Fig. 16.6** Quadcopter vehicle free-fall simulation where one motor has an unmodeled malfunction. The goal is to learn to stabilize the quadcopter under two learning approaches. The first is a passive learning approach (green) where a Koopman operator is learned online over 1 s while attempting to stabilize the quadcopter. The second is an active learning approach (blue) where a Koopman operator is generated while optimizing Fisher information, then the system switches objectives to achieve stabilization

tributed initial conditions was carried out. Figure 16.6 summarizes the experimental results. Figure 16.6a depicts timestamps from a pair of quadcopter trajectories under active and passive learning.

In Fig. 16.6b we display the results from the Monte Carlo trials. The active Koopman operator stabilized the quadcopter within 2.5 s in over 65% of trials, while the passive operator took another full second on average. Figure 16.6c shows the instantaneous information gain over the 1 s learning period for both passive and active learning approaches, where the active learning approach acquires orders of magnitude more information than the passive approach in a selected trial. Finally, Fig. 16.6d depicts the stabilization error for a particular trial with passive and active learning. The active learning trial stabilizes the quadcopter more quickly than the passive learning approach and with lower error after the stabilizing objective switch.

Through this experiment, we have shown that the Koopman operator with active learning can be used in time-critical applications. We introduced an unmodeled motor malfunction into the quadcopter dynamics and learned a Koopman operator in real time in order to stabilize the system. We showed that by first executing information

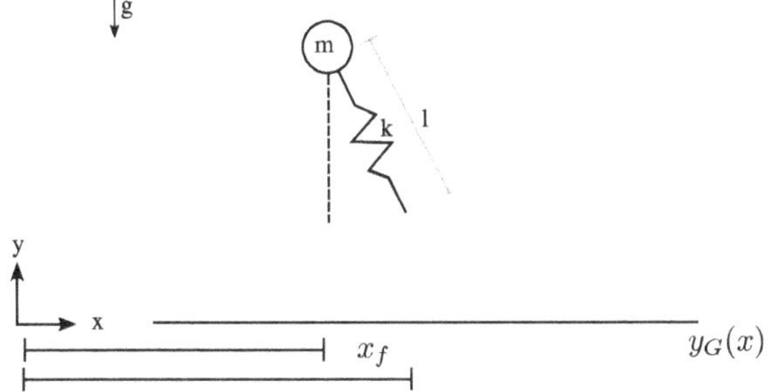


**Fig. 16.7** SLIP model configuration. The SLIP system obeys a set of hybrid dynamics corresponding to the separate stance and flight modes

maximizing trajectories prior to attempting to stabilize, we outperform passive online learning techniques.

### *16.4.3 Learning SLIP Hybrid Dynamics*

Hybrid systems can be difficult to model because of inherent discontinuities in the system dynamics. These discontinuities often cannot be represented by a single model. Experimental learning in uncertain environments in the real world may demand learning techniques that model such discontinuities. In particular, field robotics often concerns itself with modeling locomotive systems, such as bipedal or quadrupedal walkers. Walking robots deal with the discontinuities of impact at each step of their locomotive cycle, and as such must have a way to reason about them. The simplest commonly used dynamic walker model is the spring-loaded inverted pendulum (SLIP) model, shown in Fig. 16.7.

The SLIP system dynamics are split into two hybrid modes: stance and flight. The transition between these hybrid modes is described by an indicator function, often referred to as a guard equation, which specifies the set of dynamics evolving the system at a given coordinate in state space. The system states are $\mathbf{x} = [x, y, \dot{x}, \dot{y}, x_f]^T$, where $(x, y)$ indicates the position of the mass in the configuration specified in Fig. 16.7, and $x_f$ is the position of the SLIP model's foot. The model's control inputs are $\mathbf{u} = [u_s, u_f]^T$. The dynamics of the SLIP model are

$$f_{stance} = \begin{bmatrix} \dot{x} \\ \dot{y} \\ (k(l_0 - l(\mathbf{x})) + u_s)\frac{x - x_f}{ml(\mathbf{x})} \\ (k(l_0 - l(\mathbf{x})) + u_s)\frac{y}{ml(\mathbf{x})} - g \\ 0 \end{bmatrix}, \quad f_{flight} = \begin{bmatrix} \dot{x} \\ \dot{y} \\ 0 \\ -g \\ \dot{x} + u_f \end{bmatrix}, \tag{16.47}$$

where $l_0$ is the resting length of the spring, $k$ is its stiffness, and $l(\mathbf{x}) = \sqrt{(x - x_f)^2 + y^2}$ is its variable length [21]. Then, the indicator function is

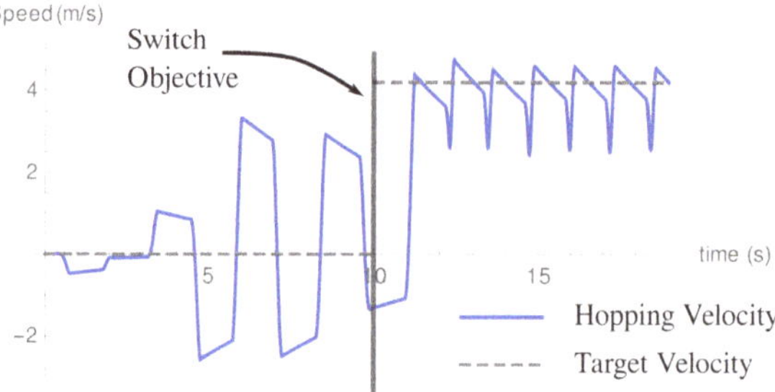


**Fig. 16.8** SLIP model objectives. First, the SLIP hopper follows an information maximizing objective for 10 s while learning a Koopman representation of its dynamics. Then, the SLIP hopper uses this model to carry out a forward trajectory

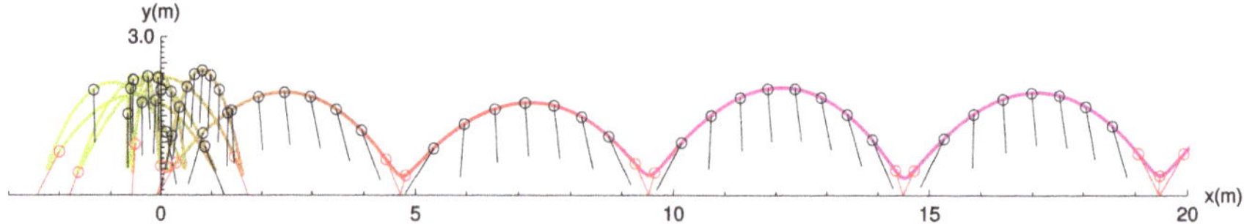


**Fig. 16.9** Active identification of simulated SLIP system dynamics. The SLIP system first obeys an information maximizing objective for 10 s, and then the objective is changed to following a forward velocity trajectory

$\Phi_{SLIP}(\mathbf{x}) = \text{sign}(1 - \frac{l_0}{l(\mathbf{x})}) \in \{-1, 1\}$. The complete hybrid dynamics are

$$f_{SLIP}(\mathbf{x}, \mathbf{u}) = \begin{cases} f_{stance}(\mathbf{x}, \mathbf{u}), & \text{if } \Phi_{SLIP}(\mathbf{x}) = -1 \\ f_{flight}(\mathbf{x}, \mathbf{u}), & \text{otherwise} \end{cases}. \tag{16.48}$$

We use the hybrid formulation of the Koopman operator as specified in Sect. 16.2.5, as well as the analytical indicator function to actively learn a hybrid Koopman operator representation of the SLIP model in simulation. For the first 10 s we have a simulated SLIP hopper follow an information maximizing trajectory, where each hybrid Koopman mode incrementally updates according to the indicator function. Following the active learning period, the objective changes and the model is made to follow a forward hopping velocity of 4.2 m/s. Results are shown in Fig. 16.9 where the hybrid Koopman model first maximizes information about its dynamics for 10 s, and then it moves forward tracking a trajectory. Figure 16.8 depicts the SLIP hopper's switch in objectives in terms of the model's forward velocity.

The Koopman operator's hybrid formulation enables modeling of discontinuous dynamical systems, which is of great importance in the field of locomotion, but also field robotics at large. Furthermore, the experiment presented an active learning framework for characterizing hybrid dynamical systems with known guard equations that could be extended to other walking robots.

### 16.4.4 Generating Finite Automata via DSS

In Sect. 16.4.3 we discussed the importance of modeling hybrid system dynamics in experimental field robotics. Although we were able to generate a hybrid Koopman model for the SLIP hopper, we required a priori specification of the system's indicator function. The indicator function is not generally known ahead of time. Thus, it is of interest to develop a systematic approach to identifying hybrid system dynamics, as well as the underlying indicator function specifying state-space boundaries between hybrid modes.

Hybrid dynamical systems are often represented as finite automata with discrete and continuous components. While the structure of the hybrid automaton is discrete, each of its nodes is a continuous dynamical system. The transitions are specified by the indicator function and guard equations. This graphical model is a compact representation of the system dynamics [19]. Dynamical system segmentation (DSS), proposed in [4], is an algorithm that generates state-space partitions of dynamical systems from data, thereby generating a hybrid automaton. For hybrid systems, DSS can identify the dynamics of each hybrid mode and generate an indicator function from data. For non-hybrid systems, DSS synthesizes a finite automaton representation of the dynamics from data. The state-space continuous dynamics are partitioned into a set of distinct dynamics governing a given region of the state-space manifold.

Given a dataset $\mathbf{X} = [\mathbf{x}_1, \ldots, \mathbf{x}_M]$ from some dynamical system and a set of basis functions $\Psi(\mathbf{x}) \in \mathbb{R}^N$, DSS synthesizes a set of Koopman operators from subsets of the dataset $\mathbf{U}_i = GenerateKoopman(\mathbf{X}_{a:b}),\ s.t.\ \mathbf{X}_{a:b} = [\mathbf{x}_a, \ldots, \mathbf{x}_b]$. The choice of method for splitting the dataset $X$ into subsets is left to the user. This set of $W$ Koopman operators $\mathcal{U} = \{\mathbf{U}_1, \ldots, \mathbf{U}_W\}$ are each a local estimate of the system dynamics for some neighborhood of the state-space manifold. However, given that this set $\mathcal{U}$ may have redundancies we are interested in distilling a minimal set of Koopman operators to represent the system dynamics. By considering each Koopman operator $\mathbf{U}_i \in \mathbb{R}^{N\times N}$ as points in $\mathbb{R}^{N^2}$ space, we can use a clustering algorithm to construct a set of exemplar operators $\overline{\mathcal{U}}$ from the set $\mathcal{U}$. We suggest nonparametric clustering algorithms, such as HDBSCAN [8], in order to avoid presupposing a number of clusters. We will consider the minimal set of operators $\overline{\mathcal{U}}$ as a node set in a finite automaton. To derive the edge set $\mathcal{E} = \{(\mathbf{U}_i, \mathbf{U}_j), \ldots\}$, we generate a support vector machine (SVM), $\Pi(\Psi(\mathbf{x}))$, using the labeled dataset $\mathbf{X}$ and determine state-space transitions between nodes [30]. This graphical model $\mathbb{G} = (\overline{\mathcal{U}}, \mathcal{E})$ is the output of DSS. Figure 16.10 depicts a notional example of DSS applied to some dynamical system, as well as the relationship between the graph and the partitioned state-space manifold.

To demonstrate the DSS algorithm's performance in identifying hybrid systems, we applied DSS to a dataset of a simulated SLIP hopper as formulated in Sect. 16.4.3 following a forward constant velocity trajectory. Figure 16.11 depicts a successful identification of the bimodal SLIP dynamics purely from data without *any* prior knowledge of the system dynamics. The data-driven SVM indicator function

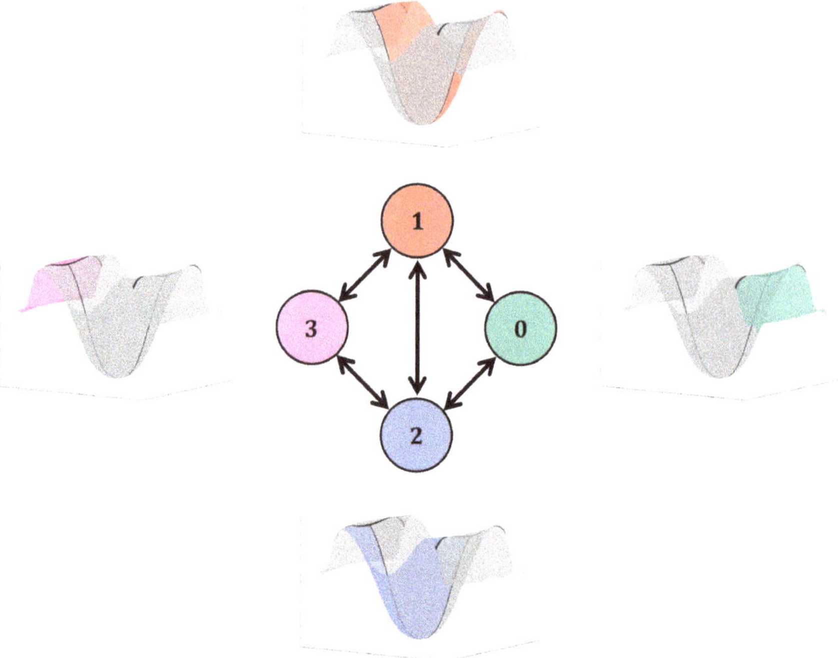


**Fig. 16.10** Segmentation of a non-hybrid dynamical system into a hybrid automaton represented as a graph. Each node in the graph is a distinct dynamical system represented by a Koopman operator governing the system dynamics over a region of the underlying state-space manifold. The partitions of the manifold are specified by an SVM model $\Pi(\Psi(x))$

$\Pi(\Psi(\mathbf{x}))$ correctly captures the SLIP system transitions between flight and stance modes.

Additionally, we segment non-hybrid systems to showcase the resulting finite descriptions of state-space continuous systems. We use the well-studied cart–pendulum inversion problem as an example. The cart–pendulum system, shown in Fig. 16.12, is actuated about its horizontal axis and has $\theta = 0$ defined at the unstable equilibrium point. We used the same nonlinear model-predictive controller as in Sect. 16.4.1, SAC, to synthesize a set of 30 trajectories with the goal of inverting and stabilizing the cart–pendulum system. We then applied DSS to this dataset and found that the pendulum inversion task could be segmented into three hybrid modes corresponding to swing-up, slow-down, and stabilization dynamics. The synthesized finite automaton resulting from DSS is shown in Fig. 16.13. Each node is accompanied by its corresponding partition of the original dataset shown in the $(\theta, \dot{\theta})$ phase plane.

Dynamical system segmentation is an extension of the hybrid dynamical formulation of the Koopman operator described in Sect. 16.2.5, and showcased in Sect. 16.4.3. DSS generates finite representations of dynamical systems from data without any prior knowledge of the system. As an experimental tool, as was shown in [4], DSS

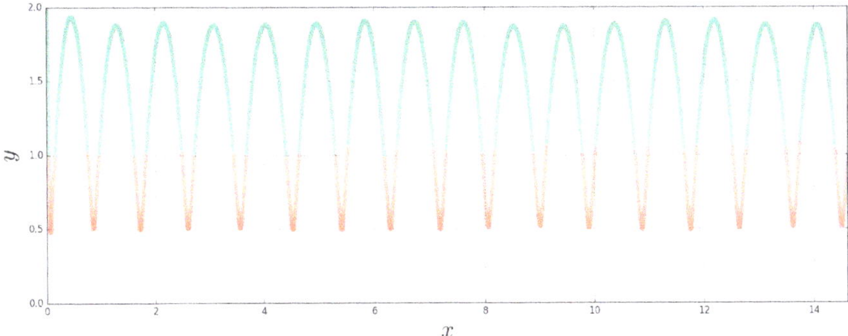


**Fig. 16.11** Dynamical system segmentation of the SLIP model system described in Sect. 16.4.3. The DSS algorithm without any a priori knowledge of the SLIP system dynamics is able to discern the two hybrid modes of flight and stance, as well as the indicator function that switches between modes. The plot shows the position of the SLIP hopper's mass color coded according to the hybrid mode it is currently in. Green corresponding to flight, orange to stance

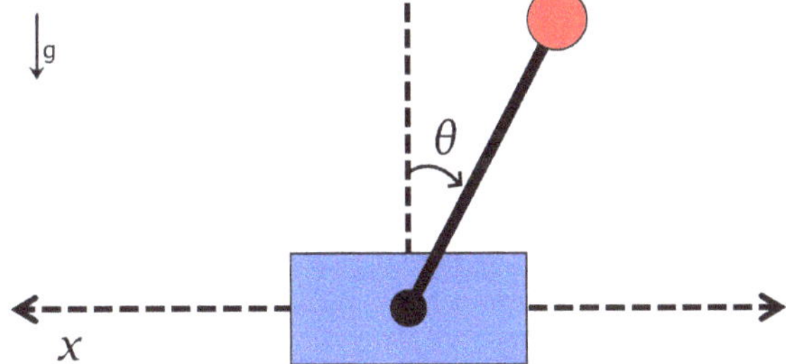


**Fig. 16.12** Cart–pendulum system used for DSS, with its unstable equilibrium defined at $\theta = 0$

is capable of discerning complex relationships from motion data and formulating them in a low-dimensional representation. Identification of discontinuous phenomena from data is of great importance in experimental settings where environmental interactions are often difficult to model, and the Koopman operator provides us with the ability to formulate algorithms to address this problem.

## 16.5 Conclusion

Experimental learning is necessary for real-time contexts where the environment and internal dynamics can change rapidly. The learning techniques we apply should reflect that these settings demand incremental learning in time-varying environments, and most machine learning techniques do not. We have proposed the EDMD-synthesized finite-dimensional Koopman operator as a technique well suited to experimental settings and compared it to similar techniques to characterize their relationships. Moreover, we provide a set of conditions that specify the connec-

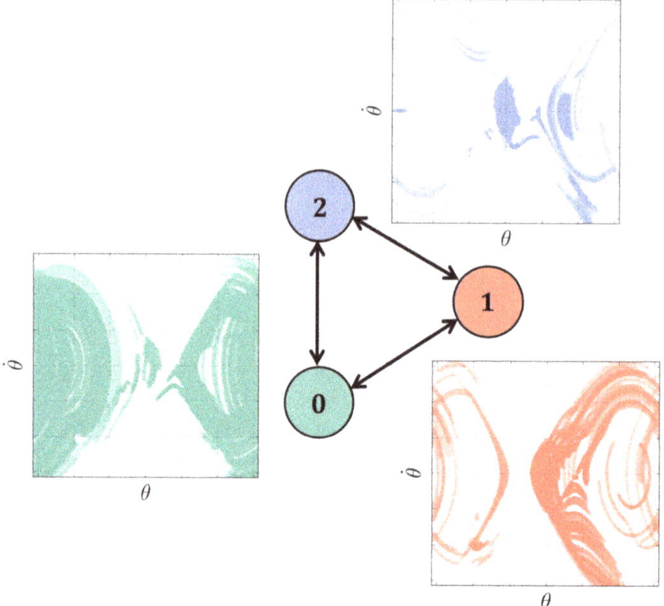


**Fig. 16.13** Application of DSS to a set of 30 control solutions to the cart–pendulum inversion problem. Each node in the generated graph represents a Koopman operator that describes the cart–pendulum nonlinear dynamics over a particular region of the state-space manifold. Mode 0 corresponds to energy pumping and swing-up, mode 1 corresponds to energy removal and slow-down, and mode 2 corresponds to stabilization

tion between EDMD and alternative learning methods. We highlight the synthesized Koopman operator's capabilities in a variety of experimental applications. Modeling real processes requires techniques capable of purposeful learning, and we propose the Koopman operator as a suitable model representation for these settings.

**Acknowledgements** This work was supported by the National Science Foundation under grant CBET-1637764. Any opinions, findings, conclusions, or recommendations expressed in this material are those of the authors and do not necessarily reflect the views of the National Science Foundation.

## References

1. Abraham, I., Murphey, T.D.: Active learning of dynamics for data-driven control using koopman operators. IEEE Trans. Robot. **35**(5), 1071–1083 (2019)
2. Abraham, I., Torre, G., Murphey, T.: Model-based control using Koopman operators. In: Robotics: Science and Systems (2017)
3. Ansari, A., Murphey, T.D.: Sequential action control: Closed-form optimal control for nonlinear and nonsmooth systems. IEEE Trans. Robot. **32** (2017)

4. Berrueta, T.A., Pervan, A., Fitzsimons, K., Murphey, T.D.: Dynamical system segmentation for information measures in motion. IEEE Robot. Autom. Lett. **4**(1), 169–176 (2019)
5. Bishop, C.: Pattern Recognition and Machine Learning. Information Science and Statistics. Springer, Berlin (2006)
6. Brunton, S., Brunton, B., Proctor, J., Kutz, J.: Koopman invariant subspaces and finite linear representations of nonlinear dynamical systems for control. PLOS ONE **11**(2), 1–19 (2016)
7. Budišić, M., Mohr, R., Mezić, I.: Applied Koopmanism. Chaos: Interdiscip. J. Nonlinear Sci. **22**(4), 047, 510 (2012)
8. Campello, R., Moulavi, D., Sander, J.: Density-based clustering based on hierarchical density estimates. Advances in Knowledge Discovery and Data Mining, pp. 160–172. Springer, Berlin (2013)
9. Chen, B.W., Abdullah, N.N.B., Park, S., Gu, Y.: Efficient multiple incremental computation for kernel ridge regression with Bayesian uncertainty modeling. Futur. Gener. Comput. Syst. **82**, 679–688 (2018)
10. Chirikjan, G.S.: Stochastic Models, Information Theory, and Lie Groups, vol. 2. Springer, Berlin (2012)
11. Dias, J., Leitao, J.: A method for computing the information matrix of stationary Gaussian processes. In: 1996 8th European Signal Processing Conference (EUSIPCO 1996), pp. 1–4 (1996)
12. Engel, Y., Mannor, S., Meir, R.: The kernel recursive least squares algorithm. IEEE Trans. Signal Process. **52**(8), 2275–2285 (2004)
13. Fan, T., Murphey, T.: Online feedback control for input-saturated robotic systems on Lie groups. In: Robotics: Science and Systems (2016)
14. Gallager, R.: Stochastic Processes: Theory for Applications. Cambridge University Press, Cambridge (2013)
15. Govindarajan, N., Arbabi, H., van Blargian, L., Matchen, T., Tegling, E., Mezić, I.: An operator-theoretic viewpoint to non-smooth dynamical systems: Koopman analysis of a hybrid pendulum. In: 2016 IEEE 55th Conference on Decision and Control (CDC), pp. 6477–6484 (2016)
16. Hamilton, J.: Time Series Analysis. Princeton University Press, Princeton (1994)
17. Heersink, B., Warren, M., Hoffmann, H.: Dynamic mode decomposition for interconnected control systems (2017). arxiv:1709.02883
18. Hemati, M.S., Williams, M.O., Rowley, C.W.: Dynamic mode decomposition for large and streaming datasets. Phys. Fluids **26**(11), 111, 701 (2014)
19. Henzinger, T.A.: The theory of hybrid automata. In: Proceedings of the 11th Annual IEEE Symposium on Logic in Computer Science, pp. 278–283 (1996)
20. Kaiser, E., Kutz, J.N., Brunton, S.L.: Data-driven discovery of Koopman eigenfunctions for control (2017). arxiv:1707.01146
21. Kalinowska, A., Fitzsimons, K., Dewald, J.P., Murphey, T.D.: Online user assessment for minimal intervention during task-based robotic assistance. In: Robotics: Science and Systems (2018)
22. Koopman, B.: Hamiltonian systems and transformation in Hilbert space. Proc. Natl. Acad. Sci. **17**(5), 315–318 (1931)
23. Korda, M., Mezić, I.: Linear predictors for nonlinear dynamical systems: Koopman operator meets model predictive control. Automatica **93**, 149–160 (2018)
24. Lee, J., Bahri, Y., Novak, R., Schoenholz, S.S., Pennington, J., Sohl-Dickstein, J.: Deep neural networks as Gaussian processes (2018). arxiv:1711.00165
25. Li, Q., Dietrich, F., Bollt, E.M., Kevrekidis, I.G.: Extended dynamic mode decomposition with dictionary learning: a data-driven adaptive spectral decomposition of the Koopman operator. Chaos: Interdiscip. J. Nonlinear Sci. **27**(10), 103, 111 (2017)
26. Mezić, I.: Analysis of fluid flows via spectral properties of the Koopman operator. Annu. Rev. Fluid Mech. **45**(1), 357–378 (2013). https://doi.org/10.1146/annurev-fluid-011212-140652
27. Murphy, K.P.: Machine Learning: A Probabilistic Perspective. The MIT Press, Cambridge (2012)

28. Neal, R.M.: Priors for infinite networks. Bayesian Learning for Neural Networks, pp. 29–53. Springer, Berlin (1996)
29. Nguyen-Tuong, D., Peters, J., Seeger, M.: Local Gaussian process regression for real time online model learning and control. In: Proceedings of the 21st International Conference on Neural Information Processing Systems, pp. 1193–1200 (2008)
30. Pedregosa, F., Varoquaux, G., Gramfort, A., Michel, V., Thirion, B., Grisel, O., Blondel, M., Prettenhofer, P., Weiss, R., Dubourg, V., Vanderplas, J., Passos, A., Cournapeau, D., Brucher, M., Perrot, M., Duchesnay, E.: Scikit-learn: machine learning in python. J. Mach. Learn. Res. **12**, 2825–2830 (2011)
31. Peitz, S., Klus, S.: Koopman operator-based model reduction for switched-system control of PDEs (2017). arXiv:1710.06759
32. Rasmussen, C.E., Williams, C.K.I.: Gaussian Processes for Machine Learning. Adaptive Computation and Machine Learning. The MIT Press, Cambridge (2005)
33. Schmid, P.J.: Dynamic mode decomposition of numerical and experimental data. J. Fluid Mech. **656**, 5–28 (2010)
34. Seal, H.L.: Studies in the history of probability and statistics. XV: the historical development of the Gauss linear model. Biometrika **54**(1/2), 1–24 (1967)
35. Shahriari, B., Swersky, K., Wang, Z., Adams, R., de Freitas, N.: Taking the human out of the loop: a review of Bayesian optimization. Proc. IEEE **104**(1), 148–175 (2016)
36. Srivastava, J., Anderson, D.: A comparison of the determinant, maximum root, and trace optimality criteria. Commun. Stat. **3**(10), 933–940 (1974)
37. Takeishi, N., Kawahara, Y., Tabei, Y., Yairi, T.: Bayesian dynamic mode decomposition. In: Proceedings of the Twenty-Sixth International Joint Conference on Artificial Intelligence, pp. 2814–2821 (2017)
38. Tu, J., Rowley, C., Luchtenburg, D., Brunton, S., Kutz, J.: On dynamic mode decomposition: theory and applications. J. Comput. Dyn. **1**, 391 (2014). https://doi.org/10.3934/jcd.2014.1.391
39. Umlauft, J., Beckers, T., Kimmel, M., Hirche, S.: Feedback linearization using Gaussian processes. In: 2017 IEEE 56th Annual Conference on Decision and Control (CDC), pp. 5249–5255 (2017)
40. Van Vaerenbergh, S., Fernandez-Bes, J., Elvira, V.: On the relationship between online Gaussian process regression and kernel least mean squares algorithms. In: 2016 IEEE International Workshop on Machine Learning for Signal Processing (2016)
41. Williams, M., Kevrekidis, I., Rowley, C.: A data-driven approximation of the Koopman operator: extending dynamic mode decomposition. J. Nonlinear Sci. **25**, 1307–1346 (2015)
42. Williams, M.O., Rowley, C.W., Kevrekidis, I.G.: A kernel-based method for data-driven Koopman spectral analysis. J. Comput. Dyn. **2**, 247–265 (2015)
43. Yeung, E., Kundu, S., Hodas, N.: Learning deep neural network representations for Koopman operators of nonlinear dynamical systems (2017). arxiv:1708.06850

# Chapter 17
# Application of Koopman-Based Control in Ultrahigh-Precision Positioning

**Saša Zelenika, Ervin Kamenar, Milan Korda and Igor Mezić**

**Abstract** Ultrahigh-precision positioning devices are of outmost importance in microsystems' technologies and precision engineering. The frictional disturbances of mechanical elements in relative motion often limit their positioning performances. If nanometric positioning precision and accuracy are aimed for, frictional disturbances have thus to be identified, modeled and compensated for via appropriate control algorithms. Suitable experimental setups are therefore employed to study the effects of frictional disturbances. The parameters related to state-of-the-art friction models are experimentally identified. Different control algorithms, such as a PID controller, a feedforward controller, and adaptive controllers, are experimentally and numerically validated and compared. It is proven that adaptive controllers enable nanometric precision positioning, but in point-to-point positioning applications can give rise to large overshoots and issues related to lengthy settling times. It is shown that these

S. Zelenika · E. Kamenar (✉)
Faculty of Engineering, University of Rijeka, Vukovarska 58, 51000 Rijeka, Croatia
e-mail: ekamenar@riteh.hr

S. Zelenika
e-mail: sasa.zelenika@riteh.hr

S. Zelenika · E. Kamenar · I. Mezić
Centre for Micro- and Nanosciences and Technologies, University of Rijeka, Radmile Matejčić 2, 51000 Rijeka, Croatia

M. Korda
Centre national de la recherche scientifique, 7 avenue du Colonel Roche, 31400 Toulouse, France
e-mail: korda@laas.fr

I. Mezić
Department of Mechanical Engineering, University of California Santa Barbara, Santa Barbara, CA 93105, USA
e-mail: mezic@ucsb.edu



A. Mauroy et al. (eds.), *The Koopman Operator in Systems and Control*, Lecture Notes in Control and Information Sciences 484,
https://doi.org/10.1007/978-3-030-35713-9_17

problems can be minimized by employing the Koopman-based model predictive control that allows simplifying the modeling burden while successfully compensating the frictional effects.

## 17.1 Introduction

Ultrahigh-precision positioning devices are of outmost importance in microsystems' technologies and precision engineering in general. In fact, these devices are often used in machine tools, optical systems, robots, automotive industry, aerospace and astrophysics, home appliances, scientific instrumentation, measurement equipment, information technologies (IT), in handling and assembly of micro- and nano-electromechanical systems (MEMS and NEMS), etc. high-precision assures in this frame, among others, better product quality and reliability, the automation of the assembly processes, interchangeability of parts, longer lifetime (i.e., lower wear and fatigue) and higher packaging density [14, 25, 26, 31–33, 43].

Compliant devices and mechanisms are a viable means to achieve ultrahigh-precision positioning, but their usage is limited by the achievable travel ranges and load capacities [10, 15, 23, 24, 34, 42]. Displacements of ultrahigh-precision mechatronics systems are therefore still generally achieved by employing sliding and rolling machine elements, such as ball- and lead-screws, ball bearings, and linear guideways, all characterized by frictional stochastic disturbances often limiting their positioning performances. In fact, friction is characterized by a stochastic nonlinear behavior, which is time-, position-, and temperature-dependent (with measured variations of friction forces sometimes even bigger than 10%) [14]. This influences significantly precision and accuracy of the positioning systems by inducing limit cycles around the reference position (i.e., hunting), stick-slip, static ("lost motion") and tracking errors, as well as large settling times, and can thus be an obstruction to effective performances of precision machines directly influencing product quality [4, 12].

If nanometric positioning precision and accuracy are needed, frictional disturbances have therefore to be identified, modeled and compensated for via appropriate control methodologies [14]. Suitable experimental setups, characterized by multiple friction sources and pre-sliding and sliding motion regimes, are thus employed to study the effects of frictional disturbances on positioning performances and to investigate compensation typologies based on different control algorithms [12, 41].

In this frame, sliding and pre-sliding regimes have to be discerned. In fact, in the sliding motion regime, friction is characterized by stiction (i.e., static friction), Coulomb friction and viscous friction, i.e., by the conventional Stribeck friction model as shown in Fig. 17.1.

On the other hand, pre-sliding, a consequence of adhesion, material transfer due to adhesion, and "ploughing" due to surface roughness, gives rise to displacements that occur even for loads smaller than those needed to overcome stiction. In this motion regime, the adhesive forces at asperity contacts are dominant and asperity junctions deform inducing an elastoplastic (i.e., similar to the effect of nonlinear

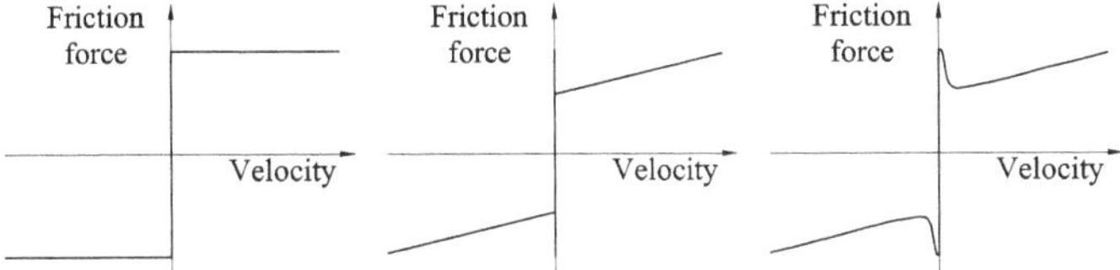


**Fig. 17.1** Conventional friction models: Coulomb model (left), Coulomb plus static plus viscous model (center) and overall Stribeck curve (right). Reproduced from [12]

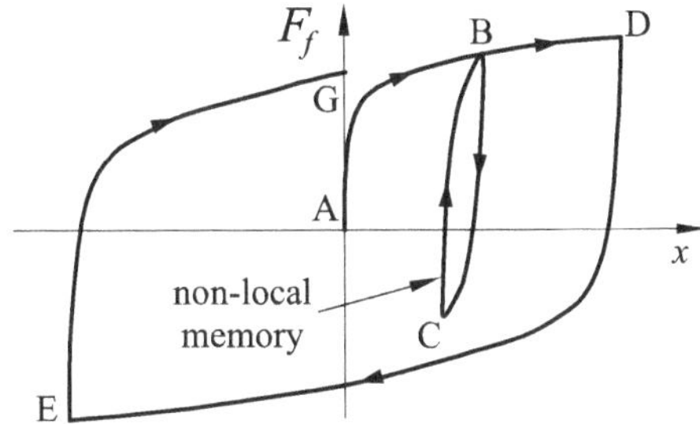


**Fig. 17.2** Frictional response in the pre-sliding motion regime

springs—though mainly plastic) hysteretic frictional disturbance. The latter causes a frictional force versus displacement dependence such that, at each displacement reversal, a closure of the inner hysteresis loop is obtained (resulting in a so-called nonlocal memory effect) and the curve of the outer loop is followed again [1, 14, 16, 35]. In this case, friction is a function of displacement rather than velocity, resulting in a substantial nonlinear displacement (Fig. 17.2). As displacement increases, the junctions will break and the friction force in sliding will become a function of velocity [1, 4, 14, 16, 29, 30, 35, 43].

The models based on the physics of the frictional behavior in the pre-sliding motion regime, which have to be related to the experimentally identified characteristic parameters of the studied phenomenon, are referred to in literature as grey-box models [38]. Among many grey-box pre-sliding friction models described in literature, extensively used for control purposes are Dahl's model [8], the LuGre model [7, 9], Hsieh's model [11], and the generalized Maxwell-slip (GMS) model [2].

The parameters related to these state-of-the-art friction models are hence experimentally identified [16] and the applicability of different control approaches, such as the proportional-integral-derivative (PID) controller, the feedforward controller, adaptive control typologies (model reference adaptive control (MRAC) and a self-tuning regulator (STR)), is experimentally assessed as well as numerically validated and compared in the MATLAB/Simulink environment. It is hence proven that adaptive controllers enable nanometric precision positioning, but in point-to-point experiments can give rise to large overshoots and longer settling times. It is shown that these problems can be minimized by employing Koopman-based model predictive control (MPC). In fact, this approach allows simplifying the modeling burden while successfully compensating the frictional effects and concurrently eliminating steady-state errors as well as substantially reducing overshoot and settling times [17].

## 17.2 Modeling and Identification of Friction

The basic characteristics of the friction models outlined and compared in this work are elucidated in this section.

### 17.2.1 *Dahl's Pre-sliding Friction Model*

Dahl's friction model, which can account only for the pre-sliding motion regime, is based on a spring-like behavior at the junction between the surfaces in relative motion that, for moderately large displacements, exhibits plastic deformations. The friction force $F_f$ in pre-sliding can here be expressed as a function of the displacement $x$ [8]:

$$\begin{aligned}\frac{dF_f}{dt} &= \frac{dF_f}{dx} \times \frac{dx}{dt} \\ &= \sigma_0 \left| 1 - \frac{F_f}{F_C} \times \text{sgn}\left(\frac{dx}{dt}\right) \right|^n \times \text{sgn}\left(1 - \frac{F_f}{F_C} \times \text{sgn}\left(\frac{dx}{dt}\right)\right) \times \frac{dx}{dt},\end{aligned} \tag{17.1}$$

where $\sigma_0$ represents the stiffness of the asperities on the surfaces in relative motion at the beginning of detectable displacements (i.e., the slope of the force vs. deflection curve for $F_f \approx 0$), $F_C$ is the value of the Coulomb friction force when slipping begins, $n$ is a parameter that determines the shape of the force versus displacement curve, while $t$ is time. The characteristic parameters of Dahl's model can be obtained according to the method described in [29]. The method is based on the minimization of the sum-of-squares of the relative errors between the measured and the modeled friction.

### 17.2.2 *The LuGre Friction Model*

In the LuGre friction model the contact asperities between the bodies in relative motion are represented as bristles whose average deformation is defined by a state variable $z$ [7, 9]:

$$\frac{dz}{dt} = \frac{dx}{dt} - z \times \frac{\sigma_0}{s(v)} \times \left| \frac{dx}{dt} \right|, \tag{17.2}$$

where $s(v)$ represents a velocity weakening curve commonly associated to the Stribeck effect at the transition between pre-sliding and sliding motion. The friction force is, in turn, defined as

$$F_f = \sigma_0 \times z + \sigma_p \times \frac{dz}{dt}, \tag{17.3}$$

where $\sigma_p$ is the viscous damping coefficient related to the pre-sliding friction state.

The parameters of the LuGre friction model can be obtained by using the same method as in the case of Dahl's model [29].

### 17.2.3 Hsieh's Pre-sliding Friction Model

Hsieh's friction model consists of a nonlinear spring module and a plastic module (Fig. 17.3a). The nonlinear spring embraces the hysteresis loop with memory and wipeout effects and it is connected in parallel with a viscous damper $C_v$ that takes into account energy dissipation. The plastic module, in turn, describes deformation effects and it encompasses creep and work hardening.

Due to the serial connection of the two modules, an external force $F$ will result in a pre-sliding displacement $x$, given by the sum of the displacement of the nonlinear spring module $x_s$ and of the deformation of the plastic module $x_p$, and an opposing pre-sliding friction force $F_f$ [11]:

$$F_f = \begin{cases} k_{1s}(x_s - x_r) + \frac{k_{2s}}{\beta}\left(1 - e^{-\beta|x_s - x_r|}\right) + \sigma_r + C_v\dot{x}_s, & \text{if } x_s \geq x_r \\ k_{1s}(x_s - x_r) + \frac{k_{2s}}{\beta}\left(e^{-\beta|x_s - x_r|} - 1\right) + \sigma_r + C_v\dot{x}_s, & \text{if } x_s < x_r, \end{cases} \tag{17.4}$$

where $k_{1s}$ and $k_{2s}$ are stiffness coefficients, $\beta$ is a positive scalar, while $\sigma_r$ and $x_r$ are, respectively, the force and the displacement at the reversal of the inner hysteresis loop (see points B and C in Fig. 17.2). The parameters $k_{1s}$, $k_{2s}$, and $\beta$ can be estimated from the experimentally obtained $F_f$ versus $x$ curve as shown in Fig. 17.3b [16].

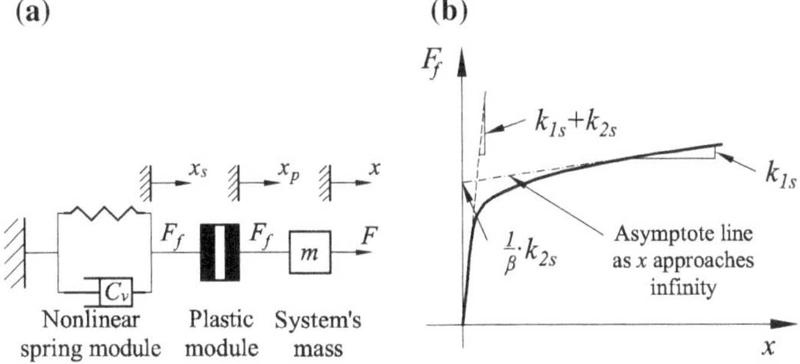


**Fig. 17.3** Schematic representation of Hsieh's friction model (**a**) and determination of its nonlinear spring parameters (**b**). Reprinted from [16] by permission of SAGE Publications

On the other hand, the plastic behavior with its work hardening $x_h$ is defined as

$$\dot{x}_h = \begin{cases} \rho \left[ \frac{\left| F_f^a \right|}{\psi} - x_h \right], & \text{if } \left( \frac{\left| F_f^a \right|}{\psi} \right) > x_h \\ 0, & \text{otherwise} \end{cases} \tag{17.5}$$

$$\dot{x}_p = \operatorname{sgn}\left(F_f\right) \dot{x}_h, \tag{17.6}$$

where $a$ and $\psi$ are positive constants related to work hardening, while $\rho$ is a positive constant related to creep. The parameters $\rho$, $a$ and $\psi$ can be estimated from experimental data assuming that the nonlinear spring has not undergone any deflection prior to the experiment [11, 16].

### 17.2.4 The GMS Friction Model

The GMS friction model is based on $N$ massless Maxwell-slip blocks connected in parallel that have all the same input—velocity $v$, and one output—the friction force $F_i$ acting on the $i$th block (Fig. 17.4a).

Two states of either hysteresis loop with nonlocal memory in pre-sliding (where $v$ is the derivative of the already defined state variable $z$) and slip with frictional lag, determine the behavior of each block. The latter depends on the stiffness $k_i$ of the block and the force threshold $W_i$ when the block starts slipping. Sliding dynamics of each block can, in turn, be represented by the Coulomb slip law (see Fig. 17.1 left); if a more accurate modeling is needed, this can be replaced by the well-known Stribeck effect, i.e., via the so-called velocity weakening effect $s(v)$ bounded on the lower end by Coulomb friction.

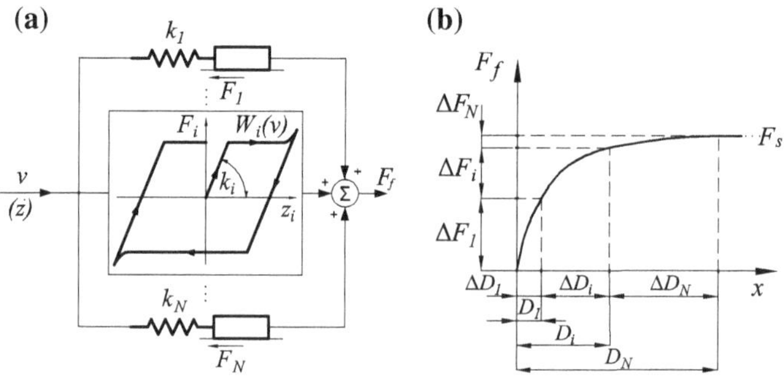


**Fig. 17.4** GMS pre-sliding model: scheme (**a**) and approximation of the experimental $F_f$ versus $x$ curve (**b**). Reprinted from [16] by permission of SAGE Publications

The parameters of the model can thus be determined by a piecewise approximation of the experimentally obtained friction force $F_f$ versus displacement $x$ data (Fig. 17.4b). This is expressed mathematically as [16]

$$k_i = K_i - K_{i+1}, \tag{17.7}$$

$$K_i = \frac{\Delta F_i}{\Delta D_i} = \frac{F_{i+1} - F_i}{D_{i+1} - D_i}. \tag{17.8}$$

Here $K_i$ is the comprehensive stiffness contribution of all the Maxwell-slip blocks that are still in the pre-sliding state in a determined region of the overall pre-sliding motion regime, while the stiffness of the last slip block is defined as $k_N = K_N$. The relative weight of each Maxwell-slip block can be expressed as

$$\alpha_i = \frac{k_i \times D_i}{F_s}, \tag{17.9}$$

where $D_i$ is the maximum deflection of the $i$th Maxwell-slip block before it starts slipping. Taking into consideration that the whole considered system will start sliding when the actuating force reaches the value of the breakaway (static friction) force $F_s$, equations (17.7)–(17.9) imply also that the sum of the frictional contributions $\alpha_i$ adds up to 1.

The state of each block is hence determined based on the following conditions:

- if $|F_i(v)| < |W_i(v)|$: the $i$th block sticks:

$$\frac{dF_i}{dt} = k_i \times v \tag{17.10}$$

- otherwise: the $i$th block slips:

$$\frac{dF_i}{dt} = \text{sgn}(v) \times C \times \left(\alpha_i - \frac{F_i}{s(v)}\right). \tag{17.11}$$

Here the positive constant number $C$ is the attraction parameter associated to frictional lag that is relevant in the characterization of the transition from the pre-sliding to the sliding motion regime [2]. In [12] it was shown that a change of $C$ in a large range of values has no major impact on system's response. This finding is substantiated by the fact that, in nanometric positioning, precision and accuracy are far more important than positioning velocity and acceleration, i.e., that generally there are no sudden dynamical and/or periodical effects that would induce frictional lag [2, 19].

The total pre-sliding friction force $F_f$ of the GMS model can thus finally be calculated as the sum of the contributions of the frictional forces of all the Maxwell-slip blocks:

$$F_f = \sum_{i=1}^{N} F_i(t). \tag{17.12}$$

It is to be noted also that the sensitivity analysis of the normalized mean square error (MSE), depending on the number of characteristic blocks of the GMS friction model, performed in [16], allowed establishing that three GMS blocks are the minimum needed to approximate the behavior of the factual precision positioning systems, six blocks allow representing excellently the real behavior, while the cost-to-benefit ratio of using more elements does not justify their usage.

Due to its comprehensiveness and simplicity, the GMS model is often used for real-time control purposes [14, 19, 36].

## 17.3 Used Experimental Setups

In order to identify the parameters of the above friction models, suitable ultrahigh-precision positioning experimental setups are needed. In [12, 14, 41, 43] these are successfully devised, designed and put together while using commercially available elements (Fig. 17.5). Given their foreseen applications, the developed devices are intended mainly for point-to-point positioning so that their positioning accuracy is more relevant than the respective velocity.

As actuators are hence used simple DC motors with planetary gear reducers allowing to reduce the output velocity of the actuator's shaft while increasing the respective output torques and motion resolutions. The motor-gear reducer assembly is characterized by several friction sources (motor components, bearings, and gears' teeth) [14, 43].

This assembly is then connected by means of elastic joints to a lead or ball screw that converts the rotation of the actuating assembly into several tens of millimeters translating motion of linear guides based on rolling elements (balls). The screws

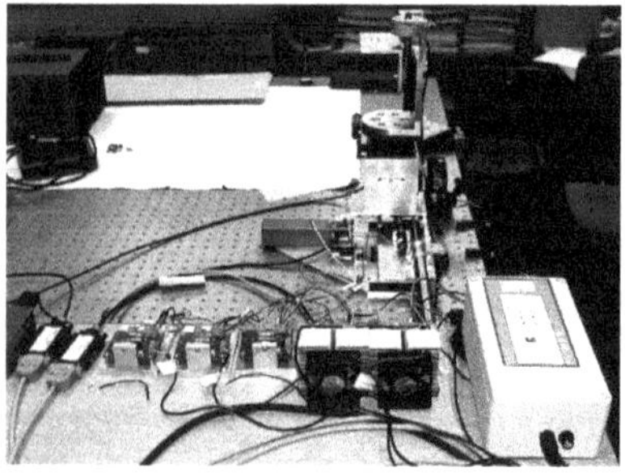

**Fig. 17.5** Considered ultrahigh-precision positioning systems. Reprinted from [12] and from [43] by permission of SAGE Publications

are supported via suitable miniature ball bearings. The screws-guides assemblies are thus characterized by several friction sources on their own. In this frame, standard mechanical elements without particular accuracy requirements are employed.

The feedback signal is, in turn, obtained by using high-resolution positioning transducers such as optical encoders (that, coupled to suitable interpolation electronic units, allow attaining resolutions down to 25 nm), linear variable differential transducers (LVDTs) or, to validate the positioning accuracy and precision with the highest possible performances, via laser interferometric systems. All the components of the positioning devices are mounted on a high-stability optical table.

The data acquisition and control systems used in the experimental setups are based on digital signal processors (DSPs) or field-programmable gate array (FPGA) modules. The respective control algorithms are programmed in the LabVIEW or the C++ environments based on MATLAB/Simulink systems' models [14, 16, 43].

The functional scheme of the used setups is shown in Fig. 17.6, whereas the simplified scheme of a setup in the MATLAB/Simulink environment is shown in Fig. 17.7.

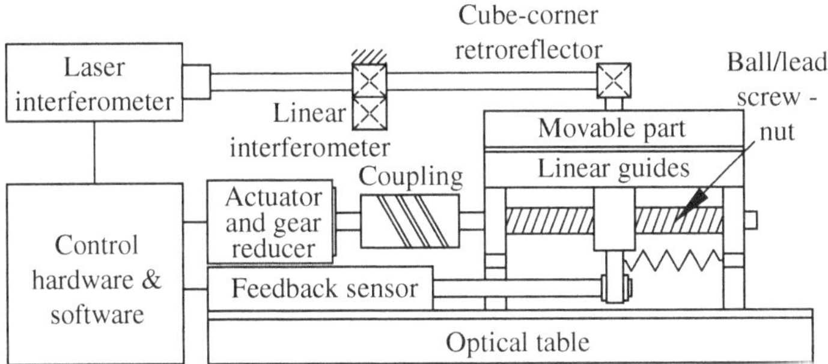


**Fig. 17.6** Functional scheme of the used experimental setups

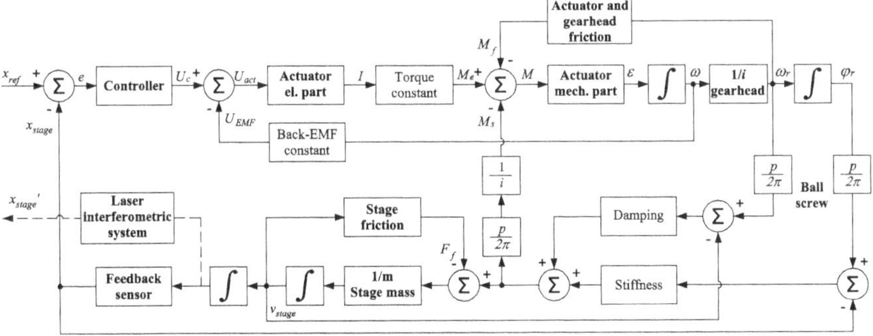


**Fig. 17.7** Simplified MATLAB/Simulink model of the positioning system. Reprinted from [14] by permission of Taylor & Francis Ltd

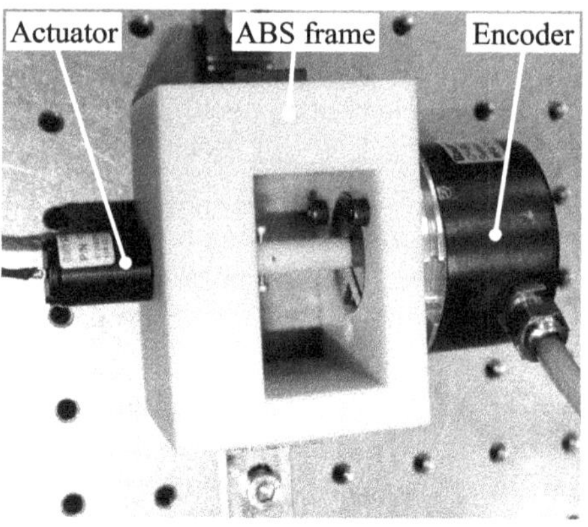


**Fig. 17.8** Experimental setup for the measurement of the frictional behavior of the rotational elements. Reproduced from [12]

In Fig. 17.7 the interconnections of all the electromechanical elements (indicated in bold), with their main physical characteristics, are depicted. The typical electromechanical dynamics of the DC actuator is inducing here, via the gear reducer, a rotation $\varphi_r$ of the screw characterized by its pitch $p$. Rotation $\varphi_r$ is hence converted, taking into account the stiffness and damping effects of the elements of the system, into a linear displacement $x_{stage}$ of the translator bearing the movable load of mass $m$. The frictional effects of the rotational elements $M_f$ and of the translating elements $F_f$ are considered as disturbances and the parameters of the respective models are to be determined so as to enable their compensation via a suitable controller.

## 17.4 Identification of the Parameters of the Friction Models

As indicated in Fig. 17.7, the friction parameters are experimentally assessed separately for the DC actuator-gear reducer assembly (i.e., the rotational elements of the considered setups) and for the linear guides inducing the translatory motion of the working part of the considered devices.

### *17.4.1 Friction Parameters of the Rotation Elements*

The friction in the actuator-gear reducer assembly is induced by the brushes–commutator contacts, the radial bearings, the point contacts of gears' teeth as well as the velocity-dependent viscous effect of the back-electromagnetic force [43]. In order to determine the respective pre-sliding behavior, a suitable experimental setup, coupling the rotation assembly to a rotational encoder that, coupled to an interpolation unit, enables a 3.8 μrad resolution, is developed as shown in Fig. 17.8 [12].

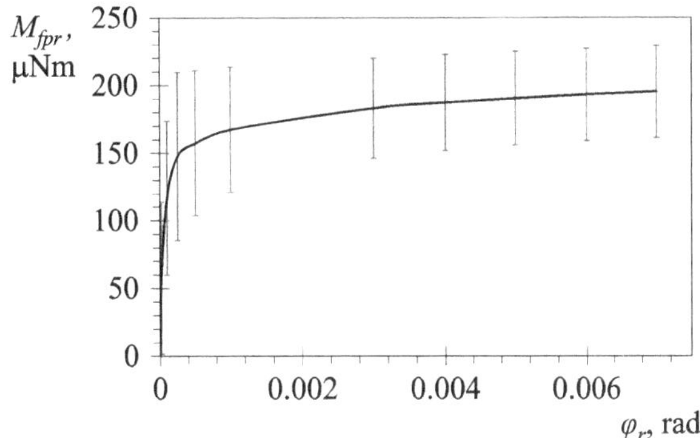


**Fig. 17.9** Pre-sliding friction of the rotational elements. Reproduced from [12]

Considering that in the used setups the inertial effects and damping are negligible, the friction torque $M_f$ is obtained by multiplying the current absorbed by the motor with the torque constant of the actuator. The average obtained results, with the respective deviations in repetitive measurements, are reported in Fig. 17.9.

In the particular case depicted in Fig. 17.9 it can be observed that a typical nonlinear pre-sliding behavior, characterized by a large time and position variability, is obtained. Pre-sliding frictional torques $M_{fpr}$ with average values of up to 200 μNm, and respective rotations at the output of the gear reducer of up to ca. 7 mrad, are obtained. The data shown in Fig. 17.9, referring to Eqs. (17.7) and (17.8), can also be used to determine the parameters of the corresponding GMS friction model [14, 16].

A similar measurement setup is used to assess the behavior of the actuator-gear reducer assemblies in the sliding regime as well. The angular velocity of the steady-state rotation at the output of the gear reducer is calculated in this case as the derivative of the readings of the encoder and is validated from the input voltage to the actuators. The friction versus angular velocity $\omega_r$ dependence is obtained by averaging a large number of measurements. The resulting frictional torque in the sliding regime $M_{fsl}$ versus $\omega_r$ is hence shown in Fig. 17.10, where, despite the different orders of magnitudes of the studied effects in considered experimental setups, the expected Stribeck curve is clearly visible. The marked viscous friction component is mainly due to the influence of the back EMF effect [14, 43].

The dispersion from average values in repetitive measurements is roughly ±6% for lower angular velocities and up to ±15% for larger ones. A careful examination of Fig. 17.10 allows noticing also a certain asymmetry depending on the direction of motion. All these facts confirm, hence, the marked stochastic nature of frictional effects in both pre-sliding and sliding motion regimes.

The remaining parameters of the GMS model, relative to its transition from pre-sliding to sliding and then in the sliding regime itself, are also determined from the measurements reported in Fig. 17.10. The Stribeck frictional behavior can hence be modeled as

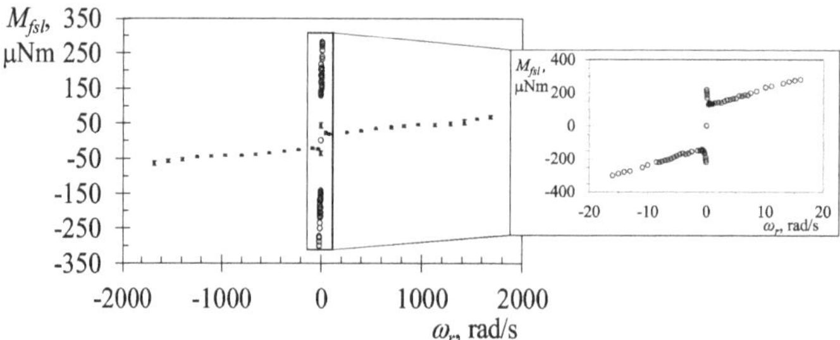


**Fig. 17.10** Friction torque of the rotational elements in the sliding regime in different experimental setups

$$M_{fsl} = \left[ M_C + (M_s - M_C) \times e^{-\left(\frac{|\omega_r|}{\omega_S}\right)^{\delta}} \right] + \sigma \times \omega_r, \qquad (17.13)$$

where $M_s$ and $M_C$ are the static and Coulomb friction torques, $\omega_S$ is the Stribeck angular velocity, $\delta$ is the Stribeck shape factor and $\sigma$ is the viscous friction coefficient [14].

### *17.4.2 Friction Parameters of the Linear Elements*

To assess the pre-sliding frictional behavior of the linear guides, elaborated experimental setups are conceived (Fig. 17.11). Tangential forces are hereby applied via precisely controlled loading to the guides, allowing load increments with resolutions down to 10 mN. Load increments are made when the system comes to an almost complete rest, since displacements on the nanometric level can be observed even after extended time periods [43]. The resulting displacements are measured via the laser interferometric systems. In order to validate the position and time variability of the measured effects, several tens of experiments are conducted using different starting points in various periods during the day. The obtained results are depicted in Fig. 17.12.

The typical nonlinear elastoplastic pre-sliding behavior with nonlocal memory can hence be observed. The narrow loops, obtained by reducing the tangential forces and increasing them again, allow evidencing that the elastic component of the overall behavior is rather small, i.e., that the overall behavior is essentially irreversible, whereas the slope (i.e., stiffness) of the elastic component is almost constant irrespective of the position on the curve where the return loop is started. The average overall pre-sliding behavior is displayed for frictional forces of up to a couple of N and respective displacements of up to even 100 µm. These values have, however, a noticeable variability depending on the point of the stage where the measurement is performed. A large dispersion (up to about ±15%) in repetitive measurements is obtained (not shown in Fig. 17.12 for clarity reasons), which is partly due to tem-

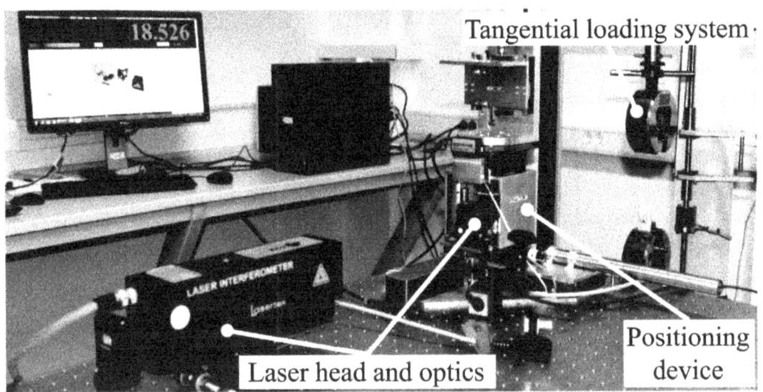


**Fig. 17.11** Setup for the measurement of the pre-sliding behavior of the linear guideways

**Fig. 17.12** Typical measured pre-sliding friction of the linear guideways

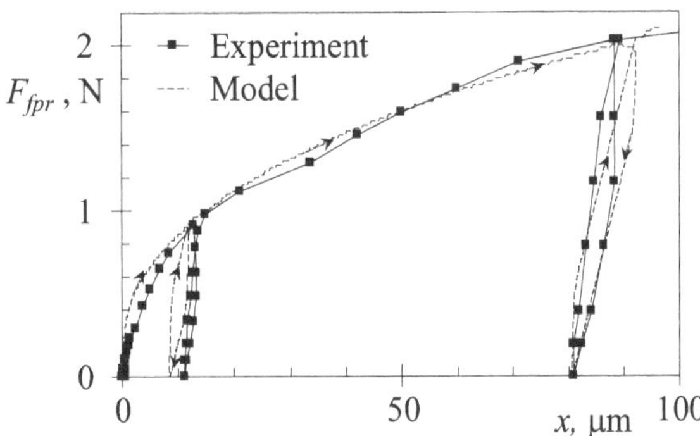


perature variations and partly to differing dwell times inducing rising static friction effects [14, 43]

The data shown in Fig. 17.12 can also be used to determine not only the parameters of the GMS model, but also of other considered friction models [14, 16]. In fact, based on the characteristic features of the considered models and the respective procedures of identifying their main parameters outlined in Sect. 17.2, the experiments allow obtaining the responses of Dahl's, the LuGre, and Hsieh's (HM) friction models, as reported in Fig. 17.13.

From Fig. 17.13 it can thus be seen that neither Dahl's nor the LuGre model allow capturing the nonlocal memory effects (see also Fig. 17.2). On the other hand, the pre-sliding behavior attained by using Hsieh's and the GMS models matches excellently experimental data, especially considering that in the repetitive measurements the data is characterized by a considerable dispersion. What is more, although Hsieh's model gives results closely matching the experimental ones, the determination of the numerous parameters of this model proves to be cumbersome, while they are physically hard to interpret. As shown in recent literature, the formulation of Hsieh's model itself is, moreover, challenging to incorporate into real-time control systems [21]. On the other hand, the GMS pre-sliding friction model not only provides results

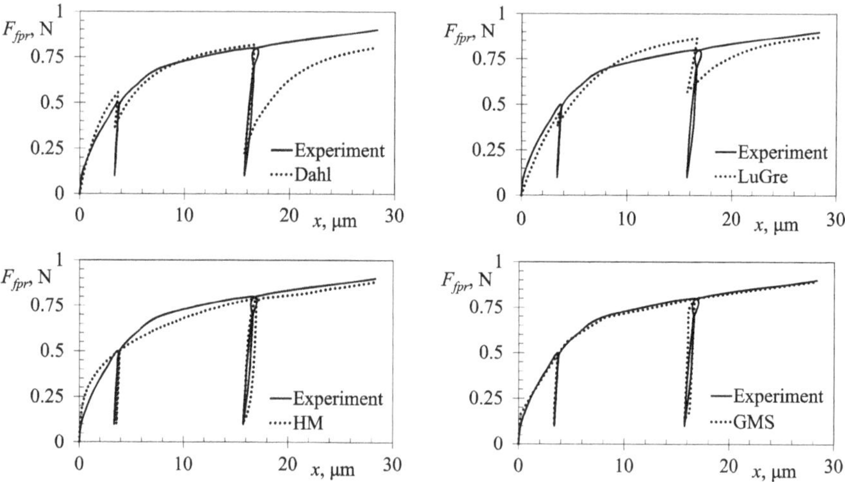


**Fig. 17.13** Comparison of the experimental and modeled pre-sliding responses for different friction models. Reprinted from [16] by permission of SAGE Publications

closely approximating the experimental ones, but it is relatively simple to implement. In fact, six Maxwell-slip blocks used in the GMS model are sufficient to keep the normalized Mean Square Error below 1%, i.e., the model approximates excellently the behavior of the real system. What is more, while Hsieh's model can be used to model pre-sliding friction only, when the motion of the considered positioning device extends also to the sliding regime, the GMS model can be used to simulate the overall behavior in both motion regimes, with a smooth transition between the description of the respective frictional disturbances [16].

It is to be noted here in particular that, due to the reduction ratios of the gear reducer and the screw, the frictional effects of the mechanical parts of the system, when reduced to the motor shaft, will amount to a frictional torque comparable to the variability of the friction present in the actuator-gear reducer assembly [14, 43]. This consideration, proven by experimental measurement on the whole system as well, confirms that the biggest frictional contribution will be that of the actuator. Moreover, due to the reduction ratios, even when the motor-gear reducer assembly overcomes stiction and enters sliding, the mechanical transmission elements are still in pre-sliding. Most significantly, ultrahigh-precision positioning will certainly happen when the linear slide is in the pre-sliding regime [14, 43]. All this implies that only the pre-sliding frictional behavior of the linear guideways is to be considered in the quest for attaining ultrahigh precision. Moreover, given its magnitude, in the development of a suitable controller, this behavior can be considered as a perturbation to the markedly stochastic frictional effects and can hence be compensated by using an appropriate control typology [14, 43].

Up to this point, a thorough overview of the peculiarities of ultrahigh-precision devices based on sliding and rolling elements is given. Suitable experimental setups are introduced and described. The methods of identifying the characteristic friction parameters of the widely used friction models are illustrated. Due to its shown advantages and its comprehensiveness, the GMS model will hence be used as the reference model for the subsequent study of the effectiveness of the various controllers in compensating the frictional disturbances. The drawn conclusions allow also establishing the relative importance of the various sources of frictional disturbances, thus providing important clues for the development and implementation of suitable control typologies aimed at achieving nanometric positioning accuracy and precision.

## 17.5 Nanopositioning Performances via Active Compensation of Friction

Ultrahigh-precision positioning in the presence of frictional disturbances is possible only when the latter are compensated by employing advanced control algorithms. In this section, based on the findings in Sect. 17.4, several control approaches are investigated, starting from those widely used in industrial practice (i.e., PID control) and proceeding with mathematically more sophisticated ones that incorporate adaptive features. The latter, although being computationally more intensive, provide several advantages in terms of the obtained performances, as will be illustrated via the factual experimental data in the following paragraphs [14, 43].

### *17.5.1 Proportional-Integral-Derivative (PID) Control*

PID controllers use in the feedback loop proportional ($K_P$), integral ($K_I$) and derivative ($K_D$) terms (i.e., PID gains) that multiply the error $e$ determined as the difference between the reference value (set point) and the measured value (process value) (Fig. 17.14).

To apply successfully PID control in the used DSP and FPGA architectures, its discrete form is to be used. The output of the PID controller $U_c(n)$ can thus be defined as [6, 13, 14]

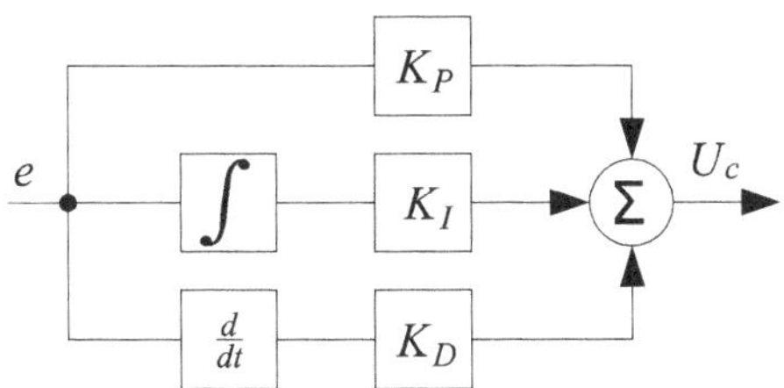


**Fig. 17.14** PID controller scheme. Reproduced from [12]

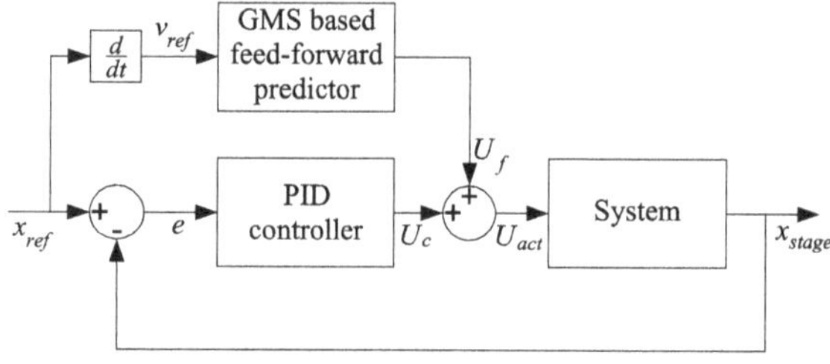


**Fig. 17.15** PID controller with a feedforward predictor

$$U_c(n) = K_\mathrm{P} \times e(n) + K_\mathrm{I} \times \sum_{k=1}^{n} e(k) + K_\mathrm{D} \times \left[x_\mathrm{stage}(n) - x_\mathrm{stage}(n-1)\right], \quad (17.14)$$

where $e(n)$ is the discrete error term, $e(k)$ is the integral error term, $k$ is the summation variable for the integral term, $n$ is the discrete-time step and $x_{stage}(n)$ and $x_{stage}(n-1)$ are the measured positions in two subsequent time steps. The integral and derivative gains are calculated based on the signal sampling period $T$, as well as the integral and derivative time constants $T_I$ and $T_D$, as [14]

$$K_I = \frac{K_P \times T}{T_I}, \quad K_D = \frac{K_P \times T_D}{T}. \quad (17.15)$$

The tuning of the PID gains is conducted in two steps: the time-domain Ziegler–Nichols method is used to achieve a rough estimate of the gains, whereas fine-tuning is performed experimentally. The obtained gain values are used to assess, via the MATLAB/Simulink model, the stability of the system proving that in all the considered cases a positive phase margin of at least 30° and a gain margin of about 40 dB are obtained. On the other hand, as expected, the PID gains optimized for a certain motion range, do not allow achieving similar accuracy for different motion amplitudes. Moreover, the PID gains determined for one position on the linear guideways can prove to be far from optimal even for the same motion amplitude at a different point along the stage and/or in different time or temperature conditions. Therefore, the PID control alone does not allow to capture the effects induced by the stochastic nature of friction [6, 13, 14].

### *17.5.2 PID Controller with Feedforward Compensation*

As suggested in [20, 35, 36, 40], the nonlinear stochastic effects related to friction can be compensated by complementing the PID control scheme with a model-based feedforward friction predictor (Fig. 17.15).

In this case, the disturbances are represented by the GMS friction models of the actuator-gear reducer assembly and of the linear guides with the values of the characteristic parameters determined via the procedure described in Sect. 17.4. Obviously, the stochastic variability of frictional effects is not taken into account in this case either. Despite this fact, this control configuration allows achieving good results for varying motion amplitudes and positions along the stage, albeit at the expense of computational complexity and the need to perform slower motions. In addition, in the considered setups it is based on the nominal values of the velocity of the mechatronics system. In fact, the algorithm needed to determine the actual velocity of the linear stage by differentiating its measured position, can give rise to overloading problems or to positioning performances limited by the reachable sampling times, which are limited by the used hardware [14].

### 17.5.3 Model Reference Adaptive Control (MRAC)

The aim of the work performed in [43] is to develop a simple control approach, where a unique control law is applied in both motion regimes. The possibility of using a model-based adaptive control approach based on pulse width modulation (PWM), developed in [39], is hence explored with the purpose of extending its applicability to the pre-sliding region. It is hence shown that, for pulse widths limited to $t_{imp} \leq 200$ ms, i.e., for high-precision (pre-sliding) displacements limited to approximately 10 μm, a quadratic relation describes excellently both numerically and experimentally the behavior of the used positioning system (Fig. 17.16) whose pre-sliding regime was described via Hsieh's model (cf. Sect. 17.2.3 and Fig. 17.12). The coefficient of proportionality $b$ between the square of the pulse width and the respective displacement, due to a significant stochastic component, can, however, be determined here only adaptively by using the model reference adaptive control (MRAC) algorithm. The resulting displacement of the system, based on the characteristic quadratic equation implemented in the regulator proposed by [39], is hence determined by a parameter adaptation algorithm (PAA). The input signals to this

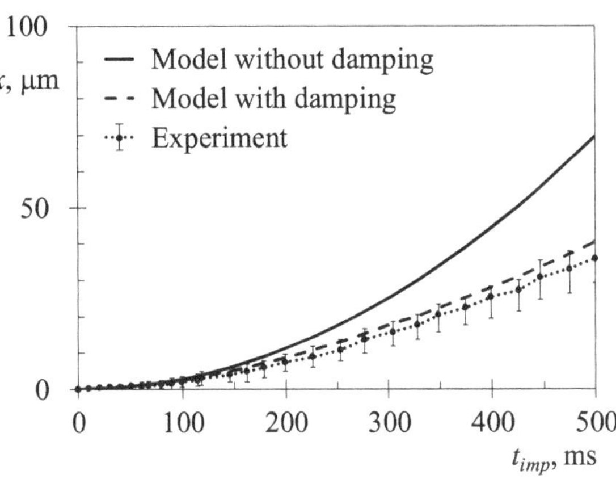


**Fig. 17.16** Displacement versus pulse width

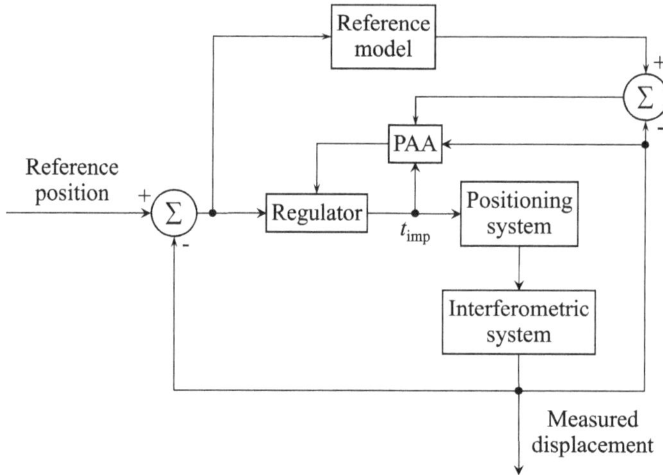


**Fig. 17.17** Implemented model reference adaptive control algorithm. Reprinted from [43] by permission of SAGE Publications

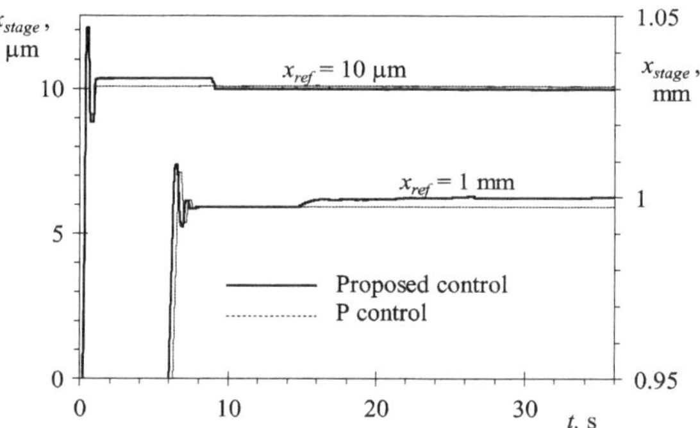


**Fig. 17.18** Measured displacements for a 10 μm and for a 1 mm reference position. Reprinted from [43] by permission of SAGE Publications

algorithm are pulse width, the measured displacement and the displacement that is obtained using a reference model in which an ideal response of the system is assumed. The block diagram of the implemented MRAC system is thus shown in Fig. 17.17 [43].

When displacement amplitudes larger than the limit of validity of the quadratic relation are needed, a simple P control suffices to bring the positioning stage within the range of validity of the MRAC, i.e., the overall positioning algorithm is structured in this case as a dual control algorithm. This approach allows obtaining accuracy and precision limited only by the resolution of the feedback system, i.e., nanometric positioning can be attained both for smaller (micrometric) and larger (millimeter) distances (Fig. 17.18).

In all the considered cases the resulting errors are hence within the interval of uncertainty of the measurements of the used laser interferometric system, i.e., the error is always within ±20 nm. However, the applicability of this approach is limited not only by the characteristic marked overshoots of the P controller, but particularly by a considerable lowering of the positioning speed, especially for longer travel ranges, when a suitable switching element between the P and the MRAC-based PWM controller has to be introduced. Furthermore, this switching element can introduce perturbations in applications with higher dynamics [43].

### *17.5.4 Self-tuning Regulator (STR)*

In the set of available adaptive nonlinear control schemes, self-tuning regulators (STR) are a viable choice for settings with stochastic changes in systems to be controlled online in digital closed-loop environments, as it is flexible and computationally easy to implement (Fig. 17.19) [5]. An STR adaptive PID control scheme, originally proposed in [22], is therefore used to control the herein considered ultrahigh-precision positioning systems [14].

The adaptation of the PID gains, based on the theory of adaptive interactions [22], is hence achieved, thus allowing the probabilistic changes of frictional effects and other eventual external disturbances to be effectively compensated [38]. Moreover, the implementation of the tuning algorithm is quite straightforward since, in the simplified form, the dependence on the model of the device can be eliminated, whereas the adaptation of the parameters of the regulator is reduced to an algorithm based on a single adaptation coefficient and the outputs of the considered subsystems [22], thus providing a significant advantage with respect to the previously described MRAC adaptive approach.

In [22] it is shown that the simplified STR scheme is stable and converges quickly for linear and nonlinear systems, as well as with or without noise and/or time delays.

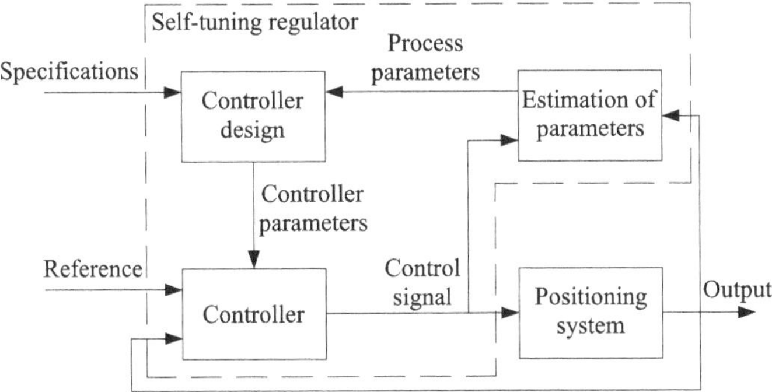


**Fig. 17.19** Structure of the STR adaptive controller

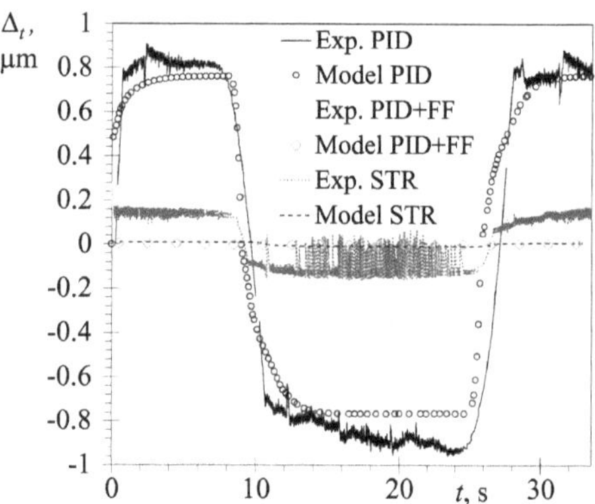


**Fig. 17.20** Tracking errors for sinusoidal motions of the ultrahigh-precision positioning device for different control approaches. Reprinted from [14] by permission of Taylor & Francis Ltd

What is more, this scheme does not require an intricate computational implementation of the frictional model, since the adaptation of the PID gains compensates effectively all the dynamical changes. This implies that this control scheme, in contrast to the considered feedforward scheme, can be efficiently implemented even when large parts of the computational capacities are devoted to real-time monitoring of the effective velocity of the system [14].

When the results obtained by using the PID, the feedforward and the STR controllers are compared in Fig. 17.20, it can be observed that the smallest tracking error is obtained by using the STR controller. In fact, when STR is used, in all the considered cases excellent positioning accuracies ($\pm$150 nm), limited basically by the limitations of the PID controller at velocity reversals, by the time constants of the used electrical elements and by the measurement uncertainty of the feedback sensor, are achieved rapidly and reliably. By tuning via the MATLAB model the values of the STR adaptation parameter so that the resulting PID gains never approach the values leading to a possible occurrence of instabilities, the used simplified STR algorithm allows thus fast convergence to high-accuracy tracking conditions. The regulator compensates efficiently the variability of frictional effects, despite the fact that it is basically a black-box model not relying at all on a physical model of friction [14].

In point-to-point positioning experiments performed on the factual experimental setup described in Sect. 17.3, the results shown in Fig. 17.21 allow evidencing that the adaptive nature of the used STR controller allows compensating efficiently all the present mechanical nonlinearities and their variability in this case as well, permitting in all the considered cases to achieve nanometric positioning precision. It is, in fact, evident that in all the considered cases the reference positions are reached. The resulting error in the MATLAB model is hence negligible, confirming its excellent physical foundation while, in the case of experimental measurements, precision, and accuracy comparable to the resolution of the used feedback sensor are always obtained [14].

Despite this advantages, that make STR results useful as a benchmark for high-precision applications, it is to be noted that in point-to-point positioning experiments

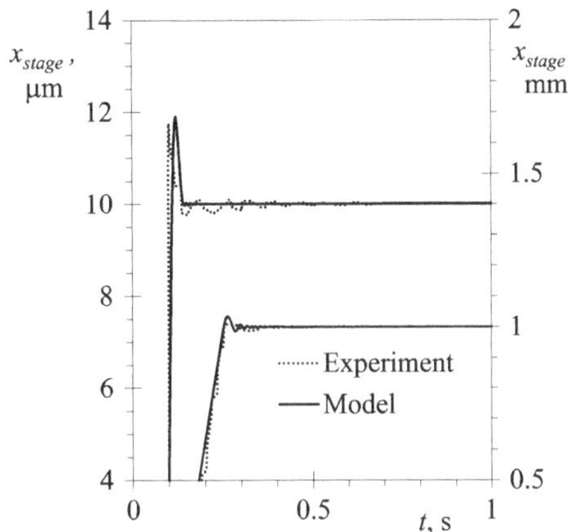


**Fig. 17.21** Point-to-point positioning for short-range (10 μm) and long-range (1 mm) motions controlled via the STR control scheme. Reprinted from [14] by permission of Taylor & Francis Ltd

STR gives rise to an overshoot of up to 20%, which could cause problems especially in the vicinity of the mechanical limits of the ultrahigh-precision positioning devices.

### 17.5.5 Koopman-Based Control

With the aim of verifying the possibility of overcoming the remaining positioning issues present when using the above controllers, the results of the positioning performances attained by employing the STR, used as a benchmark that allows achieving very good results with some remaining shortcomings, are finally compared with results attained by employing innovative optimization-based model predictive control (MPC) algorithms. Koopman-based MPC is hence employed [17].

In fact, the nonlinearities inherently present in this, as well as in many other physical systems, would often imply the need to use nonlinear controllers with potential difficulties (evident particularly in higher dimensional problems) associated with computational times, stability, and robustness (see the above treatise and Chap. 8 of this volume). On the other hand, model predictive controllers (MPCs) are an emerging class of algorithms based on an iterative open-loop optimization of the model of the considered device subject to constraints, by allowing the time frame of the behavior of the device to be extended to a finite future time horizon (prediction), that is increasingly used in industrial settings. In this frame, the Koopman operator, whose origins relate to the 1930s and the work of Bernard Koopman, allows linearized representations for highly nonlinear dynamical systems, exhibiting performance superior in terms of accuracy to existing linear predictors, such as those based on local linearization or the so-called Carleman linearization which uses polynomial basis to represent the Koopman operator, while the procedure to construct the inherent linear predictors is data-driven and quite simple, providing a global and smooth perspective on systems' behavior (i.e., the dynamical time evolution of nonlinear systems' outputs in a linear fashion), especially for longer prediction times (cf. in this regard [18, 27, 28]). Since numerical schemes, such as the extended dynamic mode decomposition (EDMD) [37], can be applied to approximate the infinite-dimensional nature of the Koopman operator with a projected finite-dimensional subspace that has the form of a linear controlled dynamical system, the obtained linear predictors can potentially be used to design controllers for nonlinear dynamical systems by using linear controller design methodologies. The latter, coupled to state-of-the-art sensors, allow using existing robust data-driven solvers, thus reducing the computational burden in real-time applications [18] (see also Chaps. 1 and 8).

Koopman operators are therefore increasingly investigated for model-based control in experimental settings such as, e.g., robotics, biology, power grids, flight and fluid dynamics, traffic, locomotion devices, or DC motor control (see [3, 18], and Chaps. 1, 8, and 16). However, a systematic comparison with well-established controllers in specific applications, needed to assess the factual applicability and advantages of Koopman-based MPC, is still in its infancy, requiring a larger set of reliability proofs and quantifications (see Chap. 1). This study aims at contributing to bridging this gap.

Koopman-based MPC is here applied to ultrahigh-precision positioning experimental setups while their frictional disturbances are once more modeled via the GMS model, where separate friction blocks are used to model the friction of the actuator-gear reducer assembly and that of the linear guideways (cf. also Fig. 17.7). A set of linear predictors is thus applied in order to estimate the behavior (i.e., the future states) of the considered nonlinear dynamical ultrahigh-precision positioning system, wherein the numerical approximations of the Koopman operator allow "lifting" the nonlinear dynamics of the considered factual device (i.e., of its state-space model) into a higher dimensional space, where its behavior can be accurately predicted by a linear system. The scheme depends on the current state and the current and future inputs to the system, i.e., it is data-driven and practically reduces to a nonlinear transformation of the data (the lifting) and a linear least squares problem in the lifted space. What is more, the obtained linear predictors were recently successfully used in the design of MPCs for nonlinear dynamical systems, with the resultant computational complexity comparable to that of MPCs for linear dynamical systems of the same size [18].

The Koopman-based procedure is therefore followed in the investigated cases to track the position of the used ultrahigh-precision positioning device. The open-loop response of the nanopositioning device, numerically modeled in the MATLAB/Simulink environment, obtained for a random input, is shown in Fig. 17.22. It can thus be inferred that the Koopman-based scheme indeed accurately follows the dynamics of the system.

The numerical model is used next to track the closed-loop responses of the used mechatronics system while controlled with the STR algorithm, used as benchmark, and via the Koopman-based MPC. The sampling time is set to 1 ms. Tracking of sinusoidal excitations comparable with the experimental and numerical values of Fig. 17.20 is hence shown in Fig. 17.23. It can thus be inferred that for both the STR and the Koopman-based controller the tracking errors are significantly reduced with

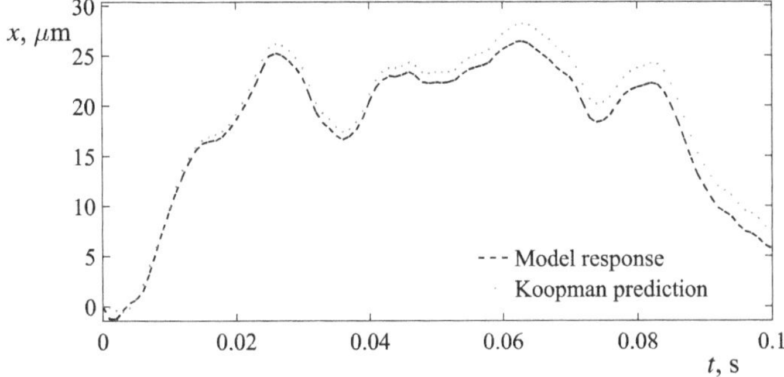


**Fig. 17.22** Validation of prediction performances of the Koopman-based scheme. Reproduced from [17]

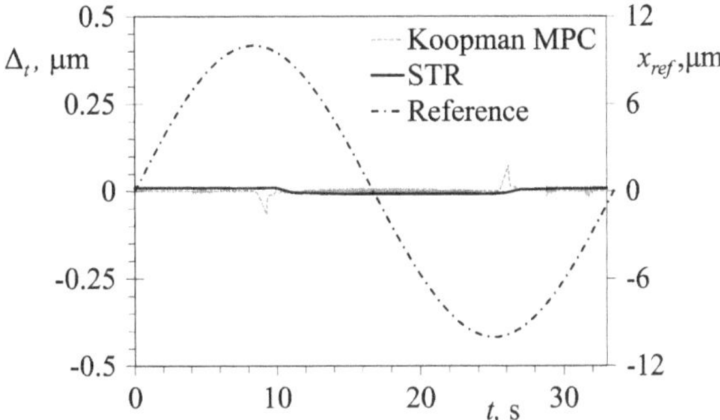


**Fig. 17.23** Tracking errors for a 10 μm sinusoidal excitation and different controllers

respect to those obtained with the other control typologies used in Fig. 17.20. It is evident, however, that STR shows a slightly more accurate tracking of the reference signal at direction reversals, whereas in these points the MPC controller, with the cost function matrices set to $Q = 0.1$, $Q_{NP} = 0.2$ and $R = 4$ [18], results in small "glitches". However, in the here performed simulations the lifting map is a simple delay embedding, so that presumably a more elaborate choice of this map could eliminate these slight irregularities [17, 18].

In the next step, point-to-point positioning is considered for short-range (10 μm) and long-range (1 mm) positioning steps. In this case, the parameters of the Koopman-based controller are optimized again, resulting in cost function matrices $Q = 0.1$, $Q_{Np} = 3.5 \times 10^{-6}$ and $R = 3.555$; once optimized, these are valid for all the considered reference position magnitudes. The used prediction horizon is set to 50 ms. The results shown in Fig. 17.24 allow thus evidencing that with both the STR and the Koopman control algorithms the steady-state error is eliminated, while the rise times are similar. Nonetheless, when using the Koopman-based MPC the overshoot is reduced to a value lower than 2% of the reference position (i.e., by ca. 10 times with respect to the STR values). What is more, Koopman-based MPC is characterized also by considerably shorter settling times [17, 18].

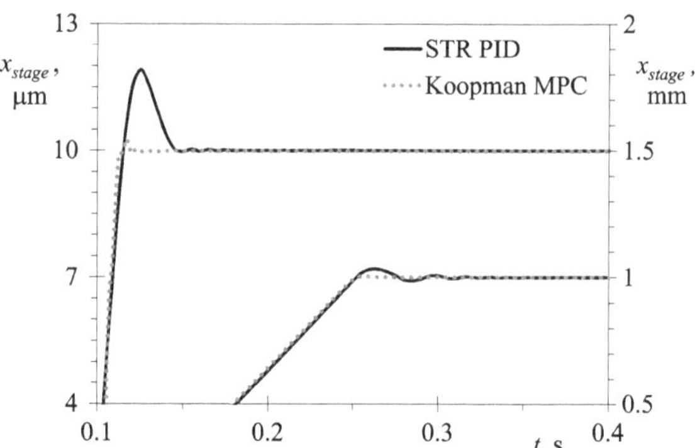


**Fig. 17.24** Validation of point-to-point responses. Reproduced from [17]

The Koopman-based MPC controller is hence proven to be a valid alternative in efficiently compensating factual frictional nonlinear stochastic disturbances in ultrahigh-precision positioning devices, allowing to attain precision and accuracy equivalent to those of the most elaborate adaptive controllers, while concurrently enabling not only to reduce overshoot and settling times, but also to simplify the overall numerical procedure [17, 18]. The goal of subsidizing with tangible data the assumed advantages of using Koopman-based MPC in dynamically controlling advanced experimental systems with inherently nonlinear features is thus achieved.

## 17.6 Conclusions and Outlook

The pre-sliding and sliding stochastic nonlinear frictional effects present in the ultrahigh-precision positioning systems, which limit their positioning precision, are thoroughly studied, modeled, and identified via elaborated experimental setups. The devised procedure of identifying the friction parameters allows establishing that, due to reduction ratios, the frictional contribution of the actuator-gear reducer assembly is the most significant one and that, even when this assembly enters the sliding motion regime, the downstream elements will still be in pre-sliding. The identification of friction parameters enables also to study different compensation approaches. It could therefore be shown that, while simpler control schemes result in marked errors and/or computational problems in real-time applications, by implementing an STR adaptive PID control scheme, the stochastic nonlinear frictional disturbances can be compensated rather efficiently, but the controller introduces quite large overshoots.

The data-driven Koopman-based MPC is therefore used, since it allows "lifting" the nonlinear system's dynamics into a higher dimensional space where its behavior can be accurately predicted via a linear system. It is hence proven that the Koopman-based approach ensures an excellent dynamical tracking of the behavior of ultrahigh-precision positioning systems with several sources of nonlinear frictional disturbances, even when it is randomly excited. The Koopman controller is numerically validated next by comparing the respective results to those attained by using the STR controller. It is concluded that, irrespective of the excitation amplitudes, the MPC virtually eliminates steady-state errors and is characterized by very small overshoots, although in some tracking applications its simplified implementation could give rise to eventual "glitches" at direction reversals.

In the follow-up of the work, the Koopman-based MPC will be used to control the factual mechatronics system, i.e., it will be implemented in the used data acquisition and control systems based on digital signal processors (DSPs) or field-programmable gate array (FPGA) modules. Given the stochastic nature of the frictional disturbances, special attention will be dedicated in this frame to the efficient real-time resolution of the quadratic optimization problem, i.e., to the lifting mapping in each time step of the closed-loop control [18]. The thus attained positioning accuracy and precision of the nanometric positioning device, measured by using a laser interferometric system, will finally be validated according to the ISO 230-2 machine tools' standard.

In the framework of the initiated collaboration of the University of Rijeka, Croatia, with the University of California Santa Barbara, Koopman-based MPC will also be applied to the dynamical control of pneumatic soft robotic devices adaptable to varying applications and environments (e.g., gripping, actuating, and/or manipulating of objects or in the field of, potentially medical, wearable devices). The Koopman operator will be used in this frame first to identify and model the inherently nonlinear input–output dependence of the behavior of the robot, and then to control it so as to achieve the desired planar positioning performances of its end effector. Special attention will also be dedicated to the optimized design of the compliant robotic structure as well as to the number and location of the used tracking sensors so as to increase the efficiency of the collection of the data needed for the real-time data-driven control of the used devices.

This and other potential applications of Koopman-based control to factual devices will contribute to the growing body of knowledge proving the advantageous applicability of this controller typology for robust, reliable, fast, and computationally efficient real-world applications.

## 17.7 Notations

| | |
|---|---|
| $a$ | positive constant (exponent) in Hsieh's model related to a work hardening, - |
| $C$ | attraction parameter in the GMS friction model, N/s, (Nm)/s |
| $C_v$ | damping coefficient of the viscous damper in Hsieh's friction model, (Ns)/m |
| $D_i$ | maximum deflection of the $i$th Maxwell-slip block before it starts to slip, m |
| $e$ | positioning error, i.e., difference between reference and measured position, m |
| $F$ | tangential force, N |
| $F_C$ | Coulomb friction force, N |
| $F_f$ | friction force, N |
| $F_{fpr}$ | pre-sliding friction force of the linear slide, N |
| $F_i$ | friction force of the $i$th Maxwell-slip block in the GMS friction model, N |
| $F_s$ | static friction force, N |
| $i$ | gear reducer ratio, - |
| $I$ | actuator's input current, A |
| $k_{1s}, k_{2s}$ | stiffness coefficients in Hsieh's friction model, N/m |
| $K_D$ | derivative gain in PID control, - |
| $K_i$ | comprehensive stiffness contribution of all Maxwell-slip blocks, N/m, (Nm)/rad |
| $K_I$ | integral gain in PID control, - |
| $K_N$ | stiffness of the last Maxwell-slip block, N/m, (Nm)/rad |
| $K_P$ | proportional gain in PID control, - |
| $m$ | overall system's mass, kg |
| $M$ | mechanical torque produced by the DC motor, Nm |
| $M_C$ | Coulomb friction torque, Nm |
| $M_e$ | overall torque generated by the DC motor, Nm |
| $M_f$ | friction torque of the actuator-gear reducer assembly, Nm |
| $M_{fpr}$ | pre-sliding friction torque of the actuator-gear reducer assembly, Nm |
| $M_{fsl}$ | sliding friction torque of the actuator-gear reducer assembly, Nm |
| $M_s$ | static friction torque, Nm |
| $n$ | force versus displacement curve shape factor in Dahl's friction model, - |
| $N$ | number of Maxwell-slip elements in the GMS model, - |
| $p$ | pitch of the ball screw, m |
| $s$ | velocity weakening curve (Stribeck effect) |
| $t$ | time, s |
| $T$ | sampling time, s |
| $T_D$ | derivative time constant of the PID control, s |
| $T_I$ | integral time constant of the PID control, s |
| $U_{act}$ | output voltage from the controller, V |
| $U_{DC}$ | input voltage to the DC motor, V |
| $U_{EMF}$ | motor back-electromagnetic force, V |
| $U_f$ | output voltage from the feedforward friction predictor, V |
| $v$ | velocity of the moving object (gross sliding velocity), m/s |
| $v_{stage}$ | velocity of the positioning stage, m/s |
| $W_i$ | force threshold where a Maxwell-slip block starts slipping, N |
| $x$ | displacement, m |
| $x_h$ | accumulated work hardening in Hsieh's friction model, m |
| $x_p$ | deformation of the plastic module in Hsieh's friction model, m |
| $x_r$ | reverse point of the hysteresis loop in Hsieh's friction model, m |
| $x_{ref}$ | reference position, m |

$x_s$ displacement of the nonlinear spring module in Hsieh's model, m
$x_{stage}$ displacement of the stage measured by the feedback sensor, m
$x_{stage}'$ displacement measured by the laser interferometer, m
$y_i$ $i$th output of the considered subsystems in the adaptive regulator, -
$z$ state variable in the friction models, -
$z_i$ state variable of the $i$th Maxwell block in the GMS model, -
$\alpha_i$ relative weight of a Maxwell-slip block, -
$\beta$ positive scalar in Hsieh's friction model, -
$\delta$ Stribeck shape factor, -
$\Delta_t$ tracking deviation, m
$\varepsilon$ angular acceleration of the DC actuator before the gear reducer, rad/s$^2$
$\mu$ friction coefficient, -
$\rho$ positive constant in Hsieh's friction model related to creep, -
$\sigma$ viscous coefficient related to the sliding motion regime, (Ns)/m
$\sigma_0$ the slope of the force versus deflection curve for $F_f = 0$, N/m
$\sigma_p$ viscous coefficient related to the pre-sliding motion regime, (Ns)/m
$\sigma_r$ friction force datum of a reverse point in Hsieh's friction model, N
$\phi$ angular position of motor's shaft before the gear reducer, rad
$\phi_r$ angular position at the output of the actuator-gear reducer assembly, rad
$\psi$ positive constant in Hsieh's friction model related to work hardening, -
$\omega$ angular velocity of the DC motor before the gear reducer, rad/s
$\omega_r$ angular velocity at the output of the actuator-gear reducer assembly, rad/s
$\omega_S$ Stribeck angular velocity, rad/s

**Acknowledgements** The work performed by S. Zelenika and E. Kamenar is made possible by using the equipment funded via the ERDF project no. RC.2.2.06-0001 "Development of Research Infrastructure at the University of Rijeka Campus - RISK" and supported by the University of Rijeka installation grant "Measuring, modeling, and compensating friction in high-precision devices: from macro- to nanometric scale" no. 17.10.2.1.02 and the University of Rijeka grant "Innovative mechatronics constructions for smart technological solutions", uniri-tehnic-18-32. I. Mezić acknowledges support from ARO-MURI grant W911NF-17-1-0306.

## References

1. Al-Bender, F., De Moerlooze, K.: Characterization and modeling of friction and wear: an overview. Sustain. Constr. Des. **2**(1), 19 (2011)
2. Al-Bender, F., Lampaert, V., Swevers, J.: The generalized Maxwell-slip model: a novel model for friction simulation and compensation. IEEE Trans. Autom. Control **50**(11), 1883–1887 (2005)
3. Arbabi, H., Korda, M., Mezic, I.: A data-driven Koopman model predictive control framework for nonlinear flows (2018). arXiv:1804.05291
4. Armstrong-Hélouvry, B., Dupont, P., De Wit, C.C.: A survey of models, analysis tools and compensation methods for the control of machines with friction. Automatica **30**(7), 1083–1138 (1994)
5. Åström, K.J., Hägglund, T.: PID Controllers: Theory, Design, and Tuning, vol. 2. Instrument Society of America, Research Triangle Park (1995)
6. Baćac, N., Slukić, V., Puškarić, M., Štih, B., Kamenar, E., Zelenika, S.: Comparison of different DC motor positioning control algorithms. In: 37th International Convention on Information

and Communication Technology, Electronics and Microelectronics (MIPRO), pp. 1654–1659 (2014)
7. Canudas-de Wit, C.: Comments on "A new model for control of systems with friction". IEEE Trans. Autom. Control **43**(8), 1189–1190 (1998)
8. Dahl, P.R.: A solid friction model. Technical report, Aerospace Corporation, El Segundo, CA (1968)
9. De Wit, C.C., Olsson, H., Astrom, K.J., Lischinsky, P.: A new model for control of systems with friction. IEEE Trans. Autom. Control **40**(3), 419–425 (1995)
10. Howell, L.L.: Compliant Mechanisms. Wiley, New York (2001)
11. Hsieh, C., Pan, Y.C.: Dynamic behavior and modelling of the pre-sliding static friction. Wear **242**(1–2), 1–17 (2000)
12. Kamenar, E.: Ultra-high precision positioning via a mechatronics approach. Ph.D. thesis, University of Rijeka, Faculty of Engineering, Rijeka, Croatia (2016)
13. Kamenar, E., Zelenika, S.: Micropositioning mechatronics system based on FPGA architecture. In: 36th International Convention on Information and Communication Technology, Electronics and Microelectronics (MIPRO), pp. 125–130 (2013)
14. Kamenar, E., Zelenika, S.: Nanometric positioning accuracy in the presence of presliding and sliding friction: modelling, identification and compensation. Mech. Based Des. Struct. Mach. **45**(1), 111–126 (2017)
15. Kamenar, E., Zelenika, S.: Characterisation of positioning performances of a mechatronics device actuated via a frictionless voice-coil actuator. In: Proceedings of the 18th EUSPEN International Conference, pp. 61–62 (2018)
16. Kamenar, E., Zelenika, S.: Issues in validation of pre-sliding friction models for ultra-high precision positioning. Proc. Inst. Mech. Eng. Part C: J. Mech. Eng. Sci. **233**(3), 997–1006 (2019)
17. Kamenar, E., Korda, M., Zelenika, S., Mezić, I., Maćešić, S.: Koopman-based model predictive control of a nanometric positioning system. In: Proceedings of the 18th EUSPEN International Conference, pp. 75–76 (2018)
18. Korda, M., Mezić, I.: Linear predictors for nonlinear dynamical systems: Koopman operator meets model predictive control. Automatica **93**, 149–160 (2018)
19. Lampaert, V., Al-Bender, F., Swevers, J.: A generalized Maxwell-slip friction model appropriate for control purposes. In: 2003 IEEE International Workshop on Workload Characterization (IEEE Cat. No. 03EX775), vol. 4, pp. 1170–1177 (2003)
20. Lampaert, V., Swevers, J., Al-Bender, F.: Comparison of model and non-model based friction compensation techniques in the neighbourhood of pre-sliding friction. In: Proceedings of the 2004 American Control Conference, vol. 2, pp. 1121–1126 (2004). https://doi.org/10.23919/acc.2004.1386722
21. Lin, C.J., Yau, H.T., Tian, Y.C.: Identification and compensation of nonlinear friction characteristics and precision control for a linear motor stage. IEEE/ASME Trans. Mechatron. **18**(4), 1385–1396 (2013)
22. Lin, F., Brandt, R.D., Saikalis, G.: Self-tuning of PID controllers by adaptive interaction. In: Proceedings of the American Control Conference, vol. 5, pp. 3676–3681 (2000)
23. Lobontiu, N.: Compliant Mechanisms: Design of Flexure Hinges. CRC Press, Boca Raton (2002)
24. Marković, K., Zelenika, S.: Optimized cross-spring pivot configurations with minimized parasitic shifts and stiffness variations investigated via nonlinear FEA. Mech. Based Des. Struct. Mach. **45**(3), 380–394 (2017)
25. McKeown, P.: The role of precision engineering in manufacturing of the future. CIRP Ann. **36**(2), 495–501 (1987)
26. McKeown, P., Corbett, J.: Ultra precision machine tools. Autonome Produktion, pp. 313–327. Springer, Berlin (2004)
27. Mezić, I.: Spectral properties of dynamical systems, model reduction and decompositions. Nonlinear Dyn. **41**(1–3), 309–325 (2005)

28. Mezić, I., Banaszuk, A.: Comparison of systems with complex behavior. Phys. D: Nonlinear Phenom. **197**(1–2), 101–133 (2004)
29. Piatkowski, T.: Dahl and LuGre dynamic friction models—the analysis of selected properties. Mech. Mach. Theory **73**, 91–100 (2014)
30. Ruderman, M., Bertram, T.: Two-state dynamic friction model with elasto-plasticity. Mech. Syst. Signal Process. **39**(1–2), 316–332 (2013)
31. Schmidt, R.M., Schitter, G., Rankers, A.: The Design of High Performance Mechatronics: High-Tech Functionality by Multidisciplinary System Integration. IOS Press, Amsterdam (2014)
32. Shore, P., Cunningham, C., DeBra, D., Evans, C., Hough, J., Gilmozzi, R., Kunzmann, H., Morantz, P., Tonnellier, X.: Precision engineering for astronomy and gravity science. CIRP Ann. **59**(2), 694–716 (2010)
33. Slocum, A.H.: Precision machine design. Soc. Manuf. Eng. (1992)
34. Smith, S.T.: Flexures: Elements of Elastic Mechanisms. CRC Press, Boca Raton (2014)
35. Swevers, J., Al-Bender, F., Ganseman, C.G., Projogo, T.: An integrated friction model structure with improved presliding behavior for accurate friction compensation. IEEE Trans. Autom. Control **45**(4), 675–686 (2000)
36. Tjahjowidodo, T., Al-Bender, F., Van Brussel, H.: Friction identification and compensation in a DC motor. In: 16th IFAC World Congress, Prague (2005)
37. Williams, M.O., Kevrekidis, I.G., Rowley, C.W.: A data-driven approximation of the Koopman operator: extending dynamic mode decomposition. J. Nonlinear Sci. **25**(6), 1307–1346 (2015)
38. Worden, K., Wong, C., Parlitz, U., Hornstein, A., Engster, D., Tjahjowidodo, T., Al-Bender, F., Rizos, D., Fassois, S.: Identification of pre-sliding and sliding friction dynamics: grey box and black-box models. Mech. Syst. Signal Process. **21**(1), 514–534 (2007)
39. Yang, S., Tomizuka, M.: Adaptive pulse width control for precise positioning under influence of stiction and Coulomb friction. In: 1987 American Control Conference, pp. 188–193. IEEE (1987)
40. Yoon, J.Y., Trumper, D.L.: Friction modeling, identification, and compensation based on friction hysteresis and Dahl resonance. Mechatronics **24**(6), 734–741 (2014)
41. Zelenika, S.: Design of high-precision positioning systems: correlation between non-linearities and control typology. Ph.D. thesis, Politecnico di Torino, Torino, Italy (1996)
42. Zelenika, S., DeBona, F.: Analytical and experimental characterisation of high-precision flexural pivots subjected to lateral loads. Precis. Eng. **26**(4), 381–388 (2002)
43. Zelenika, S., DeBona, F.: Nano-positioning using an adaptive pulse width approach. Proc. Inst. Mech. Eng. Part C. J. Mech. Eng. Sci. **223**(8), 1955–1963 (2009)

# Chapter 18
# Modeling of Advective Heat Transfer in a Practical Building Atrium via Koopman Mode Decomposition

**Yohei Kono, Yoshihiko Susuki and Takashi Hikihara**

**Abstract** This chapter develops a methodology for modeling of heat transfer dynamics in a building atrium via Koopman mode decomposition (KMD). We address the phenomenon of heat transfer due to movement of air inside an atrium, where the air slowly moves over the distance between individual rooms. The heat transfer is modeled as a heat advection equation with a vector-valued velocity coefficient and a nonlinear input term from heating, ventilation, and air-conditioning (HVAC) operations, which corresponds to a control variable. KMD is applied to the equation so that the heat transfer and HVAC operation are characterized in terms of frequencies and wavenumber vectors of Koopman modes (KMs). The velocity coefficient is then identified with a KM governing the heat transfer. The effectiveness of the modeling is demonstrated with measurement data on temperature field and HVAC operation of a practically used building. A possibility of how to use this modeling for control of heat transfer dynamics is discussed at the end of this chapter.

## 18.1 Introduction

In-building thermal dynamics occur on a wide range of scales in both space and time. For their mathematical representation, two different types of mathematical models have been used in the literature. First, lumped-parameter models are used in order to predict and control coarse-scale thermal dynamics: see, e.g., [1]—range over multiple rooms and hours. The models play an important role in the design stage

Y. Kono (✉) · T. Hikihara
Department of Electrical Engineering, Kyoto University, Kyoto, Japan
e-mail: y-kono@dove.kuee.kyoto-u.ac.jp

T. Hikihara
e-mail: hikihara.takashi.2n@kuee.kyoto-u.ac.jp

Y. Susuki
Department of Electrical and Information Systems, Osaka Prefecture University, Osaka, Japan
e-mail: susuki@eis.osakafu-u.ac.jp



A. Mauroy et al. (eds.), *The Koopman Operator in Systems and Control*,
Lecture Notes in Control and Information Sciences 484,
https://doi.org/10.1007/978-3-030-35713-9_18

of building architecture and heating, ventilation, and air-conditioning (HVAC) system. Second, distributed-parameter models have been studied for fine-scale thermal dynamics inside a room [2–4]. One of the motivations of developing the distributed-parameter models is to enhance HVAC performance including energy efficiency and thermal comfort. This is enabled by local and precise control of in-room temperature field with taking human occupancy into account. Indeed, it was reported in [5] that the energy consumption of air conditioning can be saved by changing the amount of air ventilation based on human occupancy and in [6] that thermal comfort for each occupant can be improved by creating an optimal thermal environment around it.

In this chapter, we address the thermal dynamics emerging in a building with *atrium*. Atrium is a large open space that provides light and ventilation to the interior space of a building [7]. Since the transfer of heat inside an atrium develops over the height between neighboring floors, and the time for the heat transfer is of the order of minutes [8], it can disturb the temperature inside a room connected to the atrium in the short-temporal range. For quantifying the short-temporal disturbance and exploiting it by design of dynamics, it is necessary to establish a methodology for mathematical modeling to capture the underlying heat transfer in and through the atrium.

For mathematical modeling, however, no model structure has been developed that is computationally simple and physically relevant. Conventional models of the heat transfer in atrium are based on computational fluid dynamics (CFD) analysis [9], which is computationally hard and hence unsuitable to real-time control. Also, discretized forms of such models, called *zonal models*, are used to simulate the indoor temperature with a set of coupled equations of temperature and airflow volume in the so-called subzones, into which a room is divided [10, 11]. The zonal models are computationally simple, but difficult to describe the short-temporal thermal disturbance coupled with multiscale airflows, which can develop in the large open space, because the models are constructed under a single discretization scheme. Moreover, even if a new model structure is developed, no technique for estimating the model's parameters was available due to complex responses of an HVAC system. It is widely recognized in, e.g., [12] that an HVAC system includes inevitable nonlinearities such as dead zone, hysteresis, and saturation between set point and flow rate of actuators (e.g., valves, dampers, and fans). Because it involves complex time series of the output temperature and flow rate of the HVAC systems, it is naturally pointed out that linear techniques including standard Fourier analysis and least squares do not work well for estimating model parameters of the heat transfer.

In this chapter, we propose a simple and physically relevant method for modeling the heat transfer in and through atrium. The model proposed here is based on the so-called averaging and homogenization techniques [13], which approximate the first-principle model of heat transfer as *macroscale* advection and diffusion equations. Although the model is mathematically simple, it requires prior knowledge on *microscale* air movement that is still difficult in terms of computational correctness and time. To avoid the difficulty, we propose to utilize the Koopman mode decomposition (KMD) (see [14, 15] or Chap. 1 of this book) in order to extract the macroscale thermal dynamics directly from measurement data and to estimate their

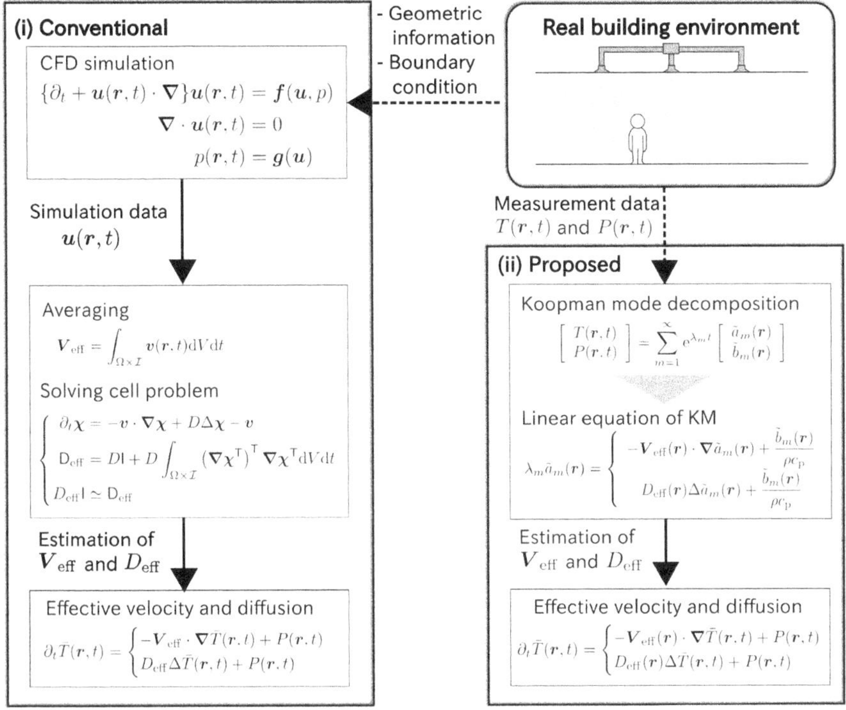


**Fig. 18.1** (i) Conventional and (ii) proposed methods for determining parameters of averaged and homogenized models. The conventional one is based on classical averaging and homogenization techniques, which need to simulate indoor airflow field **v**, compute the spatial average of **v** and solve the so-called *cell problem* [13]—a PDE of **v**, and thus causes computational burden to the modeling. On the other hand, the proposed method utilizes Koopman mode decomposition to estimate the model parameters directly from measurement data on indoor temperature $T$ and heat input $P$ from HVAC units

parameters in the linear fashion. Figure 18.1 summarizes the difference between the conventional and proposed methods for the parameter estimation. KMD has been applied in various ways to the building energy engineering: validation and zoning approximation of building energy models [16–18], improving HVAC performance via energy waste analysis [19], energy consumption monitoring of multiple buildings [20], pattern recognition, classification, and detection of HVAC faults [21, 22]. These papers investigate the long-term thermal dynamics that occur on a range of daily to yearly scales. In this chapter, we mainly develop a model for hourly-scale thermal dynamics with the KMD. Figure 18.2 exhibits thermal phenomena related to our modeling and their space- and timescales. For the modeling, we introduce a formula of linear *wave propagation* of Koopman mode (KM) that makes it possible to effectively extract a spatiotemporal component of indoor temperature and heat input from HVAC. Then, we estimate the model parameters with the KM that govern the heat transfer of our interest. It is shown that the KMD-based modeling can

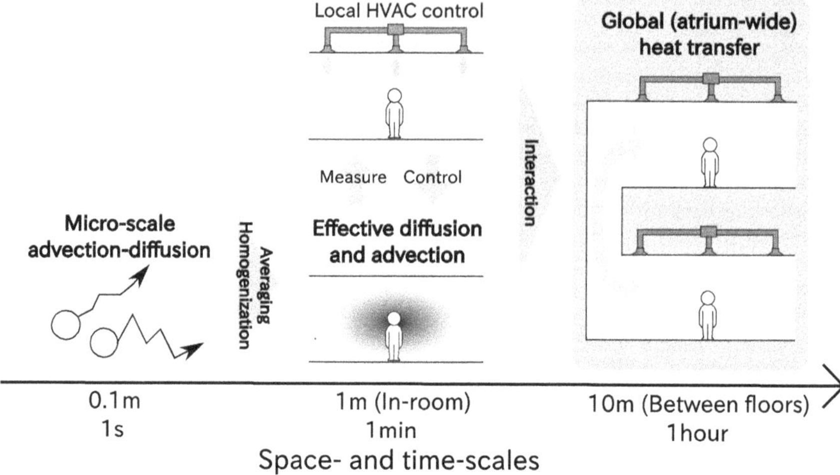


**Fig. 18.2** Space- and timescales of thermal phenomena related to the modeling of this chapter. The advection–diffusion phenomenon described by Eq. (18.1) appears on the scales of 0.1 m and 1 s, which is coarse-grained by averaging and homogenization techniques as the effective diffusion and advection. Since the spatial distance between neighboring HVAC units is several meters, and the sampling period of HVAC control is of the order of minutes, the effective diffusion and advection become relevant on these scales. Also, the effective phenomena and the associated interaction of multiple HVAC units induce global (atrium-wide) heat transfer over the height between floors and hours, which we address for the estimation of model's parameters

effectively describe the direction of heat transfer caused by a laminar airflow in an atrium. Note that the proposed modeling has been reported in the archival papers by the authors and collaborators [23–25].

The rest of this chapter is organized as follows. Section 18.2 derives mathematical models of the heat transfer dynamics applied to a building atrium. Section 18.3 illustrates the main idea of analyzing and modeling spatiotemporal behavior of indoor temperature and HVAC operation in a building directly from measurement data via KMD. Section 18.4 numerically demonstrates the main idea with measurement data on temperature field and HVAC operation of a practically used building. Section 18.5 discusses a possibility of how to use the proposed modeling for control of heat transfer dynamics in atrium. Section 18.6 is the conclusion with the summary of this chapter and comments on future prospects.

## 18.2 Mathematical Formulation

This section is devoted to mathematical models of heat transfer dynamics in a building atrium. The models are based on the simplification of the first principle model (the conservation law of energy) with the so-called averaging and homogenization techniques [13].

We start by considering a mathematical model of the time evolution of air temperature field in $d$-dimension ($d = 2$ or 3) represented by the *advection–diffusion* equation [1]:

$$\frac{\partial T(\mathbf{r}, t)}{\partial t} = L_0 T(\mathbf{r}, t) + \frac{P(\mathbf{r}, t, T)}{\rho c_\mathrm{p}}, \tag{18.1}$$
$$L_0 := -\mathbf{v}(\mathbf{r}, t) \cdot \boldsymbol{\nabla} + D\Delta, \quad \mathbf{r} \in \Omega \subset \mathbb{R}^d, t \in \mathbb{R}$$

where $T(\mathbf{r}, t)$ stands for the temperature on position $\mathbf{r}$ and time $t$, $\mathbf{v}(\mathbf{r}, t) \in \mathbb{R}^d$ for the air velocity, $D$ for the molecular heat diffusion constant, $P(\mathbf{r}, t, T)$ for the heat input per unit volume, which is possibly a nonlinear function of $T$ due to the HVAC mechanism, $\rho$ and $c_\mathrm{p}$ for the density and the specific heat at constant pressure, $\boldsymbol{\nabla}$ and $\Delta$ for the $d$-dimensional vector-differential and Laplace operators. The first term of the right-hand side of Eq. (18.1) is commonly calculated by CFD program [2, 26]. This term is tightly coupled with the *Navier–Stokes equation* [1]. Since it is mathematically and computationally hard to derive $\mathbf{v}(\mathbf{r}, t)$ from Eq. (18.1), it is impractical and unsuitable to the real-time control.

Now, we simplify Eq. (18.1) into computationally efficient and physically accurate forms. Assuming that the velocity $\mathbf{v}$ is smooth (at least $C^2$), incompressible, and periodic in time, it is possible to evaluate the long-term behavior of solutions of Eq. (18.1) via the singular perturbation theory [13]. First, we pose the following *centering condition* at every $\mathbf{r}$ and $t$:

$$\int_{V_\mathrm{c} \times I_\mathrm{c}} \mathbf{v}(\mathbf{r}, t) \mathrm{d}\mathbf{r} \mathrm{d}t = 0, \tag{18.2}$$

where $V_\mathrm{c} \subset \mathbb{R}^d$ and $I_\mathrm{c} \subset \mathbb{R}$ are the domain and interval around $\mathbf{r}$ and $t$ with constant volume $|V_\mathrm{c}|$ and length $|I_\mathrm{c}|$. If Eq. (18.2) holds, a solution of Eq. (18.1) can be approximated (coarse-grained) as a solution of *pure* diffusion equation by the *diffusive scaling*

$$\mathbf{r} \to \mathbf{r}/\varepsilon, \qquad t \to t/\varepsilon^2.$$

If Eq. (18.2) does not hold, then the approximation involves the advection term based on the *advective scaling*

$$\mathbf{r} \to \mathbf{r}/\varepsilon, \qquad t \to t/\varepsilon,$$

where $\varepsilon \ll 1$ is the parameter that scales $(\mathbf{r}, t)$ over $V_\mathrm{c} \times I_\mathrm{c}$. We apply both the two scalings to a practical building. By asymptotic expansions of $T$, $\partial/\partial t$, and $L_0$ in terms of $\varepsilon$, we derive the rescaled field $\overline{T}$ satisfying

$$\frac{\partial \overline{T}(\mathbf{r}, t)}{\partial t} = L\overline{T}(\mathbf{r}, t) + \frac{\overline{P}(\mathbf{r}, t, \overline{T})}{\rho c_{\mathrm{p}}}, \quad \mathbf{r} \in \Omega \subset \mathbb{R}^d, t \in \mathbb{R}, \tag{18.3}$$

$$L = \begin{cases} -\mathbf{V}_{\mathrm{eff}}(\mathbf{r}) \cdot \nabla & (\mathbf{r} \to \mathbf{r}/\varepsilon, t \to t/\varepsilon), \\ D_{\mathrm{eff}}(\mathbf{r}) \Delta & (\mathbf{r} \to \mathbf{r}/\varepsilon, t \to t/\varepsilon^2) \end{cases}$$

where the coefficients $\mathbf{V}_{\mathrm{eff}}(\mathbf{r}) \in \mathbb{R}^d$ and $D_{\mathrm{eff}}(\mathbf{r}) \in \mathbb{R}$ denote the *effective velocity* and *effective diffusivity* at the position $\mathbf{r}$, respectively. The coefficients quantify global behavior of $\overline{T}$ caused by the advection–diffusion phenomena developing in scales smaller than $V_{\mathrm{c}}$ and $I_{\mathrm{c}}$. These approximations are called *averaging* and *homogenization* in the field of multiscale analysis.[1] Also, $\overline{P}$ represents the heat input averaged in space and time over $V_{\mathrm{c}} \times I_{\mathrm{c}}$. In Eq. (18.3) we assume that the spatiotemporal change of $\overline{P}$ arising from HVAC units appears over $V_{\mathrm{c}} \times I_{\mathrm{c}}$ so that $\overline{P}$ does not depend on the original (fine-scale) field $T$ but depends on the rescaled field $\overline{T}$. In the case of this chapter, the assumption is reasonable due to the two reasons. The distance of neighboring HVAC units is several meters and larger than the space scales of the effective diffusion and advection (see Fig. 18.4 for details). Also, the time constants of temperature measurement system and the sampling period of HVAC control are of the order of minutes [8] so that outlet temperature and volume of HVAC units change more slowly than the effective diffusion and advection. This is why $\overline{P}$ is related to a large-scale phenomenon in space and time governed by $\overline{T}$. In the rest of this chapter, for simplicity of presentation we denote the rescaled field $\overline{T}$ by $T$ and the averaged input $\overline{P}$ by $P$.

Here, for the estimation of $\mathbf{V}_{\mathrm{eff}}$ and $D_{\mathrm{eff}}$, it is required to determine the spatiotemporal average of $\mathbf{v}(\mathbf{r}, t)$ and a solution of the *cell problem* [13]—a PDE governed by $\mathbf{v}$, which is still computationally hard. In addition, Eq. (18.3) contains a nonlinear element in the term of $P(\mathbf{r}, t, T)$. Thus, it is impossible to identify $\mathbf{V}_{\mathrm{eff}}$ and $D_{\mathrm{eff}}$ from simulation data via the standard Fourier analysis and linear least-square method. Such hard problems in averaging and identification can be resolved by the Koopman mode decomposition—a data-based method for full characterization of nonlinear spatiotemporal dynamics in the linear fashion.

## 18.3 Modeling of Heat Transfer Dynamics Based on Koopman Mode Decomposition

This section shows the main idea of analyzing and modeling the spatiotemporal evolution of heat transfer and HVAC operation via KMD. Here we introduce a new interpretation of each oscillatory pattern of KM as linear wave propagation. Note

[1] Although $\mathbf{V}_{\mathrm{eff}}$ and $D_{\mathrm{eff}}$ are *constant* in the typical averaging and homogenization frameworks, we now treat them as space dependent functions in order to represent global trend of $\mathbf{v}(\mathbf{r}, t)$ over the domain $V_{\mathrm{c}}$. The space- (and time-)dependency of $D_{\mathrm{eff}}$ and $\mathbf{V}_{\mathrm{eff}}$ is investigated in the literature: see, e.g., [27].

that a similar concept has been reported under the assumption that targets dynamical system is periodic in both space and time [28].

### 18.3.1 The Koopman Operator

First of all, we consider a formulation of the Koopman operator for the target dynamical system (18.3). The formulation is based on the celebrated papers [14, 15], in which the author interpreted spatiotemporal dynamics, precisely speaking, time evolution of a field governed by a PDE as a family of observables parameterized by geometrical positions.

Let us delineate the state space on which observables evolve in time via the Koopman operator. For simplicity of the formulation, we consider below the temperature field $T$ and the heat input $P$ as elements of the space $\mathscr{L}_2$ of square-integrable scalar-valued functions defined on a bounded space $\Omega$ in the configuration space $\mathbb{R}^d$, which is a typical Hilbert space. We also assume that an attractor of the system (18.3) exists (in a suitable sense) and its dimension is finite so that the dynamics restricted to the attractor have a finite-dimensional representation. This assumption holds for a wide class of dynamical systems such as the reaction–diffusion equations and the incompressible Navier–Stokes equations [14, 29]. For the application in this chapter, since the timescale of heat transfer addressed here is several hours, which is much larger than the order of the transient change of $T$ and $P$ due to advection, effective diffusion, and HVAC control (see Fig. 18.2), the dynamics of $T$ and $P$ are almost steady and expected to evolve almost on an attractor of Eq. (18.3). For the finite-dimensional representation, we introduce the state space $X$ corresponding to the attractor. The finite-dimensional representation is supposed to be a form of the following differential equation: for $\mathbf{x} \in X$,

$$\frac{\mathrm{d}\mathbf{x}}{\mathrm{d}t} = \mathbf{F}(\mathbf{x}), \tag{18.4}$$

where $\mathbf{F} : X \to \mathbb{T}X$ is the vector field ($\mathbb{T}X$ is the tangent bundle of $X$) and assumed to be tractable in the region of interest. Analogously to linear advection and diffusion equations without any input term [13], $\mathbf{x}$ represents the position of each *fluid parcel* described by $\mathbf{V}_{\mathrm{eff}}(\mathbf{r})$ and $D_{\mathrm{eff}}(\mathbf{r})$. Also, since $P(\mathbf{r}, t)$ depends on $T(\mathbf{r}, t)$ due to control operation of HVAC units, $\mathbf{x}$ might represent internal states of each HVAC unit that determine the control operation: e.g., valve opening and supplied air temperature. By construction, for every state $\mathbf{x} \in X$, a function $f \in \mathscr{L}_2$ (or $f(\mathbf{r})$, $\forall \mathbf{r} \in \Omega$) is assigned, which will be explicitly denoted by $f(\mathbf{x}; \mathbf{r})$, and can contain information of the temperature field $T$ and the heat input $P$ on the attractor of Eq. (18.3).

The Koopman operator is introduced for the dynamics described by (18.4). Let us denote by $\mathbf{S}^t : X \to X, t \in \mathbb{R}$, the flow (that is, one-parameter family of (non-singular) maps) induced by $\mathbf{F}(\mathbf{x})$: for $\mathbf{x} \in X$,

$$\left.\frac{\mathrm{d}\mathbf{S}^t(\mathbf{x})}{\mathrm{d}t}\right|_{t=0} = \mathbf{F}(\mathbf{x}). \tag{18.5}$$

Then, for a (Banach) space $\mathscr{F}$ of observables $f : X \to \mathbb{C}$, the Koopman operator $U^t$ is defined through the composition

$$U^t f := f \circ \mathbf{S}^t, \quad \forall f \in \mathscr{F}. \tag{18.6}$$

Based on the above, the original $T(\mathbf{r}, t)$ and $P(\mathbf{r}, t)$ can be thought of as a family of observables on $X$ parameterized by the position $\mathbf{r}$, and denoted by $T(\mathbf{x}; \mathbf{r})$ and $P(\mathbf{x}; \mathbf{r})$. From this viewpoint, we are able to describe the time evolutions of $T$ and $P$ as the action of $U^t$:

$$\begin{aligned}\begin{bmatrix} T(\mathbf{r}, t) \\ P(\mathbf{r}, t) \end{bmatrix} &:= \begin{bmatrix} T(\mathbf{x}(t); \mathbf{r}) \\ P(\mathbf{x}(t); \mathbf{r}) \end{bmatrix} = \begin{bmatrix} T(\mathbf{S}^t\mathbf{x}(0); \mathbf{r}) \\ P(\mathbf{S}^t\mathbf{x}(0); \mathbf{r}) \end{bmatrix} \\ &= \begin{bmatrix} U^t T(\mathbf{x}(0); \mathbf{r}) \\ U^t P(\mathbf{x}(0); \mathbf{r}) \end{bmatrix} = \begin{bmatrix} U^t T(\mathbf{r}, 0) \\ U^t P(\mathbf{r}, 0) \end{bmatrix}.\end{aligned} \tag{18.7}$$

Note that Eq. (18.7) is an approximation of the long-term (hourly-scale) dynamics of $T$ and $P$, and does not represent the short-term dynamics.

### 18.3.2 *Koopman Mode Decomposition and Associated Linear Wavenumber Vector*

Next, we introduce the KMD and the notion of linear wave propagation of KM. Let us denote an eigenvalue $\lambda$ and associated (normalized) eigenfunction $\phi_\lambda$ of the Koopman operator $U^t$ as follows:

$$U^t \phi_\lambda = \exp(\lambda t)\phi_\lambda, \quad \lambda \in \mathbb{C}, \quad \phi_\lambda \in \mathscr{F} \setminus \{0\}. \tag{18.8}$$

The value $\lambda$ is called the *Koopman Eigenvalue* (KE), and the number of eigenvalues is normally (countably) infinite [14]. Since the dynamical system (18.4) is introduced on the attractor, it is relevant that the flow preserves a measure defined on $X$ [30]. This implies that in a suitable choice of $\mathscr{F}$ such as $\mathscr{L}_2$, the Koopman operator $U^t$ is unitary [30, 31]. The author of [14, 15] shows that an observable $f \in \mathscr{F}$ is expanded in terms of spectral properties of $U^t$. In particular, the part of the spectral expansion in terms of Koopman eigenfunctions $\phi_\lambda(\mathbf{x})$ is called the KMD [14, 15]. If the latter expansion holds for the target $T$ and $P$, then we have

$$\begin{bmatrix} T(\mathbf{x}; \mathbf{r}) \\ P(\mathbf{x}; \mathbf{r}) \end{bmatrix} = \sum_{m=1}^{\infty} \phi_{\lambda_m}(\mathbf{x}) \begin{bmatrix} a_m(\mathbf{r}) \\ b_m(\mathbf{r}) \end{bmatrix}, \tag{18.9}$$

where $a_m(\mathbf{r})$ and $b_m(\mathbf{r})$ are the projections of $T$ and $P$ onto the subspace in $\mathscr{L}_2$ spanned by the Koopman eigenfunction $\phi_{\lambda_m}$: with the inner product $\langle\cdot,\cdot\rangle : \mathscr{F} \times \mathscr{F} \to \mathbb{C}$ in $\mathscr{L}_2$,

$$a_m(\mathbf{r}) = \langle T(\mathbf{x};\mathbf{r}), \phi_{\lambda_m}(\mathbf{x})\rangle, \tag{18.10a}$$

$$b_m(\mathbf{r}) = \langle P(\mathbf{x};\mathbf{r}), \phi_{\lambda_m}(\mathbf{x})\rangle. \tag{18.10b}$$

Therefore, we have the following decompositions of the time evolutions of $T$ and $P$:

$$\begin{aligned}\begin{bmatrix} T(\mathbf{r},t) \\ P(\mathbf{r},t) \end{bmatrix} &= \begin{bmatrix} T(\mathbf{x}(t);\mathbf{r}) \\ P(\mathbf{x}(t);\mathbf{r}) \end{bmatrix} = \sum_{m=1}^{\infty} \phi_{\lambda_m}(\mathbf{x}(t)) \begin{bmatrix} a_m(\mathbf{r}) \\ b_m(\mathbf{r}) \end{bmatrix} \\ &= \sum_{m=1}^{\infty} U^t \phi_{\lambda_m}(\mathbf{x}(0)) \begin{bmatrix} a_m(\mathbf{r}) \\ b_m(\mathbf{r}) \end{bmatrix} = \sum_{m=1}^{\infty} \exp(\lambda_m t)\phi_{\lambda_m}(\mathbf{x}(0)) \begin{bmatrix} a_m(\mathbf{r}) \\ b_m(\mathbf{r}) \end{bmatrix} \\ &=: \sum_{m=1}^{\infty} \exp(\lambda_m t) \begin{bmatrix} \tilde{a}_m(\mathbf{r}) \\ \tilde{b}_m(\mathbf{r}) \end{bmatrix}. \end{aligned} \tag{18.11}$$

In this chapter, we refer to $\tilde{a}_m(\mathbf{r}) := \phi_{\lambda_m}(\mathbf{x}(0))a_m(\mathbf{r})$ and $\tilde{b}_m(\mathbf{r}) := \phi_{\lambda_m}(\mathbf{x}(0))b_m(\mathbf{r})$ as $m$th *Koopman modes* of $T$ and $P$, respectively. Note that in Eq. (18.11), $\lambda_m$ is *common* in both the decompositions of $T$ and $P$.

Here, for the parameter estimation of Eq. (18.3), we derive PDEs in terms of $\tilde{a}_m(\mathbf{r})$ and $\tilde{b}_m(\mathbf{r})$. Assume that the KE $\lambda_m$ is distinct. By substituting Eq. (18.11) into Eq. (18.3) and equating the terms evolving in time with $\exp(\lambda_m t)$, the PDEs for $m$th KM are derived as

$$\lambda_m \tilde{a}_m(\mathbf{r}) = \begin{cases} -\mathbf{V}_{\text{eff}}(\mathbf{r}) \cdot \nabla \tilde{a}_m(\mathbf{r}) + \dfrac{\tilde{b}_m(\mathbf{r})}{\rho c_{\text{p}}} & (L = -\mathbf{V}_{\text{eff}}(\mathbf{r}) \cdot \nabla), \\ D_{\text{eff}}(\mathbf{r}) \Delta \tilde{a}_m(\mathbf{r}) + \dfrac{\tilde{b}_m(\mathbf{r})}{\rho c_{\text{p}}} & (L = D_{\text{eff}}(\mathbf{r})\Delta). \end{cases} \tag{18.12}$$

In Sect. 18.4, we provide a concrete scheme of estimating $\mathbf{V}_{\text{eff}}$ and $D_{\text{eff}}$ by a combination of Eq. (18.12) and measurement data.

In the present application of building, we sample the dynamics of the temperature field $T(\mathbf{r},t)$ and the heat input $P(\mathbf{r},t)$ at finite $M$ locations in the target space $\Omega$, denoted by $\mathbf{r}_1,\ldots,\mathbf{r}_M \in \Omega$. According to Eq. (18.11), the dynamic data are represented as

$$\begin{bmatrix} T(\mathbf{r}_1, t) \\ \vdots \\ T(\mathbf{r}_M, t) \\ P(\mathbf{r}_1, t) \\ \vdots \\ P(\mathbf{r}_M, t) \end{bmatrix} = \sum_{m=1}^{\infty} \exp(\lambda_m t) \begin{bmatrix} \tilde{a}_m(\mathbf{r}_1) \\ \vdots \\ \tilde{a}_m(\mathbf{r}_M) \\ \tilde{b}_m(\mathbf{r}_1) \\ \vdots \\ \tilde{b}_m(\mathbf{r}_M) \end{bmatrix} =: \sum_{m=1}^{\infty} \exp(\lambda_m t) \begin{bmatrix} \mathbf{A}_m \\ \mathbf{B}_m \end{bmatrix}. \tag{18.13}$$

Here, the two location-dependent vectors $\mathbf{A}_m := [\tilde{a}_m(\mathbf{r}_1) \ \ldots \ \tilde{a}_m(\mathbf{r}_M)]^\mathsf{T}$ and $\mathbf{B}_m := [\tilde{b}_m(\mathbf{r}_1) \ \ldots \ \tilde{b}_m(\mathbf{r}_M)]^\mathsf{T}$ ($\mathsf{T}$ denotes the transpose of vector) will be also called $m$th KMs.

For detailed analysis, we describe the oscillatory pattern of a single KM as a linear wave propagation. To do this, the polar forms of $\tilde{a}_m(\mathbf{r}_p)$ and $\tilde{b}_m(\mathbf{r}_p)$ for $p = 1, \ldots, M$ are introduced with the moduli $A_m(\mathbf{r}_p), B_m(\mathbf{r}_p) \in [0, \infty)$ and the arguments $\alpha_m(\mathbf{r}_p), \beta_m(\mathbf{r}_p) \in [0, 2\pi]$:

$$\begin{bmatrix} \mathbf{A}_m \\ \mathbf{B}_m \end{bmatrix} = \begin{bmatrix} A_m(\mathbf{r}_1) \exp(\mathrm{i}\alpha_m(\mathbf{r}_1)) \\ \vdots \\ A_m(\mathbf{r}_M) \exp(\mathrm{i}\alpha_m(\mathbf{r}_M)) \\ B_m(\mathbf{r}_1) \exp(\mathrm{i}\beta_m(\mathbf{r}_1)) \\ \vdots \\ B_m(\mathbf{r}_M) \exp(\mathrm{i}\beta_m(\mathbf{r}_M)) \end{bmatrix}, \tag{18.14}$$

where $\mathrm{i}$ is the imaginary unit. If the matrix $[\mathbf{r}_1 \cdots \mathbf{r}_M]$ of sampling locations is row full rank, then it is possible to uniquely determine the two constant vectors $\mathbf{k}_m, \mathbf{l}_m \in \mathbb{R}^d$ satisfying

$$\mathbf{k}_m^\mathsf{T}[\mathbf{r}_1 \cdots \mathbf{r}_M] = -[\tilde{\alpha}_m(\mathbf{r}_1) \ldots \tilde{\alpha}_m(\mathbf{r}_M)],$$
$$\mathbf{l}_m^\mathsf{T}[\mathbf{r}_1 \cdots \mathbf{r}_M] = -[\tilde{\beta}_m(\mathbf{r}_1) \ldots \tilde{\beta}_m(\mathbf{r}_M)],$$

which correspond to the direction on which *wavefront* (surface of constant phase of a KM) propagates forward in time. Hence, we show that the spatiotemporal evolution of a single KM is described as the *wave propagation*:

$$\exp(\lambda_m t) \begin{bmatrix} \mathbf{A}_m \\ \mathbf{B}_m \end{bmatrix} = \exp(\sigma_m t) \begin{bmatrix} A_m(\mathbf{r}_1) \exp\{\mathrm{i}(\omega_m t - \mathbf{k}_m^\mathsf{T}\mathbf{r}_1)\} \\ \vdots \\ A_m(\mathbf{r}_M) \exp\{\mathrm{i}(\omega_m t - \mathbf{k}_m^\mathsf{T}\mathbf{r}_M)\} \\ B_m(\mathbf{r}_1) \exp\{\mathrm{i}(\omega_m t - \mathbf{l}_m^\mathsf{T}\mathbf{r}_1)\} \\ \vdots \\ B_m(\mathbf{r}_M) \exp\{\mathrm{i}(\omega_m t - \mathbf{l}_m^\mathsf{T}\mathbf{r}_M)\} \end{bmatrix}, \tag{18.15}$$

where $\sigma_m := \mathrm{Re}[\lambda_\mathrm{m}]$ represents the amplifying/attenuating rate of $m$th KM and $\omega_m := \mathrm{Im}[\lambda_\mathrm{m}]$ the oscillation frequency which is nonzero. Equation (18.15) implies that the dynamics of observables can be represented by a spatiotemporal wave whose amplitude changes in space (sampling locations). In this chapter, we refer to $\mathbf{k}_j$ and $\mathbf{l}_j$ as $m$th *Koopman wavenumber vectors* (KWs). The KWs $\mathbf{k}_j$ and $\mathbf{l}_j$ enable us to visualize the direction of heat transfer induced by an interaction between multiple HVAC units (see Sect. 18.4.2).

## 18.4 Demonstration for a Practical Building Atrium

This section demonstrates the above modeling framework with measurement data in a practical building.

### *18.4.1 Target Building and Measurement Data*

First of all, we describe the practical building where our modeling is demonstrated: architectural geometry; spatial arrangements of HVAC units; and measurement of temperature and HVAC outlet.

The target building is the research laboratory of Azbil Corporation in Kanagawa, Japan. Figure 18.3a, b shows the photograph and cross section of the target building.

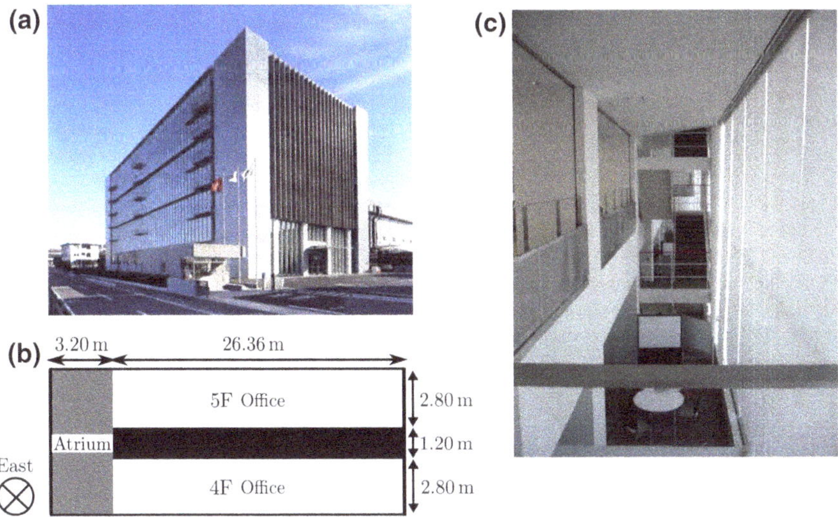


**Fig. 18.3** **a** Photograph and **b** cross section of the target building; **c** Photograph of atrium in the target building. The photographs and figure are reprinted with permission from [25]. ©2017, SICE

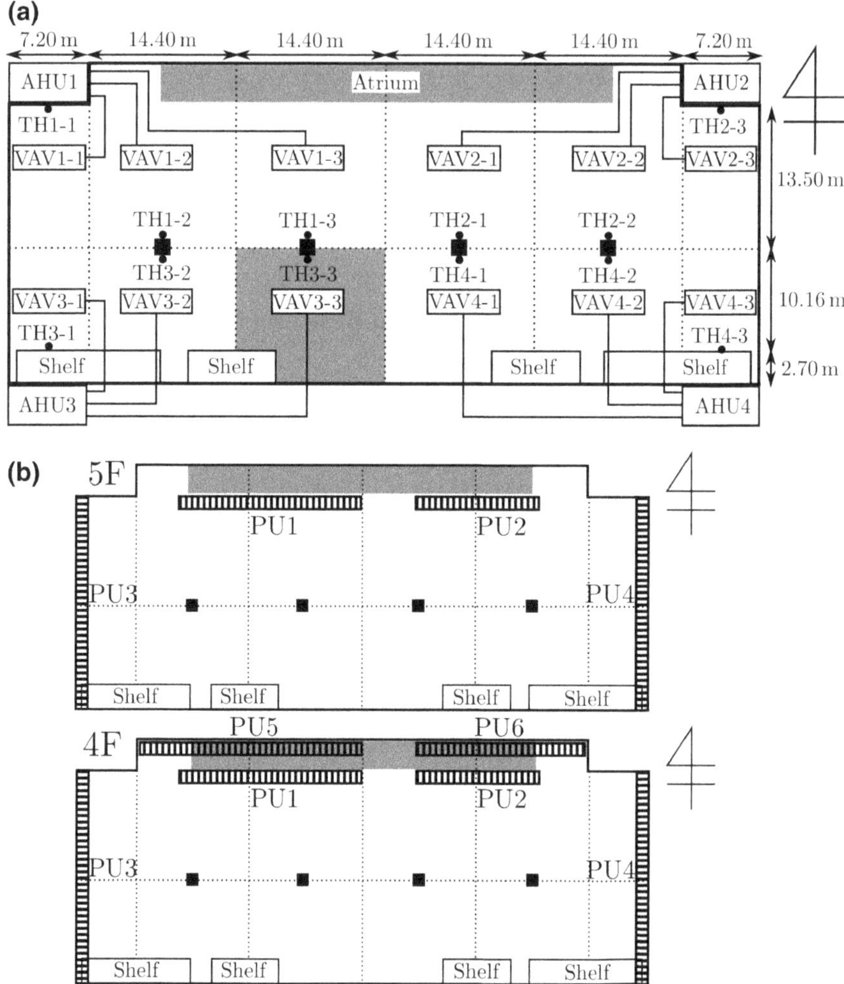


**Fig. 18.4** Floor outline with the deployment of HVAC units: **a** Outline of fourth and fifth floors of the target building. The circles stand for in-room temperature sensors and the rectangles for VAV units and AHU units. **b** Outlet ducts of HVAC units in the perimeter zone of the target building. PU1 to PU4 are located on the ceiling and generate downward airflow. On the other hand, PU5 and PU6 are located near the floor and generate upward airflow. Reprinted with permission from [25]. ©2017, SICE

The southern parts of fourth and fifth floors are used as office spaces. An atrium is located in the northern part of this building and connected to the offices, as shown in Fig. 18.3c. We address the atrium as the main target for the modeling of heat transfer dynamics.

Next, we review the spatial arrangement of HVAC units. Figure 18.4a shows floor outline of the target building, where the black and white rectangles denote pillars and shelves, respectively. There are four air handling units (AHUs) on each floor that provide heating/cooling air to the building. An AHU unit is connected to three variable air volume (VAV) units [32], which locally regulate temperature by delivering the air supply from the AHU to *local zones*, illustrated by the shaded area in Fig. 18.4a. Here, a VAV unit has multiple ducts on the ceiling for delivering the air. In the rest of this chapter, each AHU is identified by the index $n_a \in \{1, 2, 3, 4\}$, and the VAVs associated with AHU$n_a$ are called VAV$n_a$-1, VAV$n_a$-2, and VAV$n_a$-3. Moreover, as denoted by the striped rectangles in Fig. 18.4b, there are six auxiliary units for conditioning perimeter zones on the fourth floor, and four ones on the fifth floor. The four units on both the fourth and fifth floors, called perimeter unit 1 (PU1) to PU4, are located on the ceiling and generate downward airflow. Also, the two units only on the fourth floor, called PU5 and PU6, are near the floor and generate upward airflow.

Here, we introduce the measurement setting for the target building. In the building, indoor air temperature, outlet flow rate, and temperature of HVAC units were sampled by 0.1 °C or 1 m$^3$/h every 1 min. Indoor temperature was measured by sensors on the pillars, walls, or shelves. The sensors are called TH1-1 to TH4-3 and deployed as in Fig. 18.4a. Outlet flow rate was measured in the 12 VAVs on each floor; on the other hand, outlet temperature was sampled in the four AHUs. Also, for PU1 and PU2 on the fourth floor and PU1 to PU4 the on fifth floor, their outlet flow rate and outlet temperature were measured, but for the other PUs no measurement was performed.

Finally, we introduce the measurement data for our demonstration. All the data were sampled on August 1, 2014 (Friday). Figure 18.5 shows the indoor temperature on the fifth floor, and Figs. 18.6 and 18.7 the outlet flow rate of VAVs and the outlet temperature of AHUs on the fifth floor. The outlet flow rate and temperature tempo-

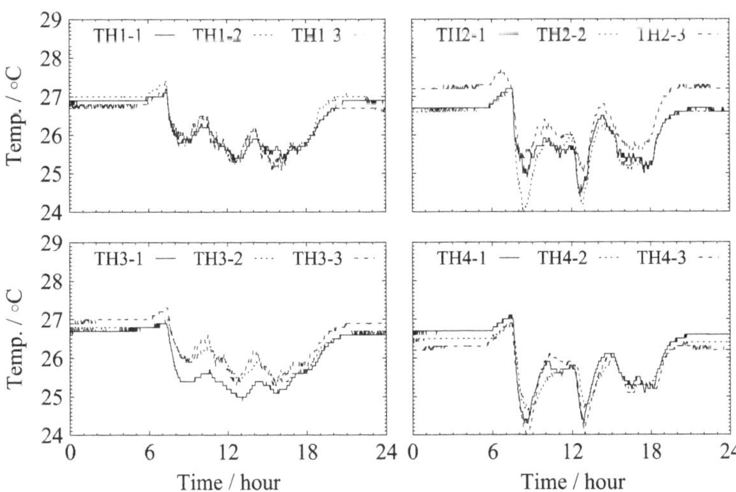


**Fig. 18.5** In-room temperature on fifth floor sampled on August 1, 2014 (Friday). Reprinted with permission from [25]. ©2017, SICE

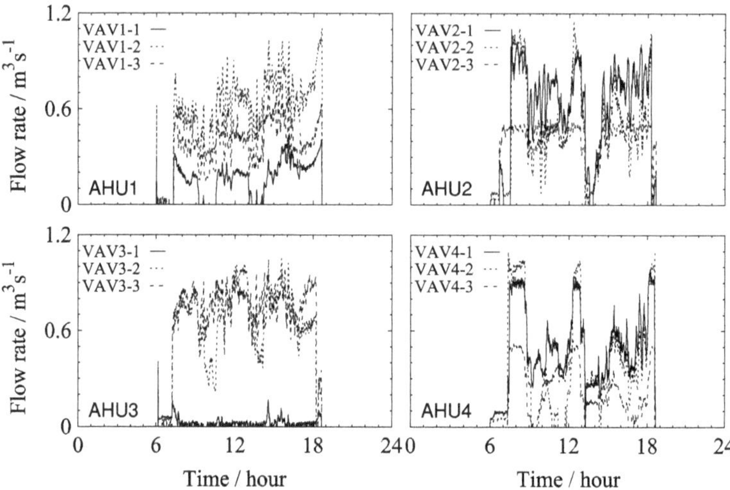


**Fig. 18.6** Outlet volume of VAVs on fifth floor sampled on August 1, 2014 (Friday). Reprinted with permission from [25]. ©2017, SICE

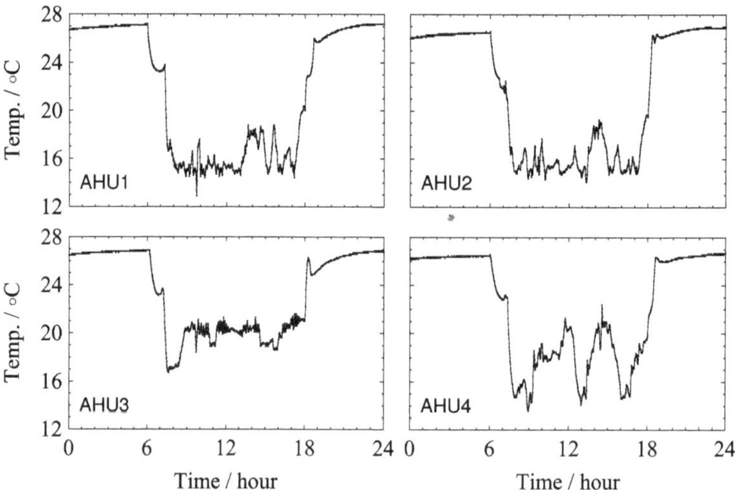


**Fig. 18.7** Outlet temperature of AHUs on fifth floor sampled on August 1, 2014 (Friday). Reprinted with permission from [25]. ©2017, SICE

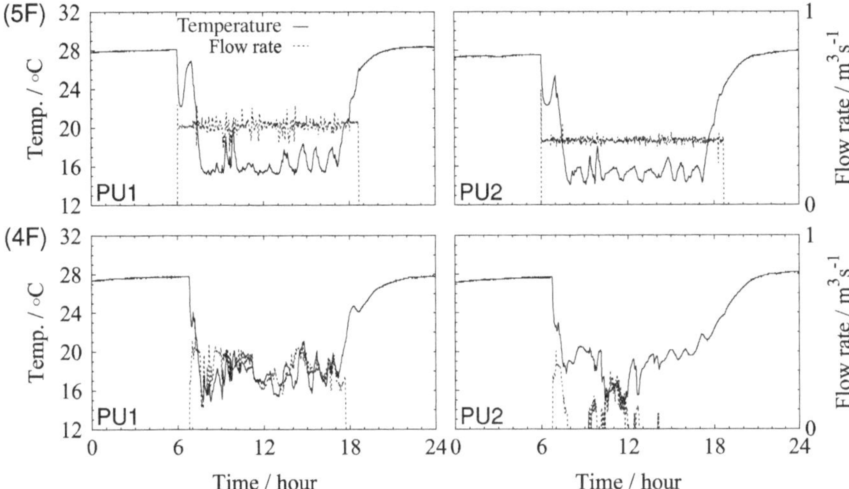


**Fig. 18.8** Outlet temperature and volume of PUs sampled on August 1, 2014 (Friday). Reprinted with permission from [25]. ©2017, SICE

rally changed with the indoor temperature, implying that VAV operations regulated the indoor temperature. For example, the indoor temperature falls at about 8:00 and surges at about 13:00. The VAVs, especially VAV2-* and 4-*, changed their outlet flow rate in response to the fall and surge of indoor temperature. Furthermore, Fig. 18.8 shows the outlet flow rate and temperature of PUs. The measurement data are different for every location in space, implying that the magnitude of thermal loads such as appliances and humans depends on the locations in the perimeter zones.

## 18.4.2 Koopman Mode Decomposition of Measurement Data

The measurement data in the last subsection are analyzed with the KMD. A variety of computation algorithms has been developed for finite-order approximation of KMD: phase averaging [14], Arnoldi-type algorithm [33], dynamic mode decomposition (DMD) [34, 35], extended DMD [36], and Prony approximation [37]. In this chapter, we use the Arnoldi-type algorithm to compute KEs and KMs directly from the measurement data. It has been reported in [38] that compared with the standard (not extended) DMD, the Arnoldi-type algorithm is suitable to spatially low-dimensional and temporally high-dimensional data, which is the case of our current analysis. For more details of the Arnoldi-type algorithm, see Chap. 1 of this book and [33].

Before the analysis, we preprocess the measurement data and derive time-series data in discrete time on the indoor temperature $T$ and heat input $P$. The original data on indoor temperature are multivariate with dimension $M$ and sampled every time period $\Delta t$, denoted by

$$\mathbf{T}[n] = [T(\mathbf{r}_1, n\Delta t), \ldots, T(\mathbf{r}_M, n\Delta t)]^{\mathsf{T}}, \quad n = 0, \ldots, N,$$

where $n$ is the discrete time and $N + 1$ corresponds to the number of sampled data. The data in this work are derived for the time interval 8:00–17:00 every 1 min, that is, $N = 540$ and $\Delta t = 1\,\text{min}$. The indoor temperature $T$ in Fig. 18.5 and the heat input $P$ are associated with the 12 zones of the fourth and fifth floors, implying $M = 24$. Here, the sampled data $\mathbf{P}[n]$ of the heat input are derived in the following manner. We assume that $P$ mainly arises from HVAC units, and that the residue of heat input from HVAC is negligible for evaluating the target heat transfer. The relevance of this assumption depends on the timescale of heat transfer dynamics which we address in this modeling. By choosing an appropriate KM for the evaluation, it will be shown that the timescale of the heat transfer dynamics is separated from the residue of heat input so that the above assumption holds (see the next paragraph in detail). Then, by using the well-known *bulk convection* formula [1], we are able to calculate the $\ell$th element $P_\ell[n]$ ($\ell = 1, \ldots, M$) of $\mathbf{P}[n]$ as follows:

$$\begin{aligned} P_\ell[n] :=& \frac{\rho c_{\mathrm{p}} V_{\mathrm{out}}(\mathbf{r}_\ell, n\Delta t)}{V_\ell} \times (T_{\mathrm{out}}(\mathbf{r}_\ell, n\Delta t) - T(\mathbf{r}_\ell, n\Delta t)) \\ &+ \frac{\rho c_{\mathrm{p}} V^{\mathrm{p}}_{\mathrm{out}}(\mathbf{r}_\ell, n\Delta t)}{V_\ell} \times (T^{\mathrm{p}}_{\mathrm{out}}(\mathbf{r}_\ell, n\Delta t) - T(\mathbf{r}_\ell, n\Delta t)), \end{aligned} \tag{18.16}$$

where $V_\ell$ is the control volume around $\mathbf{r} = \mathbf{r}_\ell$, that is, the volume of each zone of the target space (see Fig. 18.4a), $T$ the indoor temperature (Fig. 18.5), $V_{\mathrm{out}}$ and $T_{\mathrm{out}}$ the outlet flow rate and temperature of VAVs (Figs. 18.6 and 18.7), and $V^{\mathrm{p}}_{\mathrm{out}}$ and $T^{\mathrm{p}}_{\mathrm{out}}$ the outlet flow rate and temperature of PUs (Fig. 18.8).

We are now in a position to present the KMD result on the sampled data $\mathbf{T}[n]$ and $\mathbf{P}[n]$. They are decomposed with the Arnoldi-type algorithm into the following finite series:

$$\begin{bmatrix} \mathbf{T}[n] \\ \mathbf{P}[n] \end{bmatrix} = \sum_{m=1}^{N} \mu_m^n \begin{bmatrix} \mathbf{A}_m \\ \mathbf{B}_m \end{bmatrix}, \quad n = 0, \ldots, N-1, \tag{18.17}$$

where $\mu_m := \exp(\lambda_m \Delta t)$ is associated with the KE $\lambda_m$. Table 18.1 summarizes modal information in order of the magnitude of $|\mu_m|$. In the table, $\|\cdot\|$ denotes the vector norm and $T_m := 2\pi/|\mathrm{Im}[\lambda_{\mathrm{m}}]|$ the oscillation period. Note that for simplicity of the parameter estimation in Sect. 18.4.3, KMs $\mathbf{B}_m$ are divided by $\rho c_{\mathrm{p}}$. KM 7 with $\mu_m = 1.0000$ represents the bias (time-average) component of the original data. From the other modes, we choose KMs $\{16,17\}$ as *dominant* modes that will be used for the current modeling of heat transfer. The main reason for this choice is that both the norms $\|\mathbf{A}_m\|$ and $\|\mathbf{B}_m\|$ are large, hence they capture well the heat transfer and associated HVAC operation.[2] Moreover, it is naturally recognized that heat sources

[2]KMs $\{10,11\}$ also have large magnitudes of $\|\mathbf{A}_m\|$ and $\|\mathbf{B}_m\|$. However, they can be neglected because the original data contain no oscillatory component associated with the period $T_{10} = 6.490\,\text{h}$. Indeed, we applied the discrete Fourier transform (DFT) to the original data $\{T(\mathbf{r}_p, n\Delta t)|n =$

**Table 18.1** Koopman mode decomposition of the data measured on August 1, 2014 (Friday). Reproduced from [25]. ©2017, SICE

| $\{m, m+1\}$ | $\lvert\mu_m\rvert$ | $T_m$ / h | $\lVert\mathbf{A}_m\rVert$ | $\lVert\mathbf{B}_m\rVert/\rho c_\mathrm{p}$ |
|---|---|---|---|---|
| $\{1, 2\}$ | 1.0036 | 1.014 | $2.68 \times 10^{-2}$ | $3.76 \times 10^{-4}$ |
| $\{3, 4\}$ | 1.0019 | 1.466 | $8.59 \times 10^{-2}$ | $9.46 \times 10^{-4}$ |
| $\{5, 6\}$ | 1.0001 | 0.583 | $3.00 \times 10^{-2}$ | $6.92 \times 10^{-4}$ |
| 7 | 1.0000 | $\infty$ | $1.27 \times 10^{2}$ | $4.63 \times 10^{-2}$ |
| $\{8, 9\}$ | 0.9997 | 1.151 | $1.46 \times 10^{-1}$ | $1.86 \times 10^{-3}$ |
| $\{10, 11\}$ | 0.9997 | 6.490 | $4.24 \times 10^{-1}$ | $6.01 \times 10^{-3}$ |
| $\{12, 13\}$ | 0.9995 | 0.634 | $4.16 \times 10^{-2}$ | $7.30 \times 10^{-4}$ |
| $\{14, 15\}$ | 0.9992 | 0.381 | $4.83 \times 10^{-2}$ | $8.39 \times 10^{-4}$ |
| $\{\mathbf{16}, \mathbf{17}\}$ | 0.9991 | **3.999** | **1.06** | $\mathbf{7.66 \times 10^{-3}}$ |
| $\{18, 19\}$ | 0.9989 | 0.272 | $3.45 \times 10^{-2}$ | $7.38 \times 10^{-4}$ |

except for HVAC outlet—solar radiation and outdoor temperature—change more slowly than the period $T_{16} = 3.999\,\mathrm{h}$. This implies that the assumption of the above paragraph, in which we have neglected the residue of heat input from HVAC units, holds in this modeling.

Finally, we illustrate the spatial shapes of the dominant KM 16 and its KW. The space dependency of $A_{16}(\mathbf{r}_p)$ and $\alpha_{16}(\mathbf{r}_p)$ of KM $\mathbf{A}_{16}$ is shown in Fig. 18.9a-1, a-2, and that of $B_{16}(\mathbf{r}_p)/\rho c_\mathrm{p}$ and $\beta_{16}(\mathbf{r}_p)$ of $\mathbf{B}_{16}$ in Fig. 18.9b-1, b-2. The arrows in Fig. 18.9a-2, b-2 represent the KWs $\mathbf{k}_{16}$ and $\mathbf{l}_{16}$, projected onto the width–height plane. To visualize the vertical heat transfer (in the height direction), the elements of KMs are illustrated only on the four locations connected to the atrium, that is, TH1-2, TH1-3, TH2-1, and TH2-2. In Fig. 18.9a-1, b-1, both $A_{16}$ and $B_{16}$ have large amplitudes on the fourth floor, suggesting a possibility that the temperature of each floor is mainly regulated by the HVAC of the same floor. Also, we observe that the argument $\alpha_{16}$ varies between the fourth and fifth floors, and that the KW $\mathbf{k}_{16}$ is directed from the fifth floor to the fourth floor. The numerical results show that the vertical (height direction) heat transfer occurs in the atrium. Here, the spatial pattern of $\beta_{16}$ differs from the pattern of $\alpha_{16}$, and the KW $\mathbf{l}_{16}$ is in the opposite direction of $\mathbf{k}_{16}$. These results indicate that multiple HVAC units interact through the vertical heat transfer in the atrium to diminish it. It should be noted that the HVAC interaction emerges due to the local feedback mechanism of each VAV unit, not any global (i.e., atrium-wide) feedback control. For this reason, we contend that the vertical heat transfer in atrium is not fully suppressed by the HVAC operation.

$0, 1, \ldots, N\}$ for every position $\mathbf{r}_p$, and confirmed that the values of the period of dominant components derived by DFT are different from $T_{10}$: see [25].

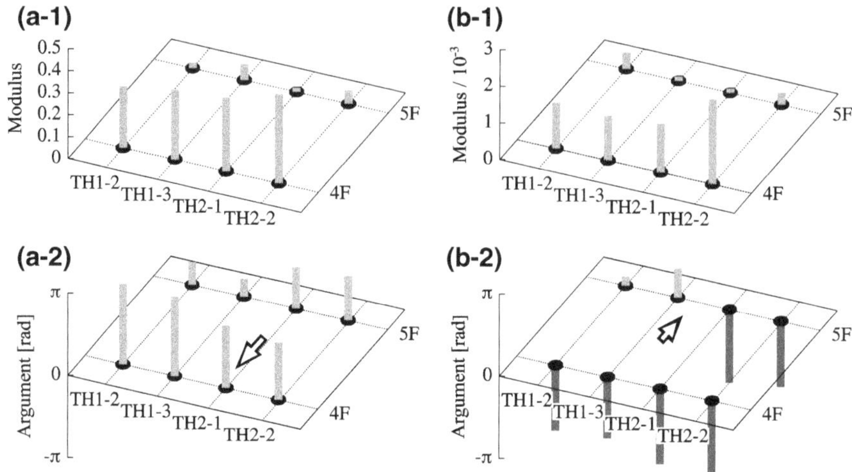


**Fig. 18.9** Modulus and argument of 16th Koopman mode in Table 18.1: **a-1**, **a-2** KM $\mathbf{A}_{16}$ of atrium temperature, where the arrow represents the direction of the calculated Koopman wavenumber vector $\mathbf{k}_{16}$, and **b-1**, **b-2** KM $\mathbf{B}_{16}$ of outlet temperature, where the arrow represents the direction of the calculated Koopman wavenumber vector $\mathbf{l}_{16}$. Reprinted with permission from [25]. ©2017, SICE

### *18.4.3 Numerical Implementation*

In this section, we describe how the parameter estimation based on Eq. (18.12) is numerically implemented to the practical building in Sect. 18.4.1. This includes the grid generation (i.e., discretization in space) to the target space and the discretization of Eq. (18.12) in time so that it is applicable to the sampled data introduced and analyzed in Sect. 18.4.2.

Before discretization, we mention the domain and heat transfer phenomena that we analyze for the parameter estimation. The atrium space studied in Fig. 18.9 corresponds to the target space where we estimate the parameters of Eq. (18.3). In order to accurately simulate the heat input from VAVs to atrium, we analyze the heat transfer in the overall domain including the atrium and neighboring office spaces of the fourth and fifth floors. Inside these offices, a lot of ducts (about six per zone shown in Fig. 18.4a) on the ceiling generate airflow. Because the spatial variance of indoor temperature is small (see Fig. 18.5), the airflow field is not affected by buoyancy of air and is mainly governed by the *nonlinear* convection effect (convection term in the Navier–Stokes equation) [1]. Thus, from the well-established phenomenology of energy cascade [39], the airflow field can be periodic in spatial scale smaller than the distance between ducts and satisfies the centering condition (18.2). Therefore, the associated heat transfer can be modeled as the diffusion phenomenon with effective diffusion $D_{\text{eff}}$. The phenomenon becomes dominant only in the horizontal direction because the height of each office is much smaller than its length and width as shown in Figs. 18.3b and 18.4a. Also, from operational experience and the fact that no HVAC

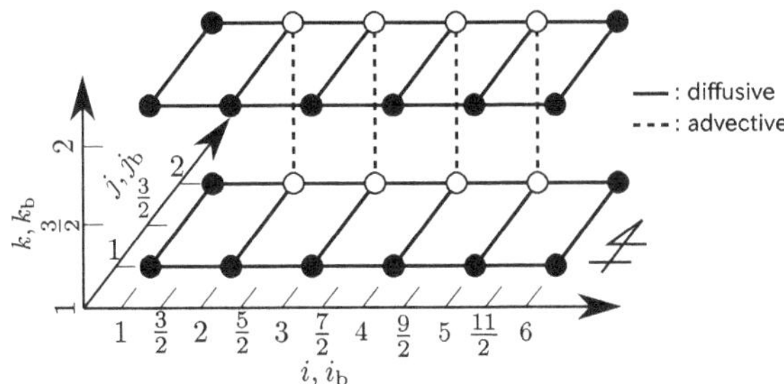


**Fig. 18.10** Nodes and branches of spatial discretization of the target building space. The *white* circles denote the nodes neighboring to the atrium, and the *black* circles the ones not neighboring to it. The branches with the *broken* lines are located on the atrium where the effective velocity governs the heat transfer. The branches with the *solid* lines are located on the floors where the heat transfer is diffusive. The walls with the Neumann-type boundary condition surround the office spaces and the atrium, and separate the fourth and fifth floors. The boundary condition is represented by omitting the branches between the sampling locations $\mathbb{I}$ and the outside, and the vertical ones through the wall, namely, not neighboring the atrium

unit is deployed inside the atrium, the perimeter units PU1 to PU4 generate a laminar airflow over the atrium, implying that the centering condition (18.2) does not hold. Therefore, the heat transfer inside the atrium can be modeled as the advection phenomenon with effective velocity $V_{\mathrm{eff}}$. The phenomenon emerges only in the vertical direction because the width of the atrium is much smaller than its length and height (see Fig. 18.3b), and the heat flux of the advection in the length direction is negligible compared with the effective diffusion.

Next, we set the spatial discretization of Eq. (18.12), that is, the grid generation. Figure 18.10 shows nodes and branches for the discretization. Here the sampling locations $\mathbf{r}_p$ on which measurement data on $T$ and $P$ are available in this work are used as the nodes given by

$$(i, j, k) \in \mathbb{I} := \{1, \ldots, 6\} \times \{1, 2\} \times \{1, 2\}.$$

Here, $i$ denotes the length direction and $j$ the width direction so that $[i, j]$ is assigned to the center of each zone shown in Fig. 18.4a. Also, $k$ shows whether on the fourth or fifth floor the node is placed. Then, KMs $\tilde{a}_m(\mathbf{r}_p)$ and $\tilde{b}_m(\mathbf{r}_p)$ are discretized into $\tilde{a}_m[i, j, k]$ and $\tilde{b}_m[i, j, k]$. From the fact that the temperature sensors are non-uniformly deployed as in Fig. 18.4a, we introduce the length step $\Delta r$ between nodes that depends on $i, j, k$. In addition, to represent the branches, we introduce the fractional indices as

$$i_{\mathrm{b}} \in \mathbb{I}_{\mathrm{b}} := \{3/2, 5/2, \ldots, 11/2\}, \ \ j_{\mathrm{b}} \in \mathbb{J}_{\mathrm{b}} := \{3/2\}, \ \ k_{\mathrm{b}} \in \mathbb{K}_{\mathrm{b}} := \{3/2\},$$

and we describe the branch dependency of $\Delta r$ and the parameters $D_{\mathrm{eff}}$ and $V_{\mathrm{eff}}$ by $\Delta r_{i_{\mathrm{b}},j,k}$, $\Delta r_{i,j_{\mathrm{b}},k}$, $\Delta r_{i,j,k_{\mathrm{b}}}$, $D_{i_{\mathrm{b}},j,k}$, $D_{i,j_{\mathrm{b}},k}$, and $V_{i,j,k_{\mathrm{b}}}$. The *white* circles in the figure denote the nodes neighboring the atrium while the *black* ones not neighboring it. The vertical branches denoted by the *broken* lines are located on the atrium, and the horizontal ones denoted by the *solid* lines are on the floors.

Also, we adopt the Neumann-type boundary condition on the walls that the spatial derivative of $T$ is zero, namely neglecting any heat input from the walls. This condition is reasonable for the period $T_{16} = 3.999\,\mathrm{h}$ of the target KM for modeling because the heat input is decayed during the period due to large thermal capacitance of the walls [1, 40]. The walls surround both the office spaces and the atrium, and separate the fourth and fifth floors (see Fig. 18.3b). In Fig. 18.10, the boundary condition for the walls is represented by omitting the branches between the sampling locations $\mathbb{I}$ and the outside, and the vertical ones through the wall, namely, not neighboring the atrium.

Finally, we describe the concrete scheme of the parameter estimation. Following the above preliminaries in this subsection, the heat flux at the horizontal branches (*solid* lines in Fig. 18.10) is modeled as the effective diffusion, and the heat flux at the vertical branches (*broken* lines) as the advection with effective velocity. Thus, by approximating the spatial derivatives in the model with first- and second- order central differences, Eq. (18.12) is rewritten in the following form (the double sign in same order):

$$\begin{aligned}\lambda_m \tilde{a}_m[i,j,k] = &\sum_{i_\mathrm{b}=i\pm 1/2} D_{i_\mathrm{b},j,k} \frac{\tilde{a}_m[i\pm 1,j,k]-\tilde{a}_m[i,j,k]}{\Delta r^2_{i_\mathrm{b},j,k}} \\ &+ \sum_{j_\mathrm{b}=j\pm 1/2} D_{i,j_\mathrm{b},k} \frac{\tilde{a}_m[i,j\pm 1,k]-\tilde{a}_m[i,j,k]}{\Delta r^2_{i,j_\mathrm{b},k}} \\ &+ \sum_{k_\mathrm{b}=\mp 1/2} V_{i,j,k_\mathrm{b}} \frac{\pm\tilde{a}_m[i,j,k\mp 1]\pm\tilde{a}_m[i,j,k]}{2\Delta r_{i,j,k_\mathrm{b}}} + \frac{\tilde{b}_m[i,j,k]}{\rho c_\mathrm{p}} \\ &((i,j)\in\{2,3,4,5\}\times\{2\}),\end{aligned} \tag{18.18}$$

and

$$\begin{aligned}\lambda_m \tilde{a}_m[i,j,k] = &\sum_{i_\mathrm{b}=i\pm 1/2} D_{i_\mathrm{b},j,k} \frac{\tilde{a}_m[i\pm 1,j,k]-\tilde{a}_m[i,j,k]}{\Delta r^2_{i_\mathrm{b},j,k}} \\ &+ \sum_{j_\mathrm{b}=j\pm 1/2} D_{i,j_\mathrm{b},k} \frac{\tilde{a}_m[i,j\pm 1,k]-\tilde{a}_m[i,j,k]}{\Delta r^2_{i,j_\mathrm{b},k}} \\ &+ \frac{\tilde{b}_m[i,j,k]}{\rho c_\mathrm{p}} \quad \text{(otherwise)},\end{aligned} \tag{18.19}$$

where the coefficients $D$ and $U$ take nonzero values for their subscripts $i_\mathrm{b} \in \mathbb{I}_\mathrm{b}$, $j_\mathrm{b} \in \mathbb{J}_\mathrm{b}$, $k_\mathrm{b} \in \mathbb{K}_\mathrm{b}$; otherwise, they take zero values to represent the Neumann-type boundary condition. Moreover, to obtain the parameter values in a physically tractable manner, we set the following constraint condition:

$$D_{i,3/2,k} = \frac{\Delta r_{i+1/2,1,k} D_{i+1/2,1,k} + \Delta r_{i-1/2,1,k} D_{i-1/2,1,k}}{2(\Delta r_{i+1/2,1,k} + \Delta r_{i-1/2,1,k})} + \frac{\Delta r_{i+1/2,2,k} D_{i+1/2,2,k} + \Delta r_{i-1/2,2,k} D_{i-1/2,2,k}}{2(\Delta r_{i+1/2,2,k} + \Delta r_{i-1/2,2,k})}, \tag{18.20}$$

for $(i, k) \in \{1, \ldots, 6\} \times \{1, 2\}$. Equation (18.20) causes from a discretized form of the gradient $\|\nabla D_{\rm eff}\|$, implying that a drastic change of $D_{\rm eff}$ in space is not allowed. We identify the parameters $D$ and $V$ by solving Eqs. (18.18)–(18.20) in terms of them.

### *18.4.4 Estimation Result*

Figure 18.11a shows the estimated effective velocity $V_{i,2,3/2}$ $(i = 2, \ldots, 5)$ for the dominant KM 16 in Fig. 18.9. A positive value of the velocity stands for the downward flow. Also note that the coefficient matrices of $V_{\rm eff}$ and $D_{\rm eff}$ in Eqs. (18.18) and (18.19) were not full rank, and that the pseudoinverse was used for the computation. In the figure we see that the values of the effective velocity are of the order of 0.1 m/s. Since the velocity of air outlet from PU1 and PU2 is about 3 m/s, the estimated velocity suggests that the PUs' operation possibly governs the heat transfer in atrium. Also, the estimated velocity is positive at all positions, implying the development of downward (in the averaged sense) air movement in the atrium. The air movement is mainly caused by the following two reasons: PU1 and PU2 output cool air to the atrium on

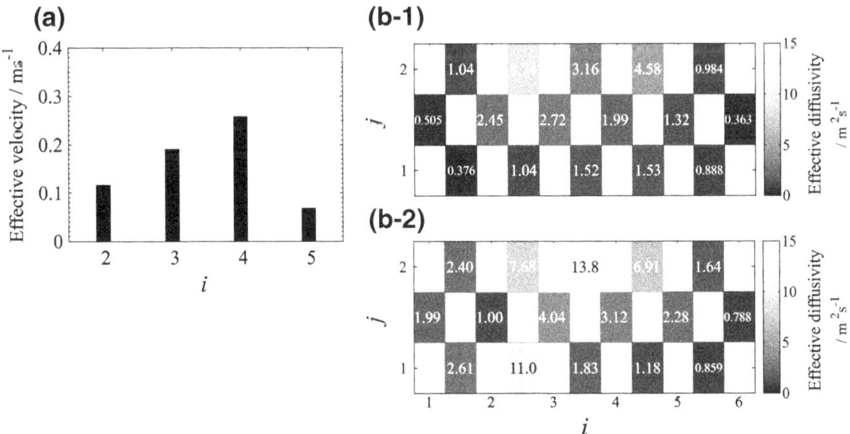


**Fig. 18.11** **a** Estimated velocity $V_{\rm eff}$ at $[i, 2, 3/2]$ $(i = 2, \ldots, 5)$. The downward flow (from the fifth floor to the fourth floor) takes a positive value of the velocity. **b-1** Estimated effective diffusivity at $[i_{\rm b}, j, k]$ $(i_{\rm b} = 3/2, 5/2, \ldots, 11/2,\ j = 1, 2)$ and $[i, 3/2, k]$ $(i = 1, \ldots, 6)$ on the fifth floor $(k = 2)$. **b-2** Estimated effective diffusivity on the fourth floor $(k = 1)$. Reprinted with permission from [25]. ©2017, SICE

the target day (August in Japan is summer: see Fig. 18.8); and the outlet direction of the PUs is downward (see Sect. 18.4.1). Moreover, the velocity takes the largest value at $i = 4$. This is consistent with the architectural geometry of the atrium. There is a stairway at $i = 4$ (see Fig. 18.3c) connecting the atrium with the upper floors. Thus, since HVAC units on the upper floors also output cool air, it possibly enhances the downward flow at $i = 4$. This shows that the estimated velocity is relevant to the target atrium. In addition, the downward flow corresponds to the direction of the KW $\mathbf{k}_{16}$ of temperature field, which implies that the use of $\mathbf{k}_{16}$ enables to describe advective heat transfer caused by a laminar airflow in the atrium.

Figure 18.11b-1 shows estimated effective diffusivity on the fifth floor ($k = 2$), and Fig. 18.11b-2 the fourth floor ($k = 1$). At most of the branches, the magnitudes of the effective diffusivity are of the order of $0.1\,\mathrm{m^2/s}$ to $1\,\mathrm{m^2/s}$. The order corresponds to effective diffusivity of in-room heat transfer governed by HVAC operation [4, 24] so that it is physically relevant.

Here, we note the space dependency of the estimated effective diffusivity. The effective diffusivity takes large values (about $10\,\mathrm{m^2/s}$) near the atrium (namely, at $j = 2$) compared with other positions. This is possibly due to the use of pseudoinverse to solve Eqs. (18.18)–(18.20). Since the effective diffusivity is a numerical approximation by pseudoinverse, it becomes singular near the atrium in order to minimize the influence of heat transfer in the atrium (at $j = 3/2$). Moreover, the effective diffusivity takes a large value at $[i_b, j, k] = [5/2, 1, 1]$, which is because of nonsteady (transient) component of HVAC operation data. In the target day, flow rate of VAV3-1 on the fourth floor transiently changed (diverged) overtime (see Fig. 7 of [25] for details). Because of this, modulus of $b_{16}(\mathbf{r}_p)$ at TH3-1 becomes greater than the other positions, which induces high effective diffusivity at the position. As above, the space dependency of the estimation result is affected by KMD and parameter estimation algorithms. Further improvement is needed to make the space-dependency more physically relevant.

## 18.5 Possible Applications to Control

Here, we discuss the applicability of the proposed modeling to control problems. The main application is global control of indoor temperature field. In buildings, local operation of HVAC systems induces their undesirable phenomena. One of the phenomena is the so-called *hunting* behavior [41] in which an undesired oscillation of indoor temperature field develops. This behavior is often observed in a building with dispersed multiple HVAC units and is caused by not only their nonlinear characteristics and oscillatory responses of thermal loads but also the dynamic thermal interaction of neighboring HVAC units. The proposed modeling is capable of representing the hunting behavior via a linear wave propagation. This idea is extended to a case of nonlinear propagation in [42]. Moreover, by using the model in a feedback loop, it becomes possible to eliminate the hunting behavior in the design and operational stages of the HVAC systems. A possible idea for this is the so-called model predictive control (MPC) [43] that has been reported for control of in-building thermal dynamics [44, 45]. Note that an MPC framework based on Koopman operator theory has been developed and applied to nonlinear ODE and PDE systems [46].

The proposed modeling is also closely related to system identification theory. If $D_{\text{eff}}$ and $\mathbf{V}_{\text{eff}}$ change in space and time due to fluctuating thermal sources or mobile objects [4], their estimation process needs to be extended in a wider frequency domain; i.e., multiple KEs and KMs (and Koopman eigenfuctions). A hint of the extension is the Koopman operator-based framework of system identification [47, 48]. With this framework, it is basically possible to identify $\mathbf{V}_{\text{eff}}(\mathbf{r})$ as a vector field of a nonlinear dynamical system with open-loop input and/or white noise, which is represented by the Kolmogorov backward equation in a space of observable functions. Unfortunately, it is still impossible to directly apply the novel framework to the system (18.3) since the equation has control (closed-loop) input $P(\mathbf{r}, t, T)$. The identification of $D_{\text{eff}}(\mathbf{r}, t)$ and $\mathbf{V}_{\text{eff}}(\mathbf{r}, t)$ requires further development of closed-loop identification of nonlinear PDE, possibly based on the Koopman operator theory.

## 18.6 Conclusion

This chapter proposed a method for modeling of heat transfer phenomena in a building atrium via Koopman mode decomposition (KMD). First of all, through scale transformation in space and time, we formulated the in-building heat transfer as simple advection and diffusion equations with effective velocity and effective diffusivity. KMD was applied to the two equations for extracting a dominant oscillatory pattern of the heat transfer, describing it as a linear wave propagation, and introducing linear equations of the oscillatory pattern for identification. Then, we demonstrated the effectiveness of the modeling with measurement data on temperature and HVAC operation of a practically used building, where the advection equation was implemented in the atrium and the diffusion equation in the office spaces. Consequently, we visualized the heat transfer in the atrium and its interaction with HVAC operation, obtained the physically relevant values of the effective velocity, and showed the capability of KMD to describe advective heat transfer caused by a laminar flow in the atrium.

As stated in Sect. 18.5, the proposed modeling is applicable to the control of in-building thermal dynamics. In addition, the modeling enables us to investigate in-building airflow structure. The proposed modeling identifies both the effective diffusion by small-scale (zero mean) airflow and the advection by large-scale (nonzero mean) one [24, 25]. Thereby, it becomes possible to evaluate from measurement data whether zero mean or nonzero mean flow appears in buildings. This contributes to developing a data-driven HVAC management technology, comprising not only control but also monitoring, fault detection, and maintenance of HVAC units/systems. We conclude this chapter with the applicability of the proposed modeling to the *practical* HVAC control and management.

**Acknowledgements** The authors appreciate Mr. Chosei Kaseda, Mr. Junnya Nishiguchi, and Mr. Kei Koga (Azbil Corporation) for the provision of the photographs and measurement data, and insightful discussion. This work was partly supported by JST, CREST #JPMJCR15K3.

## References

1. Hensen, J.L.M., Lamberts, R. (eds.): Building Performance Simulation for Design and Operation. Spon Press, London (2011)
2. Zhang, W., Hiyama, K., Kato, S., Ishida, Y.: Building energy simulation considering spatial temperature distribution for nonuniform indoor environment. Build. Environ. **63**, 89–96 (2013)
3. Osawa, S., Hara, S., Koga, K., Honda, M., Kaseda, C.: Optimal control for room air conditioning using time-space muti-scale modeling. In: Proceedings of SICE Control Division Conference (2013) (in Japanese)
4. Kono, Y., Susuki, Y., Hayashida, M., Mezić, I., Hikihara, T.: Multiscale modeling of in-room temperature distribution with human occupancy data: a practical case study. J. Build. Perform. Simul. **11**(2), 145–163 (2018)
5. Agarwal, Y., Balaji, B., Gupta, R., Lyles, J., Wei, M., Weng, T.: Occupancy-driven energy management for smart building automation. In: Proceedings of the 2nd ACM Workshop on Embedded Sensing Systems for Energy-Efficiency in Building, pp. 1–6 (2010)
6. Veselý, M., Zeiler, W.: Personalized conditioning and its impact on thermal comfort and energy performance–a review. Renew. Sustain. Energy Rev. **34**, 401–408 (2014)
7. Saxon, R.: Atrium Buildings: Development and Design. Architectural Press, London (1983)
8. Levermore, G.J.: Building Energy Management Systems: Applications to Low-energy HVAC and Natural Ventilation Control, 2nd edn. E & FN Spon, London (2000)
9. Liu, P.C., Lin, H.T., Chou, J.H.: Evaluation of buoyancy-driven ventilation in atrium buildings using computational fluid dynamics and reduced-scale air model. Build. Environ. **44**(9), 1970–1979 (2009)
10. Ren, Z., Stewart, J.: Simulating air flow and temperature distribution inside buildings using a modified version of COMIS with sub-zonal divisions. Energy Build. **35**(3), 257–271 (2003)
11. Megri, A.C., Haghighat, F.: Zonal modeling for simulating indoor environment of buildings: review, recent developments, and applications. HVAC&R Res. **13**(6), 887–905 (2007)
12. Price, C.R., Rasmussen, B.P.: Effective tuning of cascaded control loops for nonlinear HVAC systems. In: Proceedings of ASME 2015 Dynamic Systems and Control Conference, vol. 2 (2015)
13. Pavliotis, G.A., Stuart, A.M.: Multiscale Methods: Averaging and Homogenization. Springer, Berlin (2008)
14. Mezić, I.: Spectral properties of dynamical systems, model reduction and decompositions. Nonlinear Dyn. **41**(1), 309–325 (2005)
15. Mezić, I.: Analysis of fluid flows via spectral properties of the Koopman operator. Annu. Rev. Fluid Mech. **45**, 357–378 (2013)
16. Eisenhower, B., Maile, T., Fischer, M., Mezić, I.: Decomposing building system data for model validation and analysis using the Koopman operator. In: Proceedings of 4th National Conference of IBPSA-USA (SIMBUILD 2010) (2010)
17. Georgescu, M., Eisenhower, B., Mezić, I.: Creating zoning approximations to building energy models using the Koopman operator. In: Proceedings of 5th National Conference of IBPSA-USA (SimBuild2012), vol. 5, pp. 40–47 (2012)
18. Georgescu, M., Mezić, I.: Building energy modeling: a systematic approach to zoning and model reduction using Koopman mode analysis. Energy Build. **86**, 794–802 (2015)
19. Georgescu, M., Mezić, I.: Improving HVAC performance through spatiotemporal analysis of building thermal behavior. In: Proceedings of 3rd International High Performance Buildings Conference at Purdue (2014)

20. Georgescu, M., Mezić, I.: Site-level energy monitoring and analysis using Koopman operator methods. In: Proceedings of 2014 ASHRAE/IBPSA-USA Building Simulation Conference (SimBuild 2014) (2014)
21. Littooy, B., Loire, S., Georgescu, M., Mezić, I.: Pattern recognition and classification of HVAC rule-based faults in commercial buildings. In: Proceedings of 2016 IEEE International Conference on Big Data (Big Data), pp. 1412–1421 (2016)
22. Georgescu, M., Loire, S., Kasper, D., Mezić, I.: Whole-building fault detection: a scalable approach using spectral methods. In: Proceedings of the 2017 ASHRAE Winter Meeting (2017)
23. Kono, Y., Susuki, Y., Hayashida, M., Hikihara, T.: Modeling of effective heat diffusion in a building atrium via Koopman mode decomposition. In: Proceedings of 2016 International Symposium on Nonlinear Theory and its Applications (NOLTA2016), pp. 362–365 (2016)
24. Kono, Y., Susuki, Y., Hayashida, M., Hikihara, T.: Applications of Koopman mode decomposition to modeling of heat transfer dynamics in building atriums–I–effective heat diffusion by small-scale air movement. Trans. Soc. Instrum. Control Eng. **53**(2), 123–133 (2017) (in Japanese)
25. Kono, Y., Susuki, Y., Hikihara, T.: Applications of Koopman mode decomposition to modeling of heat transfer dynamics in building atriums–II–advection by large-scale air movement. Trans. Soc. Instrum. Control Eng. **53**(2), 188–197 (2017) (in Japanese)
26. Gao, N.P., Niu, J.L.: CFD study of the thermal environment around a human body: a review. Indoor Built Environ. **14**(1), 5–16 (2005)
27. Pavliotis, G.A.: Homogenization theory for advection–diffusion equations with mean flow. PhD thesis, Rensselaer Polytechnic Institute (2002)
28. Sharma, A.S., Mezić, I., McKeon, B.J.: Correspondence between Koopman mode decomposition, resolvent mode decomposition, and invariant solutions of the Navier-Stokes equations. Phys. Rev. Fluids **1**(3), 032402 (2016)
29. Temam, R.: Infinite-Dimensional Dynamical Systems in Mechanics and Physics, vol. 68. Springer Science & Business Media, New York (2012)
30. Arnol'd, V.I., Avez, A.: Ergodic Problems of Classical Mechanics. WA Benjamin, New York (1968)
31. Koopman, B.O.: Hamiltonian systems and transformations in Hilbert space. Proc. Natl. Acad. Sci. USA **17**, 315–318 (1931)
32. Okochi, G.S., Yao, Y.: A review of recent developments and technological advancements of variable-air-volume (VAV) air-conditioning systems. Renew. Sustain. Energy Rev. **59**, 784–817 (2016)
33. Rowley, C.W., Mezić, I., Bagheri, S., Schlatter, P., Henningson, D.S.: Spectral analysis of nonlinear flows. J. Fluid Mech. **641**, 115–127 (2009)
34. Schmid, P.J.: Dynamic mode decomposition of numerical and experimental data. J. Fluid Mech. **656**, 5–28 (2010)
35. Tu, J.H., Rowley, C.W., Luchtenburg, D.M., Brunton, S.L., Kutz, J.N.: On dynamic mode decomposition: theory and application. J. Comput. Dyn. **1**(2), 391–421 (2015)
36. Williams, M.O., Kevrekidis, I.G., Rowley, C.W.: A data-driven approximation of the Koopman operator: extending dynamic mode decomposition. J. Nonlinear Sci. **25**(6), 1307–1346 (2015)
37. Susuki, Y., Mezić, I.: A prony approximation of Koopman mode decomposition. In: Proceedings of 54th IEEE Conference on Decision and Control, pp. 7022–7027 (2015)
38. Raak, F., Susuki, Y., Mezić, I., Hikihara, T.: On Koopman and dynamic mode decompositions for application to dynamic data with low spatial dimension. In: Proceedings of 2016 IEEE 55th Conference on Decision and Control (CDC), pp. 6485–6491 (2016)
39. Pope, S.B.: Turbulent Flows. Cambridge University Press, Cambridge (2000)
40. Gouda, M.M., Danaher, S., Underwood, C.P.: Building thermal model reduction using nonlinear constrained optimization. Build. Environ. **37**(12), 1255–1265 (2002)
41. Chintala, R., Price, C., Rasmussen, B.: Identification and elimination of hunting behavior in HVAC systems. ASHRAE Trans. **121**, 294–305 (2015)
42. Hiramatsu, N., Susuki, Y., Ishigame, A.: An estimation of in-room temperature gradient using Koopman mode decomposition. In: Proceedings of SICE Annual Conference 2018 (SICE2018), pp. 78–82 (2018)

43. Camacho, E.F., Bordons, C.: Model Predictive Control, 2nd edn. Springer, London (2007)
44. Ma, Y., Anderson, G., Borrelli, F.: A distributed predictive control approach to building temperature regulation. In: Proceedings of the 2011 American Control Conference, pp. 2089–2094 (2011)
45. Afram, A., Janabi-Sharifi, F.: Theory and applications of HVAC control systems-a review of model predictive control (MPC). Build. Environ. **72**, 343–355 (2014)
46. Korda, M., Mezić, I.: Linear predictors for nonlinear dynamical systems: Koopman operator meets model predictive control. Automatica **93**, 149–160 (2018)
47. Mauroy, A., Goncalves, J.: Linear identification of nonlinear systems: a lifting technique based on the Koopman operator. In: Proceedings of 2016 IEEE 55th Conference on Decision and Control (CDC), pp. 6500–6505 (2016)
48. Mauroy, A., Goncalves, J.: Koopman-based lifting techniques for nonlinear systems identification (2017). arXiv:1709.02003

# Chapter 19
# Data-Driven Voltage Analysis of an Electric Power Grid via Delay Embedding and Extended Dynamic Mode Decomposition

**Yoshihiko Susuki and Kyoichi Sako**

**Abstract** Data-driven analysis of dynamic performance has attracted a lot of interest in the modern power grid with highly accurate measurement technologies such as Synchrophasor. In this chapter, we report the research effort to do this for voltage dynamics in a rudimentary model of power grids. We use the combination of the so-called delay embedding with the extended dynamic mode decomposition, which is a technique for approximating the Koopman operator directly from time-series data. This combination enables us to approximate the spectral properties of the Koopman operator through voltage–amplitude measurement of a single bus. The spectral properties clarify not only the local information on the voltage dynamics such as frequency/damping in transient responses but also global one such as the flow structure of the underlying mathematical model.

## 19.1 Introduction

Data-driven analysis of dynamic performance has attracted a lot of interest in the modern power grid. A key enabler to doing this is the progress of information, communication, and measurement technologies such as the so-called Synchrophasor: see, e.g., [7]. Synchrophasor is based on a highly accurate device for measuring voltage phasors in AC power grids. The measured data on voltage phase, which are commonly time stamped at different measurement locations in a large-scale system, are widely addressed in stability studies of rotor angle dynamics (synchronization) and frequency dynamics: see, e.g., [29]. The measured data on voltage amplitude are

Y. Susuki (✉)
Department of Electrical and Information Systems,
Osaka Prefecture University, Osaka, Japan
e-mail: susuki@eis.osakafu-u.ac.jp

K. Sako
Department of Electrical Engineering, Kyoto University, Kyoto, Japan
e-mail: kyoichi.sako@gmail.com

Nidec Corporation, Kyoto, Japan



A. Mauroy et al. (eds.), *The Koopman Operator in Systems and Control*,
Lecture Notes in Control and Information Sciences 484,
https://doi.org/10.1007/978-3-030-35713-9_19

also expected for monitoring of the so-called voltage stability in both short term and long term. For example, real-time monitoring of short-term voltage stability with Synchrophasor is reported in [6].

In this chapter, we report the research effort to do the data-driven analysis of voltage dynamics in power grids. The novel point here is to apply the data-driven spectral analysis of the Koopman operator (see, e.g., [4]). The Koopman operator is a linear but infinite-dimensional operator defined for nonlinear dynamical systems, and its spectral properties have been studied in analysis of nonlinear dynamical systems: see [14, 17] and Chap. 1 of this book. The spectral computation enables us to gain statistical and geometric insights into the nonlinear dynamical systems from time-series data, such as nonlinear oscillatory modes [21] and invariant sets/manifolds [16, 18]. This has been applied to the data-driven analysis of dynamic performance in power grids by the authors and their collaborators (see, e.g., [19, 20, 24]). This chapter is the first application of data-driven Koopman spectral analysis to voltage dynamics, to the best of the authors' knowledge, and its contents were preliminarily reported in [25, 26].

The purpose of this chapter is to conduct the data-driven Koopman spectral analysis of voltage data in a rudimentary model of power grids. In particular, we address the stable voltage dynamics of the model in terms of *eigenvalues* of the Koopman operator. The data-driven analysis is conducted with the Extended Dynamic Mode Decomposition (EDMD) [28], which is a widely recognized technique for approximating the Koopman operator directly from time-series data. EDMD requires time-series data on the full states for a target nonlinear dynamical system (mathematical model): in terms of power grids, all bus voltages, all rotor angles of generators, and so on. Since the amount of available data in a practical power grid is not enough to cover the full states of the underlying mathematical model, it is desirable to conduct EDMD through a small number of measurements. In this chapter, we use the delay embedding [27] in order to reconstruct the full-state dynamics of the rudimentary model through voltage–amplitude measurement of a single bus. The technique of delay embedding and delay coordinate has been applied to theory and numerics for spectral characterization of the Koopman operators: see, e.g., [1, 5, 9, 10, 17, 22]. We then apply EDMD to the reconstructed data and derive spectral properties of the Koopman operator. The spectral properties clarify local and global information on the rudimentary model: frequency/damping in transient responses through eigenvalues of the Koopman operator and the flow structure through associated eigenfunctions. We hence show that a single-bus measurement of voltage amplitude is capable of extracting information on the underlying nonlinear mathematical model through spectral properties of the Koopman operator, which is never accomplished with any technique based on linearization.

The rest of this chapter is organized as follows. Section 19.2 presents a summary of theoretical contents that guide us in the current data-driven analysis. Section 19.3 introduces the rudimentary model of power grids on which we focus in the analysis. Section 19.4 is the main part of this chapter in which we conduct the data-driven voltage analysis via Koopman spectral analysis. Section 19.5 concludes this chapter with several remarks.

## 19.2 Summarized Theory

This section summarizes the theoretical contents that guide us in the data-driven voltage analysis performed in this chapter. Readers are encouraged to read the introductory chapter of this book (Chap. 1).

### *19.2.1 Dynamical System and Koopman Operator*

In the following, we suppose that the dynamics of a power grid are modeled by the following ordinary differential equation on a finite-dimensional manifold $X$ with dimension $d$ ($< \infty$): for time $t \in \mathbb{R}$,

$$\frac{\mathrm{d}}{\mathrm{d}t}\mathbf{x}(t) = \mathbf{F}(\mathbf{x}(t)), \tag{19.1}$$

where $\mathbf{x} \in X$ is the state of the system dynamics, and $\mathbf{F} : X \to TX$ (tangent bundle of $X$) a nonlinear function representing the rule of how the state evolves in time. For this equation, the following finite time-$t$ map $\mathbf{S}^t$ is defined as

$$\mathbf{S}^t : X \to X;\ \mathbf{x}(0) \mapsto \mathbf{x}(t) = \mathbf{x}(0) + \int_0^t \mathbf{F}(\mathbf{x}(\tau))\mathrm{d}\tau.$$

The one-parameter group of maps, $\{\mathbf{S}^t; t \in \mathbb{R}\}$, is called the *dynamical system* or *flow*.

Here we introduce the Koopman operators for the map $\mathbf{S}^t$. To do this, the so-called *observable* $f$ is introduced as a scalar-valued function defined on $X$, namely

$$f : X \to \mathbb{C}.$$

This observable is a mathematical formulation of observation or measurement in a power grid, that is, bus voltage in this chapter. Below, we will denote by $\mathscr{F}$ a given space of observables. For an observable $f \in \mathscr{F}$, the *Koopman operator* $U^t$ for $\mathbf{S}^t$ maps $f$ into a new function as follows:

$$U^t f := f \circ \mathbf{S}^t \qquad \forall t \in \mathbb{R}.$$

That is, the Koopman operator $U^t$ describes the time $t$ evolution of observation (or measurement) along the state's dynamics. Although the model (19.1) can be nonlinear and evolve in the finite-dimensional space, the Koopman operator is *linear* but *infinite dimensional*. This type of composition operator does not rely on linearization; indeed, it captures the full information on the original nonlinear system (19.1): see [4, 15, 16].

Spectral properties of the Koopman operator are of paramount importance in applied perspectives. The pair of eigenvalue $\nu \in \mathbb{C}$ and eigenfunction $\phi_\nu \in \mathscr{F} \setminus \{0\}$ of the Koopman operator $U^t$ is defined as follows:

$$U^t \phi_\nu = \exp(\nu t)\phi_\nu \qquad \forall t \in \mathbb{R},$$

where we call $\nu$ the *Koopman eigenvalue* and $\phi_\nu$ the associated *Koopman eigenfunction*. Generally, the number of pairs of eigenvalues and eigenfunctions of the Koopman operator is not finite.

### 19.2.2 Extended Dynamic Mode Decomposition

A numerical method for approximating the Koopman operator from time-series data is proposed in [28] and called the extended dynamic mode decomposition (EDMD). Consider a finite-length series of sampled data on the state dynamics $\{\mathbf{x}_1, \mathbf{x}_2, \ldots, \mathbf{x}_{N+1}\}$ ($\mathbf{x}_i := \mathbf{x}((i-1)\Delta t) \in \mathbb{X}$), where $\Delta t$ is the sampling period, and $N+1$ corresponds to the number of samples. In this setting, the time evolution from $\mathbf{x}_i$ to $\mathbf{x}_{i+1}$ is represented by the map $\mathbf{S}^{\Delta t}$. Also, we consider a finite-number ($L$) set of observables $\{\psi_1, \psi_2, \ldots, \psi_L\}$ ($\psi_i \in \mathscr{F} \setminus \{0\}$). Thus, we define the two matrices $\mathbf{A}$ and $\mathbf{B}$ as follows:

$$\mathbf{A} := \begin{bmatrix} \psi_1(\mathbf{x}_1) & \psi_2(\mathbf{x}_1) & \cdots & \psi_L(\mathbf{x}_1) \\ \psi_1(\mathbf{x}_2) & \psi_2(\mathbf{x}_2) & \cdots & \psi_L(\mathbf{x}_2) \\ \vdots & \vdots & \ddots & \vdots \\ \psi_1(\mathbf{x}_N) & \psi_2(\mathbf{x}_N) & \cdots & \psi_L(\mathbf{x}_N) \end{bmatrix},$$

$$\mathbf{B} := \begin{bmatrix} \psi_1(\mathbf{x}_2) & \psi_2(\mathbf{x}_2) & \cdots & \psi_L(\mathbf{x}_2) \\ \psi_1(\mathbf{x}_3) & \psi_2(\mathbf{x}_3) & \cdots & \psi_L(\mathbf{x}_3) \\ \vdots & \vdots & \ddots & \vdots \\ \psi_1(\mathbf{x}_{N+1}) & \psi_2(\mathbf{x}_{N+1}) & \cdots & \psi_L(\mathbf{x}_{N+1}) \end{bmatrix}.$$

Then, the following matrix $\mathbf{U}$ is defined as a finite-dimensional approximation of the Koopman operator $U^{\Delta t}$ for the map $\mathbf{S}^{\Delta t}$:

$$\mathbf{U} := \mathbf{A}^{+}\mathbf{B},$$

where $\mathbf{A}^{+}$ denotes the Moore–Penrose matrix of $\mathbf{A}$. For each eigenvalue of $\mathbf{U}$ denoted by $\lambda \in \mathbb{C}$, $(\ln \lambda)/\Delta t$ approximates an eigenvalue of $U^{\Delta t}$. For the associated right eigenvector of $\mathbf{U}$ denoted by $\mathbf{V}_\lambda \in \mathbb{C}^L$, $\mathbf{AV}_\lambda$ and $\mathbf{BV}_\lambda$ provide datasets of values of the associated Koopman eigenfunction on the given state trajectories $\{\mathbf{x}_1, \mathbf{x}_2, \ldots, \mathbf{x}_N\}$ in the state space $X$. See [28] for the detailed description of EDMD including its theoretical foundation.

### 19.2.3 Delay Embedding with Connections to the Koopman Operator

Here we introduce the Takens embedding theorem [27], in which we use the Koopman operator $U^\tau$ for fixed $\tau \in \mathbb{R}$.

**Theorem 19.1** (Takens 1981) *Suppose $X$ is compact. For pairs $(\mathbf{S}^\tau, f)$, $\mathbf{S}^\tau : X \to X$ a smooth diffeomorphism and $f : X \to \mathbb{R}$ a smooth observable, it is a generic property that the map $\Phi_{(\mathbf{S}_\tau, f)} : X \to \mathbb{R}^{2d+1}$, defined by*

$$\begin{aligned}\Phi_{(\mathbf{S}_\tau, f)}(\mathbf{x}) &:= [f(\mathbf{x}), f(\mathbf{S}^\tau(\mathbf{x})), \ldots, f(\mathbf{S}^{2d\tau}(\mathbf{x}))]^\top \\ &= [f(\mathbf{x}), (U^\tau f)(\mathbf{x}), \ldots, ((U^\tau)^{2d} f)(\mathbf{x})]^\top \end{aligned} \tag{19.2}$$

*is an embedding; by "smooth" above we mean at least $C^2$. The symbol $\top$ denotes the transpose operation of vectors.*

The domain of $\Phi_{(\mathbf{S}_\tau, f)}$ is $X$, and its image can be a subset of the codomain $\mathbb{R}^{2d+1}$. From Theorem 19.1, there exists a homeomorphism $\varphi : X \to \Phi_{(\mathbf{S}_\tau, f)}(X) \subseteq \mathbb{R}^{2d+1}$. Thus, the *reconstructed map* $\tilde{\mathbf{S}}^\tau : \varphi(X) \to \varphi(X)$ is defined with the following commutative relation:

$$\varphi \circ \mathbf{S}^\tau = \tilde{\mathbf{S}}^\tau \circ \varphi.$$

The commutative diagram is shown in Fig. 19.1. The map $\tilde{\mathbf{S}}^\tau$ is the mathematical object constructed via the delay embedding (19.2) directly from scalar time-series data $\{y_1, y_2, \ldots\}$ ($y_i := (U_\tau^{i-1} f)(\mathbf{x})$ for fixed initial $\mathbf{x}$), to which we apply the EDMD in Sect. 19.4.

It is here valuable to introduce the connection of the above delay embedding with the Koopman operator. For this, we denote by $\tilde{U}^\tau$ the Koopman operator defined for the reconstructed map $\tilde{\mathbf{S}}^\tau$ and by $\mathscr{F}_\varphi$ the space of observables defined on $\varphi(X)$ (it becomes compact in the above theorem) on which $\tilde{U}^\tau$ acts. Here, let us define the composition operator with the homeomorphism $\varphi$, denoted by $C_\varphi : \mathscr{F}_\varphi \to \mathscr{F}$: for $g \in \mathscr{F}_\varphi$,

$$C_\varphi g = g \circ \varphi.$$

For $g \in \mathscr{F}_\varphi$, we have

$$\begin{aligned}\tilde{U}^\tau g &= g \circ \tilde{\mathbf{S}}^\tau, \\ &= g \circ (\varphi \circ \mathbf{S}^\tau \circ \varphi^{-1}), \\ (\tilde{U}^\tau g) \circ \varphi &= (g \circ \varphi) \circ \mathbf{S}^\tau, \\ C_\varphi(\tilde{U}^\tau g) &= \left(C_\varphi g\right) \circ \mathbf{S}^\tau, \\ &= U^\tau \left(C_\varphi g\right).\end{aligned}$$

Because $g$ is arbitrarily chosen from $\mathscr{F}_\varphi$, the following commutative relation of the two Koopman operators $U^\tau$ and $\tilde{U}^\tau$ holds:

$$\begin{array}{ccc} X & \xrightarrow{\mathbf{S}^\tau} & X \\ \varphi \downarrow & & \downarrow \varphi \\ \Phi_{(\mathbf{S}_\tau, f)}(X) & \xrightarrow{\tilde{\mathbf{S}}^\tau} & \Phi_{(\mathbf{S}_\tau, f)}(X) \end{array}$$

**Fig. 19.1** Commutative diagram of the original map $\mathbf{S}^\tau$ and reconstructed one $\tilde{\mathbf{S}}^\tau$ from scalar time-series data via delay embedding

$$\begin{array}{ccc} \mathscr{F} & \xrightarrow{U^\tau} & \mathscr{F} \\ C_\varphi \uparrow & & \uparrow C_\varphi \\ \mathscr{F}_\varphi & \xrightarrow{\tilde{U}^\tau} & \mathscr{F}_\varphi \end{array}$$

**Fig. 19.2** Commutative diagram of the Koopman operators $U^\tau$ for the original map and $\tilde{U}^\tau$ for the reconstructed one

$$C_\varphi \circ \tilde{U}^\tau = U^\tau \circ C_\varphi.$$

The diagram for the two Koopman operators is shown in Fig. 19.2. Note that the direction of vertical arrows used in Figs. 19.1 and 19.2 are opposite for maps (dynamics of states) and Koopman operators (dynamics of observables). This diagram implies that it is possible for us to investigate spectral properties of $U^\tau$ for the original map using $\tilde{U}^\tau$ for the reconstructed one, which is computable only with the single-bus measurement as demonstrated in Sect. 19.4. See Appendix of this chapter for a theoretical proof of unitary equivalence for the two Koopman operators $U^\tau$ and $\tilde{U}^\tau$.

## 19.3 Rudimentary Model of Voltage Dynamics

This section introduces the rudimentary model of power grids for which we analyze voltage dynamics in Sect. 19.4.

### *19.3.1 Power Grid Model*

Figure 19.3 shows the rudimentary model of two-machines power grid proposed in [8]. This model has the two synchronous machines, one of which is regarded as the slack bus, denoted by Generator 1 in Fig. 19.3. In this model, the two synchronous machines are connected to the load center that consists of induction motor and constant power load. The rudimentary model is famous because it was used in [8] for analysis of voltage collapse phenomena in terms of the theory of local bifurcation and center manifold.

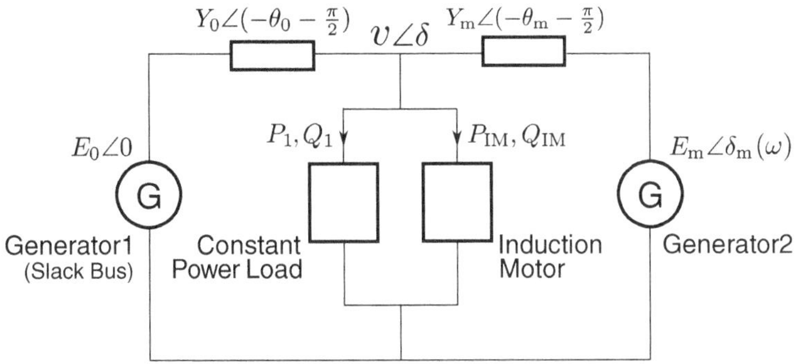


**Fig. 19.3** Rudimentary model of two-machine power grid for voltage studies. This model has been proposed in [8]

## 19.3.2 Mathematical Model

The mathematical model to represent electromechanical and voltage dynamics of the two-machines power grid was originally derived in [8]. In this chapter, by applying the standard singular perturbation technique to the original model, we derive a reduced order model with a small number of dependent variables. The variables include the angle (or argument) $\delta_{\mathrm{m}}$ of phasor in bus voltage on Generator 2, its angular frequency deviation $\omega$ to nominal, the amplitude $v$ of phasor in bus voltage at the load center, and its argument $\delta$. The reduced order model has a form of differential-algebraic equations as follows:

$$\left.\begin{aligned}
\frac{\mathrm{d}\delta_{\mathrm{m}}}{\mathrm{d}t} &= \omega, \\
\frac{\mathrm{d}\omega}{\mathrm{d}t} &= \left(\frac{1}{M}\right)\{-d_{\mathrm{m}}\omega + P_{\mathrm{m}} + E_{\mathrm{m}}^2 Y_{\mathrm{m}} \sin\theta_m \\
&\quad + E_{\mathrm{m}} Y_{\mathrm{m}} v \sin(\delta - \delta_{\mathrm{m}} - \theta_{\mathrm{m}})\}, \\
\frac{\mathrm{d}v}{\mathrm{d}t} &= \left(\frac{1}{T K_{\mathrm{qw}} K_{\mathrm{pv}}}\right)\{K_{\mathrm{pw}} K_{\mathrm{qv2}} v^2 + (K_{\mathrm{pw}} K_{\mathrm{qv}} - K_{\mathrm{qw}} K_{\mathrm{pv}}) v \\
&\quad + K_{\mathrm{qw}}(P(\delta_{\mathrm{m}}, \delta, v) - P_0 - P_1) \\
&\quad - K_{\mathrm{pw}}(Q(\delta_{\mathrm{m}}, \delta, v) - Q_0 - Q_1)\}, \\
0 &= \left(\frac{1}{K_{\mathrm{qw}}}\right)\{-K_{\mathrm{qv2}} v^2 - K_{\mathrm{qv}} v + Q(\delta_{\mathrm{m}}, \delta, v) - Q_0 - Q_1\}.
\end{aligned}\right\} \tag{19.3}$$

The setting of constant parameters for simulations of (19.3) is summarized in Table 19.1. The parameter setting is basically from [8]: also see their paper for physical meanings of the parameters. The simulations are conducted for generating data on dynamics of the two-machines power grid. It should be noted that (19.3) is regular in the following simulations of this chapter, implying that existence and uniqueness of solutions in (19.3) are guaranteed.

**Table 19.1** Simulation setting for rudimentary model of two-machines power grid in Fig. 19.3

| Symbol | Value w/unit | Value in PU |
|---|---|---|
| $E_0$ | 50 kV | 2.5 |
| $E_m$ | 20 kV | 1.0 |
| $M$ | $13.2639 \times 10^6$ J · s | 0.02652 |
| $d_m$ | – | $0.1326 \times 10^{-3}$ |
| $P_m$ | 500 MW | 1.0 |
| $P_0$ | 300 MW | 0.6 |
| $Q_0$ | 650 MW | 1.3 |
| $P_1$ | 0 MW | 0 |
| $Q_1$ | 3867 MW | 7.7356 |
| $T$ | 5 s | 5 |
| $K_{pw}$ | $0.5305 \times 10^6$ W · s | $1.0610 \times 10^{-3}$ |
| $K_{pv}$ | 7500 W/V | 0.3 |
| $K_{qw}$ | $-0.03979 \times 10^6$ W · s | $-0.7958 \times 10^{-4}$ |
| $K_{qv}$ | $-70 \times 10^3$ W/V | −2.8 |
| $K_{qv2}$ | 2.625 W/V$^2$ | 2.1 |
| $Y_0$ | 6.9989 $\Omega^{-1}$ | 5.6 |
| $-\theta_0 - \pi/2$ | −1.36 radian | – |
| $Y_m$ | 4.3752 $\Omega^{-1}$ | 3.5 |
| $-\theta_m - \pi/2$ | −1.48 radian | – |
| Base power | 500 MW | – |
| Base voltage | 20 kV | – |
| Base frequency | 377 rad | – |
| Base time | 1 s | – |

## 19.4 Numerical Experiment

In this section, we conduct a numerical experiment of the data-driven voltage analysis of the rudimentary power grid model. The main idea here is to use the delay embedding for reconstructing the state dynamics of the model from a single-bus measurement and to apply EDMD to the reconstructed dynamics as shown in Fig. 19.4. Thereby, we approximately derive spectral properties of the Koopman operator for the rudimentary model.

### *19.4.1 Setting*

First of all, we numerically solve the reduced order model (19.3) and obtain time-series data on the state dynamics $[\delta_m, \omega, v, \delta]^\top$. Equation (19.3) is numerically

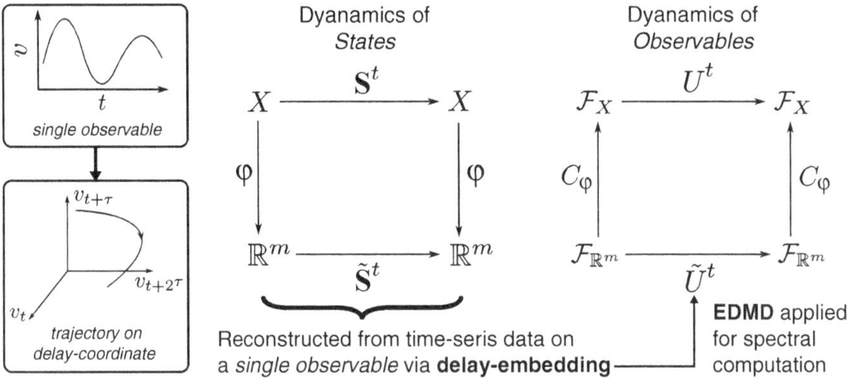


**Fig. 19.4** Main idea of the data-driven voltage analysis via delay embedding and extended dynamic mode decomposition (EDMD) where $m$ is the dimension of embedding

integrated with a mixture of Euler method and Newton–Raphson method. The initial conditions are chosen in a grid of $11^3$ points, spanned by $0.4243 \leq \delta_\mathrm{m} \leq 0.4643$, $-0.05 \leq \omega \leq 0.05$, and $0.9125 \leq v \leq 0.9725$. All the grid points are placed inside the basin of attraction of an asymptotically stable equilibrium point.[1] The time duration of the numerical integration is fixed at $\Delta t \times 20$ where $\Delta t = 0.05$s. The time-series data on the bus voltage $v(t)$ at the load center will be addressed below.

EDMD requires the finite-number set of observables. In particular, due to the combination of delay embedding, we need to define the observables in the embedded space $\mathbb{R}^{2d+1}$. In this chapter, since the dimension of the rudimentary model is known, from Theorem 19.1 we use the case of $d = 3$, that is, the embedded space is $\mathbb{R}^7$:

$$[v(t), v(t+\tau), \ldots, v(t+6\tau)]^\top, \quad \tau - 2\Delta t.$$

Note that if the dimension is unknown, then we will rely on techniques developed for estimating $d$ (see, e.g., [11]). Here, we use the thin-plate spline function that is as one type of radial basis functions and used for nonlinear time-series prediction [11]. The number $L$ of observables is equal to that of the initial conditions, i.e., $L = 11^3$. The observables $\psi_i$ $(i = 1, 2, \ldots, L)$ are as follows:

$$\psi_i(\tilde{\mathbf{x}}) = (\|\tilde{\mathbf{x}} - \tilde{\mathbf{c}}_i\|_2)^2 \ln\|\tilde{\mathbf{x}} - \tilde{\mathbf{c}}_i\|_2, \quad \tilde{\mathbf{x}} \in \mathbb{R}^7,$$

where $\|\cdot\|_2$ denotes the Euclidean norm. The constant vectors $\tilde{\mathbf{c}}_i \in \mathbb{R}^7$ are called the centers of thin-plane spline functions. The centers are fixed with the $k$-means method [3]. See [28] for the details of the choice of radial basis functions and the use of $k$-means method.

[1] In this sense, the target voltage dynamics simulated numerically are dissipative, which is different from the target system for theoretical analysis on spectral equivalence in the appendix. See the remark therein.

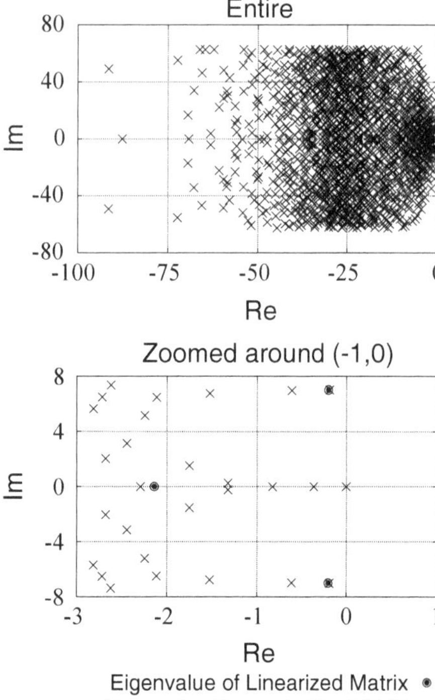


**Fig. 19.5** Distribution of Koopman eigenvalues computed with delay embedding and EDMD. The computed Koopman eigenvalues are denoted by the *black* crosses, and the eigenvalues of the linearized matrix of (19.3) by the *black* circles. All the Koopman eigenvalues are located in the left half of the complex plane, implying that the sampled voltage dynamics are stable

## *19.4.2 Results*

This section presents a set of the numerical experiment on the data-driven analysis of the rudimentary power grid model. Figure 19.5 shows the distribution of Koopman eigenvalues computed with EDMD. The Koopman eigenvalues are denoted by the *black* crosses. All the Koopman eigenvalues are located in the left half of the complex plane, implying that the sampled voltage dynamics are stable. The eigenvalues of the linearized matrix of (19.3) around the stable equilibrium point are also denoted by the *black* circles: $-0.1993 \pm i7.0118$ and $-2.1413$ (here i denotes the imaginary unit). There exist three of the Koopman eigenvalues, $-0.1852 \pm i7.0147$ and $-2.2950$, close to the linear eigenvalues because the voltage dynamics are sampled in the neighborhood of the stable equilibrium point. This suggests that the combination of delay embedding with EDMD accurately locates the linear eigenvalues in this numerical experiment, that is, local information on frequency and damping of stable voltage transients.

Next, we investigate the distributions of values of the Koopman eigenfunctions for the two Koopman eigenvalues $-0.1852 + i7.0147$ and $-2.2950$ which are close to the linear eigenvalues. Here it is impossible to visualize the distributions entirely because the observables are defined on the embedded space $\mathbb{R}^7$. Thus, we will project

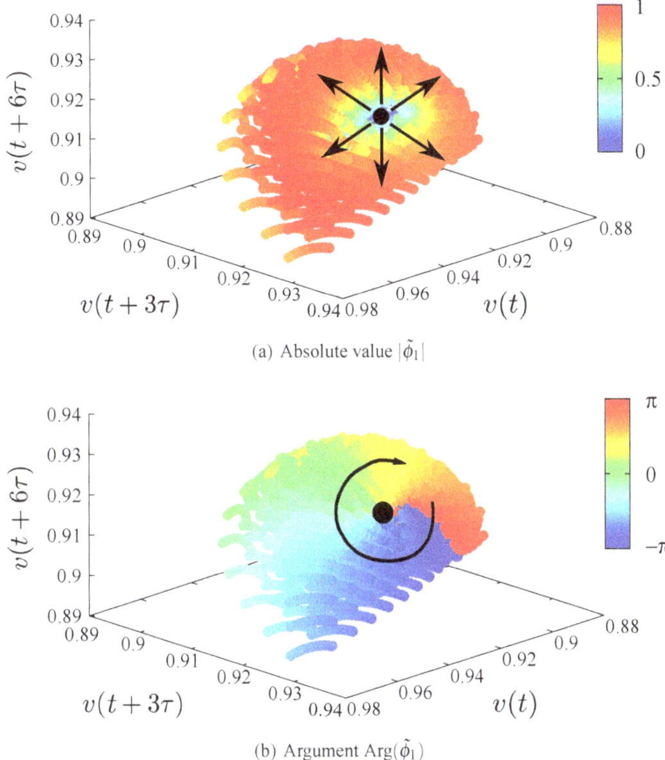


(a) Absolute value $|\tilde{\phi}_1|$

(b) Argument $\mathrm{Arg}(\tilde{\phi}_1)$

**Fig. 19.6** Distribution of values of the Koopman eigenfunction $\tilde{\phi}_1$ for the Koopman eigenvalue $-0.1852 + \mathrm{i}7.0147$. The figures are plotted for absolute value and argument, and obtained by projecting the values onto the space spanned by the three coordinates $v(t)$, $v(t + 3\tau)$, and $v(t + 6\tau)$

them onto a Euclidean space with less than three dimensions in order to look at their structure.

Figure 19.6 shows the distribution of values of the Koopman eigenfunction, denoted by $\tilde{\phi}_1$, for the Koopman eigenvalue $-0.1852 + \mathrm{i}7.0147$. Since the values are complex, we plot (a) absolute value $|\tilde{\phi}_1|$ and (b) argument $\mathrm{Arg}(\tilde{\phi}_1)$ in this figure. The level set of absolute value of a Koopman eigenfunction forms a subspace in the state space, on which the asymptotic component is well captured; this concept is developed in [13] as *isostable*. Also, the level set of argument of a Koopman eigenfunction forms a subspace in the state space and is related in [12] to the so-called *isochron*. The two figures are also plotted by projecting the values onto the space spanned by the three coordinates $v(t)$, $v(t + 3\tau)$, and $v(t + 6\tau)$. In Fig. 19.6a, the absolute values are distributed in a concentric manner and increase far from the center. In fact, the center of the concentric distribution corresponds to the stable equilibrium point (precisely, projected onto the embedded space). Also in Fig. 19.6b, the values of argument are distributed in a clockwise-rotating manner. These distributions are

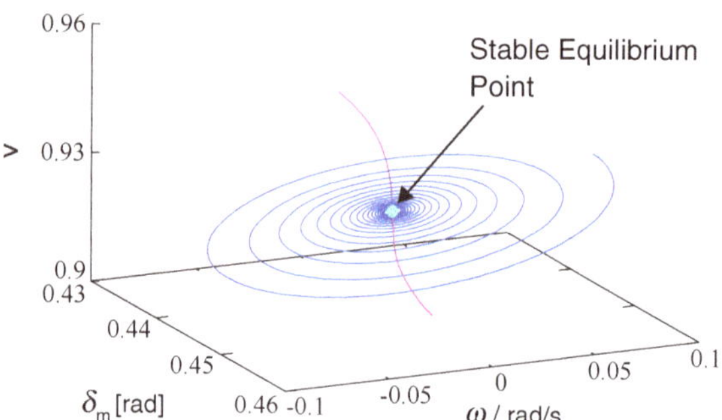


**Fig. 19.7** Flow structure of the original model (19.3) around the stable equilibrium point. This is obtained by direct computation of two distinct trajectories (oscillatory and non-oscillatory) of (19.3) converging to the stable equilibrium point

consistent with the flow structure of the original model (19.3) around the stable equilibrium point as shown in Fig. 19.7. In Fig. 19.7, we see two distinct trajectories to the stable equilibrium point, denoted by the *black* line (oscillatory) and *black* one (non-oscillatory), and the stable equilibrium point denoted by the *cyan* point. The *black* trajectory and the distributions of values in Fig. 19.6 behave in a similar manner. The computed Koopman eigenfuction $\tilde{\phi}_1$ thus captures an oscillatory response in the bus voltage around the stable equilibrium point. It should be noted that the Koopman eigenfunction of the conjugated Koopman eigenvalue $-0.1852 - \mathrm{i}7.0147$ shows a similar distribution of values.

Figure 19.8 shows the distribution of values of the Koopman eigenfunction $\tilde{\phi}_3$ for the Koopman eigenvalue $-2.2950$. The values, in this case, are real. The figure is plotted by projecting the values onto the space spanned by the two coordinates $v(t + 5\tau)$ and $v(t + 6\tau)$. We see the values change in an almost vertical direction in this figure. This structure is also consistent with the flow structure shown in Fig. 19.7: see the *black* non-oscillatory trajectory converging to the stable equilibrium point. The computed Koopman eigenfuction $\tilde{\phi}_3$ thus captures a non-oscillatory response in

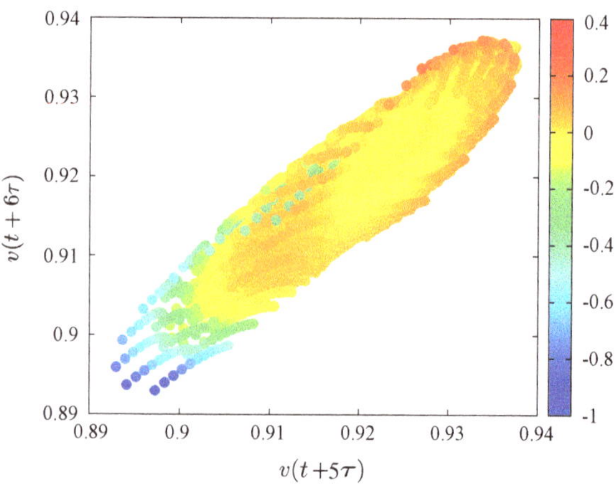


**Fig. 19.8** Distribution of values of the Koopman eigenfunction $\tilde{\phi}_3$ for the Koopman eigenvalue $-2.1413$. The values are real in this case. The figure are plotted by projecting the values onto the two-dimensional plane spanned by the two coordinates $v(t + 5\tau)$ and $v(t + 6\tau)$

the bus voltage that is often crucial to the finite-time development of voltage collapse phenomena (see, e.g., [8]).

From the Koopman eigenfunctions, we suggest that the combination of delay embedding with EDMD extracts global information on the flow structure of the underlying model directly from the single-bus measurement in this numerical experiment.

## 19.5 Concluding Remarks

We conducted the data-driven voltage analysis of the rudimentary power grid model via Koopman spectral analysis. The numerical experiment suggests that the data-driven analysis, which is a combination of delay embedding and EDMD, provides dynamical insights into the voltage dynamics of the rudimentary model directly from the single-bus measurement. The derived properties will be exploited for reduced-order modeling of voltage dynamics for monitoring the distance to instability and synthesizing a controller.

Several follow-up studies on the present work are possible. A method for appropriately choosing the coordinates like Figs. 19.6 and 19.8 is required for comparison of global information. An application of the combination to practical voltage data is also necessary for evaluating its effectiveness. The combination is theoretically founded for a class of nonlinear dynamical systems—measure-preserving systems: see Appendix and [26]. It is important to establish the theoretical foundation beyond this class of dynamical systems, especially, to dissipative ones that appear in the phenomenology in this chapter.

**Acknowledgements** The authors greatly appreciate Professor Takashi Hikihara (Kyoto University) for valuable discussion on this work and Professor Igor Mezić (University of California, Santa Barbara) for his careful reading of the draft of this chapter with valuable comments, which are cited in the remark of the appendix. During part of the work on this chapter, the authors were at Laboratory of Advanced Electrical Systems Theory, Department of Electrical Engineering, Kyoto University, Kyoto, Japan. The work is supported in part by JST-CREST program #JP-MJCR15K3 and JSPS-KAKEN #15H03964.

## Appendix

In this appendix, instead of $\mathbf{S}^\tau$, we will consider a smooth diffeomorphism (map, dynamical system) $\mathbf{S}$ defined on a probability space $(X, \mathfrak{B}_X, \mu)$, where $X$ is now a compact and metric space, $\mathfrak{B}_X$ is the Borel $\sigma$-algebra of $X$, and $\mu$ a probability measure. The diffeomorphism $\mathbf{S} : X \to X$ is called *measure-preserving* with $\mu$ if $\mu \circ \mathbf{S}^{-1} = \mu$ holds. Also, we denote by $\mathscr{F}_X$ a space of observables which correspond to scalar-valued functions defined on $X$, and we denote by $\mathscr{L}_2(X)$ the space of functions for which the second power of absolute value is integrable on $X$.

First of all, we introduce the definition of unitary equivalence of two bounded linear operators prior to the main theorem.

**Definition 19.1** Consider bounded linear operators $U$ and $\tilde{U}$, acting in a Hilbert space, with domains $\mathscr{D}$ and $\tilde{\mathscr{D}}$, respectively. Then, the linear operators are said to be *unitarily equivalent* if $CUC^{-1}f = \tilde{U}f$ holds for any $f \in \tilde{\mathscr{D}}$, where $C$ is an unitary operator.

An important fact is that the spectra of unitarily-equivalent operators have identical structures: either both are point spectra, continuous, or mixed. In particular, in the case of pure point spectrum, the eigenvalues of unitarily equivalent operators are identical and the multiplicities of corresponding eigenvalues are the same. An intuitive example of the unitary operator accomplishing the unitary equivalence is the Fourier transform. The notion of unitary equivalence is important in traditional ergodic theory to characterize properties of measure-preserving dynamical systems, called spectral invariants [2].

We now state the theorem on unitary equivalence of two Koopman operators related through the delay-embedding. We again denote by $U$ the Koopman operator defined for $\mathbf{S}$, by $\tilde{U}$ the Koopman operator defined for the reconstructed map $\tilde{\mathbf{S}}$ through the delay embedding $\Phi_{(\mathbf{S},f)}$, and by $\mathscr{F}_{\varphi(X)}$ the space of observables defined on $\varphi(X)$ through the associated homeomorphism $\varphi$, on which $\tilde{U}$ acts.

**Theorem 19.2** *Suppose that* $\mathbf{S} : X \to X$ *is measure-preserving. For a smooth observable* $f \in \mathscr{F}_X = \mathscr{L}_2(X)$, *consider the reconstructed map* $\tilde{\mathbf{S}} : \varphi(X) \to \varphi(X)$ *through the delay-embedding* $\Phi_{(\mathbf{S},f)}$. *Then, the associated Koopman operators* $U : \mathscr{F}_X \to \mathscr{F}_X$ *and* $\tilde{U} : \mathscr{F}_{\varphi(X)} \to \mathscr{F}_{\varphi(X)}$ *are unitarily equivalent.*

*Proof* Based on (19.2.3), we need to prove that the composition operator $C_\varphi : \mathscr{F}_{\varphi(X)} \to \mathscr{F}_X$ is a unitary operator:

1. $C_\varphi$ preserves the inner-product in $\mathscr{F}_{\varphi(X)} = \mathscr{L}_2(\varphi(X))$ and $\mathscr{F}_X = \mathscr{L}_2(X)$.
2. $C_\varphi$ is surjective.
3. $C_\varphi$ is linear and bounded.

By assumption, the measure $\mu$ is preserved under $\mathbf{S}$, that is,

$$\mu \circ \mathbf{S}^{-1} = \mu. \tag{19.4}$$

Then, we see that the new (pullback) measure $\mu \circ \varphi^{-1}$ is preserved under the reconstructed map $\tilde{\mathbf{S}}$ as shown below:

$$\begin{aligned} \mu \circ \mathbf{S}^{-1} &= \mu \\ \mu \circ \left(\varphi^{-1} \circ \tilde{\mathbf{S}}^{-1} \circ \varphi\right) &= \mu \\ \left(\mu \circ \varphi^{-1}\right) \circ \tilde{\mathbf{S}}^{-1} &= \mu \circ \varphi^{-1}. \end{aligned} \tag{19.5}$$

With this, we estimate the inner product of observables $g_1, g_2 \in \mathscr{F}_{\varphi(X)} = \mathscr{L}_2(\varphi(X))$ on the set $\varphi(X)$ with the pullback measure $\mu \circ \varphi^{-1}$ as follows:

$$
\begin{aligned}
(g_1, g_2)_{\mathscr{L}_2(\varphi(X))} &= \int_{\varphi(X)} \overline{g_1(\mathbf{y})} g_2(\mathbf{y}) d(\mu \circ \varphi^{-1})(\mathbf{y}) \\
&= \int_X \overline{g_1(\mathbf{x})} g_2(\varphi(\mathbf{x})) d(\mu \circ \varphi^{-1})(\varphi(\mathbf{x})) \\
&= \int_X \overline{(C_\varphi g_1)(\mathbf{x})} (C_\varphi g_2)(\mathbf{x}) d\mu(\mathbf{x}) \\
&= (C_\varphi g_1, C_\varphi g_2)_{\mathscr{L}_2(X)}
\end{aligned}
\tag{19.6}
$$

That is to say, $C_\varphi$ preserves the inner-product.

Next, we show that for the homeomorphism $\varphi$, $C_\varphi : \mathscr{F}_{\varphi(X)} \to \mathscr{F}_X$ is surjective. Assume not. Then, for a $f \in \mathscr{F}_X$, there exists no $g \in \mathscr{F}_{\varphi(X)}$ such that $C_\varphi g = f$. By focusing on the continuous inverse $\varphi^{-1}$ and making a new observable $f \circ \varphi^{-1}$ by composition, the following holds

$$
\begin{aligned}
\infty > \int_X |f(\mathbf{x})|^2 d\mu(\mathbf{x}) &= \int_{\varphi(X)} |f(\varphi^{-1}(\mathbf{y}))|^2 d\mu(\varphi^{-1}(\mathbf{y})) \\
&= \int_{\varphi(X)} |(f \circ \varphi^{-1})(\mathbf{y})|^2 d(\mu \circ \varphi^{-1})(\mathbf{y}),
\end{aligned}
\tag{19.7}
$$

implying that $f \circ \varphi^{-1}$ is an element of the space $\mathscr{F}_{\varphi(X)}$ with the pullback measure $\mu \circ \varphi^{-1}$. By applying $C_\varphi$ to this $f \circ \varphi^{-1}$, we see the contradiction. Thus, for $\varphi$ homeomorphism, $C_\varphi$ is surjective.

Finally, by definition, it is easily shown that $C_\varphi$ is linear and bounded. Hence, $C_\varphi$ is a unitary operator. □

*Remark* The theorem is proved for the case of measure-preserving dynamical systems for which the associated Koopman operators are unitary. As mentioned in the last paragraph of the conclusion, its generalization is expected to be a large class of dissipative dynamical systems. It is shown in [17] that the so-called ergodic partition of a state space, which is based on eigenspace of eigenvalue 1 of the associated Koopman operator, can be inferred from the delay embedding. It is also shown in [23] that the eigenspace of eigenvalue 1, which is computable via time average of a continuous observable, is direct to characterizing the basin of attraction in dissipative systems. These suggest that a sort of spectral equivalence of Koopman operators will hold generally.

## References

1. Arbabi, H., Mezić, I.: Ergodic theory, dynamic mode decomposition and computation of spectral properties of the Koopman operator. SIAM J. Appl. Dyn. Syst. **16**(4), 2096–2126 (2017)
2. Arnold, V.I., Avez, A.: Ergodic Problems of Classical Mechanics. Benjamin, Amsterdam (1968)
3. Bishop, C.M.: Pattern Recognition and Machine Larning. Springer, New York (2006)
4. Budišić, M., Mohr, R., Mezić, I.: Applied Koopmanism. CHAOS **22**(4), 047, 510 (2012)

5. Das, S., Giannakis, D.: Delay-coordinate maps and the spectra of Koopman operators (2017). arXiv:1706.08544v6
6. Dasgupta, S., Paramasvam, M., Vaidya, U., Ajjarrapu, V.: Real-time monitoring of short-term voltage stability using PMU data. IEEE Trans. Power Syst. **28**(4), 3702–3711 (2013)
7. De La Ree, J., Centeno, V., Thorp, J.S., Phadke, A.G.: Synchronized phasor measurement applications in power systems. IEEE Trans. Smart Grid **1**(1), 20–27 (2010)
8. Dobson, I., Chiang, H.D.: Towards a theory of voltage collapse in electric power systems. Syst. Control Lett. **13**, 253–262 (1989)
9. Giannakis, D., Slawinska, J., Zhao, Z.: Spatiotemporal feature extraction with data-driven Koopman operators. In: JMLR: Workshop and Conference Proceedings, vol. 44, pp. 103–115 (2015)
10. Kamb, M., Kaiser, E., Brunton, S.L., Kutz, J.N.: Time-delay observables for Koopman: theory and applications (2018). arXiv:1810.01479v1
11. Kantz, H., Schreiber, T.: Nonlinear Time Series Analysis, 2nd edn. Cambridge University Press, Cambridge (2003)
12. Mauroy, A., Mezić, I.: On the use of Fourier averages to compute the global isochrons of (quasi)periodic dynamics. CHAOS **22**(3), 033, 112 (2012)
13. Mauroy, A., Mezić, I., Moehlis, J.: Isostables, isochrons, and Koopman spectrum for the action-angle representation of stable fixed point dynamics. Phys. D **261**, 19–30 (2013)
14. Mezić, I.: Spectral properties of dynamical systems, model reduction and decompositions. Nonlinear Dyn. **41**, 309–325 (2005)
15. Mezić, I.: Analysis of fluid flows via spectral properties of the Koopman operator. Annu. Rev. Fluid Mech. **45**, 357–378 (2013)
16. Mezić, I.: On applications of the spectral theory of the Koopman operator in dynamical systems and control theory. In: Proceedings of 2015 IEEE 54th Annual Conference on Decision and Control, pp. 7034–7041. Osaka, Japan (2015)
17. Mezić, I., Banaszuk, A.: Comparison of systems with complex behavior. Phys. D **197**(1–2), 101–133 (2004)
18. Mezić, I., Wiggins, S.: A method for visualization of invariant sets of dynamical systems based on the ergodic partition. CHAOS **9**(1), 213–218 (1999)
19. Raak, F., Susuki, Y., Hikihara, T.: Data-driven partitioning of power networks via Koopman mode analysis. IEEE Trans. Power Syst. **31**(4), 2799–2807 (2016)
20. Raak, F., Susuki, Y., Tsuboki, K., Kato, M., Hikihara, T.: Quantifying smoothing effects of wind power via Koopman mode decomposition: a numerical test of wind speed predictions in Japan. Nonlinear Theory Its Appl., IEICE **8**(4), 342–357 (2017)
21. Rowley, C.W., Mezić, I., Bagheri, S., Schlatter, P., Henningson, D.S.: Spectral analysis of nonlinear flows. J. Fluid Mech. **641**, 115–127 (2009)
22. Susuki, Y., Mezić, I.: A Prony approximation of Koopman mode decomposition. In: Proceedings of IEEE Conference on Decision and Control, pp. 7022–7727 (2015)
23. Susuki, Y., Mezić, I.: Invariant sets in quasiperiodically forced dynamical systems (2019). arXiv:1808.08340
24. Susuki, Y., Mezić, I., Raak, F., Hikihara, T.: Applied Koopman operator theory for power systems technology. Nonlinear Theory Its Appl., IEICE **7**(4), 430–459 (2016)
25. Susuki, Y., Sako, K.: Data-based voltage analysis of power systems via delay-embedding and extended dynamic mode decomposition. IFAC-PapersOnLine **51**, 221–226 (2018)
26. Susuki, Y., Sako, K., Hikihara, T.: On the spectral equivalence of Koopman operators through delay embedding (2017). arXiv:1706.1006
27. Takens, F.: Detecting strange attractors in turbulence. In: D. Land, L.S. Young (eds.) Dynamical Systems and Turbulence. Lecture Notes in Mathematics, vol. 898. Springer, Berlin (1981)
28. Williams, M.O., Kevrekidis, I.G., Rowley, C.W.: A data-driven approximation of the Koopman operator: extending dynamic mode decomposition. J. Nonlinear Sci. **25**, 1307–1346 (2015)
29. Yan, J., Liu, C.C., Vaidya, U.: PMU-based monitoring of rotor angle dynamics. IEEE Trans. Power Syst. **26**(4), 2125–2133 (2011)

# Chapter 20
# Koopman Performance Analysis of Nonlinear Consensus Networks

**Hossein K. Mousavi, Christoforos Somarakis, Qiyu Sun and Nader Motee**

**Abstract** Spectral decomposition of dynamical systems is a popular methodology to investigate the fundamental qualitative and quantitative properties of these systems and their solutions. In this chapter, we consider a class of nonlinear cooperative protocols, which consist of multiple agents that are coupled together via an undirected state-dependent graph. We develop a representation of the system solution by decomposing the nonlinear system utilizing ideas from the Koopman operator theory and its spectral analysis. We use recent results on the extensions of the well-known Hartman theorem for hyperbolic systems to establish a connection between the original nonlinear dynamics and the linearized dynamics in terms of Koopman spectral properties. The expected value of the output energy of the nonlinear protocol, which is related to the notions of coherence and robustness in dynamical networks, is evaluated and characterized in terms of Koopman eigenvalues, eigenfunctions, and modes. Spectral representation of the performance measure enables us to develop algorithmic methods to assess the performance of this class of nonlinear dynamical networks as a function of their graph topology. Finally, we propose a scalable computational method for approximation of the components of the Koopman mode decomposition, which is necessary to evaluate the systemic performance measure of the nonlinear dynamic network.

H. K. Mousavi · C. Somarakis · N. Motee (✉)
Department of Mechanical Engineering and Mechanics,
Lehigh University, Bethlehem, PA 18015, USA
e-mail: motee@lehigh.edu

H. K. Mousavi
e-mail: mousavi@lehigh.edu

C. Somarakis
e-mail: csomarak@lehigh.edu

Q. Sun
Department of Mathematics, Orlando, FL 32816, USA
e-mail: qiyu.sun@ucf.edu



A. Mauroy et al. (eds.), *The Koopman Operator in Systems and Control*,
Lecture Notes in Control and Information Sciences 484,
https://doi.org/10.1007/978-3-030-35713-9_20

## 20.1 Introduction

The central objective in the theory of networked control systems is to address and analyze the practical challenges in implementations of real-world dynamical networks, in order to develop design algorithms with certified convergence properties [9, 19, 22, 28, 38, 52, 54, 55]. The application areas, nowadays, range from multi-robot systems [3] to social networks [23], power systems [20], metabolic pathways [11, 44, 46], and brain networks [8]. One of the inherent unappealing features of these real-world networks is the nonlinearity of the interactions among the subsystems that stem from how subsystems affect each other's dynamics [2, 4, 13, 15, 24, 38]. For example in the natural networks, physical interactions such as fluid field coupling [45], coupled biochemical reactions [46], or visual coordination [15] may result in nonlinear coupling among the subsystems.

The main focus of the existing body of literature is on stability analysis of nonlinear dynamical networks, where some of these works investigate effects of coupling topologies [20, 24], time delay [40, 41, 51] and exogenous noise [14]. The common approach to deal with the existing nonlinearities is to study linearized forms of network dynamics. There is a rich number of works devoted to performance and robustness analysis and optimal design of linear dynamical networks [5, 6, 17, 25, 29, 34, 35, 37, 39, 42, 47–50, 53]. Despite a growing need to analyze and synthesize the nonlinear dynamical networks in nonequilibrium modes of operation, consistent and systematic methods to tackle these problems are sorely missing in the literature. The main reason is that the linear network techniques, which are mainly based on eigendecomposition, cannot be applied to nonlinear systems. Recent advances in analysis of dynamical systems using Koopman operator theory have opened up a new venue to study the properties of nonlinear systems in a systematic manner [10, 26, 27, 31, 32].

In this chapter, we build upon concepts and tools from Koopman methodology to assess the performance of a class of nonlinear consensus networks. These networks are defined over an undirected state-dependent interconnection graph topology, where the control input of each agent is equal to a weighted combination of the difference between its own state and its neighbors. The expected value of the output energy of the network is adopted as the performance measure. We obtain a closed-form series representation for this quadratic performance measure and show that the value of performance measure depends on the spectra of the Koopman operator. The idea of spectral characterization of performance measure can be potentially utilized to analyze and design nonlinear networks; we refer to [37, 49, 50, 53] for the successfulness of this approach in the case of linear dynamical networks. An efficient numerical algorithm is developed to compute the value of the performance measure for a given dynamical network. Several analytical and numerical examples have been provided to highlight the usefulness of our theoretical findings.

## 20.2 Preliminaries

Consider an autonomous dynamical system given by

$$\dot{\mathbf{x}} = \mathbf{F}(\mathbf{x}) \tag{20.1}$$

with $\mathbf{F}(\mathbf{x}) : \mathbb{R}^n \to \mathbb{R}^n$ representing a $C^2$ vector field on $\mathbb{R}^n$. For the initial condition $\mathbf{x}_0 \in \mathbb{R}^n$, $\mathbf{x}(t) := \mathbf{S}(t, \mathbf{x}_0) : \mathbb{R}_+ \times \mathbb{R}^n \to \mathbb{R}^n$ is the generated flow of (20.1), which is assumed to be defined for all $t \geq 0$. We assume that $\mathbf{F}$ attains a hyperbolic stable fixed point at the origin, i.e., $\mathbf{F}(\mathbf{0}) = \mathbf{0}$. Moreover, we denote the Jacobian of $\mathbf{F}$ at the fixed point by

$$\mathbf{A} := \frac{\partial}{\partial \mathbf{x}} \mathbf{F}|_{\mathbf{x}=\mathbf{0}}, \tag{20.2}$$

which we assume to be Hurwitz; i.e., the eigenvalues of $\mathbf{A}$ have strictly negative real parts. The basin of attraction of the origin is an open neighborhood of $\mathbf{0}$ with $\Omega \subset \mathbb{R}^n$, a compact subset of this neighborhood. By definition, $\mathbf{S}(t, \mathbf{x}_0) \in \Omega$ for any $\mathbf{x}_0 \in \Omega$ and $t \geq 0$, such that $\mathbf{S}(t, \mathbf{x}_0) \to \mathbf{0}$ as $t \to +\infty$. Let us define the functional space

$$\mathscr{F} = \left\{ f \in C^1(\Omega, \mathbb{R}) : \sup_{\mathbf{x} \in \Omega} \left| f(\mathbf{x}) \right| + \sup_{\mathbf{x} \in \Omega} \left\| \nabla f(\mathbf{x}) \right\| < \infty \right\}, \tag{20.3}$$

that together with norm $|f|_{C^1} := \sup_{\mathbf{x} \in \Omega} \left| f(\mathbf{x}) \right| + \sup_{\mathbf{x} \in \Omega} \left\| \nabla f(\mathbf{x}) \right\|$, constitute a Banach space. This will be the space of observable functions on flow $\mathbf{S}(\cdot, \mathbf{x}_0)$. For fixed $t \geq 0$, the Koopman operator $U^t : \mathscr{F} \to \mathscr{F}$ associated with (20.1) is

$$(U^t f)(\mathbf{x}_0) = f \circ \mathbf{S}(t, \mathbf{x}_0). \tag{20.4}$$

For any fixed $t \geq 0$, it can be shown that $U^t$ is linear in $\mathscr{F}$. Furthermore, the collection $\{U^t\}_{t \geq 0}$ constitutes a semigroup known as *the Koopman semigroup* [10]. In the context of continuous autonomous dynamical systems, (20.4) is interpreted as the action of semigroup on observable $f \in \mathscr{F}$. The spectrum of operator $U^t$ may consist of a discrete, continuous, and residual part. The discrete part, also known as point spectrum of $U^t$, is defined as

$$\sigma_p(U^t) = \left\{ \lambda \in \mathbb{C} \,\middle|\, U^t \phi = e^{\lambda t} \phi, \text{ for some } \phi = \phi_\lambda \in \mathscr{F} \right\}. \tag{20.5}$$

Throughout this chapter $(\lambda, \phi_\lambda)$, for $\lambda \in \sigma_p(U^t)$, is called the Koopman pair of an eigenvalue with its corresponding eigenfunction. The purpose of this work is to discuss the role of the Koopman operator theory in evaluating quadratic performance measures for a class of nonlinear consensus protocols that enjoy a great interest in the field of networked control systems. More specifically, we leverage a recent extension of the Hartman's theorem for hyperbolic dynamical systems [27] to outline the pivotal role of point spectrum in approximating the output energy of nonlinear distributed cooperative algorithms.

The rest of the chapter is organized as follows. In Sect. 20.3, we will apply the extension of the Hartman's theorem in order to investigate the conditions under which one is able to express the flow $\mathbf{S}(\cdot, \mathbf{x}_0)$ of (20.1) in terms of the Koopman pairs, i.e., to write the $i$th element of $\mathbf{S}(\cdot, \mathbf{x}_0)$ as

$$\big[\mathbf{S}(t, \mathbf{x_0})\big]_i \approx \sum_{\lambda} c_\lambda^{(i)} \mathrm{e}^{\lambda t} \phi_\lambda(\mathbf{x_0}), \quad \text{for every } t \geq 0$$

for some coefficients $\mathbf{c}_\lambda = [c_\lambda^{(1)}, \ldots, c_\lambda^{(n)}]^T$. Then, each collection $\{(\lambda, \phi_\lambda, \mathbf{c}_\lambda)\}_\lambda$ will constitute a Koopman mode decomposition (KMD) [10]. We use an interesting fact about the map created by stacking specific eigenfunctions of the Koopman operator and its inverse map for dynamical systems with hyperbolic stable fixed points: polynomial approximations of the inverse map yields a Koopman mode decomposition.

Based on the results of Sect. 20.3, we proceed in Sect. 20.4 with the calculation of the performance measures for nonlinear consensus networks. The measures are expressed in series form as a function of KMD's. In addition, we discuss a number of special cases where KMD's can be explicitly calculated.

In Sect. 20.5 we describe a method to come up with a sparse approximation to the eigenfunctions of the Koopman operator. The method strongly depends on a nearly optimal fitting technique called Smolyak-Collocation projection. We use the same method to compute the approximate Koopman modes. Using the above developments, we may derive quantitative information about the stability and performance of nonlinear dynamical networks. In fact, inspired by our previous work [36], we look at the performance measure of a class of nonlinear dynamical systems and illustrate how their performance can be assessed using the spectra of the Koopman operator.

## 20.3 Koopman Mode Decomposition of System Flows

The celebrated theorem of Hartman (stated below for convenience) establishes a crucial connection between autonomous dynamical system (20.1) and the dynamics of the linearized system around the origin. A moment of reflection, initially mentioned in [27], can lay the groundwork of bridging the gap between spectral properties of the nonlinear and the linearized system around the fixed point. The aim of the present section is to conduct a rigorous discussion of these exact steps. We begin our analysis with parts adapted from the literature to keep the manuscript self-contained.

**Theorem 20.1** (Hartman's Theorem [43]) *Consider dynamical system* (20.1) *with the smoothness assumptions on* $\mathbf{F}$ *to hold and the origin to be a hyperbolic fixed point. Then there exists a* $C^1$ *diffeomorphism* $\mathbf{H}$ *of a neighborhood* $U$ *of the origin* *on an open set $\Omega' \subset \Omega$ containing the origin such that for each $\mathbf{x}_0 \in \Omega'$, there exists an open interval $I(\mathbf{x}_0) \subset \mathbb{R}_+$ containing zero such that for all $\mathbf{x_0} \in U$ and $t \in I(\mathbf{x_0})$*

$$\mathbf{H} \circ \mathbf{S}(t, \mathbf{x_0}) = \mathrm{e}^{\mathbb{A}t}\, \mathbf{H}(\mathbf{x_0}),$$

*where $\mathbf{A} = \frac{\partial}{\partial \mathbf{x}} \mathbf{F}|_{\mathbf{x}=0}$.*

*Remark 20.1* The set $I(\mathbf{x}_0)$ stands for the maximal interval of existence of the solution of system (20.1), that defines flow $\mathbf{S}(t, \mathbf{x}_0)$, for any $t \geq 0$. Evidently, $I(\mathbf{x}_0) = \mathbb{R}_+$ for all $\mathbf{x}_0$ in the basin of attraction $\Omega$.

The next result extends the theorem of Hartman to hold true over the whole the basin of attraction of the fixed point at the origin.

**Theorem 20.2** ([27]) *If $\mathbf{F}$ is $C^2$ and $\mathbb{A} = \frac{\partial}{\partial \mathbf{x}} \mathbf{F}|_{\mathbf{x}=0}$ is Hurwitz, then there exists a diffeomorphism $\boldsymbol{\alpha} : \Omega \to \mathbb{R}^n$ such that*

$$\boldsymbol{\alpha} \circ \mathbf{S}(t, \mathbf{x_0}) = \mathrm{e}^{\mathbb{A}t}\, \boldsymbol{\alpha}(\mathbf{x}_0), \tag{20.6}$$

*for all $\mathbf{x_0} \in \Omega$ and $t \geq 0$.*

Next, assuming that $\mathbb{A}$ is diagonalizable, we can write $\mathbb{A} = \mathbf{R}\Lambda\mathbf{R}^{-1}$ where $\Lambda$ is a diagonal matrix, having diagonal elements with strictly negative real parts. Let us define

$$\mathbf{H}(\mathbf{x}) := \mathbf{R}^{-1}\boldsymbol{\alpha}(\mathbf{x}). \tag{20.7}$$

Then, one may observe that

$$\mathbf{H}\left(\mathbf{S}(t, \mathbf{x}_0)\right) = \mathbf{R}^{-1}\mathrm{e}^{\mathbb{A}t}\mathbf{R}\mathbf{H}(\mathbf{x}_0) = \mathrm{e}^{\Lambda t}\mathbf{H}(\mathbf{x_0}). \tag{20.8}$$

Clearly, map $\mathbf{H} : \Omega \to \mathbb{C}^n$ is a diffeomorphism. Hence, flow of the dynamical system $\mathbf{S}(\cdot, \mathbf{x}_0)$ can be expressed as

$$\mathbf{S}(t, \mathbf{x_0}) = \mathbf{H}^{-1}\left(\mathrm{e}^{\Lambda t}\mathbf{H}(\mathbf{x_0})\right), \quad \text{for every } t \geq 0 \text{ and } \mathbf{x}_0 \in \Omega. \tag{20.9}$$

This suggests that knowledge of maps $\mathbf{H}$ and $\mathbf{H}^{-1}$ helps identify the flow of the system. In an interesting turn of events, there is an important correlation between Koopman spectrum and the eigenvalues of the Jacobian matrix at the fixed point $\mathbf{A}$.

**Theorem 20.3** *Let map $\mathbf{H}$ given in (20.7) have component-wise expression*

$$\mathbf{H} = \left[H_1, H_2, \ldots, H_n\right]^T, \tag{20.10}$$

*for $H_i : \Omega \to \mathbb{C}^n$ and $i = 1, \ldots, n$. If $\lambda_i$ is the $i$th eigenvalue of $\mathbb{A}$, then $\left(\lambda_i,\ H_i\right)$ is a pair of Koopman eigenvalue and its corresponding eigenfunction.*

*Proof* The result immediately follows after comparing the definition of the Koopman eigenfunction in (20.5) with identity (20.8). □

We take advantage of this connection to provide a Koopman mode decomposition (KMD) for dynamical systems with a stable hyperbolic fixed point. One may find the general aspects of this decomposition in [10]. In this context, extended dynamic mode decomposition (EDMD) [56] is a framework with focus on derivation of numerical estimations to Koopman operator and KMD. At first, we make two crucial remarks before coming up with the advertised decomposition.

*Polynomial Expansion of* $\mathbf{H}^{-1}$. Clearly, all elements of $\mathbf{H}^{-1}(\mathbf{x}) = \boldsymbol{\alpha}^{-1}(\mathbf{R}\,\mathbf{x})$ are continuous in $\Omega$, hence they map compact sets onto compact sets. Therefore, the domain of definition of $\mathbf{H}^{-1}$ is compact.

By virtue of the Stone–Weierstrass Theorem [16], $\mathbf{H}^{-1}(\mathbf{x})$ can be uniformly $\varepsilon$-approximated over the domain of $\mathbf{H}^{-1}$ by multivariate polynomials. Therefore, for every $\mathbf{x} \in \operatorname{dom} \mathbf{H}^{-1}$ we can write

$$\mathbf{H}^{-1}(\mathbf{x}) \overset{\varepsilon}{\approx} \sum_{\gamma \in \Gamma_\varepsilon} \mathbf{c}_\gamma^\varepsilon \, x_1^{j_1} \cdots x_n^{j_n}, \tag{20.11}$$

where $\gamma = [j_1, \ldots j_n]^T \in \mathbb{Z}_+^n$, $\overset{\varepsilon}{\approx}$ implies the approximation with maximal error of $\varepsilon$, and $\mathbf{c}_\gamma^\varepsilon = \mathbf{c}_{j_1,\ldots,j_n}^\varepsilon$ is represented using the multi-index notation. The index set $\Gamma_\varepsilon \subset \mathbb{Z}_+^n$ consists of finite number of indices based on the desired level of accuracy $\varepsilon > 0$. If map $\mathbf{H}^{-1}$ is analytic, then it admits a Maclaurin expansion with a positive radius of convergence and we can have an infinite series representation (at least in a subset of $\Omega^{-1}$) similar to (20.11).

*Remark 20.2* Not every polynomial approximation of $\mathbf{H}^{-1}$ is suitable in this chapter. It is necessary for the right-hand side of (20.11) to vanish at the origin, as $\mathbf{H}^{-1}$ does as well. This property permits a credible polynomial approximation of the output energy of (20.1) in terms of Koopman modes. Examples of polynomial expansions that can approximate $\mathbf{H}^{-1}$ under such constraints are interpolation-based methods using multi-variate polynomials of the Bernstein or Chebyshev families, with appropriate scaling of domain of $\mathbf{H}^{-1}$ [18].

*Superposition of Koopman Eigenpairs.* The closedness of the set of eigenfunctions under multiplication is an important property that is stated in the next lemma.

**Lemma 20.1** ([10]) *Let $\phi_1, \phi_2 \in \mathscr{F}$ with associated eigenvalues $\lambda_1$ and $\lambda_1$, respectively. Then $\phi_3(\mathbf{x}) := \phi_1(\mathbf{x})\phi_2(\mathbf{x}) \in \mathscr{F}$ with associated eigenvalue $\lambda_3 = \lambda_1 + \lambda_2$.*

We are ready now to formulate a KMD-based expression for flow $\mathbf{S}(\cdot, \mathbf{x}_0)$. For its exposition we consider an arbitrary but fixed ordering of the elements of $\mathbb{Z}_+^n$, with $\mathbb{Z}_+^n = \{\boldsymbol{\gamma}_1, \boldsymbol{\gamma}_2, \ldots, \boldsymbol{\gamma}_i, \ldots\}$ where $\boldsymbol{\gamma}_i = (j_1, j_2, \ldots, j_n)^T$.

**Proposition 20.1** *Let $\mathbf{A} = \frac{\partial}{\partial \mathbf{x}} \mathbf{F}(\mathbf{x})|_{\mathbf{x}=\mathbf{0}}$ be diagonalizable and Hurwitz with eigenvalues $\lambda_1, \ldots, \lambda_n$. Consider map $\mathbf{H}^{-1}(\mathbf{x})$ with elements given (20.7) for every $\mathbf{x}_0 \in \Omega$,* *where $\Omega$ is a compact set. Then, using the approximation* (20.11) *for* $\mathbf{H}^{-1}(\mathbf{x})$, *flow* $\mathbf{S}(\cdot, \mathbf{x}_0)$ *of nonlinear system* (20.1) *attains the representation*

$$\mathbf{S}(t, \mathbf{x}_0) \overset{\varepsilon}{\approx} \sum_{i \geq 1} \mathbf{c}_i \, \mathrm{e}^{\overline{\lambda}_i t} \phi_i(\mathbf{x}_0), \quad \textit{for all } t \geq 0$$

*where for the ordered vector* $\boldsymbol{\gamma}_i = [j_1, \ldots, j_n]^T \in \Gamma_\varepsilon \subset \mathbb{Z}_+^n$, *we have*

$$\overline{\lambda}_i := \sum_{k=1}^{n} j_k \lambda_k \quad \textit{and} \quad \phi_i(\mathbf{x}_0) := \prod_{k=1}^{n} H_k^{j_k}(\mathbf{x}_0). \tag{20.12}$$

*Proof* Recall the expression (20.9) that is true for every $\mathbf{x}_0 \in \Omega$. Substituting $\mathrm{e}^{\Lambda t}\mathbf{H}(\mathbf{x}_0)$ into (finite) series representation (20.11), we may write the flow of system (20.1) as

$$\mathbf{S}(t, \mathbf{x}_0) \overset{\varepsilon}{\approx} \sum_{\gamma \in \Gamma_\varepsilon} \mathbf{c}_\gamma^\varepsilon \, \left(\mathrm{e}^{\lambda_1 t} H_1(\mathbf{x}_0)\right)^{j_1} \ldots \left(\mathrm{e}^{\lambda_n t} H_n(\mathbf{x}_0)\right)^{j_n},$$

which can be reorganized to obtain

$$\mathbf{S}(t, \mathbf{x}_0) \overset{\varepsilon}{\approx} \sum_{\gamma \in \Gamma_\varepsilon} \mathbf{c}_\gamma^\varepsilon \, \mathrm{e}^{(j_1\lambda_1 + j_2\lambda_2 + \cdots + j_n\lambda_n)t} H_1^{j_1}(\mathbf{x}_0) \ldots H_n^{j_n}(\mathbf{x}_0).$$

Let us define $\overline{\lambda}_i$ and $\phi_i(\mathbf{x})$ according to (20.12). Using Lemma 20.1, we deduce that $\phi_i(\mathbf{x})$ is an eigenfunction of Koopman operator with eigenvalue $\overline{\lambda}_i$. Rewriting the flow and using the introduced notation gives us the desired representation. □

In fact, we derive the explicit decomposition introduced in Proposition 20.1 by extending the material presented in [32] or [27]. We will see that this decomposition is a necessary tool for the subsequent analysis. Before that, we recall a useful lemma, that identifies a partial differential equation to associate the Koopman pairs.

**Lemma 20.2** (See [32]) *Consider a pair of Koopman eigenvalue and its corresponding eigenfunction denoted by* $\left(\lambda, \phi_\lambda(\mathbf{x})\right)$ *associated with nonlinear dynamics* (20.1). *The pair satisfies the identity*

$$\mathbf{F}(\mathbf{x})^T \nabla \phi_\lambda(\mathbf{x}) = \lambda \phi_\lambda(\mathbf{x}). \tag{20.13}$$

## 20.4 Performance of Nonlinear Consensus Networks

The standard multi-agent setting regards a finite collection of agents is labeled as $i = 1, 2, \ldots, n$. The $i$th agent is characterized by a real-valued state $x_i$. In a consensus network with first-order dynamics, the agents update their states by communicating with their adjacent (neighboring) agents. Our focus in this work is on the class of dynamic protocols of the form

$$\dot{x}_i = \sum_{\{i,j\}\in\mathcal{E}} w_{ij}\,(x_j - x_i), \tag{20.14}$$

where $\mathcal{E}$ is the set of edges of the undirected graph of the network whose weights are symmetric and state dependent in the form of

$$w_{ij} = w_{ji} = \tilde{w}_{ij}\, g\left(|x_i - x_j|^2\right), \tag{20.15}$$

for $g : \mathbb{R}_+ \to \mathbb{R}_{++}$ a positive coupling function of the graph, and constant $\tilde{w}_{ij} > 0$. We note that such a state dependence of the couplings is motivated by a natural assumption: the remote or dissimilar agents less likely interact with each other. For instance, this is the case in the context of social networks, oscillatory networks [24], or biological networks. For this reason function $g$ is usually considered to be monotonically decreasing [15, 52]. By defining the state of the network as $\mathbf{x} := [x_1, \ldots, x_n]^T \in \mathbb{R}^n$, we may express the collective dynamics of the agents as

$$\dot{\mathbf{x}} = -\mathcal{L}_{\mathbf{x}}\,\mathbf{x}, \tag{20.16}$$

where $\mathcal{L}_{\mathbf{x}}$ is the state-dependent graph Laplacian matrix with coupling weights that vary according to (20.15). For subsequent analysis, we rely on two conditions stated right below.

**Assumption 20.4** The function $g$ is analytic and it satisfies $g(0) = 1$.

**Assumption 20.5** The graph with coupling weights $\{\tilde{w}_{ij}\}_{\{i,j\}\in\mathcal{E}}$ is connected.

Connectedness implies that there exists a linked path between any two distinct nodes $i$ and $j$ in the graph of the network. A consequence of the latter assumption is that the graph corresponding to $\mathcal{L}_{\mathbf{x}}$ remains connected and undirected for all $\mathbf{x} \in \mathbb{R}^n$ since $w_{ij} > 0$ for every $\{i, j\} \in \mathcal{E}$. The next result provides a standard sufficient condition for convergence of dynamical network (20.14) to consensus equilibrium.

**Theorem 20.6** *Let Assumptions 20.4 and 20.5 hold true. For any initial state $\mathbf{x}_0 \in \mathbb{R}^n$ the long-term dynamics satisfy*

$$\lim_{t\to\infty} \mathbf{S}(t, \mathbf{x}_0) = \overline{x}\,\mathbf{1}_n,$$

*where the average vector of the network is $\overline{x} := \frac{1}{n}\sum_{i=1}^{n} x_i(0)$. The convergence to consensus occurs exponentially fast, with a rate that depends on initial state $\mathbf{x}_0$.*

*Proof* At first, observe that

$$\max_{i,j=1,\dots,n} |x_i(t) - x_j(t)| \leq \max_{i,j=1,\dots,n} |x_i(0) - x_j(0)| \text{ for all } t \geq 0.$$

This is easily verified since for the node $i$ with the maximum initial condition $\max_i \dot{x}_i(t) \leq 0$. Similarly for the node $i$ with the minimum starting value $\min_i \dot{x}_i(t) \geq 0$. The solution $\mathbf{x}(t, \mathbf{x}_0)$ remains bounded in $\Omega_0 := [\min_i x_i(0), \max_i x_i(0)]$. Consider the Lyapunov functional $\Lambda(\mathbf{x}) = \frac{1}{2}\sum_{i \neq j} |x_i - x_j|^2$. Then for the solution $\mathbf{x}(t),\ t \geq 0$ of (20.14), we have

$$\frac{d}{dt}\Lambda(\mathbf{x}(t)) = \sum_{i \neq j} \big(x_i(t) - x_j(t)\big)\big(\dot{x}_i(t) - \dot{x}_j(t)\big) \leq -\beta(t)\Lambda(\mathbf{x}(t)),$$

where the value of $\beta(t)$ is given by

$$\beta(t) := \min_{\substack{s \in [0,t] \\ \{i,j\} \in \mathscr{E}}} w_{ij}\big(\mathbf{x}(s)\big) \geq \underline{w} \cdot \underline{g} > 0,$$

for $\underline{w} = \min_{i,j=1,\dots,n} \tilde{w}_{ij}$ and $\underline{g} = \min_{s_1, s_2 \in \Omega_0} g(|s_1 - s_2|) > 0$ [33]. By virtue of graph connectivity, the convergence to the agreement space $x_1 = x_2 = \cdots = x_n$ occurs exponentially fast. Finally, observe that $\frac{1}{n}\sum_{i=1}^{n} x_i$ is a first integral of motion to conclude about the consensus point. □

The average of $\mathbf{x}_0$ is called the consensus equilibrium of the network over the state of interest [38]. The central objective of this work is to evaluate systemic measures of performance that quantify the necessary effort the dynamical system takes to converge to consensus. We aim at leveraging the Koopman framework, developed in the previous section. The requirement for the implementation of that machinery is to have a hyperbolic and asymptotically stable fixed point. One may notice that

$$\mathbf{A} := -\mathscr{L}_{\mathbf{0}}$$

with a smallest eigenvalue in magnitude being $\lambda_1(\mathbf{A}) = 0$. Hence, the fixed point at the origin is not hyperbolic. In order to overcome this difficulty we introduce output dynamics vector $\mathbf{y}$ with elements $y_i := x_i - \frac{1}{n}\sum_{k=1}^{n} x_k$, or in matrix form, $\mathbf{y} = \mathbf{M}_n\mathbf{x}$, where $\mathbf{M}_n$ is the the centering matrix given by

$$\mathbf{M}_n := \mathbf{I}_n - \mathbf{J}_n/n \in \mathbb{R}^{n \times n},$$

where $\mathbf{J}_n$ is the square matrix of all ones. The dynamics of $\mathbf{y}$ which constitute the disagreement network associated with (20.14) are defined to pass this obstacle [38, 48]. The disagreement Laplacian matrix is

$$\mathscr{L}_d(\mathbf{x}) := \mathscr{L}_{\mathbf{x}} + \frac{\delta}{n}\mathbf{J}_n,$$

for some $\delta > 0$. The next stability result is a straightforward corollary of Theorem 20.6 and it is stated without proof.

**Corollary 20.1** *The output dynamics of* $\mathbf{y} = \mathbf{M}_n\mathbf{x}$ *of* (20.14) *satisfy*

$$\dot{\mathbf{y}} = -\mathscr{L}_d(\mathbf{y})\,\mathbf{y}, \qquad (\mathscr{N}_d)$$

*with* $\mathbf{y} = \mathbf{0}$ *is the a globally exponentially stable hyperbolic fixed point.*

The dynamics of $(\mathscr{N}_d)$ satisfy $\mathbf{y}(t, \mathbf{y_0}) = \mathbf{M}_n\,\mathbf{S}(t, \mathbf{x}_0),\ t \geq 0$. The energy of the output once weighted with a positive-definite and symmetric matrix $\mathbf{Q}$ is

$$\int_0^\infty \mathbf{y}^T(t, \mathbf{y_0})\mathbf{Q}\,\mathbf{y}(t, \mathbf{y_0})\,dt.$$

We choose the performance measure as the mean energy of the vanishing signal $\mathbf{y}$, when the state of the consensus system starts from a random initial condition $\mathbf{x}_0$. The long-term energy of the output signal $\mathbf{y}$ that converges to zero is equivalent to the energy of the state vector $\mathbf{x}$ to converge to consensus. We take this mean for uncertain initial conditions, by assuming that the initial state is a random variable $\mathbf{x}_0 : \Omega_s \to \Omega$ from the sample space $\Omega_s$, with some probability measure (e.g., a probability density function or a probability mass function). In either case, we define the performance measure as

$$\rho\,(\mathscr{L}) := \mathbb{E}_{\mathbf{x}_0}\left\{\int_0^\infty \mathbf{S}^T(t, \mathbf{x}_0)\mathbf{M}_n^T\mathbf{Q}\,\mathbf{M}_n\mathbf{S}(t, \mathbf{x}_0)\,dt\right\}. \qquad (20.17)$$

The next result establishes an analytical expression for the performance measure of $(\mathscr{N}_d)$ that reflects the contributions of the spectra of the linearized graph Laplacian and eigenfunctions of the Koopman operator.

**Theorem 20.7** (Performance Measure) *Consider the disagreement dynamics* $(\mathscr{N}_d)$ *and the associated flow* $\mathbf{S}(\cdot, \mathbf{y_0})$ *for all initial disagreements* $\mathbf{y_0}$. *Then, the performance measure* (20.17) *can be expressed as*

$$\rho(\mathscr{L}) = \sum_{i,j\geq 1} \phi_{ij}c_{ij}\frac{1}{\overline{\lambda}_i + \overline{\lambda}_j}, \qquad (20.18)$$

*where* $\{\overline{\lambda}_i\}_{i=1,2,\ldots}$ *is the sequence of Koopman eigenvalues in the KMD of* $(\mathcal{N}_d)$, *enumerated by an arbitrary numbering of* $\gamma_i = (j_2, \ldots, j_n) \in \mathbb{Z}_+^{n-1}$ *as*

$$\overline{\lambda}_i := \sum_{k=2}^{n} j_k \lambda_k \quad \text{and} \quad \phi_i(\mathbf{x}_0) := \prod_{k=2}^{n} H_k^{j_k}(\mathbf{x}_0)$$

*with* $\lambda_2, \ldots, \lambda_n$ *being the nonzero eigenvalues of* $\mathcal{L}_\mathbf{0} := \frac{\partial}{\partial \mathbf{x}} \mathcal{L}(\mathbf{x})|_{\mathbf{x}=\mathbf{0}}$. *Moreover,* $\phi_{ij} := \mathbb{E}_{\mathbf{x}_0}\{\phi_i(\mathbf{y})\phi_j(\mathbf{y})\}$ *and* $c_{ij} := \mathbf{c}_i^T Q\, \mathbf{c}_j$ *are computed in terms of Koopman eigenfunctions and modes, respectively.*

*Proof* The disagreement dynamics $(\mathcal{N}_d)$ attain a globally exponentially stable hyperbolic origin. In view of Assumptions 20.4 and 20.5, one can sort the eigenvalues of $-\mathbf{A} = \frac{\partial}{\partial \mathbf{y}} \mathcal{L}_d(\mathbf{y})|_{\mathbf{y}=\mathbf{0}}$ as $\lambda_1 < \lambda_2 \leq \cdots \leq \lambda_n$ such that $\lambda_1 = \delta$. We claim that the restriction of $\phi_1(\mathbf{x}) = H_1(\mathbf{x})$ to $\mathbf{1}^\perp$ is zero since $\phi_1(\mathbf{x}) = \mathbf{1}_n^T \mathbf{x}$. We substitute $\phi_1(\mathbf{x})$, $\mathbf{F}(\mathbf{x}) = -\mathcal{L}_d(\mathbf{x})\mathbf{x}$, and $\lambda_1 = -\delta$ into the left-hand side of (20.13) to obtain

$$\nabla^T \phi_1(\mathbf{x}) \mathbf{F}(\mathbf{x}) = \mathbf{1}_n^T (-\mathcal{L}(\mathbf{x}) - \delta \mathbf{J}_n / n)\mathbf{x},$$

which implies that

$$\nabla^T \phi_1(\mathbf{x}) \mathbf{F}(\mathbf{x}) = 0 - \delta \sum_{i=1}^{n} x_i = -\delta \times \mathbf{1}_n^T \mathbf{x} = -\lambda_1 \phi_1.$$

Therefore, $\phi_1(\mathbf{x}) = \mathbf{1}_n^T \mathbf{x}$ is in fact a Koopman eigenfunction with eigenvalue $-\delta$. We observe that for any $\mathbf{y} \in \mathbf{1}^\perp$, it holds that $\phi_1(\mathbf{y}) = 0$. Considering the restricted dynamics, $H_1(\mathbf{y}) = \phi_1(\mathbf{y}) = 0$. Hence, any Koopman eigenfunction parametrized with $\gamma_i = (j_1, j_2, \ldots, j_n)$ with $j_1 \geq 1$ is zero, because the corresponding eigenfunction is

$$\phi_i(\mathbf{x}) = \prod_{k=1}^{n} H_k^{j_k}(\mathbf{x}).$$

Now let a $\mathbf{H}^{-1}$ have the form (20.11) for $\mathbf{y}$. We consider a KMD based on Proposition 20.1. This implies that all terms related to $\lambda_1$ are canceled out of the decomposition. Thus, we can restrict the numbering of summation indices to $\mathbb{Z}_+^{n-1}$ and then write the KMD for $\mathbf{y}(\cdot, \mathbf{y}_0) = \mathbf{M}_n \mathbf{S}(\cdot, \mathbf{x}_0)$ as

$$\mathbf{y}(t, \mathbf{y}_0) = \sum_{i \geq 1} \mathbf{c}_i e^{-\overline{\lambda}_i t} \phi_i(\mathbf{y}_0), \tag{20.19}$$

where for any multi-index $\gamma_i = (j_2, \ldots, j_n) \in \mathbb{Z}_+^{n-1}$ inducing

$$\overline{\lambda}_i := \sum_{k=2}^{n} j_k \lambda_k \quad \text{and} \quad \phi_i(\mathbf{x}_0) := \prod_{k=2}^{n} H_k^{j_k}(\mathbf{x}_0).$$

The integrand of the integral in the performance measure is

$$\mathbf{y}^T(t, \mathbf{y_0})\, \mathbf{Q}\, \mathbf{y}(t, \mathbf{y_0}) = \left( \sum_{i \geq 1} \mathrm{e}^{-\overline{\lambda}_i t} \phi_i(\mathbf{y_0}) \mathbf{c}_i^T \right) \mathbf{Q} \left( \sum_{j \geq 1} \mathbf{c}_j \mathrm{e}^{-\overline{\lambda}_j t} \phi_j(\mathbf{y_0}) \right).$$

We reorganize this quadratic term as

$$\mathbf{y}^T(t, \mathbf{y_0})\, \mathbf{Q}\, \mathbf{y}(t, \mathbf{y_0}) = \sum_{i,j \geq 1} \mathrm{e}^{-(\overline{\lambda}_i + \overline{\lambda}_j)t} \phi_i(\mathbf{y_0}) \phi_j(\mathbf{y_0}) \mathbf{c}_i^T\, \mathbf{Q}\, \mathbf{c}_j.$$

The induced eigenvalues satisfy $\overline{\lambda}_i = \sum_{k=2}^{n} j_k \lambda_k > 0$, hence, $\overline{\lambda}_i + \overline{\lambda}_j > 0$ for all $i, j \geq 1$. Integrating over all times yields

$$\int_0^\infty \mathbf{y}^T(t, \mathbf{y_0}) \mathbf{Q}\, \mathbf{y}(t, \mathbf{y_0})\, dt = \sum_{i,j \geq 1} \phi_i(\mathbf{y_0}) \phi_j(\mathbf{y_0}) \frac{\mathbf{c}_i^T \mathbf{Q} \mathbf{c}_j}{\overline{\lambda}_i + \overline{\lambda}_j}.$$

The result follows by virtue of the linearity of the expected value. □

### *20.4.1 Analytic Examples*

The Koopman representation of flows in consensus networks can be derived analytically, for some special cases. In this section, we discuss a few such types of networks in the form of (20.14) where the associated Koopman modes (subsequently $\rho(\mathscr{L})$) can be calculated in a closed form.

*Example 20.1* (*Linear Consensus Network*) We evaluate the performance measure of a first-order LTI consensus network of order $n$, which has the dynamics

$$\dot{\mathbf{x}} = -\mathscr{L}\,\mathbf{x},$$

for a graph Laplacian $\mathscr{L}$ that is state independent (i.e., $g \equiv 1$) but satisfies Assumption 20.5. To use (20.18), we let $\mathbf{Q} = \mathbf{I}_n$ and choose the initial conditions such that

$$\mathbb{E}_{\mathbf{x}_0} \left\{ \mathbf{y_0} \mathbf{y_0}^T \right\} = \mathbf{I}_n.$$

We denote the eigenvalues of $\mathscr{L}$ as $\lambda_i$ for $i = 1, \ldots, n$. Based on Lemma 20.3, $\lambda_i$ has a Koopman eigenfunction $\phi_i(\mathbf{x}) = H_i(\mathbf{x})$, that is

$$\phi_i(\mathbf{x}) = \mathbf{v}_i^T \mathbf{x},$$

where $\mathbf{v}_i$ is the unit eigenvector of $\mathscr{L}$ corresponding to $\lambda_i$ (see [10, 32]). Let $\mathbf{V} = [\mathbf{v}_1|\mathbf{v}_2|\dots|\mathbf{v}_n]$ be the orthonormal matrix of eigenvectors of $\mathscr{L}$, then for the disagreement dynamics we have $\mathbf{H}(\mathbf{y}) = \mathbf{V}^T\mathbf{y}$. Since $(\mathbf{V}^T)^{-1} = (\mathbf{V}^{-1})^{-1} = \mathbf{V}$, the inverse of this map is $\mathbf{H}^{-1}(\mathbf{y}) = \mathbf{V}\mathbf{y}$. This lets us compute the components of the performance measure as follows.

$$\phi_{ij} = \mathbb{E}_{\mathbf{x}_0}\left\{\phi_i(\mathbf{y})\phi_j(\mathbf{y})\right\} = \mathbb{E}_{\mathbf{x}_0}\left\{\mathbf{v}_j^T\mathbf{y}\cdot\mathbf{v}_i^T\mathbf{y}\right\},$$

for all $i, j = 2, \dots, n$. We rearrange to obtain

$$\phi_{ij} = \mathbb{E}_{\mathbf{x}_0}\left\{\mathbf{v}_j^T\mathbf{y}\mathbf{y}^T\mathbf{v}_i\right\} = \mathbf{v}_j^T\mathbb{E}_{\mathbf{x}_0}\left\{\mathbf{y}\mathbf{y}^T\right\}\mathbf{v}_i = \mathbf{v}_j^T\mathbf{v}_i = \delta_{ij}$$

since $\mathbb{E}_{\mathbf{x}_0}\{\mathbf{y}\mathbf{y}^T\} = \mathbf{I}_n$ and $\mathbf{V}$ is orthonormal. Obviously, $\mathbf{H}^{-1}(\mathbf{y}) = \mathbf{V}\mathbf{y}$ is a exact polynomial representation, thus

$$\mathbf{c}_i = \begin{cases} \mathbf{v}_i \ \text{if } i = 2, \dots, n \\ \mathbf{0} \ \text{if } i = 1 \end{cases},$$

which allows to compute the coefficients used in the performance measure as

$$c_{ij} = \mathbf{c}_i^T\mathbf{c}_j = \begin{cases} \delta_{ij} \ \text{if } i, j = 2, \dots, n \\ 0 \ \ \text{if } i \text{ or } j = 1 \end{cases}.$$

We substitute these terms into the result in (20.18) to find

$$\rho(\mathscr{L}) = \sum_{i=2}^{n} \frac{1}{2\lambda_i}, \tag{20.20}$$

that is the $\mathscr{H}_2$-norm squared of a first-order linear consensus network [48].

It turns out that for the case when we have only two agents, we may be able to compute the eigenfunctions analytically. The next two examples highlight this fact.

*Example 20.2* Suppose that the network consists of two agents with dynamics dictated by (20.14) and weight functions

$$w_{ij} = \frac{1}{(1 + (x_i - x_j)^2)^{\alpha}}, \tag{20.21}$$

for some constant $\alpha \in \mathbb{R}_+$. Parameter $\alpha$ in (20.21) defines how localized the interactions are within the network. As $\alpha$ increases, the agents update their states mainly with respect to their closest neighbors. In fact, the particular type of link implies that magnitude of interaction between two subsystems becomes weaker as their state becomes more different. For such a consensus network, in the case of two nodes, let

$p$ and $q$ denote the states of the agents. Consequently, we can explain the interaction of these two nodes through the dynamics

$$\begin{bmatrix} \dot{p} \\ \dot{q} \end{bmatrix} = \frac{-1}{(1+(p-q)^2)^\alpha} \begin{bmatrix} 1 & -1 \\ -1 & 1 \end{bmatrix} \begin{bmatrix} p \\ q \end{bmatrix}.$$

Now, we turn into the disagreement dynamics $\mathscr{N}_d$ with $\delta = 1$,[1] whose Jacobian at the origin attains the eigevnalues $\lambda_1 = -1$ and $\lambda_2 = -2$. Based on Lemma 20.3, each eigenvalue corresponds to a Koopman eigenfunction, say $\phi_1(x)$ and $\phi_2(x)$. We restrict the dynamics to $\mathbf{1}_n$, however, for the sake of simplicity, we denote the restricted variables with the same notation (i.e., $p$ and $q$). Hence, $[p, q]^T \in \mathbf{1}^\perp$; i.e., $p + q = 0$. We already know that

$$\phi_1(\mathbf{x}) = \mathbf{1}_n^T \mathbf{x},$$

whose restriction to $\mathbf{1}^\perp$ is indeed zero. Once $\phi_2(\mathbf{x})$ is restricted to $\mathbf{1}^\perp$, we may compute it in an explicit fashion using (20.13), that is

$$\begin{bmatrix} \partial\phi_2/\partial p \\ \partial\phi_2/\partial q \end{bmatrix}^T \frac{-1}{(1+(p-q)^2)^\alpha} \begin{bmatrix} q-p \\ p-q \end{bmatrix} = -2\phi_2.$$

We let $z := p - q$ and use the chain rule to obtain

$$\frac{\partial\phi_2}{\partial p} = \frac{\partial\phi_2}{\partial z} \quad \text{and} \quad \frac{\partial\phi_2}{\partial q} = -\frac{\partial\phi_2}{\partial z}.$$

Noting that the only free variable is now $z$, we can change the partial derivatives with respect to $z$. The resulting scalar ordinary differential equation is

$$\frac{2z}{(1+z^2)^\alpha} \frac{d\phi_2}{dz} = 2\phi_2,$$

which together with $\phi_2(0) = 0$ implies that

$$\phi_2 = \pm z \exp\left( \int \frac{(1+z^2)^\alpha - 1}{z} dz \right).$$

Without loss of generality, we choose to work with the plus sign. Using the binomial series we write the numerator of the integrand as

$$h(z) := (1+z^2)^\alpha - 1 = \alpha z^2 + \frac{(\alpha)(\alpha-1)z^4}{2!} + \cdots,$$

[1] Note that choice of $\delta$ is arbitrary.

for all $|z| < 1$. We integrate the series to get

$$\int \frac{h(z)}{z} dz = \sum_{n=1}^{\infty} \frac{\alpha(\alpha-1)\dots(\alpha-n+1)z^{2n}}{2n \times n!},$$

which completes the evaluation of $\phi_2(z)$ as

$$\phi_2(z) = z \exp\left(\sum_{n=1}^{\infty} \frac{\alpha(\alpha-1)\dots(\alpha-n+1)z^{2n}}{2n \times n!}\right),$$

that is a convergent series for $|z| < 1$ since exp(.) is analytic everywhere. The identity $p + q = 0$ implies $z = 2p$, thus

$$\phi_2(p) = 2p \exp\left(\sum_{n=1}^{\infty} \frac{\alpha(\alpha-1)\dots(\alpha-n+1)2^{2n-1}p^{2n}}{n \times n!}\right).$$

Thus, the component-wise description of $\mathbf{H}(p, q)$ is

$$\mathbf{H}(p, q) = \mathbf{H}(p) = \left[\phi_2(p)\ 0\right]^T.$$

The definition of inverse of a map implies

$$\mathbf{H}^{-1}(\mathbf{H}(p, q)) = \left[p\ q\right]^T$$

so $H_1^{-1}(p, q)$ and $H_2^{-1}(p, q)$ are simply the inverse functions of $\phi_2(p)$ and $-\phi_2(p)$, respecively. These functions are locally analytic around the origin based on Lagrange inversion theorem [1]. Futhermore, one can calculate their coefficients in terms of Bell polynomials [12]. We assume that the initial value of $p$ to come from a discrete random variable that takes the values from $\{0, 0.1, 0, 2, 0.3, 0.4\}$ with the uniform probability distribution. Because the mean of each initial condition must be zero, the initial value of $q$ will be $-p$. We compute the value of $\rho(\mathcal{L})$ for $\mathbf{Q} = \mathbf{I}_n$ using the 17th order Maclaurin series of both eigenfunctions and the inverse map $\mathbf{H}^{-1}(p)$. Because the value of the performance measure may be computed from the numerical integration of the trajectory as well, we may compare the results of the analytic approximations to the KMD and the real value of performance measure. These two values for a range of $\alpha \in [0, 0.4]$ have been illustrated in Fig. 20.1, where they are in good numerical agreement. The relative error is observed to increase from zero in the case that $\alpha = 0$ to less than 0.13% for $\alpha = 0.4$.

*Remark 20.3* There is a limitation on the magnitude of the admissible initial conditions for the analysis conducted in the previous example. However, notice that this does not imply that it is a linear analysis since for any value of $\alpha$, the linearization matrix of $\mathcal{N}_d$ (with $k = 1$) in Example 20.2 is the following matrix:

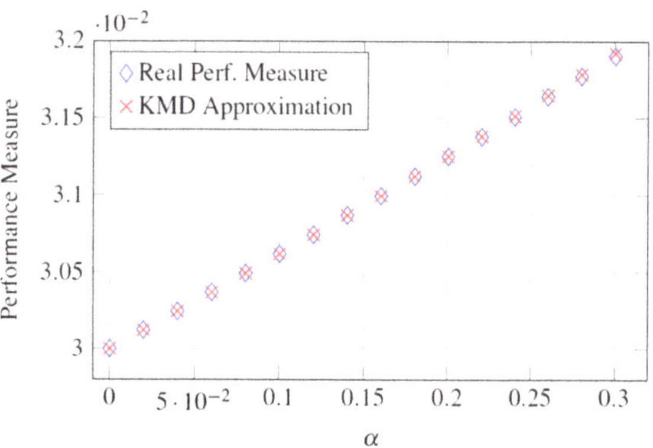


**Fig. 20.1** The performance measure of the network of two agents in Example 20.2 with the decaying parameter $\alpha$

$$\mathbf{A} = \begin{bmatrix} 1 & -1 \\ -1 & 1 \end{bmatrix} + \mathbf{J}_n.$$

This means that the value of the performance measure computed using the linearized system for each value of decaying parameter $\alpha$ is only one value.

*Example 20.3* The dynamics of oscillators have been observed to be closely related to the consensus dynamics. Kuramoto suggested a model of biological oscillation, in which each oscillator was connected to the other one; i.e., the topology was a complete graph. Instead the interactions can be limited over a certain graph [24], so the dynamics of agent $i$ can be represented as

$$\dot{x}_i = \omega_i + \sum_{\{i,j\} \in \mathscr{E}} w_{ij}(x_j - x_i), \tag{20.22}$$

where $\omega_i$ is the natural frequency and the coupling weight is

$$w_{ij} = K \frac{\sin(x_i - x_j)}{x_i - x_j}. \tag{20.23}$$

When the agents are identical (i.e., when $\omega_i = \omega$ for some $\omega$), the change of variable $x_i \to x_i - \omega t$ induces a nonlinear consensus network that is

$$\dot{x}_i = \sum_{\{i,j\} \in \mathscr{E}} w_{ij}(x_j - x_i). \tag{20.24}$$

We proceed with a procedure for computation of performance measure similar to one introduced in Example 20.2. For two identical oscillators, with phases $\theta$ and $\gamma$, the dynamics are

$$\frac{d}{dt}\begin{bmatrix}\theta\\ \gamma\end{bmatrix} = K\begin{bmatrix}\sin(\gamma-\theta)\\ \sin(\theta-\gamma)\end{bmatrix}.$$

Setting $k = 1$, the disagreement Jacobian has eigenvalues $\lambda_1 = -1$ and $\lambda_2 = -2K$, with eigenfunctions $\phi_1$ and $\phi_2$, respectively. Again we only need the restriction of $\phi_2$ to $\mathbf{1}^{\perp}$. The Eq. (20.13) for these dynamics becomes

$$\begin{bmatrix}\partial\phi_2/\partial\theta\\ \partial\phi_2/\partial\gamma\end{bmatrix}^T K\sin(\theta-\gamma)\begin{bmatrix}1\\ -1\end{bmatrix} = -2K\phi_2.$$

The new variable $z := \theta - \gamma$ creates a single ordinary differential equation that is

$$\frac{2\sin(z)}{d\phi_2}dz = 2\phi_2 \Rightarrow \frac{d\phi_2}{\phi_2} = \frac{1}{\sin(z)}dz,$$

which is integrated and manipulated to get

$$\phi_2 = \pm\exp\left(-\ln(\cot(z/2))\right) = \pm\tan(z/2),$$

where we arbitrarily choose +. The eigenfunction $\phi_2$ is locally analytic around the origin for $|z| < \pi$. Restricting to $\mathbf{1}^{\perp}$, $\theta + \gamma = 0$, so $z = 2\theta$, hence

$$\phi_2(\theta,\gamma) = \phi_2(\theta) = \tan(\theta),$$

which helps write the components of $\mathbf{H}(\theta,\gamma)$ as $\mathbf{H}(\theta,\gamma) = \begin{bmatrix}\phi_2 & 0\end{bmatrix}^T$. First component of $\mathbf{H}^{-1}(\theta)$ satisfies $H_1^{-1}(\phi_2(\theta), 0) = \theta$. Hence, $H_1^{-1}(\theta,\gamma) = \arctan(\theta)$, which is again locally analytic for $|\theta| < 1$ around zero.

We sample initial conditions from a uniform discrete random variable of 63 equally distributed initial conditions in $\theta \in [0, \pi/5]$ (and $\gamma = -\theta$) with equal distance of 0.01. We set $\mathbf{Q} = \mathbf{I}_n$ and use 17th order Maclaurin series of eigenfunctions and $\mathbf{H}^{-1}(\theta)$ to assess $\rho(\mathscr{L})$. As shown in Fig. 20.2, we alter $K \in [0.2, 4]$ and take a look at the values of the performance measure, once compared with the exact value of

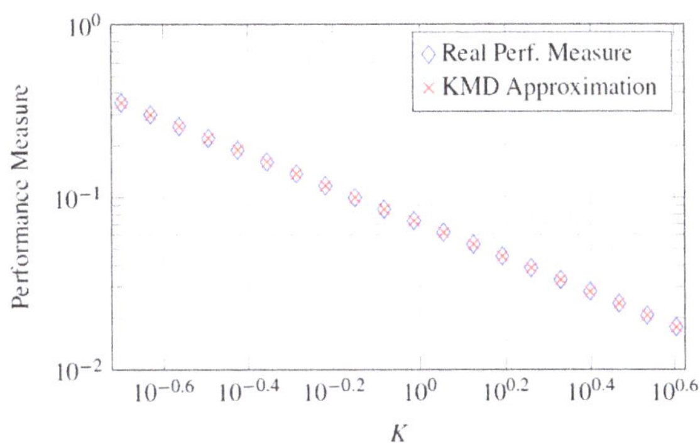


**Fig. 20.2** The performance measure of the Kuramoto model of two agents in Example 20.3 with the parameter $K$

$\rho(\mathscr{L})$ (computed with the numerical integration of the trajectories). The numerical agreement in this experiment can be measured by the relative error of the performance using the KMD approximation, which was about $3 \times 10^{-4}\%$ in the worst case.

## 20.5 Sparse Polynomial Approximations

We have observed that any polynomial approximation to the inverse map of eigenfunctions $\mathbf{H}^{-1}(\mathbf{x})$ will result in the Koopman mode decompositions. In this section, first, we detail a general sparse approximation technique for multivariate interpolation. Then, we demonstrate how we can use this technique for the map of Koopman eigenfunctions $\mathbf{H}(\mathbf{x})$ as well as the inverse $\mathbf{H}^{-1}(\mathbf{x})$.

### *20.5.1 Smolyak-Collocation Method*

To introduce the notion of sparsity for the approximation of both Koopman eigenfunctions and Koopman mode decomposition, we may use sparse functional approximation methods. The idea is that instead of searching for the approximant in the whole space of polynomials, the search is carried out over a nearly optimal sparse basis, called *Smolyak basis*. The output of the method would be a polynomial: the weighted sum of the tensor product of Chebyshev polynomials that are in the basis. Naturally, we choose the coefficients of the polynomial with respect to some error criterion. One way of doing so is collocation, where we enforce the approximant to (perhaps approximately) satisfy the governing equation of the problem at the given points of a grid, called Smolyak Sparse Grid. To describe this method, we need a few basic tools (consult [7, 30] for more details).

**Definition 20.1** (*Chebyshev Polynomials*) The sequence of the (scalar) Chebyshev polynomials of first kind $\{T_i(x)\}_{i=1,2,\ldots}$ are initialized with $T_1(x) = 1$ and $T_2(x) = x$ and recursively defined as follows.

$$T_{i+1}(x) = 2xT_i(x) - T_{i-1}(x), \text{ for } i = 2, 3, \ldots \tag{20.25}$$

Similarly, the Chebyshev polynomials of the second kind $\{U_i(x)\}_{i=1,2,\ldots}$ start with $U_1(x) = 1$ and $U_2(x) = 2x$, and then iteratively

$$U_{i+1}(x) = 2xU_i(x) - U_{i-1}(x), \text{ for } i = 2, 3, \ldots \tag{20.26}$$

We define the integer function $m(i) : \mathbb{N} \to \mathbb{N}$ with $m(1) = 1$ and for $i = 2, 3, \ldots,$ it is evaluated according to

$$m(i) := 2^{i-1} + 1. \tag{20.27}$$

We also define the sequence of sets $\{\mathscr{G}^i\}_{i=1,2,\ldots}$ wherein, $\mathscr{G}^1 = \{0\}$, and for $i = 2, 3, \ldots$, it holds that $\mathscr{G}^i = \{\zeta_1, \ldots, \zeta_i\} \subset [-1, 1]$, that is the set of the extrema of the Chebyshev polynomials with the components given by

$$\zeta_j := -\cos\left(\frac{\pi(j-1)}{i-1}\right), \quad \text{for all } j \in \{1, \ldots, i\}. \tag{20.28}$$

In the next definition, we use the multi-index notation $\mathbf{i} = (i_1, \ldots, i_n) \in \mathbb{N}^n$ inducing $|\mathbf{i}| := \sum_{k=1}^{n} i_k$.

**Definition 20.2** (*Smolyak Sparse Grid*) The Smolyak Sparse Grid of $[-1, 1]^n$ is a union of the Cartesian products of the form

$$\mathscr{H}^{n,\mu} := \bigcup_{|\mathbf{i}|=n+\mu} \left(\mathscr{G}^{m(i_1)} \times \cdots \times \mathscr{G}^{m(i_n)}\right), \tag{20.29}$$

where positive integer $\mu \in \mathbb{N}$ is the order of the grid.[2]

**Definition 20.3** (*Smolyak Approximant Polynomial*) The Smolyak approximant to a function $f : [-1, 1]^n \to \mathbb{R}$ is given by

$$\hat{f}^{n,\mu}(\mathbf{x}) := \sum_{q \le |\mathbf{i}| \le n+\mu} (-1)^{n+\mu-|\mathbf{i}|} \binom{n-1}{n+\mu-|\mathbf{i}|} \mathrm{p}^{\mathbf{i}}(\mathbf{x}) \tag{20.30}$$

with $q = \max(n, \mu + 1)$ the tensor product polynomials for each multi-index $\mathbf{i} = (i_1, \ldots, i_n)$ defined as

$$\mathrm{p}^{\mathbf{i}}(\mathbf{x}) := \sum_{l_1=1}^{m(i_1)} \cdots \sum_{l_n=1}^{m(i_n)} \theta_{l_1,\ldots,l_n} T_{l_1}(x_1) \ldots T_{l_n}(x_n), \tag{20.31}$$

where $\theta_{l_1,\ldots,l_n}$ the coefficients that are to be determined.

To find the optimal vector of coefficients of $\Theta = \mathrm{vec}(\theta_{l_1,\ldots,l_n}) \in \mathbb{R}^m$ for approximation of some function $f$ that is $C^k([-1, 1]^n)$, an error objective should be defined and minimized. One way is to consider the error function $E(f, \Theta) : C^k([-1, 1]^n) \times \mathbb{R}^n \to \mathbb{R}_+$ to be

$$\begin{aligned} E(f, \Theta) &:= \int_{[-1,1]^n} \left|f(\mathbf{x}) - f^{n,\mu}(\mathbf{x})\right|^2 \prod_{\mathbf{y} \in \mathscr{H}^{n,\mu}} \delta(\mathbf{x} - \mathbf{y})\, \mathrm{d}\mathbf{x} \\ &= \sum_{\mathbf{y} \in \mathscr{H}^{n,\mu}} \left|f(\mathbf{y}) - f^{n,\mu}(\mathbf{y})\right|^2, \end{aligned}$$

[2]One can show that grids of higher order include all grids of lower order; i.e., $\mathscr{H}^{n,\mu} \subset \mathscr{H}^{n,\mu+1}$.

where $\delta$ is the Dirac's delta function. This metric is certainly minimized (i.e., $E(f, \Theta) = 0$) if $\Theta$ is chosen such that

$$f(\mathbf{x}) = f^{n,\mu}(\mathbf{x}) \text{ for all } \mathbf{x} \in \mathscr{H}^{n,\mu}. \tag{20.32}$$

This procedure is called collocation. A pivotal property of the overall method is that size of vector of coefficients $\Theta$ and the number of interpolation points is equal; i.e., $|\mathscr{H}^{n,\mu}| = |\Theta| = M$. Therefore, enforcing equalities (20.32) constitutes of searching for the solution to $M$ equations involving $M$ unknown entries of $\Theta$. Once we evaluate the coefficients with the described scheme, the following error bound would hold, wherein the used functional norm $\|.\| : C^k([-1, 1]^n) \to \mathbb{R}_+$ is defined as

$$\|f\| = \max\left\{\left\|D^i f\right\|_\infty : i = 1, \ldots, k\right\}. \tag{20.33}$$

**Theorem 20.8** (Theorem 2 in [7]) *Suppose that function $f(\mathbf{x}) : [-1, 1]^n \to \mathbb{R}$ is $C^k([-1, 1]^n)$, together with a Smolyak approximant $\hat{f}^{n,\mu}(\mathbf{x})$ that interpolates $f$ on $\mathscr{H}^{n,\mu}$ with $|\mathscr{H}^{n,\mu}| = M$. Then, for some positive constant $c_{n,k}$, the error of the approximation is bounded according to*

$$\left\|f - \hat{f}^{n,\mu}\right\| \le c_{n,k} M^{-k} (\log M)^{(k+2)(n+1)+1}. \tag{20.34}$$

Each multi-index $\mathbf{i}$ in (20.30) induces a number of tensor product polynomials that are summed together as in (20.31). We gather the indices of all these tensor product polynomials in a set $\mathbf{L}^{n,\mu}$; i.e.,

$$\mathbf{L}^{n,\mu} := \bigcup_{q \le |\mathbf{i}| \le n+\mu} \{(l_1, \ldots, l_n) : l_j \le m(i_j)\}. \tag{20.35}$$

One can show that at the end of the day, the approximant constructed in (20.30) using the polynomials (20.31) boils down to the following simple representation

$$\hat{f}^{n,\mu}(\mathbf{x}) = \sum_{\mathbf{l}_i = (l_1^i, \ldots, l_n^i) \in \mathbf{L}^{n,\mu}} \Theta_i \mathrm{T}_i(\mathbf{x}) = \sum_{i=1}^{M} \Theta_i \mathrm{T}_i(\mathbf{x}), \tag{20.36}$$

where $M = |\mathscr{H}^{n,\mu}| = |\mathbf{L}^{n,\mu}|$, and for $\mathbf{l}_i = (l_1^i, \ldots, l_n^i) \in \mathbf{L}^{n,\mu}$, the coefficients $\Theta_i$ and polynomial terms are given by

$$\Theta_i := \theta_{l_1^i, \ldots, l_n^i}, \tag{20.37}$$

$$\mathrm{T}_i(\mathbf{x}) := T_{l_1^i}(x_1) \ldots T_{l_n^i}(x_n). \tag{20.38}$$

To compute the partial derivatives of approximation, we use the definitions of the Chebyshev polynomials to define

$$\mathrm{T}_i^j(\mathbf{x}) := \begin{cases} \mathrm{T}_i \cdot \dfrac{U_{l_j^i}(x_j)}{T_{l_j^i}(x_j)} & l_j^i = 2, \ldots, n \\ 0 & l_j^i = 1 \end{cases} \tag{20.39}$$

This lets us write the partial derivatives of $\hat{f}^{n,\mu}$ in the compact form

$$\frac{\partial \hat{f}^{n,\mu}(\mathbf{x})}{\partial x_j} = \sum_{i=1}^{M} l_j^i \Theta_i \mathrm{T}_i^j(\mathbf{x}). \tag{20.40}$$

### 20.5.2 Sparse Approximation to Eigenfunctions

We denote the approximation to Koopman eigenfunction $\phi(\mathbf{x})$ by $\hat{\phi}(\mathbf{x})$. Substituting (20.36) and (20.40) into (20.13), we get the (approximate) equality

$$\sum_{j=1}^{n} \sum_{i=1}^{M} l_j^i \Theta_i \mathrm{T}_i^j(\mathbf{x}) F_j(\mathbf{x}) \approx \lambda \sum_{i=1} M \Theta_i \mathrm{T}_i(\mathbf{x}).$$

We change the order of summations to further obtain

$$\sum_{i=1}^{M} \left( \sum_{j=1}^{n} \left( l_j^i \mathrm{T}_i^j(\mathbf{x}) F_j(\mathbf{x}) \right) - \lambda \mathrm{T}_i(\mathbf{x}) \right) \Theta_i \approx 0. \tag{20.41}$$

We define and denote the vector of coefficients $\Theta \in \mathbb{R}^M$ by $\Theta := [\Theta_1, \ldots, \Theta_M]^T$. For a point in the grid $\mathbf{x}^k \in \mathscr{H}^{n,\mu}$, we may write the left-hand side of (20.41) as

$$\mathscr{A}_k \Theta = [\mathscr{A}_{ki}]_{i=1,\ldots,n}\, \Theta,$$

where the entries of row vector $\mathscr{A}_k \in \mathbb{R}^{1 \times M}$ can be computed from

$$\mathscr{A}_{ki} := \sum_{j=1}^{n} \left( l_j^i \mathrm{T}_i^j(\mathbf{x}^k) F_j(\mathbf{x}^k) \right) - \lambda \mathrm{T}_i(\mathbf{x}^k), \text{ for all } i = 1, \ldots, M.$$

Repeating this for $M$ points in the Smolyak grid, the stacked left-hand side of all equations becomes $\mathscr{A}\Theta$, where $\mathscr{A} \in \mathbb{R}^{M \times M}$ is given by

$$\mathscr{A} := [\mathscr{A}_1^T, \ldots, \mathscr{A}_M^T]^T. \tag{20.42}$$

Ideally, $\mathscr{A}\Theta$ should be zero for an eigenfunction, however, if it is not possible, we would like to minimize an error function, which we choose to be

$$J(\Theta) = \|\mathscr{A}\Theta\|_2^2. \tag{20.43}$$

Now, because $\mathbf{A}$ may have repeated eigenvalues, we add a constraint that lets us derive multiple eigenfunctions corresponding to one eigenvalue. Consider the Koopman eigenfunction $\phi(\mathbf{x})$ with Koopman eigenvalue $\lambda$, which is the eigenvalue of the linearization matrix with a left eigenvector $\mathbf{R} = [\mathbf{r}_1, \ldots, \mathbf{r}_n] \in \mathbb{R}^{n\times n}$. We omit the index of the eigenvalues and eigenvectors in the following developments for simplicity and consider $\lambda$ to be associated with the eigenvector $\mathbf{r} \in \mathbb{R}^n$. We can show that with the fixed point at the origin,

$$\nabla\phi(\mathbf{x})|_{\mathbf{x}=\mathbf{0}} = \mathbf{r}.$$

Translating this for the approximant, for each $j = 1, \ldots, n$ we have

$$\frac{\partial\hat{\phi}(\mathbf{x})}{\partial x_j}|_{\mathbf{x}=\mathbf{0}} = \sum_{i=1}^{M} l_j^i \Theta_i \mathrm{T}_i^j(\mathbf{0}) = \mathscr{B}_j\Theta = r_j,$$

where the row vector $\mathscr{B}_j \in \mathbb{R}^{1\times M}$ has the components

$$\mathscr{B}_{ji} = l_j^i \mathrm{T}_i^j(\mathbf{0}), \quad \text{for all } \ i = 1, \ldots, M.$$

The matrix form of this equality becomes

$$\mathscr{B}\Theta = \mathbf{r}, \tag{20.44}$$

where the matrix $\mathscr{B} \in \mathbb{R}^{n\times M}$ is the result of stacking row vectors as

$$\mathscr{B} = \left[\mathscr{B}_1^T, \ldots, \mathscr{B}_n^T\right]^T. \tag{20.45}$$

Recall that our approximation requires $\hat{\phi}(\mathbf{0}) = 0$ to provide exponential convergence for the performance measure integrals (see Remark 20.2). This condition can be translated to single scalar equality

$$\mathscr{C}\Theta = 0, \tag{20.46}$$

where $\mathscr{C} \in \mathbb{R}^{1\times M}$ is the row vector with elements

$$\mathscr{C}_i := \mathrm{T}_i(\mathbf{0}) \ \text{ for all } \ i = 1, \ldots, M.$$

Now, we would like to minimize the error function defined by (20.43) while constraints (20.44) and (20.46) are satisfied. We define the optimization problem

$$\begin{aligned} \underset{\Theta \in \mathbb{R}^m}{\text{minimize}} \quad & \|\mathscr{A}\Theta\|_2^2, \\ \text{subject to} \quad & \begin{bmatrix} \mathscr{B} \\ \mathscr{C} \end{bmatrix} \Theta = \begin{bmatrix} \mathbf{r} \\ 0 \end{bmatrix}. \end{aligned} \tag{20.47}$$

This program is equivalent to a semidefinite program (SDP) and can be solved using the conventional convex programming such as CVX [21].

### 20.5.3 Sparse Approximation to Koopman Mode Decomposition

In the previous subsection, we illustrated a way to find approximations to the Koopman eigenfunctions. Thus, the components of the map $\mathbf{H}(\mathbf{x})$ can be approximated. Here, following a similar approach, we seek approximations to the components of its inverse map $\mathbf{H}^{-1}(\mathbf{x})$. The very natural equation for component $H_j^{-1}(\mathbf{x})$ is

$$H_j^{-1}(\mathbf{H}(\mathbf{x})) = x_j, \quad \text{for all} \quad j = 1, \ldots, n. \tag{20.48}$$

We only have an approximation to $\mathbf{H}(\mathbf{x})$, namely $\hat{\mathbf{H}}(\mathbf{x})$, and we need approximations to $\hat{\mathbf{H}}^{-1}(\mathbf{x})$, namely $\hat{\mathbf{H}}^{-1}(\mathbf{x})$. Hence, we consider the approximate equality

$$\hat{H}_j^{-1}(\hat{\mathbf{H}}(\mathbf{x})) \approx x_j, \quad \text{for all} \quad j = 1, \ldots, n. \tag{20.49}$$

Again, following the spirit of the collocation method, for each point of the grid $\mathbf{x}^k \in \mathscr{H}^{n,\mu}$, we enforce this equation to hold. Suppose that we have found the series approximation to each components of $\mathbf{H}(\mathbf{x})$, denoted by $\hat{\mathbf{H}}(\mathbf{x})$. Moreover, we define

$$\mathbf{z}^k := \hat{\mathbf{H}}(\mathbf{x}^k). \tag{20.50}$$

Then, we consider a Smolyak series representation for this function as

$$\hat{g}_j^{n,\mu}(\mathbf{z}^k) = \sum_{i=1}^{M} \Phi_i^j \mathrm{T}_i(\mathbf{z}^k). \tag{20.51}$$

Similar to the essence of the method that we discussed in the previous subsection, we define vector of coefficients $\Phi_1^j \in \mathbb{R}^M$ to be

$$\Phi^j := \left[\Phi_1^j, \ldots, \Phi_M^j\right].$$

Inserting (20.51) into (20.48), we get

$$\mathscr{D}_k \Phi^j = [\mathscr{D}_{ki}]_{i=1,\ldots,M}\, \Phi^j \approx x_j^k, \tag{20.52}$$

where the components of row vector $\mathscr{D}_k \in \mathbb{R}^{1\times M}$ are

$$\mathscr{D}_{ki} := \mathrm{T}_i(\mathbf{z}^k). \tag{20.53}$$

Concatenating these vectors and the right-hand side scalars for each point in the grid (i.e., $M$ points), we may write these equations as

$$\mathscr{D}\Phi^j \approx \mathbf{X}_j, \tag{20.54}$$

where matrix $\mathscr{D} \in \mathbb{R}^{M\times M}$ and vector $\mathbf{X}_j \in \mathbb{R}^M$ are given by

$$\mathscr{D} := \left[\mathscr{D}_1^T, \ldots, \mathscr{D}_M^T\right]^T, \tag{20.55}$$

$$\mathbf{X}_j := \left[x_j^1, \ldots, x_j^M\right]^T, \tag{20.56}$$

respectively. Again one hopes that (20.54) holds with a minimal error for each $j = 1, \ldots, n$. Therefore, we define the optimization problem

$$\underset{\Phi^j \in \mathbb{R}^M}{\text{minimize}} \quad \left\| \mathscr{D}\Phi^j - \mathbf{X}_j \right\|_2^2. \tag{20.57}$$

Note that matrix $\mathscr{D}$ is deliberately denoted without index $j$, because it is the same matrix for the optimization problem for each component $H_j^{-1}(\mathbf{x})$. The solution to this least-squares optimization problem is given by

$$\Phi^j = \mathscr{D}^\dagger \mathbf{X}_j \quad \text{for all} \quad j = 1, \ldots, n.$$

We should repeat this for each component of $\mathbf{H}^{-1}(\mathbf{x})$. Putting the results in a matrix gives us

$$\Phi := \left[\Phi^1, \ldots, \Phi^n\right].$$

Because the value of matrix $\mathscr{D}$ is shared between $M$ optimization problems defined by (20.57), by inspection, we find that

$$\Phi = \mathscr{D}^\dagger \mathbf{X}^T, \tag{20.58}$$

where $\mathbf{X} \in \mathbb{R}^{n\times M}$ is the matrix containing the vector of all grid points. Now, we have a polynomial approximation to $\mathbf{H}^{-1}$, which would give us a Koopman mode decomposition.

### 20.5.4 Numerical Examples

In this section, we show that we may be able to effectively estimate the performance measure of nonlinear consensus networks with more than two subsystems. Note that in all cases, the real performance measure is calculated from the numerical solution of the network output followed by numerical integration.

*Example 20.4* (*Complete Graphs*) Using the described numerical approximation method, we estimate the performance measure for the nonlinear consensus network with exponentially decaying weights defined in (20.21). The corresponding linearized graph Laplacian corresponds to the undirected unweighted complete graph; i.e.,

$$\mathbf{A} = -\mathscr{L}_{\mathscr{K}_n} = \mathbf{J}_n/n - \mathbf{I}_n.$$

We evaluate the performance measure from the KMD approximation based on the numerical integration of the solutions. The initial conditions are uniformly sampled random initial conditions from $[-1, 1]^n$. The numerical values for data using Koopman approach have been obtained by implementation of the suggested numerical method with and the results are shown in Fig. 20.3. In this example, the error in the evaluated performance measure using our numerical method is less than 2%.

*Example 20.5* (*Random Graphs*) We fix $n = 8$ and create Erdős–Rényi graphs with different edges probabilities (and consequently, different edge numbers). Then, we consider again the exponentially decaying weights given by (20.21). The performance measures from Monte Carlo simulations as well as the formula (using the method discussed in the previous sections) are also evaluated and illustrated in Fig. 20.4. The error of approximation, in this case, is less than 1.4%.

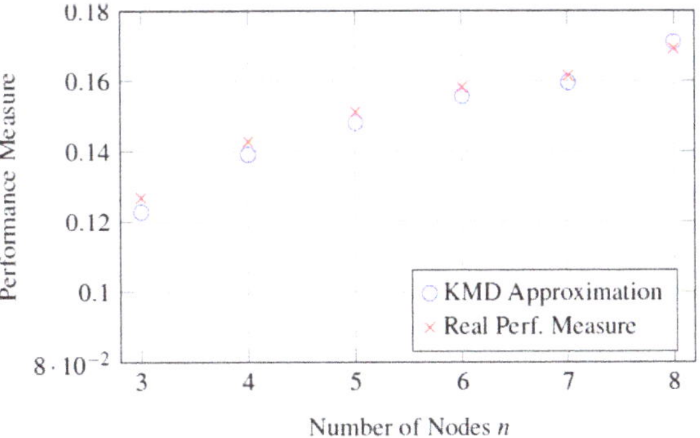


**Fig. 20.3** The performance measure of nonlinear consensus network with $\alpha = 0.25$ and the graph at the linearized Laplacian of complete graph

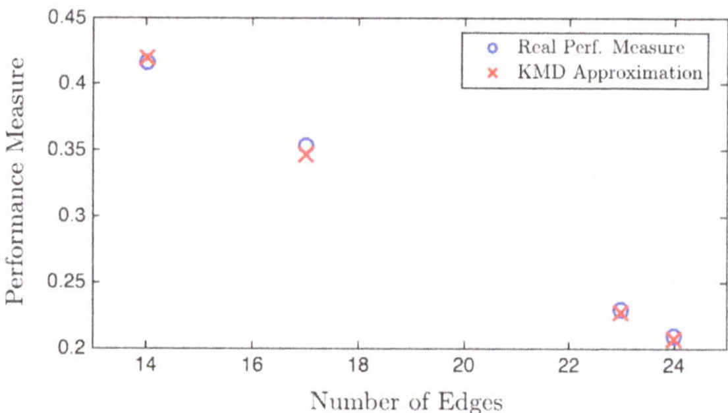


**Fig. 20.4** The performance measure of nonlinear consensus network with $n = 8$, $\alpha = 0.25$ and random graphs with different number of edges

## *20.5.5 Comparison to Extended Dynamic Mode Decomposition*

The numerical method explained in this section is related to the notion of extended dynamic mode decomposition (EDMD) [56], which has been a promising procedure for extracting information about the Koopman spectrum of the dynamical system. In EDMD, to find a KMD for the flow of the dynamical system, one should first assume a rich enough dictionary of basis functions such that hopefully, the Koopman eigenfunctions lie in their span. Then, using the snapshots from the trajectory, one may find a truncated approximation to the Koopman operator and finite number of approximations to the Koopman eigenfunctions and their corresponding eigenvalues. Then, the solution to the dynamical system is approximated as a truncated KMD using those eigenvalues and eigenfunctions.

In the current settings, we know what are the eigenvalues and eigenfunctions that are required for the representation of the flow of the nonlinear system. Hence, we do not need the first step of the EDMD for the computation of the (approximate) eigenvalues and eigenfunctions. In fact, we build the dictionary that one needs for EDMD based on the principal eigenfunctions in the map $\mathbf{H}(\mathbf{x})$.

On the other hand, the second step in both methods are connected in the spirit: in our approach, we find an approximation to map $\mathbf{H}^{-1}(\mathbf{x})$ using identity

$$\mathbf{H}^{-1}\left(\mathbf{H}(\mathbf{x})\right) = \mathbf{x}.$$

While in EDMD, the identity observable (i.e., $\mathbf{f}(\mathbf{x}) \equiv \mathbf{x}$) has to be represented in terms of the basis functions in the dictionary. Then one is allowed to write down a KMD for the system dynamics as explained before.

## 20.6 Conclusion and Discussion

Koopman mode decomposition approaches hold promise for performance analysis and synthesis of nonlinear dynamical systems, that are of interest in various disciplines of engineering and control. The vital connection between the eigenspectrum of linearized dynamics and Koopman operator provides a closed-form evaluation of the first moment of energy integral of the solutions, in terms of Koopman eigenvalues, eigenfunctions, and modes. The numerical approximation of KMD components is implemented by a scalable computational algorithm using a sparse Smolyak grid with certifiable accuracy. Future directions include, but are not limited to the following: investigation and analysis of various performance metrics in nonlinear systems as extensions of linear control systems [48]. Another research line regards dynamical systems with higher order integrators as well as a systemic performance-based network synthesis for optimal interactions among interconnected entities in the face of uncertain initial conditions or other structural parameters.

**Acknowledgements** The authors would like to thank Prof. Alex Mauroy, Prof. Igor Mezić, and the anonymous reviewer for fruitful discussions and comments that enhanced the quality of this chapter. This work is supported by NSF CAREER ECCS-1454022, AFOSR FA9550-19-1-0004 and ONR YIP N00014-16-1-2645.

## References

1. Abramowitz, M., Stegun, I.A., et al.: Handbook of mathematical functions. Appl. Math. Ser. **55**, 62 (1966)
2. Ajorlou, A., Momeni, A., Aghdam, A.G.: Sufficient conditions for the convergence of a class of nonlinear distributed consensus algorithms. Automatica **47**(3), 625–629 (2011)
3. Ali, Q., Gageik, N., Montenegro, S.: A review on distributed control of cooperating mini UAVs. Int. J. Artif. Intell. Appl. **5**(4), 1 (2014)
4. Arcak, M.: Passivity as a design tool for group coordination. IEEE Trans. Autom. Control **52**(8), 1380–1390 (2007)
5. de Badyn, M.H., Mesbahi, M.: Growing controllable networks via whiskering and submodular optimization. In: 2016 IEEE 55th Conference on Decision and Control (CDC), pp. 867–872. IEEE (2016)
6. Bamieh, B., Jovanovic, M.R., Mitra, P., Patterson, S.: Coherence in large-scale networks: Dimension-dependent limitations of local feedback. IEEE Trans. Autom. Control **57**(9), 2235–2249 (2012)
7. Barthelmann, V., Novak, E., Ritter, K.: High dimensional polynomial interpolation on sparse grids. Adv. Comput. Math. **12**(4), 273–288 (2000)
8. Bassett, D.S., Bullmore, E.: Small-world brain networks. The Neuroscientist **12**(6), 512–523 (2006)
9. Beard, R.W., Lawton, J., Hadaegh, F.Y.: A coordination architecture for spacecraft formation control. IEEE Trans. Control. Syst. Technol. **9**(6), 777–790 (2001)
10. Budišić, M., Mohr, R., Mezić, I.: Applied koopmanism chaos: an interdisciplinary. J. Nonlinear Sci. **22**(4), 047,510 (2012)
11. Buzi, G., Topcu, U., Doyle, J.C.: Quantitative nonlinear analysis of autocatalytic pathways with applications to glycolysis. In: American Control Conference (ACC), 2010, pp. 3592–3597. IEEE (2010)

12. Charalambides, C.A.: Enumerative Combinatorics. CRC Press, Boca Raton (2002)
13. Chiang, H.D., Wu, F.F., Varaiya, P.P.: A bcu method for direct analysis of power system transient stability. IEEE Trans. Power Syst. **9**(3), 1194–1208 (1994)
14. Cucker, F., Mordecki, E.: Flocking in noisy environments (2007). arXiv preprint arXiv:0706.3343
15. Cucker, F., Smale, S.: Emergent behavior in flocks. IEEE Trans. Autom. Control **52**(5), 852–862 (2007)
16. Cullen, H.F.: Introduction to General Topology. D C Heath & Co (1968)
17. Dai, R., Mehran, M.: Optimal topology design for dynamic networks. In: 2011 50th IEEE Conference on Decision Control/European Control Conference, pp. 1280–1285. IEEE (2011)
18. Derriennic, M.M.: On multivariate approximation by bernstein-type polynomials. J. Approx. Theory **45**, 155–166 (1985)
19. Dorfler, F., Bullo, F.: Synchronization and transient stability in power networks and nonuniform kuramoto oscillators. SIAM J. Control Optim. **50**(3), 1616–1642 (2012)
20. Dörfler, F., Chertkov, M., Bullo, F.: Synchronization in complex oscillator networks and smart grids. Proc. Natl. Acad. Sci. **110**(6), 2005–2010 (2013)
21. Grant, M., Boyd, S., Cvx, Y.Y.: Matlab software for disciplined convex programming, version 1.21 (2011). cvxr.com/cvx
22. Jadbabaie, A., Lin, J., Morse, A.S.: Coordination of groups of mobile autonomous agents using nearest neighbor rules. IEEE Trans. Autom. Control **48**(6), 988–1001 (2003)
23. Jadbabaie, A., Molavi, P., Sandroni, A., Tahbaz-Salehi, A.: Non-bayesian social learning. Games Econ. Behav. **76**(1), 210–225 (2012)
24. Jadbabaie, A., Motee, N., Barahona, M.: On the stability of the kuramoto model of coupled nonlinear oscillators. In: American Control Conference. Proceedings of the 2004, vol. 5, pp. 4296–4301. IEEE (2004)
25. Kim, Y., Mesbahi, M.: On maximizing the second smallest eigenvalue of a state-dependent graph laplacian. In: American Control Conference, 2005. Proceedings of the 2005, pp. 99–103. IEEE (2005)
26. Kutz, J.N., Brunton, S.L., Brunton, B.W., Proctor, J.L: Dynamic Mode Decomposition: Data-Driven Modeling of Complex Systems. OT149. SIAM (2016)
27. Lan, Y., Mezić, I.: Linearization in the large of nonlinear systems and koopman operator spectrum. Phys. D: Nonlinear Phenom. **242**(1), 42–53 (2013)
28. Leonard, N.E., Fiorelli, E.: Virtual leaders, artificial potentials and coordinated control of groups. In: Proceedings of the 40th IEEE Conference on Decision and Control, 2001, vol. 3, pp. 2968–2973. IEEE (2001)
29. Lin, F., Fardad, M., Jovanović, M.R.: Design of optimal sparse feedback gains via the alternating direction method of multipliers. IEEE Trans. Autom. Control **58**(9), 2426–2431 (2013)
30. Malin, B.A., Krueger, D., Kubler, F.: Solving the multi-country real business cycle model using a smolyak-collocation method. J. Econ. Dyn. Control **35**(2), 229–239 (2011)
31. Mauroy, A., Goncalves, J.: Koopman-based lifting techniques for nonlinear systems identification (2017). arXiv preprint arXiv:1709.02003
32. Mauroy, A., Mezić, I.: Global stability analysis using the eigenfunctions of the koopman operator. IEEE Trans. Autom. Control **61**(11), 3356–3369 (2016)
33. Mesbahi, M., Egerstedt, M.: Graph Theoretic Methods in Multiagent Networks. Princeton University Press, Princeton (2010)
34. Moghaddam, S.H., Jovanović, M.R.: An interior point method for growing connected resistive networks. In: 2015 American Control Conference, pp. 1223–1228 (2015)
35. Mousavi, H.K., Somarakis, C., Bahavarnia, M., Motee, N.: Performance bounds and optimal design of randomly switching linear consensus networks. In: American Control Conference (ACC), 2017, pp. 4347–4352. IEEE (2017)
36. Mousavi, H.K., Somarakis, C., Motee, N.: Koopman performance analysis of a class of nonlinear dynamical networks. In: 2016 IEEE 55th Conference on Decision and Control (CDC), pp. 117–122. IEEE (2016)

37. Mousavi, H.K., Somarakis, C., Motee, N.: Spectral performance analysis and design for distributed control of multi-agent systems. In: 2017 IEEE 56th Annual Conference on Decision and Control (CDC), pp. 2918–2923. IEEE (2017)
38. Olfati-Saber, R., Murray, R.M.: Consensus problems in networks of agents with switching topology and time-delays. IEEE Trans. Autom. Control **49**(9), 1520–1533 (2004)
39. Olshevsky, A., Tsitsiklis, J.N.: Convergence speed in distributed consensus and averaging. SIAM J. Control Optim. **48**(1), 33–55 (2009)
40. Papachristodoulou, A., Jadbabaie, A.: Synchonization in oscillator networks with heterogeneous delays, switching topologies and nonlinear dynamics. In: 2006 45th IEEE Conference on Decision and Control, pp. 4307–4312. IEEE (2006)
41. Papachristodoulou, A., Jadbabaie, A., Munz, U.: Effects of delay in multi-agent consensus and oscillator synchronization. IEEE Trans. Autom. Control **55**(6), 1471–1477 (2010)
42. Patterson, S., Bamieh, B.: Leader selection for optimal network coherence. In: 2010 49th IEEE Conference on Decision and Control (CDC), pp. 2692–2697. IEEE (2010)
43. Perko, L.: Differential Equations and Dynamical Systems, vol. 7. Springer Science & Business Media, Berlin (2013)
44. van Riel, N.A., Sontag, E.D.: Parameter estimation in models combining signal transduction and metabolic pathways: the dependent input approach. IEE Proc.-Syst. Biol. **153**(4), 263–274 (2006)
45. Shilong, L., Mao, S.: Aerodynamic force and flow structures of two airfoils in flapping motions. Acta Mech. Sin. **17**(4), 310–331 (2001)
46. Siami, M., Motee, N.: On existence of hard limits in autocatalytic networks and their fundamental tradeoffs. IFAC Proc. **45**(26), 294–298 (2012)
47. Siami, M., Motee, N.: Performance analysis of linear consensus networks with structured stochastic disturbance inputs. In: 2015 American Control Conference, pp. 4080–4085 (2015)
48. Siami, M., Motee, N.: Fundamental limits and tradeoffs on disturbance propagation in linear dynamical networks. IEEE Trans. Autom. Control **61**(12), 4055–4062 (2016)
49. Siami, M., Motee, N.: Growing linear dynamical networks endowed by spectral systemic performance measures. IEEE Trans. Autom. Control **63**(8) (2018)
50. Siami, M., Motee, N.: Network abstraction with guaranteed performance bounds. IEEE Trans. Autom. Control (2018)
51. Somarakis, C., Baras, J.S.: Delay-independent convergence for linear consensus networks with applications to non-linear flocking systems. In: Submitted to the 12th IFAC Workshop on Time Delay Systems. Ann Arbor, MI, USA (2015)
52. Somarakis, C., Baras, J.S.: A simple proof of the continuous time linear consensus problem with applications in non-linear flocking networks. In: 14th European Control Conference. Linz, Austria (2015)
53. Somarakis, C., Ghaedsharaf, Y., Motee, N.: Time-delay origins of fundamental tradeoffs between risk of large fluctuations and network connectivity (2018). arXiv preprint arXiv:1801.06856
54. Susuki, Y., Mezić, I.: Nonlinear koopman modes and a precursor to power system swing instabilities. IEEE Trans. Power Syst. **27**(3), 1182–1191 (2012)
55. Wang, P.K.: Navigation strategies for multiple autonomous mobile robots moving in formation. J. Field Robot. **8**(2), 177–195 (1991)
56. Williams, M.O., Kevrekidis, I.G., Rowley, C.W.: A data-driven approximation of the koopman operator: extending dynamic mode decomposition. J. Nonlinear Sci. **25**(6), 1307–1346 (2015)

# Index





A. Mauroy et al. (eds.), *The Koopman Operator in Systems and Control*,
Lecture Notes in Control and Information Sciences 484,
https://doi.org/10.1007/978-3-030-35713-9

GPSR Compliance
The European Union's (EU) General Product Safety Regulation (GPSR) is a set of rules that requires consumer products to be safe and our obligations to ensure this.

If you have any concerns about our products, you can contact us on

ProductSafety@springernature.com

In case Publisher is established outside the EU, the EU authorized representative is:

Springer Nature Customer Service Center GmbH
Europaplatz 3
69115 Heidelberg, Germany

www.ingramcontent.com/pod-product-compliance
Ingram Content Group UK Ltd.
Pitfield, Milton Keynes, MK11 3LW, UK
UKHW021008290726
14059UKWH00001BA/16

* 9 7 8 3 0 3 0 3 5 7 1 2 2 *